Y0-BSD-783

The chemistry of **the hydroxyl group**

Part 1

THE CHEMISTRY OF FUNCTIONAL GROUPS

A series of advanced treatises under the general editorship of Professor Saul Patai

The chemistry of alkenes (published in 2 volumes)
The chemistry of the carbonyl group (published in 2 volumes)
The chemistry of the ether linkage (published)
The chemistry of the amino group (published)
The chemistry of the nitro and nitroso groups (published in 2 parts)
The chemistry of the carboxylic acids and esters (published)
The chemistry of the carbon–nitrogen double bond (published)
The chemistry of amides (published)
The chemistry of the cyano group (published)
The chemistry of the hydroxyl group (published in 2 parts)

—OH

The chemistry of **the hydroxyl group**

Part 1

Edited by

SAUL PATAI

The Hebrew University, Jerusalem

1971

INTERSCIENCE PUBLISHERS

a division of John Wiley & Sons

LONDON – NEW YORK – SYDNEY – TORONTO

First published 1971 John Wiley & Sons Ltd.

Library of Congress Catalog Card No. 77–116164
ISBN 0 471 66939 3

Made and Printed in Great Britain by
Butler & Tanner Ltd, Frome and London

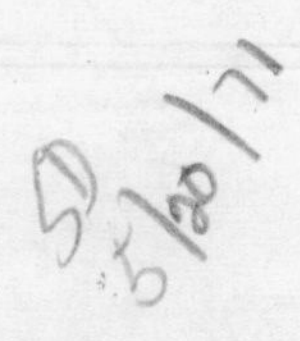

Foreword

The present volume of the series 'The Chemistry of the Functional Groups' is again organized according to the general plan as described in the Preface to the series, printed on the following pages.

Only one of the originally planned chapters did not materialize, that on 'Oxidation and Reduction of Alcohols'.

Jerusalem, March 1970 SAUL PATAI

The Chemistry of the Functional Groups

Preface to the series

The series 'The Chemistry of the Functional Groups' is planned to cover in each volume all aspects of the chemistry of one of the important functional groups in organic chemistry. The emphasis is laid on the functional group treated and on the effects which it exerts on the chemical and physical properties, primarily in the immediate vicinity of the group in question, and secondarily on the behaviour of the whole molecule. For instance, the volume *The Chemistry of the Ether Linkage* deals with reactions in which the C—O—C group is involved, as well as with the effects of the C—O—C group on the reactions of alkyl or aryl groups connected to the ether oxygen. It is the purpose of the volume to give a complete coverage of all properties and reactions of ethers in as far as these depend on the presence of the ether group, but the primary subject matter is not the whole molecule, but the C—O—C functional group.

A further restriction in the treatment of the various functional groups in these volumes is that material included in easily and generally available secondary or tertiary sources, such as Chemical Reviews, Quarterly Reviews, Organic Reactions, various 'Advances' and 'Progress' series as well as textbooks (i.e. in books which are usually found in the chemical libraries of universities and research institutes) should not, as a rule, be repeated in detail, unless it is necessary for the balanced treatment of the subject. Therefore each of the authors is asked *not* to give an encyclopaedic coverage of his subject, but to concentrate on the most important recent developments and mainly on material that has not been adequately covered by reviews or other secondary sources by the time of writing of the chapter, and to address himself to a reader who is assumed to be at a fairly advanced post-graduate level.

With these restrictions, it is realized that no plan can be devised for a volume that would give a *complete* coverage of the subject with *no* overlap between chapters, while at the same time preserving the readability of the text. The Editor set himself the goal of attaining *reasonable* coverage with *moderate* overlap, with a minimum of cross-references between the chapters of each volume. In this manner, sufficient freedom is given to each author to produce readable quasi-monographic chapters.

The general plan of each volume includes the following main sections:

(a) An introductory chapter dealing with the general and theoretical aspects of the group.

(b) One or more chapters dealing with the formation of the functional group in question, either from groups present in the molecule, or by introducing the new group directly or indirectly.

(c) Chapters describing the characterization and characteristics of the functional groups, i.e., a chapter dealing with qualitative and quantitative methods of determination including chemical and physical methods, ultraviolet, infrared, nuclear magnetic resonance, and mass spectra; a chapter dealing with activating and directive effects exerted by the group and/or a chapter on the basicity, acidity or complex-forming ability of the group (if applicable).

(d) Chapters on the reactions, transformations and rearrangements which the functional group can undergo, either alone or in conjunction with other reagents.

(e) Special topics which do not fit any of the above sections, such as photochemistry, radiation chemistry, biochemical formations and reactions. Depending on the nature of each functional group treated, these special topics may include short monographs on related functional groups on which no separate volume is planned (e.g. a chapter on 'Thioketones' is included in the volume *The Chemistry of the Carbonyl Group*, and a chapter on 'Ketenes' is included in the volume *The Chemistry of Alkenes*). In other cases, certain compounds, though containing only the functional group of the title, may have special features so as to be best treated in a separate chapter, as e.g., 'Polyethers' in *The Chemistry of The Ether Linkage*, or 'Tetraaminoethylenes' in *The Chemistry of the Amino Group*.

This plan entails that the breadth, depth and thought-provoking nature of each chapter will differ with the views and inclinations of the author and the presentation will necessarily be somewhat uneven. Moreover, a serious problem is caused by authors who deliver their manuscript late or not at all. In order to overcome this problem at least to some extent, it was decided to publish certain volumes in several parts, without giving consideration to the originally planned logical order of the chapters. If after the appearance of the originally planned parts of a volume it is found that either owing to non-delivery of chapters, or to new developments in the subject, sufficient material has accumulated for publication of an additional part, this will be done as soon as possible.

The overall plan of the volumes in the series 'The Chemistry of the Functional Groups' includes the titles listed below:

The Chemistry of the Alkenes (published in two volumes)
The Chemistry of the Carbonyl Group (published in two volumes)
The Chemistry of the Ether Linkage (published)
The Chemistry of the Amino Group (published)
The Chemistry of the Nitro and the Nitroso Group (published in two parts)
The Chemistry of Carboxylic Acids and Esters (published)
The Chemistry of the Carbon–Nitrogen Double Bond (published)
The Chemistry of the Cyano Group (published)
The Chemistry of the Amides (published)
The Chemistry of the Hydroxyl Group (published in two parts)
The Chemistry of the Carbon–Halogen Bond (in preparation)
The Chemistry of Carbonyl Halides (in preparation)
The Chemistry of the Azido Group (in preparation)
The Chemistry of the Carbon–Carbon Triple Bond
The Chemistry of Imidoates and Amidines
The Chemistry of the Thiol Group
The Chemistry of the Quinonoid Compounds
The Chemistry of the Hydrazo, Azo and Azoxy Groups
The Chemistry of the SO, —SO_2, —SO_2H *and* —SO_3H *Groups*
The Chemistry of the —OCN, —NCO and —SCN Groups
The Chemistry of the —PO_3H_2 *and Related Groups*

Advice or criticism regarding the plan and execution of this series will be welcomed by the Editor.

The publication of this series would never have started, let alone continued, without the support of many persons. First and foremost among these is Dr. Arnold Weissberger, whose reassurance and trust encouraged me to tackle this task, and who continues to help and advise me. The efficient and patient cooperation of several staff-members of the Publisher also rendered me invaluable aid (but unfortunately their code of ethics does not allow me to thank them by name). Many of my friends and colleagues in Jerusalem helped me in the solution of various major and minor matters, and my thanks are due especially to Prof. Y. Liwschitz, Dr. Z. Rappoport and Dr. J. Zabicky. Carrying out such a long-range project would be quite impossible without the non-professional but none the less essential participation and partnership of my wife.

The Hebrew University, SAUL PATAI
Jerusalem, ISRAEL

Contributing authors

R. F. W. Bader	McMaster University, Hamilton, Ontario, Canada.
I. R. L. Barker	Brighton College of Technology, Sussex, England.
R. A. Basson	Atomic Energy Board, Pretoria, South Africa.
Hans-Dieter Becker	General Electric Research and Development Center, Schenectady, New York, U.S.A.
Geoffrey W. Brown	Sir John Cass College, London, England.
Živorad Čeković	Institute of Chemistry, Technology and Metallurgy, Belgrade, Yugoslavia.
R. Graham Cooks	Kansas State University, Manhattan, Kansas, U.S.A.
S. K. Erickson	University of Leicester, Leicester, England.
H. M. Flowers	The Weizmann Institute of Science, Rehovoth, Israel.
Colin A. Fyfe	University of Guelph, Guelph, Ontario, Canada.
J. Gordon Hanna	Connecticut Agricultural Experiment Station, New Haven, Connecticut, U.S.A.
D. A. R. Happer	University of Canterbury, Christchurch, New Zealand.
Alpo Kankaanperä	University of Turku, Turku, Finland.
H. Knözinger	Institute of Physical Chemistry, University of Munich, Germany.
Henning Lund	University of Aarhus, 8000 Aarhus C, Denmark.
Elliot N. Marvell	Oregon State University, Corvallis, Oregon, U.S.A.
Mihailo Lj. Mihailović	University of Belgrade, Belgrade, Yugoslavia
Juhani Murto	University of Helsinki, Helsinki, Finland.
S. Oae	Osaka City University, Osaka, Japan.

Kalevi Pihlaja — University of Turku, Turku, Finland.

C. H. Rochester — Nottingham University, Nottingham, England.

Pentti Salomaa — University of Turku, Turku, Finland.

D. F. Sangster — A.A.E.C. Research Establishment, Sutherland, New South Wales, Australia.

J. Schädelin — Justus-Liebig Universität, 63 Giessen, Germany.

U. Schmeling — Justus-Liebig Universität, 63 Giessen, Germany.

H.-H. Schott — Justus-Liebig Universität, 63 Giessen, Germany.

Sidney Siggia — University of Massachusetts, Amherst, Massachusetts, U.S.A.

Hj. Staudinger — Justus-Liebig Universität, 63 Giessen, Germany.

Thomas R. Stengle — University of Massachusetts, Amherst, Massachusetts, U.S.A.

S. Tamagaki — Osaka City University, Osaka, Japan.

V. Ullrich — Justus-Liebig Universität, 63 Giessen, Germany.

J. Vaughan — University of Canterbury, Christchurch, New Zealand.

William Whalley — Oregon State University, Corvallis, Oregon, U.S.A.

Contents

CHAPTER 1

Theoretical aspects of the chemistry of the hydroxyl group

R. F. W. Bader

Department of Chemistry, McMaster University, Hamilton, Ontario, Canada

I. INTRODUCTION

A. A Density Approach to Chemical Binding

This chapter is concerned with the quantum mechanical prediction and interpretation of the properties of the hydroxyl group as a radical, as an ion, either positively or negatively charged, or as a functional group. The theoretical description of the electronic structure is given in terms of molecular orbital theory and all quantitative results are based on self-consistent field (SCF) calculations at or near the Hartree–Fock limit. The results of such calculations are known to yield good representations of molecular charge distributions and in this chapter the attempt is made to relate the chemistry of the hydroxyl group directly to the properties of the spatial distribution of electronic charge in the molecule.

The molecular charge distribution describes the manner in which the electronic charge is distributed throughout real space. Thus the properties of a system may be given a direct *physical* description and interpretation when they are related to the charge distribution. A discussion of the properties in terms of the spatial details of the wavefunction ψ does not yield a direct physical picture because of the multidimensional nature of the wavefunction for a many-electron system.

The function ψ for an N-electron system is a function of the space and spin coordinates of all the electrons. The instantaneous *simultaneous* probability of each electron being in some particular small region of space with a given spin is given by the product

$$\psi^*\psi \, d\tau_1 \, d\tau_2 \ldots d\tau_i \ldots d\tau_N$$

where $d\tau_i$ denotes both an infinitesimal volume element in the space of electron i and a definite spin component either α or β. The integration of this product over *all* the spins (thereby changing each $d\tau_i$ into a spatial volume element dx_i) and over the spatial coordinates of all the electrons but one, say electron 1,

$$dx_1 \int \psi^*\psi \, dx_2 \, dx_3 \ldots dx_N \tag{I-1}$$

yields a function which describes the probability of finding one of the electrons in some particular region of its cartesian space; i.e., it yields a single-electron probability distribution in three-dimensional space. Since all N electrons are equivalent and indistinguishable, a consequence of the antisymmetry requirements imposed on ψ by the Pauli exclusion principle, the *total* probability of finding negative charge in a given region of space is N times the one-electron probability given in equation (I-1)

$$dx_1 \, N \int \psi^*\psi \, dx_2 \, dx_3 \ldots dx_N \tag{I-2}$$

The three-dimensional distribution function $\rho(x)$

$$\rho(x) = N \int \psi^*\psi \, dx_2 \, dx_3 \ldots dx_N \tag{I-3}$$

is the total electronic charge density or total electronic charge distribution, a function in three-dimensional space.

The charge distribution $\rho(x)$ determines all of the electrical moments of the system (dipole, quadrupole, etc., and fields and field gradients at the positions of the nuclei). The charge distribution also determines the 'size' and 'shape' of a molecule in its nonbonded interactions with other systems, and is responsible for the scattering observed in X-ray and electron diffraction experiments. Electron and X-ray diffraction results provide in principle a method for the experimental determination of molecular charge distributions. Reviews of such attempts, mainly in the field of X-ray crystallography, have been given by Brill[1] and by O'Connell, Rae and Maslen[2]. Kohl and Bartell[3,4] have reported electron diffraction results for small-angle scattering on gas phase molecules which suggest that electron scattering techniques now in development should be able to provide information about the charge distribution of the bonding electrons.

Hohenberg and Kohn[5] have presented a theorem which shows that in principle even the energy of the system may be expressed as a function of the charge distribution. The function $\rho(x)$ therefore contains all the information necessary for a complete physical description of a system.

The increase in our understanding and prediction of chemical phenomena may be related to the increase in our understanding of how the electronic charge is distributed in a molecule and of how these distributions change during a chemical reaction. Concepts such as ionic-covalent character and electronegativity[6] are outstanding examples of earlier attempts to determine empirically how the total charge is partitioned between the atomic components of a molecular system. With the advent of relatively good quantum mechanical calculations, we are now able to determine and relate chemical phenomena to the actual distribution of charge throughout three-dimensional space.

In addition to its use in the direct calculation of physical properties, the charge distribution may be analysed in terms of the total amount of charge which is found in different regions of space, for example, the amount of charge in the 'binding region' between the nuclei, or in the regions normally pictured as occupied by 'lone pairs'. Related to the total density maps are the *density difference* maps which are obtained by subtracting the density distributions of the constituent atoms from the total molecular density. Such maps provide a picture of the redistribution of charge which results in the formation of a chemical bond. The density difference maps show patterns which are characteristic of and distinct for limiting types of bonds ionic and covalent[7]. The density and density-difference maps can therefore serve as the basis for definitions of distinct bond types.

The physical picture provided by the charge distribution may be carried even further through the use of the Hellmann–Feynman theorem which relates in a rigorous manner the forces acting on the nuclei to the distribution of charge in the molecule[8, 9]. Because of the essentially classical nature of the connexion between the forces and the electronic charge distribution, a study of the forces exerted on the nuclei can provide a physical basis for the interpretation of chemical binding.

Since the results which follow lean so heavily on Hartree–Fock wavefunctions, a brief discussion of the ultimate accuracy and limitations of these functions is given in the following section.

B. Hartree–Fock Wavefunctions

A Hartree–Fock wavefunction is by definition the best possible single determinantal wavefunction for a system. Such a wavefunction consists of an antisymmetrized product of one-electron functions, the orbitals. The antisymmetrization of the wavefunction is a necessary consequence of the Pauli principle and has the result of correlating the motions of electrons with identical spins. The motions of electrons with different spins are, however, completely uncorrelated because the probability function for all such pairs of electrons is given simply as a product of the individual probabilities, i.e., the electrons act independently. For this reason the difference between the true energy of the system and that predicted by the Hartree–Fock wavefunction is called the correlation energy[10]. (There is also a correction for the neglected relativistic effects but these are very small for atoms in the first two rows of the Periodic Table.)

The orbitals are obtained as solutions to the Hartree–Fock equations*

$$\left[-\tfrac{1}{2}\nabla_1^{\,2} - \sum_\alpha \frac{Z_\alpha}{r_{1\alpha}}\right]\mu_i(x_1) + \left[\sum \int \mu_j^*(x_2)\frac{1}{r_{12}}\mu_j(x_2)\,\mathrm{d}\tau_2\right]\mu_i(x_1) - \sum_j \left[\int \mu_j(x_2)\frac{1}{r_{12}}\mu_i(x_2)\,\mathrm{d}\tau_2\right]\mu_j(x_1) = \varepsilon_i\mu_i(x_1) \qquad \text{(I-4)}$$

there being one such equation for each spin-orbital μ_i, i.e., a space orbital ϕ_i multiplied by an α or β spin function. For a closed-shell system of N electrons there will be $N/2$ occupied and distinct space orbitals. The summations in equation (I-4) are over all N of the occupied spin-orbitals.

The first bracketed term in equation (I-4) represents the kinetic

* Atomic units are used throughout this chapter: length, 1 au $= a_0 = 0{\cdot}52917$ Å; energy, 1 au $= e^2/a_0 = 6{\cdot}2771 \times 10^2$ kcal/mole; force, 1 au $= e^2/a_0^{\,2} = 8{\cdot}2377 \times 10^{-3}$ dynes; charge density, 1 au $= e/a_0^{\,3} = 6{\cdot}749\ e^-/\text{Å}^3$.

energy and the potential energy in the field of the nuclei of an electron in the orbital μ_i. The summation $\sum_j \mu_j^*(x_2)\mu_j(x_2)$ in the second term, called the coulomb term, represents the *total* electronic charge density at each point in space, and the integral of this quantity over the operator $1/r_{12}$ gives the repulsive field exerted on the electron in μ_i by the total charge distribution (including a contribution from the electron in μ_i). The third bracketed term, the exchange term, arises from the antisymmetry conditions imposed on the wavefunction, and is different from zero only for those spin-orbitals possessing the same spin as μ_i. This term removes all contributions to the total repulsive field experienced by the electron in μ_i at the position x_1 from other electrons with the same spin as that of the electron in μ_i. The exchange charge density when integrated over all space yields one electronic charge and hence its presence in equation (I-4) decreases the total number of electronic charges exerting a repulsive force on the electron in μ_i by unity. The exchange term may, therefore, be interpreted as providing a correction to the coulomb term which includes a contribution from the electron in μ_i exerting a repulsive force on itself[11].

Equation (I-4) replaces the actual instantaneous repulsions between pairs of electrons by an average interaction, one which describes each electron separately interacting with the average field of the remaining electrons. The exchange term effectively correlates the motions of electrons with parallel spins by removing from the immediate vicinity of a given electron all charge density arising from electrons with similar spin. The same electron, however, experiences only the average field exerted by electrons with opposite spin, and this is the origin of the so-called correlation error in the total energy.

The total energy of the system in terms of the orbital energies ε_i is

$$E = \sum \varepsilon_i - \sum_{i<j} \{[ii|jj] - [ij|ji]\} \tag{I-5}$$

where the terms in square brackets represent the coulomb and exchange integrals respectively.

Equation (I-4) for the one-electron orbital energies is derived by demanding that the functions μ_i give the *lowest possible energy* for the system. This particular set of orbitals, the Hartree–Fock orbitals, provide the best one-electron approximation to the system. A set of N equations of the form I-4 are too involved to solve directly, since the solution of each equation demands a knowledge of the solutions for the remaining $(N - 1)$ equations. This is a consequence of the fact that the average field exerted by the remaining $(N - 1)$ electrons as expressed by coulomb and exchange terms is known only when all the remaining μ_i are known.

To overcome this difficulty Roothaan[12] has devised a self-consistent field method for solving the Hartree–Fock equations for a system based on the expansion of each μ_i in terms of a linear combination of much simpler functions[13]. The set of simpler functions, called the basis set, is finite in number and usually consists of Slater or Gaussian type atomic orbitals centred on the various nuclei in the molecule.

The SCF equations of Roothaan may be solved for a basis set containing only $\sim N$ distinct atomic orbitals for an N-electron problem. While the molecular orbitals obtained from such a minimal size basis set are self-consistent, they are poor approximations to the true Hartree–Fock orbitals. Only by including a large number of atomic orbitals in the basis set can the expansion be made flexible enough adequately to describe the Hartree–Fock molecular orbitals. Ideally the Hartree–Fock result represents the limiting case of an expansion in terms of an infinite basis set. However, experience has shown that the Hartree–Fock limit may be reached for all practical purposes using basis sets of reasonable size. For example, the basis set required to approximate the Hartree–Fock orbitals for the OH radical[14] to an accuracy of about 0·001 consisted of 24 Slater type orbitals (STO's) with the following composition: centred on oxygen, two $1s$, two $2s$, four $2p\sigma$ and four $2p\pi$, two $3d\sigma$ and one $3d\pi$, one $4f\sigma$ and one $4f\pi$; and centred on hydrogen, two $1s$, one $2s$, one $2p\sigma$ and one $2p\pi$.

Our primary use of the Hartree–Fock results will be to obtain molecular charge distributions. With regard to this use, there is a very important theorem which can be proved for a Hartree–Fock wavefunction. The theorem itself is due to Brillouin[15] and, as a consequence of this theorem, we can show that the charge density and its dependent properties obtained from a Hartree–Fock wavefunction are correct up to the second-order. (The interested reader is referred to Ref. 16 for a discussion of this theorem which is relevant to the calculation of charge distributions.) Thus we may expect the Hartree–Fock charge distribution (a one-electron property) and the properties determined by the charge distribution to be relatively insensitive to the correlation error (a property of the two-electron probability distribution) inherent in the Hartree–Fock wavefunction. To test this assumption we have listed in Table I-1 the

TABLE I-1. A comparison of experimental and Hartree–Fock results[a].

Molecule	Term	μ (Debyes)		Forces on nuclei		D_e[b] (e.v.)		R_e (au)		ω_e (cm^{-1})	
		Calc.	Exp.	on proton	on heavy nucleus	Calc.	Exp.	Calc.	Exp.	Calc.	Exp.
LiH	$X^1\Sigma^+$	−6·002	−5·882	−0·003	−0·002	1·49	2·52	3·034	3·015	1433	1406
CH	$X^2\Pi_r$	1·570	1·46	−0·014	−0·053	2·47	3·65	2·086	2·124	3053	2869
OH	$X^2\Pi_i$	1·780	1·660	−0·024	−0·074	3·03	4·63	1·795	1·8343	4062	3735
HF	$X^1\Sigma^+$	1·942	1·8195	−0·026	−0·075	4·38	6·12	1·696	1·7328	4469	4139
NaH	$X^1\Sigma^+$	−6·962	—	+0·0002	−0·582	0·932	2·3	3·617	3·566	1187	1172
HCl	$X^1\Sigma^+$	1·197	1·12	−0·0108	+0·1833	3·48	4·616	2·389	2·4087	3181	2990
H_2O	X^1A_1	1·955	1·85	+0·002	+0·080	6·80	9·49				

[a] The calculated and experimental data for the diatomic hydrides are from Ref. 14 and for H_2O from Ref. 17.

[b] These are 'rationalized' dissociation energies calculated by subtracting the Hartree–Fock estimate of the molecular energy from the Hartree–Fock estimate of the energy of the atoms. By this method a large fraction of the correlation error in D_e is cancelled. Only the *change* in the correlation energy in passing from the molecule to the dissociated atoms remains in the estimate of D_e.

values of the dipole moments and forces acting on the nuclei calculated from Hartree–Fock wavefunctions for a number of hydride molecules. In addition to the one-electron properties just cited we have also included calculated values of some energy quantities and spectroscopic constants. The values of the one-electron properties are indeed in good agreement with experiment, the dipole moments exhibiting an average error of about 0·12 Debyes and the forces on the nuclei which should be zero at the equilibrium bond length, indicating only slight departures from electrostatic equilibrium. Brillouin's theorem holds strictly only for closed-shell molecules but the calculated properties of the open-shell systems included in Table I-1, OH and CH, do not exhibit any sudden deterioration in quality.

The electronic contribution to the dipole moment is determined primarily by the spatial details of the charge distribution in its outer regions while the forces are most sensitive to the properties of the charge distribution in regions close to the nuclei. While the dipole moment and the forces offer tests of different moments of the charge distribution, they are still averages over the complete distribution. A test of the accuracy of the actual spatial distribution of charge predicted by a Hartree–Fock wavefunction can be made only by comparing the distribution with one obtained from a more extended calculation. Such comparisons have been made for H_2 and Li_2 [18], using the wavefunctions of Das and Wahl[19] which yield a large fraction of the correlation energy. It was concluded that no noticeable error is introduced when a Hartree–Fock density distribution is used to portray the molecular charge distribution. In the case of Li_2 a plot of the difference density distribution between the extended and the Hartree–Fock results yielded no values greater than 1×10^{-4} au, a number smaller by a factor of 20 than the outer contours used to determine the nonbonded sizes of molecules.

The data in Table I-1 indicate that while the correlation error is appreciable for the energy (a quantity directly determined by the two-electron probability distribution) it is much less significant for the one-electron charge distribution and its dependent properties.

II. A STUDY OF THE O–H BOND IN OH^+, $OH^{\cdot}$ AND OH^-

A. The Molecular Charge Distributions

The introductory discussion of the electronic structure of the hydroxyl group will be concerned with the molecular charge distributions and mechanism of binding of the proton in the diatomic species OH^+, $OH^\cdot$ and OH^-.

The molecular orbital configurations and ground state symmetries for these molecules are

$$OH^+ \quad 1\sigma^2 2\sigma^2 3\sigma^2 1\pi^2;\ {}^3\Sigma^- \quad ({}^1\Delta,\ {}^1\Sigma^+)$$
$$OH^\cdot \quad 1\sigma^2 2\sigma^2 3\sigma^2 1\pi^3;\ {}^2\Pi$$
$$OH^- \quad 1\sigma^2 2\sigma^2 3\sigma^2 1\pi^4;\ {}^1\Sigma^+$$

The hydroxide ion possesses a closed-shell electronic structure and hence a singlet, symmetric ground state. The radical $OH^\cdot$ possesses both orbital and spin degeneracy. The half-filled π configuration in OH^+ results in three distinct states, ${}^3\Sigma^-$, ${}^1\Delta$ and ${}^1\Sigma^+$. To first-order the energies of these states differ because of different contributions from the repulsive energies between the electrons. The state of highest multiplicity, the ${}^3\Sigma^-$ state, is lowest in energy. Certain energy values and other physical characteristics of these molecules are listed in Table **II-1**. The bond strengths, with respect to the appropriate separated atom or ion states, decrease in the order OH^-, $OH^\cdot$, OH^+, and the bond lengths increase in the same order.

TABLE II-1. Physical properties of OH^+, $OH^\cdot$ and OH^- *.

AH	Term	R_e (au)	D_e (ev)	Electron affinity (ev)	Ionization potential (ev)	ω_e (cm^{-1})
OH^+	$X^3\Sigma^-$	1·944	(>4·4)			
$OH^\cdot$	$X^2\Pi$	1·8342	4·63	1·83	13·36	3735·2
OH^-	$X^1\Sigma^+$	(1·81)	(>3·48)			(3773)

* Data from P. E. Cade. *J. Chem. Phys.*, **47**, 2390 (1967).

The total molecular charge distributions for the OH species are shown in Figure 1 in the form of contour maps in the plane of the nuclei*. Bader, Keaveny and Cade[20] have presented an interpretation of the binding in the first-row neutral diatomic hydrides based upon the molecular charge distribution and the forces which it exerts on the nuclei. On this basis the binding in LiH is classified as ionic and that in BH → HF as covalent. The binding in BeH is

* The Hartree–Fock wavefunctions used to calculate the charge densities were obtained from P. E. Cade and W. Huo, *J. Chem. Phys.*, **47**, 614 (1967) for $OH(X^2\Pi_i)$, $CH(X^2\Pi_r)$ and $LiH(X^1\Sigma^+)$; unpublished wavefunctions of P. E. Cade were employed for $OH^-(X^1\Sigma^+)$ and $OH^+({}^3\Sigma^-)$.

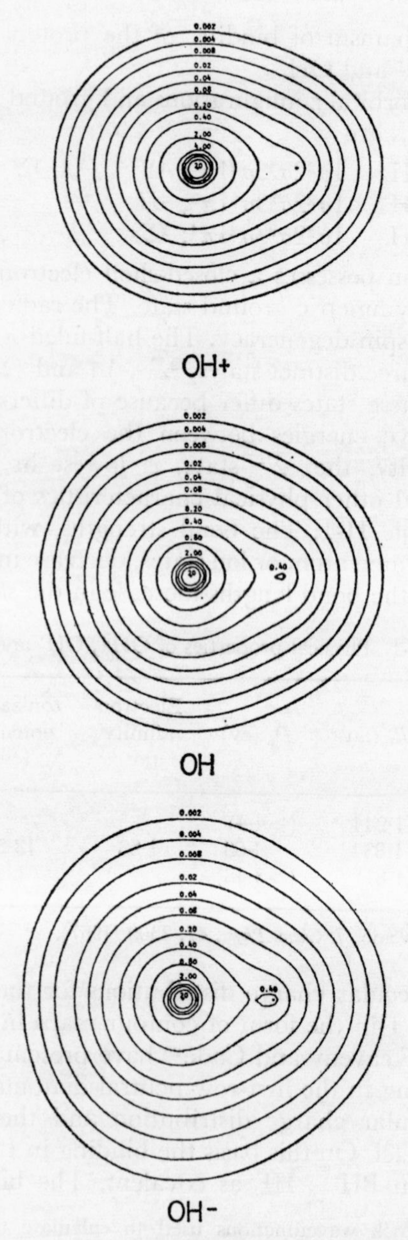

Figure 1. Contour maps of the total molecular charge distributions of diatomic hydrides in their electronic ground states. The proton is represented by the dot on the right-hand side in each contour map.

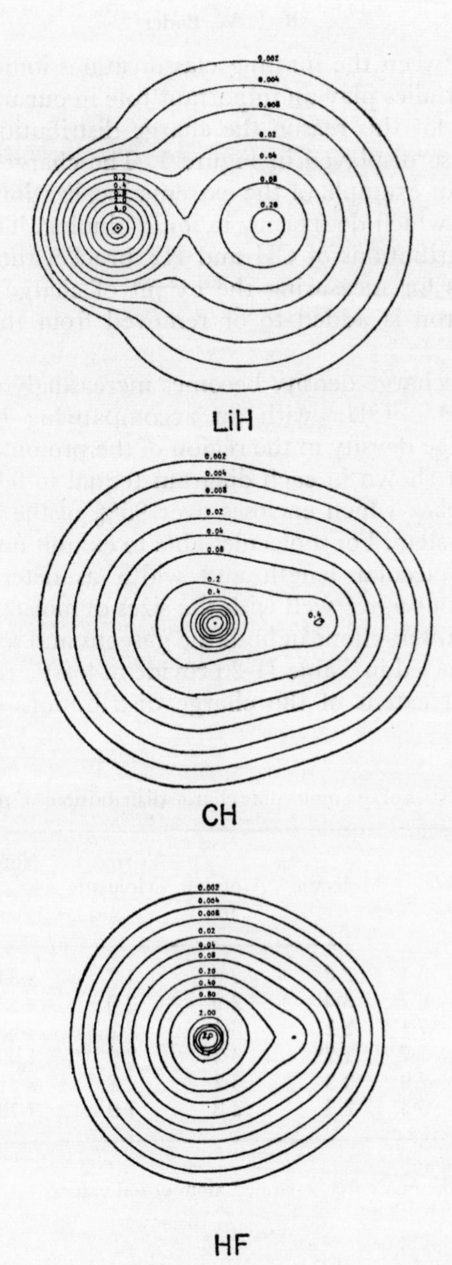

Figure 1 continued

transitional between the limiting classifications ionic and covalent. Comparative studies play an important role in our understanding of chemistry and for this reason the charge distributions of LiH, CH and HF are also displayed in Figure 1. The charge distribution in LiH provides an example of the extreme localization of the valence charge density which does occur in ionic binding. The properties of the charge distributions of CH and HF bracket those of $OH^{\cdot}$ and provide gauges for measuring the extent of charge reorganization when an electron is added to or removed from the neutral $OH^{\cdot}$ species.

The valence charge density becomes increasingly diffuse through the series $OH^{+} \rightarrow OH^{-}$ with an accompanying increase in the amount of charge density in the region of the proton. The outermost density contour shown in each diagram (equal to 0·002 au) defines a volume in space which encloses over 95% of the total electronic charge of the system. For molecules able to exhibit nonbonded interactions, the molecular length and width as determined by this particular contour agree well with the sizes of molecules determined by nonbonded interactions in both the gaseous and solid states[7]. The lengths (L) reported in Table II-2 provide at least a relative measure of the size and extent of the charge distributions of the hydride molecules.

TABLE II-2. Properties of the molecular charge distributions of diatomic hydrides[a]

AH	μ (Debyes)	L	r_A Molecule	r_A Atom or ion	r_H Molecule	Nonbonded charges on[b] A	Nonbonded charges on[b] H
OH^{+}		6·4	2·8	2·6	1·7	3·84 (3·5)	0·24
$OH^{\cdot}$	1·780	6·7	3·0	2·9	2·0	4·23 (4·0)	0·36
OH^{-}		7·1	3·1	3·1	2·2	4·68 (4·5)	0·51
LiH	−6·002	7·7	1·7[c]	3·2	2·9	1·09 (1·5)	0·71
CH	1·570	7·9	3·5	3·2	2·3	3·21 (3·0)	0·49
HF	1·942	6·3	2·7	2·8	1·9	4·72 (4·5)	0·30

[a] All lengths are in au.
[b] Values in parentheses are free separated atom or ion values.
[c] The value of r_{Li^+} is 1·8 au.

The distance measured from either nucleus along the bond axis to the outermost (0·002) contour provides a measure of the nonbonded radius of the 'atom' in the molecule. The contribution of the nonbonded density on both H and O to the length of the molecule increases as the number of electrons is increased. The value of r_H for the free atom is 2·5 au. Relative to this value the nonbonded radius of H is considerably decreased in all three molecules. The removal of an electron from $OH^\cdot$ causes an overall tightening of the charge distribution and a shift in the values of the nonbonded radii towards values characteristic of HF. In fact the amount of charge density in the region of the proton in OH^+ is less than in HF. Similarly the addition of an electron to $OH^\cdot$ results in an expansion of the charge distribution and in a set of nonbonded radii close in value to those for the preceding neutral hydride NH ($r_N = 3{\cdot}2$, $r_H = 2{\cdot}1$ au).

The electron population of any spatial region may be obtained by integrating the charge density over the corresponding restricted volume in space. Table II-2 lists such populations for the nonbonded regions of A and H, the nonbonded regions being defined by the volume of space on the nonbonded side of a plane perpendicular to the bond axis, and passing through the A or H nucleus. The nonbonded charge on oxygen exceeds that of the parent oxygen ion or atom in each of the three molecules. The nonbonded population on hydrogen is essentially unchanged from the free atom value in OH^- but decreased from this value in $OH^\cdot$ and still more so in OH^+.

The changes in the nonbonded charges on O and H upon ionization of or electron attachment to OH again reflect a shift in the properties of the charge distribution towards those characteristic of HF or NH respectively. The increase in the nonbonded charge on O and its decrease for H (compared to O^+ and H) in OH^+ indicate the presence of a greater degree of charge polarization than is found in HF. While the total nonbonded charge on O in OH^- is similar in value to that on F in the isoelectronic molecule HF, the nonbonded charges on H differ greatly in the two cases. OH^- falls into sequence when one compares the values of the nonbonded charges in the molecule with those of the parent species O^- and H. Such a comparison shows that the valence charge density is democratically delocalized over both nuclei in a manner similar to that found for the less polar central member of the hydride series CH.

A more detailed view of the differences in the charge distributions of $OH^\cdot$ and OH^- is given in Figure 2. The Figure portrays a density difference map obtained by subtracting the molecular charge density of $OH^\cdot$ from that of OH^- calculated at the equilibrium bond length of $OH^\cdot$. The map thus shows the instantaneous change in the charge density when an electron is captured by the $OH^\cdot$ molecule. (The density difference map for the ionization of $OH^{\cdot}$ is similar in all its spatial features to Figure 2 but the signs of the contours are reversed and their magnitudes larger than those for the electron attachment.)

In the orbital approximation the added electron enters the 1π molecular orbital. A glance at Figure 5 indicates that the 1π orbital

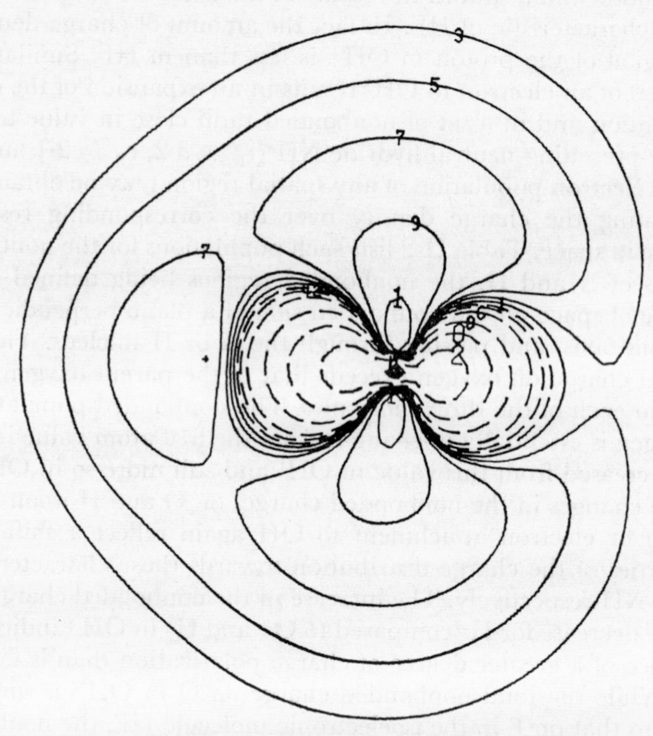

FIGURE 2. A density difference map showing the change in the molecular charge distribution when an electron is captured by the neutral OH species. Solid contour lines (odd numbered contours) denote an increase in charge density, dashed lines (even numbered contours) denote a decrease. The key relating the values of the contours to their numerical labels is given in Table II-7. The proton is on the left-hand side in this diagram.

density in all three of the OH species is highly localized on the O nucleus with a distribution very close to that of a $2p\pi$ atomic orbital on oxygen. In view of this, the enhancement of the charge density in the region of the proton in OH^- is surprising. In addition, there is a region of charge deficiency along the bond axis in the region of the oxygen nucleus corresponding to a *loss* of almost 0·1 charges to other regions when the electron is added to $OH^{\bullet}$. If the molecular orbitals remained unchanged during ionization or electron attachment (an assumption which is frequently employed) Figure 2 would reduce to a density plot of the 1π molecular orbital. The pattern of charge shift actually obtained indicates that all the orbitals undergo a substantial change when the number of electrons or the electron configuration is altered. Thus, in spite of the fact that the 1π orbital is atomic-like in both $OH^{\bullet}$ and OH^- (with zero density at the proton), the presence of the extra electron leads to changes in the σ orbitals of the system via the coupling provided by the exchange and coulomb integrals in the Hartree–Fock equations. The use of rigid or virtual orbitals will thus lead to incorrect assumptions regarding the changes in the charge distribution caused by the addition, removal or excitation of an electron[21].

One further point of interest regarding the total charge distribution is its insensitivity to changes in the spin multiplicity when these changes correspond to transitions between different electronic states arising from the same configuration[21]. For example, the molecular charge distributions for the three states of OH^+, $^3\Sigma^-$, $^1\Delta$ and $^1\Sigma^+$, all of which arise from the same $1\pi^2$ open-shell configuration, are indistinguishable to the accuracy to which they are portrayed in the present article. It should be stressed that completely separate and distinct Hartree–Fock calculations are made for each state. Thus the differences in the chemistry observed for such states of different multiplicity cannot be accounted for in terms of differences in their charge distributions. Instead, the difference in their chemistry must be related to the ability of the systems with nonzero spin to induce a spin polarization in the reacting system and in this manner follow a different reaction coordinate.

A detailed picture of the net reorganization of the charge density of the separated atoms accompanying the formation of a molecule may be obtained by subtracting the superimposed densities of the component (undistorted) atoms separated by R_e from the molecular charge density, also evaluated at $R = R_e$. This density difference distribution when illustrated in the form of a contour plot in the plane of the nuclei will be designated $\Delta\rho_{SA}(x, y)$. The density distribution which results from the superposition or overlap of the undistorted atomic densities does not place sufficient charge density in the 'binding region' to balance the forces of nuclear repulsion[22]. The regions of charge increase in the density difference maps are, therefore, the regions to which charge is transferred relative to the separated atoms to obtain a state of electrostatic equilibrium and a stable chemical bond. In this sense the charge density differences may be interpreted as pictures of the 'bond density'.

It is natural to use the location of this charge increase relative to the positions of the nuclei to characterize the bond[7]. Thus, if the density difference map exhibits a region of net accumulation of negative charge symmetrically placed between and behind the nuclei, as is the case for the homonuclear diatomic molecules, the bond is classified as covalent. If, at the other extreme, the net accumulation of negative charge is distinctly localized in the region of only one of the nuclei, as exemplified in LiH or LiF, the bond is classified as ionic. The mutually shared charge density binds the nuclei in the covalent case, while in the ionic case they are bound by the density increase localized on one nucleus.

The density difference maps for the OH species and LiF, CH and HF are illustrated in Figure 3. The contours in these maps represent the increase or decrease in the amount of charge density present in the molecules relative to the distribution obtained by the overlap of the undistorted atom or ion densities. The principal features of the $\Delta\rho_{SA}(x,y)$ maps are similar for all three of the OH species. There is an accumulation of charge density in both the bonded and nonbonded regions of the oxygen which is concentrated along the internuclear axis. The charge increase on the bonded side encompasses the proton. These charge accumulations are a result of a charge removal from the region behind the proton and from a belt-like region perpendicular to the bond at the position of the oxygen nucleus. The concentration of charge density along the axis and its removal from a torus-like region perpendicular to the axis represents a *quadrupole polarization*. The same type of quadrupolar polarization is present in the regions of the F and C nuclei in HF, CH and LiF. The simple *dipolar polarization* depicted in the $\Delta\rho_{SA}(x,y)$ maps in the vicinity of the proton or the Li, a deficiency of charge density on one side and an accumulation on the other, is typical of the charge rearrangements found for atoms which employ principally *s* orbitals in their binding. However, *for atoms which employ principally* p-*type orbitals, the reorganization of the charge density accompanying bond formation is quadrupolar in character*, regardless of the bond type[7, 20, 23].

The chemically important feature of the quadrupolar polarization is that it results in a charge increase in the antibinding region, the region normally ascribed to lone pair or unshared electron density. This polarization accounts for the increase in electronic charge found in the nonbonded regions of the A nuclei. Note that the region of charge removal in the vicinity of the heavy nuclei is largely confined to the binding region.

The $\Delta\rho_{SA}$ map for OH^- is most similar to that for CH as judged by a comparison of the extent of charge removal from the nonbonded region of the proton, of the positioning and extent of the charge increase in the binding region relative to the position of the proton and of the spatial extent of the torus-like region of charge removal from the region of the oxygen or carbon. The tightening of the charge distribution accompanying the ionization of $OH^{\bullet}$ to yield OH^+ results in a shift of the characteristics of the $\Delta\rho_{SA}$ map towards those of HF.

The nonbonded radius of the Li in LiF (and in LiH) is the same as for a Li^+ ion. Since the valence density of the Li atom is extremely diffuse, only a single negative contour appears in the $\Delta\rho_{SA}$ map to signify its essentially complete transfer to the F. The slight accumulation of charge density on the nonbonded side of the Li nucleus is the result of a polarization of the $1s$ core density, the significance of which is discussed below. It has been previously[24] shown that a plot of just the sigma density increase around F in LiF (that is, from 1σ, 2σ, 3σ and 4σ molecular orbitals) is almost coincidental to a single occupied $2p\sigma$ density on F. Thus the $\Delta\rho_{SA}$ pattern obtained in LiF can be viewed as the equivalent of filling a $2p\sigma$ orbital vacancy and characteristic of the ionic case. This limiting pattern is most closely approached here by HF and OH^+. None of the hydrides illustrated in Figure 3, however, attains the ionic limit. Instead the proton, unlike the Li nucleus in LiF, is encompassed by the density increase on A, one which in the hydrides may be associated with the *partial* filling of an asymmetrically distorted $2p\sigma$ orbital on A. The density increase binding the nuclei is thus shared by both nuclei, and the binding in these molecules is therefore covalent. The extent and details of sharing the charge increase, however, change markedly through the series. In CH and OH^- the density increase in the binding region is a maximum at the proton and results from a sharing of density centred on both nuclei. The remaining members of the series, OH, OH^+ and HF, give $\Delta\rho_{SA}$ diagrams which progressively give the appearance of an unsymmetrical $2p\sigma$ atomic orbital centred on A with a proton embedded at its extremity.

There is another feature of the $\Delta\rho_{SA}$ maps which indicates that the extent of charge transfer is not as great in the hydrides, OH^+ and HF for example, as in the ionic cases of LiH or LiF. The extreme

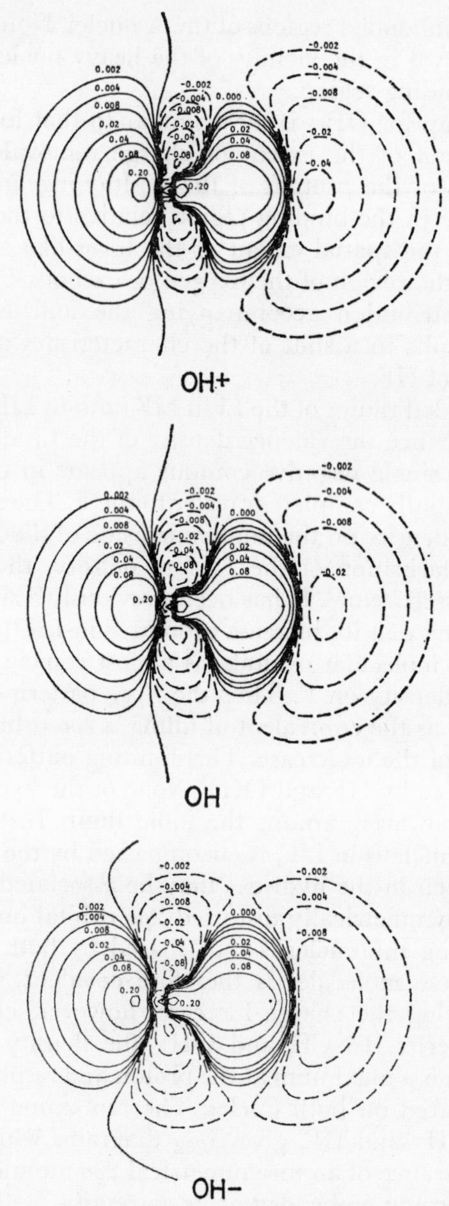

FIGURE 3. Contour maps of the density difference distributions $\Delta\rho_{\mathrm{SA}}$ (molecular minus atomic) for diatomic hydrides and LiF in their ground electronic states. The atomic densities of the A nuclei used in the construction of these maps correspond to a configuration which places a single electron in their $2p\sigma$ orbital.

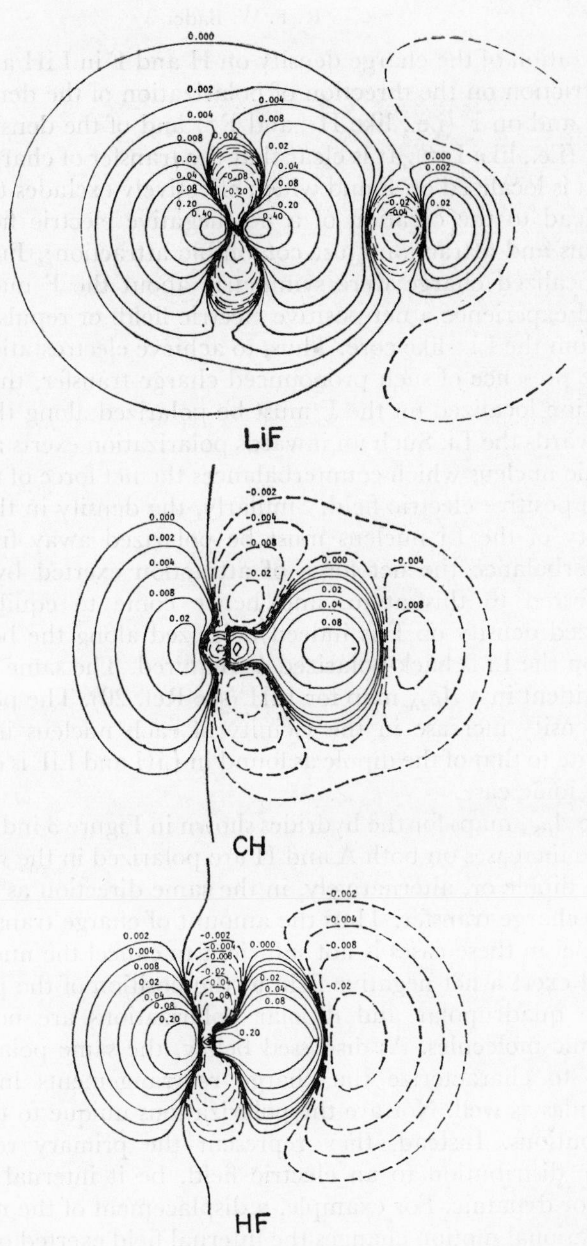

Figure 3 continued

The remaining $2p$ electrons are averaged over the $2p\pi$ orbitals. The proton is on the right-hand side in these maps.

localization of the charge density on H and F in LiH and LiF places a restriction on the direction of polarization of the density localized on H and on F (i.e., like H^- and F^-) and of the density remaining on Li (i.e., like Li^+). It is clear that the transfer of charge to a region which is localized on F and which effectively excludes the Li nucleus will lead to the creation of a net negative electric field at the Li nucleus and attraction (just coulombic attraction). Furthermore, if the localized charge were symmetric about the F nucleus, then it would experience a net positive electric field, or repulsion, originating from the Li^+-like core. Thus, to achieve electrostatic equilibrium in the presence of such pronounced charge transfer, the density distribution localized on the F must be polarized along the bond, that is, towards the Li. Such an inwards polarization exerts a force on the anionic nucleus which counterbalances the net force of repulsion due to the positive electric field. Similarly, the density in the immediate vicinity of the Li nucleus must be polarized away from the F to counterbalance the net force of attraction exerted by the density transferred to this atom, and hence come to equilibrium. The localized density on F is indeed polarized along the bond axis and that on the Li is back-polarized as required. The same polarizations are evident in a $\Delta\rho_{SA}$ map for LiH (see Ref. 20). The polarization of the density increase in the vicinity of each nucleus in a direction opposite to that of the dipole as found in LiH and LiF is characteristic of the ionic case.

The $\Delta\rho_{SA}$ maps for the hydrides shown in Figure 3 indicate that the charge increases on both A and H are polarized in the *same* direction as the dipole or, alternatively, in the same direction as the direction of the charge transfer. Thus the amount of charge transferred to the A nuclei in these cases is not sufficient to cancel the nuclear field on A and exert a net negative field at the position of the proton.

The quadrupolar and dipolar polarizations are not unique to diatomic molecules. As discussed below, the same polarizations are found to characterize the charge rearrangements in polyatomic molecules as well. Nor are the polarizations unique to the $\Delta\rho_{SA}(x, y)$ distributions. Instead, they represent the primary response of a charge distribution to an electric field, be it internal or external, static or dynamic. For example, a displacement of the nuclei during a vibrational motion changes the internal field exerted on the charge distribution causing it to change or 'relax'. Figure 4 illustrates that a bond extension of the O–H radical *diminishes* both the quadrupole polarization of the charge density in the vicinity of the O nucleus and the dipolar polarization in the region of the proton. A corresponding density difference map for a bond contraction is the same as that shown in Figure 4 with all signs reversed. Thus, as the bond is contracted the polarizations are *enhanced* and charge density is removed from the belt-like region perpendicular to the bond axis and concentrated along the axis.

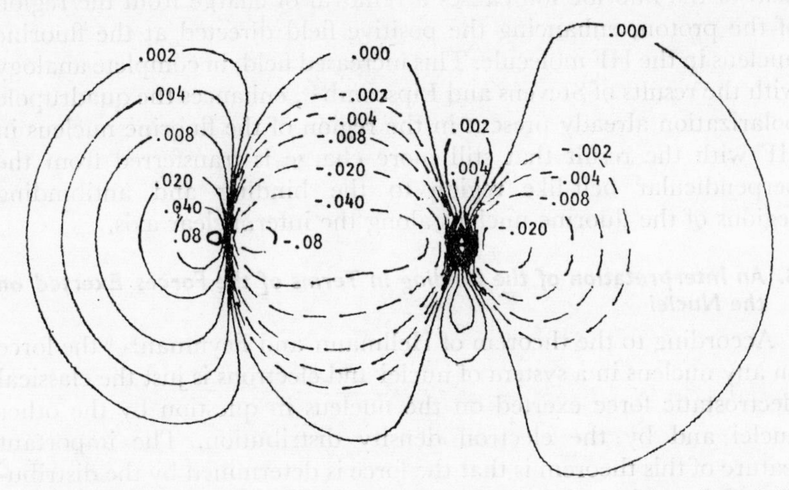

FIGURE 4. A density difference map showing the change or relaxation of the charge density in the OH molecule when the bond length is increased by 0·166 au. The oxygen nucleus is held stationary. For a bond contraction the algebraic signs of the contours are reversed.

Stevens and Lipscomb[25] have presented density difference maps which illustrate the change in a molecular charge density caused by an externally applied field. The density difference maps given by these authors

$$\Delta\rho(x, y) = \rho(\text{molecule in field}) - \rho(\text{unperturbed molecule})$$

illustrate the same polarizations of the charge density as are found in the $\Delta\rho_{SA}(x, y)$ maps. For example, the results of Stevens and Lipscomb for the hydrogen fluoride molecule show that when the direction of the positive field is from the proton to the fluorine, charge is removed from a torus-like region perpendicular to the axis at the position of the F nucleus and transferred to both the binding and antibinding regions of the F nucleus along the internuclear axis. Thus the polarizations evident in the $\Delta\rho_{SA}(x, y)$ maps, which show the response of the atomic charge densities to a field resulting from the close approach of nuclei, are identical in form with the response of a system to an externally applied field.

The same polarizations are also evident in the approach of two molecules in a chemical reaction. For example, the change in the charge distributions of the HF molecule and the F^- ion as they approach one another to form the $(FHF)^-$ ion[26] indicate that the field of the fluoride ion causes a removal of charge from the region of the proton, enhancing the positive field directed at the fluorine nucleus in the HF molecule. This increased field, in complete analogy with the results of Stevens and Lipscomb[25], enhances the quadrupole polarization already present in the region of the fluorine nucleus in HF with the result that still more charge is transferred from the perpendicular belt-like region to the binding and antibinding regions of the fluorine nucleus along the internuclear axis.

B. *An Interpretation of the Binding in Terms of the Forces Exerted on the Nuclei*

According to the theorem of Hellmann and Feynman[8, 9] the force on any nucleus in a system of nuclei and electrons is just the classical electrostatic force exerted on the nucleus in question by the other nuclei and by the electron density distribution. The important feature of this theorem is that the force is determined by the distribution of charge in real three-dimensional space, an observable property of the system. It is for this reason that a discussion of the binding in a molecule in terms of the forces exerted on the nuclei may be given a classical interpretation.

The theorem itself is easily derived. The X-component of the force on nucleus A in a system with fixed nuclei is given by

$$F_{X_A} = \partial E/\partial X_A = (\partial/\partial X_A)\langle\psi^* |H| \psi\rangle$$
$$= \langle\psi^* | \partial H/\partial X_A | \psi\rangle + \langle\partial\psi^*/\partial X_A | H|\psi\rangle + \langle\psi^* |H| \partial\psi/\partial X_A\rangle \quad \text{(II-1)}$$

For the exact wavefunction

$$H\psi = E\psi$$

and the two final terms on the right-hand side of equation (II-1) may be reduced to

$$E\frac{\partial}{\partial X_A}\int \psi^*\psi \, d\tau = 0$$

Hence

$$F_{X_A} = \langle\psi \,|\partial H/\partial X_A|\, \psi\rangle \quad \text{(II-2)}$$

When ψ and the electronic coordinates are expressed in terms of a space-fixed coordinate system a further simplification occurs. Under these conditions the only terms in the Hamiltonian operator H which depend on the nuclear coordinates are the internuclear separations R_{AB} and the nuclear-electron attraction terms. For example in a diatomic molecule AB

$$\partial H/\partial X_A = \frac{\partial}{\partial X_A}\left(\frac{Z_A Z_B}{R_{AB}} - Z_A \sum_i \frac{1}{r_{iA}}\right) = \frac{Z_A Z_B}{R_{AB}^{\ 2}} - Z_A \sum_i \frac{\cos \theta_{Ai}}{r_{iA}^{\ 2}}$$

where θ_{Ai} and r_{Ai} are polar coordinates centred on nucleus A defining the position of electron i. The only term involving the electronic coordinates, the last term, is a one-electron operator, and thus a knowledge of the full N-electron probability distribution as given by

$$\psi^* \psi \, dx_1 \, dx_2 \ldots dx_N$$

is unnecessary. Instead it is necessary to have only the probability distribution for a single electron multiplied by N, i.e., the molecular charge density

$$\rho(x) = N \int \psi^* \psi \, dx_2 \ldots dx_N \qquad \text{(I-3)}$$

Thus

$$F_{X_A} = \frac{Z_A Z_B}{R_{AB}^{\ 2}} - Z_A \int \frac{\cos \theta_A}{r_A^{\ 2}} \rho(x) \, dx \qquad \text{(II-3)}$$

The Hellmann–Feynman theorem holds for the exact wavefunction and a certain class of approximate functions (those which have been fully optimized with respect to the nuclear coordinates) which includes the Hartree–Fock function[27]. In the Hartree–Fock case the electron density or charge distribution assumes a particularly simple form. Since the molecular orbitals form an orthogonal set of functions

$$\rho(x_\mu) = \sum_i N_i \phi_i^*(x_\mu) \phi_i(x_\mu)$$

where N_i is the occupation number of the molecular orbital $\phi_i(x_\mu)$. The total electronic contribution to the force can therefore be equated to a sum of orbital contributions. For interpretative purposes it is convenient to go one step further and rewrite equation (II-3) as

$$F_{XA} = (Z_A/R^2)\,[Z_B - \sum_i f_{iA}] \qquad \text{(II-4)}$$

where f_{iA} is the force exerted on nucleus A by the charge density in the ith molecular orbital multiplied by R^2

$$f_{iA} = R^2 N_i \int \phi_i^*(x_\mu) \frac{\cos \theta_{\mu A}}{r_{\mu A}^{\ 2}} \phi_i(x_\mu) \, dx \qquad \text{(II-5)}$$

The f_{iA} may be either attractive or repulsive and thus their values can be used as a quantitative gauge of the binding or antibinding characteristics of the ith molecular orbital using a significant reference standard. The reference standard is based on the contributions to the force on A as $R \to \infty$, where $F_A = 0$; that is, the reference state is that of the component separated atoms. Clearly, at large R the unperturbed atom A possesses a centre of symmetry and exerts a zero net force on nucleus A. One may interpret the vanishing of the force at large R as resulting from each electron on B screening one of the nuclear charges on B from nucleus A. Thus the limiting value at $R \to \infty$ of the sum of the f_{iA} values for the force on nucleus A is the total electronic charge on atom B and

$$\sum_i f_{iA}{}^{(\infty)} = \sum_l N_l = Z_B \qquad \text{(II-6)}$$

where the sum over l refers to a sum over the *atomic* orbitals on B.

The f_{iA} have the dimensions of electronic charge. Each f_{iA} is numerically equal to the number of point charges which, when placed at the B nucleus, exert the same field at the A nucleus as does the density in the ith molecular orbital. The electronic contribution to the force on the A nucleus at any value of R may, therefore, be equated to an effective number of charges situated at the B nucleus, this number being the sum of the partial forces. At R_e the system is in electrostatic equilibrium, $F_A = 0$ and again one obtains the condition

$$\sum_i f_{iA}(R_e) = Z_B$$

At intermediate internuclear distances the sum of the effective charges exceeds Z_B, corresponding to a net force of attraction, and for large values of R it reduces to the number of electronic charges which correlate with the separated B nucleus, e.g. equation (II-6). This suggests that the limiting value of each individual f_{iA} should be taken as the number of electrons in the ith molecular orbital which correlate with the B atom for large values of R, N_{iB}

$$f_{iA}(R \to \infty) = N_{iB} \quad (= 0, 1 \text{ or } 2)$$

The partial forces provide an absolute measure of the binding ability of an orbital density in terms of the number of point charges at the B nucleus which produce a field at A equivalent to that exerted by the actual density distribution. A measure of the binding ability of a molecular-orbital charge distribution relative to the separated atoms as the reference standard is given by a comparison of the value of $f_{iA}(R_e)$ with N_{iB}. This compares the charge equivalent (in terms of a number of charges on B) of the electric field exerted by a pair of electrons in the molecule with the charge equivalent of the field exerted by the ancestral pair of electrons in the separated atoms. This latter number is simply the number of electrons which correlate with B, since the electrons which correlate with A exert no field at the A nucleus as $R \to \infty$. In general $f_{iA}(R_e)$ may be greater than, equal to or less than N_{iB} leading to the three definitions listed below:

$$
\begin{aligned}
f_{iA}(R_e) &> N_{iB} \quad && \text{binding MO} \\
f_{iA}(R_e) &\sim N_{iB} \quad && \text{nonbinding MO} \\
f_{iA}(R_e) &< N_{iB} \quad && \text{antibinding MO.}
\end{aligned}
$$

To allow for a more detailed understanding of the variations in the f_{iA} or f_{iH} values, each is expressed in terms of the separate contributions which arise from the atomic populations on A and H and the overlap population. These separate contributions to the f_is are easily determined since the basis set in the expansions of the present wavefunctions consists of Slater-type atomic functions centred on both A and H. Thus equation (II-5) is written as

$$f_{iA} = [f_{iA}^{(AA)} + f_{iA}^{(AH)} + f_{iA}^{(HH)}] \qquad \text{(II-7)}$$

for the A nucleus in A–H, and as

$$f_{iH} = [f_{iH}^{(HH)} + f_{iH}^{(AH)} + f_{iH}^{(AA)}] \qquad \text{(II-8)}$$

for the proton. For example, $f_{iA}^{(AA)}$ ($\equiv$ atomic force) denotes the contribution to the partial force on nucleus A from the atomic charge population on A; $f_{iA}^{(AH)}$ ($\equiv$ overlap force) is the corresponding contribution from the overlap charge density, and $f_{iA}^{(HH)}$ ($\equiv$ screening force) is the contribution to the partial force on the A nucleus from the atomic charge density centred on the proton. The screening force is a measure of the electronic shielding of the proton from the nucleus A by the electrons situated on H. The screening force provides the sole contribution to the f_i values for large values of R, i.e.,

$$f_{iA}(\infty) = f_{iA}^{(HH)}(\infty) = N_{iH} \quad \text{and} \quad f_{iH}(\infty) = f_{iH}^{(AA)}(\infty) = N_{iA}$$

The atomic, overlap, and screening contributions to the partial forces provide more information than do the population figures themselves. As important as the amount of charge in determining the binding in a molecule is the exact disposition of the charge, its polarization and whether it is diffuse or concentrated. There are certain limiting cases for which the screening contribution to a

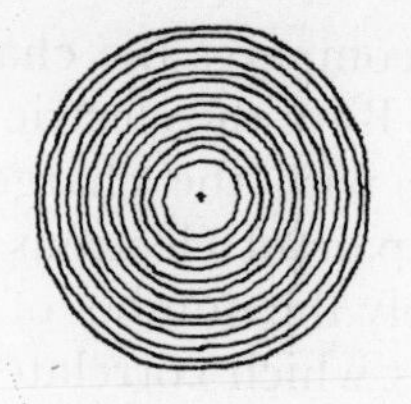

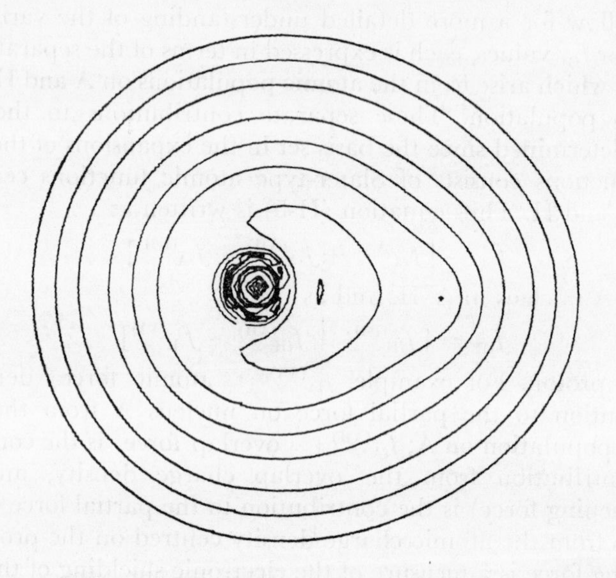

FIGURE 5. Contour plots of the molecular orbital charge densities of the OH^- ion. The values of the contours are obtained by numbering them consecutively starting with the outer contour and using the key given in Table II-7. There is a near circular node encompassing the closely spaced contours centred on the oxygen nucleus in the 2σ orbital density. The 3σ and 1π densities possess nodes which are nearly perpendicular to and along the OH bond axis respectively.

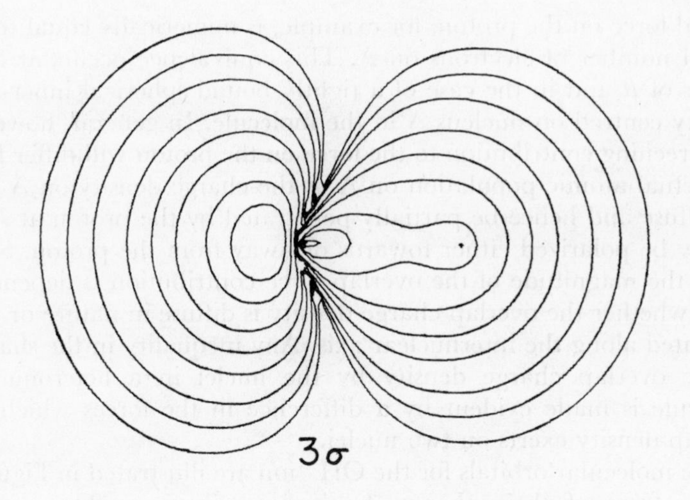

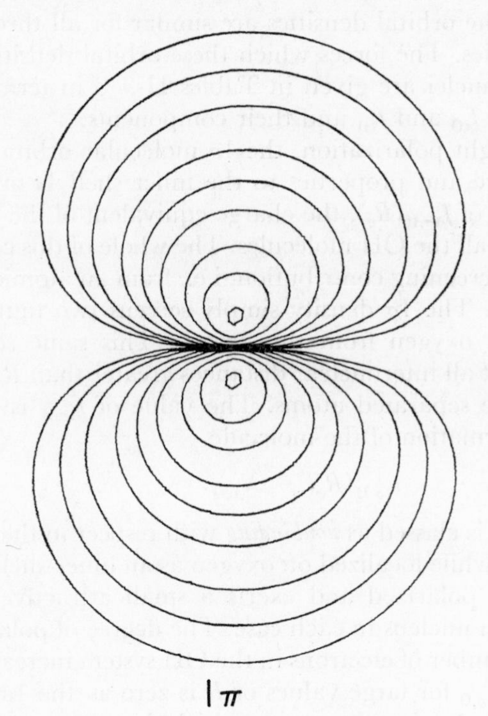

1π

Figure 5 continued

partial force on the proton, for example, is numerically equal to the actual number of electrons on A. This equivalence occurs at large values of R and in the case of a tightly bound spherical inner-shell density centred on nucleus A in the molecule. In general, however, the screening contribution to the force on the proton will differ from the actual atomic population on A as the charge density on A may be diffuse and hence be partially penetrated by the proton at R_e or it may be polarized either towards or away from the proton. Similarly, the magnitude of the overlap force contribution is dependent upon whether the overlap charge density is diffuse in nature or concentrated along the internuclear axis. Any inequality in the sharing of the overlap charge density by the nuclei in a heteronuclear molecule is made evident by a difference in the forces which the overlap density exerts on two nuclei.

The molecular orbitals for the OH^- ion are illustrated in Figure 5 in the form of their charge density contributions. The general characteristics of the orbital densities are similar for all three of the OH diatomic species. The forces which these orbital densities exert on the O and H nuclei are given in Tables II-3–6 in terms of the charge equivalents $f_{i\mathrm{O}}$ and $f_{i\mathrm{H}}$ and their components.

Aside from a slight polarization, the 1σ molecular orbital is very close in appearance and properties to the inner shell $1s$ orbital on oxygen. The value of $f_{1\sigma,\mathrm{H}}(R_e)$, the charge equivalent of the force on the proton, is 2 for all the OH molecules. The whole of this contribution arises from a screening contribution, i.e., from an atomic charge density on oxygen. The 1σ density simply screens two units of the nuclear charge on oxygen from the proton. This same screening effect is obtained at all internuclear distances greater than R_e including the case of the separated atoms. The value of $f_{1\sigma,\mathrm{H}}$ is left unchanged by the formation of the molecule,

$$f_{1\sigma,\mathrm{H}}(R_e) = N_{1s\mathrm{O}}$$

and the 1σ density is classed as *nonbinding* with respect to the proton.

The 1σ density, while localized on oxygen as an inner-shell atomic density, is slightly polarized and exerts a small attractive atomic force on the oxygen nucleus in each case. The degree of polarization decreases as the number of electrons in the OH system increases. The limiting value of $f_{1\sigma,\mathrm{O}}$ for large values of R is zero as the 1σ density correlates with a $1s$ density on oxygen, which does not exert a force on the oxygen nucleus. The 1σ charge density is therefore, slightly *binding* for the oxygen nuclei.

TABLE II-3. Forces exerted on the nuclei by the 1σ density.

	Forces on the proton					Forces on the O nuclei			
	$f_{1\sigma,H}$	Atomic	Overlap	Screening		$f_{1\sigma,O}$	Atomic	Overlap	Screening
OH+	2·001	0·000	0·000	2·001	OH+	0·291	0·290	0·001	0·000
OH·	2·000	0·000	0·000	2·000	OH·	0·247	0·245	0·002	0·000
OH−	2·000	0·000	0·000	2·000	OH−	0·214	0·213	0·001	0·000

TABLE II-4. Forces exerted on the nuclei by the 2σ density.

	Forces on the proton					Forces on the O nuclei			
	$f_{2\sigma,H}$	Atomic	Overlap	Screening		$f_{2\sigma,O}$	Atomic	Overlap	Screening
OH+	2·428	0·026	0·222	2·180	OH+	0·967	0·825	0·131	0·011
OH·	2·317	0·040	0·383	1·894	OH·	0·905	0·581	0·292	0·032
OH−	2·214	0·047	0·423	1·744	OH−	0·862	0·481	0·339	0·043

TABLE II-5. Forces exerted on the nuclei by the 3σ density.

	Forces on the proton					Forces on the O nuclei			
	$f_{3\sigma,H}$	Atomic	Overlap	Screening		$f_{3\sigma,O}$	Atomic	Overlap	Screening
OH+	2·094	0·099	0·578	1·417	OH+	−0·377	−1·341	0·832	0·132
OH·	1·684	0·068	0·475	1·141	OH·	−0·344	−1·430	0·915	0·169
OH−	1·385	0·066	0·380	0·939	OH−	−0·305	−1·473	0·959	0·209

TABLE II-6. Forces exerted on the nuclei by the 1π density.

	Forces on the proton					Forces on the O nuclei			
	$f_{1\pi,H}$	Atomic	Overlap	Screening		$f_{1\pi,O}$	Atomic	Overlap	Screening
OH+	1·558	0·001	0·024	1·533	OH+	0·166	0·141	0·024	0·001
OH·	2·084	0·001	0·044	2·039	OH·	0·226	0·180	0·045	0·001
OH−	2·413	0·001	0·058	2·354	OH−	0·242	0·177	0·063	0·002

The ancestral relationship of the 2σ molecular orbital to a $2s$ atomic orbital on oxygen is readily discernible in the form of its charge density contours. The lack of contours encircling only the proton indicates that the distribution of the 2σ charge density is determined primarily by the field of the oxygen nucleus. The density is however, strongly perturbed by the proton and charge density is accumulated in the region between the nuclei. The localization of the 2σ charge density in the region of the oxygen nucleus decreases through the series in the order OH^+, $OH^{\cdot}$, OH^-.

The 2σ molecular orbital correlates with the doubly occupied $2s$ orbital on oxygen. For large internuclear separations the correlated $2s$ density will exert no force on the oxygen nucleus and a screening force on the proton equivalent to that of two negative charges: $f_{2\sigma,O}(\infty) = 0$ and $f_{2\sigma,H}(\infty) = 2$. Thus when compared to the separated atoms, the 2σ charge density is *binding* for both the proton and the oxygen in all three molecules. The binding of the proton, which is measured by the amount by which $f_{2\sigma,H}$ exceeds 2, is primarily the result of the force exerted by the overlap density in OH^- and $OH^{\cdot}$. The decrease in the overlap contribution and the increase in the screening contributions to $f_{2\sigma,H}$ through the series from OH^- to OH^+ indicate that the 2σ charge density becomes increasingly contracted towards the oxygen as the total number of electrons in the system decreases. The 2σ atomic population on oxygen in OH^+ which is necessarily less than 2, is strongly polarized towards the proton with the result that the force which it exerts on the proton is equivalent to placing $\sim 2{\cdot}2$ electronic charges at the position of the oxygen nucleus.

The binding of the oxygen nucleus by the 2σ density, like that of the proton, decreases through the series from OH^+ to OH^-. The binding of the oxygen nucleus in OH^+ is primarily the result of the atomic population on oxygen being polarized towards the proton, while in OH^- the atomic and overlap contributions are almost equally important.

The 3σ orbital density resembles a $2p\sigma$ atomic orbital on oxygen with the lobe on the bonded side of the nucleus strongly contracted along the internuclear axis. However, in the immediate vicinity of the oxygen nucleus there is a larger amount of charge density accumulated in the nonbonded than in the bonded lobe. The 3σ orbital densities for $OH^{\cdot}$ and OH^+ are similar to that for OH^-, but, as for the 2σ density, become progressively more contracted towards the oxygen nucleus as the total number of electrons in the system decreases.

The 3σ orbital correlates with the singly occupied H $1s$ and O $2p\sigma$ atomic orbitals. In the limit of large internuclear distances the values of both $f_{3\sigma,\mathrm{H}}$ and $f_{3\sigma,\mathrm{O}}$ approach unity as the correlated atomic densities screen one nuclear charge on each nucleus. The values of $f_{3\sigma,\mathrm{H}}(R_e)$ indicate that the 3σ density is *binding* with respect to the proton and becomes progressively more so in the order OH^-, $OH^{\bullet}$, OH^+. For example, in OH^+ the force exerted on the proton by the 3σ density is equivalent to placing $\sim 2{\cdot}1$ electronic charges at the oxygen nucleus as opposed to the separated atom equivalent of one electronic charge. The number of charges which are effective in binding the proton is doubled in the formation of the 3σ orbital in OH^+. In both OH^+ and $OH^{\bullet}$ the proton is bound primarily by the 3σ charge density, while in OH^- the 2σ and 3σ charge densities are comparable in this respect.

The 3σ charge density exerts an *antibinding* force on the oxygen nucleus in spite of a large overlap contribution because of an even larger negative atomic force term. The negative values for $f_{3\sigma,\mathrm{O}}$ indicate that the 3σ density is antibinding in the absolute sense as it exerts a force which tends to pull the oxygen nucleus away from the proton. This pattern of overlap and atomic force contributions is characteristic of any orbital charge density which involves a significant $p\sigma$ component. It is the increase in the 3σ density on the oxygen and its extreme back-polarized form which are responsible for the characteristic pattern of the $\Delta\rho_{\mathrm{SA}}$ maps and for the increase in non-bonded charge densities of the oxygen atoms in the OH species.

The screening of the proton by the 3σ charge density is uniformly low throughout the series reflecting the relative localization of the 3σ charge density on the oxygen nuclei.

The molecules OH^+ to OH^- possess two to four π electrons, respectively. The 1π orbital correlates with the $2p\pi$ orbitals on the oxygen, and it is evident from Figure 5 that the 1π molecular orbital retains its basic atomic orbital character. The 1π orbital density is in each case centred on the oxygen with contours characteristic of a $2p\pi$ atomic density slightly polarized in the direction of the proton.

The 1π density screens two to four nuclear charges on oxygen from the proton in the separated atom case. Thus

$$f_{1\pi,\mathrm{H}}(\infty) = N_{p\pi}$$
$$f_{1\pi,\mathrm{O}}(\infty) = 0$$

The values of $f_{1\pi,\mathrm{H}}(R_e)$ listed in Table II-6 are less than the orbital occupation number in each case. The 1π density is, therefore, *antibinding* with respect to the proton in the relative sense that in the molecule it does not screen an equivalent number of nuclear charges on the oxygen. This antibinding effect is a direct consequence of the π density being concentrated around the internuclear axis, rather than along it (where it has a node). The small value of the overlap and atomic force contributions to $f_{1\pi,\mathrm{H}}$ illustrate that no significant π bond is present in these molecules and the 1π molecular densities are best described as inwardly polarized atomic densities on the oxygen nuclei.

The binding–antibinding properties exhibited by the molecular orbitals in the OH species are characteristic of Hartree–Fock orbitals regardless of the system in which they are found, if they have either a common correlated atomic orbital or a common major orbital component they exhibit similar binding properties. For example, a Hartree–Fock orbital which correlates with a $2s$ atomic orbital on the most electronegative atom in a molecule [the 2σ orbital in the hydrides AH ($\mathrm{A} = \mathrm{B} \rightarrow \mathrm{F}$), the $2\sigma_g$ orbital in homonuclear diatomics, or the 3σ orbital in BeO, BF, CO or LiF] is always *binding* for both nuclei. The molecular orbital density in the region of the nucleus with which it correlates is *polarized into the bond* and exerts an attractive force on this nucleus. In addition, the overlap charge density exerts almost equal forces on both nuclei in the heteropolar examples. These are the binding characteristics of an orbital which correlates with $2s$ atomic orbital on the most electronegative atom in the molecule whether the bond is covalent, polar or ionic.

Similarly a Hartree–Fock orbital which exhibits (or correlates partially with) a $2p\sigma$ component on a given nucleus (the 3σ orbital in AH, $\mathrm{A}=\mathrm{C} \rightarrow \mathrm{F}$, the $3\sigma_g$ in homonuclear diatomics or the 4 orbital in BF, CO, LiF and BeO) is strongly polarized into the antibinding region of that nucleus and exerts an antibinding force on it. It is the polarization associated with such an orbital which is responsible for the charge increase in the lone pair or nonbonded region of the charge distribution. The forces exerted by the overlap charge density in these same orbitals are, for the heteronuclear cases, approximately twice as large for the nucleus on which the $2p\sigma$ component is centred as they are for the second nucleus.

The Hartree–Fock π molecular orbitals, whether they are delocalized as in CO or strongly localized on a single nucleus in OH, are inwardly polarized and exert nearly equal overlap forces on both nuclei.

TABLE II-7. Key to density and density difference maps.

Density maps		Density difference maps	
Contour No.	Value of contour (in au)	Contour No.	Value of contour (in au)
1	0·002	1	0·000
2	0·004	2	−0·002
3	0·008	3	0·002
4	0·02	4	−0·004
5	0·04	5	0·004
6	0·08	6	−0·008
7	0·20	7	0·008
8	0·40	8	−0·02
9	0·80	9	0·02
0	2·00	0	−0·04
1	4·00	1	0·04
2	8·00	2	−0·08
3	20·00	3	0·08
4	40·00	4	−0·20
		5	0·20
		6	−0·40
		7	0·40
		8	−0·80
		9	0·80

III. THE MOLECULAR CHARGE DISTRIBUTIONS OF CH_3OH AND CH_2FOH

This section presents a discussion of the total molecular charge distributions and their orbital components for the polyatomic systems methanol and fluoromethanol. The wavefunctions for these polyatomic systems, which are close to Hartree–Fock accuracy, were obtained by Csizmadia, Tel and Wolfe[28] using a basis set of Gaussian-type atomic orbitals (GTO's) centred on the nuclei. The basis set consisted of fifty-six orbitals for methanol and seventy-two for fluoromethanol. Contracted basis functions were formed by taking linear combinations of the GTO's on the various centres as suggested by Huzinaga[29]. The twenty GTO's centred on oxygen, for example, were combined to give ten contracted basis functions. The energy minimization in the SCF calculation is obtained by varying only the linear coefficients of the contracted sets of basis orbitals, the composition of each contracted set remaining fixed. The use of contracted sets of orbitals makes feasible (in terms of computing time)

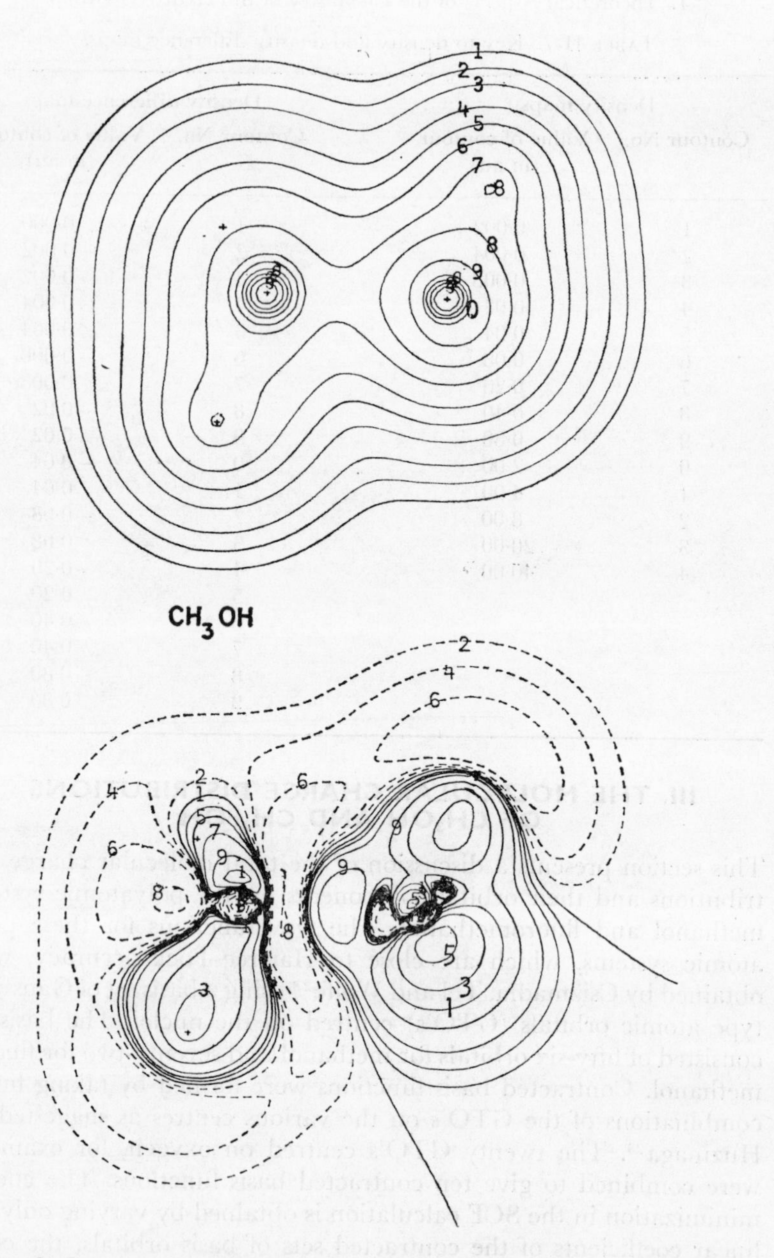

FIGURE 6. Contour maps of total molecular charge distribution in methanol in the staggered conformation. The plot on the left is in the plane of the HCOH nuclei and that on the right is in the plane perpendicular to this. Beneath each

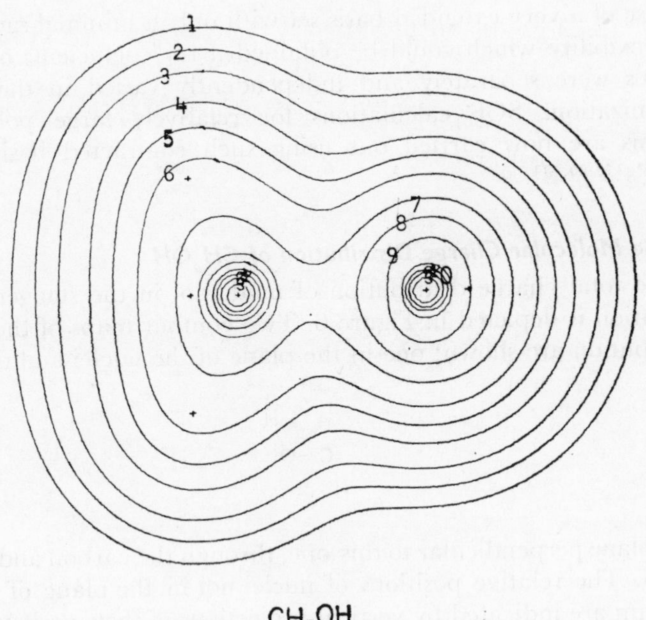

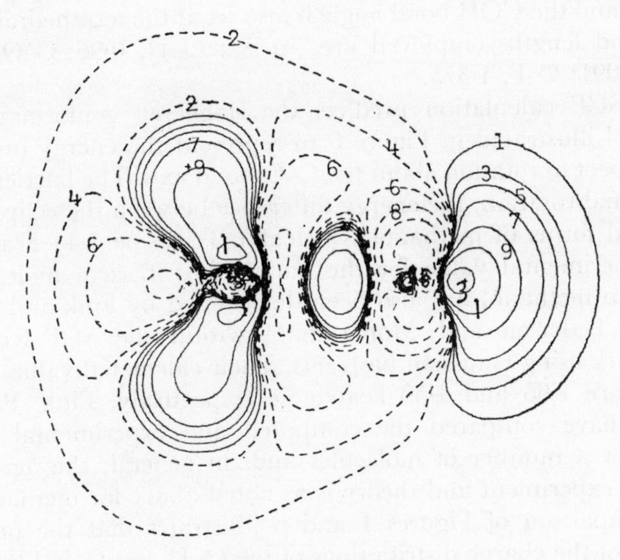

Figure 6 continued

total density plot is the corresponding $\Delta\rho_{SA}$ map. The key for the values of the contours of the total and density difference maps is given in Table II-7.

the use of a very extended basis set with only a minimal sacrifice in the flexibility which could be obtained if the coefficients of all the GTO's were separately and independently varied in the energy minimization. SCF calculations for relatively large polyatomic systems are now carried out using such contracted basis sets of GTO's[17, 30, 31].

A. The Molecular Charge Distribution of CH_3OH

The total charge distribution of methanol in the staggered conformation is depicted in Figure 6. Two contour maps of the charge distribution are shown; one in the plane of the nuclei and the other

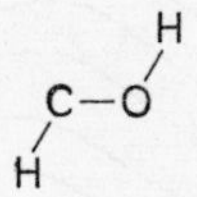

H
C—O
H

in a plane perpendicular to this one, through the carbon and oxygen nuclei. The relative positions of nuclei not in the plane of a given diagram are indicated by vertical projections of their positions on to the plane in question. A tetrahedral geometry is assumed about the carbon, and the COH bond angle is also set at the tetrahedral value. The bond lengths employed are (in Å); O–H, 0·96; C–O, 1·428; C–H, 1·091; C–F, 1·375.

The SCF calculation predicts the staggered conformation for methanol illustrated in Figure 6 to represent the energy minimum with respect to rotation about the C–O bond axis. The barrier height for internal rotation, the energy difference between the eclipsed and staggered forms of methanol is calculated[28] to be 1·44 kcal/mole. The experimental value for the barrier is 1·07 kcal/mole[32]. The barrier in methanol has also been determined by Fink and Allen[33] and by Pedersen and Morokuma[34] within the SCF–Roothaan framework using Gaussian basis sets. Their calculated values for the barrier are 1·06 and 1·59 kcal/mole respectively. Fink, Pan and Allen[35] have compared the computed and experimental barrier values for a number of molecules and, in general, the agreement between experiment and theory is as noted above for methanol.

A comparison of Figures 1 and 6 illustrates that the principal features of the charge distributions of the O–H and C–H bond fragments in methanol are remarkably similar to the molecular charge distributions of the corresponding diatomic species OH($^2\Pi$) and

$CH(^2\Pi)$. The nonbonded radii on hydrogen and oxygen in methanol are identical with the values found in the OH molecule while the nonbonded radii of carbon and of the hydrogen in the CH bond differ from those of the CH molecule by only 0·1 au, the nonbonded charge density being slightly more contracted in the molecular fragment than in the diatomic molecule. The complete outer envelopes of the charge densities of the fragments in methanol are similar in all respects to those for the diatomic species. Thus the shape and size of the charge distribution in methanol can be predicted from the appropriate bond lengths and bond angles together with the nonbonded radii and general shapes of the charge distributions of the CH and OH diatomic species.

The nonbonded charge density on oxygen exhibits a pronounced polarization whose direction undergoes a continuous change in each of the planes obtained by a rotation about the C–O bond axis. In the plane of the H–C–O–H nuclei the nonbonded charge density on oxygen is concentrated along an axis which bisects the COH bond angle. In the plane perpendicular to this, the charge density centred on oxygen is concentrated along a line perpendicular to the C–O bond axis.

The polarization in the plane containing the four nuclei is particularly evident in the density difference map (molecular density minus the overlapped atomic distributions) also shown in Figure 6. Because of its tetrahedral environment the atomic density of carbon has been sphericalized in the construction of this $\Delta\rho_{SA}$ map. The atomic density on oxygen corresponds to the configuration $1s^2\ 2s^2\ 2p_x\ 2p_y^2\ 2p_z$ where the y-axis is perpendicular to the plane containing the four nuclei. This results in spherical contours for the oxygen atom charge distribution in this plane and corresponds to the valence bond description of the two unpaired electrons on oxygen forming single bonds with the carbon and hydrogen atoms. The $\Delta\rho_{SA}$ map indicates that the charge distribution of the oxygen atom, which is initially spherical in this plane, is strongly polarized along the line which bisects the COH bond angle, into both its bonded and nonbonded regions, but particularly into the latter. The charge density is not accumulated directly along either the O–H or O–C bond. There is a region of charge removal at the oxygen which is perpendicular to the principal line of polarization. Thus the $\Delta\rho_{SA}(X)$ map exhibits a quadrupolar polarization in the region of the oxygen similar to that found in the diatomic molecules. In the COH system, however, the polarization is with respect to an axis which bisects the directions of the C–O and O–H bond directions. The same quadrupolar polarization is found in a $\Delta\rho_{SA}(X)$ map for the water molecule in the plane of the nuclei[36]. In this case the polarization is directed along the C_2 symmetry axis which bisects the HOH bond angle and the charge removal occurs from a belt-like region perpendicular to this axis at the position of the oxygen nucleus. The quadrupolar polarization also persists in the perpendicular plane of the methanol system.

The charge density in the region of the carbon also exhibits a quadrupolar polarization. In this case the charge accumulation is understandably concentrated in a belt-like region perpendicular to the C–O bond axis to encompass the protons while the region of charge removal occurs along the axis, to the extent of causing a partial depletion of the atomic densities between the carbon and oxygen nuclei. A comparison of the $\Delta\rho_{SA}(X)$ map for methanol with that for the CO molecule[23] indicates that even the sigma bond charge density (which is the only density to contribute to the charge density on the C–O axis) is greatly reduced in the polyatomic system. In contrast to this, the $\Delta\rho_{SA}(X)$ map indicates that the extent of charge accumulation between the C and H and the O and H nuclei in methanol is very similar to that found in the density difference maps for the corresponding diatomic species.

The molecular orbital charge densities for methanol in the staggered conformation are illustrated in Figure 7. They are numbered in order of increasing energy. This particular configuration possesses a plane of symmetry (the one containing the H–C–O–H nuclei) and every molecular orbital must be either symmetric or antisymmetric with respect to it. For this reason density contour maps of the antisymmetric orbitals, numbers six and nine (which have a node in the plane of symmetry), are illustrated in the plane perpendicular to the symmetry plane.

The molecular orbitals numbers one and two are 1*s* atomic-like orbitals centred on the oxygen and carbon nuclei respectively and hence are similar to the 1σ orbitals in CH or OH or to the 1σ and 2σ orbitals in CO. The major component of molecular orbital number three is from the 2*s* orbital on oxygen. Thus it strongly resembles the 2σ orbital in the diatomic hydrides (see Figure 5) or the 3σ orbital in CO. In all three cases the orbital density is strongly polarized towards the nuclei bonded to the oxygen as indicated by the contour with the shape of a half-moon. A similar pattern of contours appears again in molecular orbital number four, this time localized on the

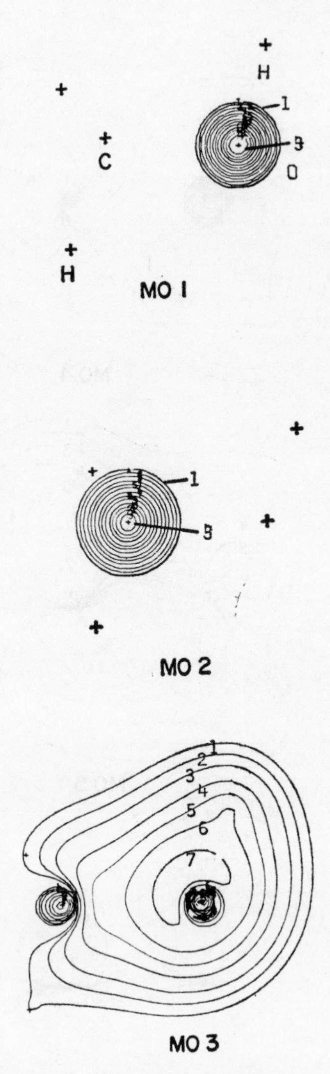

FIGURE 7. Contour plots of the molecular orbital charge densities for methanol in the staggered configuration. The maps for orbitals six and nine (overleaf) are shown in a plane perpendicular to plane containing the HCOH nuclei.

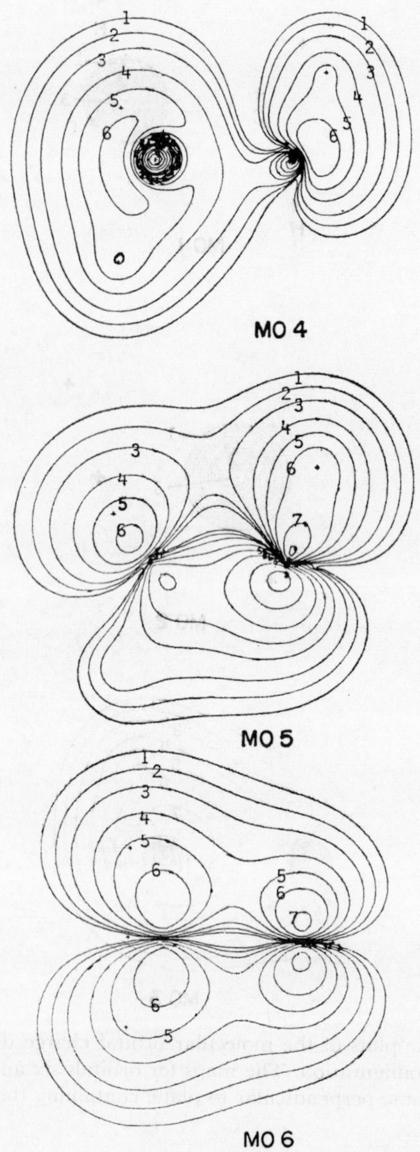

Figure 7 continued

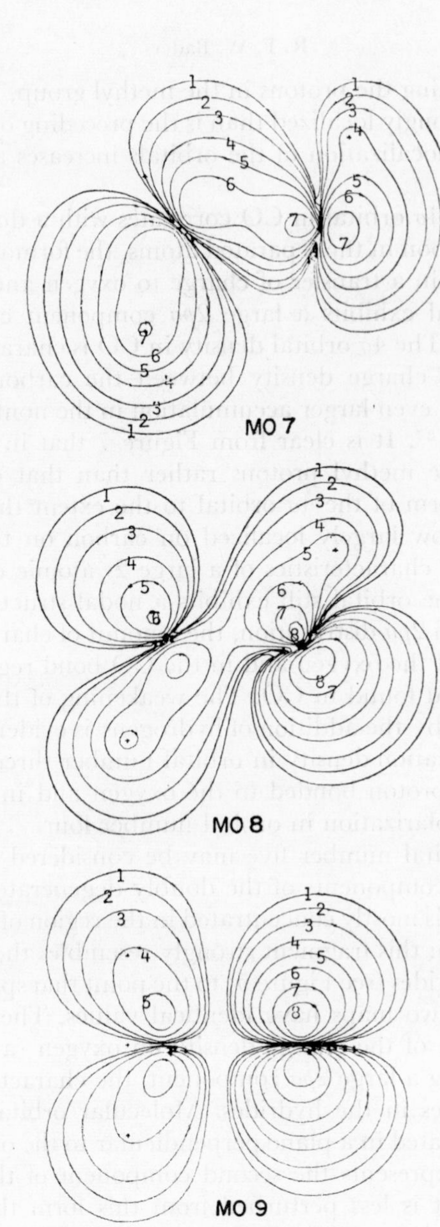

Figure 7 continued

carbon and binding the protons in the methyl group. This orbital is, however, less strongly localized than is the preceding one. In general, the extent of delocalization of the orbitals increases as their energy increases.

Although the 4σ orbital in CO correlates with a doubly occupied $2s$ orbital on carbon in the separated atoms, the formation of the CO molecule results in a transfer of charge to oxygen and the resulting molecular orbital exhibits a large $2p\sigma$ component centred on the oxygen nucleus. The 4σ orbital density in CO is characterized by an accumulation of charge density between the carbon and oxygen nuclei and by an even larger accumulation in the nonbonded region of the oxygen[23, 37]. It is clear from Figure 7 that in methanol the field of the three methyl protons rather than that of the oxygen dominates the form of the 4σ orbital to the extent that the density distribution is now largely localized on carbon on the side of the protons with the characteristics of a large $2s$ atomic component on carbon. While the orbital still exhibits a nodal structure at oxygen characteristic of a $2p\sigma$ distribution, the amount of charge in the antibinding region of the oxygen and in the CO bond region is greatly reduced from that found in CO. The weakening of the sigma bond structure in CO by the addition of hydrogens is evident in the sharing of the polarization density in orbital number three between the carbon and the proton bonded to the oxygen and in the complete reversal of the polarization in orbital number four.

Molecular orbital number five may be considered to be derived from one of the components of the doubly degenerate π orbitals in CO. The density is mostly concentrated in the region of the OH bond and the density in this fragment strongly resembles the 3σ orbital of the diatomic hydrides (see Figure 5) to the point that spatially related contours in the two maps have identical values. There is a strong back-polarization of the charge density on oxygen (away from the proton) indicating a large $2p\sigma$ component, the characteristic feature of the 3σ densities in the hydrides. Molecular orbital number six which is concentrated in a plane perpendicular to the one containing the OH bond, represents the second component of the bonding π orbital in CO. It is less perturbed from this form than is orbital number five but shows a larger concentration of charge density on carbon than is found in CO[37].

The 5σ orbital is the orbital of highest energy occupied in the ground state of CO. In this diatomic species the orbital is largely concentrated in the nonbonded region of the carbon (characteristic of an orbital with a large $p\sigma$ component on one centre) and exhibits two nodes perpendicular to the bond axis at the positions of the carbon and oxygen nuclei. This same nodal structure is evident in molecular orbital number seven of methanol but the charge density in this case is strongly delocalized over the entire system and contributes to the bonding of the protons on oxygen and carbon as well as to the bonding between the two heavy nuclei.

There are four more electrons in methanol than in CO and the two final molecular orbitals in methanol, particularly orbital number nine, are closely related to the doubly degenerate 2π antibonding orbital of CO. The 2π orbital is unoccupied in the ground state of CO. Since the 1π orbital in CO is heavily localized on the oxygen, the antibonding 2π orbital is concentrated in the region of the carbon. In methanol, however, the presence of the three methyl protons reverses this behaviour. The bonding π-like orbital, orbital number six, is more democratically shared and slightly favours the carbon. Consequently, the second π-like orbital in methanol, orbital number nine, is localized to a considerable extent in the region of oxygen.

Csizmadia et al[28] have also determined the wavefunction and molecular energy of the methoxide ion. The value predicted for the proton affinity of the methoxide ion using the molecular energies of CH_3OH and CH_3O^- is -420 kcal/mole. Hopkinson et al[38] have found the correlation between experimental and calculated proton affinities to be excellent when extensive basis sets are employed in the determination of the calculated values. These authors noted that both the experimental and calculated proton affinities fall into definite groups characterized only by their charge, e.g., dinegative ions have proton affinities between -500 and -700 kcal/mole; all mononegative species between -320 and -450 kcal/mole and the neutral species between -70 and -220 kcal/mole. The value for the methoxide ion falls within the range of values for the mononegative ions.

B. The Charge Distribution in CH_2FOH

The total molecular charge distribution for fluoromethanol is illustrated in the plane of the

```
F
 \
  C—O
     \
      H
```

nuclei in Figure 8. The charge density in the region of the oxygen

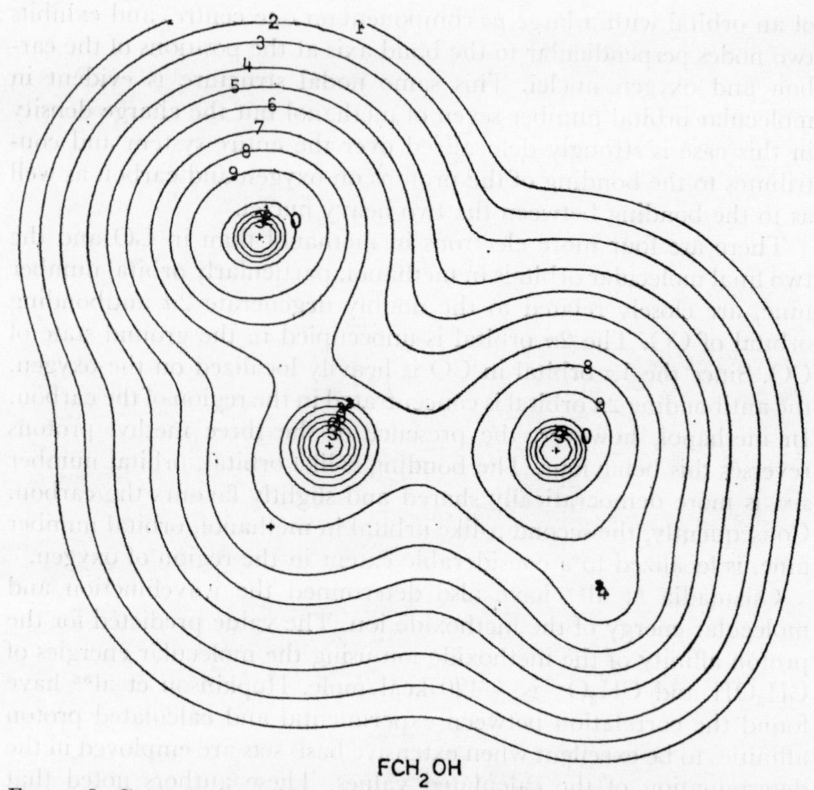

FIGURE 8. Contour plots of the total density and density difference distributions for the fluoromethanol molecule.

exhibits the same polarization as in methanol and the nonbonded radii of the OH bond fragment are essentially unchanged in value. The nonbonded charge density on carbon is slightly more contracted than it is in methanol, all the outer contours being displaced in closer to the carbon nucleus by approximately one-tenth of an au. The nonbonded radius on fluorine is 2·8 au, the same as that found in the CF diatomic molecule and close to the value of 2·7 in HF.

A much more detailed comparison of the effect which fluoro substitution has on the charge distribution of methanol may be obtained by comparing the $\Delta\rho_{SA}$ maps for CH_3OH and CH_2FOH (Figures 6 and 8). The inductive effect of the fluorine on the CO bond is very evident in such a comparison. The contours defining the region of charge increase between the C and O nuclei are increased in magnitude and extent, while those defining the charge deficit are similarly

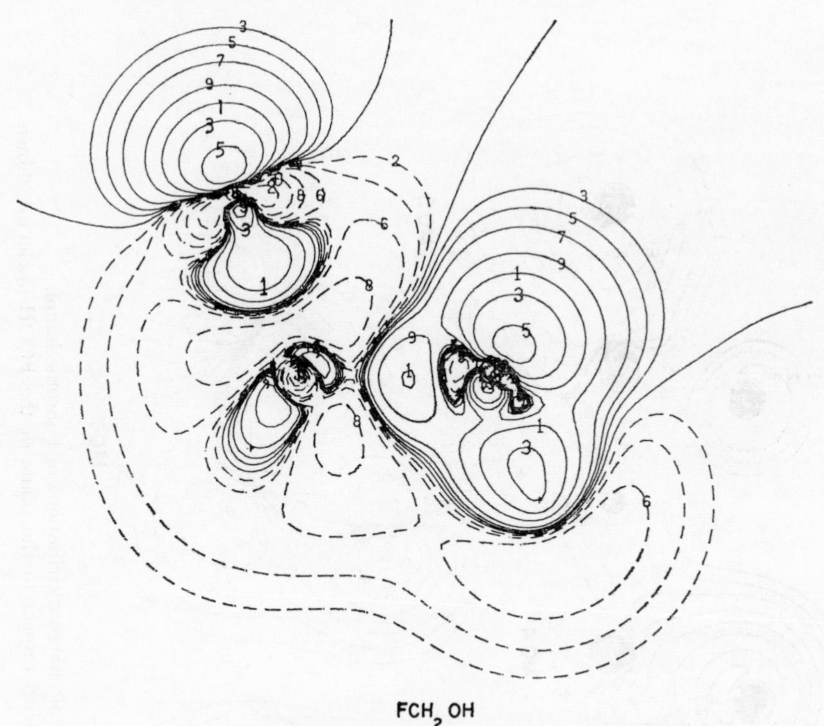

FCH_2OH

Figure 8 continued

decreased. A detailed comparison of the two $\Delta\rho_{SA}$ maps indicates that the whole of the charge increase in the vicinity of the oxygen nucleus and the proton is shifted slightly towards the carbon in CH_2FOH compared to CH_3OH. The charge increase at the position of the proton in the OH bond is slightly decreased. Aside from these effects the pattern of charge increase and decrease in the vicinity of the oxygen and hydrogen is the same as is found in CH_3OH, with a strong quadrupole polarization along the axis bisecting the COH bond angle. The regions of charge increase in the immediate vicinity of the carbon nucleus are directed along the axis which bisects the FCO bond angle. The pattern of the density difference map for the C–F bond fragment is very similar to that obtained for the CF diatomic molecule. Both maps exhibit a similar region of charge deficit in the binding region adjacent to the carbon nucleus, indicating a considerable degree of charge transfer to the region of the fluorine. The fluorine exhibits a quadrupole polarization typical of diatomic molecules.

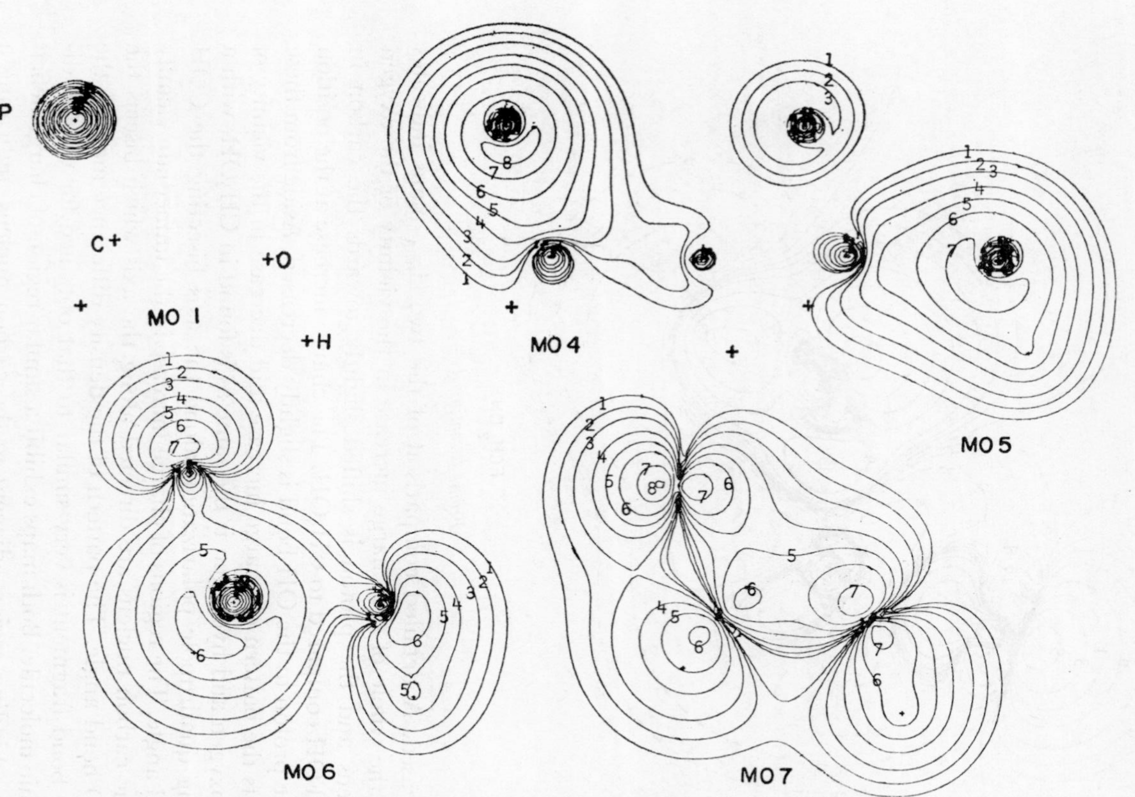

FIGURE 9. Contour plots of the molecular orbital charge distributions of fluoromethanol. Only the densities of the orbitals symmetric with respect to the plane of the FCOH nuclei are shown.

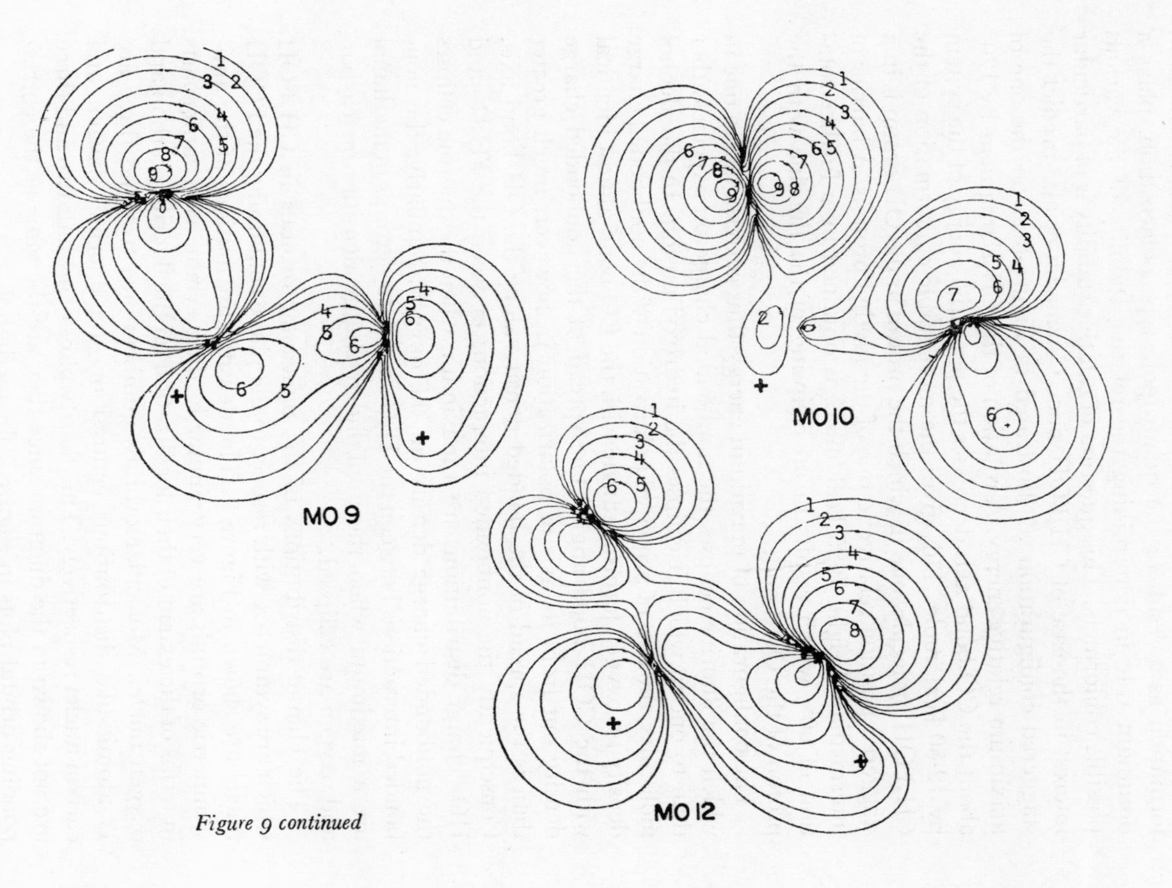

Figure 9 continued

The large increase in and expansion of the charge density into the nonbonded regions of the heavy nuclei, which occurs on bond formation as a result of the quadrupole type polarization, plays a dominant role in determining the relative stability of the various possible conformers. This question of relative stability is of particular interest in the case of CH_2FOH since Csizmadia et al[28] predict the staggered configuration as illustrated in Figure 8 to be the one of maximum relative energy. A rotation of the CH_2F fragment by 120° about the CO bond axis decreases the potential energy of the system by 12·60 kcal/mole. This gives the most stable conformation of the CH_2FOH molecule, one in which the proton in the OH group is in a staggered position relative to one of the methyl protons. A barrier of intermediate height (8·25 kcal/mole) is obtained by a further rotation of 60°, which results in an eclipsing of the fluorine with the proton of the OH bond.

The configuration of maximum energy thus corresponds to one in which the fluorine eclipses the nonbonded charge on oxygen rather than to one in which it eclipses the hydroxyl proton. In the alcohol molecules, the axis of principal polarization of the nonbonded charge density on oxygen forms an angle with the CO bond almost identical with the COH bond angle. The spread of the nonbonded charge density out from its axis of polarization is, however, much greater than is the spread of the bonded density along the OH bond axis. Consequently the nonbonded interactions between the FH_2C– and HO– charge distributions are a maximum when the fluorine eclipses the nonbonded charge density on the oxygen. Similarly, the nonbonded interactions between the H_3C– and HO– groups in methanol are a maximum when the nonbonded charge densities on carbon and oxygen are eclipsed.

The charge distributions of the molecular orbitals in CH_2FOH which are symmetric with respect to the plane containing the FCOH nuclei are shown in Figure 9. The density distributions of the antisymmetric orbitals are very similar to their symmetric counterparts in terms of the extent of their localization on the fluorine, carbon and oxygen nuclei. Molecular orbitals numbers one, two and three are $1s$ atomic-like distributions centred on the fluorine, oxygen and carbon nuclei respectively. The density plots of orbitals two and three are not shown in the diagram since they are the same as the corresponding orbital plots in Figure 7 for methanol.

Molecular orbital number four is, aside from a small tail to the oxygen, similar to the 3σ orbital in the diatomic CF species. The contours exhibit the shape characteristic of a strong 2*s*– component, in this case centred on fluorine and polarized towards the carbon. Orbitals five and six are primarily polarized 2*s* components on oxygen and carbon respectively, and, with the exception of the tails extending to fluorine, resemble very closely the corresponding orbitals, numbers three and four, in methanol.

Similarly, orbitals seven and eight resemble the π-like bonding orbitals five and six of methanol, but the distributions in CH_2FOH encompass the fluorine as well. The antibonding π-like pair of orbitals, numbers ten and eleven, are concentrated almost exclusively on the fluorine and oxygen nuclei, and more so on the former than on the latter nucleus. The final pair of orbitals, twelve and thirteen, are concentrated in the region of the oxygen.

IV. ACKNOWLEDGMENTS

The author is grateful to Dr. I. G. Csizmadia for making the wavefunctions for CH_3OH and CH_2FOH available prior to publication. He also wishes to thank Mr. G. Runtz for calculating the molecular and orbital charge distributions for these same molecules.

V. REFERENCES

1. R. Brill, in *Solid State Physics* (Ed. F. Seitz and D. Turnbull), Academic Press Inc., New York, 1967, pp. 1–35.
2. A. M. O'Connell, A. I. M. Rae and E. N. Maslen, *Acta Cryst.*, **21**, 208 (1966).
3. D. A. Kohl and L. S. Bartell, *J. Chem. Phys.*, **51**, 2891 (1969).
4. D. A. Kohl and L. S. Bartell, *J. Chem. Phys.*, **51**, 2896 (1969).
5. P. Hohenberg and W. Kohn, *Phys. Rev.*, **136**, B864 (1964).
6. L. Pauling, *The Nature of the Chemical Bond*, 3rd ed., Cornell University Press, Ithaca, New York, 1960.
7. R. F. W. Bader, W. H. Henneker and P. E. Cade, *J. Chem. Phys.*, **46**, 3341 (1967).
8. H. Hellmann, *Einfuhrung in die Quanten Chemie*, Franz Deuticke, Leipzig, Germany, 1937, pp. 285 ff.
9. R. P. Feynman, *Phys. Rev.*, **56**, 340 (1939).
10. E. Clementi, *J. Chem. Phys.*, **38**, 2248 (1963).
11. J. C. Slater, *Quantum Theory of Molecules and Solids*, Vol. 1, McGraw-Hill Book Co. Inc., New York, N.Y., 1963, pp. 93 ff.
12. C. C. J. Roothaan, *Rev. Mod. Phys.*, **23**, 69 (1951).
13. For a discussion of the SCF method see C. A. Coulson and E. Theal, in *The Chemistry of Alkenes*, Vol. 1 (Ed. S. Patai), Interscience Publishers, London, 1964, pp. 69 ff.
14. P. E. Cade and W. M. Huo, *J. Chem. Phys.*, **47**, 614, 649 (1967).

15. L. Brillouin, *Actualities Sc. Ind.*, Nos. 71, 159, 160 (1933–1934).
16. C. W. Kern and M. Karplus, *J. Chem. Phys.*, **40**, 1374 (1964).
17. D. Neumann and J. W. Moskowitz, *J. Chem. Phys.*, **49**, 2056 (1968).
18. R. F. W. Bader and A. K. Chandra, *Can. J. Chem.*, **46**, 953 (1968).
19. G. Das and A. C. Wahl, *J. Chem. Phys.*, **44**, 87 (1966); G. Das, *J. Chem. Phys.*, **46**, 1568 (1967).
20. R. F. W. Bader, I. Keaveny and P. E. Cade, *J. Chem. Phys.*, **47**, 3381 (1967).
21. P. E. Cade, R. F. W. Bader and J. Pelletier, *The Effect of Excitation, Ionization and Electron Attachment on the Molecular Charge Distribution*, to be published.
22. R. F. W. Bader, *J. Am. Chem. Soc.*, **86**, 5070 (1964).
23. R. F. W. Bader and A. D. Bandrauk, *J. Chem. Phys.*, **49**, 1653 (1968).
24. R. F. W. Bader and W. H. Henneker, *J. Am. Chem. Soc.*, **87**, 3063 (1965).
25. R. M. Stevens and W. N. Lipscomb, *J. Chem. Phys.*, **41**, 184, 3710 (1964).
26. R. F. W. Bader and G. Runtz, unpublished results.
27. A. C. Hurley, in *Molecular Orbitals in Chemistry, Physics, and Biology* (Ed. P. and O. Löwdin and B. Pullman), Academic Press Inc., New York, 1964, pp. 161–191.
28. I. G. Csizmadia, L. M. Tel and S. Wolfe, private communication (to be published).
29. S. Huzinaga and Y. Sakai, *J. Chem. Phys.*, **50**, 1371 (1969).
30. C. D. Ritchie and H. F. King, *J. Chem. Phys.*, **47**, 564 (1967).
31. E. Clementi, *J. Chem. Phys.*, **47**, 2323, 4485 (1967).
32. E. V. Ivash and D. M. Dennison, *J. Chem. Phys.*, **21**, 1804 (1953).
33. W. H. Fink and L. C. Allen, *J. Chem. Phys.*, **46**, 2261 (1967).
34. L. Pedersen and K. Morokuma, *J. Chem. Phys.*, **46**, 3941 (1967).
35. W. H. Fink, D. C. Pan and L. C. Allen, *J. Chem. Phys.*, **47**, 895 (1967).
36. R. F. W. Bader, G. Runtz and I. T. Keaveny, *Can. J. Chem.*, **47**, 2308 (1969).
37. W. M. Huo, *J. Chem. Phys.*, **43**, 624 (1965).
38. A. C. Hopkinson, N. K. Holbrook, K. Yates and I. G. Csizmadia, *J. Chem. Phys.*, **49**, 3596 (1968).

CHAPTER 2

Nucleophilic attack by hydroxide and alkoxide ions

COLIN A. FYFE

University of Guelph, Guelph, Ontario, Canada

I. INTRODUCTION

In reactions involving nucleophilic attack by hydroxide $(OH)^-$ and alkoxide $(OR)^-$ ions, a critical feature is the rather obvious observation that the former contains a second ionizable proton. The loss of this proton can give rise to further reaction after the initial nucleophilic attack by the hydroxide ion. This possibility does not arise in the case of the analogous alkoxide ions, and most of the general kinetic data on the nucleophilic attack by these two species come from investigations involving alkoxide ions.

The further reaction referred to above can manifest itself either in the form of side reactions, or in some cases, as will be shown in later sections, by complete further reaction giving rise to products entirely different from those in the corresponding alkoxide reactions. In these cases, the reactions involving attack by the simpler alkoxide ions provide excellent models for the initial products of the attack by hydroxide ion.

The reactions involving attack by these ions represent an enormous field of study, and any review must be restricted to some extent in its choice of subject material. The purpose of the present review is to present the reactions resulting from the attack of hydroxide and alkoxide ions on the range of carbon skeletons represented by (**A**—**E**) below*.

* The major omissions in this approach are the attacks on functional groups. The most important of these, attack on the carbonyl grouping, has recently been reviewed[1].

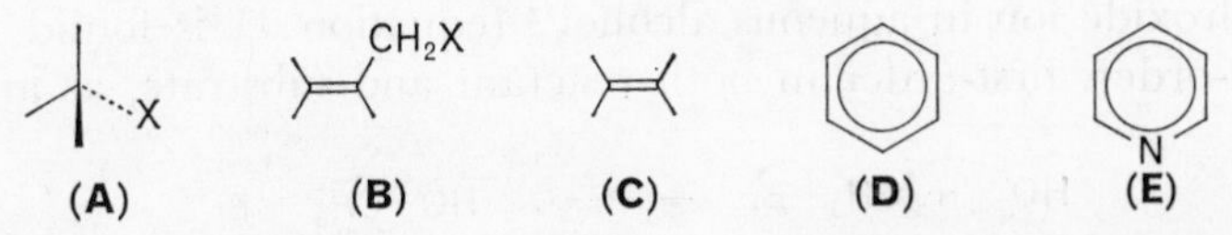

It is hoped to illustrate the change in the reactions of the ions with these substrates and also the general trend in the mechanism of their substitution reactions which changes from a synchronous one-step reaction to a two-step reaction involving an intermediate. Both of these changes are determined by the presence of π-electron systems, and the activating groups which these require for nucleophilic attack to occur.

Substitution reactions at saturated carbon atoms have been extensively investigated and reviewed. Because of this, these particular alkoxylation and hydroxylation reactions are presented in outline only, for comparison with those in later sections. In general, the emphasis in the chapter is placed strongly on mechanistic aspects; papers of a preparative nature are normally referred to only where they are relevant to a discussion of the mechanism. (This is not a severe restriction, as most of the reactions are rather obvious from a preparative point of view, involving either substitution of an activated group, or addition at a specifically activated position.)

One important aspect which is not considered here, as it does not usually affect the overall mechanisms which are the main subject of the chapter, but which is common to all the reactions, is the role of the solvent—in particular the use of dipolar aprotic solvents as reaction media. These generally enhance the rate of substitution reactions involving an ionic species and a neutral substrate which involve a transition state containing both moieties (which includes all the reactions considered here), and are of both preparative and theoretical interest. This aspect has been treated in reviews by Parker[2–4], Reichard[5] and in other publications[6–11].

Finally, the particular reactions and mechanisms which are presented here are only part of a much larger field of study, and references which are review articles or which contain a large amount of general data are marked with an asterisk (*) in the reference section to facilitate additional reading.

II. ATTACK AT SATURATED CARBON

A. Bimolecular Reactions Involving Hydroxide and Alkoxide Ions

Kinetically, the displacement of bromide ion from methyl bromide by hydroxide ion in aqueous alcohol[12] (equation 1) is found to be second-order, first-order in both reactant and substrate, as are the

$$HO^- + CH_3{-}Br \longrightarrow HO{-}CH_3 + Br^- \quad (1)$$

reactions of simple alkyl bromides with ethoxide ions in ethanol solution[13]. These are typical of the reactions of simple *n*-alkyl halides with these two ionic species.

The reaction is considered [14–17] to involve attack by the ion on the carbon atom from the side opposite the group to be displaced (equation 2). The reaction begins with the attachment of the nucleo-

$$RO^- + \;\rangle C{-}X \longrightarrow RO \cdots C \cdots X \longrightarrow RO{-}C\langle \; + X^- \quad (2)$$

Transition state

phile to the smaller lobe of the sp^3 orbital by which the group to be displaced is attached to the carbon atom and proceeds through a transition state, where both nucleophiles are attached to the opposite lobes of a p orbital while the remaining orbitals are sp^2 hybridized and lie in a plane at right angles to these groups. It terminates by the re-establishment of sp^3 hybridization of the carbon orbitals with the attacking nucleophile now attached to the carbon.

It is implicit in the above equation (2) that there is inversion of the configuration of the attached groups in the product. If the reaction is carried out with an optically active substrate, one should obtain the enantiomer from the reaction. Thus, treatment of the optically active α-phenylethyl chloride with ethoxide ion in ethanol yields[18] almost pure, optically active inverted product (equation 3). Similar results are found for other substitutions involving the highly reactive alkoxide or hydroxide ions.

$$EtO^- + C(Ph)(H)(CH_3){-}Cl \longrightarrow EtO{-}C(Ph)(H)(CH_3) + Cl^- \quad (3)$$

In general, the S_N2 mechanism is favoured for these strongly basic ions. One particular case of the S_N2 mechanism involving alkoxide ions which is of synthetic importance is the base-catalysed formation of epoxides from chlorohydrins[19]. The reaction is considered to involve an internal S_N2 displacement of chloride ions by initially formed alkoxide ion (equation 4).

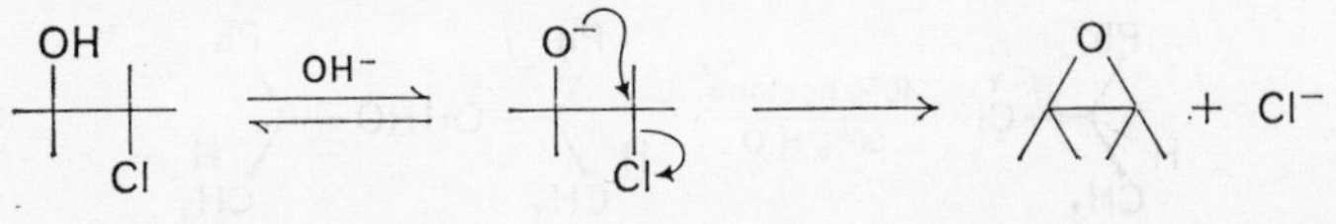

(4)

B. Unimolecular Reactions

Although, as indicated above, the S_N2 type of reaction is favoured for displacements involving the highly reactive alkoxide or hydroxide ions, under certain conditions substitution can occur by a unimolecular mechanism to give products which have arisen from formal replacement of X by OR or OH. This mechanism is favoured for solvolysis reactions involving actual attack by OH_2 and ROH. The reaction under these conditions is considered to involve an initial ionization to yield a carbonium ion (**1**) which is then attacked by the nucleophilic solvent (equation 5). Because the conditions favour-

(5)

able for this mode of reaction are often those involving solvolysis, kinetic evidence cannot be used to determine the molecularity of the reaction, and this decision must be made from other considerations. Very important is the stereochemistry of the reaction. Thus, implicit in the formation of the carbonium ion **1** in equation (5) is the idea that, since the carbonium ion is planar, nucleophilic attack can occur from either side to give a product which is a racemic mixture, irrespective of the stereochemistry of the reactant (equation 6).

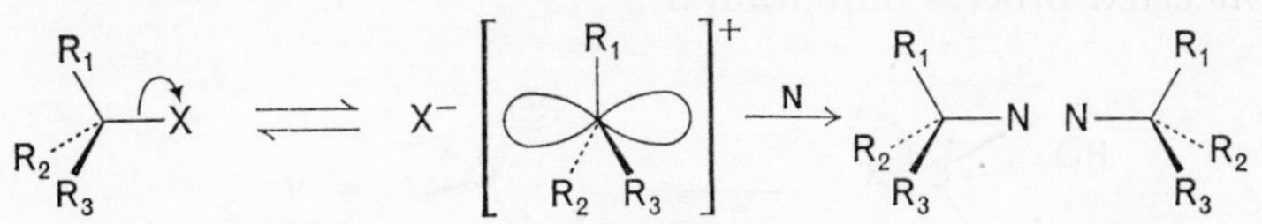

(6)

Indeed, when optically active α-phenylethyl chloride reacts under solvolysis conditions to give the alcohol, the product is almost* 100% racemized[18] (equation 7).

$$\text{Ph(H)(CH}_3\text{)C–Cl} \xrightarrow[60\%\ H_2O]{40\%\ \text{acetone}} \underset{49\%}{\text{Ph(H)(CH}_3\text{)C–OH}} + \underset{51\%}{\text{HO–C(Ph)(H)(CH}_3\text{)}} \qquad (7)$$

Thus the overall substitution by hydroxide and alkoxide ions at saturated carbon atoms can be considered as proceeding via two possible extreme modes of substitution.

III. ATTACK ON ALLYLIC SYSTEMS

In both structure and reactions[20, 21] the allylic system 2 occupies a place intermediate between saturated and alkene systems. In some reactions, both the point of attack and the mechanism are those

$$\overset{\gamma}{C}=\overset{\beta}{C}–\overset{\alpha}{C}H_2–X$$

(2)

discussed above for reactions at saturated carbon atoms, and the main point of interest is how the presence of the C=C bond affects the rates of the reactions. In other reactions, however, attack can occur at the alkene double bond to give quite different products.

A. Bimolecular Substitution Reactions

1. Involving attack at the α-carbon atom

This mode of reaction is exactly analogous to the S_N2 reactions of saturated systems, the nucleophile attacking the saturated carbon atom at the side remote from the substituent X which is displaced in a concerted process (equation 8).

$$RO^- + C=C–C–X \longrightarrow C=C–C–OR + X^- \qquad (8)$$

* Absolutely unambiguous results are rare. For a discussion of the complicating features see Refs. 14–17.

The most significant feature of the reaction is the effect of the alkene moiety on the rate of the reaction compared with that of the corresponding aliphatic substrate. In general, much faster reactions are observed, the increase in rate being reflected (Table 1) in a decrease in the activation energy of the reaction. This is considered

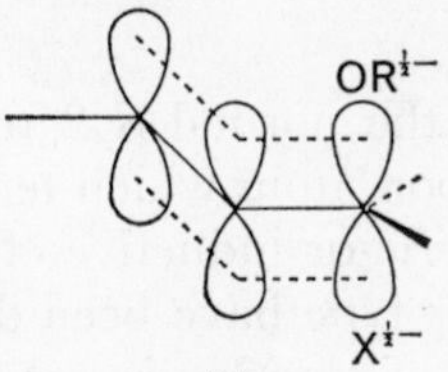

$OR^{\delta-}$

$X^{\delta-}$

(3)

to be due to stabilization of the transition state by overlap of the π orbitals of the double bond with the p orbital formed on the α-carbon atom in the transition state (**3**). The conjugation over the three atoms which is now possible lowers the overall energy of the system. The effect of substituents on the reactivity of these compounds has been reviewed[21].

TABLE 1. Relative rates and activation energies for the reaction of allylic halides and the corresponding saturated halides with ethoxide ion in ethanol solution at 44·6°C[a].

Substrates	$\frac{k_2(\text{allyl})}{k_2(\text{alkyl})}$	E_a (kcal mole^{-1})
$CH_2{=}CHCH_2Cl$	~27	20·6
$CH_3CH_2CH_2Cl$		21·1
$CH_3CH{=}CHCH_2Cl$	~88	20·5
$CH_3CH_2CH_2CH_2Cl$		21·2
$CH_2{=}C(CH_3)CH_2Cl$	~340	20·5
$CH_3CH(CH_3)CH_2Cl$		22·4

[a] C. A. Vernon, *J. Chem. Soc.*, 4462 (1954).

2. Involving attack on the alkene moiety

A second possible mode of bimolecular substitution is available in allyl systems because of the double bond. This attack can take place at the γ-carbon atom (equation 9) to give, for any substituted alkene which is unsymmetrical with respect to the transition state, a different product from that described in section I above. This mode of displacement is usually called S_N2' and, in general, will

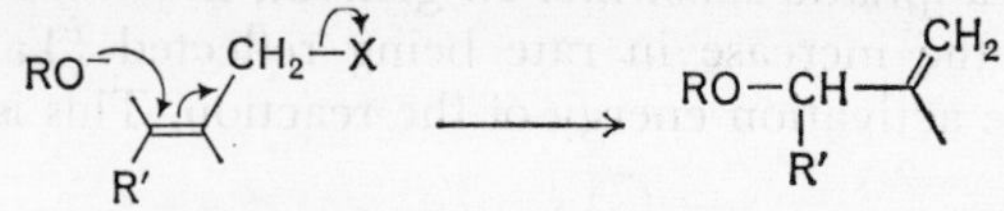

$$RO^- \; CH_2{-}X \; R' \longrightarrow RO{-}CH(R'){-}C(=CH_2) \qquad (9)$$

be promoted relative to the normal S_N2 reaction where there are substituents on the α-carbon atom which tend to inhibit the normal S_N2 reaction by either steric or inductive effects.

Several reactions of this type have been documented for alkoxide ions, although this mode of reaction is not as favoured for charged as for uncharged nucleophiles.

Thus, although α-methylallyl chloride with ethoxide ion in ethanol yields only a small quantity of the abnormal product[22], *t*-butylallyl chloride forms substantial amounts of 'abnormal' product with both ethoxide[23] and phenoxide ions[24], and α,α-dichloroallyl chloride gives exclusively the abnormal product with ethoxide ion[25].

The reaction is considered[21] to proceed in an analogous manner to the S_N2 mechanism, i.e., via a one-step mechanism involving synchronous bond breaking and formation with a transition state such as **4** where there is stabilization from the overlap of the (β) *p*-orbital with the (α) and (γ) 'pseudo' *p*-orbitals. However, the ex-

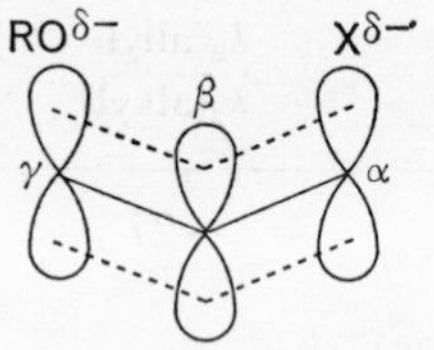

$RO^{\delta-}$ $X^{\delta-}$ β γ α

(4)

perimental evidence on this point is not completely conclusive. Bordwell and co-workers[26] have shown that the rates of reaction of 3-halomethylbenzothiophene 1,1-dioxides with thiourea in methanol were very dependent on the halogen atom, and had the same ratio of halogen activities as normal S_N2 reactions with the same nucleophile. However, although this eliminates mechanisms for this reaction in which the carbon–halogen bond fission is not the rate-determining step, it is equally in accord with a two-step mechanism where the second step is rate-determining (some cases of aromatic substitution are known where this is so). However unlikely this might be considered for relatively unactivated substrates, the mechanism could be altered by the presence of strongly electronegative substituents on C_β (e.g. CN, NO_2, Ph, NO_2) which would not only facilitate attack on C_γ, but would also tend to stabilize the possible resultant ion.

B. Unimolecular Substitution Reactions

The first step in allylic unimolecular reactions is similar to that in the aliphatic series (section II) in so far as it involves ionization to yield a carbonium ion (**5**) (equation 10).

$$\text{(allylic substrate, } \gamma, \beta, \alpha\text{–X)} \longrightarrow [\text{allyl cation}]^+ + X^- \quad (10)$$

(**5**)

However, the carbonium ion formed is now mesomeric by virtue of the initial double bond, and attack by the second nucleophile can now occur at either end to give, in the case of an unsymmetrical carbonium ion, different products. For example, the solvolysis of $(CH_3)_2C{=}CHCH_2Cl$ in acetic acid–silver acetate yields the two possible products (equation 11). That the isomeric $(CH_3)_2CClCH{=}CH_2$

$$(CH_3)_2C{=}CHCH_2Cl \xrightarrow[CH_3COOAg]{CH_3COOH} \underset{55\%}{(CH_3)_2C{=}CHCH_2OH} + \underset{45\%}{(CH_3)_2C(OH){-}CH{=}CH_2} \quad (11)$$

which should yield the same intermediate carbonium ion, in fact gives the same mixture of products[27] is good evidence for the S_N1 mechanism. Electron-releasing substituents at the α and γ positions stabilize the intermediate (**5**) and lower the activation energy, thereby increasing the rate[28]. The effect of monosubstitution by an alkyl group in these positions in allyl chloride increases the rate by a factor of (2–5) $\times 10^3$. Two alkyl substituents are roughly twice as effective as one in a given position. A phenyl substituent is somewhat more activating than a methyl group[21].

However, most of the data pertinent to this reaction come from solvolysis reactions such as that described above, and, in these cases, kinetic evidence for the molecularity of the reaction is missing. As the solvent is one of the reactants, 'first-order' kinetics will be shown by both unimolecular and bimolecular reactions. Care must therefore be taken in the interpretation of the data in these systems, and the molecularity of the reactions decided from other factors. A complete discussion of these is given by DeWolfe and Young[21].

A particularly interesting case of the above reaction is the solvolysis of allylic alcohols. In the case where the attack is at the end of the carbonium ion where the initial ionization took place, the starting material is regenerated. If, however, attack takes place at the other end of the carbonium ion, the net result is isomerization of the allylic alcohol (equation 12). Although the reaction is acid-catalysed, it comes within the scope of the present chapter because of the nucleophilic attack of OH_2 which actually represents the isomeriza-

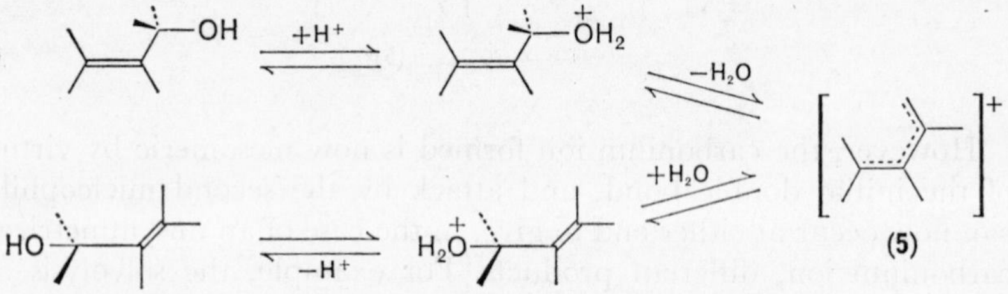

OH $+H^+$ $\overset{+}{O}H_2$ $-H_2O$ $+H_2O$ HO $-H^+$ $H_2\overset{+}{O}$ (5)

Isomerization (12)

tion process. The reaction can also be described in terms of an S_Ni mechanism involving a cyclic intermediate of type **6**.

Although the reaction is again of the solvolysis type, and inferences as to the molecularity of the reaction must be made from sources other than kinetic ones, the evidence favours the intermediacy of a carbonium ion such as **5** or **6**, and not a concerted bimolecular process.

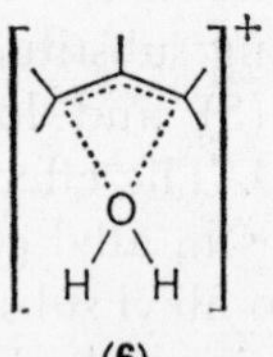

O H H

(6)

In some cases, a decision between the S_N1 and the intramolecular S_Ni reaction can be made by using ^{18}O-labelled alcohol where, if the reaction is of the S_Ni type, the ^{18}O should be retained in the product. In the isomerization of α-phenylallyl alcohol in acidic 60% aqueous dioxane[29] and in 40% dioxane–aqueous perchloric acid[30], the rearranged alcohol contains little of the ^{18}O of the starting alcohol, and the mechanism is considered to be S_N1.

However, a similar study[31] of the isomerization of *cis*- and *trans*-5-methyl-2-cyclohexene showed that although the *trans*-isomer reacted by the S_N1 mechanism, the *cis*-isomer isomerized mainly by the intramolecular S_Ni mechanism.

IV. ATTACK ON CARBON–CARBON MULTIPLE BONDS

A. General Comments on C═C Bonds

The very high electron density in the double bond system of ethylenes makes direct nucleophilic attack unfavourable, unless there are one or more electronegative groups present which can lower the electron density at the carbon atoms. Common activating groups are NO_2, CN, COR.

For such electronegative groups, the resonance structures of the molecule are:

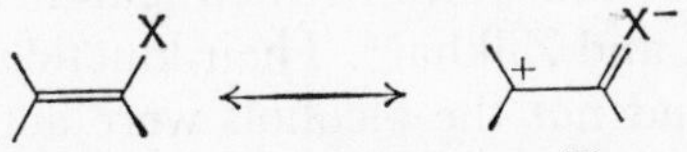

$$\text{>C=C<X} \longleftrightarrow \text{>}\overset{+}{C}\text{–C=X}^{-}$$

(7)

The contribution of structure **7** depends on the nature of X, the net effect being the lowering of the electron density at the carbon atom β to the electronegative group, and nucleophilic attack will normally take place at this position. In general, the attack of alkoxide and hydroxide ions on C═C bonds can be rationalized[32, 33] in terms of a general first equilibrium reaction in which an anion **(8)** is formed by the attack of the nucleophile on the β-carbon atom of the double bond (equation 13). According to the nucleophile and

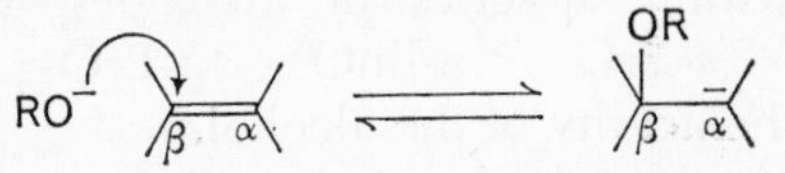

$$RO^{-} + \text{>}C_{\beta}\text{=}C_{\alpha}\text{<} \rightleftharpoons \text{>}C_{\beta}(OR)\text{–}\overset{-}{C}_{\alpha}\text{<} \qquad (13)$$

(8)

the conditions, the further reaction of **8** can give rise to a whole variety of products. For hydroxide and alkoxide ions, reactions involving addition, decomposition and substitution are the most important and these will be discussed separately. As will be seen below, it is in some cases critical whether the nucleophile is hydroxide or alkoxide ion, the presence of the ionizable hydrogen on the former giving rise to a quite different reaction path, although the initial reaction is similar in both cases.

B. Attack on C=C Bonds

1. Attack by alkoxide ions: addition of alcohols

In 1905, Meisenheimer[34] found that sodium methoxide or sodium ethoxide added instantaneously and at room temperature to β-nitrostyrene with the formation of an alkoxy derivative (**9**).

$$C_6H_5CH{=}CHNO_2 \xrightarrow[CH_3ONa]{CH_3OH} C_6H_5CH(OCH_3){-}CH_2NO_2$$

(9)

This is found to be a general reaction of arylnitroalkenes[35] and other activated double bonds[33].

The kinetics of the cyanoethylation reaction between cyanoethylene and various alkoxides in their parent alcohols have been investigated by Feit and Zilkha[36]. Their kinetic analysis showed that the alkoxide ions and not the alcohols were acting as nucleophiles. The reaction was first-order in both cyanoethylene and alkoxide ion and proceeded according to the scheme below (equation 14). The

$$RO^- + C{=}C(CN) \rightleftharpoons RO{-}C{-}\bar{C}(CN) \xrightarrow{H^+} RO{-}C{-}C(H)(CN) \qquad (14)$$

(10)

first step in the reaction, the formation of the intermediate **10**, is a particular case of reaction (13). Feit and Zilkha also found that the rate of the reaction was independent of the counterion for a given alkoxide, and within a series of alkoxides, was in the order $OMe^- < OEt^- < n\text{-}PrO^- < n\text{-}BuO^- < i\text{-}PrO^-$, reactivity inversely proportional to the acidity of the alcohol.

The same relative reactivities of alkoxides was found by Ferry and McQuillin[37] for their reaction with $CH_2{=}CHCOCH_3$. The butenone was formed in an initial, very fast reaction of the methobromide of 4-dimethylaminobutan-2-one with base (equation 15a). The rate-determining step (equation 15b) is second-order, first-order in both base and substrate, and is thought to proceed in an analogous manner to equation (14) above. A similar mechanism (equation 16) was proposed by Crowell and co-workers[38] for the basic methanolysis

$$Me_3\overset{+}{N}CH_2CH_2COCH_3 \xrightarrow[\text{fast}]{OR^-} CH{=}CHCOCH_3 \qquad (15a)$$

$$CH_2{=}CHCOCH_3 \underset{}{\overset{ROH}{\rightleftharpoons}} RO{-}CH_2CH_2COCH_3 \qquad (15b)$$

of dibenzoylethylene. They found that a linear Hammett relation

$$PhCOCH{=}CHCOPh \overset{OCH_3^-}{\rightleftharpoons} PhCO\overset{-}{C}H{-}CH(OCH_3){-}COPh \xrightarrow[\text{fast}]{HOCH_3} PhCOCH_2{-}CH(OCH_3){-}COPh \qquad (16)$$

existed between the rates and the sum of the σ-values of the *para*-substituents in both rings.

Although the stereochemical course of the reaction has not been investigated in detail, one can envisage different products depending on the lifetime of the intermediate ion **9**. Thus, if the addition of H^+ was very fast, giving a short lifetime to the intermediate, the addition of the alcohol would approximate to a concerted reaction, and one would obtain a product in which the elements of the alcohol were orientated *trans*. If, however, the intermediate ion **9** had a finite lifetime longer than the time for rotation about the C_α–C_β bond, one would obtain a range of configurations.

2. Attack by hydroxide ion or H_2O: cleavage reactions

Attack by OH^- to yield a simple hydration of the double bond (equation 17a) is in accord with the attack by alkoxide ions discussed in the previous section. However, the product (**11**) formed in

$$>C{=}C< \xrightarrow[HOH]{HO^-} >C(OH){-}CH< \quad \textbf{(11)} \qquad (17a)$$

$$\underset{\textbf{(11)}}{>C(OH){-}CH<} \xrightarrow{OH^-} >C{=}O + H{-}CH< \qquad (17b)$$

this case differs from the corresponding alkoxy product in that it contains an ionizable hydrogen and, in general, further reaction occurs (equation 17b), giving cleavage to the corresponding carbonyl and active methylene compounds. There is no reported case of cleavage caused by alkoxides alone. Steps (17a) and (17b) are essentially the reverse equilibria of the condensation reactions commonly used to prepare alkenes[39].

Reactions corresponding to step (17b) above can be investigated independently of the initial stages (equation 17a) in the above reaction mechanism[40, 41]. Thus, Westheimer and Cohen[40] studied the dealdolization of diacetone alcohol and found the reaction to be base-catalysed and to follow the reaction equation (18), with the rate-determining step the decomposition of the ion **12**.

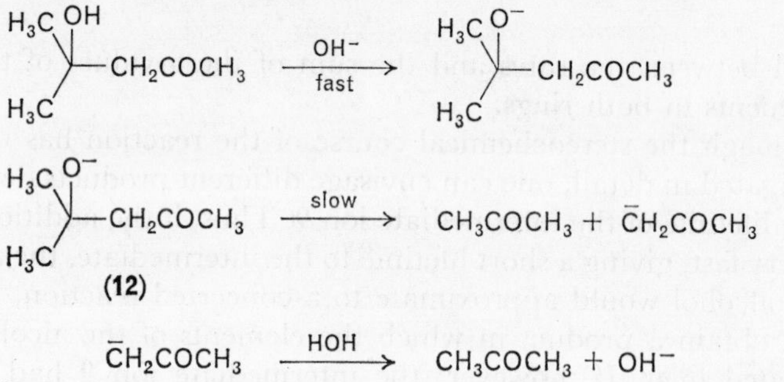

$$(H_3C)_2C(OH)-CH_2COCH_3 \xrightarrow[\text{fast}]{OH^-} (H_3C)_2C(O^-)-CH_2COCH_3$$

$$\underset{(\mathbf{12})}{(H_3C)_2C(O^-)-CH_2COCH_3} \xrightarrow{\text{slow}} CH_3COCH_3 + \bar{C}H_2COCH_3 \qquad (18)$$

$$\bar{C}H_2COCH_3 \xrightarrow{HOH} CH_3COCH_3 + OH^-$$

In contrast, Rondestvedt and Rowley[41] found that, in the cleavage of β-hydroxy acids and esters of general formula **13**, the results were best correlated in terms of a concerted mechanism in which the breaking of the C_α–C_β bond in **13** was essentially simultaneous with the breaking of the O–H bond (equation 19), there being no significant build-up of an intermediate corresponding to **12**. The

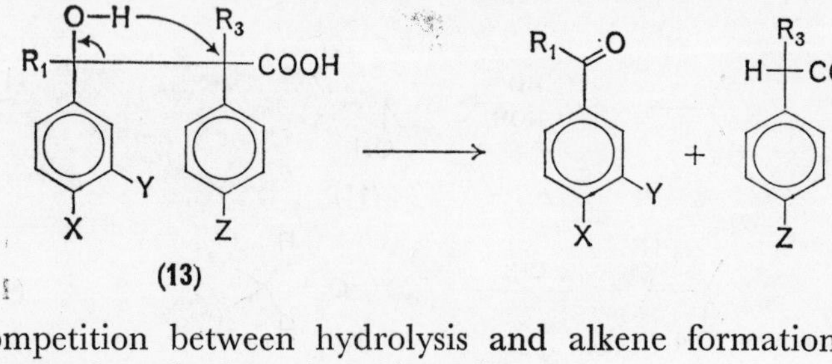

$$\underset{(\mathbf{13})}{R_1C(OH)(C_6H_3\text{-}X,Y)-C(R_3)(C_6H_4\text{-}Z)COOH} \longrightarrow R_1C(=O)(C_6H_3\text{-}X,Y) + H-C(R_3)(C_6H_4\text{-}Z)COOH \qquad (19)$$

competition between hydrolysis and alkene formation in similar systems has also been investigated[47].

Attack by OH_2 (20a)

Attack by OH^- (20b)

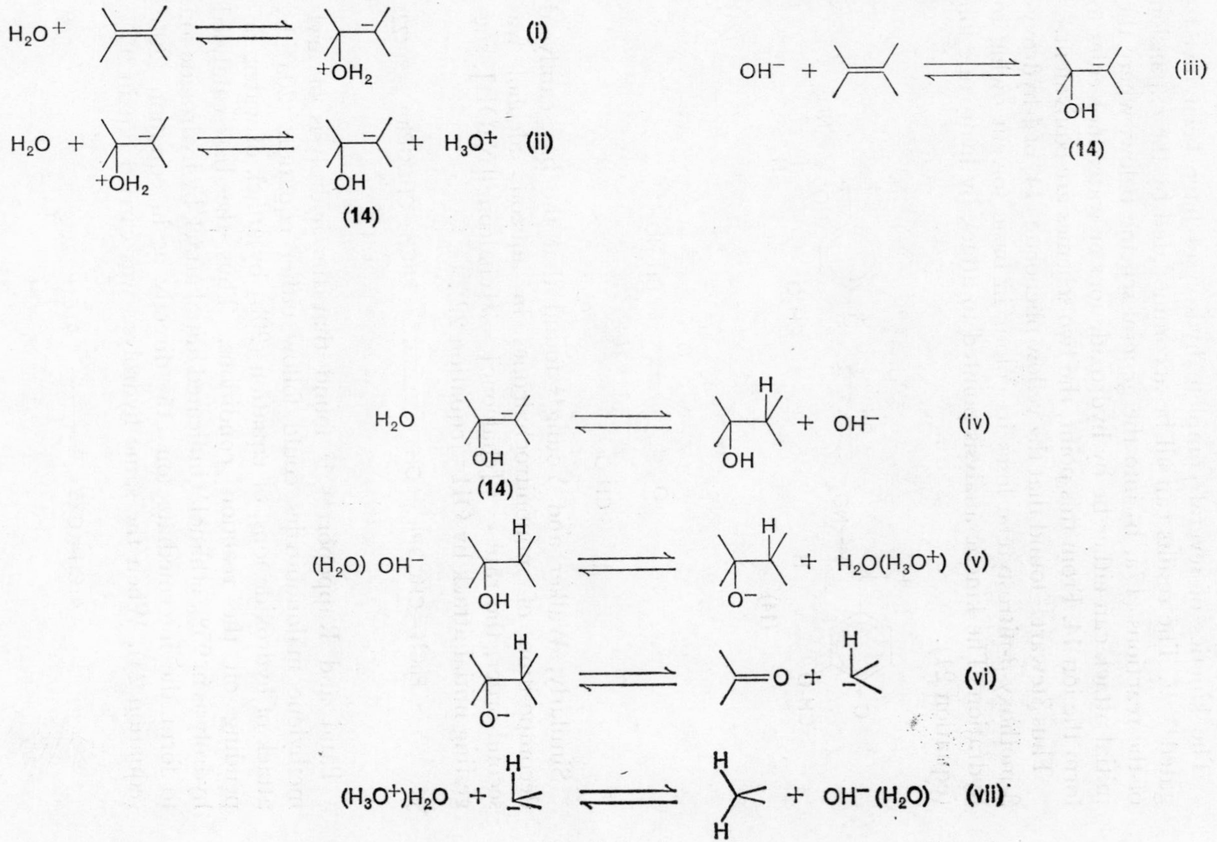

H_2O^+ + alkene ⇌ carbanion with $^+OH_2$ (i)

H_2O + ($^+OH_2$ carbanion) ⇌ (14) (OH carbanion) + H_3O^+ (ii)

OH^- + alkene ⇌ (14) (OH carbanion) (iii)

H_2O + (14) ⇌ (H, OH adduct) + OH^- (iv)

$(H_2O)\ OH^-$ + (H, OH adduct) ⇌ (H, O^- adduct) + $H_2O(H_3O^+)$ (v)

(H, O^- adduct) ⇌ C=O + (H carbanion) (vi)

$(H_3O^+)H_2O$ + (H carbanion) ⇌ (H, H product) + $OH^-\ (H_2O)$ (vii)

The kinetics of several complete hydrolyses have been investigated[42–47]. The results can all be accommodated by the expansion of the reactions (17a, b) into the general scheme below, where the initial attack can either be by hydroxide ions or water molecules to form the ion **14**. From this point, the two schemes are equivalent.

Thus Stewart[42] found that the yellow phenolate (**14**) of 4-hydroxy-3-methoxy-β-nitrostyrene loses its colour in basic solvent owing to hydration. The kinetic analysis pointed to attack by hydroxide ion (equation 21).

(14) $\xrightleftharpoons{OH^-}$ $\bar{O}$–$C_6H_3(OCH_3)$–CH(OH)–$\bar{C}H$–NO_2 → $\bar{O}$–$C_6H_3(OCH_3)$–CHO + CH_3NO_2 (21)

Similarly, Walker and Young[43] found that the base-catalysed decomposition of mononitrochalcones in aqueous alcohol was second-order, the rate $= k'$[chalcone] $= k$[chalcone][NaOH], suggesting initial attack by OH^- (equation 22).

$$PhCH{=}CHCOPh + OH^- \rightleftharpoons PhCH(OH){-}\bar{C}HCOPh \qquad (22)$$

Patai and Rappoport[44, 45] found that the hydrolysis of arylmethylene malononitriles could follow either equation (20a) by attack of hydroxide ions, or equation (20b) by attack of water, depending on the reaction conditions. Thus the base-catalysed hydrolysis in 95% ethanol[44] indicated initial attack by hydroxide ion to form the intermediate ion (**15**) directly as in equation (20b) (equation 23). When the same hydrolysis was carried out in 95%

$$ArCH{=}CXY \longrightarrow ArCH(OH){-}\bar{C}XY \quad \textbf{(15)} \qquad (23)$$

ethanol in the absence of base, the reaction scheme followed was equation (20a), the rate-determining step being the relatively slow initial attack by H_2O to give the ion **16** which then reacted to give **15** as in equations (24a, b). Evidence for this reaction scheme was

$$\mathrm{ArCH{=}CXY + H_2O \rightleftharpoons ArCH(\overset{+}{O}H_2){-}\bar{C}XY} \quad \mathbf{(16)} \tag{24a}$$

$$\mathrm{ArCH(\overset{+}{O}H_2){-}\bar{C}XY + H_2O \rightleftharpoons ArCH(OH){-}\bar{C}XY + H_3O^+} \quad \mathbf{(15)} \tag{24b}$$

that the rate depression by acids was dependent[45] on the concentration of the substrate.

A similar dependence of mechanism on the reaction conditions was found by Crowell and Francis[46] in the hydrolysis of substituted β-nitrostyrenes in aqueous solution. At pH 0·8—6·0, they found two consecutive, pseudo-first-order reactions, the first of which showed general base catalysis, and the second of which was pH-dependent, in agreement with equations (25a, b), involving attack by H_2O.

$$\mathrm{ArCH{=}CHNO_2 + OH_2 \rightleftharpoons ArCH(\overset{+}{O}H_2){-}\bar{C}HNO_2} \tag{25a}$$

$$\mathrm{ArCH(\overset{+}{O}H_2){-}\bar{C}HNO_2 + OH_2 \rightleftharpoons ArCH(OH){-}\bar{C}HNO_2 + H_3O^+} \quad \mathbf{(17)} \tag{25b}$$

At higher pH, the cleavage occurred at an enhanced rate, and at pH 11, the primary step changed, attack being by OH^- to form the ion (**17**) directly.

3. Substitution reactions

If there is a labile group attached directly to the double bond, a replacement reaction can occur after the initial attack by the anion (equation 26), by an 'addition–elimination' mechanism.

There are several other routes which can give the substitution

$$\mathrm{{>}C{=}C(X){<} + {}^-OR \rightleftharpoons {>}\bar{C}{-}C(OR)(X){<}\ \mathbf{(18)} \longrightarrow {>}C{=}C(OR){<}\ \mathbf{(19)} + X^-} \tag{26}$$

Addition–Elimination

product (**19**). The most important of these is the 'elimination–addition' mechanism shown in equation (27) in which an acetylene is formed initially which then adds the elements of the alcohol (see section IV.C) to give **19**.

$$\text{H(X)C=C} \xrightleftharpoons{-HX} \cdot - \equiv - \xrightarrow{ROH} \text{H(OR)C=C} \cdot \quad (27)$$

Elimination–Addition **(19)**

The reactions (26) and (27) differ in several significant aspects which can, in principle, be used to distinguish between them. Thus in route (26) there should be no deuterium exchange with the solvent, but (27) should give complete exchange. Stereochemically, one might also hope to differentiate between the two reactions: (27) will give no relationship between the stereochemistry of the two isomers, as both *cis*- and *trans*-isomers will give the same intermediate acetylene. In equation (26), however, there is the possibility that the stereochemistry of the starting material will influence that of the product: thus if the ion **18** has a relatively long lifetime, there will be a distribution of the isomeric products depending on their thermodynamic stability, but if **18** is very short lived, approaching in the limit a concerted reaction, then one might expect a reaction where the stereochemistry of the reactant is retained in the product. However, it is seldom that a clear-cut distinction can be made between the two schemes (which may, of course, occur simultaneously), owing to complicating features such as the possible *cis–trans*-isomerization and further addition to the double bond (section B.1 above) which can obscure differences in stereochemistry.

In general, mechanism (26) will be favoured for compounds with a low electron density on the β-carbon atom to provide the initial attack. Mechanism (27) will be especially favoured in *cis*-isomers where there is a favourable *trans*-disposition of the elements of HX (where X is the leaving group) to promote the elimination reaction. For a given substrate, the occurrence of the elimination–addition mechanism will depend on the proton basicity of the nucleophile. Since alkoxide ions are highly basic with respect to protons, in many instances, reaction (27) will compete with reaction (26).

Thus, although the *cis*- and *trans*-isomers of β-chlorocrotonate undergo substitution by thiophenoxides with retention of configuration, only the *trans*-isomer is obtained in substitution by alkoxides[48].

However, both the *cis*- and *trans*-isomers of β-chloro-α-cyanoethylene react with alkoxides to give products of the same stereochemistry as the starting materials[49], and the reaction can be formulated as in equation (28) with the restriction that the intermediate ion (**20**) must have a very short lifetime, or that the reaction proceeds in a concerted manner.

$$\text{H(NC)C=CH(Cl)} + {}^-\text{OR} \rightleftharpoons \text{H(NC)C}^-\text{–CH(RO)(Cl)}\ (\mathbf{20}) \longrightarrow \text{H(NC)C=CH(OR)} \quad (28)$$

The competition between the two possible mechanisms for substitution has been investigated in detail for the reactions of arylsulphonylhaloethylenes (**21**) with methoxide ion by Modena and co-workers[50–54].

$$RC_6H_4SO_2CR^1{=}CR^2X \quad (\mathbf{21})$$

$$ArSO_2\bar{C}{=}CRX \underset{}{\overset{1}{\rightleftharpoons}} ArSO_2CH{=}CRX + OCH_3^- \text{ (either isomer)}$$

$$ArSO_2\bar{C}{=}CRX \xrightarrow{2} ArSO_2C{\equiv}CR + X^- \xrightarrow{3} (ArSO_2)HC{=}CH(OCH_3)$$

$$ArSO_2CH{=}CRX + OCH_3^- \xrightarrow{4} ArSO_2\bar{C}H{-}C(OCH_3)(X){-}R \xrightarrow{5} ArSO_2CH{=}CR(OCH_3) + X^- \quad (29)$$

Their results, some of which are summarized in Table 2, are discussed[54] in terms of the reaction scheme (29). The addition–elimination mechanism 1, 4, 5, under the restrictions discussed above, could give retention of configuration of both isomers, whereas both isomers would give the *cis*-product by the elimination–addition mechanism (steps 2 and 3 in equation 29), as both would give the acetylene and the stereochemistry of step 3 is known[50, 54] to give the *cis*-product. There is retention of configuration in the *trans*-isomers, indicating that these react via 1, 4, 5, and the lack of any element effect in two series of *trans*-compounds (Table 2) suggests that the breaking of the C–Hal bond is not rate-determining, in accordance with $k_4 < k_5$. A critical case is compound IV (Table 2) where α-elimination is not possible and reaction must be by addition–elimination. There is now no difference, either in rates or in activation energies, between the four isomers.

The *cis*-isomers behave quite differently, showing both greatly enhanced rates of reaction and also very strong element effects (Table 2). The large differences in the rates of reaction between compound IV (*cis*), where no elimination is possible, and the other

TABLE 2. Rate coefficients $k_2 \times 10^3$ ($l\ m^{-1}\ sec^{-1}$) of reactions of substituted arylsulphonylethylenes (**21**) with methoxide ion in methanol solution

R	R¹	R²	Temp.	X = Br *cis*	X = Br *trans*	X = Cl *cis*	X = Cl *trans*	Ref.
p-NO_2 I	H	H	0°C	—	—	280	168	50
p-Me II	H	H	0°C	1780	5·35	9·6	6·4	51
H III	H	H	0°C	2600	8·8	18	10·5	51
p-NO_2 IV	Me	H	(25°C)	5·05	5·26	5·40	6·22	52
p-NO_2	H	Me	(25°C)	7850	—	71·9	—	53

TABLE 3. Kinetic and thermodynamic parameters for the elimination–addition (subscript e) and addition–elimination (subscript a) reactions of *cis* arylsulphonyl chloroethylenes[55].

RC_6H_4—SO_2CH=CHCl with methoxide ion in methanol[a].

R	Temp.	k_e/k_a	E_e (kcal/mole)	E_a (kcal/mole)
H	0	1·1	24	17
	13	2·2	—	—
	25	3·0	—	—
p-CH_3	0	0·8	—	—
	13	1·0	24	20
	25	1·4	—	—

[a] The estimated errors are ca 10% for the rate coefficients and ±1 kcal/mole for the activation energies.

cis-isomers suggests the occurrence of the competing addition–elimination reaction (steps 2 and 3 in equation 29) which would be favoured in the case of the *cis*-compounds by a facile *trans*-elimination of HX.

Retention of configuration in the case of the *cis*-isomers is ambiguous, as both routes would give the same stereochemistry (equation 29). The exact relationship between the two mechanisms in the case of the *cis*-isomers has recently been studied in detail[55]. Direct evidence was found for the occurrence of the elimination–addition mechanism (steps 2 and 3) in the detection of the intermediate acetylene, both by isolation in cases of suitable kinetics, and by infrared spectroscopy in others. In a detailed examination of the kinetics, it was found possible to separate the kinetic parameters for the two processes. These are summarized in Table 3. The activation energy for the elimination–addition mechanism is larger than for the addition–elimination mechanism and the elimination mechanism becomes more important at higher temperatures. It was found to be significant at room temperature in all the cases studied.

An extreme case of the difference in the reaction of *cis*- and *trans*-isomers, due to the tendency of the *cis*-isomer to react by elimination, is found in the reactions of the isomers of 4-nitro-β-bromostyrene[56, 57, 58a] and 2,4-dinitro-β-bromostyrene[58a] with alkoxides. The *cis*-isomers react by a *trans*-elimination of HBr to form the corresponding acetylenes (equation 30). It has been suggested that

$$RC_6H_4CH{=}CHBr \xrightarrow{OCH_3^-} RC_6H_4C{\equiv}CH + HOCH_3 + Br^- \quad (30)$$

TABLE 4. Rate coefficients and thermodynamic parameters for the reaction of bromostyrenes of general formula ArCX=CHBr with methoxide ion in MeOH solution[58a] [a].

Compound	Ar	X	Temp. (°C)	$k \times 10^3$ ($l\,m^{-1}\,sec^{-1}$)	E^a (kcal mole^{-1})	$\Delta S_{25}^{\ddagger}$ (e.u.)
I *cis*	4-nitro	H	25	0·71	25	+8·8
II *cis*	4-nitro	D	25	0·32	—	—
III *trans*	4-nitro	H	78·25	0·97	25·1	−2·8
III *trans*	4-nitro	D	78·25	0·98	—	—
IV *cis*	2,4-dinitro	D	25	1070	19·9	+6·3
V *trans*	2,4-dinitro	H	25	720	18·9	−7·1

[a] Similar results are obtained from the corresponding chloro-compounds[58b].

the elimination is by a concerted mechanism[59]. However, the *trans*-isomers react much more slowly to form the acetals as indicated in equation (31). The addition of the OCH_3 groups to the β-carbon

$$RC_6H_4CH{=}CHBr \xrightarrow[CH_3OH]{^-OCH_3} RC_6H_4CH_2CH(OCH_3)_2 + Br^- \quad (31)$$

atom is in accord with the occurrence of the addition mechanism with subsequent addition of MeOH as in equation (32).

$$\mathrm{Ar(H)C{=}C(H)Br} \xrightarrow{+OCH_3^-} \mathrm{Ar(H)\bar{C}{-}CH(OCH_3)Br} \xrightarrow{-Br^-} \mathrm{Ar(H)C{=}C(H)OCH_3} \xrightarrow{CH_3OH} \mathrm{ArCH_2{-}CH(OCH_3)_2} \quad (32)$$

That equation (30) was not important in the case of the *trans*-isomers was shown[58a] by studies on the α-deuterated compounds (Table 4). There is no isotope effect, and no exchange of deuterium with the solvent, eliminating equation (30). The isotope effect shown by the *cis*-isomers is in accord with the reaction proceeding completely by elimination. By contrast, nucleophiles which are much less hydrogen-basic react with both the *cis*- and *trans*-isomers of these systems to give products with retention of configuration, indicating an addition–elimination mechanism for both isomers[58, 60]. Very similar results have recently been reported for the reactions of the corresponding chloro compounds[58b].

1,1-*diphenylethylenes*. The possibility of α-hydrogen abstraction is removed in the case of 2-halo-1,1-diphenylethylenes, and these would seem to be ideally suited for kinetic investigations of alkene nucleophilic substitution reactions. However, 1,1-diphenyl-2-chloro- or -2-bromoethylenes react with basic alcoholic solutions to give, in addition to substitution of the halogen by alkoxide ions, or, in some cases, instead of this substitution, rearrangement by α-elimination to give diphenylacetylenes as in equation (33) (Fritsch rearrange-

$$\mathrm{Ar_2C{=}C(H)Hal} \xrightarrow{OCH_3^-} \underset{\mathbf{(21)}}{\mathrm{Ar{-}C{\equiv}C{-}Ar}} + CH_3OH + Hal^- \quad (33)$$

ment[61, 62]). When *t*-BuO⁻ is used as base, only α-elimination occurs[63], while other alkoxides give mixtures of the two mechanisms[64]. Replacement of the α-hydrogen with a methyl group does not give a clean substitution reaction[65].

However, some suitably substituted derivatives do give clean substitution reactions. Thus, Silversmith and Smith[66] found that 1,1-diphenyl-2-fluoroethylene reacted with OEt^- in ethanol to give 1,1-diphenyl-2-ethoxethylene. The reaction was first-order with respect to both reactants, and the authors concluded that the results were consistent with an addition–elimination mechanism in which the first step was the formation of an ionic intermediate [(equation 34), with Ar = C_6H_5 and Hal = F].

$$Ar_2C{=}CH(Hal) + {}^{-}OEt \rightleftharpoons Ar_2\bar{C}{-}CH(OEt)(Hal) \xrightarrow{\text{either}} Ar_2CH{-}CH(OEt)(Hal)$$

$$\xrightarrow{\text{or}} Ar_2C{=}CH(OEt) \qquad (34)$$

Similarly, Beltrame and co-workers[67] found that *p*-nitro substitution in the phenyl rings gave simple alkoxydehalogenation of the chloro- and bromo-derivatives. The reactions were again first-order with respect to each reactant, and the results can be accommodated in equation (34) above (with Ar = *p*-$NO_2C_6H_4$ and Hal = Cl, Br). The rates of reaction were $10^6 \times$ those of the unsubstituted compounds, and a correlation between the rates of reaction for a series of *p*,*p*-disubstituted derivatives and the sum of the Hammett parameters for the two substituents was found. A discussion of the theoretical approach of these authors to this reaction will be given in section IV.D.

C. Attack on C≡C Bonds: Addition of Alcohols

The addition of alcohols and phenols to acetylenes can be catalysed by both alkoxides[68] and tertiary amines[69]. As with alkenes, reaction is greatly facilitated by the presence of electron-withdrawing substituents. The addition is generally[68, 70] considered to proceed by a *trans* mechanism.

Thus Miller[68] found that the reaction of phenylacetylene with sodium methoxide in methanol solution gave only one isomer, considered to be the *cis*-form. The mechanism (35) (Ar = Ph) was suggested to account for this.

Similarly, Modena and co-workers[71, 72] have shown that the

$$\text{Ar—C≡C—H} \xrightarrow{OCH_3^-} \text{Ar(}^-\text{)C=C(OCH}_3\text{)(H)} \xrightarrow{CH_3OH} \text{Ar(H)C=C(OCH}_3\text{)(H)} + OCH_3^- \quad (35)$$

methoxide-catalysed addition of methanol to *p*-tolylsulphonylacetylene gives the *cis*-2-alkoxy-1-*p*-tolysulphonylethylene [equation (35), Ar = *p*-tolyl·SO_2].

The reaction, measured by the rate of disappearance of the acetylene by measurement of the C≡C infrared stretching frequency, was found to obey a first-order kinetic equation, as the alkoxide concentration does not change with time, but the rate depends on the first power of the alkoxide concentration as required by equation (35). As judged by the second-order rate constants for the reaction, ethoxide ion reacts faster than methoxide ion. The activation energies for the addition of MeOH and EtOH are very similar: E_a(MeOH) = 16·4 kcal mole^{-1}, E_a(EtOH) = 17·6 kcal mole^{-1}.

Although substitution of the acetylene hydrogen by a methyl group does not alter the kinetics of the reaction significantly[73], it alters the course of the reaction[74], giving 2-methoxy-3-phenylsulphonylpropene (**22**) as the kinetic product which then isomerizes to *trans*-2-methoxy-1-phenylsulphonylpropene (**23**) as in equation (36).

$$PhSO_2C{\equiv}CCH_3 \longrightarrow PhSO_2CH_2C(OCH_3){=}CH_2 \ (\mathbf{22}) \longrightarrow (PhSO_2)(H)C{=}C(CH_3)(OCH_3) \ (\mathbf{23}) \quad (36)$$

The addition of alcohols to disubstituted acetylenes has been investigated by Winterfeldt and co-workers[69, 70, 75, 76]. In general, the rule of *trans*-addition is borne out. Thus, addition of alcohols to acetylene dicarboxylate gives mainly the compound **24**, although additions at high temperatures lead to the *cis*-derivative. With dicyanoacetylene, however, the *cis*-isomers predominate, even although the reactions were performed at room temperature.

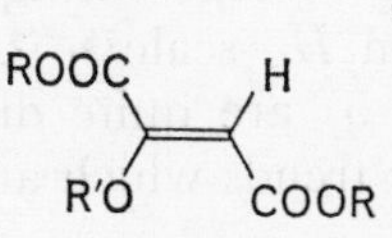

ROOC H
R'O COOR
(24)

In general, alkoxide-catalysed additions to triple bonds will be complicated by the possibility of the base-catalysed isomerization of the resulting alkene(s). Arguments regarding the stereochemistry of the reaction which are based on product analysis will be more valid when the thermodynamically less stable alkene predominates.

D. *Investigations Involving Intermediates*

In 1885, Friedlander[77] found that *p*-nitro-*β*-nitrostyrene formed a complex which he formulated as **25** when it was treated with an alcoholic solution of potassium hydroxide (equation 37). The salt decomposed rapidly in the free state yielding the original components. This compound does not seem to have been investigated further, but recently intermediates corresponding to those postulated

$$p\text{-}NO_2C_6H_4CH{=}CHNO_2 \xrightarrow[EtOH]{OEt^-K^+} p\text{-}NO_2C_6H_4\underset{OEt}{\underset{|}{C}}H{-}CH{=}NOO^-K^+ \quad (37)$$

(25)

TABLE 5. Wavelengths (mμ) and extinction coefficients (l m^{-1} cm^{-1}) of the positions of maximum absorption of substituted α-cyanostilbenes in basic DMSO–ethanol mixtures[78].

Stilbene	λ_{max}	ε
I 4-nitro	547	42,600
II 3′-chloro-4-nitro	547	43,700
III 4,4′-dinitro	553	42,100
IV 4-cyano	402	47,400
V 3′-chloro-4-cyano	402	47,200
VI 4′-nitro-4-cyano	393	44,000
VII 3-cyano	363	28,800

as the product of initial attack by alkoxide ions o. general formula **9** have been completely characterized in activated stilbene systems[78–80].

Thus, the interaction between alkoxide ions and a large number of substituted stilbenes in DMSO–methanol mixtures has been investigated by Stewart and Kroeger using u.v. spectroscopy for the purpose of establishing an H_- scale[78, 79]. The u.v.–visible spectra of these solutions (Table 5) are quite different from those of the parent stilbenes and show trends which are indicative of the nature of the species formed.

Of the possible reactions (38–40), it was considered that[48], under

$$\text{Ar}^{\beta}\text{—CH=C(CN)Ar}^{\alpha} + \text{OCH}_3^- \rightleftharpoons \text{Ar}^{\beta}(\text{C}^-\text{=C})(\text{CN})\text{Ar}^{\alpha}\ (\mathbf{26}) + \text{CH}_3\text{OH} \quad (38)$$

$$\text{Ar}^{\beta}\text{—CH=C(CN)Ar}^{\alpha} + \text{OCH}_3^- \rightleftharpoons \text{Ar}^{\beta}\text{C(OCH}_3\text{)(H)—C}^-\text{(CN)Ar}^{\alpha}\ (\mathbf{27}) \quad (39)$$

$$\text{Ar}^{\beta}\text{C(OCH}_3\text{)(H)—C}^-\text{(CN)Ar}^{\alpha}\ (\mathbf{27}) + \text{CH}_3\text{OH} \rightleftharpoons \text{Ar}^{\beta}\text{—CH(OCH}_3\text{)—CH(CN)Ar}^{\alpha} \quad (40)$$

the conditions of the experiment, the contribution from equation (40) would not be significant. From the similarity in shape and position between the u.v. spectra of the species formed from α-cyano-4-nitrostilbene and the anion of 4-nitrobenzyl cyanide, it was concluded that the colour-producing species was **27** and not **26** which could be formed by proton abstraction. Further support for the colour-producing reaction being one of methoxide addition, rather than proton abstraction, comes from the general effect of substituents in the two rings, both on the position of the u.v.–visible maximum[78] (Table 5) and the equilibrium constant for the reaction[79]. In general, substituents in the β-phenyl ring (I—III and IV—VI) have much less effect than substituents in the α-phenyl ring (I, IV, VII; II, V; III, VI) as would be expected from structure **27**, whereas **26** would give exactly the opposite trend.

The interactions of anhydrous sodium methoxide with a series of α-cyano-4-nitro-4′-X-stilbenes have also been studied by n.m.r.[80], and the intermediates formed in these cases characterized unambiguously. Because of the lack of abstractable hydrogens in the system, reaction (40) cannot occur.

The n.m.r. spectra of the products of reactions corresponding to equations (38) and (39) for α-cyano-4-nitro-4′-X-stilbenes would be expected to be quite different and diagnostic: Thus the product (**28**) one would expect to have an n.m.r. spectrum consisting of a 'normal' aromatic spectrum for the β-phenyl ring and a spectrum closely corresponding to the spectrum of the anion of 4-nitrophenylacetonitrile (**30**) for the α-phenyl ring, but showing some variation with both nucleophile and substituents. There should be a large change in the chemical shift of H_β from an alkene hydrogen at low field to a hydrogen attached to an sp^3 carbon atom at higher field, whose chemical shift would be dependent on the nature of the nucleophile.

The spectrum of compound **29** would be quite different, a critical feature being the complete lack of an absorption corresponding to the abstracted alkene proton H_β. If hydrolysis occurred due to small traces of water, then the α-phenyl ring should show exactly the spectrum of the anion of 4-nitrophenylacetonitrile (**30**) with the methylene hydrogen at high field, and the β-phenyl ring the spectrum of 4-X-benzaldehyde, with the aldehydic proton at low fields and independent of the nucleophile.

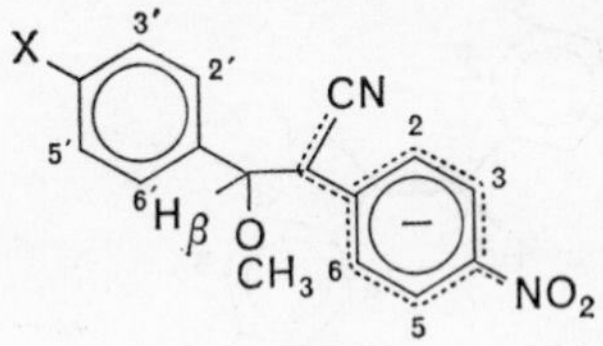

(28) 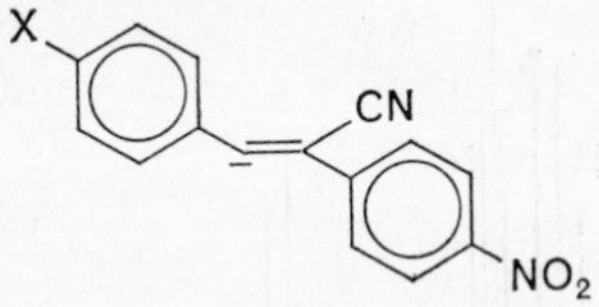 (29) 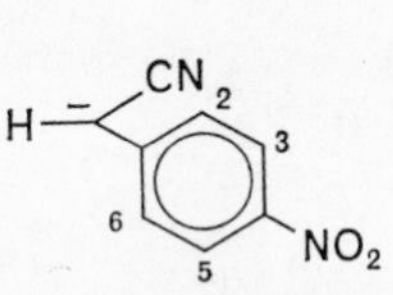 (30)

The addition of anhydrous sodium methoxide to a DMSO solution of α-cyano-4-nitro-4′-methoxystilbene gives an intense violet-pink coloration (λ_{max} 550 mμ) and causes the broadening and eventual disappearance of the spectrum of the stilbene and the appearance of a new spectrum at higher fields (Figure 1). The new spectrum consists of an AA′XX′ pattern centred at $\delta = -6{\cdot}95$ (rel. intens. 4), two multiplets centred at $\delta = -7{\cdot}34$ (doublet, rel. intens. 2) and $\delta = -6{\cdot}38$ (approximately a triplet with further splitting, rel. intens. 2), and a sharp singlet at $\delta = -5{\cdot}01$ (rel. intens. 1). The spectrum is consistent with the formation of the intermediate **28** (X = OCH_3), as shown in Figure 1.

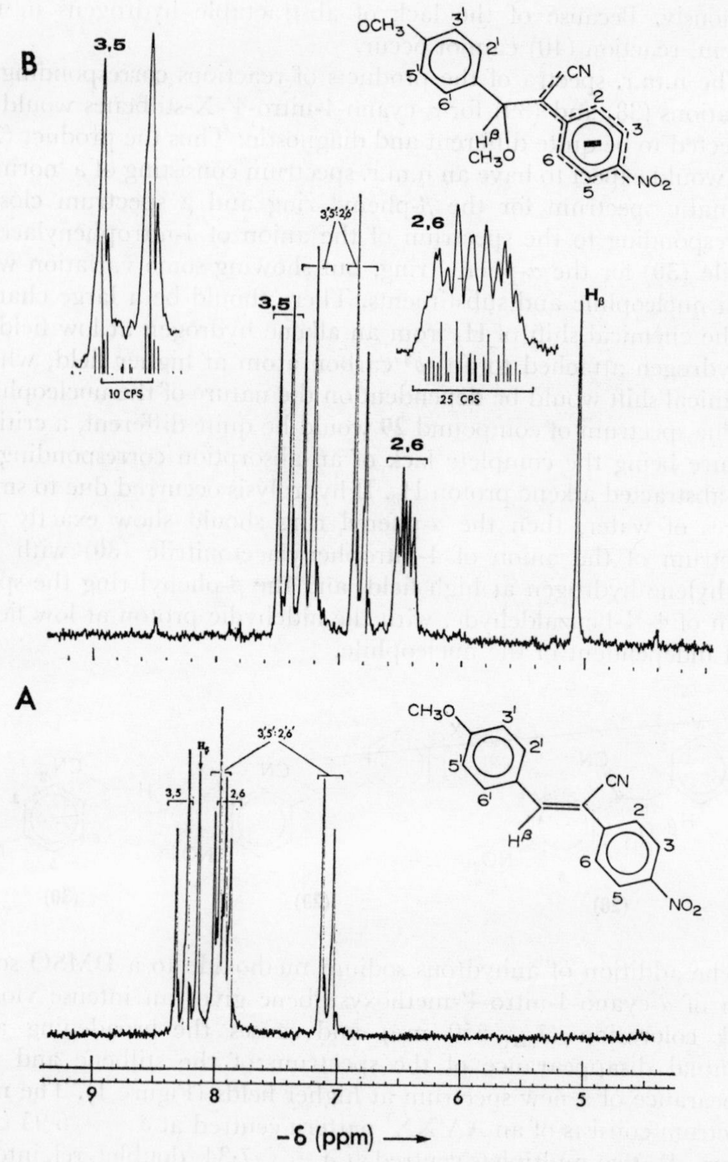

FIGURE 1. 100 MHz n.m.r. spectra of (A) X-cyano-4-nitro-4′-methoxystilbene in DMSO solution and (B) this solution after the addition of anhydrous sodium methoxide.

It is inconsistent with either hydrolysis or the occurrence of reaction (38). Thus the AA'XX' pattern can be assigned to the four hydrogens on the β-phenyl ring and the singlet at $\delta = -5{\cdot}01$ to the H_β proton. The shifts of the other two multiplets correspond very closely to those observed for the ring hydrogens of 4-nitrophenylacetonitrile anion (Table 6), although the multiplet structures are different, and can be assigned to the four hydrogens of the α-phenyl ring. That they do not form a simple AA'XX' system suggests that, as in the case of the 4-nitrophenylacetonitrile anion[81], there is restricted rotation around the α-phenyl–C_α bond, the complexity of the resonance at $\delta = -6{\cdot}38$ arising because $J_{23} \sim \gamma_0\delta_{2,6}$. This, and the very large changes in the chemical shifts of the α-phenyl resonances, suggest very substantial delocalization of the negative charge into the α-phenyl ring, so that the anion is probably more properly formulated as in **28**, rather than with the negative charge localized on the α-carbon atom as is usually done. This is thought to be the case for all the compounds studied. Such a delocalization would be favoured by a planar rather than a tetrahedral configuration at the α-carbon atom, and it is possible that this is the case for stilbenes with very electronegative groups in the 4-position.

The sharpness of the H_β proton would suggest that there is free rotation round the C_α–C_β bond as required for such intermediates to take part in the *cis–trans* isomerization of alkenes. The solutions are quite stable for several days.

Similar conclusions can be drawn regarding the structures of the intermediates formed from a whole series of α-cyano-4-nitro-4'-X-stilbenes. The n.m.r. parameters (Table 6) are in close agreement with those described above, but show small differences which rule out the formation of **30**, as identical spectra for the α-phenyl ring would be obtained in this case. In general, the complexes seem to be quite stable (at least in solution). The n.m.r. spectra reveal that, whereas there is free rotation about the β-phenyl–C_β and C_α–C_β bonds, there is restricted rotation around the α-phenyl–C_α bond in all the complexes and delocalization of the negative charge into the α-phenyl ring, suggesting perhaps the favouring of a planar configuration at this carbon atom. Although the above effects will, in general, be very dependent on the nature and position of substituents in the α-phenyl ring, it is thought that the results obtained from these more activated substrates may be of somewhat general applicability.

Both the *cis*- and *trans*-isomers of α-cyano-4-nitrostilbene give rise to the same n.m.r. spectrum on treatment with base.

TABLE 6. Chemical shift data ($-\delta$, ppm), coupling constants (cps) and chemical shift differences (cps) from the 100 MHz n.m.r. spectra of compounds of general formula **28** and **30** in DMSO solution[a].

X	H_β	$H_{2\,6}$[b]	$\gamma_0\delta_{2,6}$	$J_{2,6}$	$J_{2,3}$ $J_{5,6}$	$J_{3,5}$	$\gamma_0\delta_{3,5}$	$H_{3,5}$[b]	β-Phenyl ring
28 OCH_3	5·01	6·38	5·0 ± 0·5	2·5 ± 0·4	9·5 ± 0·5	1·8 ± 0·3	0·0 + 0·5	7·34	AA′XX′ system
28 CH_3	5·09	6·44	8·5 ± 0·5	2·1 ± 0·3	9·5 ± 0·5	1·8 ± 0·3	0·0 + 0·5	7·45	AA′BB′ system
28 NO_2[c]	5·36	6·56	18.0 ± 0·5	—	10·0 ± 0·5	—	—	7·50	AA′XX′ system
28 H	5·16	6·51	9·5 ± 0·3	2·3 ± 0·3	10·0 ± 0·3	1·8 ± 0·3	0·5 ± 0·4	7·47	Unresolved multiplet
30 —	—	6·27	19·0 ± 0·3	2·3 ± 0·2	9·0 ± 0·2	2·6 ± 0·2	13·0 ± 0·3	7·39	—

[a] C. A. Fyfe, *Can. J. Chem.*, **47,** 2331 (1969).
[b] Denotes multiplet centre.
[c] Owing to radical-exchange broadening, only the major splittings were properly resolved.

E. Theoretical Approaches

In 1953, Gold[82] discussed alkene nucleophilic substitution reactions in terms of the possible intermediates or transition states **31** and **32.** In **31**, attack by the nucleophile occurs in the plane of the alkene so that in the transition state, the groups C,Y,R and X are all coplanar. Such a mode of displacement would give a change of configuration, and more recent work has eliminated this. The second possibility (**32**) where the attacked carbon atom assumes a

(31) (32)

tetrahedral configuration, is more in accord with the experimental evidence. **32** could range from approximating to the transition state in a one-step mechanism to being a relatively stable carbanion.

In fact, the analogies between aromatic and alkene nucleophilic substitutions are very strong. The activation energies and the effect of substituents in both cases are very similar, and during the reaction the attacked carbon atom must change its configuration from sp^2 to sp^3, either in a transition state or intermediate complex (equation 41). In both cases, the existence of stable intermediates from suitably activated substrates can be demonstrated.

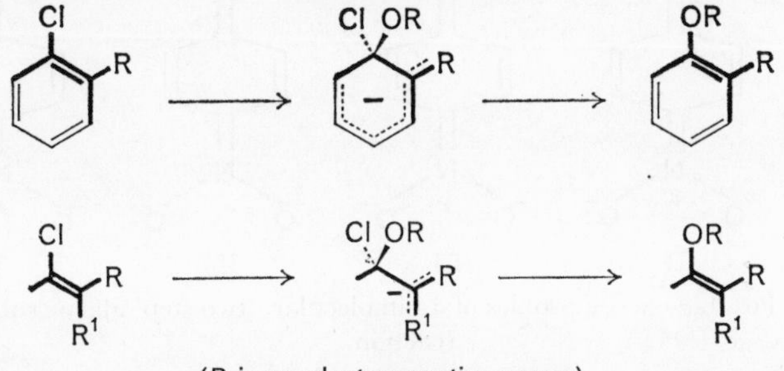

(41a)

(41b)

(R is an electronegative group)

It should be possible directly to extend the semi-quantitative approach of Miller[203] for aromatic substitution (see section V), using the same data, to, for example, the alkoxy dehalogenation of 2-(*p*-nitrophenyl)-1-haloethylenes. The transition states and intermediate complexes could be represented as I—V in Figure 2. By analogy with the aromatic system (see on), these would give rise in the general case to curves of the types A—C in Figure 2. Thus a useful qualitative guide could be got as to the activation energies and also to the life-time of the intermediate complex, and hence the occurrence of essentially concerted reaction and possible retention of stereochemistry (curve C) or formation of the thermodynamically more stable isomer (curve A).

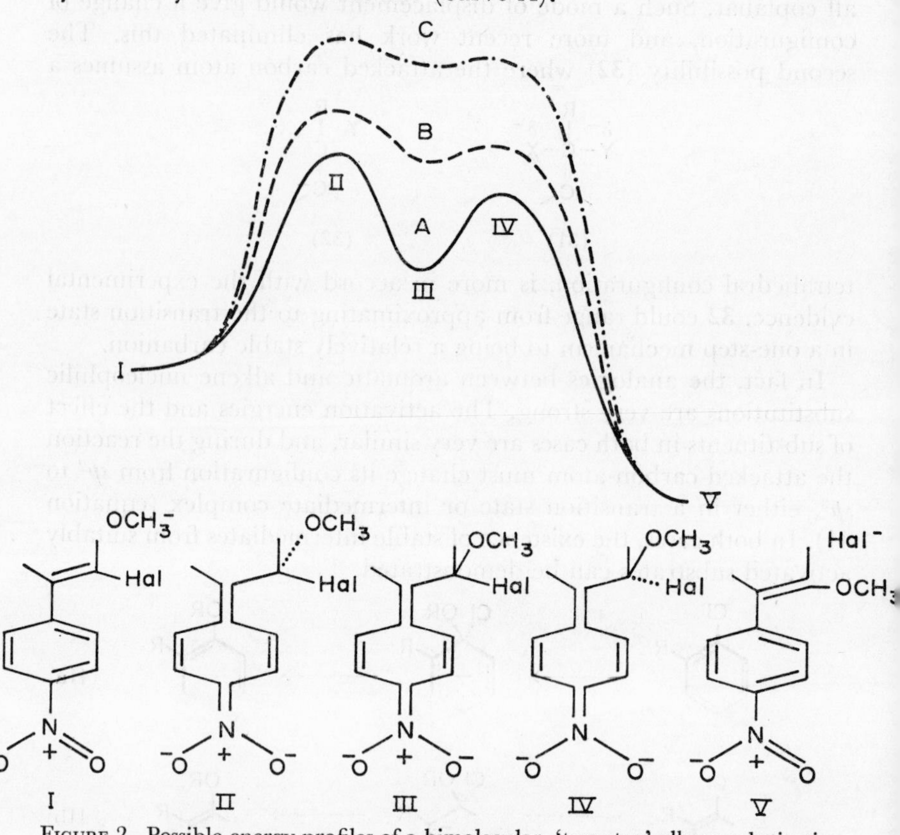

FIGURE 2. Possible energy profiles of a bimolecular, 'two-step' alkene substitution reaction.

Although there is this close analogy with aromatic substitutions, there has been little attempt to accommodate alkene substitutions within the general theoretical approaches to aromatic substitutions. Recently, however, Simonetta and co-workers[67] have extended their delocalization approach (see section V) to the alkoxy dehalogenation of 1,1-di-*p*-nitrophenyl-2-haloethylenes. They considered the attacking alkoxide group to approach from above the plane of the alkene and to form a transition state of type **33** by attack at the α-carbon atom.

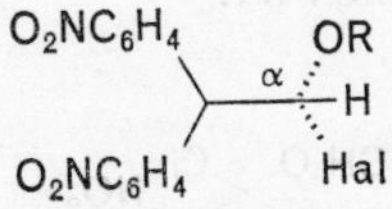

$$\begin{array}{c} O_2NC_6H_4 \quad\quad OR \\ \quad \alpha \\ \quad\quad\quad -H \\ O_2NC_6H_4 \quad\quad Hal \end{array}$$

(33)

The leaving and attacking groups were considered together as a 'pseudo-atom' bonded to the residue by a 'quasi-σ' bond to the α-carbon atom and by interaction with the π-electrons of the residue via a 'quasi-π' orbital and two electrons just as in the aromatic case (see on). On this model, differences in π-electron energies in the initial substrates and the transition states (**33**) were calculated. A correlation was obtained between experimental ΔG values and calculated $\Delta E\pi$ values, but the authors considered that the fact that the parameters which had to be chosen for the pseudo-atom were different from the corresponding ones in the aromatic and aza-aromatic series gave rise to doubts as to the applicability of the method to alkene substitutions.

V. ATTACK ON AROMATIC SYSTEMS

A. General Comments

Because of the high electron density of the aromatic system, as with alkenes, nucleophilic substitution usually occurs only where there are one or more suitable electronegative groups (e.g. NO_2, CN, COR) oriented *ortho* and/or *para* to the position of substitution. Thus, although the Dow process for the direct conversion of chlorobenzene into phenol by Na_2CO_3 requires copper as catalyst and temperatures in excess of 300°C, *ortho*- and *para*-nitrohalobenzenes undergo relatively facile reaction with base.

In general, the reactions are kinetically second-order, first-order with respect to both nucleophile and substrate, and are usually discussed[83–88] in terms of the 'two-stage' mechanism introduced by Bunnett[83, 85] (equation 42). In this scheme, the displacement of, for example, chloride ion from 2,4-dinitrochlorobenzene by methoxide ion is thought to proceed via an intermediate (**34**) of finite stability. In **34** both leaving and attacking groups are covalently bonded to the carbon atom at the point of attack which is now of sp^3 hybridization, and the negative charge from the attacking alkoxide ion is delocalized round the residue of the aromatic ring and on to the two nitro groups. The energy profile of the reaction will have a minimum corresponding to the complex **34.**

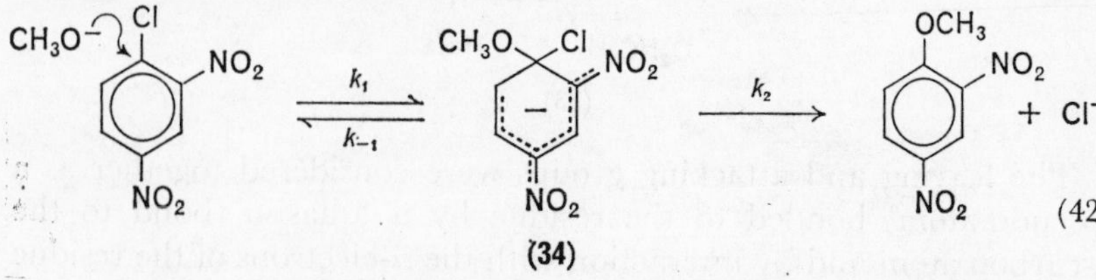

(42)

However, an alternative mechanism has been considered[89–91] (equation 43). In this, the reaction proceeds in one step, with synchronous bond making and breaking in an analogous manner to the aliphatic S_N2 reaction discussed in section II. There is *no* intermediate in this mechanism, and no dip in the energy profile of the reaction*.

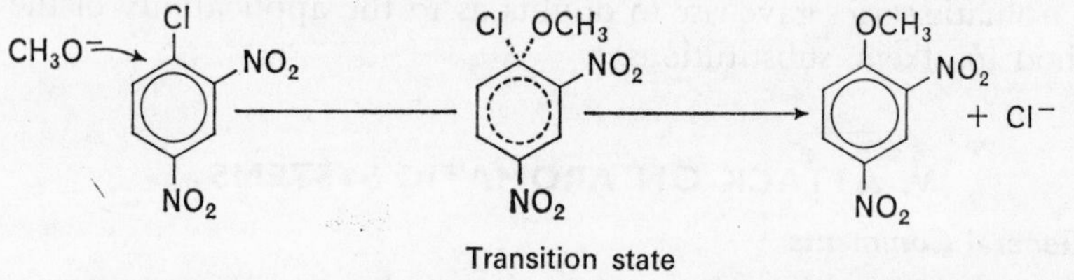

(43)

Transition state

There are several general pieces of kinetic evidence in favour of the two-step mechanism: (i) In reactions where the fission of a carbon–halogen bond is a rate-determining step, the order of the rates of reaction is $F < Cl < Br \sim I$. In many aromatic substitution reactions involving halogens, the observed order is $F \gg Cl \sim Br \sim I$[85]. This can be accounted for in terms of the two-stage mechanism with $k_2 < k_1$ in equation (42), but is difficult to explain on the basis of equation (43). The reverse order[89, 92] can also occur, and can also be accommodated in the two-stage mechanism with $k_2 > k_1$ in equation (42).

(ii) It is found[85] that the rates of reaction of a whole series of 1-X-2,4-dinitrobenzenes with piperidine are within a factor of five of one another, i.e., independent of the strength of the C–X bond.

* The structure assigned to the transition state in equation (43) should not be taken too literally as it is not entirely clear what type of structure is envisaged by those who have suggested this mode of reaction.

This again can be explained in terms of equation (42) with $k_2 < k_1$, but like (i) above is difficult to interpret in terms of the 'one-step' reaction scheme.

(iii) Very convincing kinetic evidence for the existence of an intermediate is the general base catalysis observed under certain conditions when amines are the nucleophiles. In terms of the intermediate complex theory (equation 44), the intermediate **35** will have a labile proton on the nitrogen atom. Any process which can facilitate the loss of this proton will increase the rate of conversion of **35** into products, as NR_2 is a much poorer leaving group than $^{+}NHR_2$. In the presence of a base, there will be the 'normal' conversion of the intermediate into products, and the faster 'base-catalysed' conversion, the overall effect being an increase in the reaction rate.

$$\text{1-F-2,4-}(NO_2)_2C_6H_3 \underset{k_{-1}}{\overset{k_1}{\rightleftharpoons}} \mathbf{(35)} \xrightarrow[\text{uncat.}]{k_2} \text{1-}NR_2\text{-2,4-}(NO_2)_2C_6H_3 \qquad (44)$$

(35) → products: base catalysed, $k_{B.C.} > k_2$

Since the catalysis involves the decomposition of the intermediate into products, it should only be observed in systems where the decomposition of the intermediate is the rate-determining step, i.e., $k_1 > k_2$, and will therefore be favoured by the presence of relatively poor leaving groups, and its occurrence and magnitude will vary as these groups are changed. In fact, base catalysis has been observed[93-97] in a variety of systems, and the general trends predicted[93] by the two-stage mechanism are found. Particularly incisive is the observation of a variable $^{16}O/^{18}O$ isotope effect in the substitution of 2,4-dinitrophenyl phenyl ether by piperidine, when catalysed by varying concentrations of hydroxide ion[97]. The general mechanism for the base catalysis is considered[94, 95] to involve a fast proton transfer, followed by a rate-determining removal of the leaving group by the moiety BH^{+} [98].

Although any one particular case of catalysis could be accounted for in terms of the one-step theory by the postulate that the base intervened in the transition state, the differences observed with the variation of substrate and nucleophile cannot reasonably be accommodated in this fashion, and the above observations provide the strongest kinetic evidence for the 'two-step' mechanism.

However, the cases which can kinetically be unambiguously assigned to the bimolecular mechanism are relatively few (and must obviously involve amines), and the general acceptance of the 'two-step' mechanism is to some extent due to its more flexible character, and to the inferences which can be drawn from other studies (see on).

B. *Substitutions Involving Alkoxide and Hydroxide Ions*

1. 'Activated' substrates

Displacement reactions involving attack by alkoxide and hydroxide at activated centres have been widely investigated. Typical of these are the displacement reactions involving halogens, some illustrative examples of which are given in Table 7. Activation by substituents is generally in the order *para* > *ortho* when alkoxide or hydroxide ions are the nucleophiles, both when judged by rates of reaction and by activation energies. The main effect of the substituents is seen in the very large lowering of the activation energy from that of the unactivated substrate.

The activating effects are not however additive (Table 7). Thus the introduction of a second nitro group lowers the activation energy by a much smaller amount than the introduction of a single group in either the *ortho* or *para* position.

The relative activating effects of various substituents have been found[83, 85] to be in the order

$$-N_2^+ > -NO_2 > -SO_2CH_3 > -NMe_3^+ > -CN > -CF_3$$

In general, the effects of substituents will be shown in the ground state, the transition state, and in the intermediate complex, all of which may determine the kinetic pattern of the reaction. Correlations can be obtained with the variation of one or more substituents within a series of closely related compounds, which may allow inferences to be drawn regarding the mechanism. For example, Norman

Cl NO₂ NO₂
(36)

Cl NO₂ Cl
(37)

and co-workers[99] found that in the bimolecular reactions of both 1-chloro-2,4-dinitrobenzene (**36**) and 1,4-dichloro-2-nitrobenzene (**37**) with a series of substituted phenoxide ions, the rates correlated with the σ-values of the phenoxide ion substituents, suggesting that in the transition state the phenoxide ion is appreciably bound. The parallel nature of the correlations for the two series suggested[99] that the transition states for the two were similar, even though the reactivities were very different.

TABLE 7. Second-order rate coefficients ($l\ m^{-1}\ sec^{-1}$) and activation energies ($kcal\ mole^{-1}$) for the reactions of alkoxide ions with some nitro-activated halobenzenes.

Substrate	Conditions	k_2	E_a	Ref.
Chlorobenzene	OMe^-|MeOH (200°)	—	40·0	105b
2-chloronitrobenzene	OEt^-| EtOH (90°)	$3{\cdot}97 \times 10^{-4}$	22·2	218
4-chloronitrobenzene	OEt^-| EtOH (90°)	$9{\cdot}63 \times 10^{-4}$	20·1	218
4-chloronitrobenzene	OMe^-|MeOH (0°)	$8{\cdot}9 \times 10^{-9}$	24·0	204
4-chloro-1,3-dinitrobenzene	OMe^-|MeOH (0°)	$2{\cdot}0 \times 10^{-3}$	17·5	204
Fluorobenzene	OMe^-|MeOH (200°)	—	34·9	[a],
4-fluoronitrobenzene	OMe^-|MeOH (0°)	$6{\cdot}26 \times 10^{-6}$	21·2	[a], 207
4-fluoro-1,3-dinitrobenzene	OMe^-|MeOH (0°)	$1{\cdot}74 \times 10^{0}$	13·5	[a], 207

[a] B. A. Botto, M. Liveris and J. Miller, *J. Chem. Soc.*, 750 (1956).

As shown above, groups *ortho* and *para* to nitro groups undergo facile displacement by hydroxide and alkoxide ions. This is true of nitro groups, and the *ortho*- and *para*-isomers of dinitrobenzene as well as higher homologues are subject to ready displacement of one group by base. In accord with the general characteristics of nitro-activated substrates, the relative rates are *para* > *ortho*, and the reaction proceeds faster with ethoxide than with methoxide ions[100].

In substitutions involving hydroxide ions[100c, 101–104], the product obtained is not the phenol, but the phenate ion. A typical example is the reaction of trinitroanisole with hydroxide ion. Gold and Rochester[104] examined its rate of decomposition in weakly basic phosphate buffer solutions and considered that the kinetics were consistent with the rate-determining step being the attachment of the hydroxide ion to form the intermediate **38** (equation 45). There are two pathways by which the phenate ion can now be formed from the complex. In general, it will be very difficult to distinguish between them.

$$\text{O}_2\text{N}\text{-}C_6H_2(OCH_3)(NO_2)_2 \rightleftharpoons \underset{(38)}{[\text{HO, OCH}_3\text{ adduct}]^-} \xrightarrow{-H^+} [{}^-\text{O, OCH}_3\text{ adduct}]$$
$$\downarrow -CH_3O^- \qquad\qquad \downarrow -CH_3O^- \qquad (45)$$
$$\text{2,4,6-}(NO_2)_3C_6H_2OH \xrightarrow{-H^+} \text{2,4,6-}(NO_2)_3C_6H_2O^-$$

2. Non-activated substrates

Discussion of nucleophilic aromatic substitution normally centres on the reactions of the so-called 'activated' substrates discussed above, where the activating groups are orientated *ortho* and/or *para* to the reaction site, and can stabilize excess negative charge in the ring in both transition state and intermediate complex. However, if 'activation' is judged by the rates and activation energies of the reaction, then there is considerable activation in substrates where the activating groups are orientated *meta* to the site of reaction. Thus, although the rate of reaction of *m*-nitrofluorobenzene is approximately 10^{-4} that of the *ortho*- or *para*-isomers, there is a considerable increase compared with the unactivated substrate[22]. If there are two nitro groups present (to give 1-fluoro-3,5-dinitrobenzene), then the decrease in activation energy (cf. fluorobenzene) is equal to that effected by a nitro group in the *ortho*- or *para*-position. Since there can be no direct delocalization of charge into the nitro groups in this displacement, an intermediate of finite stability is unlikely and reaction probably proceeds in one step. However, these reactions can be considered within a general formalism* involving an intermediate corresponding to those of the two-stage mechanism (equation 46). If the reaction involved the intermediate, then a considerably smaller activation energy might be required than if the reaction proceeded directly from the aromatic substrate. The

* Of which the reaction of trinitrobenzene with methoxide ion is a special case.

$$\text{3,5-}(O_2N)(NO_2)C_6H_3F \underset{}{\overset{OR^-}{\rightleftharpoons}} \text{(39)} \xrightarrow{-F^-} \text{3,5-}(O_2N)(NO_2)C_6H_3OR \qquad (46)$$

(39)

reaction between trinitrobenzene and methoxide ion (see below) where there is some evidence of the intermediacy of a complex corresponding to **39** is a special case of this general scheme where the displaced group is a nitro group. In fact, intermediates corresponding to **39** for a variety of halogen substituents can be observed in basic DMSO solution[106], and it may be possible to investigate their possible role in the reaction mechanism.

C. Investigation of Attack by OR^- and OH^- leading to the Formation of Intermediates

Under certain conditions, attack by hydroxide or alkoxide ions can lead to the formation of intermediates in large enough concentration for their detection, characterization and possibly isolation[88, 107]. This is achieved if the two-stage mechanism is stopped at the intermediate complex, effectively making the substitution reaction an equilibrium reaction involving the reactants and intermediate complex. The two common methods are (a) by the direct attack of alkoxide or hydroxide ion on an 'unsubstituted' nitroaromatic, where further reaction of the intermediate involves the energetically unfavourable elimination of H^- and the reaction effectively stops at this point and (b) by attack of alkoxide ion on a nitroanisole so that the complex is symmetrical with respect to both the attached groups, and any subsequent reaction must lead to the formation of the original reactants. Equilibria established under conditions (a) or (b) above, are then altered in favour of the complex by the use of dipolar aprotic solvents such as dimethyl sulphoxide or dimethylformamide.

I. Attack on nitro- and polynitro-aromatics

a. Attack on 1,3,5-*trinitrobenzene**. The interaction between TNB and alkoxides was investigated by several authors[108–111] before the end of the last century. In 1895 Lobry de Bruyn and Van Leent[108c] described the isolation of a solid complex by the action of KOH on

* Hereafter referred to as TNB.

a methanolic solution of TNB. Similar observations were made by other workers[109, 110]. De Bruyn showed that it was formed only in the presence of alcohol and had the empirical formula $C_6H_3(NO_2)_3.CH_3OK$. The formulation **40** was suggested by analogy

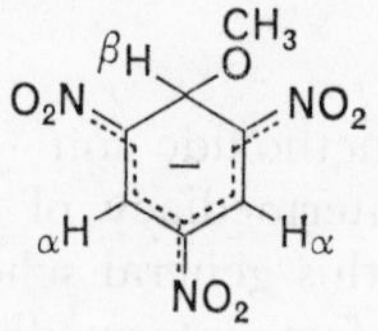

(40) 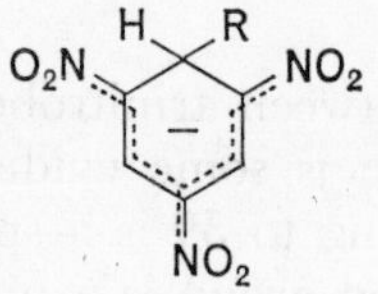 (41)

with the suggestions of Meisenheimer for the corresponding complex with trinitroanisole (see on). More recently, the visible[112] and infrared[113] spectra of this complex have been shown to correspond closely to those of the dialkoxy Meisenheimer compounds. A very elegant confirmation that the compound has structure **40** was given by Crampton and Gold[114]. The three equivalent protons in TNB give rise to a single sharp absorption at $\delta = -9{\cdot}21$ in DMSO solution, but the isolated product from the action of methanolic KOH on TNB shows two absorptions; a doublet at $\delta = -8{\cdot}42$ of relative intensity 2 ($J \sim 1$ cps) and a triplet at $\delta = -6{\cdot}14$ or relative intensity 1. These are in excellent agreement with those expected from H_α and H_β respectively in **40**. The large upfield shift in the resonance ascribed to H_β is caused by the change from sp^2 to sp^3 hybridization in the ring carbon atom, and chemical shifts in the range $\delta = -6{\cdot}0$ to $-6{\cdot}5$ are characteristic of this environment[107]. A product analogous to **40** is formed by the action of ethoxide ion on TNB[115]. Solutions of **40** undergo very facile exchange reactions with ketones[115, 116], amines [117] and aliphatic nitro compounds[118] to form the corresponding compounds **41** (R = CXYCOR′, NR′R″, $CXYNO_2$ respectively) by treatment of a DMSO solution of **40** with the reacting species. N.m.r. measurements have also been made on the 'in situ' formation of **40**. If increasing amounts of methoxide are added to a solution of TNB in DMSO, the n.m.r. spectrum shows first the formation of **40**[115, 119, 120], followed by changes at higher methoxide concentrations which may be indicative[115, 120] of the formation of **42**.

Quantitative spectrophotometric studies have been made on the interaction between TNB and alkoxides in alcoholic solution[121, 122]. Gold and Rochester found a value of $K = 15\,l\,m^{-1}$ for the equili-

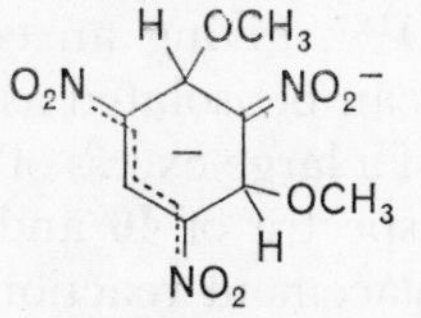

(42)

brium constant at 28°. Caldin and co-workers[122] studied the rate of formation of the ethoxy analogue of **40** and the rate of its decomposition by acids between −70° and −100°C. They found the activation energies for the forward and back reactions to be very similar, indicating considerable stability for the complex. At high ratios of alkoxide to TNB it has been suggested that complexes of stoichiometry 2 : 1 and 3 : 1 are formed[123, 124]. Solutions of **40** in methanol are known to be unstable[108, 121, 124–126], a slow irreversible reaction yielding mainly 3,5-dinitroanisole and nitrite ion taking place. The reaction is first-order in the complex **40**[121] but this does not necessarily mean that it is an intermediate in the reaction. The reaction would seem to proceed through a *transition state* (**43**). Such a

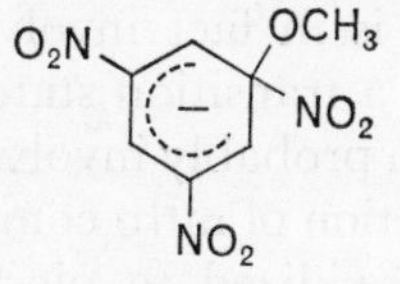

(43)
Transition state

species would have a very low stability, as there could be no delocalization of the negative charge by the nitro groups by a mesomeric effect, and it must be emphasized that *no* evidence has been found for a stable species corresponding to **43** in *any* system.

Attack by hydroxide ion on TNB in DMSO solution yields

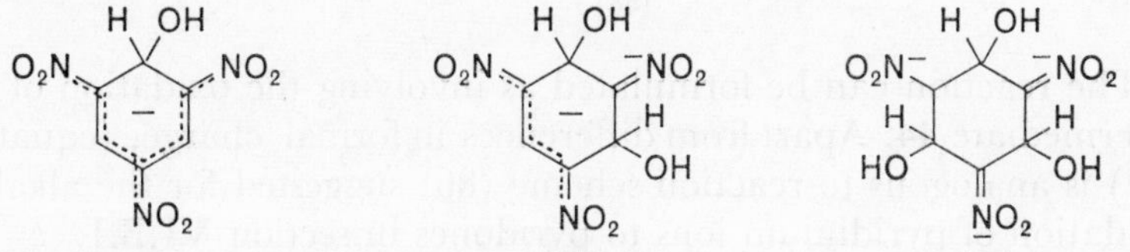

(44) **(45)** **(46)**

44[119, 127]. The n.m.r. spectrum of **44** is very similar to that of **40**, showing absorptions of $\delta = -8{\cdot}2_0$ (rel. intens. 2) and $\delta = -6{\cdot}1_5$ (rel. intens. 1). Compound **44** can also be made by treating a DMSO solution of **40** with H_2O[127], giving an equilibrium mixture of **40** and **44***. Compound **44** can be isolated as its sodium salt in quantitative yield by addition of a large excess of H_2O to a DMSO solution of **40**. The u.v. and i.r. spectra of **40** and **44** are very similar, and **44** undergoes all the replacement reactions described above for **40**.

The formation of **44** by the action of aqueous alkali on TNB has been investigated spectrophotometrically by several workers[128–134]. General agreement is found with an equilibrium constant of $K \sim 2{\cdot}7$ l m^{-1} when determined spectrophotometrically, but this differs greatly from the value of 347 l m^{-1} found by a polarographic method[135]. At high ratios of hydroxide ion to TNB, it has been suggested that complexes **45** and **46** are formed[131, 135]. Again, solutions of complex **44** are unstable, slowly yielding 3,5-dinitrophenol and nitrite ion with traces of 3,5,3′,5′-tetranitroazoxybenzene. Gold and Rochester[129, 130] found the reaction to be light-sensitive, the apparent quantum yield increasing with the concentration of hydroxide ion and the rate being dependent on the wavelength of light used, the excitation causing reaction being at the longest wavelength absorption of the complex **44**. These observations suggest strongly that **44** is in fact involved in the reaction which presumably proceeds via a transition state analogous to **43**.

Another reaction which probably involves intermediates analogous to **44** is the alkaline oxidation of nitro compounds to nitrophenols[136]. Thus trinitrobenzene is oxidized to picric acid when boiled with alkaline potassium ferrocyanide (equation 47).

(47)

The reaction can be formulated as involving the oxidation of the intermediate **44**. Apart from differences in formal charge, (equation (47) is analogous to reaction scheme (86) suggested for the alkaline oxidation of pyridinium ions to pyridones in section VI.B.1.

b. Attack on other nitroaromatics. In basic DMSO solution, 1,3-dinitrobenzene undergoes isotopic exchange, mainly in the 2-position. The

* In fact, the spectrum of **44** formed by reaction of **40** with traces of H_2O in DMSO solution was wrongly assigned a structure involving attack by the DMSO[115].

solutions also show a characteristic red colouration. From a detailed examination of the kinetics of the methoxide-induced tritium exchange in DMSO solution, Crampton and Gold[137] concluded that the colour-producing species was not **47** formed by proton abstrac-

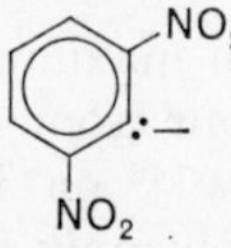

(47) 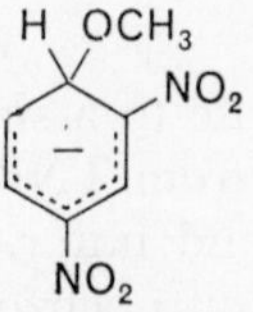 **(48)** 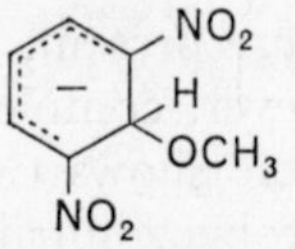 **(49)**

tion, a possibility considered by Pollitt and Saunders[138], but was due to a complex of dinitrobenzene which was inactive in the tritium exchange. These properties would be expected from the adduct **48**[137]. The formulation of the adduct as **48** rather than **49**, where attack has taken place between the two nitro groups, is suggested by the close correspondence of the visible spectrum (λ_{max} = 576 mμ) compared to that of the corresponding compound from 2,6-dinitroanisole (λ_{max} = 584 mμ). However, several workers have tried without success to measure the n.m.r. spectrum of **48**. The difficulty is at least partly caused by the extreme ease with which anion-radicals are formed in this system[139]. These can exchange with the unreacted dinitrobenzene and through this possibly affect the spectrum of the adduct. This effect is greatly reduced when the dinitrobenzene is present in relatively low concentration, so that as large a proportion as possible is present in the form of the adduct[106]. The n.m.r. spectrum is consistent with the formulation **48** with H_1, $\delta = -8{\cdot}36$; H_2, $\delta = -5{\cdot}43$; H_4, $\delta = -5{\cdot}35$, $J_{2,3} = 10$ cps and $J_{3,4} = 4{\cdot}5$ cps, in good agreement with the corresponding acetone adduct which has recently been reported[140], but it is inconsistent with the formulation **49** which would show only the three multiplets of an AB_2X spectrum.

Investigations have also been carried out on 1,3-dinitrobenzene systems which contain an electronegative group in the 5-position, making them more analogous to the case of 1,3,5-trinitrobenzene. However, there are now two possibilities for attack, namely **50** and **51**. Pollitt and Saunders[141] noted that the visible absorption spectra of a number of 1,3-dinitro-5-X-benzenes in DMSO solution in the presence of certain nucleophiles showed a series of two absorption maxima, and concluded that species analogous to both **50** and

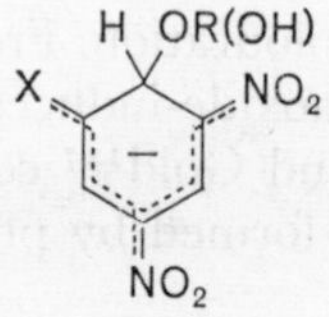

(50) 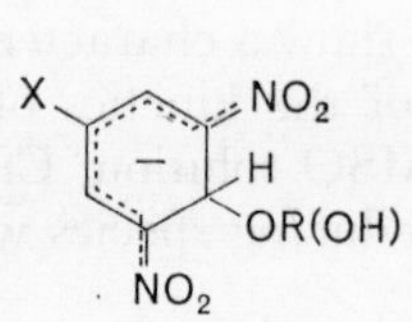 (51)

51 were present, each giving rise to a single absorption maximum. However, complex **40** formed from TNB where only one isomer is possible shows two maxima, and n.m.r. measurements [142] on 3,5-dinitrobenzonitrile and 3,5-dinitrobenzotrifluoride show only absorptions which can be attributed to complexes **50** (X = CN and X = CF_3 respectively) (Table 8). A similar situation is found with

TABLE 8. N.m.r. chemical shift parameters ($-\delta$, ppm) for the ring proton absorptions of complexes of general formula **50** in DMSO solution.

X	OR	H_α	H_β	H_γ
NO_2[114]	OCH_3	8·42	8·42	6·14
[127]	OH	8·20	8·20	6·15
CN[142]	OCH_3	8·08	7·35	5·40
	OH	8·21	7·52	5·50
CF_3[142]	OCH_3	8·02	7·13	5·44
	OH	8·15	7·31	too weak
H[106] ($\delta = -5{\cdot}43$)	OCH_3	8·36	6·96	5·35

3,5-dinitropyridine (section VI). However, although the non-observation of species rules out the possibility that they contribute to the absorption it does not exclude the possibility that they are formed in the initial attack and then isomerize very quickly to the more stable isomers (**50**).

N.m.r. studies suggest that stable adducts are also formed in nitropolycyclic aromatics[143]. Thus 9-nitroanthracene adds one equivalent of methoxide ion in DMSO solution to form the adduct **52**, and 1,3-dinitronaphthalene similarly gives **53**. The adducts **52**

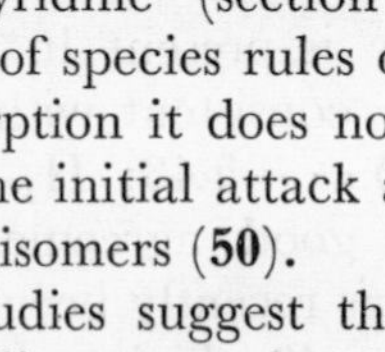

(52) 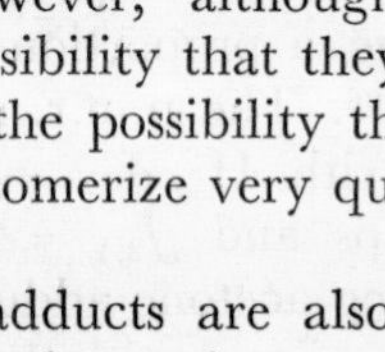 (53)

and **53** again undergo all the replacement reactions described in section IV.C.1a for the methoxy-adduct of TNB.

c. Attack on 4,6-*dinitrobenzofuroxan*. The formation of salts by the action of hydroxides on 4,5-dinitrobenzofuroxan (**54**) was described by Drost[144] in 1894, and investigated by several early workers[144–147]. Drost considered that a compound such as **55** was formed by proton abstraction (a similar suggestion had been made in the case of TNB[109]). Jackson and Earle[153a] suggested a structure of type **56** or **57** involving attack by the base. Recently, the simultaneous publication of results from three different research groups has unambiguously determined the structure of the species formed[148–150]. Thus no deuterium incorporation was found [148–150] on treating the

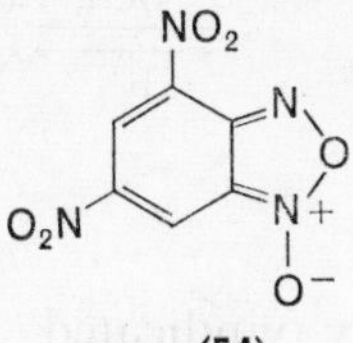

(**54**) 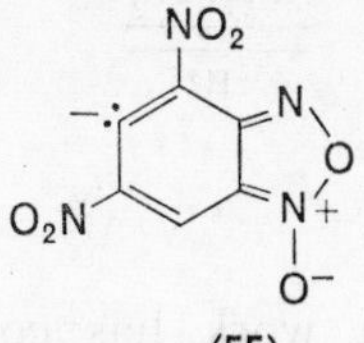 (**55**)

potassium salt with D_2SO_4, eliminating structures of type **55**, and no ^{18}O exchange with $^{18}OH_2$ [150], eliminating structures of type **58** where the heterocyclic ring has opened. The n.m.r. spectrum[148–150] of the salt formed by hydroxide addition shows doublet absorptions

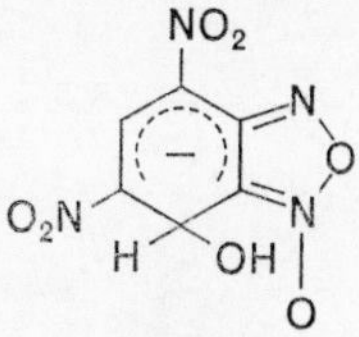

(**56**) 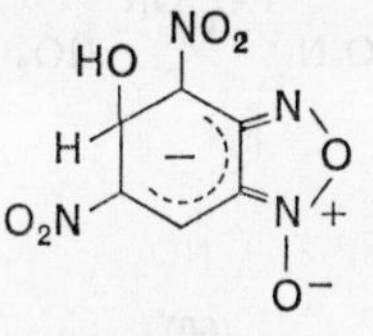 (**57**) 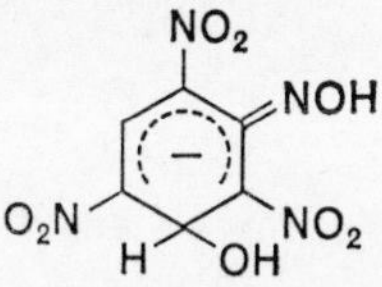 (**58**)

at $\delta = -8{\cdot}73$ and $\delta = -5{\cdot}94$ in agreement with structure **56** *or* **57**, but does not differentiate between them. However, Brown and Keys[148] synthesized the 4,6-dinitrobenzofuroxan with 50% deuterium in position 5. The n.m.r. spectrum of the hydroxide salt from this substrate shows a decrease in the low field absorption, hence H_5 is not attached to the sp^3 carbon atom, and the salt must have structure **56**.

2. Attack on nitro- and polynitro-anisoles

a. Attack on trinitroanisole and homologues.* The interaction between

*Referred to as TNA.

TNA and alkoxide ions is one of the most amenable systems and has been the subject of intensive study from the early investigations of Meisenheimer[151] and Jackson and co-workers[152, 153] up to the present day. In an early definitive study[151], Meisenheimer isolated red solid salts by treating trinitroanisole with potassium ethoxide and trinitrophenetole with potassium methoxide. Both salts gave the same mixture of trinitroanisole and trinitrophenetole on treatment with mineral acid, and Meisenheimer concluded that they were in fact identical and had structure **59***, representing the reaction as (48).

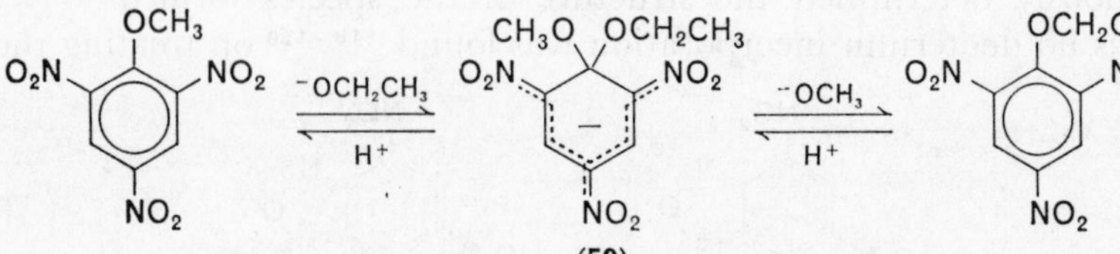

$$\text{2,4,6-}(NO_2)_3C_6H_2OCH_3 \underset{H^+}{\overset{^-OCH_2CH_3}{\rightleftharpoons}} \mathbf{(59)}\ [CH_3O,\ OCH_2CH_3] \underset{H^+}{\overset{^-OCH_3}{\rightleftharpoons}} \text{2,4,6-}(NO_2)_3C_6H_2OCH_2CH_3 \qquad (48)$$

More recent work has completely vindicated Meisenheimer's formulation, and complexes of this type are often referred to as 'Meisenheimer complexes' (or adducts, or salts). Compounds of the general type **60** have several characteristic properties. The absorption spectra show two maxima in the visible region at ca 420 mμ

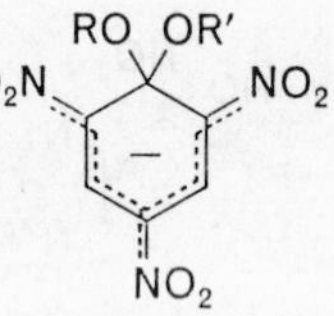

(60) [RO, OR′ adduct of 2,4,6-trinitrocyclohexadienide]

and 500 mμ due to the trinitrocyclopentadienide residue[154–156] (Table 9, Figure 3). There is little variation within a series, and the absorption may be considered characteristic of these systems. In fact, the spectra of the different compounds are so similar that the observation of identical u.v. spectra from the two adducts originally isolated by Meisenheimer is not proof that they are identical, as has been suggested, since the other possibility, that an equimolar mixture of **60** ($R = R' = CH_3$) and **60** ($R = R' = Et$) is formed, would in fact give rise to an identical absorption spectrum (Table 9).

* Although formulated slightly differently with the charge localized on one nitro group.

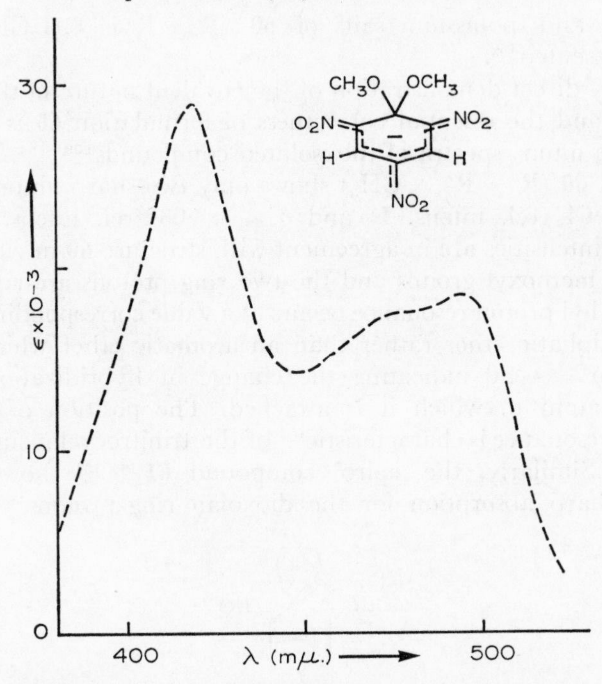

FIGURE 3. U.v.–visible absorption spectrum of compound **60** (OR = OR′ = OCH_3) in acetone solution.

The solid complexes also show characteristic i.r. spectra[157, 158] which are quite different from those of the parent ethers, and are in agreement with the general formulation **60**. Thus, the increased negative charge on the nitro groups leads to a lowering of the N–O stretching frequencies from 1552 cm^{-1} to 1513 or 1489 cm^{-1} on complex

TABLE 9. U.v. spectral parameters of compounds of general formula **60** in acetone solution.

OR	OR′	λ_1 (mμ)	ε_1 (l m^{-1} cm^{-1})	λ_2 (mμ)	ε_2 (l m^{-1} cm^{-1})	Ref.
OCH_3	OCH_3	420	29,000	494	19,000	169
OCH_3	OCH_2CH_3	421	29,000	495	19,000	169
OCH_2CH_3	OCH_2CH_3	422	30,000	495	19,000	169
OCH_2CH_2O		414	30,500	490	22,200	160

formation. There is also a strong broad absorption between 1040 cm^{-1} and 1225 cm^{-1}. Data on the crystal structures of the caesium and potassium salts of **60** (R = R′ = CH_2CH_3) have been presented[193].

A very direct demonstration of the covalent nature of these compounds and the essential correctness of formulation **60** is obtained from the n.m.r. spectra of the isolated compounds[159–161]. Thus the complex **60** (R = R′ = CH_3) shows only two sharp absorptions at $\delta = -8{\cdot}64$ (rel. intens. 1) and $\delta = -3{\cdot}03$ (rel. intens. 3). The relative intensities are in agreement with structure **60** in which both the two methoxyl groups and the two ring protons are equivalent. The methyl proton resonance occurs at a value corresponding to that of an aliphatic ether rather than an aromatic ether which would absorb $\delta = -4{\cdot}0$ indicating the change in hybridization at the carbon atom to which it is attached. The position of the ring proton resonance is characteristic[107] of the trinitrocyclopentadienide system. Similarly, the 'spiro' compound **61**[160, 162] shows only a single sharp absorption for the dioxolan ring protons[160] indicat-

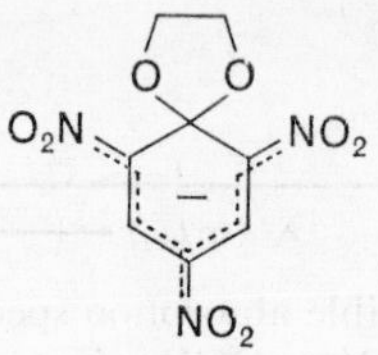

O O
O_2N NO_2
NO_2
(61)

ing the equivalence expected from structure **61**. Many investigations by n.m.r. have also been carried out on complexes generated in situ[115, 116, 119, 143, 159, 161–166] from the parent picryl ethers. The most incisive are those of Servis[164, 165]. The addition of sodium methoxide to a solution of TNA in dimethyl sulphoxide causes the disappearance of the TNA resonances and the appearance of two doublets of equal intensity at $\delta = -6{\cdot}17$ and $\delta = -8{\cdot}42$ ($J \sim 2$ cps) and a sharp resonance at $\delta = -8{\cdot}90$ (Figure 4). With time, the single resonance increases in intensity at the expense of the two doublet absorptions until these eventually disappear. The single resonance corresponds to the ring proton absorption in **62**, and Servis accounted for the spectral changes in terms of initial attack by methoxide ion at the unsubstituted 3-position to yield **63** which then isomerizes to give the thermodynamically more stable isomer (**62**) as in equation (49). The resonance at $\delta = -6{\cdot}17$ ascribed to H_β in **62** occurs at a position characteristic of protons attached to sp^3 hybridized carbon atoms in these systems. Since then, similar

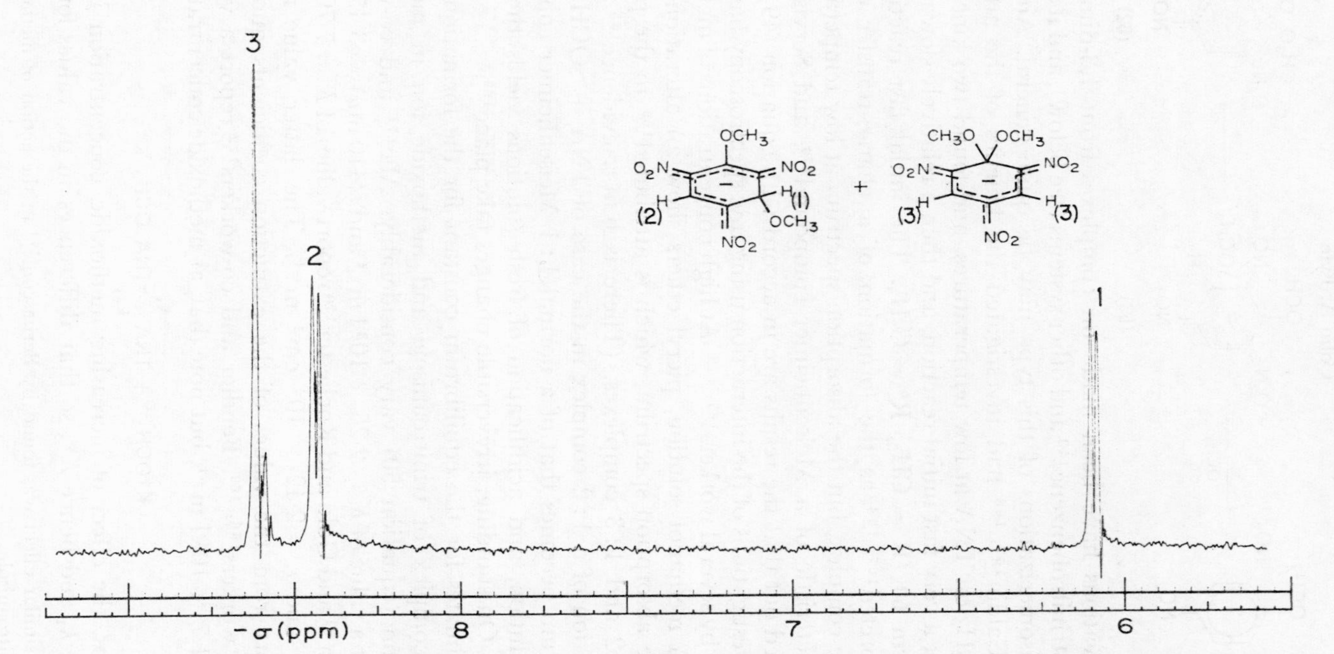

FIGURE 4. 100 MHz n.m.r. spectrum of a 1M solution of TNA in DMSO, 3 minutes after the addition of anhydrous sodium methoxide.

$$\text{2,4,6-}(NO_2)_3C_6H_2OCH_3 \xrightarrow{^-OCH_3} \textbf{(63)} \longrightarrow \textbf{(62)} \qquad (49)$$

observations have been made on complexes from 1,3-dimethoxy-2,4,6-trinitrobenzene[143] and other systems (see below), and it appears that isomerizations of this type may be quite general. Ainscough and Caldin[167, 168] had investigated the kinetics of the attack by $^-OC_2H_5$ on TNA at low temperatures, and found two kinetic processes; a very fast initial reaction, and then a relatively slow reaction to form **60** (R = CH_3, R′ = C_2H_5. The initial fast reaction was interpreted[167, 168] as the formation of a charge-transfer or ion–dipole complex, but the absorption spectrum at low temperatures is characteristic of a Meisenheimer compound[170], and Servis[164] has pointed out that the results are in accord with equation (49).

Investigations of the interaction using u.v. spectroscopy have been made by several workers[169–172]. At high concentrations of methoxide ion in methanol solution, picryl ethers show an alteration in the visible absorption spectrum, which is attributed[169] to the presence of 1 : 2 and 1 : 3 complexes. (There is n.m.r. evidence[165] for the formation of a 1 : 2 complex in the case of TNA + $^-OCH_3$.) The spectrum becomes that of a normal 1 : 1 Meisenheimer compound on dilution, and acidification of fresh solutions yields the picryl ether. On standing, irreversible changes take place[169].

Estimates for the equilibrium constant for the formation of the 1 : 1 complex of trinitroanisole and methoxide ion in methanol solution (equation 50) vary considerably. Abe[171] and co-workers report a value of $K = 2{\cdot}26 \times 10^3$ l m^{-1} and $\varepsilon(410\ \text{m}\mu) = 3{\cdot}45 \times 10^4$ cm^2 m^{-1} and Gold and Rochester[170] report values of $K = 7{\cdot}70 \times 10^3$ l m^{-1} and $\varepsilon = 2{\cdot}42 \times 10^4$ cm^2 m^{-1}. The latter value receives support from the values of the extinction coefficients quoted by other workers[166, 169]. Fendler and co-workers[172] report a value of $K = 1{\cdot}7 \times 10^4$ l m^{-1}, but note that, at methoxide concentrations of

$$K^+OCH_3^- + TNA \underset{k_{-1}}{\overset{k_1}{\rightleftharpoons}} TNA.OCH_3^-K^+ \qquad (50)$$

10^{-2} M, the effect of increasing methoxide concentration is to increase k_1 and hence K*, so that differences in the values found by

* A similar effect was found by Bernasconi[173] in the action of methoxide on 2,4-dinitroanisole.

various workers may be due to differences in concentration ranges. All the values are, however, very much larger than the value found for the attack of methoxide ion on trinitrobenzene (section IV.1a above). In this system, the isomerization equilibrium found by Servis for TNA cannot occur, and the much larger equilibrium constant in the latter system may be due to the measurement being of two consecutive equilibria. The equilibrium has also been investigated by measurement of the methoxide-catalysed loss of ^{13}C from 2,4,6-$(NO_2)_3C_6H_2O^{13}CH_3$ [174]. The rate-determining step in the isotopic exchange was found to be the unimolecular heterolysis of the complex, rather than the bimolecular formation of it, indicating that the complex is more stable than the starting ether, in agreement with the thermodynamic results from

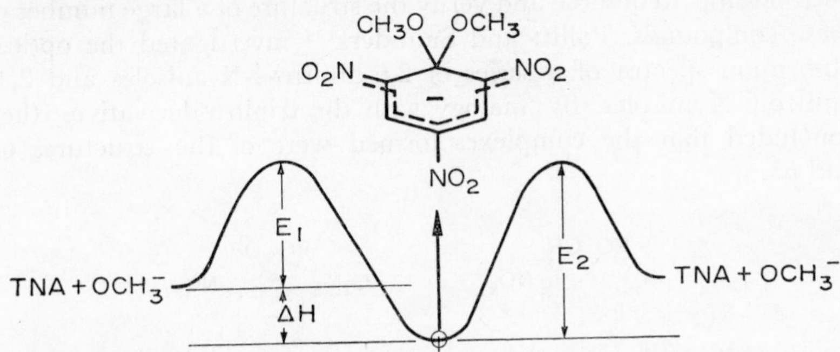

FIGURE 5. Energy profile for the reaction of TNA with methoxide ion in methanol solution.

optical measurements. (The technique of isotopic labelling has also been employed by Gitis and co-workers[175], but no equilibrium or rate measurements were made.)

There is general agreement between the thermodynamic parameters obtained from a whole variety of methods (Table 10, Figure 5). The main feature is that the complex formed is in fact not only stable, but more stable than the original substrate, by 2—7 kcal/mole.

Solutions of trinitroanisole in the presence of methoxide are unstable when irradiated at the wavelength of the absorption of the 1 : 1 complex, reacting by loss of a nitro group[176].

b. Attack on dinitro- and substituted dinitroanisoles. From both intuitive and theoretical[177] points of view, complexes from dinitroanisoles might be expected to be less stable than the trinitro-analogues. Although this has been verified experimentally, in fact it has still

TABLE 10. Thermodynamic parameters (kcal mole^{-1}) for the attack of methoxide ion on TNA in methanol solution (Figure 5) as determined by various methods.

System	Method	E_1	E_2	ΔH	Ref.
TNA + OEt^-	low T. kin. (u.v.)	13 kcal	—	—	167, 168
TNA + OCH_3^-	u.v.	10·0	12·0	−2·0	171
TNA + OCH_3^-	$O^{13}CH_3$	12[a]	19·4	−7·15	174
TNA + OCH_3^-	u.v.	13·5	19·0	−5·5	172
TNA + OCH_3^-	Miller	14·0	16·0	−2·0	203

[a] Not determined directly; value by subtraction.

been possible to observe and verify the structure of a large number of these compounds. Pollitt and Saunders[141] investigated the optical absorption spectra of a series of 2,6-dinitro-4-X-anisoles and 2,4-dinitro-6-X-anisoles. By analogy with the trinitro-derivatives, they concluded that the complexes formed were of the structures **64** and **65**.

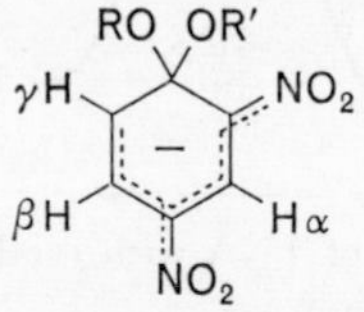

(**64**) 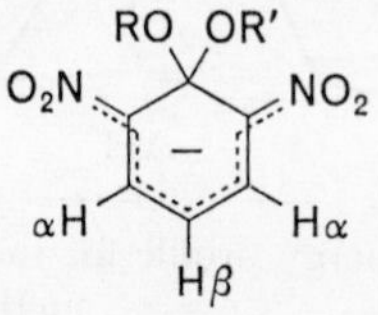 (**65**)

There is also a close correspondence between the absorptions of the adduct of 2,4-dinitroanisole and methoxide ion of suggested structure **64** and that of 1-(2′-hydroxyethoxy)-2,4-dinitrobenzene where it is thought[141] that the spiro compound **66** is formed. Measure-

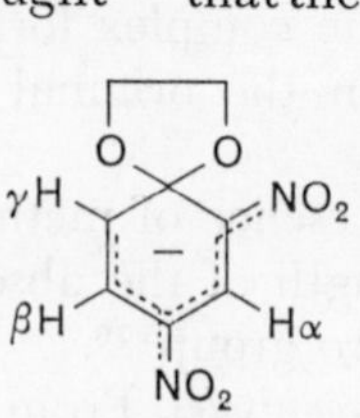

(**66**) 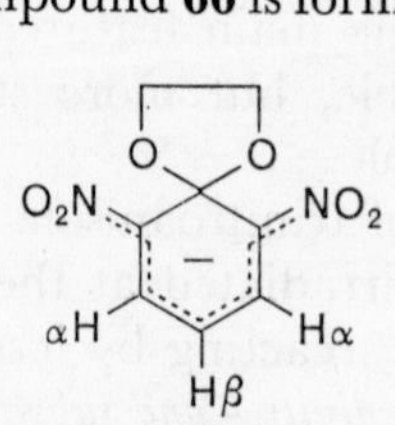 (**67**)

ments of the n.m.r. spectrum of **64** generated in situ in DMSO solution by the addition of sodium methoxide fully support the structure suggested by Pollitt and Saunders. The ring proton absorptions occur at $\delta = -8{\cdot}52$, H_α; $\delta = -7{\cdot}90$, H_β; $\delta = -5{\cdot}36$, H_γ. There is (Table 11) a close similarity between the different homo-

TABLE 11. N.m.r. absorptions ($-\delta$, ppm) of ring protons of σ-complexes of dinitrophenyl ethers in DMSO solution.

Structure	OR	OR′	H_α	H_β	H_γ	Ref.
2,4-dinitroanisole	OCH_3	—	8·75	8·57	7·64	
64	OCH_3	OCH_3	8·68	7·26	5·09	107, 143, 119
64	OCH_3	OCH_2CH_3	8·70	7·20	5·10	107, 143, 119
64	OCH_2CH_3	OCH_2CH_3	8·68	7·19	5·10	107, 143
66	OCH_2CH_2O		8·52	6·90	5·36	107, 143, 180, 181
2,6-dinitroanisole	OCH_3	—	8·31	7·58	—	
65	OCH_3	OCH_3	7·93	5·02	—	107, 143
65	OCH_3	OCH_2CH_3	7·84	4·98	—	107, 143
67	OCH_2CH_2O		7·67	5·09	—	107, 143, 180, 181

logues of **64** and the spiro compound **66,** confirming the assignment.

Similarly, complex **65** can be generated by the addition of methoxide ion to DMSO solutions of 2,6-dinitroanisole. The ring protons give rise to an AX_2 system consistent with formulation **65.** The resonances occur at $\delta = -5{\cdot}0_2$, H_β and $\delta = -7{\cdot}93$, H_α, with $J_{H_\alpha - H_\beta} = 8$ cps, the large upfield shift in H_γ occurring because it is *meta* to both nitro groups. Again (Table 11), the spectra of homologues of **65** are very similar to those of the spiro compound **67,** giving added proof of the structure. In both cases, there is eventual decomposition to the dinitrophenate ion (this is true to some extent for the picryl compounds also). This may be due to reaction with trace amounts of H_2O in the DMSO solvent, but there is growing evidence[188] that phenate ion formation can occur by an S_N2 displacement on the methoxyl carbon atom (equation 51), so that at

$$Ar{-}O(CH_3) + {}^{-}OR \longrightarrow ArO^- + CH_3OR \qquad (51)$$

least traces of phenate ions might be expected to be present in *all* solutions containing intermediates of these types*. Complexes of

* An analogous displacement involving an amine is an alternative explanation of the results of Servis[120] on the action of triethylamine on TNA.

type **64** have also been isolated as solid compounds[178, 179]. Originally they were obtained in a mixture with the phenate ion and were considered unstable[178], but recently Griffin and co-workers[179] have isolated them in a pure form, and demonstrated that they show considerable stability. This stability has been estimated quantitatively by several workers from kinetic investigations. The results are given in Table 12 and illustrated in Figure 6. Although experimentally difficult to estimate an accurate value of the equilibrium

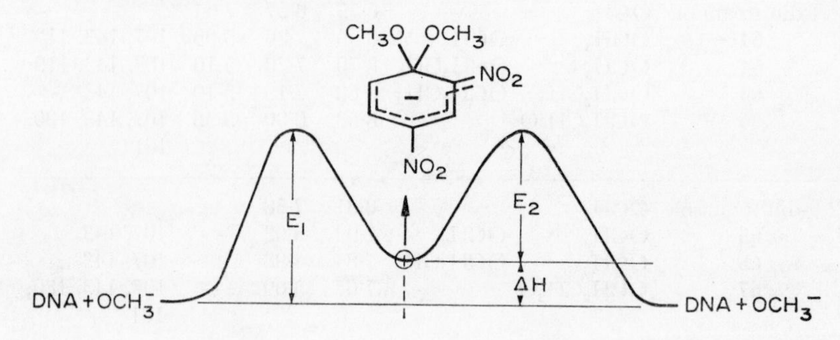

FIGURE 6. Energy profile for the reaction of 2,6-dinitroanisole with methoxide ion in methanol solution.

TABLE 12. Thermodynamic parameters (kcal/mole) for the attack of methoxide ion on 2,4-dinitroanisole in methanol solution as determined by various methods.

Method	E_1	E_2	ΔH (kcal/mole)	Ref.
^{13}C Isotopic exchange	(16·8)	—	—	174
Temp. jump.	(16·1—16·8)	(11·20—13·0)[a]	(5·6—3·1)[a]	173
Miller	19·5	12·5	7·0	203

[a] Value dependent on OCH_3^- concentration.

constant for complex formation[179], it is thought to be several thousand times less than that for the corresponding complex from trinitroanisole. Fendler[174] found that the rate of complex formation was the rate-determining step in the methoxyl exchange of ^{13}C-labelled 2,4-dinitroanisole, indicating that, unlike the case of trinitroanisole, the complex was less stable than the substrate. The activation energy (E_1) for this process was found to be 16·8 kcal/mole. Using the 'temperature-jump' technique, Bernasconi[173] found a value of 11·8 ± 0·5 kcal/mole for the decomposition (E_2) of the complex, compared with a value of 12·5 kcal/mole predicted by Miller (see below).

Several workers have investigated the complex formation of 2-cyano-4,6-dinitroanisole (**68**)[182–184] and 4-cyano-2,6-dinitroanisole (**69**)[166, 183, 184].

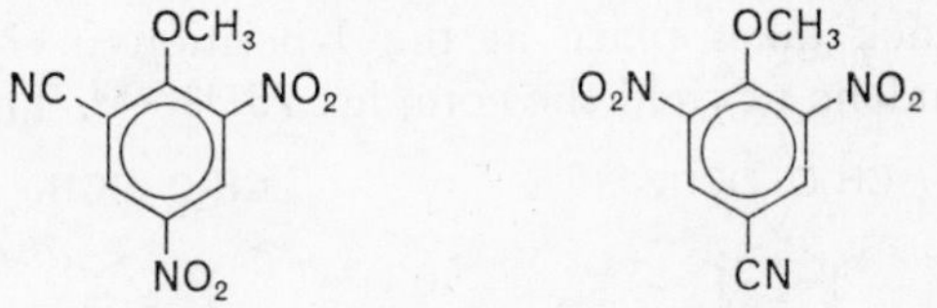

(68) **(69)**

The n.m.r. spectrum of the stable isolated complex formed from 2-cyano-4,6-dinitroanisole indicated that it had structure **70**[182–184], but Griffin and co-workers[183, 184] showed that a rearrangement similar to that found by Servis in the case of trinitroanisole took place, the thermodynamically less stable isomer (**71**) being formed first which then rearranged to **70** (equation 52). They found a

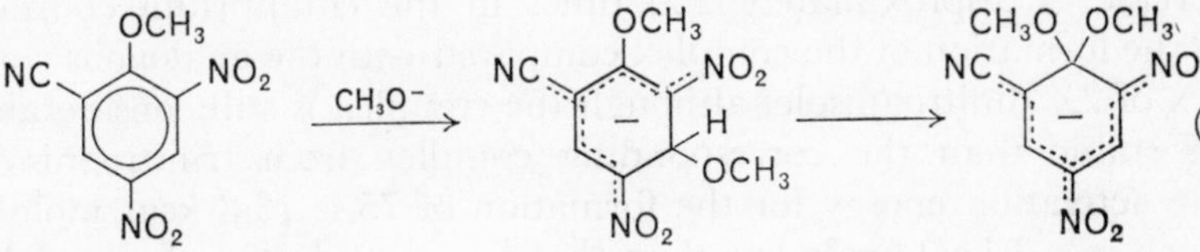

$$\mathbf{(68)} \xrightarrow{CH_3O^-} \mathbf{(71)} \longrightarrow \mathbf{(70)} \qquad (52)$$

similar rearrangement in the case of 4-cyano-2,6-dinitroanisole (equation 53), and were also able to isolate the complexes **70** and **72** as stable solids.

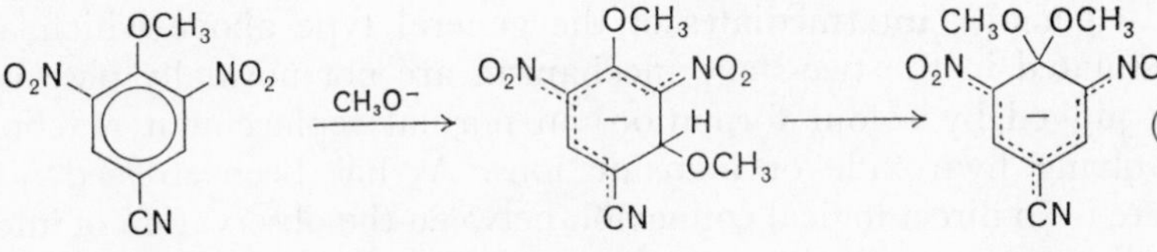

$$\mathbf{(69)} \xrightarrow{CH_3O^-} \mathbf{(73)} \longrightarrow \mathbf{(72)} \qquad (53)$$

In a detailed examination of the kinetics of formation of **70** and **72** compared with the corresponding complex **62** from trinitroanisole, Griffin and co-workers[184] found that the replacement of a nitro by a cyano group in trinitroanisole caused a 6·5-fold decrease for replacement in the 2-position. These parallel exactly the relative activating effects found for *ortho-* and *para-*nitro groups in the methoxydechlorination of nitrochlorobenzenes[189]. N.m.r. investigations[143] of the addition of methoxide ion to a solution of 9-nitro-10-methoxyanthracene confirm the suggestion of Meisenheimer[151] that attack takes place at the 10-position to yield the complex **74**. Similarly, attack takes place at the 1-position in 1-methoxy-2,4-dinitronaphthalene to yield the complex **75**[143, 186]. The correspond-

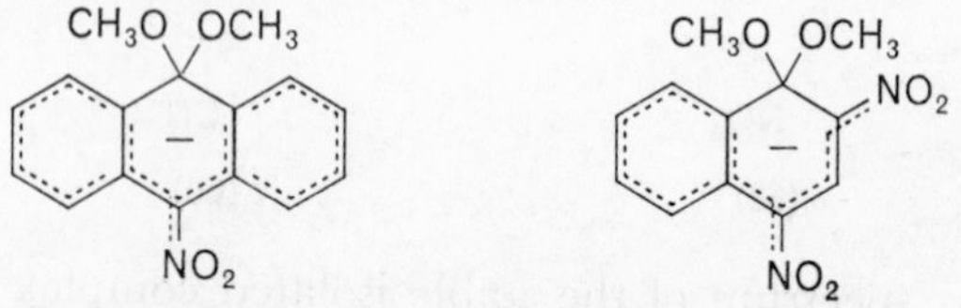

CH_3O OCH_3 NO_2 **(74)**

CH_3O OCH_3 NO_2 NO_2 **(75)**

ing spiro compounds to **74** [187] and **75** [186, 187] can also be made. Measurements on the kinetics of the formation of **75** in methanol solution indicate that the complex is more stable than the anisole substrate[186]. The presence of the second aromatic ring causes an increase of approximately 250 times in the equilibrium constant for the formation of the complex compared with the analogous complex of 2,4-dinitroanisole, although the complex is still considerably less stable than the corresponding complex from trinitroanisole. The activation energy for the formation of **75** is 13·8 kcal/mole[186] which is ~3 kcal/mole less than that found[173, 174] for the complex from 2,4-dinitroanisole and reflects exactly the trend found for methoxy-dechlorinations of nitrochlorobenzenes and naphthalenes[189].

3. Relation of the observation of intermediates to the mechanism of substitution reactions

In general, intermediates of the general type above which are postulated in the two-stage mechanism are not normally observed (as judged by colour formation) in normal replacement reactions involving hydroxide or alkoxide ions. As has been stressed[87, 88], there is no direct logical connexion between the observation of intermediates in equilibrium processes as described above and their possible participation in kinetically controlled substitution reactions, and any inferences must be made with caution. However, even if one were able to detect the intermediates in these reactions, and determine kinetically that the reaction was first-order with respect to the intermediate, the concentration of the intermediate could be re-expressed in terms of those of the reactants using the first equilibrium, and the participation of the intermediate in the reaction would always be an inference (however reasonable).

The general acceptance of the two-stage mechanism has been effected to some extent by the observation of intermediates of the same structural type and electronic configuration as those postulated in the S_NAr2 mechanism. This 'association' of the two has become somewhat more valid in recent years with the observation and characterization of intermediates from less activated substrates as discussed above, and the parallels found in the trends shown by the activation energies for complex formation and replacement reactions for similar changes in substrate. Also the work of Miller (see below) has included both processes in a general treatment with considerable success.

One promising possibility in the future may be the observation of intermediates in low concentrations and short lifetimes during reactions by use of pulsed n.m.r. and flow systems.

D. Theoretical Approaches

Several authors[190–200, 244] have performed molecular orbital calculations to obtain parameters which can be related to the structure and reactivity of the transition states and intermediate complexes of these substitution reactions. Most calculations employ the 'Wheland model'[190] for the intermediate **76** where the structure of the intermediate is that discussed in section C above.

The carbon atom at the point of attack is removed from the calculation, only the residue (shown in dark lines) and any groups attached to it contributing to the π-electron energy of the system.

(76)

Caveng and Zollinger[191] have employed this model and method of calculation to calculate π-electron densities in a whole series of intermediates formed from polynitroanisoles and the parent substrates, and have compared their results with those of the n.m.r. studies discussed in section C. Their calculations suggest that the negative charge in these complexes is mostly localized on the nitro groups, and that there is relatively little increase in the π-electron density at the ring carbon atoms, a slight *decrease* being found in some cases.

Very recently, however, Hosoya, Hosoya and Nagakura[244] have disagreed with both the results and the method of calculation used in this study. They consider that nitro compounds are beyond the limit of the HMO treatment, and have calculated the π-electron structures of the 1,3,5-trinitro-, 1,3-dinitro-, 1,5-dinitro-, 2,4-dinitro-, and 3-nitro-pentadienyl anions by the 'method of composite molecules'. In this method, the electronic structure of the substituted pentadienyl anion is considered in terms of charge-transfer interaction between the MOs of the pentadienyl anion and those of the nitro groups. A 'Pariser–Parr–Pople' type self-consistent field molecular orbital calculation with configuration interaction was also made for the 1,3,5-trinitropentadienyl anion for purposes of comparison. In contrast to the results of Caveng and Zollinger[191], they found that both methods predicted a net negative charge in the ring in the 1,3,5-trinitropentadienyl complex. The electronic transitions of the anions were considered as due to charge-transfer from the pentadienyl group to the nitro group, and the calculated values of both the transition energies and the extinction coefficients were in good agreement with available experimental values.

One very important aspect considered by Caveng and Zollinger, but often neglected in calculations, is the possibility of distortion of the *ortho*-nitro groups from the plane of the ring, thus lessening their capacity for delocalizing the negative charge. In fact, molecular models indicate that steric interactions[166] can occur, and this will affect the substrate, transition states and intermediate complex. The non-coplanarity of the nitro groups with the ring can in fact be introduced very simply into the HMO calculation, and Caveng and Zollinger[191] performed their calculations for rotations of the nitro group from the plane of the ring of 0°, 30° and 60° in cases where distortion was possible.

Direct estimates of this distortion can now be made in the picryl series from the recently published X-ray structures of trinitrophenetole[192] and its ethoxy complex (**62**)[193]. In the case of the adduct **62** no significant distortions of the two *ortho*-nitro groups were found[193], and steric interaction in the case of the intermediate must be minimal. However, very considerable steric interactions were indicated in trinitrophenetole, the two *ortho*-nitro groups being rotated by 30° and 60° from the plane of the aromatic ring[192]. Since the structure of the transition state will lie somewhere between these two extremes the possibility of distortion by steric interactions in this case will remain, to some extent at least, an unknown factor in calculations.

Correlation of parameters from molecular orbital calculations with activation energies from kinetic measurements is not so direct, however, as the MO calculations are based on a Wheland type structure which represents an *intermediate*, and the kinetic parameters depend on the energy of a *transition state*. The justification for this approach will depend on how closely the structure of the transition state resembles that of the intermediate complex*. The activation energy calculated (E_{MO}) is related to the difference in atom localization energies (A_n) of the substrate and intermediate (equation 54). The term Δ includes differences in solvation energies of the

$$E_{MO} = A_n\beta_{cc} + \Delta \tag{54}$$

substrate and intermediate. Equation (54) also assumes that ΔS is a constant.

Murto[194] found reasonable agreement using this method of calculation with the experimentally determined activation energies for a series of methoxydehalogenations of nitrohaloaromatics. He also found a correlation between the logarithm of the rate constants and the electron densities at the position of substitution, indicating the importance of the charge density as a rate-determining factor in the reactions considered. Abe[177] has used the same model introducing a 'reaction parameter' by which the values of the resonance integrals can be changed during the course of substitution. His results suggest that Meisenheimer intermediates are more easily formed, and are more stable, with increasing number of nitro groups, in agreement with the trend found experimentally (section C). However, where direct comparisons can be made with experimental results, the agreement is only qualitative.

An alternative to the Wheland representation has been given by Simonetta and Carra [195, 196] as an extension of their treatment of the Fritsch rearrangement reaction[195]. In this representation, the sp^3 hybridized carbon atom and its substituents are not removed from the calculation, and the leaving and attacking groups (e.g. Cl and $-OCH_3$) are considered together as a 'pseudo atom'. The orbitals of the oxygen and chlorine atoms of these groups can be combined

* Some guide to the validity of this approximation is given in the calculations of Miller (see below).

together to give a bonding orbital, resembling a σ orbital, which is symmetric with respect to the plane of the ring (equation 55a),

$$\psi_\sigma = (1/\sqrt{2})(\psi_{Cl} + \psi_O) \tag{55a}$$

and an antibonding orbital which is antisymmetric with respect to the plane of the ring, resembling a Π orbital ψ_π (equation 55b).

$$\psi_\pi = (1/\sqrt{2})(\psi_{Cl} - \psi_O) \tag{55b}$$

The 'sigma' orbital ψ_σ can combine with the sp^2 orbital of the attached carbon atom (Figure 7a) and the 'π' orbital with its p_z orbital (Figure 7b). This model is much more flexible than the

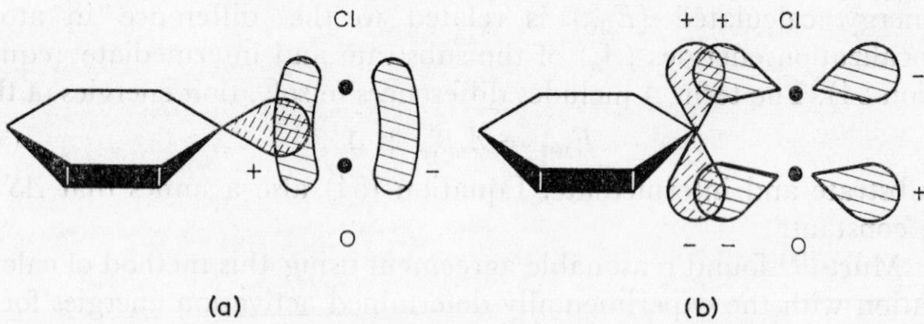

FIGURE 7. Bonding and antibonding molecular orbitals of the 'pseudo atom' in an aromatic nucleophilic substitution reaction.

Wheland one, in that the total bond order between the pseudo atom and the bonded carbon can be greater than 1. In fact the parameters for the pseudo atom are evaluated from some experimental results, and, in this sense, the calculations may be thought of as applying to a transition state, although it may be difficult to obtain a physical idea of this from the parameters found. Carra and Simonetta have applied this method to methoxy- and amino-dehalogenation reactions of benzene and naphthalene derivatives[196], and have obtained excellent correlations between experimental activation energies and calculated values of ΔE_π. The agreement is considerably better than that obtained for the same reactions using either π-electron densities, or the Wheland intermediate. Some idea of the relation of the transition state to the intermediate complex is given in the calculations of Nagakura[198]. From a consideration of the relative electron affinities and ionization potentials of substrate and reagents[197], Nagakura formulated the substitution process within the framework of the general theory of charge-transfer interactions introduced by Mulliken[201]. Thus the reaction is seen as a progressive transfer of charge from the reagent to the substrate, and a larger contribution of the charge-transfer form $\psi_{R\text{-}S^-}$ to the total wavefunction ψ_{TOT}

$$\psi_{TOT} = a\,\psi_{R^-,S} + b\,\psi_{R\text{-}S^-} \tag{56}$$

(equation 56). In this context, in the initial state, $a \gg b$, and in the intermediate complex, $b \gg a$. The transition state is thought of as the point where there is almost equal contribution from both $\psi_{R^-,S}$ and $\psi_{R\text{-}S^-}$ to the total wavefunction[202], and can be considered as the point where the whole system transfers from the no-bond to the charge-transfer structure. (A similar treatment has been given by Brown[199, 200] for electrophilic aromatic substitution reactions.)

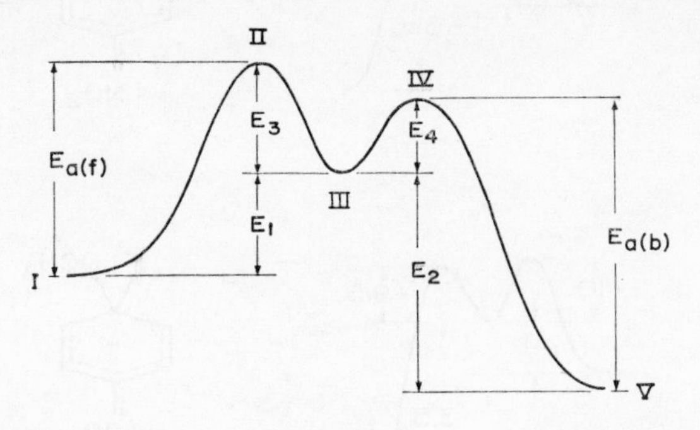

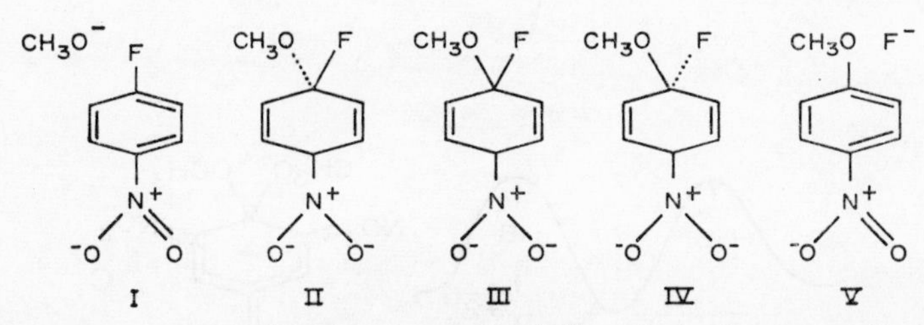

FIGURE 8. Energy for a bimolecular nucleophilic substitution proceeding via a 'two-step' mechanism.

A different approach, and one which is of very wide applicability, is that developed by Miller and co-workers[203–207]. The reaction (e.g. between methoxide ion and *p*-nitrofluorobenzene below) is thought of as proceeding according to the two-stage mechanism via structures I—V (Figure 8) where III is the intermediate complex in the Wheland representation. The energy levels of the reactants and products relative to the intermediate complex (E_1 and E_2 in Figure 8) are calculated by taking into account changes in bonding, electron affinity, solvation and delocalization energies in the dissociation of the intermediate complex to either the reactants or products[203].

Where possible, known thermodynamic values are used for these terms, but some must be approximated to, and others estimated. (See Ref.[203] for a full discussion of the approximations involved.) At this stage in the calculation one has obtained the same information as from the HMO calculations based on the Wheland model for the intermediate.

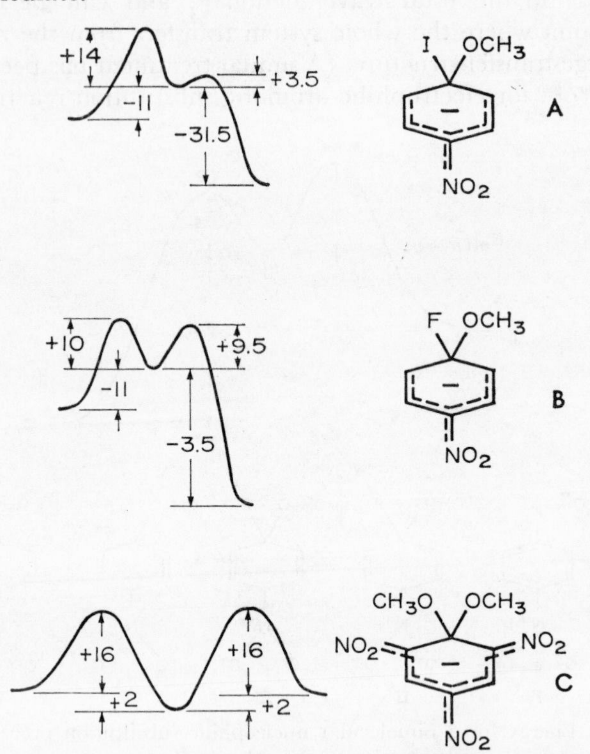

FIGURE 9. Energy profiles for aromatic nucleophilic substitution reactions involving the intermediates shown; from the calculations of J. Miller[203].

However, as discussed above, the kinetically measured activation energies are $E_{a(f)}$ or $E_{a(b)}$ in Figure 8. Miller[203] has calculated the activation energies E_3 and E_4 for the activation of the intermediate complex to the transition states for its decomposition by employing a semi-empirical correlation between % bond-dissociation energies and the exothermicity of the dissociation reaction*. Thus E_3 will

* Although semi-empirical, the correlation is not arbitrary. If one accepts a connexion between activation energies and reaction thermicity, the application in the range encountered in these reactions does not involve large errors.

be a percentage of the C–OCH_3 bond energy depending on the energy difference E_1, and E_4 a percentage of the C–F energy depending on the energy difference E_2. In this way, the kinetic activation energies $E_{a(f)}$ and $E_{a(b)}$ can be calculated. The approach is quite generally applicable to all substitution reactions for which the relevant bond-dissociation energies are known, and it would be quite reasonable to justify the method solely on the basis of being a general qualitative interpretation. However, in spite of the approximations involved, the agreement between calculated and experimental activation energies is *extremely* good (e.g. Table 13). Some representative schemes are shown in Figure 9. In no case is the Wheland intermediate a very good representation of the energy of the transition state. Of special note is the case of the stable complex (case C), discussed in section C. The method of Miller provides a strong direct link between these observable intermediates, and the general kinetic case.

The generally very good agreement with experimental activation energies provides justification not only for the method of calculation, but also for the model of the two-stage mechanism involved in it.

TABLE 13. Comparison of experimental activation energies with values calculated by the method of Miller[203] for the methoxydehalogenation of some activated halobenzenes[a].

Substrate	E_a(calc.) (kcal)	E_a(exp.) (kcal)	Ref.
1-F-4-nitrobenzene	21	21	203
1-Cl-4-nitrobenzene	24	24	203
1-I-4-nitrobenzene	25	25	203
1-F-2,4-dinitrobenzene	15	13·5	203
1-Cl-2,4-dinitrobenzene	16·8	18	194
1-I-2,4-dinitrobenzene	19	19	203
1-F-2,4,6-trinitrobenzene	8	10·5	194
1-Cl-2,4,6-trinitrobenzene	11·6	13·5	194

[a] See also Tables 11, 12.

VI. ATTACK ON AZA-AROMATIC SYSTEMS

A. Pyridine and Homologues

I. Kinetics and mechanism

a. General. Nucleophilic substitution reactions in aza-aromatic systems proceed much more readily than in the corresponding benzene analogues owing to the activating effect of the ring nitrogen atom. For example, 2- and 4-halopyridines react readily with alkoxides to yield the corresponding alkoxypyridines[208], and with hydroxide ion to yield the corresponding pyridones[209].

In general, second-order kinetics (first-order in both alkoxide and substrate) are found, and the reactions are usually discussed[210, 211] in terms of the intermediate complex theory introduced by Bunnett for aromatic substitution (section V). Thus the reaction of 4-chloropyridine would be represented as in equation (57).

$$\text{4-ClC}_5\text{H}_4\text{N} + {}^-\text{OCH}_3 \rightleftharpoons [\text{4-Cl-4-OCH}_3\text{ complex}]^- \longrightarrow \text{4-CH}_3\text{OC}_5\text{H}_4\text{N} + \text{Cl}^- \quad (57)$$

However, despite the extensive investigations of Chapman and other workers, much of the internal kinetic evidence for the S_NAr2 mechanism proposed for aromatic systems is missing in the heterocyclic series, and discussions on nucleophilic heterocyclic substitutions [210, 211] have relied heavily on mechanistic evidence from the benzenoid series. There is also a lack of kinetic data on the reaction of alkoxides with simple aza-substrates.

The discussion is restricted to consideration of the simpler nitrogen heterocycles, particularly pyridines, to make comparison with earlier sections as direct as possible (the facile hydration of polyaza-heterocycles has been reviewed[212, 213]). The main points of interest in the kinetic investigations, apart from the establishing of the bimolecular nature of the reaction, have been the quantitative investigation of the activating effect of the aza-nitrogen, both at various positions in the ring system and compared with the commonly used nitro activating group in the benzenoid series.

b. Activating effect of the aza group. The activating effect of the aza-nitrogen has been investigated in simple pyridine systems of several authors[214–216, 218–222]. Some relevant data[214–217] are given in Table 14. In general, the activating effect in the 4-position is somewhat larger than that in the 2-position, giving larger rates of reaction and smaller activation energies for the replacement of halogens by alkoxide ions. The activation energies for halogen replacement in unactivated aromatic hydrocarbons are considered[105, 214, 216] to be in excess of 30 kcal/mole, and the activating effect of the aza-nitrogen relative to the unsubstituted hydrocarbon is reflected in the lower activation energies observed in these cases, the introduction of an azine nitrogen lowering the activation energy by roughly 10 kcal/mole. A direct comparison of the rates of reaction can be made in the case of 2-chloro-naphthalene where the introduction of an aza-nitrogen in the 1-position (to give 2-chloroquinoline) increases the rate of reaction with methoxide ion by a factor of $6{\cdot}9 \times 10^9$ and the introduction of an aza-nitrogen in the 3-position (to give 3-chloroisoquinoline) by a factor of $1{\cdot}3 \times 10^5$. Other, less direct, comparisons can be made by comparing the effect of the insertion of an aza-nitrogen at various positions in a substrate already activated by an aza or nitro group[218–222]. In general, the aza group gives a very large increase in reactivity compared with the corresponding hydrocarbon. The activation at various positions is in the order *para* > *ortho* > *meta*.

TABLE 14. Kinetic and thermodynamic parameters for the alkoxydechlorination of chloropyridines.

Substituent	Conditions	$k \times 10^6$ ($l\ m^{-1} sec^{-1}$)	E_a (kcal $mole^{-1}$)	ΔS (e.u.)	Ref.
2-Cl	OMe^-, MeOH, 50°	$3{\cdot}31 \times 10^{-2}$	28·9	−5·3	217
3-Cl	OMe^-, MeOH, 50°	$1{\cdot}09 \times 10^{-5}$	32·9	−9·2	217
4-Cl	OMe^-, MeOH, 50°	$8{\cdot}9 \times 10^{-1}$	25·2	−10·4	217
2-Cl	OEt^-, EtOH, 20°	$2{\cdot}2 \times 10^{-3}$	26·8	−9·2	214, 215
4-Cl	OEt^-, EtOH, 20°	$8{\cdot}7 \times 10^{-2}$	20·9	−22·3	214, 216

The activating effect of the aza-nitrogen is often compared with the effect of the nitro group in benzenoid systems, and the two are very similar both in the preferred activation of positions *ortho* and *para* to the group and also in the magnitude of their effect.

Direct comparisons have been made in some simple systems by several authors[223–225]. The relative ratios for alkoxydechlorination in the simplest systems are shown below, where X = N or C–NO_2.

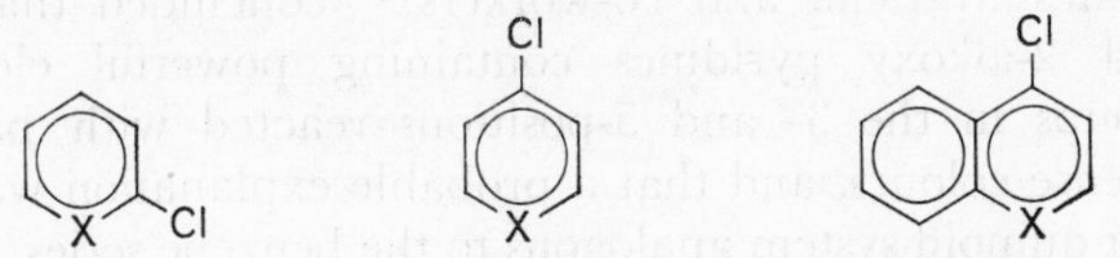

approx. ratio	(2-chloro, X at 1)	(4-chloro, X at 1)	(4-chloro bicyclic, X at 1)
$\frac{k_{2\ nitro}}{k_{aza}}$	10	15·1	8·4
conditions	90°/OEt^-	60°/OEt^-	60°/OEt^-
reference	214, 224a	214, 224b	214, 224c

A large amount of data is also available in systems containing more than one activating group[225]. Although the exact relation between the two groups will depend on both the substrate and nucleophile, some generalizations can be made. Firstly, the difference between the two groups is often within an order of magnitude which is a small difference compared with the very large effect of either relative to the unsubstituted aromatic hydrocarbon. Secondly, in reactions with alkoxide ions, the relative activating effect of NO_2 compared with N is larger in the *ortho* than in the *para* position. These observations are in substantial agreement with the early qualitative study of Mangini and Frenguelli[223].

2. Evidence for intermediate complexes

The detection of intermediates in the nucleophilic additions of alkoxides to activated substrates in the benzenoid series is considered an important piece of evidence in favour of the more general applicability of the two-stage mechanism for attack by alkoxides and hydroxides (section V).

The large activating effect of the aza group compared both with the unsubstituted hydrocarbon and with the C–NO_2 grouping (section VI.A.1) could suggest that the stabilization of some intermediates in the pyridine series might be high enough to allow for their detection.

In the conversion of 2-chloro-3-cyano-6-methyl-5-nitropyridine (**77**) into the corresponding alkoxy compound (**78**), Mariella and Hyalik[226] noted that an intense colour was produced. A similar observation was made by Fanta and Stein[227] in the treatment of 2-chloro-3-cyano-5-nitropyridine with methoxide ion. In further

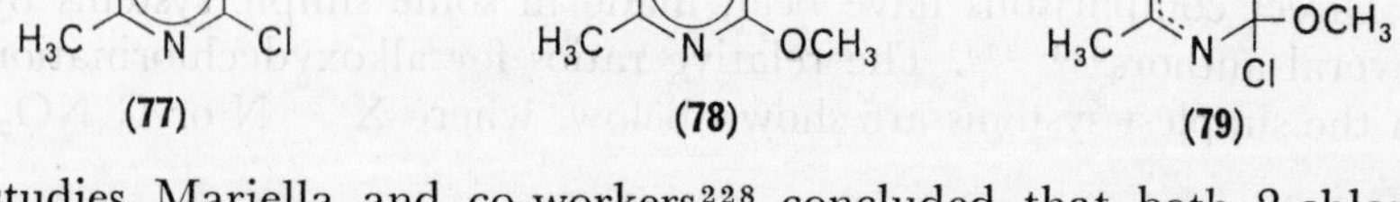

O_2N CN H_3C N Cl **(77)** O_2N CN H_3C N OCH_3 **(78)** O_2N CN − H_3C N OCH_3 Cl **(79)**

studies Mariella and co-workers[228] concluded that both 2-chloro and 2-alkoxy pyridines containing powerful electron-attracting groups in the 3- and 5-positions reacted with bases to produce intense colours, and that a probable explanation was the formation of a quinoid system analogous to the benzene series, for example, **79**, for the colour-producing species formed in the interaction of **77** with base. Although more recent research in the benzene series might suggest alternative explanations for some of their observations, the essence of their conclusions is undoubtably correct, and more recently intermediates of this type have been characterized by the n.m.r.[166, 229, 230]. The n.m.r. and u.v. spectral characteristics of these systems are collected in Table 15.

TABLE 15. N.m.r. and u.v. spectral parameters of pyridine σ-intermediates in DMSO solution.

Structure	OR	X	$H_\alpha(-\delta)$	$H_\beta(-\delta)$	$H_\gamma(-\delta)$	λ_{max} (mμ)	Ref.
80	OCH_3	H	6·08	8·30	8·62	487	229
82a or **82b**[a]	OCH_3	NMe_2	6·05	8·20	—	448	229
	OCH_3	OCH_3	5·99	8·59	—	455	230
83	OCH_2CH_2O	H	—	8·35[b]	—	462	229
84	OCH_3	H	—	8·78	—	455	230, 166

[a] Assignment can be made in terms of either structure.
[b] Multiplet centre.

Thus the addition of sodium methoxide to a solution of 3,5-dinitropyridine in DMSO gives a bright red colouration and causes the disappearance of the 3,5-dinitropyridine resonances at 0·27 and 0·86τ and the appearance of three new resonances of equal intensity at 1·38, 1·70 and 13·92τ ascribable to the intermediate **80** ($OR = OCH_3$)[229]. The shift to higher fields has been found to be characteristic of the cyclopentadienide system, and the resonance at 3·92τ ascribed to the hydrogen atom on the sp^3 carbon atom is in good agreement with those found in benzenoid aromatic systems (section V). There is no evidence for the formation of the second possible isomer (**81**) ($OR = OCH_3$) and the solutions are quite stable with time. The corresponding hydroxyl compound can be made either directly by the addition of KOH, or by solvolysis of **80** ($R = OCH_3$). Both of these compounds undergo replacement reactions with diethylamine and acetone[229]. When there is a substitu-

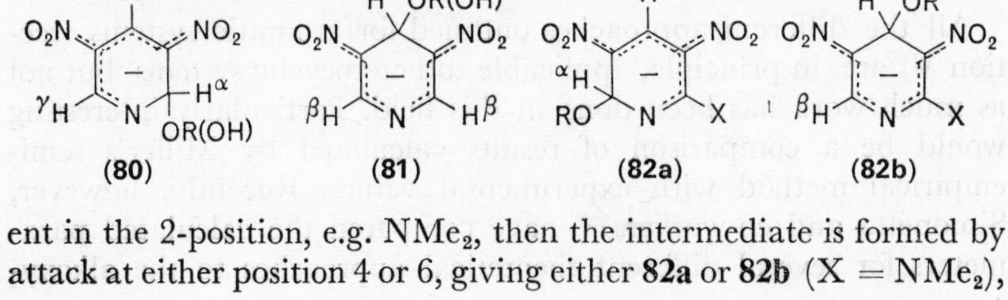

(80) **(81)** **(82a)** **(82b)**

ent in the 2-position, e.g. NMe_2, then the intermediate is formed by attack at either position 4 or 6, giving either **82a** or **82b** ($X = NMe_2$), but it is not possible to distinguish between these two structures on the basis of the n.m.r. spectrum alone[229]. Similarly, Illuminati and

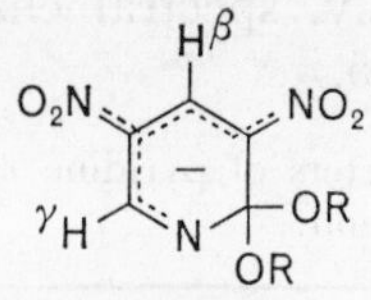

(83) 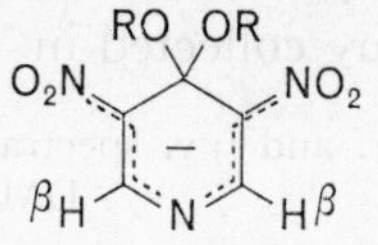 (84)

Stegel[230] found that the initial product of the action of base on 2-methoxy-3,5-dinitropyridine was a complex of the general type **82** (X = OCH_3, OR = OCH_3). Again it is impossible to establish the point of attack from the n.m.r. spectrum alone. Further possible reaction to form the dimethoxy intermediate **83** (OR = OCH_3) was obscured by a demethylation reaction. A complex of this type is, however, formed by the action of base on 2-(2′-hydroxyethoxy)-3,5-dinitropyridine where the complex **83** (OR, OR=OCH_2CH_2O)[229] is formed. The dialkoxy intermediate **84** (OR = OCH_3) is formed by the action of methoxide on 4-methoxy-3,5-dinitropyridine[166, 230]. It is not clear whether this is the initial point of attack or not. The complex can be isolated[166, 230] and shows the general characteristics expected of a structure of this type (section V).

Intermediates have also been found in the pyrimidine series[230, 231]. Illuminati and Stegel[230] reported that attack by methoxide ion on the 2- and the 4-methoxy-5-nitropyrimidines occurs at a CH position, but it is not clear whether rearrangement occurs to the more stable dialkoxy intermediates or not.

Although the extension of the above observations of intermediates in highly activated systems to a consideration of the general applicability of the two-stage mechanism must be treated with the same reservations as in the benzenoid series, there is nevertheless a strong inference that this may be the case, at least in activated systems.

3. Theoretical approaches

All the different approaches outlined for aromatic systems (section V) are, in principle, applicable to heterocyclic systems, but not as much work has been done in this field. Particularly interesting would be a comparison of results calculated by Miller's semi-empirical method with experimental values. Recently, however, Simonetta and co-workers[232] have compared the calculated parameters for several different theoretical approaches to the alkoxydechlorination of a large number of halo–aza compounds with experimental activation energies.

The methods used were the isolated molecule, localization and their own delocalization approach as described in section V. The different parameters used were the π-electron density at the carbon atom at the site of attack, the Wheland localization energy, and the difference in π-electron energy in the transition and initial states. When the activation energies were correlated, the delocalization method gave the best results and the isolated molecule the poorest. When the free energy changes were used, the delocalization method again gave the best fit, though not as good as with the activation energies. Again, the lack of suitable kinetic data for the alkoxydehalogenation of simpler aza aromatics is a serious limitation.

B. *Pyridinium Ions*

1. Kinetics and mechanism

A ready replacement of halogen by alkoxide occurs in the 2- and 4-positions of 1-alkylpyridinium salts[217, 233]. Kinetically, the reactions are first-order in both alkoxide and substrate and are thought of as proceeding according to equation (58), involving an intermediate complex (**85**). The activating effect of the quaternary

Cl, N+, CH_3 $\xrightarrow{^-OR}$ Cl OR, N, CH_3 (**85**) $\xrightarrow{-Cl^-}$ OR, N+, CH_3 + Cl^- (58)

nitrogen can be seen in a comparison of the rates of alkoxydechlorination of isomeric chloropyridines and chloropyridinium ions[217, 233] (Table 16). In fact the activating effect of the quaternary nitrogen is so much greater than the aza-nitrogen that a direct comparison with a common alkoxide is not possible, the rates of reaction of the 2- and 4-chloropyridinium ions with methoxide being too fast to measure, even at $-15°C$. Some idea of the difference can be obtained from the fact that even the rates of reaction of the pyridinium ions with the *p*-nitrophenoxide ion are still much larger than those of the corresponding pyridines with the very much stronger methoxide ion. The greater activating effect of the quaternary nitrogen, which has been estimated at $\times 10^7$ to $\times 10^{13}$ depending on position, is also reflected in very much lower activation energies. The activating effect is in fact larger than that of the NO_2 group, the kinetics of the reaction of 2- and 4-chloronitrobenzene with methoxide ion being quite measurable at 50°.

In the pyridinium ions, the relative activation at different positions is $2 > 4 \gg 3$. The activation energies for the 2- and 4-positions are very similar, and the reversal in relative rates compared with the pyridines is mainly due to a much higher activation energy for the 2-chloropyridinium ion.

TABLE 16. Kinetic parameters for the alkoxydechlorination of isomeric chloropyridines and chloropyridinium ions in methanol solution at 50°C[217, 233].

Isomer	Reagent	Pyridium ion $k \times 10^6$ $(l\,m^{-1}\,sec^{-1})$	Pyridium ion E_a (kcal/mole)	Pyridine $k \times 10^6$ $(l\,m^{-1}\,sec^{-1})$	Pyridine E_a (kcal/mole)
2-Cl	$-OCH_3$	very fast		$3{\cdot}3 \times 10^{-2}$	28·9
	$-OC_6H_4NO_2$	$1{\cdot}39 \times 10^7$	18·6		
3-Cl	$-OCH_3$	ca. 1×10^2		$1{\cdot}09 \times 10^{-5}$	32·9
	$-OC_6H_4NO_2$	0·28	30·2		
4-Cl	$-OCH_3$	very fast		0·89	25·2
	$-OC_6H_4NO_2$	$4{\cdot}6 \times 10^5$	17·6		

Pyridone formation proceeds readily by the replacement of alkoxy, halogen and cyano groups in the 2- and 4-positions in pyridinium salts[234]. The reaction is thought to proceed as shown in equation (59)

$$\text{1-methyl-2-cyanopyridinium} \underset{}{\overset{^{-}OH}{\rightleftharpoons}} \text{1-methyl-2-cyano-2-hydroxy-1,2-dihydropyridine} \xrightarrow{-HCN} \text{1-methyl-2-pyridone} + HCN \quad (59)$$

by a scheme analogous to equation (58) above, except that the final product is now a neutral species.

Pyridone formation also proceeds readily by the action of hydroxide ion on pyridinium ions without a readily replaceable group if the reaction is carried out in the presence of an oxidant[235–237]. Commonly used is potassium ferricyanide[238]. Although the pyridinium hydroxide salts can be prepared by the action of moist silver oxide on the pyridinium halides, they are normally generated in situ by performing the reaction in alkaline solution. A very important feature of the reaction is its specificity to the 2-position, no report ever having been made of the formation of a 4(γ)-pyridone. The mechanism proposed[239] for the reaction (equation 60) is analogous

$$\text{N}^+\text{–CH}_3 \quad \text{OH}^- \rightleftharpoons \underset{\textbf{(86)}}{\text{N–CH}_3\text{ (H, OH)}} \xrightarrow{[\text{O}]} \text{N–CH}_3\text{ (=O)} \qquad (60)$$

to those suggested for the replacement of labile groups (58 and 59) and requires the formation of a 'pseudo'- or 'carbinol'-base (**86**) which is then oxidized to the pyridone. This mechanism is analogous to that suggested for TNB (equation 47). However, the mechanism does not really give a complete picture of the reaction, as it does not account for the absence of the 4-pyridone while the activating effect of the quaternary nitrogen is of the same magnitude in both the 2- and 4-positions. The explanation may possibly lie in the actual mode of oxidation (see also below). Pyridinium ions substituted in the 3-position can give rise to isomeric products. Nicotinamide methiodide[240] gives both the 2- and 6-pyridones, as does 3-ethylpyridine[241], but nicotinic acid yields only the 6-isomer[242].

2. Investigations of intermediates

The equilibrium between a pyridinium hydroxide or alkoxide and a pseudo-base formed by attack of the ion in the 2- or 4-position of the pyridinium ring represents the first step in equations (58—60) above. In the case of unsubstituted pyridinium salts (equation 60), further reaction requires elimination of the very high energy hydride ion, which will be energetically unfavourable, and, in the absence of an oxidant, the reaction will not proceed further than the initial equilibrium. Investigation of this equilibrium should provide a model for the other two reactions.

Such an equilibrium gives rise to a neutral species whose presence should be detectable by conductivity measurements. This approach has been used[243, 236], but no change in conductivity which might indicate the formation of complexes corresponding to **86** was found on basification of pyridinium salts with hydroxide. It was concluded that the oxidation must proceed through a very small quantity of pseudo-base.

However, an equilibrium of this type will be very dependent on the solvating properties and dielectric constant of the solvent, and pseudo-base formation can be observed with alkoxide ions in DMSO solution[106]. The pseudo-bases are quite stable under the conditions of the experiment, and the isomers observed are very dependent on substitution in the ring.

Thus the addition of anhydrous sodium methoxide to a DMSO solution of pyridinium methiodide causes the disappearance of the resonances of the pyridinium ring, and the appearance of two new multiplets at much higher fields, consisting of a doublet $\delta = -5{\cdot}84$ and a multiplet $\delta = -4{\cdot}34$. The latter can be resolved from the overlapping absorption of the methyl group of the pyridinium salt by the use of *N*-trideuteromethylpyridinium iodide. These absorptions can be assigned to the 2,6 and 3,4,5 protons in the adduct **87**.

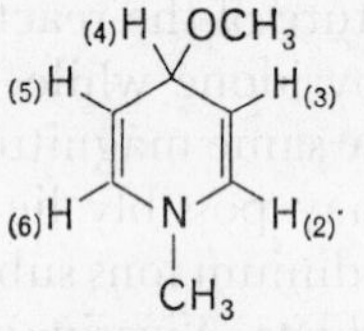

(87) 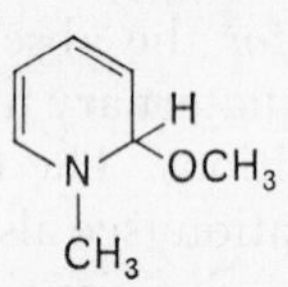 **(88)**

There are no resonances at any stage attributable to the isomeric complex **88**.

Similarly, the addition of methoxide to 3-cyanopyridinium methiodide yields the adduct (**89**) formed by attack in the 4-position and a small amount of another isomer.

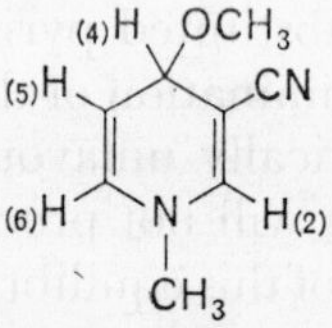

(89)

The relationship between adducts formed by attack at the 2- and at the 4-positions can be seen from the results on 3,5-dichloropyridinium methiodide. Addition of methoxide to this pyridinium salt in DMSO gives rise initially to three resonances of equal intensity at $\delta = -6{\cdot}79$, $\delta = -6{\cdot}51$, $\delta = -5{\cdot}45$ attributable to $H_{(6)}$, $H_{(4)}$ and $H_{(2)}$ respectively in **90**, and two other less intense resonances at

(90) **(91)**

$\delta = -6{\cdot}86$ (rel. intens. 2) and $\delta = -4{\cdot}94$ (rel. intens. 1) corresponding to $H_{(2,6)}$ and $H_{(4)}$ in the small amount of the isomeric complex **91** present. With time the composition of the mixture changes in favour of the thermodynamically more stable 4-isomer (Figure 10).

The above observations suggest that the 'pseudo-base' adducts of

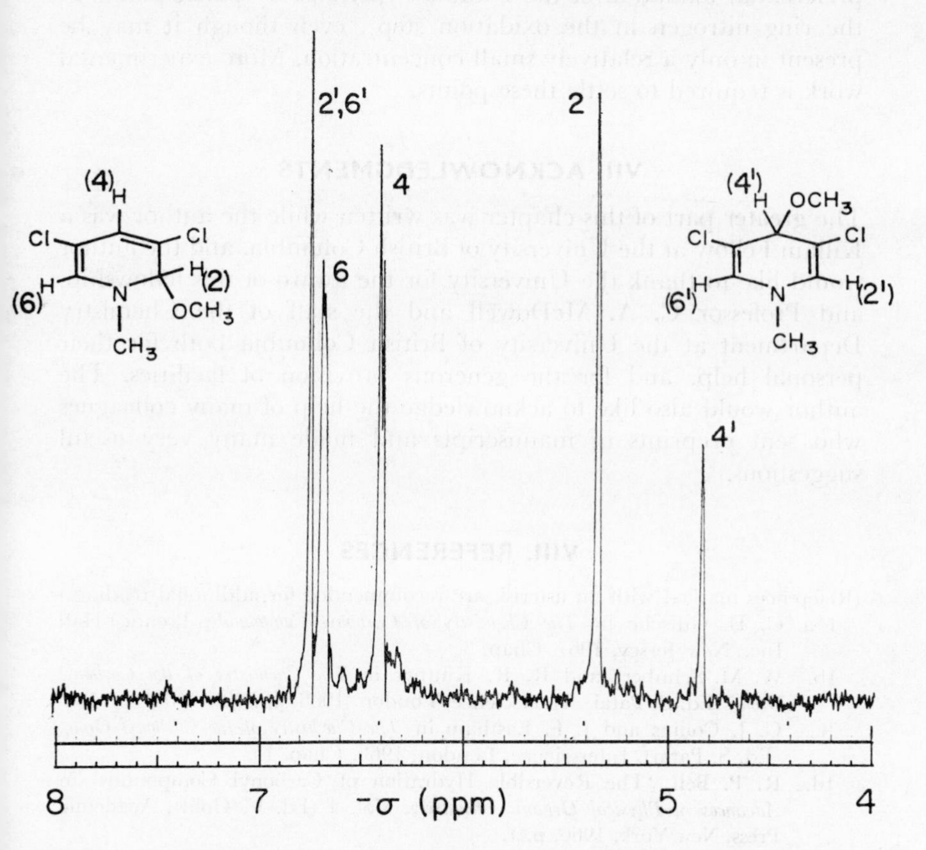

FIGURE 10. 100 MHz n.m.r. spectrum of a solution of 3,5-dichloropyridinium methiodide plus anhydrous sodium methoxide after 30 minutes.

this general type are quite stable species and can be produced quantitatively, at least in a very basic solvent. However, the factors governing which isomer is formed in a given case are not so clear. In the case of 3,5-dichloropyridinium methiodide it seems clear that it is the 2-isomer which is formed initially and then rearranges to the thermodynamically more stable 4-isomer. One could not argue from this, however, that this is the general case, as the factors which give the 2-isomer its stability (electronegative groups in the 3- and 5-positions) could also cause the attack in this position in the first place. The experimental non-observation of 4-pyridones is in agreement with either the very fast oxidation of a preferentially formed 2-adduct before it can rearrange to the 4-isomer, or the very specific preferential oxidation of the 2-adduct (perhaps by participation of the ring nitrogen in the oxidation step), even though it may be present in only a relatively small concentration. More experimental work is required to settle these points.

VII. ACKNOWLEDGMENTS

The greater part of this chapter was written while the author was a Killam Fellow at the University of British Columbia, and the author would like to thank the University for the award of this fellowship, and Professor C. A. McDowell and the staff of the Chemistry Department at the University of British Columbia both for their personal help, and for the generous provision of facilities. The author would also like to acknowledge the help of many colleagues who sent preprints of manuscripts and made many very useful suggestions.

VIII. REFERENCES

(References marked with an asterisk are recommended for additional reading.)

1*a. C. D. Gutsche in *The Chemistry of Carbonyl Compounds*, Prentice-Hall Inc., New Jersey, 1967, Chap. 3.

1b. W. M. Schubert and R. R. Kintner in *The Chemistry of the Carbonyl Group* (Ed. S. Patai), Interscience, London, 1966, pp. 706–710.

1c. C. J. Collins and J. F. Eastham in *The Chemistry of the Carbonyl Group* (Ed. S. Patai), Interscience, London, 1966, Chap. 15.

1d. R. P. Bell, 'The Reversible Hydration of Carbonyl Compounds' in *Advances in Physical Organic Chemistry*, Vol. 4 (Ed. V. Gold), Academic Press, New York, 1966, p. 1.

2*. A. J. Parker, *Quart. Rev.* (*London*), **16**, 163 (1962).

3*. A. J. Parker, *Adv. in Org. Chem.*, **5**, 1, (1965).

4*. A. J. Parker, *Advances in Physical Organic Chemistry*, Vol. 5 (Ed. V. Gold), Academic Press, New York, 1967, p. 173.
5*. C. Reichard, *Angew. Chem., Intern. Ed., Engl.*, **4**, 29 (1965).
6*. C. Agami, *Bull. Soc. Chim. France*, 1029 (1965).
7*. D. Martin, A. Weise and H-J. Niclas, *Angew Chem., Intern. Ed., Engl.*, **6**, 318 (1967).
8*. *D.M.S.O., Reaction Medium and Reactant*, Crown Zellerbach Corp. (1962).
9*. *Hexamethylphosphoramide, properties and uses*, Piernefitte Chimie (1966).
10*. L. Robert, 'Hexamethylphosphoramide, properties and uses', *Chim. Ind.*, Paris, **97**, 337 (1967).
11*. H. Normant, 'Hexamethylphosphoramide', *Angew. Chem., Intern. Ed., Engl.*, **6**, 1046 (1967).
12. E. D. Hughes and C. K. Ingold, *J. Chem. Soc.*, 244 (1935).
13. L. C. Bateman, K. A. Cooper, E. D. Hughes and C. K. Ingold, *J. Chem. Soc.*, 925 (1940).
I. Dostrovsky and E. D. Hughes, *J. Chem. Soc.*, 157 (1946).
14*. C. K. Ingold, *Structure and Mechanism in Organic Chemistry*, Cornell University Press, Ithaca, New York, 1953.
15*. C. A. Bunton, *Nucleophilic Substitution at a Saturated Carbon Atom*, Elsevier, New York, 1963.
16*. A. Streitweiser Jr., *Solvolytic Displacement Reactions*, McGraw-Hill, New York, 1962.
17*. W. H. Saunders Jr., *Ionic Aliphatic Reactions*, Prentice-Hall Inc., New Jersey, 1965.
18. E. D. Hughes, C. K. Ingold and A. D. Scott, *J. Chem. Soc.*, 1201 (1937).
19. P. Ballinger and F. A. Long, *J. Am. Chem. Soc.*, **81**, 2347 (1959).
C. G. Swain, A. D. Ketley and R. F. W. Bader, *J. Am. Chem. Soc.*, **81**, 2353 (1959).
20*. R. H. DeWolfe and W. G. Young, *Chem. Rev.*, **56**, 753 (1956).
21*. R. H. DeWolfe and W. G. Young, *The Chemistry of the Alkenes* (Ed. S. Patai), Interscience, London, 1964, p. 681.
22. E. D. Hughes, *Trans. Faraday Soc.*, **34**, 185 (1938).
23. P. B. D. de la Mare, E. D. Hughes, P. C. Merriman, L. Pichat and C. A. Vernon, *J. Chem. Soc.*, 2563 (1958).
24. P. B. D. de la Mare and C. A. Vernon, *J. Chem. Soc.*, 3331 (1952).
25. P. B. D. de la Mare and C. A. Vernon, *J. Chem. Soc.*, 3628 (1952).
26. F. G. Bordwell, P. E. Sokol and J. D. Spainhour, *J. Am. Chem. Soc.*, **82**, 2881 (1960).
27. I. N. Nazarov and I. N. Azerbaev, *Zur. Obs. Khim.*, **18**, 414 (1948).
28. C. A. Vernon, *J. Chem. Soc.*, 3628 (1952).
29. C. A. Bunton and Y. Pocker, *Chem. Ind.* (*London*), 1516 (1958).
30. H. L. Goering and R. E. Dilgren, *J. Am. Chem. Soc.*, **81**, 2556 (1959); *J. Am. Chem. Soc.*, **82**, 5744 (1960).
31. H. L. Goering and R. R. Josephson, *J. Am. Chem. Soc.*, **83**, 2585 (1961); *J. Am. Chem. Soc.*, **84**, 2779 (1962).
32. S. Patai and Z. Rappoport, *J. Chem. Soc.*, 377 (1962).
33*. S. Patai and Z. Rappoport in *The Chemistry of the Alkenes* (Ed. S. Patai), Interscience, London 1964, Chap. 8.
34. J. Meisenheimer and F. Heim, *Ber.*, **38**, 467 (1905).

35a. J. Thiele and S. Haechkel, *Liebigs Ann. Chem.*, **325,** 8 (1902); *Liebig Ann. Chem.*, **325,** 15 (1902).
35b. B. Flurscheim, *J. Prakt. Chem.*, **66,** 16 (1902).
35c. J. Meisenheimer and L. Jochelson, *Liebigs Ann. Chem.*, **355,** 293 (1907).
35d. A. Lambert, C. W. Scalfe and A. Wilder-Smith, *J. Chem. Soc.*, 1474 (1947).
35e. W. Seagers and P. Elving, *J. Am. Chem. Soc.*, **71,** 2947 (1949).
36. B.-A. Feit and A. Zilkha, *J. Org. Chem.*, **28,** 406 (1963).
37. N. Ferry and F. J. McQuillin, *J. Chem. Soc.*, 103 (1962).
38. T. I. Crowell, G. C. Helsley, R. E. Lutz and W. L. Scott, *J. Am. Chem. Soc.*, **85,** 443 (1963).
39*. T. I. Crowell, in *The Chemistry of the Alkenes* (Ed. S. Patai), Interscience, London, 1964, Chap. 4.
40. F. H. Westheimer and H. Cohen, *J. Am. Chem. Soc.*, **60,** 90 (1938).
41. C. S. Rondestvedt Jr. and M. E. Rowley, *J. Am. Chem. Soc.*, **78,** 3804 (1956).
42. R. Stewart, *J. Am. Chem. Soc.*, **74,** 4531 (1952).
43. E. A. Walker and J. R. Young, *J. Chem. Soc.*, 2045 (1957).
44. S. Patai and Z. Rappoport, *J. Chem. Soc.*, 392 (1962).
45. S. Patai and Z. Rappoport, *J. Chem. Soc.*, 383 (1962).
46. T. I. Crowell and A. W. Francis, *J. Am. Chem. Soc.*, **83,** 591 (1961).
47. D. S. Noyce, W. A. Pryer and A. H. Bottini, *J. Am. Chem. Soc.*, **77,** 1402 (1955).
48. D. E. Jones, R. O. Morris, C. A. Vernon and R. F. M. White, *J. Chem. Soc.*, 2349 (1960).
49. F. Scotti and E. J. Frazza, *J. Org. Chem.*, **29,** 1800 (1964).
50. L. Maioli and G. Modena, *Gazz. Chim. Ital.*, **89,** 854 (1959).
51. A. Campagni, G. Modena and P. E. Todesco, *Gazz. Chim. Ital.*, **90,** 694 (1960).
52. G. Modena, F. Taddei and P. E. Todesco, *Ric. Sci.*, **30,** 6, 894 (1960).
53. L. Maioli, G. Modena and P. E. Todesco, *Boll. Sci. Fac. Chim. Ind. Bologna*, **18,** 66 (1960).
54. S. Ghersetti, G. Modena, P. E. Todesco and P. Vivarelli, **91,** *Gazz. Chim. Ital.*, **91,** 620 (1961).
55. L. Di. Nunno, G. Modena and G. Scorrano, *J. Chem. Soc.* **(B)** 1186 (1966).
56. S. J. Cristol and W. P. Norris, *J. Am. Chem. Soc.*, **76,** 3005 (1954).
57. S. J. Cristol and W. P. Norris, *J. Am. Chem. Soc.*, **76,** 4558 (1954).
58a. G. Marchese, G. Modena and S. Naso, *Tetrahedron*, **24,** 663 (1968).
58b. G. Marchese, G. Modena and S. Naso, *J. Chem. Soc.* **(B)** 958 (1968).
59. G. Marchese, G. Modena and S. Naso, *Chem. Commun.*, 492 (1966).
60. S. I. Miller and P. K. Yonan, *J. Am. Chem. Soc.*, **79,** 5931 (1957).
61. P. Fritsch, W. P. Buttenberg and J. Wiechell, *Ann. Chem.*, **179,** 319, 324, 337 (1894).
62. M. Simonetta and S. Carrà, *Tetrahedron*, **19, Suppl. 2,** 467 (1963).
63a. J. G. Pritchard and A. A. Bothner-By, *J. Phys. Chem.*, **64,** 1271 (1960).
63b. W. M. Jones and R. Damico, *J. Am. Chem. Soc.*, **85,** 2273 (1963).
64. P. Beltrame and G. Favini, *Gazz. Chim. Ital.*, **93,** 757 (1963).
P. Beltrame and S. Carrà, *Gazz. Chim. Ital.*, **91,** 889 (1961).
65. P. Beltrame, S. Carrà, P. Macchi and M. Simonetta, *J. Chem. Soc.*, 4386 (1964).

P. Beltrame, D. Pitea, A. Marzo and M. Simonetta, *J. Chem. Soc.* **(B)**, 71 (1967).
66. E. F. Silversmith and D. Smith, *J. Org. Chem.*, **23**, 427 (1958).
67. P. Beltrame, P. L. Beltrame, O. Sighinolfi and M. Simonetta, *J. Chem. Soc.* **(B)**, 1103 (1967).
68. S. I. Miller, *J. Am. Chem. Soc.*, **78**, 6091 (1956).
69. E. Winterfeldt, *Chem. Ber.*, **97**, 1952 (1964).
70*. E. Winterfeldt, *Angew. Chem., Intern. Ed. Engl., B*, **6**, 423 (1967).
71. S. Ghersetti, G. Modena, P. E. Todesco and P. Vivarelli, *Gazz. Chim. Ital.*, **91**, 620 (1961).
72. L. Maioli and G. Modena, *Gazz. Chim. Ital.*, **89**, 854 (1959).
73. L. Di. Nunno, G. Modena and G. Scorrano, *J. Chem. Soc.* **(B)**, 1186 (1966).
74. C. J. M. Stirling, *J. Chem. Soc.*, 5863 (1964).
75. E. Winterfeldt and H. Preuss, *Chem. Ber.*, **99**, 450 (1966).
76. E. Winterfeldt, W. Krohn and H. Preuss, *Chem. Ber.*, **99**, 2572 (1966).
77. P. Friedlander and J. Mahly, *Liebigs Ann. Chem.*, **229**, 224 (1885).
78. D. J. Kroeger and R. Stewart, *Can. J. Chem.*, **45**, 2163 (1967).
79. R. Stewart and D. J. Kroeger, *Can. J. Chem.*, **45**, 2173 (1967).
80. C. A. Fyfe, *Can. J. Chem.*, **47**, 3331 (1969).
81. M. R. Crampton, *J. Chem. Soc.* **(B)**, 85 (1967).
82. V. Gold, *J. Chem. Soc.*, 1430 (1951).
83*. J. F. Bunnett and R. E. Zahler, *Chem. Rev.*, **49**, 273 (1951).
84*. J. Miller, *Rev. Pure Appl. Chem.* (*Australia*), **1**, 171 (1951).
85* J. F. Bunnett, *Quart. Rev.* (*London*), **12**, 1, (1958).
86*. J. Sauer and R. Huisgen, *Angew. Chem.*, **72**, 294 (1960).
87*. S. D. Ross, *Progress in Physical Organic Chemistry*, Vol. 1, Interscience, New York, 1963, p. 31.
88*. E. Buncel, A. R. Norris and K. E. Russell, *Quart. Rev.* (*London*), **1**, 123 (1968).
89. G. S. Hammond and L. R. Parks, *J. Am. Chem. Soc.*, **77**, 340 (1955).
90. N. B. Chapman and D. A. Russell-Hill, *J. Chem. Soc.*, 1563 (1956).
91. R. E. Parker and T. O. Read, *J. Chem. Soc.*, 3149 (1962).
92. J. Cortier, P. J. C. Fierens, M. Bilon and A. Halleux, *Bull. Soc. Chim. Belges*, **64**, 709 (1955).
93. J. F. Bunnett and J. J. Randall, *J. Am. Chem. Soc.*, **80**, 6020 (1958).
94. A. J. Kirby and W. P. Jencks, *J. Am. Chem. Soc.*, **87**, 3217 (1965).
95. J. F. Bunnett and R. H. Garst, *J. Am. Chem. Soc.*, **87**, 3879 (1965).
96. J. F. Bunnett and C. Bernasconi, *J. Am. Chem. Soc.*, **87**, 5209 (1965).
97. C. R. Hart and A. N. Bourns, *Tetrahedron Letters*, 2995 (1966).
98. See Reference 88, pp. 136–138 for a fuller discussion of this point.
99. J. R. Knowles, R. O. C. Norman and J. H. Prosser, *Proc. Chem. Soc.*, 341 (1961).
100a. C. A. Lobry de Bruyn and A. Steger, *Rec. Trav. Chim.*, **18**, 9 (1899), 41 (1899); *Z. Physik. Chem.*, **49**, 333 (1904).
100b. A. Steger, *Rec. Trav. Chim.*, **18**, 13 (1899); *Z. Physik. Chem.*, **49**, 329 (1904).
100c. E. Tomila and J. Murto, *Acta Chem. Scand.*, **16**, 53 (1962).
101. J. Murto, *Acta Chem. Scand.*, **18**, 1029 (1964), 1043 (1964).
102. J. Murto, *Acta Chem. Scand.*, **20**, 303, 310 (1966).

103. J. Murto and M. L. Murto, *Acta Chem. Scand.*, **20**, 297 (1966).
104. V. Gold and C. H. Rochester, *J. Chem. Soc.*, 1710 (1964).
105a. A. E. Parlath and A. J. Leffler, *Aromatic Fluorine Compounds*, Reinhold, New York, 1962, p. 326.
105b. J. Miller and W. Kai-Yan, *J. Chem. Soc.*, 3492 (1963).
106. C. A. Fyfe, unpublished results.
107*. R. Foster and C. A. Fyfe, *Rev. Pure Appl. Chem.*, **16**, 61 (1966).
108a. C. A. Lobry de Bruyn, *Rec. Trav. Chim.*, **9**, 198, 208 (1890).
108b. C. A. Lobry de Bruyn, *Rec. Trav. Chim.*, **13**, 106, 109 (1894).
108c. C. A. Lobry de Bruyn and Van Leent, *Rec. Trav. Chim.*, **14**, 89, 150 (1895).
108d. C. A. Lobry de Bruyn and Van Leent, *Rec. Trav. Chim.*, **23**, 26, 47 (1904).
109. P. Hepp, *Ann. Chem.*, **215**, 316 (1882).
110. V. Meyer, *Ber.*, **29**, 848 (1896).
111. A. Angeli, *Gazz. Chim. Ital.*, **27**, II, 366 (1897).
112. R. Foster, *Nature*, **183**, 1042 (1959).
113. R. Foster and R. K. Mackie, *J. Chem. Soc.*, 3796 (1963).
114. M. R. Crampton and V. Gold, *J. Chem. Soc.*, 4293 (1964).
115. R. Foster and C. A. Fyfe, *Tetrahedron*, **21**, 8363 (1965).
116. R. Foster and C. A. Fyfe, *J. Chem. Soc.* **(B)**, 53, (1966).
117. R. Foster and C. A. Fyfe, *Tetrahedron*, **22**, 1831 (1966).
118. C. A. Fyfe, *Can. J. Chem.*, **46**, 3047 (1968).
119. M. R. Crampton and V. Gold, *Chem. Commun.*, 256 (1965).
120. K. L. Servis, *J. Am. Chem. Soc.*, **89**, 1508 (1967).
121. V. Gold and C. H. Rochester, *J. Chem. Soc.*, 1692 (1964).
122a. E. F. Caldin and G. Long, *Proc. Roy. Soc.*, **A228**, 263 (1955).
122b. J. B. Ainscough and E. F. Caldin, *J. Chem. Soc.*, 2540 (1956).
123. M. Busch and W. Kögel, *Ber.*, **43**, 1549 (1910).
124. V. Gold and C. H. Rochester, *J. Chem. Soc.*, 1704 (1964).
125. A. F. Holleman and F. E. van Halften, *Rec. Trav. Chim.*, **40**, 67 (1921).
126. F. Reverdin, *Org. Syn.*, **7**, 28 (1927).
127. R. Foster, C. A. Fyfe and M. I. Foreman, *Tetrahedron Letters*, 1521 (1969).
128. V. Gold and C. H. Rochester, *J. Chem. Soc.*, 1710 (1964).
129. V. Gold and C. H. Rochester, *J. Chem. Soc.*, 1717 (1964).
130. V. Gold and C. H. Rochester, *J. Chem. Soc.*, 403 (1960).
131. F. Cuta and J. Pisecky, *Collection Czech. Chem. Commun.*, **23**, 628 (1958).
132. T. Abe, *Bull. Chem. Soc. Japan*, **32**, 339 (1959).
133. J. Eisenbrand and H. V. Halban, *Z. Physik. Chem.*, **A146**, 30, 101, 111 (1930).
134. E. Salm, *Z. Physik. Chem.*, **57**, 471 (1906).
135. L. Holleck and G. Pernet, *Z. Electrochem.*, **59**, 114 (1955); **60**, 463 (1955).
136. P. Hepp, *Ber.*, **13**, 2346 (1880); *Ann. Chem.*, **215**, 344 (1882).
137. M. R. Crampton and V. Gold, *J. Chem. Soc.* **(B)**, 498 (1966).
138. R. J. Pollitt and B. C. Saunders, *Proc. Chem. Soc.*, 176 (1962).
139. G. A. Russell, E. G. Janzen and E. T. Strom, *J. Am. Chem. Soc.*, **86**, 1807 (1964).
140. R. Foster and C. A. Fyfe, *Chem. Commun.*, 1219 (1967).
141. R. J. Pollitt and B. C. Saunders, *J. Chem. Soc.*, 4615 (1965).

142. R. Foster and M. I. Foreman, *Can. J. Chem.*, **47**, 729 (1969).
143. R. Foster, C. A. Fyfe, P. H. Emslie and M. I. Foreman, *Tetrahedron*, **23**, 227 (1967).
144. P. Drost, *Ann. Chem.*, **307**, 49 (1899); **313**, 299 (1900).
145. T. L. Zincke and P. L. Schwartz, *Ann. Chem.*, **307**, 32 (1899).
146. A. G. Green and F. M. Rowe, *J. Chem. Soc.*, **103**, 2023 (1913).
147. R. J. Graughran, J. P. Picard and J. V. R. Kaufman, *J. Am. Chem. Soc.*, **76**, 2233 (1954).
148. N. E. Brown and C. T. Keys, *J. Org. Chem.*, 2452 (1965).
149. A. J. Boulton and D. P. Clifford, *J. Chem. Soc.*, 5414 (1965).
150. N. P. Norris and J. Osmundsen, *J. Org. Chem.*, **30**, 2407 (1965).
151. J. Meisenheimer, *Ann. Chem.*, **323**, 205 (1902).
152. C. L. Jackson and F. H. Gazzolo, *Am. Chem. J.*, **23**, 376 (1900).
153a. C. L. Jackson and R. B. Earle, *Am. Chem. J.*, **29**, 89 (1903).
153b. C. L. Jackson and W. F. Boos, *Am. Chem. J.*, **20**, 444 (1898).
154. R. Foster, *Nature*, **176**, 746 (1955).
155. V. Gold and C. H. Rochester, *J. Chem. Soc.*, 1687 (1964).
156. S. S. Gitis and A. I. Glaz, *J. Gen. Chem. USSR*, **27**, 1960 (1957).
157. R. Foster and D. Ll. Hammick, *J. Chem. Soc.*, 2153 (1954).
158. L. K. Dyall, *J. Chem. Soc.*, 5160 (1960).
159. M. R. Crampton and V. Gold, *J. Chem. Soc.*, 4293 (1964).
160. R. Foster, C. A. Fyfe and J. W. Morris, *Rec. Trav. Chim.*, **84**, 516 (1965).
161. R. Foster and C. A. Fyfe, *Tetrahedron*, **21**, 3363 (1965).
162. J. Murto, *Suomen Kemistilehti*, **B38**, 255 (1965).
163. M. R. Crampton and V. Gold, *J. Chem. Soc.*, **(B)**, 893 (1966).
164. K. L. Servis, *J. Am. Chem. Soc.*, **87**, 5495 (1965).
165. K. L. Servis, *J. Am. Chem. Soc.*, **89**, 1508 (1967).
166. J. E. Dickeson, L. K. Dyall and V. A. Pickles, *Aust. J. Chem.*, **21**, 1267 (1968).
167. J. B. Ainscough and E. F. Caldin, *J. Chem. Soc.*, 2528 (1956).
168. E. F. Caldin, *J. Chem. Soc.*, 3345 (1959).
169. R. Foster and R. K. Mackie, *J. Chem. Soc.*, 3796 (1963).
170. V. Gold and C. H. Rochester, *J. Chem. Soc.*, 1687 (1964).
171. T. Abe, T. Kumai and J. Arai, *Bull. Chem. Soc. Japan*, **38**, 1526 (1965).
172. J. H. Fendler, E. J. Fendler and C. E. Griffin, *J. Org. Chem.*, **34**, 689 (1969). **33**, 4141 (1968).
173. C. F. Bernasconi, *J. Am. Chem. Soc.*, **90**, 4982 (1968).
174. J. H. Fendler, *J. Am. Chem. Soc.*, **88**, 1237 (1966).
175. S. S. Gitis, I. P. Gragerov and A. I. Glaz, *Zur. Obs. Khim.*, **32**, 2803 (1962); **32**, 2761 (English ed.).
176. V. Gold and C. H. Rochester, *J. Chem. Soc.*, 1704 (1964).
177. T. Abe, *Bull. Chem. Soc. Japan*, **37**, 508 (1964).
178. S. Nagakura, *Tetrahedron*, **19**, **Suppl. 2**, 361 (1963).
179. W. E. Byrne, E. J. Fendler, J. H. Fendler and C. E. Griffin, *J. Org. Chem.*, **32**, 2506 (1967).
180. E. J. Fendler, J. H. Fendler, W. E. Byrne and C. E. Griffin, *J. Org. Chem.*, **33**, 4141 (1968).
181. C. E. Griffin, E. J. Fendler and W. E. Byrne, *Tetrahedron Letters*, 4473 (1967).

182. P. Caveng, P. B. Fisher, E. Heilbonner, A. L. Miller and H. Zollinger, *Helv. Chim. Acta*, **50**, 848 (1967).
183. E. J. Fendler, J. H. Fendler and C. E. Griffin, *Tetrahedron Letters*, 5631 (1968).
184. J. H. Fendler, E. J. Fendler and C. E. Griffin, *J. Org. Chem.*, **34**, 689 (1969).
185. E. J. Fendler, J. H. Fendler, W. E. Byrne and C. E. Griffin, *J. Org. Chem.*, **33**, 4141 (1968).
186. J. H. Fendler, E. J. Fendler, W. E. Byrne and C. E. Griffin, *J. Org. Chem.*, **33**, 977 (1968).
187. R. Foster and C. A. Fyfe, unpublished results.
188a. V. A. Sokolenko, *Organic Reactivity*, Vol. II, Issue 1, Tartu, April 1965, p. 208.
188b. Reference 230, p. 4171.
188c. J. F. Bunnett and R. H. Garst, *J. Org. Chem.*, **33**, 2320 (1968).
189a. J. F. Bunnett and R. J. Morath, *J. Am. Chem. Soc.*, **77**, 5501 (1955).
189b. W. Greizerstein and J. A. Brieux, *J. Am. Chem. Soc.*, **84**, 1032 (1962).
190. G. W. Wheland, *J. Am. Chem. Soc.*, **64**, 900 (1942).
191. P. Caveng and H. Zollinger, *Helv. Chim. Acta*, **50**, 866 (1967).
192a. C. M. Gramaccioli, R. Destro and M. Simonetta, *Chem. Commun.*, 331 (1967).
192b. C. M. Gramaccioli, R. Destro and M. Simonetta, *Acta Cryst.*, **B.24**, 129 (1968).
193a. R. Destro, C. M. Gramaccioli, A. Mugnoli and M. Simonetta, *Tetrahedron Letters*, 2611 (1965).
193b. R. Destro, C. M. Gramaccioli and M. Simonetta, *Acta Cryst.*, **B.24**, 1369 (1968).
194. J. Murto, *Suomen Kemistilehti*, **B**, 246 (1965).
195. M. Simonetta and J. Carra, *Tetrahedron*, **19, Suppl. 2**, 467 (1963).
196. S. Carra, M. Raimondi and M. Simonetta, *Tetrahedron*, **22**, 2673 (1966).
197a. S. Nagakura and J. Tanaka, *Bull. Chem. Soc. Japan*, **32**, 734 (1959).
197b. S. Nagakura, *Chem. Chem. Ind.* (*Japan*), **15**, 617 (1962).
198. S. Nagakura, *Tetrahedron*, **19, Suppl. 2**, 361 (1963).
199. R. D. Brown, *J. Chem. Soc.*, 2224 (1959).
200. R. D. Brown, *J. Chem. Soc.*, 2232 (1959).
201. R. S. Mulliken, *J. Phys. Chem.*, **56**, 801 (1952); *J. Am. Chem. Soc.*, **74**, 811 (1952).
202. See appendix II, p. 375 in Ref. 198 for a fuller discussion of this point.
203. J. Miller, *J. Am. Chem. Soc.*, **85**, 1628 (1963).
204. J. Miller and K. W. Wong, *Aust. J. Chem.*, **18**, 117 (1965).
205. J. Miller and K. W. Wong, *J. Chem. Soc.*, 5454 (1965).
206. D. L. Hill, K. C. Ho and J. Miller, *J. Chem. Soc.* **(B)**, 299 (1966).
207. K. C. Ho, J. Miller and K. W. Wong, *J. Chem. Soc.* **(B)**, 310 (1966).
208*. H. E. Mertel in *Pyridine and Derivatives Part Two* (Ed. E. Klingsberg), Interscience, New York, 1961, pp. 349–350.
209*. H. E. Mertel in *Pyridine and Derivatives Part Two* (Ed. E. Klingsberg), Interscience, New York, 1961, pp. 351–352.
210*. G. Illuminati, 'Nucleophilic Heteroaromatic Substitution', *Advances in Heterocyclic Chemistry*, **3**, 285 (1963).
211*. R. G. Shepherd and J. L. Frederick, 'Reactivity of Azines with Nucleophiles', *Advances in Heterocyclic Chemistry*, **4**, 145 (1965).

212*. A. Albert and W. L. F. Armarego, 'Covalent Hydration in Nitrogen-Containing Heteroaromatic Compounds: I Qualitative Aspects', *Advances in Heterocyclic Chemistry*, **4**, 1 (1965).
213*. D. D. Perrin, 'Covalent Hydration in Nitrogen Heteroaromatic Compounds: II Quantitative Aspects', *Advances in Heterocyclic Chemistry*, **4**, 43 (1965).
214. N. B. Chapman and D. Q. Russel-Hill, *J. Chem. Soc.*, 1563 (1956).
215. N. B. Chapman and D. Q. Russel-Hill, *Chem. Ind.* (*London*), 1298 (1954).
216. N. B. Chapman, *Chem. Soc.* (*London*), *Spec. Publ.* **No. 3,** 155–167 (1955).
217. M. Liveris and J. Miller, *J. Chem. Soc.*, 3486 (1963).
218. N. B. Chapman, R. E. Parker and P. W. Soanes, *J. Chem. Soc.*, 2109 (1954).
219. N. B. Chapman and C. W. Rees, *J. Chem. Soc.*, 1190 (1954).
220. K. R. Brower, W. P. Samuels, J. W. Way and E. D. Amstutz, *J. Org. Chem.*, **18**, 1648 (1953).
221. K. R. Brower, J. W. Way, W. P. Samuels and E. D. Amstutz, *J. Org. Chem.*, **19**, 1830 (1954).
222a. G. Illuminati and G. Marino, *Chem. Ind.* (*London*), 1287 (1963).
222b. G. Illuminati and G. Marino, *Tetrahedron Letters*, 1055 (1963).
223. A. Mangini and B. Frenguelli, *Gazz. Chim. Ital.*, **69**, 86 (1939).
224a. J. Miller and V. A. Williams, *J. Chem. Soc.*, 1475 (1953).
224b. C. W. L. Bevan, *J. Chem. Soc.*, 2340 (1951).
225. Complete compilation of kinetic data are given in References 210 and 211.
226. R. P. Mariella and A. J. Havlik, *J. Am. Chem. Soc.*, **74**, 1915 (1952).
227. P. E. Fanta and R. A. Stein, *J. Am. Chem. Soc.*, **77**, 1045 (1955).
228. R. P. Mariella, J. J. Callahan and A. O. Jibril, *J. Org. Chem.*, **20**, 1721 (1955).
229. C. A. Fyfe, *Tetrahedron Letters*, **6**, 659 (1968).
230. G. Illuminati and F. Stegel, *Tetrahedron Letters*, 39, 4169 (1968).
231. M. E. C. Biffin, private communication.
232. P. Beltrame, P. L. Beltrame and M. Simonetta, *Tetrahedron*, **24**, 3043 (1967).
233. M. Liveris and J. Miller, *Aust. J. Chem.*, **11**, 297 (1958).
234a. J. P. Wibault, *Rec. Trav. Chim.*, **58**, 1100 (1939).
234b. F. Krohnke and W. Heffe, *Ber.*, **70**, 864 (1937).
234c. R. I. Ellin, *J. Am. Chem. Soc.*, **80**, 6588 (1958).
235a. H. Decker, *Ber.*, **25**, 443 (1892).
235b. H. Decker, *J. Prakt. Chem.*, **47**, 28 (1893).
235c H. Decker and A. Kaufmann, *J. Prakt. Chem.*, **84**, 425 (1911).
236. Hantzsch and Kalb, *Ber.*, **32**, 3109 (1899).
237. R. G. Fargher and R. Furness, *J. Chem. Soc.*, **107**, 690 (1915).
238a. E. A. Prill and S. M. McElvain, *Org. Syn. Coll.*, Vol. II, 419 (1943).
238b. B. S. Thyagarayan, *Chem. Rev.*, **58**, 439 (1958).
239. T. W. J. Taylor and W. Baker, Sidgwick's *Organic Chemistry of Nitrogen*, Oxford Univ. Press, 1942, p. 524.
240. M. E. Pullman and S. P. Colowick, *J. Biol. Chem.*, **206**, 121 (1954).
241. S. Sugasawa and M. Kirisawa, *Pharm. Bull.* (*Tokyo*), **4**, 139 (1956).
242. H. L. Bradlow and C. A. Wanderwerf, *J. Org. Chem.*, **16**, 73 (1951).
243. J. G. Aston and P. A. Laselle, *J. Am. Chem. Soc.*, **56**, 426 (1934).
244. H. Hosoya, S. Hosoya and S. Nagakura, *Theoret. Chim. Acta* (*Berl.*), **12**, 117 (1968).

CHAPTER 3

Free radical and electrophilic hydroxylation

D. F. Sangster

A.A.E.C. Research Establishment, Sutherland, New South Wales, Australia.

I. GENERAL

Some important indirect methods of introducing a hydroxyl group into a molecule involve attack by an electrophilic reagent as one of many steps in the synthesis. However, in this chapter only the direct hydroxylation of organic compounds by free radical and electrophilic reagents is considered. In common with many electrophilic reagents, the reactions with aromatic compounds have been the most extensively and intensively investigated.

The attacking reagents are uncharged or positively charged hydroxyl or hydroperoxyl entities[1a] and they may be free or complexed to a metal ion. They may be produced chemically by reaction between two reagents, photochemically by ultraviolet light or radiolytically by high energy ionizing radiation.

In recent years, the use of flow systems to produce a stationary state concentration of short-lived species in the cavity of electron paramagnetic resonance equipment has led to a greater understanding of the action of the chemical reagents. In many cases the short-lived transient intermediate species have been characterized. At the same time, application of pulse radiolysis techniques has enabled the early radiolytic species to be identified and their reactions to be followed. The development of sensitive separation and identification techniques, has made possible the identification of the products.

Direct electrophilic hydroxylation is not favoured as a synthetic method. In unsaturated compounds, free radicals can initiate side reactions such as polymerization or hydrogen abstraction. In aromatic compounds, the phenolic products are further attacked—particularly at the activated *ortho*- and *para*-positions—giving secondary products.

The reaction mechanism must therefore be studied under conditions where a small fraction of the original material is converted into products. The use of dilute solutions is a further simplification. These conditions are not satisfactory for practical synthetic methods.

The attacking species is known and well characterized for radiolytic hydroxylation by high energy ionizing radiation and for photolytic hydroxylation by ultraviolet light. On the other hand, rather complex interactions are involved for the chemical hydroxylating reagents. Much more is known about radiolytic hydroxylation so this method will be considered first*.

* The impact of recent research in radiation chemistry on the study of organic reaction mechanisms has been reviewed by Fendler and Fendler[323].

Quantities will, in general, be expressed in S.I. (Système International) units as well as the more familiar units. I.U.P.A.C. recommended abbreviations are used.

1 eV (electron volt) = 0·16 aJ (attojoule = 10^{-18} J)

mole/litre is used in preference to M to denote solution concentration.
1 nm (nanometre) = 1 mμ (millimicron) = 10 Å (Ångström unit)
Molar decadic (logarithm base 10) extinction coefficient (ε) is derived from the equation $\varepsilon.l.c = \lg(I_0/I)$, where l is the optical path length and c is the concentration of solute reducing light of intensity I_0 to intensity I. The units of ε are 10 $M^{-1}\ cm^{-1}$ = 1 m^2/mole.

lg = common logarithm or $\log_{10}$
s = second

Second-order rate constant, k, is in units, $M^{-1}\ s^{-1}$ or litre $mole^{-1}\ s^{-1}$ $\equiv dm^3\ mole^{-1}\ s^{-1}$.

II. RADIOLYTIC HYDROXYLATION

A. Introduction

Hydroxyl products result when aqueous solutions of certain organic solutes are exposed to high energy ionizing radiation—γ-rays, X-rays, β-particles, α-particles, etc.

The following[1] will assist readers not familiar with some of the radiation chemistry terms used in this section.

The extent of chemical reaction in any system exposed to ionizing radiation depends on the dose—the energy absorbed by unit weight of the material. A dose of one *rad* corresponds to the absorption of 100 erg per gramme of material (6·242 × 10^{13} eV/g or 10^{-2} J/kg).

The amount of material changed, of reactive species produced, or of product separated is usually expressed as the radiation chemical yield or *G*-value—the number of molecules altered for every 100 eV (16 aJ) of energy absorbed. It follows that one Megarad (1 Mrad or 1000 krad, which is equivalent to 2·4 calories/g) makes a change of $G \times 1{\cdot}04 \times 10^{-3}$ mole/kg irradiated. G lies between 0·1 and 10 for many reactions.

Many products are formed within a millisecond. Those which are present a few minutes after the end of the irradiation are called 'prompt products'. Those that appear subsequently without further exposure to ionizing radiation are called 'post-irradiation products'. A wavy arrow is often used to indicate a radiation chemical reaction.

It should be noted that some of the literature references are to reports, etc., produced by or on behalf of Atomic Energy Commissions. Most of these are listed in *Chemical Abstracts* or *Nuclear Science Abstracts* and are available from libraries of corresponding authorities.

I. The reactive species

To date the main task in aqueous radiation chemistry has been to determine what are the primary species formed during the radiolysis of water and how much of each of these species is produced for a given input of energy. This task has proved to be a very difficult one, and even now that the main features appear to be clear, some details are incomplete.

The primary species are the hydrated electron, the hydrogen atom, the hydroxyl radical, molecular hydrogen, molecular hydrogen peroxide and the hydronium ion.

$$H_2O \rightsquigarrow e^-_{aq},\ H^{\cdot},\ OH^{\cdot},\ H_2,\ H_2O_2,\ H_3O^+ \quad (1)$$

Radical species will generally be indicated by a dot ($R^{\cdot}$).

In Table 1 the experimental radiation chemical yields (*G*-values) are given. The amounts of the reducing radical species (e_{aq} and $H^{\cdot}$) available in acid or in alkaline solution are greater than in neutral solution.

TABLE 1. Radiation chemical yields (*G*-values) for primary species in aqueous solution.

	G-value (molecules per 100 eV)		
pH	0–2	4–11	13–14
e_{aq}	} 3·65	2·7	3·1
$H^{\cdot}$		0·55	0·5
$OH^{\cdot}$	2·95	2·8	2·9
H_2	0·45	0·45	0·45
H_2O_2	0·8	0·7	0·7

Accounts of these species and of their reactions are given in general textbooks on radiation chemistry[1,2]. Some of the older texts[3–5] give excellent accounts of earlier theories and of experimental results but were written before the discovery of the solvated electron[6]. The interpretations advanced in these texts must be modified accordingly.

The primary radiation chemical species undergo reaction with solutes (sometimes called 'scavengers') to give the observed products. Thus, except in concentrated solution, the radiation does not act directly on the solute. The action is indirect via the solvent water.

While the essential nature and amount of the primary species was being established, and especially during recent years, radiation chemists have investigated the reactions of these species.

2. The hydroxyl radical

It has been shown that the primary oxidizing species formed in the radiolysis of aqueous solutions is uncharged[7] and is identical with that formed in the photolysis of hydrogen peroxide[8]. Hydroxyl groups have been found in the polymer produced by irradiation of acrylonitrile in aqueous solution[9]. The electron paramagnetic resonance spectrum of hydroxyl radicals has been found in irradiated ice[10]. It seems certain that the chemical entity involved is the hydroxyl radical ($OH^{\cdot}$).

The radiation chemical yield of hydroxyl radicals can be almost doubled by dissolving nitrous oxide in the solution before irradiation. The hydrated electrons which might otherwise interfere with the course of hydroxyl radical reactions are converted and then participate in those reactions.

$$e_{aq} + N_2O \longrightarrow N_2 + OH^{\cdot} + OH^- \quad (2)$$

Similarly, in hydrogen peroxide solutions the reducing primary radical species (e_{aq} and $H^{\cdot}$) may be converted into oxidizing radicals.

$$H_2O_2 + e_{aq},\ H \longrightarrow OH^{\cdot} + OH^-,\ H_2O \quad (3)$$

In these ways and under suitable conditions, ionizing radiation is a 'clean' way of making hydroxyl radicals. Their reactions may be studied with relative freedom from interfering species.

3. The hydroperoxyl radical

The hydroperoxyl radicals ($HO_2^{\cdot}$) or their ionized form ($O_2^{\cdot -}$) are secondary products formed by reaction between hydrogen atoms or hydrated electrons and molecular oxygen dissolved in water which is exposed to air or an atmosphere of oxygen.

$$H^{\cdot} + O_2 \longrightarrow HO_2^{\cdot} \quad (4)$$

$$e_{aq} + O_2 \longrightarrow O_2^{\cdot -} \quad (5)$$

$$HO_2^{\cdot} \longrightarrow H^+ + O_2^{\cdot -} \qquad pK_a = 4{\cdot}4 \quad (6)$$

Hydroperoxyl radicals are also formed by reaction between hydroxyl radicals and hydrogen peroxide. This reaction can become important when solutions of hydrogen peroxide are irradiated.

$$H_2O_2 + OH^{\cdot} \longrightarrow HO_2^{\cdot} + H_2O \quad (7)$$

Although their identity has been established[11-14], the reactions of radiolytic hydroperoxyl radicals with organic solutes have not been extensively studied. In general, they are much less reactive than hydroxyl radicals and often react with each other to give a radiation chemical yield of hydrogen peroxide greater than the primary molecular yield ($G = 0{\cdot}7$).

$$2\ HO_2^{\cdot} \longrightarrow H_2O_2 + O_2 \qquad (8)$$

They must not be overlooked in any discussion of hydroxylating species.

4. Pulse radiolysis

Since much of our detailed knowledge of the early radiation chemical reactions comes from the application of pulse radiolysis, a brief outline of this technique will be given.

Pulse radiolysis[14-16] is the radiation chemical analogue of flash photolysis. Briefly, high energy electrons in a very short pulse (microseconds or nanoseconds) at over a million volts are fired from an electron accelerator into the chemical system under study. The chemical species present after the pulse are followed by detection methods with fast response times.

Although polarographic, conductometric and electron spin resonance detectors have been used, the ones most frequently used are spectrophotometric. A beam of light is passed through the chemical system and into a monochromator. The changes in intensity, as registered by a photoelectric cell, are displayed on a cathode ray tube. In this way, the appearance and change in concentration of an absorbing species can be followed with time. Also, the absorption spectrum of such a species can be found. Thus its kinetic behaviour can be followed and some idea of its nature or identity can be obtained.

B. *Hydroxyl Radical Reactions*

Hydroxyl radicals have been shown to undergo four different types of reaction with organic compounds—addition, hydrogen abstraction, electron transfer and displacement reactions.

I. Benzenoid compounds

a. Early studies. Weiss, Stein and co-workers[17-24] studied the radiation chemistry of aqueous solutions of benzene[17, 18], benzoic acid[17, 22], nitrobenzene[20, 21] and chlorobenzene[24]. One result was to demonstrate that the oxidizing species was the hydroxyl radical in that its reactions resembled those of the reactive entity in Fenton's reagent (ferrous ion and hydrogen peroxide—see section IV.B.). It was proposed that hydroxyl radicals abstracted ring hydrogens from the solute aromatic molecules to give the corresponding phenyl radicals ($XC_6H_4^{\bullet}$). Association of two of these radicals gave substituted biphenyls which had also been isolated from Fenton's reaction. Alternatively, each phenyl radical was thought to add another hydroxyl radical yielding the range of isomeric hydroxy compounds (phenols) which were found among the products of the radiolysis. Subsequent workers[25–56], using improved methods of product separation and identification, made more refined measurements, irradiating aqueous solutions of various aromatic compounds, but mostly benzene.

It was clear that in aerated solutions the yield of phenolic products was greater than half the $G_{OH^{\bullet}}$ value. (It is now known that hydroxyl radicals react so rapidly with the solute that under the conditions used it is extremely unlikely that the short-lived radicals, which are present at very low steady state concentrations, will react with a second hydroxyl radical as required by the proposed mechanism.) Further, oxygen isotopic studies showed that all the incorporated oxygen came from radiolysis of water. It was therefore proposed by several workers[26, 32, 33, 41], particularly Russian workers[34–40, 47, 52], that the $OH^{\bullet}$ radical must first add to the aromatic ring in the same way as many other organic substituting reagents which give a benzenium structure. Acceptance of this mechanism had to await its proof[57, 58] by pulse radiolysis techniques.

b. The definitive experiment. Using pulse radiolysis techniques, Dorfman, Taub and Bühler[58] were able to show in a strikingly direct way that the hydroxyl radical added to the aromatic ring of benzene

$$C_6H_6 + OH^{\bullet} \longrightarrow C_6H_6(H)(OH)^{\bullet} \quad \textbf{(1)} \tag{9}$$

(equation 9) to give an absorption spectrum with a maximum at 313 nm corresponding to the hydroxycyclohexadienyl radical **1**. The identity of **1** was further established by showing that the rate constant for reaction (9) was greater than that for any hydrogen abstraction reaction. This would have given $C_6H_5^{\bullet}$. Furthermore, one would expect hydrogen abstraction by hydroxyl radicals from fully deuterated benzene to take place at about one third the rate of that from C_6H_6. Experimentally, no such kinetic isotope effect was observed when C_6D_6 in aqueous solution was irradiated.

The fate of this radical depended on whether molecular oxygen was present in the solution. In deaerated solution, **1** disappeared by second-order processes giving dimers which slowly decomposed to biphenyl ($G = 0{\cdot}3$ molecules/100 eV), phenol ($G = 0{\cdot}23$) and other products (equations 10 and 11).

$$2\ HOC_6H_6^{\bullet} \longrightarrow \text{dimers} \longrightarrow C_6H_5C_6H_5 + 2\ H_2O \qquad (10)$$

$$2\ HOC_6H_6^{\bullet} \longrightarrow C_6H_5OH + C_6H_7OH \qquad (11)$$

The adduct $C_6H_7^{\bullet}$, which had been formed by reaction between benzene and the hydrogen atom, also reacted with **1** (equation 12).

$$HOC_6H_6^{\bullet} + C_6H_7^{\bullet} \longrightarrow 2\ C_6H_6 + H_2O \qquad (12)$$

Also mentioned was the possibility of **1** reacting with solute benzene to give dimer hydroxylated products but these were not sought by analysis.

In aerated solution, **1** reacted (equation 13) with oxygen giving a spectrum with a maximum at a lower wavelength which could be attributed to the peroxy derivative **2**. The final product was

$$\text{(H, OH)}C_6H_5^{\bullet} + O_2 \longrightarrow \text{(H, OH)}C_6H_5O_2^{\bullet}\ \textbf{(2)} \qquad (13)$$

phenol, with an immediate radiation yield of $G = 1{\cdot}9$ (equation 14). Fifteen weeks after removal from the radiation field, G (phenol) was found to be 2·3. The additional amount was thought to have been formed from the decomposition of unstable products such as peroxides.

$$HOC_6H_6O_2^{\bullet} \longrightarrow HOC_6H_5 + HO_2^{\bullet} \qquad (14)$$

$$2\ HO_2^{\bullet} \longrightarrow H_2O_2 + O_2 \qquad (8)$$

At high pulse intensities, **2** disappeared by second-order processes in competition with reaction (14), presumably giving peroxides or hydroperoxides:

$$2\ HOC_6H_6O_2^{\bullet} \longrightarrow \text{peroxides} \qquad (15)$$

$$HOC_6H_6O_2^{\bullet} + HO_2^{\bullet} \longrightarrow \text{hydroperoxides} \qquad (16)$$

The yield of phenol decreased with increasing pulse intensity, i.e., with increasing concentration of **2**. High concentration favoured second-order reactions (equations 15 and 16) rather than reaction (14). At the highest pulse intensity used, where the maximum concentration of **2** was 9×10^{-5} mole/litre, G(phenol) was 0·19. Post-irradiation production of phenol with $G = 0{\cdot}3$–$0{\cdot}5$ was observed.

c. Transient species. Spectra similar to that of **1** were obtained for substituted **1** formed following the pulse radiolysis of aqueous solutions of chlorobenzene, bromobenzene, iodobenzene, toluene, phenol[58], benzoic acid[59], benzoate[60], nitrobenzene[61], etc.[62].

Although the transient species can be followed for but a fraction of a millisecond, their absorption spectra resemble closely[60] those of the carbonium ions formed when aromatic compounds are ring-protonated by concentrated sulphuric acid or by Lewis acid–hydrogen halide complexes.

The molar decadic extinction coefficient has been estimated in several cases. For **1** it is 3500 $\mathrm{M}^{-1}\,\mathrm{cm}^{-1}$ (350 m^2/mole)[58]. For the benzoate–$\mathrm{O}^{-\bullet}$ adduct at pH 13 it is 3100 [60].

d. Kinetics of OH addition. Because the hydroxyl radical absorbs weakly[63–66] in the short ultraviolet region ($\varepsilon = 530\ \mathrm{M}^{-1}\,\mathrm{cm}^{-1}$ or 53 m^2/mole at about 230 nm with another broad maximum below 200 nm[66]), it is not possible to determine second-order rate constants for its reactions with solutes by following its rate of disappearance as can be done for reactions of the hydrated electron[6, 14, 16]. Nevertheless, a considerable body of data is available[67] from the application of competition methods[68–74] and from following the appearance of the transient adducts **1** [59, 75–77].

The rate constants for the reactions expressed by equation (17) have been reported by Neta and Dorfman[77] to range between

$$C_6H_5X + OH^{\bullet} \longrightarrow [C_6H_5X(OH)]^{\bullet}\ (\text{OH ortho/meta/para to X}) \quad \text{or} \quad [C_6H_5X(OH)]^{\bullet}\ (\text{OH at the ipso carbon bearing X})\ \ \mathbf{(3)} \qquad (17)$$

$1{\cdot}4 \pm 0{\cdot}3 \times 10^{10}\ \mathrm{M}^{-1}\,\mathrm{s}^{-1}$ ($\mathrm{dm}^3\,\mathrm{mole}^{-1}\,\mathrm{s}^{-1}$) for phenol ($X = OH$)[75] and $3{\cdot}2 \pm 0{\cdot}4 \times 10^{9}\ \mathrm{M}^{-1}\,\mathrm{s}^{-1}$ for nitrobenzene ($X = NO_2$). An independent determination[76] of the rate of formation of the nitrobenzene–hydroxyl radical adduct gave a value $4{\cdot}7 \pm 0{\cdot}5 \times 10^{9}\ \mathrm{M}^{-1}\,\mathrm{s}^{-1}$. Further, the values determined[77] for benzoate ($X = COO^-$) show a slight downward trend with increasing benzoate concentration. Further study is needed.

Using a series of rate constants obtained by competition methods for reaction between hydroxyl radicals and substituted benzenes, Anbar, Meyerstein and Neta[72] showed that there was a reasonably good correlation using Hammett's[78] $\sigma\rho$ relationship for electrophilic substitution (equation 18).

$$\lg[k_{(OH+C_6H_5X)}/k_{(OH+C_6H_6)}] = \sigma\rho \tag{18}$$

σ denotes the Hammett substituent constant and ρ the reaction constant. Since data on the isomeric yields were not available for the whole range of compounds measured, it was not possible to evaluate the reaction rate constants for attack at positions *meta* or *para* to the substituent. As a near approach to the problem the overall reaction rate constant was used in equation (18) in conjunction with both σ_{para} and σ_{meta} values for deactivating substituents and σ_{para} values only for other substituents. (It has been established that there is little or no attack *meta* to *ortho-para* directing substituents as would be expected from such an electrophilic reagent as the hydroxyl radical.)

The value found for the series of substituted benzenes, $\rho = -0{\cdot}41$, was the same as that found when rate constants for $OH^{\bullet}$ reaction with *para*-substituted benzoate ions were plotted in a similar fashion. The σ values in disubstituted benzenes are therefore additive. This value of ρ is much lower than that for most other electrophilic reagents and this is attributed to the high reactivity of hydroxyl radicals, which makes them less susceptible to the directing influences of substituent groups and, incidentally, also makes them less selective. Using the absolute rate constants determined from pulse radiolysis, Neta and Dorfman[77] obtained a value $\rho = -0{\cdot}5$ in good agreement with that obtained from competition rate data.

The total reaction rate constants which have been used in these linear free energy correlations include contributions from $OH^{\bullet}$ reaction with the side chain, elimination of substituent groups and addition at the *ortho* position as well as the additions at *meta*- and *para*-positions. The partial rate constants for additions at each of these two positions should be separated from the overall reaction rate constant in order to test whether they reflect the electron distribution in the aromatic compound as expressed by the σ functions. The chemistry of the side reactions which affect the yields of isomeric products must be studied before this can be done. On the data now available[81, 85, 126] it would appear that the electron distribution in the ring is perturbed on the approach of the strong dipole of the hydroxyl radical.

e. Spectral correlations. The spectra of a number of **3** structures have been determined by pulse radiolysis techniques. Chutny[79] has shown that there are good correlations between these spectra and those of the corresponding aromatic compounds. The relative bathochromic shift of the primary band absorption maximum caused by substitution can be expressed as $(\nu_{C_6H_6OH} - \nu_{C_6H_5OHX})/(\nu_{C_6H_6OH})$. ν is the frequency of the absorption maximum. This was compared with the relative shift due to the same substituent in benzene. This ratio was found to lie between 0·71 and 1·12 and to be on the average 0·92. An exception was the benzoate–$OH^{\bullet}$ adduct (**3** where X = COO^{-}) for which the ratio was 0·48 and this was attributed to the influence of the negative charge on the substituent group.

The constancy indicates that the transient species have similar structure and the correlations may be used to predict absorption spectra of unknown species. Neta and Dorfman[77] have determined the spectra of other $OH^{\bullet}$ adducts, including those derived from some disubstituted benzenes, and have found fair agreement with Chutny's results.

Since it is probable that the hydroxyl group is attached to the cyclohexadienyl structure by sigma bonds (see section II.B.1.h) the observed spectrum in each case is an envelope of the contributions of the possible isomeric structures and of the various fragments remaining from side chain and substituent group elimination.

f. Reactions of the transient species. The rates of reaction of **3** with oxygen (equation 19) and with itself (equation 20) have been

$$XC_6H_5OH^{\bullet} + O_2 \longrightarrow \underset{\textbf{(4)}}{XC_6H_5(OH)O_2^{\bullet}} \tag{19}$$

$$2\,XC_6H_5OH^{\bullet} \longrightarrow \text{products} \tag{20}$$

measured by Cercek[80] for various substituents X. He found that, except for the adducts with benzene, toluene and ethylbenzene, the linear free energy relationship[77] of equation (21) held for reaction with oxygen.

$$\lg \frac{k_{(HOC_6H_5X^{\bullet}+O_2)}}{k_{(HOC_6H_6^{\bullet}+O_2)}} = \sigma\rho + C \tag{21}$$

$$\rho = -1{\cdot}0 \quad \text{and} \quad C = -1{\cdot}4$$

A corresponding relationship held for $k_{(2HOC_6H_5X^{\bullet})}$ where $\rho = -0{\cdot}75$ and $C = 0{\cdot}3$. Activation energies for the second-order decay (equation 20) were also determined. The activation energy (kcal $mole^{-1}$) was found equal to $(5\sigma + 2{\cdot}5)$ for electron-withdrawing substituents and $(7\sigma + 5{\cdot}5)$ for electron-donating substituents.

This difference can be explained by assuming that electron-donating substituents give rise to a greater dipole repulsion between the two aromatic rings in the activated complex. Once again, the three aromatic hydrocarbon adducts did not correlate well but lay between the two groups. Since it seems unlikely that these react in a way fundamentally different from that of the other adducts, an explanation was sought in entropy effects. These may be caused by the absence of a lone pair of electrons in the side chains $-CH_3$ and $-C_2H_5$. This would inhibit hydrogen bonding with the water molecules. The result is that these adducts are twenty times more likely to react with oxygen and one fourth as likely to undergo biradical dimerization or disproportionation as would be expected from their σ constants.

Once again, such linear free energy correlations can be used to draw together and explain data and to predict rate constants. They could be refined if it were possible to separate the contributions from individual isomers. It is known for instance that of the three adducts formed during radiolysis of aqueous nitrobenzene solutions, that giving finally *o*-nitrophenol behaves quite differently from that giving *meta*- and *para*-products[81].

g. Formation of hydroxyl products. In deaerated solutions of benzene, the radiolysis products contain in addition to biphenyl (equation 10), bicyclohexadienyl compounds[58] ($G \sim 1$)[36] formed by diradical reactions (equations 22 and 23).

$$2\,HOC_6H_6^{\cdot} \longrightarrow (HOC_6H_6)_2 \qquad (22)$$

$$HOC_6H_6^{\cdot} + C_6H_7^{\cdot} \longrightarrow HOC_6H_6C_6H_7 \qquad (23)$$

At low dose rates, i.e., in steady radiolysis where the radical concentrations are much lower than in pulse radiolysis, **1** might react with the benzene substrate giving hydroxyphenylation products[58]. It should be noted that these diradical reactions are speculative as there is no report of a complete analysis of the radiolytic products from any aqueous aromatic solutions. Sitharamarao (quoted in Ref. 82) has found dihydroxybiphenyl ($G = 0{\cdot}8$) from irradiation of deaerated salicylate solution at pH 6·3 and no other hydroxylation products. In acid and alkaline solutions he found other products—dihydroxy-benzoic acids, benzoic acid, phenol and catechol[68].

Cercek[80] has considered the possible structures of **3** and concludes that one need only consider the structures shown overleaf.
Structure **7** is not applicable when X is an electron-donating substituent. Reactions between these species would be expected to give

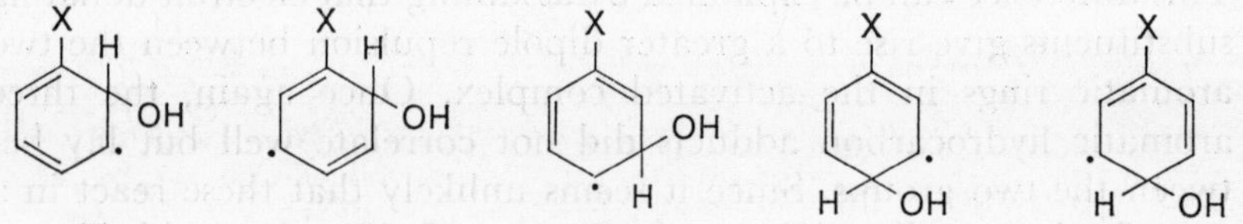

(5) (6) (7)

products containing two aromatic rings joined together predominantly at the *meta* position relative to the original substituent.

In aerated solution of most aromatic solutes, reactions are thought[58] to follow equations (17), (19) and (24), giving a phenol. The amount

$$XC_6H_5(OH)O_2^{\cdot} \longrightarrow XC_6H_4OH + HO_2^{\cdot} \tag{24}$$

of hydrogen peroxide found ($G = 2{\cdot}2$) in neutral solution can be accounted for by such a scheme in conjunction with equation (8) but more detailed evidence is lacking.

When $X = NO_2$, the peroxy derivatives of structures **5** and **6** are stabilized against unimolecular decomposition (equation 24) by forming a six-membered ring by hydrogen bonding between the out-of-plane hydroxyl radical in the *ortho* position and an oxygen of the nitro group[81]. At moderately high dose rates (10 krad per minute and higher) the diradical reaction (equation 25) predominates[81, 83–85].

$$2\,O_2NC_6H_5(OH)O_2^{\cdot} \longrightarrow O_2NC_6H_4OH + O_2NC_6H_5 + H_2O + 2\,O_2 \tag{25}$$

The yield of *o*-nitrophenol is half that at very low dose rates or that in the presence of 0·2 mmole/litre dichromate ion which presumably acts as a one-electron oxidizing agent.

Land and Ebert[75] showed that, when the substrate is a phenol, **3** ($X = OH$) is able to eliminate water in a unimolecular reaction, giving a phenoxyl radical, $C_6H_5O^{\bullet}$. A special case is hydroquinone which gives a semiquinone radical $HOC_6H_4O^{\bullet}$ [86]. Fendler and Gasowski[85] suggest that this may be a general reaction leading to the elimination of substituent groups, NO_2, Cl, OMe from the ring.

Cercek and Ebert[87] have shown that the hydroxyl radical does not attack the ring of *p*-nitrophenol. Since G(2-hydroxy-4-nitrophenol) $= 2{\cdot}95$ in good agreement with G_{OH}, one would not have suspected from steady radiolysis studies[88, 89] that anything unusual was happening.

The nitro group is strongly electron-withdrawing and is assisted by the phenolic hydroxyl group which is an electron donor to *ortho*- and *para*-positions. Hence the nitro group in this compound is more than usually electron-rich and is attacked by the hydroxyl radical in preference to the ring.

$$HOC_6H_4NO_2 + OH^{\bullet} \longrightarrow \underset{(8)}{HOC_6H_4NO_3H^{\bullet}} \quad (26)$$

8 was found to have an absorption maximum at 295 nm whereas a maximum would be expected at about 420 nm if the ring were attacked to give a substituted **3** structure[87]. **8** is a weak acid ($pK_a = 5{\cdot}3$) and ionizes in neutral solution giving $HOC_6H_4NO_3^{-\bullet}$. In deaerated solution this species undergoes an interesting first-order rearrangement (equation 27) whereby the hydroxyl group is transferred to the aromatic ring giving a species absorbing at 400 nm. The rate constant for this rearrangement (k_{27}) was found to be $14 \pm 2\ s^{-1}$.

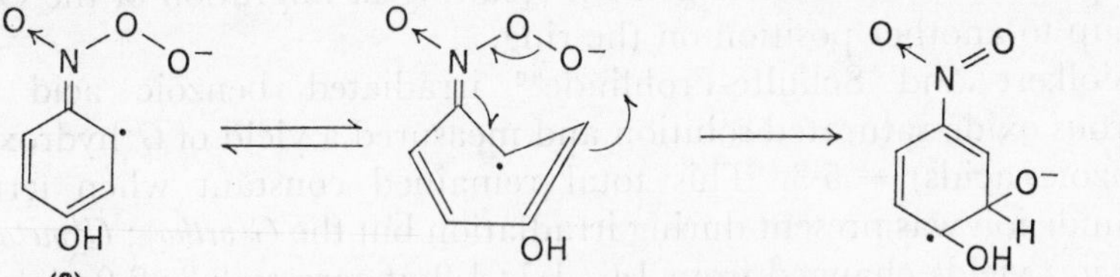

(27)

(8)

A cyclohexadienyl structure would enable the OH group to migrate from the nitro group of the *p*-nitrophenol–$O^{-\bullet}$ adduct to the 2-position without having to pass via the electron-deficient 3-position. In the benzvalene form positions 2, 3 and 5 can be depicted as equidistant from position 4. A cyclohexadienyl intermediate which provides a six-membered cyclic transition state for the oxygen transfer would therefore appear more likely than a benzvalene intermediate.

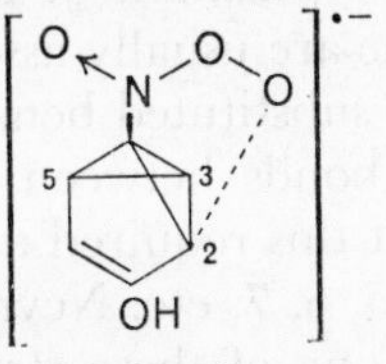

Two of these species disproportionate to give *p*-nitrophenol and the hydroxylated product. In aerated solution, **8** also disappeared

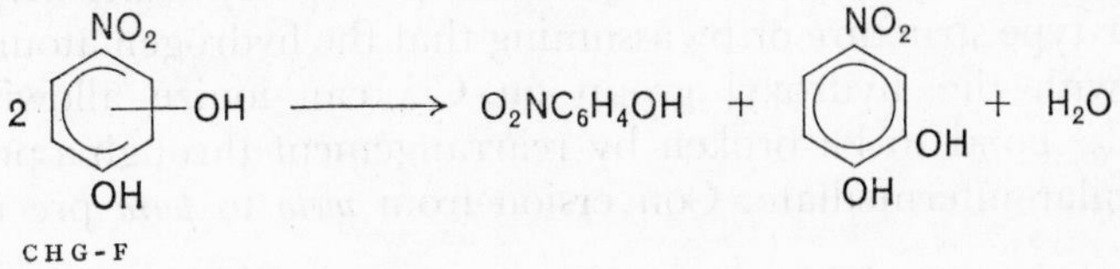

$$2\ HOC_6H_4NO_2(OH) \longrightarrow O_2NC_6H_4OH + NO_2C_6H_3(OH)_2 + H_2O \quad (28)$$

by a first-order process but 160 times faster than by reaction (27). Each hydroxyl radical gave one molecule of hydroxylated product[88].

Other examples have been found of compounds whose aromatic ring is not attacked by hydroxyl radical or only to a minor extent. In the case of nitrosobenzene[90] and phenylhydroxylamine[91] there is no subsequent transfer to the ring and no hydroxylation. Compounds in which the side chain is hydroxylated will be considered in section II.B.2.b.

h. Isomeric yields. The yields of possible isomeric products following aromatic substitution reactions do not necessarily reflect faithfully the proportion of reaction at the respective positions of the aromatic ring. Apart from various side reactions such as have been noted for the hydroxyl adducts with nitrobenzene and phenol, there is always the possibility of a rearrangement resulting in migration of the OH group to another position on the ring.

Volkert and Schulte-Frohlinde[92] irradiated benzoic acid in nitrous oxide-saturated solution and measured a yield of G(hydroxybenzoic acids) = 5·3. This total remained constant when ferricyanide ion was present during irradiation but the G(*ortho*) : G(*meta*) : G(*para*) yields changed from 1·6 : 1·7 : 1·9 at zero to 2·3 : 2·0 : 1·1 at 2 mmole/litre of ferricyanide ion and above. Allowing for enhancement by the addition of N_2O, the ratio of yields is comparable with that of 0·74 : 0·42 : 0·33 found in aerated solution[43, 51]. The change in the relative proportions of the three isomers must be due to tautomerism between the possible structures. It was suggested[92] that the $OH^{\bullet}$ radical might form a charge-transfer complex as has been observed for Cl-atoms[93]. This seems unlikely as does the possibility of **3** (X = COOH) remaining, or reverting to, a π-complex. There is the possibility of a prismane or benzvalene structure being formed[94–98] although these are usually associated with some form of excitation[97] in the case of substituted benzenes. Cercek[80] discounted the possibility of formal bonds between non-neighbouring carbon atoms on the grounds that this required a greater amount of energy than the other structures **5**, **6**, **7**, etc. Nevertheless, **3** is already non-planar and formation of one of these structures may be easier for cyclohexadienyl radicals than for benzenes. Rearrangement from *ortho* adduct to *para* precursor can be explained by either invoking a Dewar-type structure or by assuming that the hydrogen atom associated with the hydroxyl group on $C_{(2)}$ can ionize allowing the $C_{(1)}$–$C_{(2)}$ bond to be broken by rearrangement through a prismane or similar intermediate. Conversion from *meta* to *para* precursor is

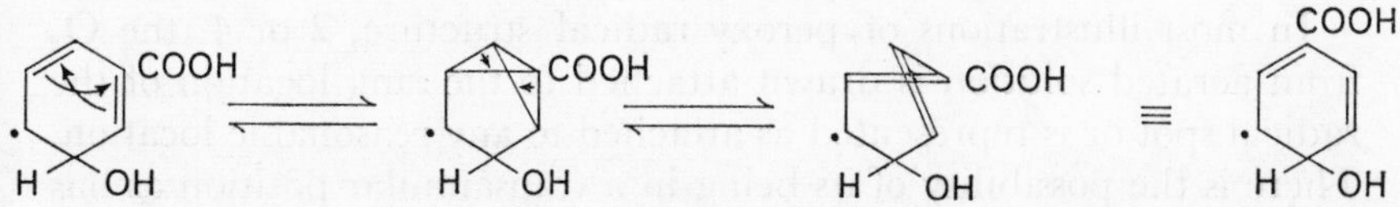

(28a)

readily explained via a benzvalene intermediate. The phenomenon may not be general. There appears to be no rearrangement in the case of the nitrobenzene–OH$^\bullet$ adduct. Here the *ortho* adduct is relatively long-lived[81] and would be expected to have time to rearrange before reacting with oxygen[80] if such rearrangement were favoured.

It seems that, with a sufficiently high concentration of oxidizing agent present—ferricyanide ion[92], dichromate ion[81], molecular oxygen, etc.—there are fewer uncertainties. The isomeric ratios determined in aerated solution are usually considered to reflect the position of attack on the ring[99, 100]. By careful choice of the right mixture of nitrous oxide and oxygen, it is possible to eliminate interference by the hydrated electrons, to increase the available amount of hydroxyl radicals and to obtain oxidized product ratios reflecting the initial positions attacked. Further, $O_2^{-\bullet}$, H_2O_2, and other species which might interfere are not introduced in significant amounts. This gives a clean source of hydroxyl radicals indeed.

Until recently, there has been some doubt whether nitrous oxide and hydrated electrons do give hydroxyl radicals on reaction (equation 2). There is no physical evidence that $N_2O^{-\bullet}$ has a finite existence but $O^{-\bullet}$ might have to react with H_2O to give $OH^\bullet$ before reacting with many solutes. Nakken, Brustad and Hansen[101] have measured the *G*-values for the yields of isomeric hydroxy acids from the radiolysis of benzoate solution and of the 3-hydroxy and 5-hydroxy derivatives from anthranilic acid. They found differences between oxygen, nitrogen and nitrous oxide-saturated solutions and conclude that nitrous oxide on reaction with hydrated electron gives a species similar to, but not identical with, the hydroxyl radical. A complete chemical study and a full product analysis might be necessary to decide this point. The author has carried out some competition experiments with *para*-aminobenzoic acid and has been unable to find evidence for a separate species[102].

Some unexpected results were obtained by Nakken and coworkers[101] when hydrogen peroxide was added before irradiation. Hydrogen peroxide also increases the amount of available hydroxyl radical (equation 3) but introduces other reactions (equation 7). A more complete chemical study is required.

In most illustrations of peroxy radical structure, **2** or **4**, the O_2 from aerated solution is drawn attached to the ring location of the radical spot or is represented as attached to any reasonable location. There is the possibility of its being in a transannular position as has often been found in photochemical attachments.

Zhikharev and Vysotskaya[103], using oxygen-18 isotope, found that the water was the source of most of the oxygen in the phenol formed by radiolysis of aqueous benzene solution[52]. A small amount originated from the dissolved oxygen, but, since the dose used was a high one for mechanistic studies (2·1 Mrad), it is probable that all, or very nearly all, the oxygen was incorporated in the phenol via hydroxyl radical attack and originated from the solvent water.

p-Nitrosodimethylaniline has a strong absorption band with a maximum at 440 nm. The effect of added solutes on the bleaching of this absorption has been used in the determination of a wide range of hydroxyl radical–solute reaction rate constants using competition methods [69, 72, 73, 104–106]. One of the main products of $OH^{\cdot}$ reaction with this compound is the nitro derivative[106] as would be expected from studies with nitrosobenzene[90] but some hydroxy isomers could also be formed[106].

In the early days of radiation chemistry the radiation bleaching of aqueous solutions of very many dyes was investigated in the search for a radiation dosimeter[3, 107, 108]. Pulse and steady state radiolysis studies have continued[109–118] but although hydroxylation of the aromatic ring is probably involved in some of these, the actual products have not been identified for certain.

In addition to the studies already described, benzene[119–122], benzoate[123] or benzoic acid[124], *p*-aminobenzoic acid[125], anisole[126], fluorobenzene[127, 128] and compounds related to tyrosine (*p*-hydroxyphenylalanine)[129–131] have been investigated recently in aerated aqueous solution and the isomeric yields determined at the low to moderate dose rates of steady state radiolysis. In some cases, dose rate, pH and other effects have been studied in order to arrive at a better understanding of detailed reaction mechanisms. The yields in aerated or oxygenated solutions probably reflect the ratio of attack at the possible isomeric positions, and hence partial rate constants can be calculated. In deaerated solutions or those containing nitrous oxide (but no oxygen) different ratios are found. These may be attributed to differences between the rates of diradical reactions (equations 22 and 23) for the different isomeric forms of **3**.

The general reaction scheme outlined above and the exceptions described explain most of the phenomena observed in the radiolytic hydroxylation of aromatic biochemical compounds[129-131].

i. Effect of additives. If ferrous ion is added to benzene solution before irradiation the G-value for phenol production is increased to 6 [34, 36, 45] or, under favourable conditions, to 14 [132-133a]. This indicates a chain reaction involving reduction of **2** (equation 13) by ferrous ion to a hydroperoxyl derivative $HOOC_6H_6OH$ which is further reduced to $OC_6H_6OH^{\cdot}$ (**9**). This can then react (equation 29) giving phenol and more **1** to continue the chain.

$$\underset{(\mathbf{9})}{OC_6H_6OH^{\cdot}} + C_6H_6 \longrightarrow \underset{(\mathbf{1})}{C_6H_6OH^{\cdot}} + C_6H_5OH \qquad (29)$$

$HO_2^{\cdot}$ and H_2O_2 are also made available as oxidizing agents by reaction with ferrous ion. $G(Fe^{3+})$ was found to be 65. Substantial post-irradiation production of ferric ion, phenol and mucondialdehyde was found. In the presence of ferrous ion, the isomeric ratio $G(ortho) : G(meta) : G(para)$ hydroxy benzoic acids was 1·9 : 3·1 : 1·6 (compare section II.B.1.h) in irradiated benzoic acid solution[134].

These high yields of phenol from benzene have attracted attention to the industrial possibilities of the process because a chain reaction often has potential for increasing yields even further. Studies have been made in the presence of ferrous ion or cupric ion or inorganic oxides and at high temperatures (up to 200°C or so)[135-152]. It is thought that at high temperatures above 130°C, a chain reaction sets in. Also $HO_2^{\cdot}$ becomes effective as a hydroxylating agent even in the absence of metal ions. G(phenol) is more than 30. This has a bearing on the autocatalytic decomposition of aqueous solutions of benzene or toluene at high temperatures[149-152].

Irradiation of naphthalene[153] produced predominantly 1-naphthol with an apparent $G = 1{\cdot}44$ up to 130°C. Beyond 140°C the yield of 1-naphthol decreased whereas that of 2-naphthol increased. Once again it was thought $HO_2^{\cdot}$ became an effective hydroxylating agent beyond 140°C, thus giving a different distribution of the isomeric phenolic products.

Irradiation of benzene dissolved in aqueous solution containing 0·5—1·0 mole/litre nitrate ion gave nitration as well as hydroxylation[154, 155]. At pH 2, G(*o*- and *p*-nitrophenol) totalled 1·5 and G(phenol) = 1·5. At higher pH lower yields were found. Higher yields were found when oxygen was excluded. Similar effects have been found in radiolysis of other organic compounds in nitrate solutions[125, 156]. No detailed mechanism has been proposed for this hydroxylation–nitration process. The absence of *m*-nitrophenol in the products suggests that phenol is formed first and is then nitrated.

j. Other conditions. In alkaline solutions the hydroxyl radical is ionized (equation 30)[157–159].

$$OH^{\bullet} + OH^- \rightleftharpoons O^{-\bullet} + H_2O \quad (pK = 11{\cdot}9) \tag{30}$$

Very little study has been made of the reactions of $O^{-\bullet}$ with organic compounds[158] but it appears to be relatively unreactive towards benzoate[102], chlorobenzene[47] or benzene[103]. It combines with oxygen to give $O_3^{-\bullet}$ [12, 160, 161].

Irradiation of a gaseous mixture of benzene and oxygen gave phenol but no chemical mechanism was established[162]. Hydroxyl radicals are well known in gas phase radiolysis but little is known about their reactions with organic compounds. A novel form of hydroxylation occurred when a mixture of benzene and nitrous oxide was irradiated[163]. The nitrous oxide captured electrons to give $O^{-\bullet}$ which attacked the benzene to give phenol with $G =$ 23·6.

k. Conclusions. To summarize, the early steps in the reaction between hydroxyl radicals and benzenoid compounds are fairly well understood. The final reactions in solutions containing an oxidizing agent such as oxygen, particularly those reactions going from **4** to the stable phenolic product, remain to be elucidated. The most probable course might be that represented by equations (31)—(36), as an alternative to equation (24). The disappearance of $HO_2^{\bullet}$ (or $O_2^{-\bullet}$) by reaction (8) is slow enough to enable a reasonable concentration to be built up.

$$\underset{(\mathbf{4})}{XC_6H_5(OH)O_2^{\bullet}} + O_2^{-\bullet} \longrightarrow \underset{(\mathbf{10})}{XC_6H_5(OH)O_2^-} + O_2 \tag{31}$$

This is similar to the reaction proposed by Daniels, Scholes and Weiss[32] and Loeff and Stein[121] except that in equation (31) the reducing agent is not $H^{\bullet}$ or e_{aq} but the much more probable $HO_2^{\bullet}$ or $O_2^{-\bullet}$ formed by their reaction with oxygen (equations 4 and 5). Although $HO_2^{\bullet}$ and $O_2^{-\bullet}$ are stoichiometrically identical, they may react at different rates with **4** or with its protonated form. Compound **10** or its protonated (un-ionized) form might react in different ways depending on what its substituent group X is and according to conditions. Equations similar to (32) and (33) can be written for the ionized form (**10**). Equation (32) is the general one giving equivalent amounts of the phenol. The hydrogens can be considered as in the *cis* configuration. In the *trans* configuration, equation (33)

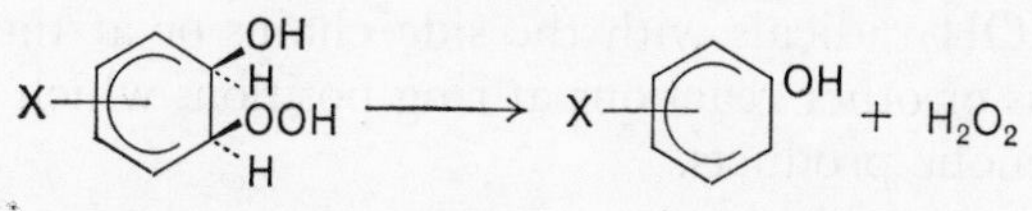

$$X\text{-}C_6H_5(OH)(H)(OOH)(H) \longrightarrow X\text{-}C_6H_4OH + H_2O_2 \quad (32)$$

$$X\text{-}C_6H_5(OH)(H)(H)(OOH) \longrightarrow X\text{-}C_4H_4(CHO)_2 + H_2O \quad (33)$$

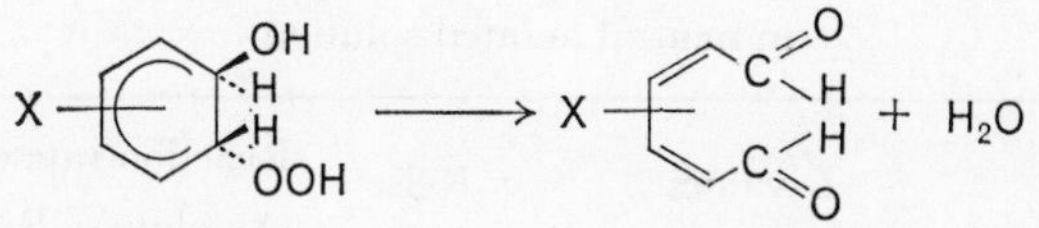

gives mucondialdehyde which has been found in irradiated benzene solutions[121]. $G = 1{\cdot}5$ in acid and $0{\cdot}8$ in neutral solutions but see Ref.163a.

These equations cannot apply to radiolysis in the presence of a mixture of N_2O and O_2 of such proportions that almost all the electrons are captured by the N_2O. Here there is no $O_2^{-\bullet}$ available for reduction according to equation (31). At both high and low dose rates nitrobenzene gives the expected yield of *o*-nitrophenol[102]. This indicates that $O_2^{-\bullet}$ is not necessary for the reaction to proceed in that particular system.

In the presence of a one-electron oxidizing agent (Ox), phenols are formed (equation 34 and, less probably, equation 35).

$$XC_6H_5(OH)O_2^{\bullet} + Ox \longrightarrow XC_6H_4OH + HOx + O_2 \quad (34)$$

$$XC_6H_5(OH)O_2H + Ox \longrightarrow XC_6H_4OH + HOx + HO_2^{\bullet} \quad (35)$$

In the case of the precursors of *o*-nitrophenol, an alternative to equation (25) is equation (36).

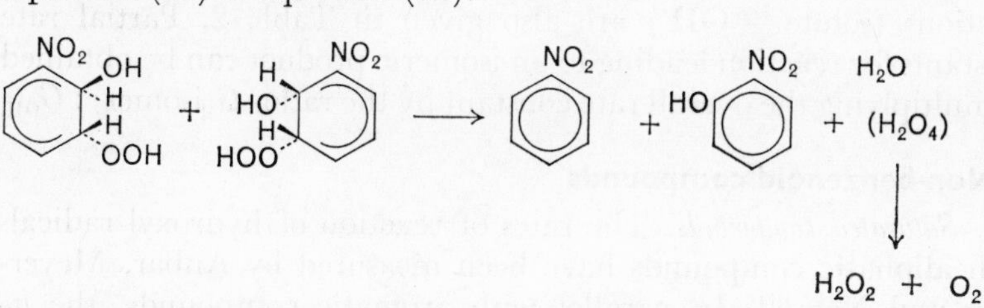

$$2\ NO_2C_6H_5(OH)(H)(H)(OOH) \longrightarrow C_6H_5NO_2 + HO\text{-}C_6H_4NO_2 + H_2O + (H_2O_4) \quad (36)$$

$$(H_2O_4) \longrightarrow H_2O_2 + O_2$$

Table 2 lists values for the hydroxylation products from radiolysis of aerated aqueous solutions of selected aromatic solutes. The isomeric distributions can be considered as reflecting most nearly the attack by OH radicals at the respective ring positions. However, account must be taken of side reactions following $OH^{\bullet}$ addition to the ring, such as that producing mucondialdehyde ($G = 0{\cdot}8$) from benzene[121]. The differences in each case between G_{OH} which is $2{\cdot}8$ (Table 1) and the total G (isomeric phenolic products) are due to reactions of OH radicals with the side chains or at the substituted carbon atoms or other reactions at ring positions which do not yield isomeric phenolic products.

TABLE 2. Isomeric yields and reaction rate constants for radiolytic hydroxylation in neutral aerated solution.

Solute	*G*-values			Refs.	Reaction rate constants		
	o	*m*	*p*		Absolute $l\ mole^{-1}\ s^{-1}$	Relative	Refs.
Benzene		2·3		58	$7{\cdot}8 \times 10^9$		77
		1·9		121		1·1	102
Benzoate	0·67	0·37	0·37	48	6·0		77
	0·62	0·32	0·26	123		1	68, 102
	0·49	0·31	0·39	101			
	0·66	0·34	0·31	102			
Nitrobenzene	1·03	0·48	0·51	81	4·7		76
					3·2		77
						0·5	102
Anisole	0·19	0	0·32	126	12		77
Fluorobenzene	0	0·2	0·32	128			
Chlorobenzene						1·1	68

Values for overall absolute and relative rate constants for the reactions (solute + OH˙) are also given in Table 2. Partial rate constants for reaction leading to an isomeric product can be obtained by multiplying the overall rate constant by the ratio $G(\text{isomer}) : G_{OH}$.

2. Non-benzenoid compounds

a. Saturated compounds. The rates of reaction of hydroxyl radicals with aliphatic compounds have been measured by Anbar, Meyerstein and Neta[104]. In parallel with aromatic compounds, the $\sigma\rho$ correlation shows that the attack is electrophilic. The predominant reaction is hydrogen abstraction which leads to dimerization or peroxy, hydroperoxy or aldehyde compounds. In the case of aerated aqueous solutions of acetic acid a small yield ($G = 0{\cdot}1$) of glycollic acid, $CH_2(OH)COOH$, has been found[164]. The radiolysis of solutions of amines gave oximes[165] with a *G*-value about 0·4. Aqueous solutions of cyclohexane gave cyclohexanol and cyclohexanone[165a].

b. Unsaturated compounds. With organic chemical compounds containing a double bond, $OH^{\bullet}$ addition reactions predominate over the slower hydrogen abstraction reactions. For aqueous solutions of ethylene[166–175] the following reaction scheme (equations 37–40) has been proposed for deaerated solutions[170, 172, 174].

$$CH_2{=}CH_2 + OH^{\bullet} \rightarrow HOCH_2CH_2^{\bullet} \quad (37)$$

$$2\,HOCH_2CH_2^{\bullet} \rightarrow CH_3CHO + C_2H_5OH \quad (38)$$

$$HOCH_2CH_2^{\bullet} + C_2H_4 \rightarrow HOC_4H_8^{\bullet} \quad (39)$$

$$2\,HOC_4H_8^{\bullet} \rightarrow C_3H_7CHO + C_4H_9OH \quad (40)$$

$G(C_2H_5OH)$ and $G(C_4H_9OH)$ were both about 0·5.

The radiolytic hydrogen atoms also add to ethylene to give ethyl radicals[174] which do not result in any hydroxylation products.

$$CH_2{=}CH_2 + H^{\bullet} \rightarrow CH_3CH_2^{\bullet} \quad (41)$$

Some of the radicals were able to initiate a polymerization chain.

$$HOCH_2CH_2^{\bullet} + nC_2H_4 \rightarrow HOCH_2(CH_2)_{2n+1}^{\bullet} \quad (42)$$

Reaction (43) also occurred to a small extent.

$$OH^{\bullet} + C_2H_4 \rightarrow C_2H_3^{\bullet} + H_2O \quad (43)$$

In aerated solution[171, 173]

$$HOCH_2CH_2^{\bullet} + O_2 \rightarrow \underset{\mathbf{(11)}}{HOCH_2CH_2O_2^{\bullet}} \quad (44)$$

$$CH_2{=}CH^{\bullet} + O_2 \rightarrow CH_2CHO_2^{\bullet} \quad (45)$$

$$CH_3CH_2^{\bullet} + O_2 \rightarrow CH_3CH_2O_2^{\bullet} \quad (46)$$

and $HO_2^{\bullet}$ also appeared able to add on to ethylene.

$$CH_2{=}CH_2 + HO_2^{\bullet} \rightarrow HOOCH_2CH_2^{\bullet} \xrightarrow{O_2} \underset{\mathbf{(12)}}{HOOCH_2CH_2O_2^{\bullet}} \quad (47)$$

Compound **11** corresponds to **2**, which is formed in the case of benzene. Interactions between the four peroxy compounds gave the observed product yields: G(glycollaldehyde) = 2·4, G(formaldehyde) = 2, G(acetaldehyde) = 1 and G(hydrogen peroxide) = 2·6. At very low doses G(glycollaldehyde) was about 30 and the reason for this is unknown.

The early parts of the reaction scheme have been confirmed by pulse radiolysis measurements[171, 175]. The reaction rate constants measured are $k_{37} = 1 \times 10^9\ \text{M}^{-1}\,\text{s}^{-1}$ (l mole^{-1} s^{-1}), $k_{38} = 6{\cdot}3 \times 10^8$ and $k_{44} = 6{\cdot}6 \times 10^9$. The last value is about ten times that measured for the corresponding reaction for benzene[58, 80] (equation 13).

In parallel with aromatic solutes, ferrous ion was found to increase the product yield—in this case, formaldehyde and acetaldehyde. The system is complex but the species, $HOCH_2CH_2O^{\bullet}$ which corresponds to **10** for aromatic solutes has been proposed[174] as formed by a similar reaction. Another possibility is that ferric ion is present in two complexed forms which react differently[171].

$$HOCH_2CH_2^{\cdot} + FeSO_4^{+} \longrightarrow CH_3CHO + Fe^{2+} + HSO_4^{-} \quad (48)$$

$$HOCH_2CH_2^{\cdot} + Fe(OH)^{2+} \longrightarrow HOCH_2CH_2OH + Fe^{2+} \quad (49)$$

Various other unsaturated compounds have been investigated. Propylene and allyl alcohol have been found to give mainly organic peroxides incorporating a hydroxyl group[176, 177]. Acrylamide also adds $OH^{\cdot}$ at the double bond[178]. Styrene and α-methylstyrene are attacked mainly at the aromatic ring but 20—40% and 15—30% respectively of the $OH^{\cdot}$ radical adds on at the side chain double bond[175, 179].

Acetylene in aerated aqueous solution gave glyoxal in yields indicating a chain reaction under some conditions[180].

The double bonds of olefinic acids are hydroxylated[181].

c. Biochemical compounds. The rates of reaction of hydroxyl radicals with a number of organic chemical compounds found in living organisms have been measured[67, 70, 71, 74].

(1) *DNA bases*

Identification of the site of radiation damage to the cell nucleus has led to the study of the radiolytic behaviour of the components of DNA. The pyrimidine and purine bases have been found to undergo hydroxylation[182–183a] in aqueous solution.

Pulse radiolysis techniques have now been applied in these studies[184–188a] and some of the reaction rate constants are known.

In neutral solution thymine (pK 9·8) is in the undissociated form. Hydroxyl radicals add on at the 6-position (equation 50)[189, 190], although there is also the possibility of addition at position 5 (the C atom to which the CH_3 group is attached)[186].

$$\text{thymine (HN–CO–C(CH}_3\text{)=CH–NH–CO ring)} + OH^{\cdot} \longrightarrow \text{HN–CO–}\dot{C}(CH_3)\text{–CH(OH)–NH–CO ring} \quad (50)$$

The overall rate for the (thymine + $OH^{\cdot}$) reaction has been measured[186] as $7{\cdot}4 \times 10^{9}$ l mole^{-1} s^{-1}. Even allowing for other reactions, this is high for attack at one position only and comparable to the rate constant for (benzene + $OH^{\cdot}$) where all π electrons are available. This double bond is a chromophoric group so its destruction is easily followed in either pulse radiolysis or steady radiolysis studies $G(-\text{double bond}) = 1{\cdot}9$.

Oxygen, if present, then adds on at the 5-position. The radical is reduced by $O_2^{-\bullet}$.

$$\xrightarrow{+O_2} \qquad \xrightarrow[-O_2]{+O_2^{-\bullet}} \qquad \xrightarrow{+H^+} \qquad (51)$$

(13)

In this case the 6-hydroxy-5-hydroperoxythymine (**13**) is stable enough to be isolated[177]. $G = 1{\cdot}05$. **13** is formed with a lower yield below pH 5. Perhaps $HO_2^{\bullet}$ (pK 4·4) is not as effective as $O_2^{-\bullet}$ as a reducing agent or, alternatively, one of the intermediates is less stable in acid solution.

In neutral solution, $OH^{\bullet}$ adds to either the 5- or the 6-position at the double bond of uracil. In aerated solution, uracil and dimethyluracil also give hydroxyhydroperoxides. However, the cytosine hydroperoxide, though detectable at low pH, decomposes in neutral solution and is not stable enough to be isolated. This suggests that a hydroperoxide of this form may be an intermediate between the aromatic solute–hydroxyl radical–oxygen adduct (**4**) and the final product, a phenol. In the case of aromatic solutes, as for cytosine, the hydroperoxides might be unstable and decompose giving hydrogen peroxide and hydroxylated product. Between cytosine and uracil there may be compounds of structure like **13** which have a gradation of stabilities. Neutral aerated irradiated solutions of pyrimidine bases on acidification give 5,6 diols probably from the decomposition of **13**.

At pH 11, thymine is in the singly ionized form and the site of attack is changing to the methyl group itself (equation 52), despite there being two conjugated double bonds in the compound. $G(-5{,}6\ \text{double bond}) = 1{\cdot}3$. Since uracil has not a methyl group there is no corresponding change in its reactions as the pH is increased.

$$\text{HN-C(=O)-C(CH}_3\text{)=CH-N=C(-O}^-\text{)} \xrightarrow{+\text{OH}^{\bullet}} \text{HN-C(=O)-C(CH}_2^{\bullet}\text{)=CH-N=C(-O}^-\text{)} \qquad (52)$$

At still higher pH, the ionization of the hydroxyl radical to $O^{-\bullet}$ is manifest and there is no longer any electrophilic hydroxylation.

At the highest pH studied, thymine is doubly ionized ($pK_2 > 13$) and the pyrimidine ring is aromatic. The reaction is hydrogen abstraction from the methyl group and the identified hydroxylation products are not formed directly.

The purine bases adenine and guanine appear to be more resistant to radiation than are the pyrimidine bases[182, 183]. Hydroxyl radicals appear to add at the 4,5 central double bond but no hydroxylation products have been found.

In the double helix conformation the DNA bases are protected somewhat from attack[191]. At physiological pH, the major reaction is $OH^{\bullet}$ attack on the pentose part of DNA.

(2) *Steroids, vitamins, etc.*

The effects of radiation on aerated aqueous solutions of other biochemicals have been summarized by Swallow[3]. In some of the earlier investigations no attempt was made to distinguish between the reactions and interactions of the various primary radiolytic species. The action, if any, of the hydroxyl radical, alone and without interfering reactions, is usually apparent in dilute aerated solution at low doses up to 30 krad. In the light of subsequent investigations with simpler model compounds, one must doubt the validity of reaction schemes in which the solute–$OH^{\bullet}$ adducts react with $HO_2^{\bullet}$. It is far more likely that they will first react with oxygen which is present at higher concentrations.

3. Use in synthesis

There are very few examples of the radiolytic hydroxyl radical being used in organic synthesis. Merger and Grässlin[192, 192a] have reported the synthesis of the hitherto unknown 1,2,3,4-tetrahydroxynitrobenzene with $G = 1{\cdot}41$ by irradiation of aqueous 4- or 5-nitropyrogallol. Davison[193] prepared the *o*-, *m*- and *p*-hydroxybenzoic acids from benzoic ^{14}C-acid. The *meta* derivative, in particular, is more difficult to prepare by conventional methods.

C. *Hydroperoxyl Radical Reactions*

1. General

The radiolytic production, spectra and subsequent reactions of the hydroperoxyl radical, $HO_2^{\cdot}$, and its ionized form, $O_2^{-\cdot}$ (pK_a 4·4) have been characterized (see section II.A.3) [11–13, 194, 195]. It is almost always formed in dilute aerated aqueous solutions and its rate of disappearance is low ($k_{HO_2\cdot+HO_2\cdot} = 0{\cdot}7 \times 10^6$, $k_{HO_2\cdot+O_2^-\cdot} = 3{\cdot}0 \times 10^7$ and $k_{O_2^-\cdot+O_2^-\cdot} = 1{\cdot}2 \times 10^7$ l mole^{-1} s^{-1}) [195] compared with some other reactions. Therefore it can reach significant concentrations, and figures in many reaction schemes described in section II.B particularly in reaction with other radicals. Virtually no studies have been made of its reactions with organic solutes.

2. Hydroquinones

In general, $HO_2^{\cdot}$ abstracts hydrogen from the hydroxy group of a hydroquinone and no hydroxylation results[196, 197]. Hydroxylation of the ring has been observed on irradiation of 1,2,4-trihydroxybenzene, *p*-hydroquinone, toluhydroquinone and monochloroquinone[198, 199]. Such hydroxylation did not occur in the absence of oxygen but it was not established that ring attack was initiated by $HO_2^{\cdot}$. More recently, Bielski and Allen[199] have found that $HO_2^{\cdot}$ adds to the ring of semiquinone radicals, which have been formed either by hydrogen abstraction or by $OH^{\cdot}$ attack on the ring followed by water elimination[86].

$HO_2^{\cdot}$; $-H_2O$; acid; **(14)** (53)

The *G*-value for production of **14** depended on the oxidation reduction potential of the hydroquinone. Where R = Cl, *G*(**14**) = 1·41.

Phenols probably undergo a similar reaction but the product remains in the *o*-quinone form.

3. Ethylene

Basson and du Plessis consider that $HO_2^{\cdot}$ radicals add to ethylene (equation 47)[173].

Compound **12** forms a ring peroxide which, in acid solution, is converted into glycollaldehyde.

$$2HOOCH_2CH_2O_2^{\bullet} \xrightarrow[-H_2O_2]{-O_2} 2\ \underset{O\!-\!\!-\!O}{\overset{CH_2-CH_2}{|\qquad|}} \longrightarrow 2\ \underset{OH}{\overset{CH_2CHO}{|}} \qquad (54)$$

4. Dyes

Schulte-Frohlinde and co-workers[200] have shown that some dyes are bleached by $HO_2^{\bullet}$ radicals. No product analysis was carried out.

III. PHOTOLYTIC HYDROXYLATION

A. Production

If an aqueous solution of hydrogen peroxide is exposed to light of wavelength 370 nm or less, hydroxyl radicals are formed[201, 202].

$$H_2O_2 + h\nu \longrightarrow 2\ OH^{\bullet} \qquad (55)$$

Hochanadel[8] showed that the species formed was identical with the radiolytic primary oxidizing radical and was $OH^{\bullet}$ uncontaminated by other species. Quantum yields (molecules changed per quantum of light absorbed) as high as 80 have been reported[202]. Obviously a chain reaction is involved. At higher light intensities the quantum yield falls to a steady 1·0—1·4 [203]. Further reaction of $OH^{\bullet}$ with hydrogen peroxides produces $HO_2^{\bullet}$ (equation 7) so in that sense it is a mixed system[204]. At wavelengths less than 242 nm the photodissociation of water becomes energetically possible (equation 56)[205].

$$H_2O + h\nu \longrightarrow H^{\bullet} + OH^{\bullet} \qquad (56)$$

The hydroxyl radicals produced by photolysis can hydroxylate benzenoid, olefinic and heterocyclic compounds. In some cases the hydroxyl radicals can be prepared free of other interfering species. Sometimes the organic solute is raised to an excited state by the light and can then react with water adding, for example, H and OH groups to a double bond[183a]. The final product is often a hydroxyl compound but, because a free radical mechanism is not involved, this type of hydroxylation reaction will not be considered in this chapter.

Another system involves photoexcited electron transfer in the $Fe^{3+}OH^-$ complex in aqueous solutions of ferric ion. It was considered that free hydroxyl radicals were produced[206, 207] (equation 57).

$$Fe^{3+}OH^- \xrightarrow{h\nu} Fe^{2+}OH \longrightarrow Fe^{2+} + OH^{\bullet} \qquad (57)$$

Similarly the hydroxy ion, OH^-, on illumination releases an electron leaving a hydroxyl radical (equation 58).

$$OH^- \xrightarrow{h\nu} e_{aq} + OH^{\bullet} \qquad (58)$$

B. Benzenoid Compounds

Compounds related to naphthalene were irradiated in aqueous hydrogen peroxide solution with ultraviolet light by Boyland and Sims[208]. Phenols were the main products found. Benzoic acid, for example, gave the three isomeric hydroxybenzoic acids. The *ortho* : *meta* : *para* ratio was 2 : 1 : 1 approximately.

Loeff and Stein[48] studied the photodecomposition of hydrogen peroxide solutions at various concentrations with benzene as a solute. Phenol, mucondialdehyde and pyrocatechol were found to be the major products.

Jefcoate, Lindsay Smith and Norman have recently investigated the photolysis of hydrogen peroxide–toluene and hydrogen peroxide–benzene mixtures[208a].

Norman and Radda[209] found that anisole under ultraviolet irradiation gave *ortho* : *meta* : *para* derivatives in the ratio 84 : 0 : 16 whereas fluorobenzene gave the ratio 37 : 18 : 45.

Omura and Matsuura[210] found that ultraviolet-irradiated mixtures of phenols and hydrogen peroxide in aqueous solutions gave predominantly *o*-dihydroxy compounds. Smaller quantities of *para*-derivatives were also formed but there was no substitution *meta* to the existing OH group. *p*-Carboxy- and *p*-methoxy-phenols gave hydroquinones in addition to the usual catechol derivatives.

The percentage conversions of starting materials were rather too high in these experiments for firm conclusions to be drawn regarding the detailed mechanisms. The proposed attack of a second hydroxyl radical on each phenoxyl radical ($XPhO^{\bullet}$) to give the phenolic product cannot be considered the general mode of formation of the phenols.

Pacifici and Straley[210a] have observed hydroxylation of the aromatic nuclei of polyesters exposed to ultraviolet light.

C. Unsaturated Compounds

Milas and co-workers[201] studied the photolysis of mixtures of unsaturated compounds and hydrogen peroxide. Allyl alcohol gave glycerol, crotonic acid gave dihydroxybutyric acid, maleic acid or diethyl maleate gave mesotartaric acid and ethylene gave diethylene glycol or, in oxygenated solution, aldehydes[210b].

Kraljic[73] has used the bleaching of *p*-nitrosodimethylaniline[69] by the photolytic hydroxyl radical as a basis for determining relative rate constants for reaction of solutes with $OH^{\bullet}$. The agreement with rate constants measured for radiolytic $OH^{\bullet}$ is said to be satisfactory. Fluorescein is also bleached[210c].

The results obtained in all these cases can be explained readily in terms of the reactions described for the radiolytic hydroxyl radical (equations 17, 19, 24, etc.).

The photolysis of hydrogen peroxide has been used as a source of hydroxyl radicals for electron paramagnetic resonance studies[211, 212]. Because of the technical difficulties encountered when liquid water is introduced into the e.p.r. cavity, many of these studies have been conducted in frozen aqueous solution at low temperatures or with strong hydrogen peroxide added to an organic liquid. Ultraviolet light is directed into the cavity itself. Although product determination does not normally form part of such investigations, valuable information regarding intermediate species is obtained. Strangely, allyl alcohol[211–212a] does not add hydroxyl radicals but undergoes hydrogen abstraction giving $^{\bullet}CH_2CHCHOH$. Oxygen, if present, adds to the radicals giving peroxy radicals.

D. Irradiation of Complexes

Bates, Evans and Uri[206] showed that 300—400 nm ultraviolet irradiation of ferric complexes such as $Fe^{3+}OH^-$, $Fe^{3+}Cl^-$ and $Fe^{3+}F^-$ gave a species which hydroxylated aromatic compounds. Benzoic acid gave the *o*-, *m*-, and *p*-hydroxybenzoic acids in the statistical ratio 2 : 2 : 1 [207]. Saldick and Allen[213] showed that the hydroxylating species was definitely the free radical $OH^{\bullet}$ and not an activated complex, such as $(Fe^{3+}OH)$. With benzoic acid present as substrate the products were the hydroxybenzoic acids[207] which were further attacked on continued exposure[213].

Benzene solutions were studied by Baxendale and Magee[214]. They found that Fe^{3+} ions could be replaced by Cu^{2+} ions.

Solutions containing $(UO_2{}^{2+}H_2O)$ and $(Ce^{4+}\,OH^-)$ have been reported by Stein and Weiss[215] as giving free $OH^{\bullet}$ capable of hydroxylating aromatics. Richardson[216] has investigated aqueous Ce(IV) solutions. Yandell and Stranks[217] have shown that the action of light on $(Tl^{3+}OH^-)$ solutions liberates hydroxyl radicals. This list is by no means exhaustive.

Hydroxyl radicals have also been reported[218] as being formed by the action of the radical cation of 9,10-anthraquinone-2-sulphonate (A). The radical cation is formed from a photo-excited state (equation 59).

$$A^* + A \longrightarrow A^{\bullet -} + A^{\bullet +} \xrightarrow{OH^-} A + OH^{\bullet} \qquad (59)$$

E. Gas Phase

Photolysis of water vapour and of hydrogen peroxide vapour also gives hydroxyl radicals but very little is known about their reactions[219–221a].

F. Conclusions

To summarize, hydroxyl radicals produced by the action of ultraviolet light on water, hydrogen peroxide or metal–hydroxy complexes appear capable of hydroxylating aromatic or olefinic compounds in the same way as do radiolytic hydroxyl radicals. Very little work has been done on these systems. The better experimental techniques now available and the better understanding of the general reactions suggest that considerable progress would ensue from a renewed attack on the problems of the system.

IV. CHEMICAL FREE RADICAL HYDROXYLATION

A. Introduction

The study of chemical methods of generating radicals capable of hydroxylating organic substrates has received considerable impetus from the similarities of the products found to those formed during metabolic hydroxylation in biological processes. The reactive entities in vivo are one-electron oxidizing agents and demonstrate the same electrophilic character combined with lack of selectivity attributed to free radicals. Hydroxyl and perhydroxyl radicals have both been considered possibilities.

There are broadly two types of systems[222, 223]. The first, based on hydrogen peroxide together with a metal ion of variable valency, is typified by Fenton's reagent (ferrous ion and H_2O_2), titanous ion and H_2O_2 and Hamilton's system (ferric ion, catechol and H_2O_2).

$$M^{n+} + H_2O_2 \longrightarrow M^{(n+1)+} + OH^- + OH^{\cdot} \tag{60}$$

The second type is based on oxygen and is typified by Udenfriend's reagent[224, 225]. In the presence of ferrous ion, EDTA (ethylenediaminetetraacetic acid) and ascorbic acid, molecular oxygen is able to hydroxylate many compounds. Initially, it was thought that hydrogen peroxide is formed first as an intermediate. Evidence now shows that this is not so.

Udenfriend's reagent behaves as a 'mixed-function oxidase'[226] and is therefore considered to resemble closely in many details the processes obtaining in biological systems. Expressed simply—a mixed function oxidase is able to reduce molecular oxygen and to convert it into a form so that one atom of each molecule of oxygen is reduced and the other appears in the product in a new hydroxyl group (in the cases we will consider).

Two further reactions (equations 7 and 61) can give rise to $HO_2^{\bullet}$ radical.

$$M^{(n+1)+} + H_2O_2 \longrightarrow M^{n+} + H^+ + HO_2^{\bullet} \quad (61)$$

The evidence for the radical nature of the active species and reviews of their reactions are given in papers by Norman and co-workers[222–222b] and by Staudinger and co-workers[223]. It should be pointed out that not all authors agree that these species are the free $OH^{\bullet}$ or $HO_2^{\bullet}$ radicals. Some consider $OH^{\bullet}$ and $HO_2^{\bullet}$ are complexed to the metal ions.

Each of these reagent types is now considered in turn followed by peracids and the other methods of hydroxylation which have been used.

B. Fenton's Reagent

1. General

Fenton's reagent, a mixture of ferrous ion and hydrogen peroxide, has been known since 1894[227]. Haber and Weiss[228] proposed a series of reactions in which the hydroxyl radical was the reactive species. Baxendale and co-workers[229–231] and Kolthoff and Medalia[232, 233] proposed some modifications to the original scheme but $OH^{\bullet}$, and to a minor extent $HO_2^{\bullet}$, remained as essential features (equations 62—65).

$$Fe^{2+} + H_2O_2 \longrightarrow Fe^{3+} + OH^{\bullet} + OH^- \quad (62)$$

$$OH^{\bullet} + Fe^{2+} \longrightarrow OH^- + Fe^{3+} \quad (63)$$

$OH^{\bullet}$ and hydrogen peroxide react giving $HO_2^{\bullet}$ (equation 7).

$$Fe^{3+} + H_2O_2 \longrightarrow HO_2^{\bullet} + H^+ + Fe^{2+} \quad (64)$$

$$HO_2^{\bullet} + Fe^{3+} \longrightarrow O_2 + H^+ + Fe^{2+} \quad (65)$$

The kinetics of the system were recently investigated by Grinstead[234]. The arguments for the radical nature of the reagent are summarized by Norman and Lindsay Smith[222, 235].

By adding EDTA[236] or other chelating agents[235, 237, 238] to the system, the ferric ion is complexed. This means that the reagent may then be used over a far greater pH range. Kraljic[73] has found that the radiolytic $OH^{\bullet}$ radical reacts about as readily with EDTA as it does with benzene and other aromatics. This could complicate the reaction scheme considerably. Ascorbic acid[236, 239] also increases the yield presumably by reducing the ferric ions to ferrous ions. Ascorbic acid reacts with $OH^{\bullet}$ radical. The rate constant at pH 1 is listed[67, 70] as $7{\cdot}2 \times 10^{9}$ l mole^{-1} s^{-1}.

The relative rates of reaction of Fenton's $OH^{\bullet}$ with a variety of solutes were measured by Merz and Waters[240–242]. In the light of present knowledge, the values they obtained for aromatic (non-chain reaction) compounds must be doubled to allow for a modification to their reaction scheme[68]. Uri[202] criticized Merz and Waters' treatment on other grounds. Kraljic[73] has used *p*-nitrosodimethylaniline to evaluate some relative rate constants. Norman and Radda[209] have also measured the relative rates of some reactions by a competitive method.

2. Benzenoid compounds

Hydroxylation of the aromatic ring by Fenton's reagent was found for benzene[222, 235, 242, 245], nitrobenzene[20, 242, 244], benzoic acid or benzoate[236, 239, 242, 246, 247], benzamide[242], phenylacetic acid[242], dimethylaniline[242], chlorobenzene[24, 235, 242, 244], fluorobenzene[222, 244], phenol[23, 246], anisole[244], *p*-cresol[248], naphthalene[207], toluene (also some hydrogen abstraction from the side chain)[235, 244] and acetanilide[223, 236].

The course of reactions is considered to be the same as for the radiolytic $OH^{\bullet}$ radical and equation (7), followed by equation (19), has been applied under aerated conditions. Since there is a rather higher concentration of radical **3** than in steady radiolysis and a greater likelihood of anaerobic conditions, there is a greater chance of biphenyls (equation 10) being formed. Lindsay Smith and Norman[235] showed that **3** was oxidized by ferric ion to phenol (equation 66).

$$XC_6H_5OH^{\bullet} + Fe^{3+} \longrightarrow XC_6H_4OH + Fe^{2+} + H^{+} \qquad (66)$$

A lower amount of phenol was formed if the ferric ion was removed by complexing it with fluoride ion or by replacing Fe^{2+} with Ti^{3+} which gives Ti^{4+}, a weaker oxidizing agent than Fe^{3+}. More phenol was formed if an excess of ferric was added. Alternatively, **3** reacts with oxygen (equation 19). Another possibility advanced was reaction with a second $OH^{\bullet}$ radical. On competition kinetic grounds, this can be discounted for small conversions of starting material.

Although no isotope effect could be detected in the formation of **3** there was an isotope effect when chlorobenzene was present. Less phenol was formed from deuterated benzene than from protobenzene. This was explained by a crossed disproportionation between the two adducts present in the solution (equations 67 and 68—compare equation 11).

$$C_6D_6OH^{\cdot} + ClC_6H_5OH^{\cdot} \longrightarrow C_6D_5OH + ClC_6H_5DOH \quad (67)$$

$$C_6D_6OH^{\cdot} + ClC_6H_5OH^{\cdot} \longrightarrow ClC_6H_4OH + C_6D_6HOH \quad (68)$$

Reaction (67) will be slower than for the corresponding proto compound because a deuterium atom must be transferred.

3. Effect of additives

Staudinger and co-workers[223] related the greater effectiveness of certain metal ions (Cu^+ or V^{3+} in place of Fe^{2+}) with their redox potential and the speed of their reactions with molecular oxygen. EDTA, then, increases the phenol yield because it lowers the Fe^{2+}/Fe^{3+} redox potential. Ascorbic acid reduces the oxidized Fe^{3+}. Other reducing agents, such as ene-diols, would also be effective provided that they were not radical traps. These authors also gave consideration to the role of $HO_2^{\cdot}$ radicals. It was not appreciated at that time that these are considerably more stable and less reactive than are $OH^{\cdot}$ radicals.

Grinstead[247] has suggested $OC_6H_4COO^{-\cdot}$ as an intermediate in the attack of Fenton's reagent on salicylate ion. This corresponds to the phenoxyl radical found in the pulse radiolysis of phenols (section II.B.g)[75, 86]. Consequently the subsequent reactions with either O_2, H_2O_2 or $OH^{\cdot}$ to give the products 2,3- and 2,5-dihydroxybenzoic acids cannot be taken as necessarily applicable to the action of Fenton's reagent on aromatics in general.

4. Olefinic compounds

Baxendale, Evans and Park[229] showed that the Fenton's radical added on to either side of the double bond of acrylonitrile and methyl methacrylate. Because of its industrial application as a redox initiator for olefin polymerization, there was considerable interest in Fenton's reagent in about the year 1950. Most of these investigations do not throw much light on its behaviour as a hydroxylating agent.

5. Heterocyclic compounds

The literature on the action of Fenton's reagent on heterocyclic compounds has been reviewed by Norman and Radda[249].

Hydroxylation of furans has been investigated recently[253a].

Scholes and Weiss[250] studied the degradation of nucleic acids and

DNA bases by Fenton's reagent. Breslow and Lukens[239] showed that quinoline gave 3-hydroxyquinoline. Cier, Nofre and coworkers[251–253] used Fenton's reagent modified by addition of pyrophosphate to complex the ferrous ion, to study purine and pyrimidine bases. In recent years emphasis has shifted to Udenfriend's and similar reagents as more nearly resembling enzymic action.

6. Fenton-type reagents

Almost any metal ion of variable valency may be used in place of ferrous ion in a Fenton-type reagent and much the same series of reactions will be found. One of these, titanous ion, has been so useful that sufficient investigations have been done on it to merit separate treatment. Ceric ions will also be dealt with separately because the reactive species behaves quite differently from that in Fenton's reagent.

Other metal ions which have been studied in a mixture with hydrogen peroxide are listed by Walling[254]. The energetics of $OH^{\bullet}$ radical production is discussed by Uri[202]. An interesting one is cupric ion[222b, 244, 246] which increases the radiation chemical yield of hydroxylated products (section II.B.1.i). Ferrocyanide has also been used[254a].

The osmium tetroxide-catalysed addition of a hydroxyl group on each side of the double bond of an olefin (Milas reaction) has recently been reinvestigated by Norton and White[255]. The main product is the glycol.

C. Titanous System

1. General

Titanous ion reacts with hydrogen peroxide by a one-electron reaction (equation 69) and has been reported as proceeding more readily than the corresponding reaction of ferrous ion[223, 256] (Fenton's reagent).

$$Ti^{3+} + H_2O_2 \longrightarrow Ti^{4+} + OH^- + OH^{\bullet} \qquad (69)$$

Dixon and Norman[257, 258] found this sytem superior to the Fenton's ferrous system in flow experiments whereby the radicals could be observed in the cavity of an electron spin resonance spectrometer less than 0·02 second after the reactants were mixed. This technique has since been used extensively on many chemical compounds. Generally studies have been confined to e.s.r. spectra measurements. Being in a liquid system much greater resolution can be obtained than in a solid. (Ultraviolet irradiation of a low temperature glass of the compound mixed with hydrogen peroxide is another way of trapping transient radical species so that they can be studied.) From such spectra the structure of the radicals formed can be deduced and quantitative ideas on the distribution of the unpaired electron can be obtained[222b].

2. Benzenoid compounds

Using such a flow system with benzene as the substrate Dixon and Norman[259] were able to show unequivocally the presence of **1** formed by reaction (9) and to resolve the question of the primary step in the action of hydroxyl radical on the aromatic ring. Thermodynamic calculations had indicated that the elimination of water from $XC_6H_5OH^\bullet$ was quite probable. They demonstrated convincingly that this adduct was, in fact, reasonably stable. This left no doubts as to the validity of Dorfman, Taub and Bühler's[58] conclusions arrived at from pulse radiolysis studies (section II.B.1.b).

Dixon and Norman[258] found that phenol gave the phenoxyl radical, $PhO^\bullet$, and concluded that hydrogen abstraction from the OH group was more facile than from CH. Pulse radiolysis studies[75] have shown that in a two-stage reaction the hydroxyl radical adds to the ring and then water is eliminated (section II.B.1.g).

Lindsay Smith and Norman[235] showed that the titanous system and Fenton's reagent behaved similarly in the hydroxylation of aromatic compounds such as fluorobenzene and chlorobenzene, and was unaffected by the addition of EDTA. Armstrong and Humphreys[260] found that the titanous system gave a bigger yield of radicals from reaction with amino acids but otherwise behaved similarly. For both reagents, addition of EDTA made no difference.

The radicals formed by attachment of $OH^\bullet$ to a number of aromatic compounds have been characterized[208a, 259, 261–263].

3. Olefinic compounds

Dixon and Norman[258] showed that allyl alcohol gave a spectrum corresponding mainly to $HOCH_2CH^\bullet CH_2OH$ and a weaker one, corresponding to $HOCH_2CH(OH)CH_2^\bullet$. This is in accordance with the greater reactivity of the unsubstituted methylene group in an olefin $CH_2{=}CHX$, as compared with the substituted carbon atom. Smith and co-workers[264, 265] observed the same phenomenon also for acrylate esters. Oximes add an $OH^\bullet$ radical to the carbon atom of the C=N bond[266, 267] but lose a hydrogen atom to give a nitroxide.

Maleic acid, fumaric acid and crotonic acid add $OH^{\bullet}$ at the double bond[268]. The reagent demonstrates electrophilic character[268a].

4. Heterocyclic compounds

Pyridine does not react with this reagent[266].

Ormerod and Singh[269] found that the titanous reagent did not react with purines but attacked pyrimidine bases to give results in good agreement with radiation chemical results. At pH 1, $OH^{\bullet}$ added to the 6-position of thymine six times more readily than to the 5-position. Surprisingly, with EDTA present at pH 2, there is no addition at position 6 and the concentration of radicals with OH in position 5 is enhanced $2\frac{1}{2}$ times. Myers and co-workers[186] have also studied this system.

Furans are hydroxylated at position 5 or 4[253a].

D. Hamilton's System

Hamilton and co-workers[269–274], during an investigation of Udenfriend's reagent, tried the effect of replacing ascorbic acid by other ene-diols. They found that catalytic amounts of catechol, (10^{-4} mole/litre) or other 1,2-dihydroxy- or 1,4-dihydroxy-aromatics and of ferric ion or less efficiently cupric ion enabled H_2O_2 to hydroxylate aromatic compounds.

Using competition kinetic methods the ratio of the rate constants anisole : benzene : chlorobenzene : nitrobenzene was found to be 1·4 : 1 : 0·6 : 0·6 which agrees fairly well with that for the radiolytic $OH^{\bullet}$ radical. The isomeric distributions do not agree so well.

The scheme on the following page was proposed where HX is perchloric acid and C_6H_5Y is the substrate which is hydroxylated to HOC_6H_4Y. The reagent is regenerated.

The system has been shown to be capable of hydroxylating a number of aromatic compounds. An attractive feature is the low concentration of catechol or other catalyst and of ferric iron that is necessary. This introduces less interpretative complications than are found in some other systems.

E. Udenfriend's Reagent

1. Mechanism

The best known of the systems that employ an electron donor to make dissolved molecular oxygen available as a hydroxylating agent is that proposed by Udenfriend and co-workers[224, 225]. The reagent

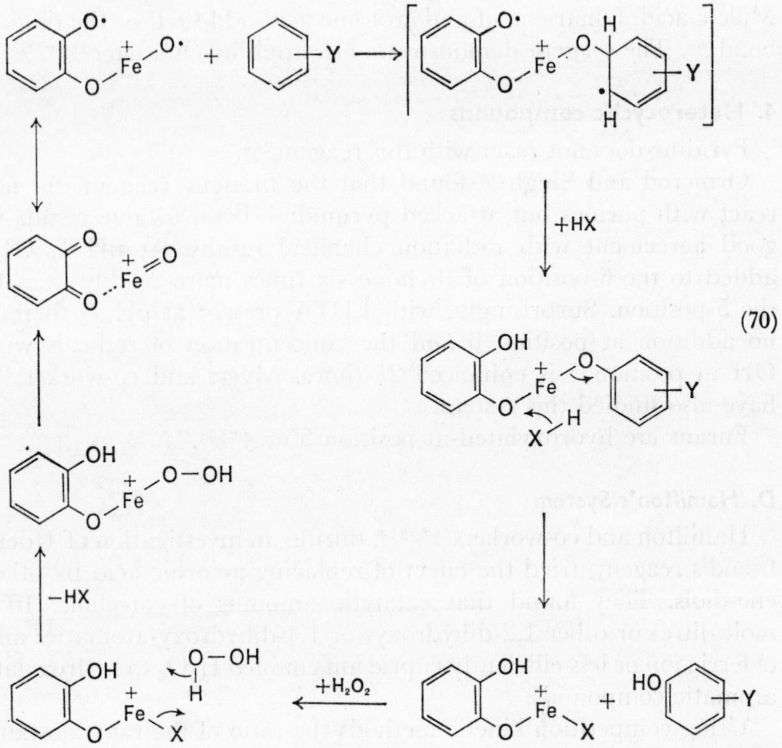

$$(70)$$

consists of ferrous ion, EDTA, ascorbic acid and oxygen. It was suggested that the oxygen was first reduced to hydrogen peroxide which then formed hydroxyl radicals. Norman and Radda[209] showed that the distribution of isomers following attack on anisole and chlorobenzene was different from that of Fenton's reagent. $HO_2^{\cdot}$ had been proposed as the radical in Udenfriend's reagent but by itself this reagent is not reactive enough. The attacking species is more selective than the radiolytic hydroxyl radical and is also electrophilic. In all, it is considered to be a better model for biological processes than any of the other systems.

Norman and Lindsay Smith[222] reported that the system is complex, and they have experienced difficulties in unravelling the mechanism. The molecular oxygen adds to the ferrous ion to give an ion, $Fe^{2+}O_2$, which is much more effective if EDTA is present. This acts as a bridge for electron transfer between the ascorbic acid and the aromatic ring and back again, being reduced itself in the process (equation 71).

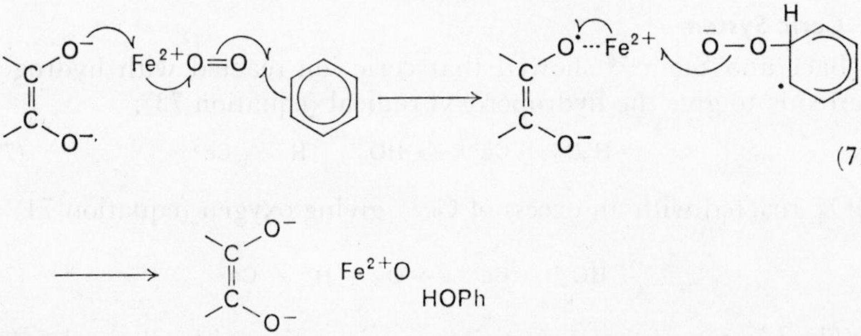

(71)

(72)

In high concentrations of metal ion, $Fe^{2+}O_2Fe^{2+}$ is formed and this gives a preponderance of *meta*-substitution, presumably because Fe is first attached to the activated *ortho*- or *para*-position and thereby the *meta*-position is made more accessible to nucleophilic attack by oxygen (equation 72). Fluorobenzene gives a high proportion of *meta* substituent even at low ferrous ion concentrations. Catechol and quinol are also found.

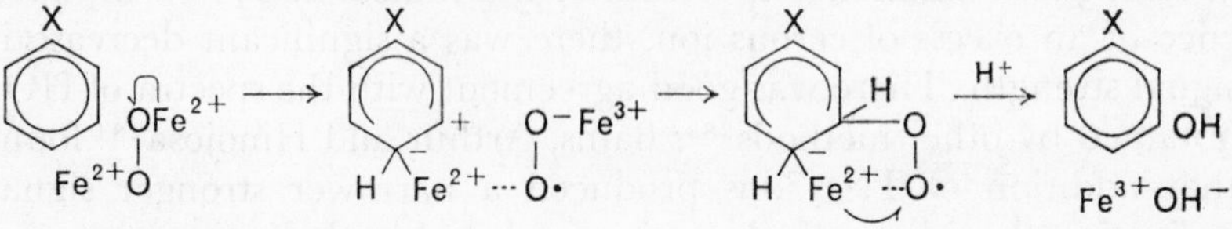

2. Systems

Some alternative complexing agents, but not all of them, and other metal ions such as titanous and copper ions[275–276a] can be used in place of ferrous ion, and other reversible electron donors can be used in place of ascorbic acid. An ascorbic acid to ferrous ion ratio of about six is an optimum, but hydroxylation will proceed very slowly even in the absence of an electron donor.

These studies have a bearing on the mechanisms obtaining in the autoxidation of aromatic compounds and, in particular, that of phenol. They are therefore of some technological significance.

Among the compounds studied with Udenfriend's reagent are a number of biochemicals and these are often hydroxylated if they are benzenoid, olefinic or heterocyclic compounds[276b, 276c].

Smith and Hays[277] have recently compared the effect of X-radiation on uracil with that of ascorbic acid–$FeSO_4$. They found many similarities.

F. Ceric System

Baer and Stein[278] showed that ceric ion reacted with hydrogen peroxide to give the hydroperoxyl radical (equation 73).

$$H_2O_2 + Ce^{4+} \longrightarrow HO_2^{\cdot} + H^+ + Ce^{3+} \quad (73)$$

$HO_2^{\cdot}$ reacted with an excess of Ce^{4+} giving oxygen (equation 74)

$$HO_2^{\cdot} + Ce^{4+} \longrightarrow O_2 + H^+ + Ce^{3+} \quad (74)$$

The kinetics of these reactions were investigated by Baxendale[279], Sigler and Masters[280], and Czapski, Bielski and Sutin[281]. They found that equation (73) was reversible. Anbar[282] proposed that an intermediate complex Ce^{III}–OOH was formed and existed for a finite time.

Using a flow system Saito and Bielski[283] determined the electron paramagnetic resonance spectrum of the radical $HO_2^{\cdot}$. In the presence of an excess of cerous ion, there was a significant decrease in signal strength. There was good agreement with the spectra of $HO_2^{\cdot}$ obtained by other methods[283]. Bains, Arthur and Hinojosa[284] found that addition of Ti^{4+} ions produced a narrower stronger signal, indicating that the radical species produced in the ceric system was less stable than its titanic analogue.

The radical species does not appear to hydroxylate benzene[235] and is generally not very reactive. In these respects it resembles the radiolytic hydroperoxyl radical.

G. Peracids

I. General

Peracids can undergo either homolysis giving free hydroxyl radicals:

$$RCO\text{–}O\text{–}OH \longrightarrow RCOO^{\cdot} + OH^{\cdot} \quad (75)$$

or heterolysis giving hydroxyl cations:

$$RCO\text{–}O\text{–}OH \longrightarrow RCOO^- + OH^+ \quad (76)$$

Both of these species appear to be capable of acting as electrophilic hydroxylating agents.

Uri[202] calculated that there should be a small amount of OH^+ present in equilibrium in hydrogen peroxide solution:

$$H_2O_2 + H^+ \rightleftharpoons H_2O + OH^+ \quad (77)$$

Derbyshire and Waters[285] chose mesitylene to demonstrate this. Mesitylene on hydroxylation with hydrogen peroxide in acetic acid–sulphuric acid mixtures gave mesitol. In this compound the positions activated by the OH group were blocked against further attack by more reagent so complications due to secondary reactions were minimized.

$$\text{mesitylene} + OH^+ \longrightarrow [\text{CH}_3\text{-substituted arenium ion (H, OH)}]^+ \xrightarrow{-H^+} \text{mesitol} \quad (78)$$

A Lewis acid such as boron trifluoride may be used in place of the acetic acid–sulphuric acid mixture[286].

2. Pernitrous acid

Pernitrous acid is formed when nitrous acid and hydrogen peroxide are mixed (equation 79). It is unstable.

$$H_2O_2 + HNO_2 \xrightarrow{-H_2O} ONOOH \longrightarrow ONO^{\bullet} + OH^{\bullet} \quad (79)$$

Halfpenny and Robinson[287, 288] concluded that the OH$^{\bullet}$ radical added almost always *ortho* or *para* to any existing substituent group in the benzene ring. This was followed by nitration in the *meta*-position giving either 1-hydroxy-2-nitro or 2-nitro-3-hydroxy derivatives. The incoming groups occupied adjacent positions in all the products found, probably because in **1** the unpaired electron was localized in that position (equation 80). Compound **15**, which is

$$C_6H_5X \xrightarrow{+OH^{\bullet}} [\text{X, OH, H, H radical}] \xrightarrow{+NO_2} [\text{X, OH, H, NO}_2\text{, H}] \quad (80)$$

not unlike **4**, was thought to decompose by elimination of either water, nitrous acid or hydrogen or the substituent as HX.

The products found from benzene are nitrobenzene, *o*-nitrophenol and smaller quantities of biphenyl, *p*-dinitrobenzene, phenol, other nitrophenols and some tar. Several other benzoid compounds were investigated but the presence of tar and dimers among the products casts doubt on the validity of deductions from the isomeric distributions of the products. It is always possible that the pernitrous acid does not dissociate until it reacts with the aromatic compound.

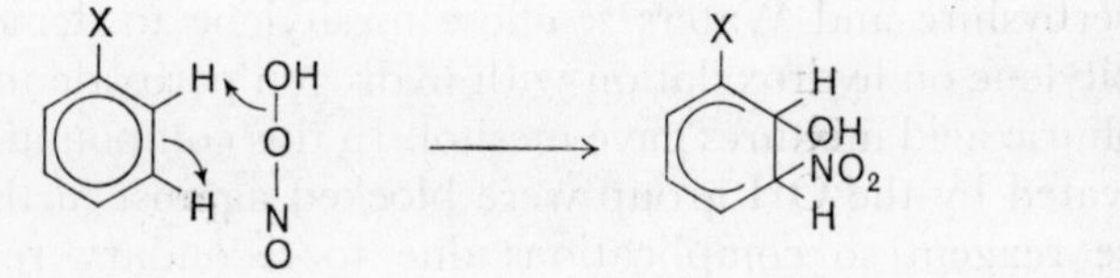

(81)

3. Trifluoroperacetic acid

a. General. Derbyshire and Waters[285] suggested that other peracids might give OH^+. The hydroxylating action of perbenzoic and peracetic acids was known but owing to secondary reactions giving quinones this action was not always recognized. In trifluoroperacetic acid (**16**) the fluorine atoms attract electrons from the O–O bond thus facilitating the fission of this bond[289]:

$$\overset{\delta-}{CF_3COO}\text{–}\overset{\delta+}{OH} \longrightarrow CF_3COO^- + OH^+ \quad (82)$$
(16)

Once again decomposition does not necessarily[290] take place giving free OH^+. The reagent may act as in equation (83).

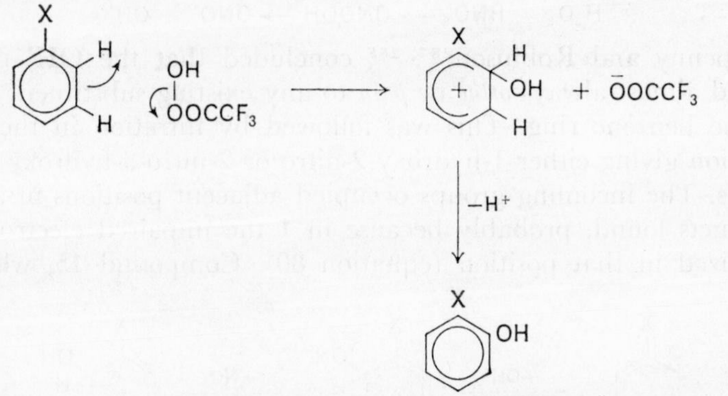

(83)

Evidence that **16** may decompose in this way is given by the reaction[291] with tetramethylethylene (equation 84).

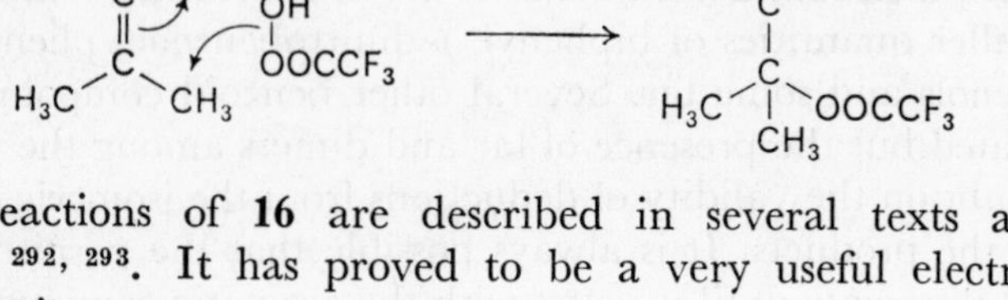

(84)

The reactions of **16** are described in several texts and reviews[222, 292, 293]. It has proved to be a very useful electrophilic hydroxylating agent.

b. Benzenoid compounds. The reaction with some aromatics has been studied by Davidson and Norman[294]. The isomeric distributions obtained were quite different from those obtained with Fenton's reagent. More importantly, published values for radiolytic hydroxyl radical reactions[126, 128] show quite a different distribution. The conditions are not strictly comparable but there seems little doubt that the attacking species is not the hydroxyl radical[294a]. The reagent can be considered as an electrophile of low selectivity but more selective and less reactive than the OH˙ radical.

The presence of a substituent nitro group in an aromatic compound does appear to inhibit the attack of **16**; although, since considerable amounts of tar were formed, the evidence is not clear[291]. In radiolytic hydroxylation nitro compounds are attacked quite readily (section II.B).

Oxidative cyclization can also be the result of attack by **16** (equation 85)[294].

H_2 C OOCCF$_3$ OH → (+) H OH —$-H^+$→ OH (85)

Attack at a ring carbon on which there is already a substituent sometimes induces methyl migration by a Wagner–Meerwein rearrangement. Prehnitene gives a variety of products including isodurenol[291] (equation 86).

—CF_3CO_3H→ OH (86)

NHAc D —CF_3CO_3H→ NHAc (+) H HO D —$-H^+$→ NHAc H D O (87)

$-D^+$; $-D^+$, $+H^+$; $+D^+$, $-H^+$ → NHAc H OH; NHAc D OH

An analogous rearrangement has been observed for deuterated acetanilide[295] (equation 87). 7·5% of the product is that deuterated in the 3-position. It is suggested that hydrogen migration might be a common reaction for phenolic cations.

c. Effect of additives. Usually hydrogen peroxide and trifluoroacetic acid are mixed to prepare **16** in situ. The effectiveness of **16** may be increased by carrying out the reaction in methylene chloride. Also effective is addition of boron trifluoride, a Lewis acid, which has been used by Hart, Buehler and Waring[291] to obtain high yields of products. The Lewis acid coordinates with one of the oxygens not used in the hydroxylation and thus facilitates decomposition. Iodine in conjunction with **16** has been used by Hey and co-workers[296] to hydroxylate steroids.

4. Other peracids

Hydroxylation has been reported following the action of peracetic acid, perbenzoic acid and persuccinic acid[297, 297a]. The presence of acid assists reaction[293] (equation 88). Smith and Fox[268] found that

$$Y{-}C_6H_5 + H{-}O{-}O{-}R \;(H^+) \longrightarrow Y{-}\overset{\bullet}{C_6H_5^+}(OH)(H) \xrightarrow[-H^+]{-ROH} YC_6H_4OH \qquad (88)$$

Ti^{3+} and peracetic produce a species which reacts like the $OH^{\bullet}$ radical.

Inorganic peroxy acids such as persulphuric acid and peroxychromic acid have been observed to effect hydroxylation[298] but hydroxyl radicals are not necessarily involved.

5. Peresters

Diisopropyl peroxydicarbonate has been shown by Kovacic and Morneweck[243] and by Kovacic and Kurz[244] to give a reasonably good yield (about 50%) in the hydroxylation of aromatic compounds to produce phenols. Reaction was carried out in the presence of a Friedel–Crafts catalyst and gave essentially no undesirable side reactions. This is a considerable advantage over those methods involving hydroxyl radicals. The initial electrophilic attack was thought to be by an oxonium cation to give a phenol ester–aluminium chloride complex which resists further attack by the reactant. On hydrolysis this yields the phenol.

H. Other Systems

There are other systems in which the action of hydroxyl radicals has been postulated in order to explain the results obtained.

Benzene is converted into *o*-nitrophenol by the Baudisch reaction (hydrogen peroxide, hydroxylamine hydrochloride and cupric ion). The first step involves hydroxylation[299, 300].

In the analogous system morpholine, copper salt and oxygen, phenols are oxidized to quinones[301]. Since known scavengers have no effect, $OH^{\bullet}$ radicals are most probably not involved.

Heckner, Landsberg and Dalchau[302] concluded that alkaline permanganate oxidized toluic acid and malonic acid through an intermediate $OH^{\bullet}$ radical which existed in equilibrium in solution[303] (equation 89).

$$MnO_4^- + OH^- \rightleftharpoons MnO_4^{2-} + OH^{\bullet} \tag{89}$$

Relative reaction rates were determined and agreed well with pulse radiolysis values[77, 90].

Powdered silica has some OH groups on the surface. If an organic compound is ground up with it, some hydroxylation results[304]. It would be interesting to know the isomeric distribution for the product from some monosubstituted benzenes.

An electrical discharge in water vapour produces $OH^{\bullet}$ radicals[305]. This should be a worthwhile source of such radicals. If an electrodeless discharge is used then no contaminants or catalytic metals are introduced. Hydroxyl radicals are also produced from hydrogen peroxide vapour[305a].

Hydrogen atoms are readily produced in the gas phase. By reaction with oxygen, $HO_2^{\bullet}$ radicals are produced[306].

Hydrogen peroxide can be homolysed by pyrolysis. Hydroxyl radicals result.

Ultrasonic waves can break up water, and one of the products is the $OH^{\bullet}$ radical[321, 322].

V. COMPARISON OF REAGENTS

A. Identity of Species

Except where there is clear evidence to the contrary, the hydroxylating species produced by chemical reagents have been called hydroxyl radicals in each case throughout this chapter. Nevertheless, since the hypothesis was first advanced in 1934 there has been speculation whether the hydroxyl radical is, in fact, formed by

Fenton's reagent. This speculation continues today and is also applied to the other reagents.

It was hoped that the e.s.r. spectra of the species would resolve these questions, but either the spectra are not exactly what would be expected on theoretical grounds for the free radicals or the radicals are perturbed in some ways by the components of the reagent system. Methods are now available[307] for finding the e.s.r. spectra of short-lived radiolytic species so these points may soon be resolved.

There is now reasonable agreement that the radiolytic and photolytic species are free $OH^{\cdot}$ radicals and are electrophilic reagents. However, only recently have techniques been available to study them without interference from other species and other reactions. The chemically produced species are undoubtedly similar in their behaviour. Some investigators have suggested that the reactive species may be complexes containing metal ions. Identification of a reactant with a known species can only follow when both kinetic data and product distribution are seen to vary together over a range of changing conditions.

Frequently one finds a categorical statement in the chemical or biological literature that one or other of the chemically generated species is known to be the hydroxyl radical. On tracing back one finds that such statements are based on the authority of a hypothesis put forward before the hydroxyl radical had been discovered for certain and its properties and reactions investigated. Furthermore, the necessary experimental equipment and techniques had not been developed at that time. At the present time, there is insufficient evidence to decide which, if any, of the chemically produced species are free $OH^{\cdot}$ radicals. Some of the considerations that must be taken into account will be summarized. In many cases conditions are different and the chemistry not understood.

B. Some Discrepancies

Livingston and Zeldes[211, 212] found that photolysis of allyl alcohol–H_2O_2 mixtures gave a spectrum quite different from that obtained by Dixon and Norman[259] using the flow system. Atkins and Symons[308] pointed out that since $OH^{\cdot}$ reacts so readily with H_2O_2 giving $HO_2^{\cdot}$, a more stable radical, it is difficult to be sure what species one is studying in ultraviolet-irradiated hydrogen peroxide solutions. Similarly, they consider $HO_2^{\cdot}$ or possibly $(TiOO^{\cdot})^{4+}$ to be the active species formed from titanous salts and hydrogen peroxide.

From product yields using the competitive method, Norman and Radda[209] have found relative rates for Fenton's radical in the ratio anisole: benzene: chlorobenzene: nitrobenzene of 6·35:1:0·55:0·14. This should be compared with the radiolytic values 1·5:1:1:0·4 [68, 77].

Myers and co-workers[186] found that the radiolytic $OH^{\bullet}$ added to carbon atom 6 of thymine whereas the titanous radical also added to position 5 on the other side of the double bond. Ormerod and Singh[269] noted a difference in site of attack between pH 1 and 2.

There is plenty of evidence that ferrous ion forms complexes with many anions[309, 310]. This influences its rate of reaction with H_2O_2. There appears to be no reason why $OH^{\bullet}$ should not remain complexed to the iron. Shiga[311] found that Fenton's reagent attacked the hydrogens on the ω carbon atoms of alcohols whereas the titanous reagent attacked the α position. The radiolytic and photolytic $OH^{\bullet}$ radical attack the α position. He concludes that Fenton's radical may be a complex, such as Fe–EDTA–H_2O_2. In a subsequent publication Shiga and co-workers[312] distinguished between the electrophilic titanous reagent and nucleophilic (sic) Fenton's reagent. On the other hand Smith and Wood[313] found that by varying the concentrations of the reactants it was possible to obtain radicals corresponding to hydrogen atom abstraction from all sites of the alkyl part of alcohols.

Staudinger[223] found different isomeric ratios for the hydroxylation products resulting from the action of Fenton's reagent and titanous system respectively on acetanilide. The reactions with cellulose are different for the two reagents[313a].

Chiang and co-workers[315] and Armstrong and Humphreys[260] consider that in the titanous system a complex $[Ti(H_2O)_4OH(substrate)]^{4+}$ is formed. To react, a substrate must be able to form this complex. If this is so, the dependence of yields on pH should differ from that in the radiolytic system. Norman[314] states that it is now apparent that the titanous reagent is far more complex than was at first thought.

Turkevich and co-workers[315], Fischer[316], Florin, Sicilio and Wall[317] and Mickewich[318], have considered the e.s.r. spectra of the titanous radical as well as its growth and decay. They conclude that it is some form of complex, perhaps $(TiOO)^{3+}$ which is essentially a complexed $O_2^{-\bullet}$. Such a complex does not necessarily behave like $O_2^{-\bullet}$.

The difficulty in finding an e.s.r. spectrum for $OH^{\bullet}$ radical in Fenton's system may be due to its very short life vis-à-vis dimerization ($k_{OH+OH} = 5 \times 10^9$ l mole^{-1}s^{-1}). When scavengers are present the time scale is very short in terms of most experimental techniques and thus the species is not readily detectable. During pulse radiolysis experiments in the presence of 1 mmole l^{-1} benzoate all hydroxyl radical reactions are more than 99% complete in $\approx$1 microsecond[60].

Sicilio, Florin and Wall[319] and Takakura and Rånby[320] were able to resolve two peaks in the e.s.r. spectrum of the titanous reagent. These were attributed to $HO^{\bullet}$ and $HO_2^{\bullet}$ and both were thought to be coordinated with Ti(IV) ions and possibly other species.

Bains, Arthur and Hinojosa[284] mixed H_2O_2 with pairs of metal ions (Fe^{2+}, Ti^{3+} and Ce^{4+}). From the changes of the e.s.r. spectra with relative concentrations they deduced that Ti^{4+} forms stable complexes with the radical species generated in these systems but Fe^{3+} and Ce^{3+} do not.

There are sufficient discrepancies between results for doubts to arise. Some of these discrepancies are due to making comparisons between experiments conducted under quite different conditions. For example, in most comparative tables one finds listed the pioneering radiation chemical results of Weiss, Stein and co-workers[17–24] obtained twenty years ago; both experimental and interpretative techniques have advanced greatly since then. It is interesting that only recently did new information appear, published by three independent groups[68, 76, 85, 90], on the aqueous radiation chemistry of solute nitrobenzene—a key compound in the study of directive effects of substituents in the benzene ring. It is quite apparent that a greater understanding of the chemistry of the systems is necessary before valid comparisons can be made.

The systems are complicated—much more complicated than the simpler radiolytic or photolytic ones—and only very recently has the understanding of these model systems progressed to a point of reasonably universal agreement. The species are very reactive, so their existence is transitory and special methods must be used to detect and identify them and to follow their reactions. They react to give other very reactive species, and several reaction steps may occur before the formation of a stable product which can be separated. Often there is a chain reaction so that results are difficult to replicate and there is uncertainty regarding what reaction is affected by a given change in experimental conditions. Further, these difficulties are compounded by the variety of the substances added in the practical systems. These substances may increase the yields of desired products but do not necessarily increase the chances of being able to sort out the mechanistics of even the basic processes. Without this knowledge the practical system is unapproachable.

C. Important Factors

Radiation chemical studies have demonstrated the importance of oxygen, low conversions, solute concentration and dose rate in some cases[321]. Addition of ferrous or copper ions introduces a chain reaction and increased yields result. Radiation chemical studies have an advantage over photochemical studies in that, within limits and certainly for dilute solutions, the rate of production and amount of hydroxyl radicals can be controlled with fair accuracy while other conditions are varied over a wide range. It should be possible to derive kinetic data for the interaction of many of the species. These data can be used, by means of computer programmes, to calculate product yields which can be compared with experimental values. Only when an evaluation has been made of a reactant's kinetic behaviour as well as the isomeric distribution of its products, can its identity be established with any certainty.

In the meantime, even without this knowledge, the practical chemist can still make use of these reagents to effect electrophilic hydroxylation in a single step.

VI. REFERENCES

1. J. H. O'Donnell and D. F. Sangster, *Principles of Radiation Chemistry*, Edward Arnold, London, 1970.

1a. *Gmelins Handbook of Inorganic Chemistry*, System No. 3, Oxygen, Section 8, *Hydroxyl–Perhydroxyl–Hydrogenozonide–Higher Hydrogen Peroxides*, Verlag Chemie G.M.B.H., Weinheim (1969).

2. P. Ausloos (Ed.), *Fundamental Processes in Radiation Chemistry*, John Wiley and Sons, New York, 1959.

3. A. J. Swallow, *Radiation Chemistry of Organic Compounds*, Pergamon Press, Oxford, 1960.

4. A. O. Allen, *The Radiation Chemistry of Water and Aqueous Solution*, Van Nostrand, Princeton, New Jersey, 1961.

5. J. W. T. Spinks and R. J. Woods, *An Introduction to Radiation Chemistry*, John Wiley and Sons, New York, 1964.

6. For references, see E. J. Hart and M. Anbar, *The Hydrated Electron*, John Wiley and Sons, New York, 1969.

7. A. Hummell and A. O. Allen, *Radiation Res.*, **17**, 302 (1962).

8. C. J. Hochanadel, *Radiation Res.*, **17**, 286 (1962).

9. H. A. Dewhurst, quoted by E. Collinson and F. S. Dainton, *Discussions Faraday Soc.*, **12,** 212 (1952).
10. M. S. Matheson and B. Smaller, *J. Chem. Phys.*, **23,** 521 (1955).
11. G. Czapski and B. H. J. Bielski, *J. Phys. Chem.*, **67,** 2180 (1963).
12. G. Czapski and L. M. Dorfman, *J. Phys. Chem.*, **68,** 1169 (1964).
13. K. Schmidt, *Z. Naturforsch.*, **16B,** 206 (1961).
14. L. M. Dorfman and M. S. Matheson in *Progress in Reaction Kinetics*, Vol. 3, (Ed. G. Porter), Pergamon Press, Oxford, 1965, Chap. 6, pp. 237–301.
15. L. M. Dorfman, *Science*, **141,** 493 (1963).
16. L. M. Dorfman and M. S. Matheson, *Pulse Radiolysis*, M.I.T. Press, Cambridge, 1969.
17. G. Stein and J. Weiss, *J. Chem. Soc.*, 3245 (1949).
18. G. Stein and J. Weiss, *J. Chem. Soc.*, 3254 (1949).
19. M. J. Day and G. Stein, *Nature*, **164,** 671 (1949).
20. H. Loebl, G. Stein and J. Weiss, *J. Chem. Soc.*, 2074 (1949).
21. H. Loebl, G. Stein and J. Weiss, *J. Chem. Soc.*, 2704 (1950).
22. H. Loebl, G. Stein and J. Weiss, *J. Chem. Soc.*, 405 (1951).
23. G. Stein and J. Weiss, *J. Chem. Soc.*, 3265 (1951).
24. G. R. A. Johnson, G. Stein and J. Weiss, *J. Chem. Soc.*, 3275 (1951).
25. J. Wright, *Discussions Faraday Soc.*, **12,** 60 (1952).
26. *Discussions Faraday Soc.*, **12,** 284–288 (1952).
27. T. J. Sworski, *J. Chem. Phys.*, **20,** 1817 (1952).
28. G. R. Freeman, A. B. Van Cleave and J. W. T. Spinks, *Can. J. Chem.*, **31,** 448 (1953).
29. T. J. Sworski, *Radiation Res.*, 1, 231 (1954).
30. W. A. Selke, A. Czikh and J. Dempsey, A.E.C. NYO 3330 (1954).
31. J. H. Baxendale and D. Smithies, *J. Chem. Phys.*, **23,** 604 (1955).
32. M. Daniels, G. Scholes and J. Weiss, *J. Chem. Soc.*, 832 (1956).
33. E. Collinson and A. J. Swallow, *Chem. Rev.*, **56,** 505 (1956).
34. M. A. Proskurnin and E. V. Barelko, *Symp. Radiation Chem. Acad. Sci., USSR*, 99 (1955).
35. E. V. Barelko, L. I. Kartasheva and M. A. Proskurnin, *Dokl. Akad. Nauk SSSR*, **116,** 74 (1957).
36. E. V. Barelko, L. I. Kartasheva, P. D. Novikov and M. A. Proskurnin, *Proceedings of first All-Union Conference on Radiation Chemistry, Moscow, 1957*, Consultants Bureau Inc., New York, 1959, p. 81.
37. M. A. Khenokh and E. M. Lapinskaia, *Proceedings of first All-Union Conference on Radiation Chemistry, Moscow, 1957*, Consultants Bureau Inc., New York, 1959, p. 167.
38. P. V. Phung and M. Burton, *Radiation Res.*, **7,** 199 (1957).
39. M. A. Proskurnin and Y. M. Kolotyrkin, *Proc. 2nd International Conference on Peaceful Uses of Atomic Energy, Geneva*, Vol. 29, 1958, p. 52.
40. M. A. Proskurnin, E. V. Barelko and L. I. Kartasheva, *Dokl. Akad. Nauk SSSR*, **121,** 671 (1958).
41. J. Weiss in *Actions Chimiques et Biologiques des Radiations*, Vol. 4 (Ed. M. Haïssinsky), Masson et Cie, Paris, 1958, p. 42.
42. W. A. Armstrong and D. W. Grant, *Nature*, **182,** 747 (1958).
43. A. M. Downes, *Australian. J. Chem.*, **11,** 154 (1958).
44. J. Goodman and J. Steigman, *J. Phys. Chem.*, **62,** 1020 (1958).

45. J. H. Baxendale and D. Smithies, *J. Chem. Soc.*, **779**, (1959).
46. K. C. Kurien, P. V. Phung and M. Burton, *Radiation Res.*, **11**, 283 (1959).
47. N. P. Krushinskaya and M. A. Proskurnin, *Russ. J. Phys. Chem.*, **33**, 237 (1959).
48. I. Loeff and G. Stein, *Nature*, **184**, 901 (1959).
49. T. Yumoto, Y. Bono and T. Matsuda, *Nagoya Kogyo Gijutsu Skikensho Hokoku*, 8, 296 (1959) in *Chem. Abstr.*, **57**, 1782b (1962).
50. A. Sugimori and G. Tsuchihashi, *Bull. Chem. Soc. Japan*, **33**, 713 (1960).
51. W. D. Armstrong, B. A. Black and D. W. Grant, *J. Chem. Phys.*, **64**, 1415 (1960).
52. L. I. Kartasheva, Z. S. Bulanovskaya, E. V. Barelko, Ya. M. Varshavskii and M. A. Proskurnin, *Dokl. Akad. Nauk SSSR*, **136**, 143 (1961).
53. A. Sakumoto and G. Tsuchihashi, *Bull. Chem. Soc. Japan*, **34**, 660 (1961).
54. A. Sakumoto and G. Tsuchihashi, *Bull. Chem. Soc. Japan*, **34**, 663 (1961).
55. M. A. Bertolaccini-Manzitti, P. L. Bertolaccini and L. Pucini, *Energia Nucleare*, 8, 445 (1961) in *Chem. Abstr.*, **56**, 3055c (1962).
56. K. Sugimoto, W. Ando and S. Oae, *Bull. Chem. Soc. Japan*, **36**, 124 (1963).
57. L. M. Dorfman, R. E. Bühler and I. A. Taub, *J. Chem. Phys.*, **36**, 549 (1962).
58. L. M. Dorfman, I. A. Taub and R. E. Bühler, *J. Chem. Phys.*, **36**, 3051 (1962).
59. L. M. Dorfman, I. A. Taub and D. A. Harter, *J. Phys. Chem.*, **41**, 2954 (1964).
60. D. F. Sangster, *J. Phys. Chem.*, **70**, 1712 (1966).
61. K.-D. Asmus, A. Wigger and A. Henglein, *Ber. Bunsenges. Physik. Chem.*, **70**, 862 (1966).
62. E. J. Land in *Progress in Reaction Kinetics*, Vol. 3 (Ed. G. Porter), Pergamon Press, Oxford, 1965, p. 369.
63. J. K. Thomas, J. Rabani, M. S. Matheson, E. J. Hart and S. Gordon, *J. Phys. Chem.*, **70**, 2409 (1966).
64. J. K. Thomas, *Trans. Faraday Soc.*, **61**, 702 (1965).
65. D. M. Brown, F. S. Dainton, D. C. Walker and J. P. Keene in *Pulse Radiolysis* (Eds. M. Ebert, J. P. Keene, A. J. Swallow and J. H. Baxendale), Academic Press, New York, 1965, p. 221.
66. P. Pagsberg, H. Christensen, J. Rabani, G. Nilsson, J. Fenger and S. O. Nielsen, *J. Phys. Chem.*, **73**, 1029 (1969).
67. M. Anbar and P. Neta, *Intern. J. Appl. Radiation Isotopes*, **18**, 493 (1967).
68. R. W. Matthews and D. F. Sangster, *J. Phys. Chem.*, **69**, 1938 (1965).
69. I. Kraljic and C. N. Trumbore, *J. Am. Chem. Soc.*, **87**, 2547 (1965).
70. G. E. Adams, J. W. Boag, J. Currant and B. D. Michael in *Pulse Radiolysis* (Eds. M. Ebert, J. P. Keene, A. J. Swallow and J. H. Baxendale), Academic Press, New York, 1965, p. 131.
71. G. Scholes, P. Shaw, R. L. Willson and M. Ebert in *Pulse Radiolysis* (Eds. M. Ebert, J. P. Keene, A. J. Swallow and J. H. Baxendale), Academic Press, New York, 1965, p. 151.
72. M. Anbar, D. Meyerstein and P. Neta, *J. Phys. Chem.*, **70**, 2660 (1966).
73. I. Kraljic in *The Chemistry of Ionization and Excitation* (Eds. G. R. A. Johnson and G. Scholes), Taylor and Francis Ltd., London, 1967, p. 303.
74. G. Scholes and R. L. Willson, *Trans. Faraday Soc.*, **63**, 2983 (1967).
75. E. J. Land and M. Ebert, *Trans. Faraday Soc.*, **63**, 1181 (1967).
76. K.-D. Asmus, B. Cercek, M. Ebert, A. Henglein and A. Wigger, *Trans. Faraday Soc.*, **63**, 2435 (1967).

77. P. Neta and L. M. Dorfman in *Radiation Chemistry*, Vol. 1 (Ed. E. J. Hart), Advances in Chemistry Series 81, 1968, p. 222.
78. L. M. Stock and H. C. Brown in *Advances in Physical Organic Chemistry*, Vol. 1 (Ed. V. Gold), Academic Press, London, 1963, p. 35.
79. B. Chutny, *Nature*, **213**, 593 (1967).
80. B. Cercek, *J. Phys. Chem.*, **72**, 3832 (1968).
81. R. W. Matthews and D. F. Sangster, *J. Phys. Chem.*, **71**, 4056 (1967).
82. C. B. Amphlett, G. E. Adams and B. D. Michael in *Radiation Chemistrv*, Vol. 1 (Ed. E. J. Hart), Advances in Chemistry Series 81, 1968, p. 231.
83. C. Carvaja, C. Farnia and E. Vianello, *Electrochim. Acta*, **11**, 919 (1966).
84. K.–D. Asmus, B. Cercek, M. Ebert, A. Henglein and A. Wigger, *Trans. Faraday Soc.*, **63**, 2435 (1967).
85. J. H. Fendler and G. L. Gasowski, *J. Org. Chem.*, **33**, 1865 (1968).
86. G. E. Adams and B. D. Michael, *Trans. Faraday Soc.*, **63**, 1171 (1967).
87. B. Cercek and M. Ebert in *Radiation Chemistry*, Vol. 1 (Ed. E. J. Hart), Advances in Chemistry Series 81, 1968, p. 210.
88. D. Grässlin, F. Merger, D. Schulte-Frohlinde and O. Volkert, *Z. Physik. Chem. NF*, **51**, 84 (1966).
89. O. Volkert, G. Termens and D. Schulte-Frohlinde, *Z. Physik. Chem. NF*, **56**, 261 (1967).
90. K.–D. Asmus, G. Beck, A. Henglein and A. Wigger, *Ber. Bunsenges. Phys. Chem.*, **70**, 869 (1966).
91. A. Wigger, A. Henglein and K.–D. Asmus, *Ber. Bunsenges. Phys. Chem.*, **71**, 513 (1967).
92. O. Volkert and D. Schulte-Frohlinde, *Tetrahedron Letters*, 2151 (1968).
93. R. E. Buhler and M. Ebert, *Nature*, **214**, 1220 (1967).
94. E. E. van Tamelen and S. P. Pappas, *J. Am. Chem. Soc.*, **84**, 3789 (1962).
95. K. E. Wilzbach and L. Kaplan, *J. Am. Chem. Soc.*, **86**, 2307 (1964).
96. A. W. Burgstahler, P.–L. Chien and M. O. Abdel-Rahman, *J. Am. Chem. Soc.*, **86**, 5286 (1964).
97. K. R. Jennings, *Z. Naturforsch.*, **22a**, 454 (1967).
98. I. Jano and Y. Mori, *Chem. Phys. Lett.*, **2**, 185 (1968).
99. M. Eberhardt and E. L. Eliel, *J. Org. Chem.*, **27**, 2289 (1962).
100. J. Cazes, *Dissertation Abstr.*, **24**, 3538 (1964); *J. Am. Chem. Soc.*, **84**, 4152 (1962).
101. K. F. Nakken, T. Brustad and A. K. Hansen in *Radiation Chemistry*, Vol. 1 (Ed. E. J. Hart), Advances in Chemistry Series 81, 1968, p. 251.
102. D. F. Sangster, unpublished results.
103. V. S. Zhikharev and N. A. Vysotskaya, *Russ. J. Phys. Chem.*, **42**, 192 (1968).
104. M. Anbar, D. Meyerstein and P. Neta, *J. Chem. Soc.* **(B)**, 742 (1966).
105. S. Shah, C. N. Trumbore, B. Giessner and W. Park in *Radiation Chemistry*, Vol. 1 (Ed. E. J. Hart), Advances in Chemistry Series 81, 1968, p. 321.
106. F. S. Dainton and B. Wiseall, *Trans. Faraday Soc.*, **64**, 694 (1968).
107. E. Collinson and A. J. Swallow, *Quart. Rev.*, **9**, 311 (1955).
108. E. Collinson and A. J. Swallow, *Chem. Rev.*, **56**, 471 (1956).
109. N. Rakintzis, W. Kunz and D. Schulte-Frohlinde, *Z. Physik. Chem. NF*, **34**, 51 (1962).
110. N. Th. Rakintzis, G. Marketos and A. P. Konstas, *Z. Physik. Chem. NF*, **35**, 234 (1962).

111. A. A. Denio, Thesis in *Nucl. Sci. Abstr.*, **18**, 12249, (1964).
112. D. G. Marketos and N. Th. Rakintzis, *Z. Physik. Chem. NF*, **44**, 270 (1965).
113. D. G. Marketos and N. Th. Rakintzis, *Z. Physik. Chem. NF*, **44**, 285 (1965).
114. J. P. Keene, E. J. Land and A. J. Swallow in *Pulse Radiolysis* (Eds. M. Ebert, J. P. Keene, A. J. Swallow and J. H. Baxendale), Academic Press, London, 1965, p. 227.
115. L. I. Grossweiner, A. F. Rodde, G. Sandberg and J. Chrysochoos, *Nature*, **210**, 1154 (1966).
116. D. R. Kalkwarf, BNWL-SA-1840 (1968). Battelle-Northwest Laboratory, Richland, Wash., U.S.A.
117. A. F. Rodde Jr and L. I. Grossweiner, *J. Phys. Chem.*, **72**, 3337 (1968).
118. L. I. Grossweiner in *Radiation Chemistry*, Vol. 1 (Ed. E. J. Hart), Advances in Chemistry Series 81, 1968, p. 309.
119. H. C. Christensen, AE-142 (1964). Aktiebolaget Atomenergi, Stockholm, Sweden.
120. M. Tsuda, *Bull. Chem. Soc. Japan*, **36**, 1582 (1963).
121. I. Loeff and G. Stein, *J. Chem. Soc.*, 2623 (1963).
122. L. I. Kartasheva and A. K. Pikaev, *Zh. Fiz. Khim.*, **41**, 2855 (1967).
123. I. Loeff and A. J. Swallow, *J. Phys. Chem.*, **68**, 2470 (1964).
124. R. Wander, P. Neta and L. M. Dorfman, *J. Phys. Chem.*, **72**, 2946 (1968).
125. K. F. Nakken, *Radiation Res.*, **21**, 446 (1964).
126. J. H. Fendler and G. L. Gasowski, *J. Org. Chem.*, **33**, 2755 (1968).
127. I. Loeff, L. M. Revetti and G. Stein, *Nature*, **204**, 1300 (1964).
128. J. H. Fendler, RRL-2310-231, p. 13 (1968). Radiation Research Laboratories, Carnegie-Mellon Institute, Pittsburgh, Pa., U.S.A.
129. G. A. Brodskaya and V. A. Sharpatyi, *Zh. Fiz. Khim.*, **41**, 2850 (1967).
130. A. Ohara and K. Toda, *J. Radiation Res.* (*Japan*), **8**, 45 (1967) in *Nucl. Sci. Abstr.* (*Japan*), **7**, 05934 (1968).
131. J. Chrysochoos, *Radiation Res.*, **33**, 465 (1968).
132. L. I. Kartasheva and A. K. Pikaev, *Dokl. Akad. Nauk SSSR*, **163**, 764 (1965).
133. L. I. Kartasheva and A. K. Pikaev, *High Energy Chem.*, **1**, 18 (1967).
133a. L. I. Kartasheva and A. K. Pikaev, *Int. J. Radiat. Phys. Chem.*, **1**, 243 (1969).
134. J. Geisselsoder, M. J. Kingkade and J. S. Laughlin, *Radiation Res.*, **20**, 263 (1963).
135. M. A. Proskurnin and Y. M. Kolotyrkin, *Proc. Second International Conference on Peaceful Uses of Atomic Energy*, Geneva, 1958, **29**, 52.
136. E. J. Henley, J. Goodman and I. Tang, *Trans. Am. Nucl. Soc.*, **3**, 387 (1960).
137. H. Hotta and A. Terakawa, *Bull. Chem. Soc. Japan*, **33**, 335 (1960).
138. H. Hotta and N. Suzuki, *Bull. Chem. Soc. Japan*, **36**, 717 (1963).
139. H. Hotta, A. Terakawa, K. Shimada and N. Suzuki, *Bull. Chem. Soc. Japan*, **36**, 721 (1963).
140. H. Hotta, N. Suzuki and A. Terakawa, *Bull. Chem. Soc. Japan*, **36**, 1255 (1963).
141. N. Suzuki and H. Hotta, *Bull. Chem. Soc. Japan*, **37**, 244 (1963).
142. K. Shimada, N. Suzuki, N. Itatani and H. Hotta, *Bull. Chem. Soc. Japan*, **37**, 1143 (1964).
143. H. Hotta, N. Suzuki, N. Itatani and K. Shimada, *Bull. Chem. Soc. Japan*, **37**, 1147 (1964).

144. H. C. Christensen, AE-192 (1965). Aktiebolaget Atomenergi, Stockholm, Sweden.
145. H. C. Christensen, AE-193 (1965). Aktiebolaget Atomenergi, Stockholm, Sweden.
146. H. C. Christensen, *Nukleonik*, **8,** 121 (1966).
147. H. C. Christensen, *Nukleonik*, **8,** 124 (1966).
148. T.–c. Hung, *Bull. Inst. Chem., Acad. Sinica*, **No. 14** (1967) in *Nucl. Sci. Abstr.*, **22**, 23072 (1968).
149. N. Suzuki and H. Hotta, *Bull. Chem. Soc. Japan*, **37**, 244 (1964).
150. H. Hotta, N. Suzuki and T. Abe, *Bull. Chem. Soc. Japan*, **39**, 417 (1966).
151. N. Suzuki and H. Hotta, *Bull. Chem. Soc. Japan*, **40,** 1361 (1967).
152. H. Hotta and N. Suzuki, *Bull. Chem. Soc. Japan*, **41**, 1537 (1968).
153. I. Balakrishnan and M. P. Reddy, *J. Phys. Chem.*, **72,** 4609 (1968).
154. K. Sugimoto, W. Ando and S. Oae, *Bull. Chem. Soc. Japan*, **36,** 124 (1963).
155. A. I. Chernova and V. D. Orekhov, *Kinetics and Catalysis*, **7,** 49 (1966).
156. R. W. Matthews and D. F. Sangster, unpublished results.
157. J. Rabani and M. S. Matheson, *J. Am. Chem. Soc.*, **86,** 3175 (1964).
158. E. Hayon, *Trans. Faraday Soc.*, **61,** 734 (1965).
159. J. M. Weeks and J. Rabani, *J. Phys. Chem.*, **70,** 2100 (1966).
160. G. E. Adams, J. W. Boag and B. D. Michael, *Nature*, **205,** 898 (1965).
161. W. D. Felix, B. L. Gall and L. M. Dorfman, *J. Phys. Chem.*, **71,** 384 (1967).
162. J. Errera and V. Henri, *J. Phys. Radium*, **7,** 225 (1926).
163. S. J. Rzad and J. M. Warman, *J. Phys. Chem.*, **72,** 3013 (1968).
163a. T. K. K. Srinvasan, I. Balakrishnan and M. P. Reddy, *J. Am. Chem. Soc.*, **73,** 2071 (1969).
164. W. M. Garrison, H. R. Haymond, W. Bennett and S. Cole, *J. Chem. Phys.*, **25,** 1282 (1956).
165. G. G. Jayson, G. Scholes and J. Weiss, *J. Chem. Soc.*, 2594 (1955).
165a. R. C. Ashline and R. L. von Berg, *Am. Inst. Chem. Eng. J.*, **15,** 387 (1969).
166. E. J. Henley and J. P. Schwartz, *J. Am. Chem. Soc.*, **77,** 3167 (1955).
167. E. J. Henley, W. S. Schiffries and N. F. Barr, *Am. Inst. Chem. Eng. J.*, **2,** 211 (1956).
168. P. G. Clay, G. R. A. Johnson and J. Weiss, *Proc. Chem. Soc.*, 96 (1957).
169. P. G. Clay, G. R. A. Johnson and J. Weiss, *J. Chem. Soc.*, 2175 (1958).
170. C. F. Cullis, J. M. Francis and A. J. Swallow, *Proc. Roy. Soc.*, **A287,** 15 (1965).
171. C. F. Cullis, J. M. Francis, T. Raef and A. J. Swallow, *Proc. Roy. Soc.*, **A300,** 443 (1967).
172. R. A. Basson and T. A. du Plessis, *Chem. Commun.*, 775 (1967).
173. R. A. Basson and T. A. du Plessis, *Radiation Res.*, **33,** 183 (1968).
174. R. A. Basson and T. A. du Plessis, *Radiation Res.*, **36,** 14 (1968).
175. A. J. Swallow in *Radiation Chemistry*, Vol. 2 (Ed. E. J. Hart), Advances in Chemistry Series 82, 1968, p. 499.
176. P. G. Clay, J. Weiss and J. Whiston, *Proc. Chem. Soc.*, 125 (1959).
177. G. Scholes and J. Weiss, *Nature*, **185,** 305 (1960).
178. K. W. Chambers, E. Collinson, F. S. Dainton, W. A. Seddon and F. Wilkinson, *Trans. Faraday Soc.*, **63,** 1699 (1967).
179. C. Schneider and A. J. Swallow, *J. Polymer Sci.*, **B4,** 277 (1966).
180. P. G. Clay, G. R. A. Johnson and J. Weiss, *J. Phys. Chem.*, **63,** 862 (1959).
181. Y. Le Roux, H. Nayer and C. Nofre, *Bull. Soc. Chim. France*, 2003 (1967).

182. G. Scholes in *Progress in Biophysics and Molecular Biology*, Vol. 13 (Eds. J. A. V. Butler, H. E. Huxley and R. E. Zirkle), Pergamon Press, London, 1963, p. 59.
183. J. J. Weiss in *Progress in Nucleic Acid Research and Molecular Biology*, Vol. 3 (Eds. J. N. Davidson and W. E. Cohn), Academic Press, New York, 1964, p. 103.
183a. E. Fahr, *Angew. Chem.* (*Intern. Ed.*), **8,** 578 (1969).
184. G. Scholes, P. Shaw and R. L. Willson in *Pulse Radiolysis* (Eds. M. Ebert, J. P. Keene, A. J. Swallow and J. H. Baxendale), Academic Press, New York, 1965, p. 151.
185. G. Scholes and R. L. Willson, *Trans. Faraday Soc.*, **63,** 2983 (1967).
186. L. S. Myers Jr., M. L. Hollis and L. M. Theard in *Radiation Chemistry*, Vol. 1 (Ed. E. J. Hart), Advances in Chemistry Series 81, 1968, p. 345.
187. C. L. Greenstock, M. Ng and J. W. Hunt in *Radiation Chemistry*, Vol. 1 (Ed. E. J. Hart), Advances in Chemistry Series 81, 1968, p. 397.
188. R. M. Danziger, E. Hayon and M. E. Langmuir, *J. Phys. Chem.*, **72,** 3842 (1968).
188a. G. Scholes, R. L. Willson and M. Ebert, *Chem. Commun.*, 17 (1969).
189. L. S. Myers Jr., J. F. Ward, W. T. Tsukamoto, D. E. Holmes and J. R. Julca, *Science*, **148,** 1234 (1965).
190. L. S. Myers Jr., J. F. Ward, W. T. Tsukamoto and D. E. Holmes, *Nature*, **208,** 1086 (1965).
191. J. F. Ward and M. M. Urist, *Intern. J. Radiation Biol.*, **12,** 209 (1967).
192. F. Merger and D. Grässlin, *Angew. Chem.* (*Intern. Ed. Engl.*), **3,** 640 (1964).
192a. F. Merger and D. Grässlin, *German Pat.*, 1,228,258. [*Chem. Abstr.*, **66,** 28526 (1967).]
193. A. Davison, A.A.E.C. TM-422 (1968). Australian Atomic Energy Commission, Sutherland, N.S.W., Australia.
194. B. H. J. Bielski and A. O. Allen in *Proceedings of the Second Tihany Symposium on Radiation Chemistry*, Akadémiai Kiadó, Budapest (1967), p. 81.
195. B. H. J. Bielski and H. A. Schwarz, *J. Phys. Chem.*, **72,** 3836 (1968).
196. C. Vermeil and L. Salomon, *Compt. Rend.*, **249,** 268 (1959).
197. B. H. J. Przybielski-Bielski and R. R. Becker, *J. Am. Chem. Soc.*, **82,** 2164 (1960).
198. G. W. Black and B. H. J. Bielski, *J. Phys. Chem.*, **66,** 1203 (1962).
199. B. H. J. Bielski and A. O. Allen, unpublished results.
200. N. Rakintzis, E. Papaconstantinou and D. Schulte-Frohlinde, *Z. Physik. Chem. NF*, **44,** 257 (1965).
201. N. A. Milas, P. F. Kurz and W. P. Anslow, *J. Am. Chem. Soc.*, **59,** 543 (1937).
202. N. Uri, *Chem. Rev.*, **50,** 375 (1952).
203. J. H. Baxendale and J. A. Wilson, *Trans. Faraday Soc.*, **53,** 344 (1957).
204. J. P. Hunt and H. Taube, *J. Am. Chem. Soc.*, **74,** 5999 (1952).
205. J. G. Calvert and J. N. Pitts Jr., *Photochemistry*, John Wiley and Sons Inc., New York, 1966, p. 200.
206. H. G. C. Bates, H. G. Evans and N. Uri, *Nature*, **166,** 869 (1950).
207. H. G. C. Bates and N. Uri, *J. Am. Chem. Soc.*, **75,** 2754 (1953).
208. E. Boyland and P. Sims, *J. Chem. Soc.*, 2967 (1953).
208a. C. R. E. Jefcoate, J. R. Lindsay Smith and R. O. C. Norman, *J. Chem. Soc.* (**B**), 1013 (1969).

209. R. O. C. Norman and G. K. Radda, *Proc. Chem. Soc.*, 138 (1962).
210. K. Omura and T. Matsuura, *Tetrahedron*, **24**, 3475 (1968).
210a. J. G. Pacifici and J. M. Straley, *J. Polymer. Sci., Part B, Polymer. Lett.*, **7**, 7 (1969).
210b. M. Ahmad and P. G. Clay, *Chem. Commun.*, 60 (1969).
210c. K. J. Youtsey and L. I. Grossweiner, *J. Phys. Chem.*, **73**, 447 (1969).
211. R. Livingston and H. Zeldes, *J. Chem. Phys.*, **44**, 1245 (1966).
212. R. Livingston and H. Zeldes, *J. Am. Chem. Soc.*, **88**, 4333 (1966).
212a. T. Ichikawa and K. Kuwata, *Bull. Chem. Soc. Japan*, **42**, 2208 (1969).
213. J. Saldick and A. O. Allen, *J. Am. Chem. Soc.*, **77**, 1388 (1955).
214. J. H. Baxendale and J. Magee, *Trans. Faraday Soc.*, **51**, 205 (1955).
215. G. Stein and J. Weiss, *Nature*, **166**, 1104 (1950).
216. W. H. Richardson in *Oxidation in Organic Chemistry* (Ed. K. B. Wiberg), Academic Press, New York, 1965, p. 274.
217. D. R. Stranks and J. R. Yandell in 'Exchange Reactions', *International Atomic Energy Agency Proceedings Series*, Vienna, 1965, p. 83.
218. G. O. Phillips, N. W. Worthington, J. F. McKellar and R. R. Sharpe, *Chem. Commun.*, 835 (1967).
219. N. Basco in *Free Radicals in Inorganic Chemistry* (Ed. R. F. Gould), Advances in Chemistry Series 36 (1962), p. 26.
220. D. H. Volman in *Advances in Photochemistry* (Eds. W. A. Noyes Jr., G. S. Hammond and J. N. Pitts Jr.), Interscience Publishers, New York, 1963, p. 63.
221. A. Y.–M. Ung and R. A. Back, *Can. J. Chem.*, **42**, 753 (1964).
221a. L. J. Stief and V. J. DeCarlo, *J. Chem. Phys.*, **50**, 1234 (1969).
222. R. O. C. Norman and J. R. Lindsay Smith in *Oxidases and Related Redox Systems*, Vol. 1 (Eds. T. E. King, H. S. Mason and M. Morrison), John Wiley and Sons Inc., New York, 1965, p. 131.
222a. R. O. C. Norman and B. C. Gilbert, in *Advances in Physical Organic Chemistrv*, Vol. 5 (Ed. V. Gold), Academic, London, 1967, p. 53.
222b. R. O. C. Norman and P. R. West, *J. Chem. Soc.* (**B**), 389 (1969).
223. H. Staudinger, B. Kerekjártó, V. Ullrich and Z. Zubrzycki in *Oxidases and Related Redox Systems*, Vol. 2 (Eds. T. E. King, H. S. Mason and M. Morrison), John Wiley and Sons Inc., New York, 1965, p. 815.
224. S. Udenfriend, C. T. Clark, J. Axelrod and B. Brodie , *J. Biol. Chem.*, **208**, 731 (1954).
225. B. Brodie, J. Axelrod, P. A. Shore and S. Udenfriend, *J. Biol. Chem.*, **208**, 741 (1954).
226. H. S. Mason in *Advances in Enzymology*, Vol. 19 (Ed. F. F. Nord), Interscience Publishers Inc., New York, 1957, p. 79.
227. H. J. H. Fenton, *Chem. Soc.* (*London*), **65**, 899 (1894).
228. F. Haber and J. Weiss, *Proc. Roy. Soc.*, **A147**, 332 (1934).
229. J. H. Baxendale, M. G. Evans and G. S. Park, *Trans. Faradav Soc.*, **42**, 155 (1946).
230. W. G. Barb, J. H. Baxendale, P. George and K. R. Hargrave, *Trans. Faraday Soc.*, **47**, 462 (1951).
231. References listed by C. Walling, *Free Radicals in Solution*, John Wiley and Sons Inc., New York, 1957, p. 567.
232. I. M. Kolthoff and A. I. Medalia, *J. Am. Chem. Soc.*, **71**, 3777 (1949).

233. I. M. Kolthoff and A. I. Medalia, *J. Am. Chem. Soc.*, **71,** 3784 (1949).
234. R. R. Grinstead, *J. Am. Chem. Soc.*, **82,** 3464 (1960).
235. J. R. Lindsay Smith and R. O. C. Norman, *J. Chem. Soc.*, 2897 (1963).
236. A. Cier, C. Nofre, M. Ranc and A. Lefier, *Bull. Soc. Chim. France*, 1523 (1959).
237. J. H. Wang, *J. Am. Chem. Soc.*, **77,** 822 (1955).
238. J. H. Wang, *J. Am. Chem. Soc.*, **77,** 4715 (1955).
239. R. Breslow and L. N. Lukens, *J. Biol. Chem.*, **235,** 292 (1960).
240. J. H. Merz and W. A. Waters, *Discussions Faraday Soc.*, **2,** 179 (1947).
241. J. H. Merz and W. A. Waters, *J. Chem. Soc.*, S15 (1949).
242. J. H. Merz and W. A. Waters, *J. Chem. Soc.*, 2427 (1949).
243. P. Kovacic and S. T. Morneweck, *J. Am. Chem. Soc.*, **87,** 1566 (1965).
244. P. Kovacic and M. E. Kurz, *J. Am. Chem. Soc.*, **87,** 4811 (1965); *J. Org. Chem.*, **31,** 2011 (1966); *J. Org. Chem.*, **31,** 2459 (1966).
245. J. H. Baxendale and J. Magee, *Discussions Faraday Soc.*, **14,** 160 (1953).
246. J. O. Konecny, *J. Am. Chem. Soc.*, **76,** 4993 (1954).
247. R. R. Grinstead, *J. Am. Chem. Soc.*, **82,** 3472 (1960).
248. S. J. Cosgrove and W. A. Waters, *J. Chem. Soc.*, 1726 (1951).
249. R. O. C. Norman and G. K. Radda in *Advances in Heterocyclic Chemistry*, Vol 2 (Ed. A. R. Katritzky), Academic Press, New York, 1963, p. 163.
250. G. Scholes and J. Weiss, *Biochem. J.*, **53,** 567 (1953).
251. C. Nofre, A. Cier, C. Michan-Sancet and J. Parnet, *Compt. Rend.*, **251,** 811 (1960).
252. C. Nofre, A. Lefier and A. Cier, *Compt. Rend.*, **253,** 687 (1961).
253. A. Cier, A. Lefier, M. A. Ravier and C. Nofre, *Compt. Rend.*, **254,** 504 (1962).
253a. T. Shiga and A. Isomoto, *J. Am. Chem. Soc.*, **73,** 1139 (1969).
254. C. Walling, *Free Radicals in Solution*, John Wiley and Sons Inc., New York, 1957, Chap. 11.
254a. S. Tobinga, Japan Pat. 21,709 (1963).
255. C. J. Norton and R. E. White in *Selective Oxidation Processes* (Ed. R. F. Gould), Advances in Chemistry Series 51, 1965 p. 10.
256. A. E. Cahill and H. Taube, *J. Am. Chem. Soc.*, **74,** 2312 (1952).
257. W. T. Dixon and R. O. C. Norman, *Nature*, **196,** 891 (1962).
258. W. T. Dixon and R. O. C. Norman, *J. Chem. Soc.*, 3119 (1963).
259. W. T. Dixon and R. O. C. Norman, *Proc. Chem. Soc.*, 97 (1963).
260. W. A. Armstrong and W. G. Humphreys, *Can. J. Chem.*, **45,** 2589 (1967).
261. W. T. Dixon and R. O. C. Norman, *J. Chem. Soc.*, 4857 (1964).
262. R. O. C. Norman and R. J. Pritchett, *J. Chem. Soc.* (**B**), 926 (1967).
263. C. R. E. Jefcoate and R. O. C. Norman, *J. Chem. Soc.* (**B**), 48 (1968).
264. P. Smith and P. B. Wood, *Can. J. Chem.*, **45,** 649 (1967).
265. P. Smith, J. T. Pearson, P. B. Wood and T. C. Smith, *J. Chem. Phys.*, **43,** 1535 (1965).
266. W. T. Dixon, R. O. C. Norman and A. L. Buley, *J. Chem. Soc.*, 3625 (1964).
267. J. Q. Adams, *J. Am. Chem. Soc.*, **89,** 6022 (1967).
268. P. Smith and W. M. Fox, *Can. J. Chem.*, **47,** in press (1969).
268a. W. E. Griffith, G. F. Longster, J. Myatt and P. F. Todd, *J. Chem. Soc.* (**B**), 530 (1967).
269. M. G. Ormerod and B. B. Singh, *Int. J. Radiation Biol.*, **10,** 533 (1966).

270. G. A. Hamilton, R. J. Workman and L. Woo, *J. Am. Chem. Soc.*, **86**, 3390 (1964).
271. G. A. Hamilton, *J. Am. Chem. Soc.*, **86**, 3391 (1964).
272. G. A. Hamilton and J. P. Friedman, *J. Am. Chem. Soc.*, **85**, 1008 (1963).
273. G. A. Hamilton, J. P. Friedman and P. M. Campbell, *J. Am. Chem. Soc.*, **88**, 5266 (1966).
274. G. A. Hamilton, J. W. Hanifin and J. P. Friedman, *J. Am. Chem. Soc.*, **88**, 5269 (1966).
275. M. M. T. Khan and A. E. Martell, *J. Am. Chem. Soc.*, **89**, 4176 (1967).
276. M. M. T. Khan and A. E. Martell, *J. Am. Chem. Soc.*, **89**, 7104 (1967).
276a. Y. Kurimura and H. Kuriyama, *Bull. Chem. Soc. Japan*, **42**, 2238 (1969).
276b. J. Hurych, *Hoppe-Seyler's Z. Physiol. Chem.*, **348**, 426 (1967).
276c. E. Eich and H. Rochelmeyer, *Pharm. Acta Helv.*, **41**, 109 (1966).
277. K. C. Smith and J. E. Hays, *Radiation Res.*, **33**, 129 (1968).
278. S. Baer and G. Stein, *J. Chem. Soc.*, 3176 (1953).
279. J. H. Baxendale, *Chem. Soc. (London), Spec. Publ.* **No. 1**, 40 (1954).
280. P. B. Sigler and B. J. Masters, *J. Am. Chem. Soc.*, **79**, 6353 (1957).
281. G. Czapski, B. H. J. Bielski and N. Sutin, *J. Phys. Chem.*, **67**, 201 (1963).
282. M. Anbar, *J. Am. Chem. Soc.*, **83**, 2031 (1961).
283. E. Saito and B. H. J. Bielski, *J. Am. Chem. Soc.*, **83**, 4467 (1961).
B. H. J. Bielski and E. Saito, *J. Phys. Chem.*, **66**, 2266 (1962).
284. M. S. Bains, J. C. Arthur and O. Hinojosa, *J. Phys. Chem.*, **72**, 2250 (1968).
285. D. H. Derbyshire and W. A. Waters, *Nature*, **165**, 401 (1950).
286. J. D. McClure and P. H. Williams, *J. Org. Chem.*, **27**, 24 (1962).
287. E. Halfpenny and P. L. Robinson, *J. Chem. Soc.*, 928 (1952).
288. E. Halfpenny and P. L. Robinson, *J. Chem. Soc.*, 939 (1952).
289. R. D. Chambers, P. Goggin and W. K. R. Musgrave, *J. Chem. Soc.*, 1804 (1959).
290. C. A. Bunton, T. A. Lewis and D. R. Llewellyn, *J. Chem. Soc.*, 1226 (1956).
291. H. Hart, C. A. Buehler and A. J. Waring in *Selective Oxidation Processes* (Ed. R. F. Gould), Advances in Chemistry Series 51, 1965, p. 1.
292. R. O. C. Norman and R. Taylor, *Electrophilic Substitution in Benzenoid Compounds*, Elsevier Publishing Co., Amsterdam, 1965, Chap. 4, p. 110.
293. R. O. C. Norman, *Principles of Organic Synthesis*, Methuen and Co. Ltd., London, 1968, p. 392.
294. A. J. Davidson and R. O. C. Norman, *J. Chem. Soc.*, 5404 (1964).
294a. D. M. Jerina, G. Guroff and J. Daly, *Arch. Biochem. Biophys.*, **124**, 612 (1968).
295. D. M. Jerina, J. W. Daly, W. Landis, B. Witkop and S. Udenfriend, *J. Am. Chem. Soc.*, **89**, 3347 (1967).
296. D. G. Hey, G. D. Meakins and M. W. Pemberton, *J. Chem. Soc.* (**C**), 1331 (1966).
297. L. F. Fieser and M. Fieser, *Reagents for Organic Synthesis*, J. Wiley and Sons Inc., New York, 1967, pp. 785–796, 820.
297a. N. P. Emel'yanov and D. V. Lopatik, *Dokl. Akad. Nauk Beloruss, SSR*, **12**, 718 (1968). [*Chem. Abstr.*, **69**, 105,] 961 (1968).
298. O. C. Dermer and M. T. Edmison, *Chem. Rev.*, **57**, 77 (1957).
299. J. Konecny, *J. Am. Chem. Soc.*, **77**, 5748 (1955).
300. K. Maruyama, I. Tanimoto and R. Goto, *Tetrahedron Letters*, 5889 (1966).
301. W. Brackman and E. Havinga, *Rec. Trav. Chim.*, **74**, 1070 (1955).

302. K. H. Heckner, R. Landsberg and S. Dalchau, *Ber. Bunsenges. Phys. Chem.*, **72**, 649 (1968).
303. K. A. K. Lott and M. C. R. Symons, *Discussions Faraday Soc.*, **29**, 205 (1960).
304. P. J. Schofield, B. J. Ralph and J. H. Green, *J. Phys. Chem.*, **68**, 472 (1964).
305. S. N. Foner and R. L. Hudson in *Free Radicals in Inorganic Chemistry* (Ed. R. F. Gould), Advances in Chemistry Series 36 (1962), p. 34. L. I. Avramenko and R. V. Kolesnikova in *Advances in Photochemistry*, Vol. 2 (Eds. W. A. Noyes Jr., G. S. Hammond and J. N. Pitts), Interscience Publishers, New York, 1964, p. 25.
305a. J. N. Herak and W. Gordy, *Proc. Natl. Acad. Sci. U.S.*, **56**, 1354 (1966).
306. L. I. Avramenko, L. M. Evlashkina and R. V. Kolesnikova, *Bull. Acad. Sci. USSR, Div. Chem. Sci. (Engl. Transl.)*, 252 (1967) in *Chem. Abstr.*, **67**, 32135x (1968).
307. B. Smaller, J. R. Remko and E. C. Avery, *J. Chem. Phys.*, **48**, 5174 (1968).
308. P. W. Atkins and M. C. R. Symons, *The Structure of Inorganic Radicals*, Elsevier Publishing Co., Amsterdam, 1967, Chap. 6, p. 105.
309. G. C. Jayson, D. A. Stirling and A. J. Swallow, *Chem. Commun.*, 931 (1967).
310. C. F. Wells and M. A. Salam, *J. Chem. Soc.* (**A**), 308 (1968).
311. T. Shiga, *J. Phys. Chem.*, **69**, 3805 (1965).
312. T. Shiga, A. Boukhors and P. Douzou, *J. Phys. Chem.*, **71**, 4264 (1967).
313. P. Smith and P. B. Wood, unpublished data.
313a. J. C. Arthur, O. Hinojosa and M. S. Bains, *J. Appl. Polymer Sci.*, **12**, 1411 (1968).
314. R. O. C. Norman, *Proc. Roy. Soc.*, **A302**, 315 (1968).
315. Y. S. Chiang, J. Craddock, D. Mickewich and J. Turkevich, *J. Phys. Chem.*, **70**, 3509 (1966).
316. H. Fischer, *Ber. Bunsenges. Phys. Chem.*, **71**, 685 (1967).
317. R. E. Florin, F. Sicilio and L. A. Wall, *J. Phys. Chem.*, **72**, 3154 (1968).
318. D. J. Mickewich, *Dissertation Abstr.*, **29**, 141-B (1968).
319. F. Sicilio, R. E. Florin and L. A. Wall, *J. Phys. Chem.*, **70**, 47 (1966).
320. K. Takakura and B. Rånby, *J. Phys. Chem.*, **72**, 164 (1968).
321. A. Weissler, *Nature*, **193**, 1070 (1962).
322. M. Anbar and I. Pecht, *J. Phys. Chem.*, **71**, 1246 (1967).
323. E. J. Fendler and J. H. Fendler in *Progress in Physical Organic Chemistry*, Vol. 7 (Eds. S. G. Cohen, A. Streitwieser Jr. and R. W. Taft), Interscience Publishers, New York, 1969, p. 229.

CHAPTER 4

Formation of hydroxyl groups via oxymetallation, oxidation, and reduction

I. R. L. Barker

Brighton College of Technology, Sussex, England

I. INTRODUCTION: OXYMETALLATION AND OXIDATION

The first parts of this review are concerned with reactions which introduce the hydroxyl group into organic compounds and in which at least one major stage is oxidative. Knowledge of mechanisms of oxidation has increased markedly in recent years and excellent texts giving extensive coverage are available (e.g., see References 1—4). The subject matter of these occasionally coincides with material in this chapter and it is for this reason that the topics here have received consideration in proportion to the extent to which they are inadequately treated in secondary sources elsewhere. Additional considerations have been the actual or potential synthetic utility and theoretical importance of the reactions.

A. The Formation of OH Groups at Saturated Carbon

The most direct method of inserting a hydroxyl group at saturated carbon is by autoxidation. This is essentially a homolytic chain reaction between triplet ground state, molecular oxygen and the substrate preferably in the presence of an initiator to counter the poor radical properties of molecular oxygen. Hydroperoxides are normally formed in the first instance but these are easily reduced, e.g., by lithium aluminium hydride, to alcohols. Although this appears to constitute a most general method of forming alcohols by oxidation it is limited in synthetic use to those examples where the desired reaction site is also the most reactive towards radicals. Alkanes are much less active than allylic and benzylic positions and reactivity generally parallels bond dissociation energies. Radicals which promote autoxidation may be generated by thermal or photochemical disruption of an initiator, photochemically via a 'sensitizer', e.g., benzophenone, or by direct homolysis of bonds in the substrate by high energy irradiation. The area is generally well served with reviews of the synthetic scope and mechanisms of the reactions (e.g., see References 5—7). Asymmetric synthesis has been observed in the autoxidation of DL-3-*p*-menthene in the presence of the optically active catalysts manganese D(−)- and -L(+)-mandelate[8].

Many carbon acids, e.g., tri-*p*-nitrophenylmethane, are quite inert towards attack by molecular oxygen but react readily in the form of their carbanions to yield hydroperoxides. Examples of these and the proposed mechanisms appear in section III.B.

Photoinitiated autoxidation of hydrocarbons proceeds via a homolytic chain process; quite different in its characteristics is the dye-photosensitized oxygenation of olefins to allylic hydroperoxides and cyclic conjugated dienes to *endo*-peroxides. These latter oxygenations find analogy in those effected by singlet oxygen, and the evidence is probably conclusive that the dye-photosensitized oxygenations also proceed via oxygen in this form. The evidence for this important development is reviewed in section III.A.

Of considerable interest are the autoxidations of organometal compounds, particularly boron and aluminium trialkyls, which find technical use in the manufacture of aliphatic alcohols and of peroxides as polymerization catalysts. The mechanistic features of these reactions appear in section III.C.

Formally resembling the autoxidation is the oxidation of trialkylboranes with hydrogen peroxide, alkyl hydroperoxides and peroxy acids. Trialkylboranes, which may be formed by hydroboration of alkenes, are oxidized by alkaline hydrogen peroxide to alcohols which correspond to an anti-Markownikoff hydration of the alkene[9, 10]. This very useful procedure continues to be extended, and amongst the developments one may note its use for the preparation of optically active alcohols with H–D asymmetry[11] and the stereoselective syntheses of alcohols via cyclic hydroboration of dienes with 2,3-dimethyl-2-butylborane[12].

Peresters undergo a copper-catalysed reaction of homolytic type with alkenes to introduce an ester group at the allyl position[13], e.g., *t*-butyl perbenzoate when heated with cyclohexene and a catalytic amount of cuprous bromide produces 73% of 3-benzoyloxycyclohexene. The mechanism of these reactions is probably as follows[14–16], e.g.. for cyclohexene:

$$PhCOO-OCMe_3 + Cu^I \longrightarrow PhCOOCu^{II} + Me_3CO^{\bullet} \quad (1)$$

$$Me_3CO^{\bullet} + \text{cyclohexene} \longrightarrow Me_3COH + \text{cyclohexenyl radical} \quad (2)$$

$$\text{cyclohexenyl radical} + PhCOOCu^{II} \longrightarrow Cu^I + \text{3-(OCOPh)cyclohexene} \quad (3)$$

There is evidence that equation (3) involves a carbonium ion[15]. Unsymmetrical alkenes yield mixtures of isomeric products which may result from a mesomeric allylic radical, terminal alkenes produce mostly an unrearranged product. Appropriate reagents are able to effect the conversion of cumene into 3-acetoxycumene in 30% yield[16] and tetralin into l-benzoyloxytetrahydronaphthalene in 15% yield[17]. The reagents are also able to introduce an ester group into other compounds containing active carbon–hydrogen bonds, e.g., ethers and thioethers[13]. The method offers a convenient route for the synthesis in particular of allyl alcohols.

Chromium compounds are capable of oxidizing tertiary alkanes to tertiary alcohols and alkenes to allylic alcohols but usually in poor yield. Further oxidation results in carbonyl compounds and carboxylic acids and the method has limited synthetic value. The mechanism of the chromic acid oxidation of triaryl alkanes to carbinols probably involves either the homolytic route (equation 4) or the cyclic transition state (equation 5):

$$Ar_3CH + Cr^{VI} \longrightarrow Ar_3C^{\cdot} + Cr^{V} \tag{4}$$

$$Ar_3C{-}H \; + \; (HO)_2CrO_2 \longrightarrow Ar_3C{-}O{-}Cr^{IV} \tag{5}$$

the fate of the radical in equation (4) is probably conversion into a chromium ester. Oxidation with chromium compounds has been reviewed recently[18–20], and the possible use of these in synthesis, e.g., of *t*-butyl chromate[21] and chromyl chloride[22], continues to be explored.

The limitations in the use of potassium permanganate for the oxidation of hydrocarbons to alcohols are similar to those of chromic acid. Further oxidation of the alcohols to carbonyl compounds and carboxylic acids is frequent, and an additional drawback is the low solubility of the oxidant in many organic solvents. The use of triphenylmethylarsonium permanganate which is soluble in chloroform has been described[23]. Potassium permanganate will effect the oxidation of tertiary hydrocarbon groups to carbinol as in branched chain carboxylic acids and in arylalkanes, and finds here synthetic use. The suggested mechanism for these oxidations involves permanganate ion abstraction of a hydrogen atom from the tertiary position giving a radical pair in a solvent cage. Recombination within the cage produces an alkyl hypomanganate:

$$R_3CH + MnO_4^- \longrightarrow [R_3C^{\cdot}MnO_4H^-] \longrightarrow R_3COMnO_3H^- \tag{6}$$

The ester may decompose by several routes; these mechanisms and other features have been reviewed recently[24].

Lead tetraacetate is able to effect acetoxylation at carbon–hydrogen bonds and thus affords a possible stage to the overall introduction of a hydroxyl group. The synthetic use of the method is limited by the occurrence of additional reactions and the reagent's ability to attack readily only those carbon–hydrogen bonds which are adjacent to carbonyl, phenyl, alkenyl or ether groups. The reaction with carbonyl compounds probably proceeds via a heterolytic mechanism involving the decomposition of an intermediate ester of the enol with lead tetraacetate:

$$>C{=}CH{-}OH + Pb(OAc)_4 \longrightarrow \begin{matrix} HC{-}O{-}Pb(OAc)_2 + HOAc \\ \| \quad\quad | \\ C \quad\quad OAc \end{matrix} \qquad (7)$$

$$\longrightarrow\ >C(OAc)CHO + Pb(OAc)_2$$

The mechanism of the acetoxylation of ethers, and allylic and benzylic acetoxylations is not yet clear. There is evidence of homolytic routes but the reactions do not appear to be accompanied by products arising from dimerization or chain transfer of radicals, although dimers have been characterized in the boron trifluoride-catalysed oxidation of benzene derivatives[25]. Lead tetraacetate whilst fairly inactive towards benzene will effect several reactions including acetoxylation of certain aromatic and heteroaromatic compounds. The reactions of this reagent have been reviewed recently[26]. Of considerable interest is the report of the use of lead tetra(trifluoracetate) which effects the trifluoroacetoxylation of hydrocarbons such as benzene and heptane. Trifluoroacetoxylation may be followed by hydrolysis and the four compounds investigated gave alcohols in $45 \pm 10\%$ yields[27]. The mechanism is not yet known.

Apart from lead tetraacetate, the acetates of Hg(II) and Tl(III), together with Pd(II) salts and selenium dioxide in acetic acid, also effect allylic acetoxylation. In so far as acetoxylation may be regarded as a route to the oxidative introduction of a hydroxyl group these reactions are described in sections II.A—D.

B. The Formation of OH Groups at Unsaturated Carbon

The direct introduction of a hydroxyl group on olefinic carbon by autoxidation is very restricted in scope and at present of little value in synthesis. The limitations are, first, that removal of allylic hydrogen is energetically preferred to the homolytic addition of oxygen and, secondly, if oxygen does add then it results in a peroxy radical which is more likely in these circumstances to add to more olefin, producing after further repetitions a polyperoxide, than it is to abstract a hydrogen atom to become a hydroperoxide. It has been observed that it is those olefins which polymerize readily which also autoxidize and that this process of autoxidation becomes equivalent to copolymerization[5]. At elevated temperatures, the addition of oxygen to many olefins increases at the expense of allylic attack and becomes a competing process. The dye-photosensitized addition of oxygen to conjugated dienes is reviewed in section III.A.

The compounds chromic acid, chromyl acetate and chloride find little use in the synthesis of alcohols from alkenes. Chromyl acetate may oxidize alkenes to epoxides but in most cases several products are formed with the epoxide in small proportion[18].

The dihydroxylation of alkenes by aqueous permanganate in basic solution is well known. The oxidation also takes place readily with periodate and catalytic amounts of permanganate in neutral solution (the Lemieux and von Rudloff reagent). Permanganate is the oxidant and is continuously regenerated by periodate[24]. Similar in mechanism and product are dihydroxylations with osmium tetroxide[28, 29]. Whereas these two methods produce *cis* glycols, the Prevost reagent—a solution of iodine in carbon tetrachloride together with an equivalent of silver acetate or benzoate—under anhydrous conditions yields the diacyl derivative of a *trans*-glycol via neighbouring acetoxy participation[30, 31]. A comparison of the mechanisms of acetoxylation by the Prevost reagent and lead tetraacetate has been reported recently[32].

The epoxidation of alkenes with peracids offers another useful route to the synthesis of *trans*-1,2-diols[13]. The formation of the epoxide is a stereospecifically *cis*-addition, the subsequent ring opening usually proceeds with inversion of configuration at the carbon atom attacked resulting in an overall *trans*-addition to the double bond[33]. Epoxides are easily converted by lithium aluminium hydride into monohydric alcohols, the reaction is of S_N2 type and the rigid epoxides of multi-ring systems such as steroids yield axial alcohols.

Lead tetraacetate reacts with alkenes in a variety of ways and yields some 1,2-diacetoxy derivative. The mechanism of the reaction is not known, the addition does not appear to be stereospecific and the reaction has little preparative value at present[26]. The reactions of alkenes with compounds of Hg(II), Tl(III), and Pd(II) are dealt with in sections II.A—C.

II. THE INTRODUCTION OF OH GROUPS VIA CERTAIN OXYMETALLATION AND ACETOXYLATION PROCEDURES

A. Mercury(II) *Salts*

Alkenes undergo electrophilic addition reactions with mercuric salts under mild conditions in solvents with nucleophilic activity to produce organomercury adducts of the type:

$$>C{=}C< + Hg(OAc)_2 \begin{cases} \xrightarrow{ROH} RO{-}\overset{|}{\underset{|}{C}}{-}\overset{|}{\underset{|}{C}}{-}HgOAc \\ \xrightarrow{H_2O} HO{-}\overset{|}{\underset{|}{C}}{-}\overset{|}{\underset{|}{C}}{-}HgOAc \\ \xrightarrow{AcOH} AcO{-}\overset{|}{\underset{|}{C}}{-}\overset{|}{\underset{|}{C}}{-}HgOAc \end{cases} \quad (8)$$

These reactions are not new and earlier work has been reviewed[34, 35]. Whereas many alkenes may be induced to react in the manner shown by equation (8), a change in reaction conditions, particularly temperature or solvent, often leads to the formation of other products. For example, in acetic acid as solvent Hg(II) acetate oxidizes cyclohexene to the allylic acetate and, in a water-suspension, to the allylic alcohol and cyclopentyl aldehyde[36].

The formation of the adduct by electrophilic addition does not appear to proceed by free carbonium ions or those involving neighbouring carbon participation. Both norbornene[37] (**1**) and norbornadiene[38] (**3**) react to give unrearranged *cis*-2,3-*exo*-mercuration products (**2** and **4**), although **4** undergoes $HgCl_2$-catalysed isomerization to the nortricyclenic oxymercurial **5** [38]:

$$(\mathbf{1}) \xrightarrow[NaX]{Hg(OAc)_2} (\mathbf{2})\ [\text{OAc, HgX}] \quad (9)$$

$$(\mathbf{3}) \xrightarrow[NaX]{Hg(OAc)_2} (\mathbf{4})\ [\text{OAc, HgX}] \xrightarrow{HgCl_2} (\mathbf{5})\ [\text{XHg, OAc}] \quad (10)$$

Similarly, both *endo*- and *exo*-dicyclopentadienes undergo oxymercuration in water and methanol without rearrangement to give *exo,cis*-addition products[39]. The results imply that the formation of initial products is under kinetic control but that lengthy reaction times may yield products of greater thermodynamic stability arising from a consecutive rearrangement of carbonium ions formed by heterolysis of the Hg–carbon bond[38, 40].

The stereochemistry of this addition reaction depends upon the structural features of the alkene taking part. There are two main divisions: the hydroxymercuration of acyclic and monocyclic alkenes is a *trans*-addition, whereas that of bicyclic alkenes in which the double bond is sterically hindered to *endo*-attack is a *cis*-addition.

The *trans*-addition of Hg(II) acetate in aqueous solution is considered to proceed via a Hg(II)–olefin π-complex (mercurinium ion) which is then *trans*-solvolysed:

$$\mathrm{>C{=}C<} + \mathrm{Hg(OAc)_2} \xrightarrow[\text{slow}]{} \mathrm{OAc^-} + \mathrm{>C{-}C<}\,(\overset{+}{\triangleright}\mathrm{HgOAc}) \xrightarrow[\text{fast}]{\mathrm{H_2O}} \mathrm{HO{-}\overset{|}{C}{-}\overset{|}{C}{-}HgOAc} + \mathrm{H^+} \tag{11}$$

Evidence for this scheme rests on the kinetics of formation of the adduct, the observed stereochemistry of the products, and kinetic studies of dehydroxymercuration which is the acid-catalysed reversion of hydroxymercuration[34, 35, 41–44]. A relevant illustration of the latter is that α-2-methoxycyclohexylmercury(II) chloride, in which the substituents may assume a *trans*-diaxial conformation, undergoes perchloric acid-catalysed dehydroxymercuration ca 10^6 times faster than the β-diastereoisomer[35]. Recently a *trans*-solvolysed mercurinium intermediate has been used to rationalize the occurrence of partial asymmetric synthesis in the methoxymercuration of α,β-unsaturated esters, namely, (−)-menthyl crotonate, cinnamate and β-methyl cinnamate[45].

The symmetrical structure assigned to the mercurinium intermediate has been disputed by Halpern and Tinker[46] who report the kinetics of the hydroxymercuration of many acyclic alkenes and cyclohexene with Hg(II) perchlorate in aqueous perchloric acid solution. They observed that the rate constants for hydroxymercuration of eight alkenes give a good log k versus σ^* plot of slope $\rho^* = -3{\cdot}3$. Consequently they suggest that the rate-determining step is the formation of an intermediate which has a high degree of positive charge localization, approaching carbonium ion character, on the carbon atom adjacent to the substituent R:

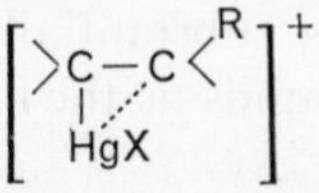

$$\left[>C-C\overset{R}{<} \atop \;\;\; | \cdot \cdot \; HgX \right]^{+}$$

There is an interesting comparison here with the mercuric acetate cleavage of substituted phenylcyclopropanes to yield an adduct, which has $\rho^+ = -3{\cdot}2$, and may involve an intermediate of similar structure[47]. In contrast, evidence of a σ-bridged mercurinium ion has been reported for the methoxymercuration of allenes[48]. The direct demonstration of kinetic nonparticipation of solvent molecule has yet to be made. The unsymmetrical π-complex has been favoured by other authors (e.g., see Reference 37) to account for features of the reaction and it lends itself readily to an explanation of the observation that the oxymercuration–demercuration procedure leads to a Markownikoff hydration of hydrocarbon alkenes.

It is the synthetic utility of this feature which has been returned to by H. C. Brown and co-workers[49, 51]. They have shown that oxymercuration combined with reduction of the oxymercurial adduct by sodium borohydride in situ provides a convenient, mild method to achieve Markownikoff hydration of carbon–carbon double bonds without rearrangement. The method complements the anti-Markownikoff hydration effected by the hydroboration–oxidation of alkenes and is considered superior to the procedure whereby the *exo*-alcohol of a bicyclic alkene is obtained via epoxidation and metal hydride reduction. A representative selection of alkenes have been hydrated in good yield by this method. The hydration of bicyclic alkenes takes place on the least hindered side resulting in *exo*-alcohols, e.g., 2-methylene-norbornane (**6**) yields 2-methyl-*exo*-norbornanol (**7**) in 99·5% yield:

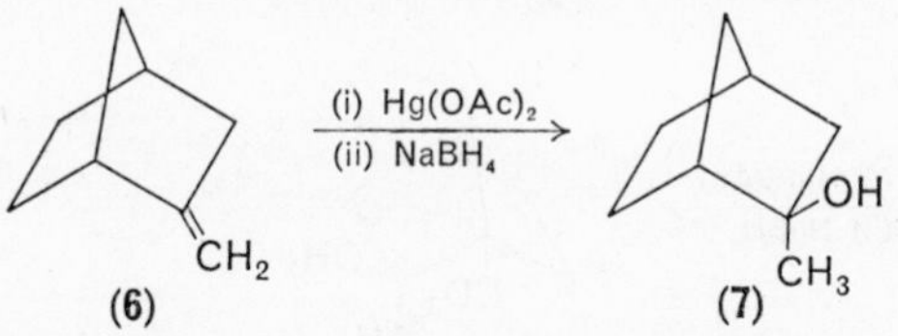

$$(\mathbf{6})\ \text{(CH}_2\text{)} \xrightarrow[\text{(ii) NaBH}_4]{\text{(i) Hg(OAc)}_2} (\mathbf{7})\ \text{(OH, CH}_3\text{)} \qquad (12)$$

In a series of molecules whose structures offer increasing steric hindrance to *endo*-approach there is an increasing preference for the formation of the *exo*-alcohol. The effect is similar to that observed in the lithium aluminium hydride reduction of hindered ketones[50]. The oxymercuration–demercuration procedure shows a high degree of stereoselectivity in effecting substantially *exo*-hydration of norbornene (**8**) and related compounds as the following examples illustrate (total yields $\geqslant 84\%$)[51]:

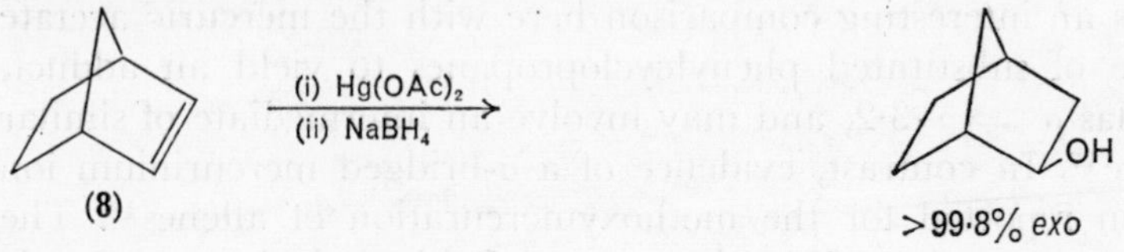

(i) $Hg(OAc)_2$
(ii) $NaBH_4$ (13) 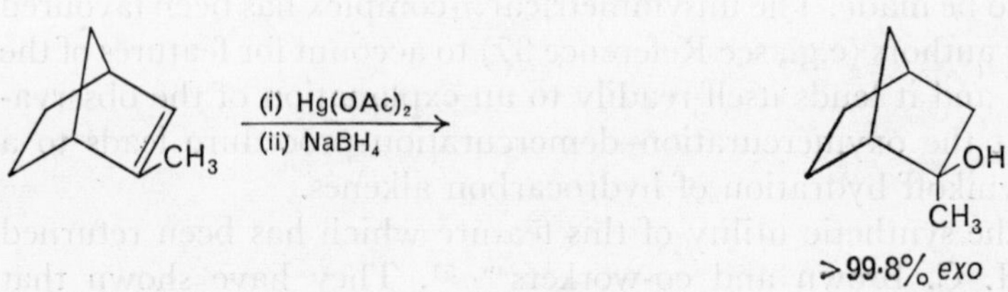
(**8**) OH
>99·8% *exo*

(i) $Hg(OAc)_2$
(ii) $NaBH_4$ (14) 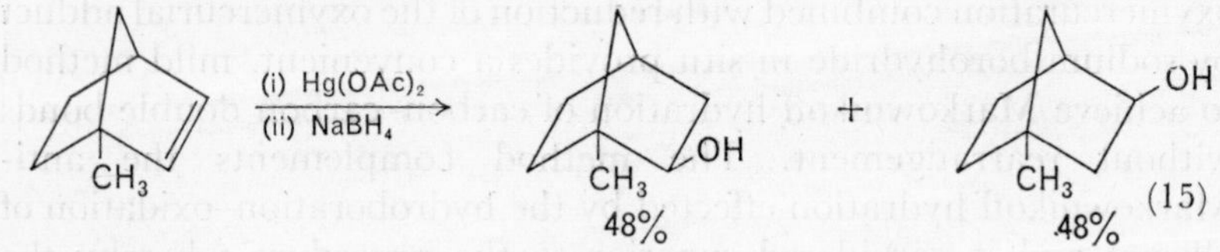
CH_3 OH CH_3
>99·8% *exo*

(i) $Hg(OAc)_2$
(ii) $NaBH_4$ + OH
CH_3 CH_3 OH CH_3 (15) 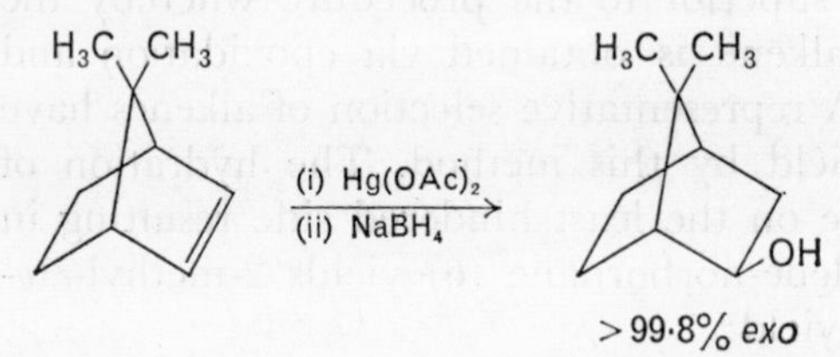
48% 48%

H_3C CH_3 H_3C CH_3
(i) $Hg(OAc)_2$
(ii) $NaBH_4$ (16) 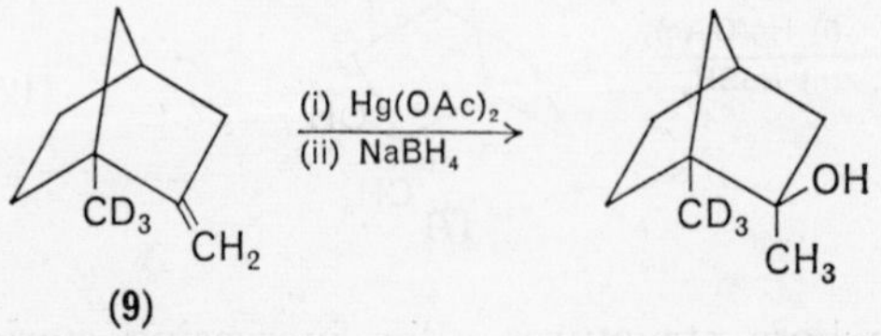
OH
>99·8% *exo*

(i) $Hg(OAc)_2$
(ii) $NaBH_4$ (17)
CD_3 CH_2 CD_3 OH CH_3
(**9**)

In common with the oxymercurations mentioned earlier, these proceed without skeletal rearrangements, or without scrambling as indicated particularly by equations (15) and (17).

Whereas the mechanism of *trans*-hydration of a mercurinium-type intermediate is reasonable for alkenes free from steric-controlling effects, the structure of the transition state leading to *cis*-addition is not yet clear. As yet there are no rate measurements for the hydroxymercuration of bicyclic olefins; however, the results of Brown and coworkers[49, 51] indicate that *cis*-oxymercuration is not handicapped by slower reaction rates than *trans*-oxymercuration. That of norbornene (**8**) is amongst the fastest and that of 7,7-dimethylnorbornene (**9**) is considerably slower.

cis-Additions commonly arise from cyclic intermediates, and in the case of hydroxymercuration this would require the replacement of acetoxy from the mercurinium intermediate by hydroxyl. The reaction could then be represented by:

$$\gt C{-}C(R)\lt \;[Hg{-}OH]\;(\mathbf{10}) \xrightarrow{H_2O} \gt C(HOHg){-}C(R)(OH)\lt + H^+ \qquad (18)$$

Structure **10** would also be present during *trans*-addition to unhindered olefins but here the intramolecular transfer of hydroxyl would have to compete with intermolecular rearwards attack by solvent. This model has been developed by Traylor[52] who envisages competitive S_N2 and S_Ni processes. Traylor has also suggested that ring strain rather than steric hindrance may be responsible for the *cis*-oxymercuration of norbornene via a mercurinium ion which itself is more reactive because of rigidity and strain than in the case of cyclohexene. In this connexion he has demonstrated that bicyclo-[2,2,2]octene undergoes both *cis*- and *trans*-oxymercuration.

Stereochemically anomalous reaction products may arise when substituents near to the carbon–carbon double bond undergo prior coordination with mercury. There are a number of examples in the literature[35] and Sung Moon and Waxman[53] have adapted the observations of Henbest and Nicholls[54] for 4-substituted cyclohexenes, to the stereospecific synthesis of *trans*-1,3-diols of six-, seven- and eight-membered rings by the oxymercuration–demercuration of cycloalk-2-en-1-ols.

Certain areas of the subject seem to be particularly beset by conflicting stereochemical claims, e.g., Jensen and Miller[55] report that the oxymercuration of 5-norbornene-2-*endo* carboxylic acid (**11**) with Hg(II) acetate yields 94% of the α-5-chloromercurilactone **12** as a consequence of *trans*-addition and not the mixture of *endo*- and *exo*-compounds reported earlier by other authors[56]. This example is also of interest in being one of a number in which a *trans*-intramolecular nucleophilic attack on the mercurinium intermediate takes place:

(i) $Hg(OAc)_2$ in MeOH
(ii) aq. NaCl

OCOH (11) · OC—O (12) HgCl (19)

The reaction of Hg(II) acetate in acetic acid with olefins at elevated temperatures commonly effects allylic acetoxylation[57-59] from which alcohols may be obtained by hydrolysis. The acetoxylations and ring contractions of cyclic olefins strongly resemble those produced by Pb(IV) acetate, and it is probable that similar mechanisms occur. A feature of the Pb(IV) acetate oxidations is the frequency of Wagner–Meerwein rearrangements[60]. Not all reactions of Pb(IV) acetate with olefins, however, are ascribed a heterolytic route. For example, the reagent reacts with styrene to give several products, one of these, PhCH(OAc)Et, is considered to result from a radical chain reaction, and another, $PhCH(OAc)CH_2OAc$, by two concurrent reactions of homolytic and heterolytic character[60, 61]. In addition, of considerable interest is the report that molecular oxygen inhibits the benzylic acetoxylation of toluene by scavenging the short-lived free radical intermediates such as $PhCH_2^{\bullet}$ or $Pb(OAc)_3^{\bullet}$, occurring in a radical chain process[62].

Whereas the oxymercuration adduct has been suggested by some authors as an intermediate in the acetoxylation reaction, Winstein and co-workers have implied the possibility of the direct formation of an allylic mercurial, envisaged mechanistically as of S_Ei' (equation 20) or S_E2' (equation 21) type[63]:

>C(H)–C=C< + AcO–HgOAc → >C=C–C(HgOAc)< + AcOH (20)

AcO^- H–C–C=C< + ^+HgOAc → >C=C–C(HgOAc)< + AcOH (21)

These processes may be regarded generally as the reversion of the reported S_Ei' or S_E2' demercurations of butenylmercuric acetate in acetic acid which yield 99·5% of the secondary allylic acetate. The demercuration is slow at 25° but greatly increased in the presence of Hg(II) acetate and the following S_Ei' (equation 22) and S_E2' (equation 23) schemes have been suggested[64]:

$$CH_3HC{=}CH{-}CH_2{-}Hg{-}OAc \xrightarrow{-Hg^0} CH_3CH(OAc){-}CH{=}CH_2 \quad (22)$$

$$CH_3HC{=}CH{-}CH_2{-}HgOAc + AcO{-}HgOAc \xrightarrow{-Hg_2(OAc)_2} CH_3CH(OAc){-}CH{=}CH_2 \quad (23)$$

Examination of butenylmercuric acetate by n.m.r. has shown that under conditions of rapid allylic equilibration (induced by HgX_2 salts) the equilibrium is far on the side of the primary butenyl structure as in the case of the analogous Grignard and PdCl compounds[65]. In terms of acetoxylation, this implies that the secondary acetate may be anticipated to be the product from the allylic oxidation of both 1- and 2-butene as any secondary mercurials initially formed isomerize rapidly to the primary structure. The expectation has been borne out by an examination of the acetoxylation of a number of 1- and 2-alkenes (C_5—C_8) with Hg(II) acetate in acetic acid at 75° [63]. However, the proportion of secondary allylic acetate diminishes with reaction time particularly with 2-olefins owing to $Hg(OAc)_2$-catalysed isomerization of the allylic acetates; olefin isomerization also occurs. The exclusive formation of secondary allylic acetates in these particular examples is, therefore, based essentially on the unique demercuration process and is not consistent with normal carbonium ion or free radical behaviour. The general applicability of the rationalization is not known and it may not apply to olefins of widely different structural type.

A completely different approach to the problem has been made by Wiberg and Nielsen[66] who investigated the stereochemistry of the allylic acetoxylation of a number of cyclic olefins with Hg(II) acetate at elevated temperatures for prolonged periods. They consider that the results, e.g., the formation of racemic carvotanacetol acetate from (+)-carvomenthene, require the intermediacy of a symmetrical, allylic carbonium ion. This arises from the heterolysis of the Hg–carbon bond of an allylmercuric acetate which is itself formed via electrophilic addition and elimination of a proton from the mercurinium-type intermediate.

B. *Thallium*(III) *Salts*

Although the reactions have been less well investigated it is apparent that Tl(III) salts react with olefins in a manner similar to that of Hg(II) salts and Pb(IV) acetate. Acyclic olefins are oxidized to glycols and carbonyl compounds by Tl(III) salts in aqueous solution[67] and the kinetics of oxidation of certain of these have been examined[68]. Cyclohexene in acetic acid solution is oxidized to the following products[36, 67, 69–71]:

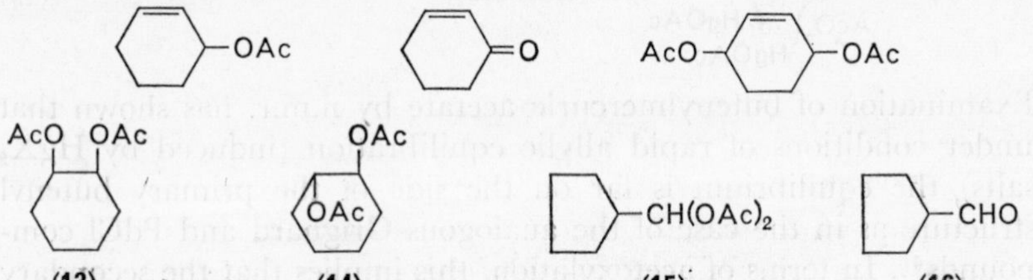

OAc AcO OAc OAc O OAc AcO OAc CH(OAc)$_2$ CHO

(13) (14) (15) (16)

and the glycol monoacetates of **13** and **14**; products **13**—**16** are often formed in high proportion. There is little work reported on the stereochemistry of the cyclohexane diacetates (**13** and **14**) but one study has shown that mainly the *trans*-isomer is formed in dry acetic acid and mainly the *cis*-isomer in the moist solvent[36]. Both isomers are considered to have a common precursor in the *trans*-oxythallation adduct which undergoes metal–carbon bond heterolysis with acetoxy participation to yield an acetoxonium ion. Solvolyses in which the stereochemistry of the product depends upon the medium and particularly the presence of water are a feature of acetoxonium compounds[36]. The ring-contracted products are presumably also the consequence of carbonium ions but on this occasion with neighbouring carbon participation.

Features such as the multi-component nature of the product, the degree of incidence of allylic attack and presence of ring-contracted products suggests that oxidations with Tl(III) salts stand nearer to the behaviour of those of Pb(IV) acetate than those of Hg(II) salts. Both Hg(II) and Pd(II) acetates oxidize cyclohexene to the allylic acetate exclusively[36]. Comparisons should be treated with caution as solvent and temperature have an important role in deciding what products form. However, it is probable that carbonium ions resulting from metal–carbon bond cleavage have a more prominent part in the reactions of Tl(III) salts and Pb(IV) acetate with olefins than do Hg(II) salts. The mechanisms of allylic oxidation and acetoxylation have received little direct attention.

Oxythallation adducts have been isolated from styrene, *o*-allylphenol[71], norbornene and norbornadiene[38]. For the latter two compounds, confirmation of expected *cis,exo*-oxythallation was obtained by an application of the nuclear Overhauser effect in their double resonance n.m.r. spectra; the single resonance n.m.r. spectra are complicated by large proton–thallium coupling constants[72]. On treatment with sodium borohydride in ether–methanol both the norbornene and norbornadiene oxythallation adducts (**17** and **19** respectively) undergo reductive deoxythallation to regenerate the parent olefin exclusively. On the other hand, the norbornene adduct (**17**) is reduced by sodium amalgam in aqueous suspension to a high yield of *exo*-norborneol (**18**). Similar treatment of the norbornadiene adduct **19** and reacetylation of the product leads to a high yield (ca 95%) of the acetates of *exo*-5-norborneol (**20**) (ca 85%) and nortricyclanol (**21**) (ca 15%)[38]:

Tl(OAc)$_3$ / NaBH$_4$; OAc, Tl(OAc)$_2$ **(17)** ; Na—Hg ; OH **(18)** (24)

Tl(OAc)$_3$ / NaBH$_4$; OAc, Tl(OAc)$_2$ **(19)** ; Na—Hg ; OH **(20)** (25)

+ OH **(21)**

The oxythallation adducts of norbornene and norbornadiene undergo rearrangements analogous to that of the oxymercuration adduct of norbornadiene noted earlier. However, there appear to be differences in that a greater variety of products is obtained fairly rapidly on acetolysis at room temperature and identical products are also obtained by direct treatment of norbornene and norbornadiene with Tl(III) acetate in acetic acid[38]—some examples follow:

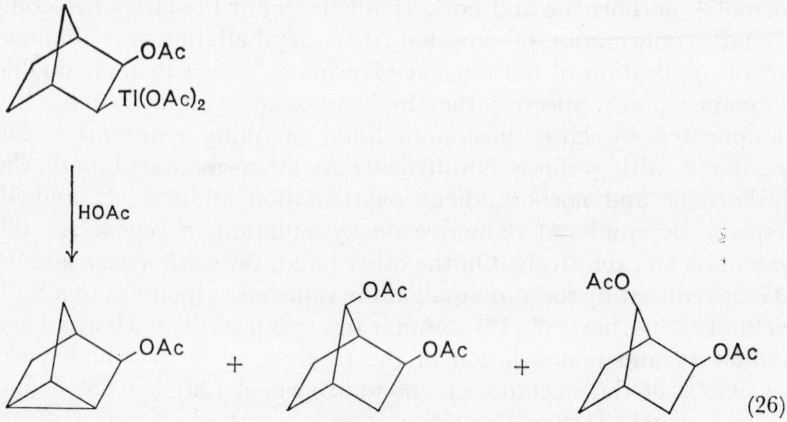

OAc, $Tl(OAc)_2$ —HOAc→ OAc + OAc, OAc + AcO, OAc (26)

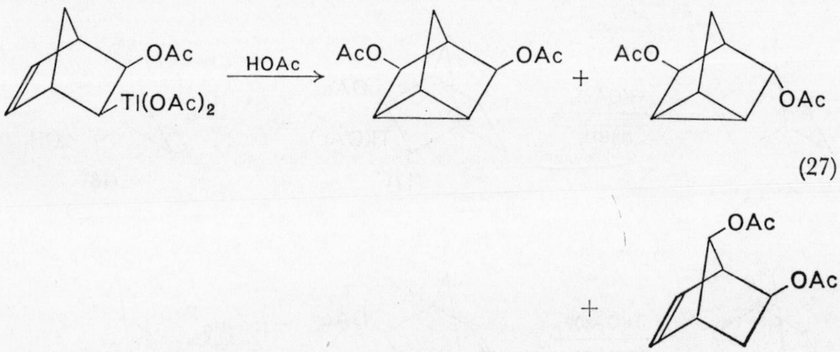

OAc, $Tl(OAc)_2$ —HOAc→ AcO, OAc + AcO, OAc (27)

+ OAc, OAc

The ease of Wagner–Meerwein rearrangement accompanying dethallation in acetic acid illustrates the importance of distinguishing whether the formation of products has been kinetically or thermodynamically controlled.

The effect of structure upon rate of reaction in the aqueous Tl(III) oxidation of simple olefins qualitatively parallels that observed in the acid-catalysed hydration of olefins[68, 73]. This observation has been offered as further evidence of carbonium ion character for the activated complex involved in oxythallation. The difference in degree of carbonium ion character of the reactive intermediates in oxythallation and oxymercuration has been ascribed to the effect of the higher charge of thallium reducing the size of the *d*-orbitals and hence the *d–p* overlap necessary for strong π-bonding. The decomposition of the mercury intermediate by nucleophiles is envisaged as essentially S_N2 (one might add also S_Ni) whereas that of the thallium intermediate is S_N1 [68, 73]. However, the unequivocal demonstration of the stereochemistry of oxythallation has yet to be reported.

Oxythallation combined with reductive dethallation under aqueous conditions does not yet appear to have been exploited in the manner of the mercury analogue for the synthesis of alcohols from olefins.

C. Palladium(II) Salts

The oxidation of olefins with Pd(II) chloride in aqueous solution is the basis of a most useful technical preparation of carbonyl compounds, particularly when Cu(II) chloride is used as a redox system to make the reaction catalytic in Pd(II) [74, 75]. 1- and 2-Olefins are oxidized to methyl ketones in high yield and in certain respects oxidations with Pd(II) resemble those of Tl(III) [68, 73, 76], in which saturated ketones and glycols are produced, more than those of Hg(II) [77, 78] in which, e.g., propene gives mainly acrolein. On the other hand, the small dependence of rate of oxidation to carbonyl compounds by Pd(II) upon olefin structure is unlike that of Hg(II) or Tl(III) oxidations and suggests little carbonium ion character in the transition state. This feature has been rationalized by a concerted, nonpolar, four-centre addition mechanism leading to an oxypalladation adduct [79].

Oxidation of olefins with Pd(II) salts in the presence of acetic acid produces vinyl and allyl acetates in a manner generally similar to Hg(II), Tl(III) and Pb(IV) acetate acetoxylations [80], [81]. The reaction of Pd(II) with olefins is considered to proceed via a preliminary PdX_2–olefin complex to an oxypalladation adduct [80–86]. Such adducts have not yet been characterized and have at the moment the status of reactive intermediates. Questions prompted by results gained in related fields as to the kinetics and stereochemistry of their formation are generally unanswered. Winstein and co-workers have studied the acetoxylation of acyclic 1- and 2-olefins (C_3—C_5) with Pd(II) acetate at 25° in acetic acid [80] and compared the results with those obtained with Hg(II) acetate [63] mentioned earlier. There is a striking contrast in that 1-olefins give mainly enol acetate while 2-olefins give mainly allylic acetate. In addition, the small amount of allylic acetate formed from a 1-olefin is mainly primary, whereas the allylic acetate from a 2-olefin is mainly secondary. The proportion of enol to allyl acetate is sensitive to solvent. Product distributions were measured early in the reaction to obtain values close to kinetic control proportions as it was observed during lengthy reaction periods that product isomerization occurred. The data were rationalized by a sequence in which an oxypalladation adduct with preferred Markownikoff orientation eliminates the elements HPdOAc with the preferred formation of an allylic acetate. The oxypalladation adduct of a 1-olefin with Markownikoff orientation can lead only to an enol acetate[80].

The detailed manner of the decomposition of the oxypalladation adducts to give products is not clear. One suggestion, based on isotopic labelling, is that the Pd–carbon bond heterolyses to a carbonium ion which rearranges and solvolyses, or loses proton[83]:

$$CH_2{=}CH_2 \longrightarrow PdX_2 \xrightarrow[-DX]{MeOD} CH_2OMe{-}CH_2PdX \xrightarrow{-PdX^-} H{-}CHOMe{-}CH_2^+ \xrightarrow{MeOD} CH_3CH(OMe)_2 \quad (28)$$

and

$$CH_2{=}CH_2 \longrightarrow PdX_2 \xrightarrow[-DX]{AcOD} CH_2OAc{-}CH_2PdX \xrightarrow{-PdX^-} H{-}CHOAc{-}CH_2^+ \quad (29)$$

$$\longrightarrow CH_3{}^+CHOAc \begin{cases} \xrightarrow{AcOD} CH_3CH(OAc)_2 + D^+ \\ \xrightarrow{-H^+} CH_2{=}CHOAc \end{cases}$$

The absence of deuterium in the saturated products precludes the possibility of their having been formed by addition of solvent molecule to a vinyl intermediate. An acceptable mechanism also has to preclude enolic intermediates which tautomerize to products. A variation of the mechanism is a Pd-assisted hydride shift followed by, or synchronously with, proton loss from an adjacent carbon atom. Differences between this and the carbonium ion mechanism outlined above are in the timing of the stages and the required *cis*-arrangements of the palladium group and the hydrogen atom involved in assisted hydride transfer. This mechanism, also incorporating the suggestion of an adduct with *cis*-orientation, has been used by Haszeldine and co-workers[86] to interpret the acetoxylation of cyclohexene with Pd(II) chloride in acetic acid (containing sodium acetate at 20°) which produced cyclohex-2-enyl acetate (76%) and cyclohex-3-enyl acetate (24%). These authors have also demonstrated that the π-allylic complex di-μ-chloro-bis(cyclohexenyl) dipalladium(II) is not a precursor in the acetoxylation of cyclohexene.

A new variation on the use of Pd(II) and Cu(II) combinations has been reported by Henry[81]. Whereas Cu(II) alone has no oxidative activity towards olefins, when present in high concentration with Pd(II), the reaction of Pd(II) takes a different course. Little vinyl acetate is formed and the product consists of saturated compounds, e.g., from ethylene, 2-chloroethyl acetate and ethylene glycol mono- and di-acetate; homologues of ethylene undergo similar reactions. *cis*- and *trans*-2-Butene give comparable quantities of both 2,3- and 1,3-isomers and this is ascribed to a prior rearrangement of the oxypalladation adduct. The results have been rationalized by a mechanism in which Cu(II) chloride successfully competes with elimination of HPdCl from the oxypalladation adduct by reacting directly with it, but the precise nature of this reaction has yet to be made clear. As with the mentioned results of Winstein and co-workers[80], the product distributions in this new reaction are best explained by preferential Markownikoff oxypalladation[81].

Norbornene (**22**) is oxidized in acetic acid solution by a mixture of Pd(II) and Cu(II) chlorides, with the latter in high concentration, to 50—84% yields of *exo*-2-chloro-*syn*-7-acetoxynorbornane (**25**). This was converted into the difficultly accessible *syn*-7-norbornenol (**26**). The formation of **25** was explained as due to heterolysis of the Pd–carbon bond in the oxypalladation adduct (**23**) to yield a norbornyl cation (**24**) which undergoes rearrangement and then nucleophilic attack as set out in equation (30)[87]. In view of the report by Henry[81] mentioned earlier it is interesting that here no special function was ascribed to Cu(II) chloride.

The quantity and range of results reported so far in these acetoxylation studies by Pd(II) do not permit a general mechanism to be established. The importance of reaction conditions such as temperature and solvent, and the structure of the olefin is such that inferences drawn from results gained under one set of conditions may not be exactly applicable to another.

D. Acetoxylation with Selenium Dioxide in Acetic Acid

Selenium dioxide is best known for its ability to oxidize ketones via enol selenite esters to α,β-diketones[3]. The reagent when used in acetic acid medium will also effect allylic acetoxylation. These reactions have been less fully studied than the ketone oxidations, and the mechanism is far from clear. It was formerly thought to be homolytic and involve hydrogen atom abstraction from the alkene in the manner of allylic bromination with *N*-bromo amides and autoxidation[88], but more recently heterolytic mechanisms have been suggested[66, 89]. A feature of these and related allylic acetoxylations by metal acetates are the relatively high temperatures and lengthy reaction periods required[66] in comparison with those for the formation of oxymetallation adducts with, e.g., Hg(II) and Tl(III) salts. The question as to whether the allylic acetoxylations proceed via these adducts is not yet settled.

The competitive allylic acetoxylation of nuclear substituted 1,3-diphenylpropenes (**27**) by SeO_2 in organic solvents of differing acidity has led Schaeffer and Horvath to suggest an electrophilic attack by SeO_2 or $HSeO_2^+$ on the carbon–carbon double bond as the initial stage of the reaction[89]. The similar oxidation of 3-deuterio-1,3-diphenylpropene gave an allylic acetate which retained about three-quarters of the original deuterium and this corresponds to a kinetic isotope effect of $k_H/k_D = 3{\cdot}1$ at 115°, thus suggesting carbon–hydrogen bond breakage during the rate-determining stage[89]. In addition, the distribution of deuterium in the product indicated that positions 1 and 3 become equivalent. The results were rationalized by the following mechanism which involves a selenite ester(**28**):

$$PhCH_2CH{=}CHPh + HSeO_2^+ \longrightarrow PhCH{=}CHCH(OSeOH)Ph + H^+ \quad (31)$$

(27) .... (28)

$$(28) \longrightarrow [PhCHCHCHPh]^+ + HSeO_2^- \quad (32)$$

(29)

$$(29) + HOAc \longrightarrow PhCH(OAc)CH{=}CHPh + H^+ \quad (33)$$

Stages (32) and (33) are described here as an S_N1 solvolysis. An alternative path in solvents of lower ionizing power or higher nucleophilicity, or where the internal structure of the selenite ester is less favourable to ionization, would be that of S_N2 or S_N2' solvolysis. Reactions proceeding by these latter routes could provide the basis of an explanation of the observed formation of optically active products[66, 89]. For example, (+)-carvomenthene (**30**) is oxidized in aqueous ethanol by SeO_2 to (+)-*p*-menth-6-en-2-one (**31**) with ca 50% retention of configuration suggesting a mainly S_N2' solvolysis by water[66].

Whereas some authors describe the reaction path in terms of the formation and decomposition of a selenite ester by analogy with ketone oxidation, Wiberg and Nielsen suggest a selenic acid as intermediate, by analogy with oxymetallation[66]. According to this scheme, the oxidation of (+)-carvomenthene (**30**) is formulated as the following S_N2' process:

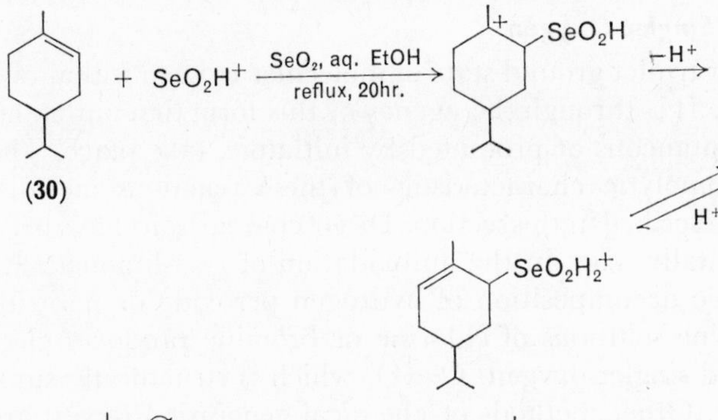

(30) + SeO_2H^+ —[SeO_2, aq. EtOH, reflux, 20hr.]→ (cation, SeO_2H) —[$-H^+$]→ (alkene, SeO_2H) (34)

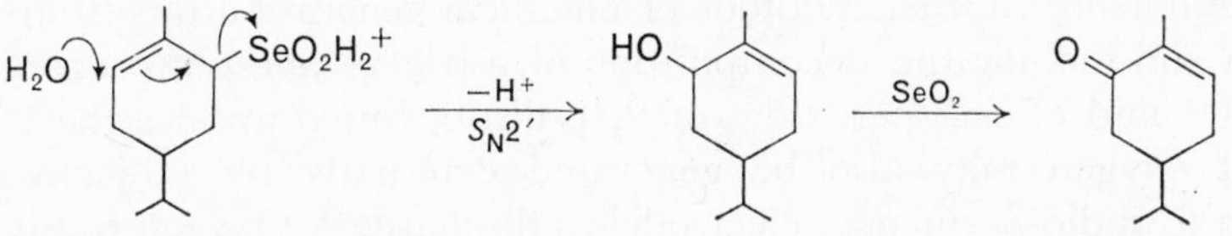

⇌ H^+ (alkene, $SeO_2H_2^+$)

H_2O + (alkene, $SeO_2H_2^+$) —[$-H^+$, S_N2']→ (HO, alkene) —[SeO_2]→ (31) (35)

The alternative S_N2 path leads to stereochemical inversion:

$$\text{(cyclohexenyl–}SeO_2H_2^+\text{, attacked by }OH_2) \xrightarrow[S_N2]{-H^+} \text{(cyclohexenyl–OH)} \quad \text{etc.} \tag{36}$$

The intermediacy of carbonium ions in these reactions involving cyclic olefins is not yet established and it may be significant that the bicyclic olefin α-pinene oxidizes without rearrangement or cleavage[66].

Unlike other olefins examined, cyclohexene reacts with SeO_2 in acetic acid to give mainly a selenium-containing adduct which yields 3-acetoxycyclohexene and selenium on pyrolysis. The reaction has been carried out using cyclohexene with ^{13}C-labelling of both carbon atoms of the double bond; analysis of the product showed that the acetoxy group was attached to one of the original olefinic carbon atoms to the extent of 90%[66]. This result requires either the S_N2 mechanism or an olefin-forming elimination of an oxyselenation adduct. It is confusing and perhaps indicative of the present state of this area to find that (+)-4-methylcyclohexene is reported to be oxidized to 5-methyl-2-cyclohexen-2-yl acetate with the same stereochemical features as (+)-carvomenthene and unlike cyclohexene[66].

III. THE INTRODUCTION OF OH GROUPS VIA CERTAIN REACTIONS WITH MOLECULAR OXYGEN

A. Oxidation via Singlet Oxygen

The normal or triplet ground state of molecular oxygen is that of a biradical ·O–O·. It is through the agency of this form that autoxidations, either spontaneous or promoted by initiators, take place. The products and homolytic characteristics of these reactions are dissimilar to those described in this section. Direct comparisons have been made experimentally, e.g., in the autoxidation of (+)-limonene[90].

The heterolytic decomposition of hydrogen peroxide or peracids with, e.g., alkaline solutions of chlorine or bromine produces electronically excited singlet oxygen, O═O, which is structurally similar to ethylene[91]. Other methods of chemical generation in situ are known and include the decomposition of a triaryl phosphite–ozone adduct[92] and of *endo*-peroxides of 9,10-disubstituted anthracenes[93]. Singlet oxygen may also be generated externally by subjecting oxygen to radio-frequency, electrodeless discharge[94]. Oxygen resulting from all these processes has one or other of the excited singlet states, $^1\Delta g$ or $^1\Sigma g^+$, whose energies are ca 22·5 and 37·5 kcal/mole respectively, above that of triplet ground state ($^3\Sigma g^-$) oxygen. Chemiluminescence occurs when the excited singlet states of oxygen emit radiation on returning to the ground state, this reversion may involve the bimolecular process:

$$2\ O_2(^1\Delta g) \longrightarrow 2\ O_2(^3\Sigma g^-) + h\nu$$

and other species [95–97]. In the case of the singlet oxygen formed by reaction between chlorine and hydrogen peroxide in alkaline solution, red light of wavelength 635 mμ is observed[96].

The singlet oxygen generated by these and other methods is of considerable theoretical and synthetic interest as an oxidant of olefins[98]. Alkenes yield allylic peroxides with double bond migration, and cyclic conjugated dienes yield *endo*-peroxides, from which alcohols may be obtained by a variety of reducing agents. A selection of illustrations from the literature is given in Table 1. The products of these oxidations are identical with those resulting from dye-photosensitized autoxidations, and high yields may be expected when high quantum yields in the photosensitized autoxidations are obtained. There is growing evidence that photosensitized autoxidation probably proceeds via singlet oxygen.

Two mechanisms have been suggested for the dye-photosensitized autoxidations of dienes and olefins which are consistent with the products, kinetics and energy considerations. These mechanisms differ in steps (39) and (40) of the following scheme, where superscripts refer to electronic spin states, Sens = sensitizer, and A is a substrate[91]:

$$\text{Sens} \xrightarrow{h\nu} {}^1\text{Sens} \qquad (37)$$

$$^1\text{Sens} \longrightarrow {}^3\text{Sens} \qquad (38)$$

$$^3\text{Sens} + {}^3O_2 \longrightarrow {}^{\cdot}\text{SensOO}^{\cdot} \qquad (39a)$$

$$^{\cdot}\text{SensOO}^{\cdot} + A \longrightarrow AO_2 + \text{Sens} \qquad (40a)$$

$$^3\text{Sens} + {}^3O_2 \longrightarrow \text{Sens} + {}^1O_2 \qquad (39b)$$

$$^1O_2 + A \longrightarrow AO_2 \qquad (40b)$$

Whereas path b was not previously favoured, the recent demonstrations of similarities in oxidations with independently generated singlet oxygen and dye-photosensitized autoxidation have now made path b more probable. Foote and co-workers[101] find that the product distributions are indistinguishable in the oxidation of three olefins both by the Rose Bengal-photosensitized reaction and with singlet oxygen from hydrogen peroxide–hypochlorite mixtures. Further evidence for the intermediacy of singlet oxygen was gained by Kopecky and

TABLE 1. Oxidations by singlet oxygen

Reaction	Method*	Reference
(cyclohexadiene) → (endo-peroxide, O–O)	a b	91, 98 92
(α-terpinene) → (ascaridole, O–O)	d	94
R, R (9,10-disubstituted anthracene) → R, R (endo-peroxide, O–O)	a d e	99 94 99
Me–(furan, O)–Me —MeOH→ Me, HO_2, O, Me, OMe	a	91, 98, 99
Ph, Ph, Ph, Ph (tetraphenylfuran, O) → Ph, Ph, PhCO, COPh	a c	91, 98 93
H_3C, H_3C C=C CH_3, CH_3 → H_2C=C(H_3C)–C(CH_3)(CH_3)(HO_2)	a b c	91, 98 92 93
(octalin) → HO_2 (octalin hydroperoxide)	a	98
CH_3, CH_3 —(ii) $LiAlH_4$→ CH_3, H_2C, OH (36%) + CH_3, H_3C, OH (18%) + CH_3, H_3C, OH (34%) (*cis* and *trans*)	a	100

* *Methods:* a, H_2O_2–NaOCl/NaOBr. b, Triphenyl phosphite–ozone adduct. c, 9,10-Diphenylanthracene *endo*-peroxide. d, Radio-frequency discharge on oxygen. e, Alkaline peracid.

Reich[102] who compared the reactivity sequences of a number of olefins towards established radical and electrophilic reagents with that for Methylene Blue-photosensitized oxidation. The results indicated that the reactive intermediate in the photosensitized oxidation is electrophilic, i.e., singlet oxygen. It was also shown, to the same end, that the relative rates of the photosensitized oxidation of pairs of olefins do not vary significantly with different sensitizers, indicating a common intermediate which could not include the sensitizer[102, 103]. Singlet oxygen is probably involved in, e.g., the dye-photosensitized oxidation of tetra-*O*-methylpurpurogallin[104], of caryophyllenes[105, 106], and of diacetylfilicinic acid[107].

The dye-sensitized photoxygenation of cholest-4-en-3β-ol (**32**) is particularly interesting as two products, an epoxy ketone (**33**) and an enone (**34**) are formed in proportions varying from 30 : 1 to 1 : 5 depending upon the sensitizer present:

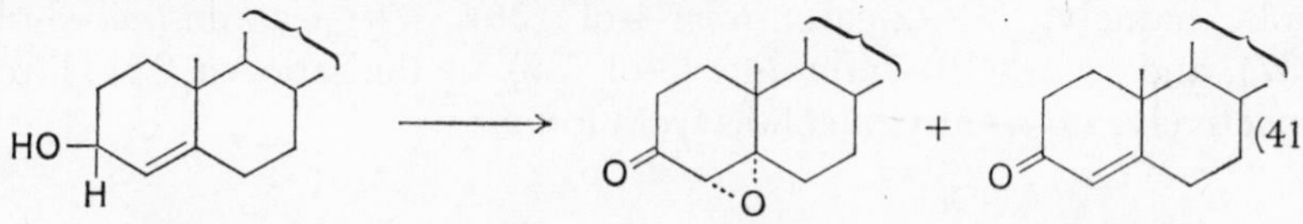

(41)

(32) **(33)** **(34)**

A correlation is reported between these product ratios and the energies of quenching the excited triplet state of the sensitizers by molecular oxygen[108, 109]. The product ratio **34**:**33** increases with increasing energy of the triplet state sensitizer, and this has been interpreted as due to the greater proportion of $^1\Sigma g^+$ oxygen formed in comparison with $^1\Delta g$ oxygen. The species $^1\Sigma g^+$ oxygen in effecting **32** $\rightarrow$ **34** has, therefore, a different chemical activity from $^1\Delta g$ oxygen which effects **32** $\rightarrow$ **33** [108, 109]. This suggestion may find an explanation in the observation that $^1\Sigma g^+$ has two antiparallel electrons in separate orbitals and possibly resembles the triplet ground state ($^3\Sigma g^-$) in its chemical properties[98]. The reaction **32** $\rightarrow$ **34** could proceed via homolytic removal of allylic hydrogen.

The mechanism of oxygenation of monoolefins by singlet oxygen from any source is not yet agreed. Sharp has suggested that an intermediate per-epoxide is formed which rearranges with a synchronized proton shift to yield an allylic peroxide[110]:

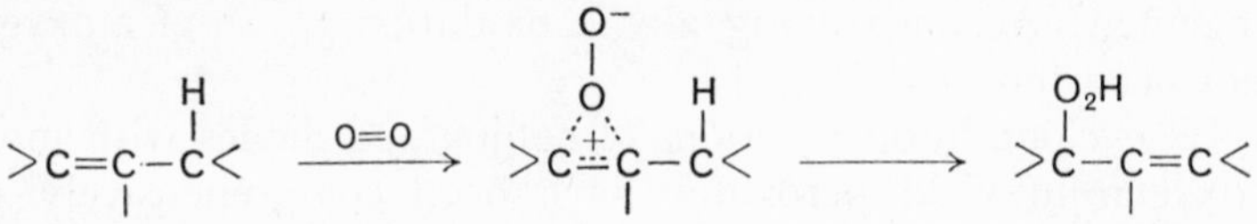

$$>C=C(-)-C(H)< \xrightarrow{O=O} >C\overset{+}{\underset{O-O^-}{\doteq}}C(-)-C(H)< \longrightarrow >C(O_2H)-C(-)=C< \quad (42)$$

whereas Nickon and Bagli have suggested a cyclic process without commitment to timing of bond breaking and making[111]:

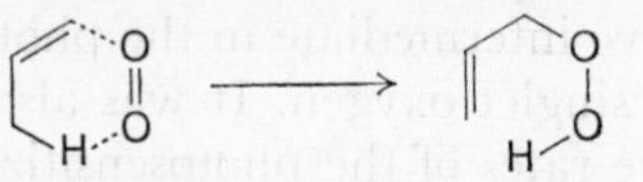

(43)

The formation of epoxy ketones from cyclohexenoid systems is stereospecifically *cis* in that the C–H bond cleaved and the C–O bond formed have a *cis* relationship (e.g. **32** → **33**). Hydroperoxides may be intermediates in the formation of these epoxy ketones[112]. Investigations with cyclic and semicyclic olefins have shown that the dye-photosensitized oxygenations are highly sterically controlled[113–116]. For example, the allylic peroxidation of (+)-3-carene (**35**) followed by reduction led to three optically active alcohols, namely, (−)-2-caren-*trans*-4-ol (**36**), (+)-4-caren-*trans*-3-ol, (**37**), and (−)-4(10)-caren-*trans*-3-ol (**38**) in the ratio ca 2 : 1 : 1 respectively; no *cis* allyl alcohols were found:

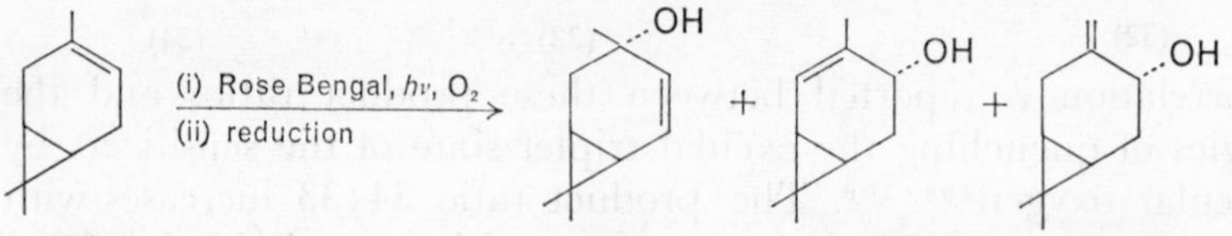

(i) Rose Bengal, $h\nu$, O_2
(ii) reduction

(35) (36) (37) (38) (44)

As had been previously observed with α-pinene, the oxygen molecule attacked that side of the olefin which was free from the steric screening effect of the isopropylidene group[113]. There is a marked dependence of rate of Methylene Blue-photosensitized oxygenation on the structure of the olefin, e.g., 2,3-dimethyl-2-butene is oxidized 5500 times faster than cyclohexene[102], suggesting considerable carbonium ion character in the transition state. This latter point is in greater accord with Sharp's proposal than a concerted form of Nickon and Bagli's. Stepwise making and breaking of bonds in the cyclic mechanism would differ essentially from Sharp's mechanism only in the molecularity of the proton transfer. Wagner–Meerwein rearrangements accompanying allylic oxidation by singlet oxygen have not been reported.

The 1,4-cycloaddition reactions of conjugated dienes with molecular oxygen to yield peroxides have been comprehensively reviewed[117]. These reactions, involving oxygen as dienophile, generally have to be performed photochemically. The dye-photosensitized peroxidation may now be interpreted in terms of singlet oxygen species thus becoming mechanistically equivalent to the thermal formation of Diels–Alder adducts in carbocycloaddition. Like the Diels–Alder reactions, the dye-photosensitized formation of *endo*-peroxides is reversible and mention has already been made of the use of 9,10-diphenylanthracene *endo*-peroxide to effect oxidation through the agency of singlet oxygen[93]. The reversion of *endo*-peroxide to hydrocarbon and oxygen is frequently accompanied by luminescence whose spectrum is often similar to that of the fluorescence of the hydrocarbon. It is possible that the origin of this radiation is the return of singlet oxygen to the triplet state[96].

The use of chemically generated singlet oxygen in organic synthesis is at present restricted because of the wastage of reagents due to unconsumed singlet oxygen, particularly with the less reactive oxygen acceptors. Likewise with dye-photosensitized oxidations, long reaction periods are often required. In connexion with the latter observation, Forbes and Griffiths report that the solvent system carbon disulphide, methanol (or ethanol) and ether (ca 14:1:1·5 v/v) produces comparable or better yields in a shorter time than the single solvents methanol or isopropanol, or a mixture of methylene chloride and benzene[118]. The role they ascribe to carbon disulphide is that of an efficient promoter of singlet-triplet intersystem crossing thus leading to a higher concentration of triplet sensitizer.

B. The Autoxidation of Carbanions

The enol forms of ketones react readily with oxygen in many solvents to give α-hydroperoxy ketones which may frequently be isolated. Keto forms are much less active and the conditions required to oxygenate these are usually sufficiently drastic to cleave the carbon skeleton of the products[119]. Although α-hydroperoxy ketones may be decomposed by base, the autoxidation of $\alpha\beta$-unsaturated ketones of the cyperone series to their γ-hydroxy derivatives and certain rotenone derivatives to the α-ketols in the presence of aqueous base have been accomplished[119]. Barton and co-workers[119, 120] have found that even in the presence of a strong base, namely reaction with oxygen in a *t*-butanol solution containing potassium *t*-butoxide, many steroidal 20-ketones (**39** and **41**) without a substituent at 17 or 21 are converted into reasonable yields of the 17α-hydroperoxides

(**40** and **42**) which may be reduced in good yield to 17α-alcohols. Some examples follow:

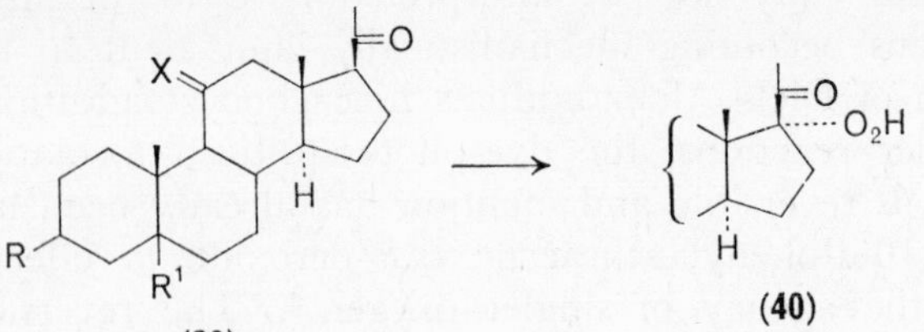

(45)

(39) (40)

(46)

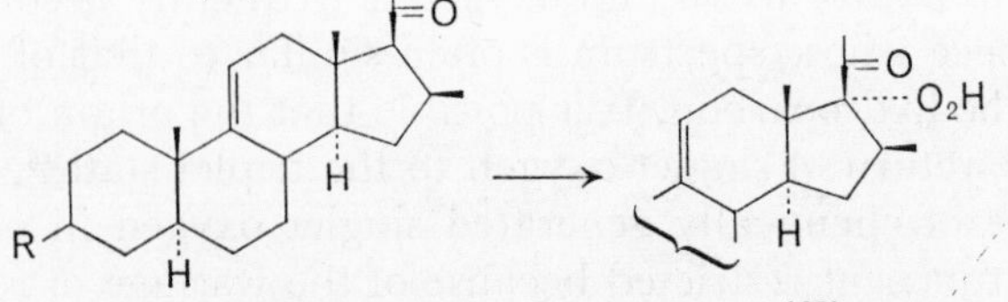

(41) (42)

R = α/β OAc, OH; R′ = α/β H; X = H_2, O.

This method has been used to convert canthaxanthin (**43**) into the corresponding bisdiosphenol, astacene (**44**) [121], and in the elucidation of the structure of limonin[122]. Both examples involve the conversion of a cyclic ketone into an enolized α-diketone probably via the hydroperoxide.

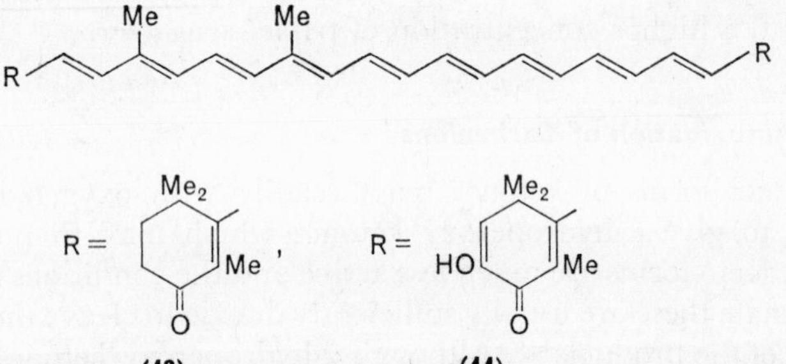

(43) (44)

Alkoxides are stronger bases in aprotic solvents, and autoxidations in the presence of potassium *t*-butoxide in an ethylene glycol–dimethyl ether solution of structurally simpler ketones and esters rapidly give high yields of α-hydroperoxides at low temperatures without significant degradation of the products[123]. Similar results with esters of diaryl- and (arylalkyl)aryl-acetic acids have been achieved with benzyltrimethylammonium hydroxide in pyridine solution[124].

The function of the base in all these autoxidations has a ready explanation, since it is the carbanion derived from the substrate which is the active entity and attacked by molecular oxygen.

Certain solvents, e.g., dimethyl sulphoxide (DMSO), markedly increase the basicity of potassium *t*-butoxide[125, 126], and it has been possible to autoxidize normally unreactive hydrocarbons via their carbanions in these systems. DMSO has a pK_a of ca 31 [127, 128], and it is itself susceptible to autoxidation to form dimethyl sulphone or methanesulphonic acid[129], or it may appear in the product via a carbanion reaction, consequently the solvent is usually diluted with *t*-butanol. This mixed solvent does not effectively ionize the weaker hydrocarbon acids and has led to the introduction of the more effective and less oxidizable diphenyl sulphoxide[130] and hexamethylphosphoramide (HMPA)[131]. The latter solvent is particularly unreactive towards base, as shown by the absence of detectable base-catalysed proton exchange, and towards oxygen[131]. It has been possible to autoxidize the weakly acidic toluene and related compounds through their carbanions in this solvent to the corresponding carboxylic acids with moderate yields[131].

The synthetic utility of the autoxidation of hydrocarbon anions in these systems to hydroperoxides or alcohols has not yet been exploited even to the limited extent of that of carbonyl compounds. There are several reports of hydrocarbon autoxidations, e.g., of xanthene[129], diphenylmethane[130], alkylated benzenes, *p*-cymene, tetralin[131], fluorene[129, 132], picolines[133], 9,10-dihydroanthracene[129, 134] and 1- and 3-arylpropenes[134], but in these cases the products isolated are mainly carboxylic acids or ketones although the reactions presumably proceed via the hydroperoxides. Limitations to applications in the synthesis of alcohols include acidity of the hydrocarbon, further oxidation of initial carbinol products, and the known facile base-catalysed dehydration of primary and secondary hydroperoxides to carbonyl compounds[135]. Hydrocarbon anion autoxidations in which hydroperoxides and alcohols are among the products are few and include those of 9-alkylfluorenes (**45**), 2,3-diphenylindene (**46**) [132], triphenylmethane (**47**), diphenylmethane (**48**) [129, 136, 137] and fulvenes (**49** and **50**) [138].

The mechanisms of carbanion autoxidations have received considerable attention particularly by G. A. Russell and co-workers[136]. Kinetic measurements show that the formation of the carbanion is the rate-determining stage in the oxidation of certain triarylmethane types[139]. These oxidations are independent of oxygen pressure above

$$\text{(45)} \xrightarrow[\text{C}_5\text{H}_5\text{N, 0}^\circ]{\text{O}_2\text{, PhCH}_2\text{N}^+\text{Me}_3\text{OH}^-} \text{R, O}_2\text{H} + \text{R, OH} \quad (47)$$

R H (45)

R = Me, Et, Bu, Ph, $PhCH_2$, $CH_3C_6H_4CH_2$, $CH_3OC_6H_4CH_2$

$$\text{(46)} \xrightarrow[\text{C}_5\text{H}_5\text{N, }-40^\circ]{\text{O}_2\text{, PhCH}_2\text{N}^+\text{Me}_3\text{OH}^-} \text{(Ph, Ph, O)} + \text{(Ph, O}_2\text{H, Ph)} \quad (48)$$

$$\underset{(47)}{\text{Ph}_3\text{CH}} \xrightarrow[\text{DMSO, }t\text{-BuOH, 25}^\circ]{\text{O}_2\text{, }t\text{-BuOK}} \text{Ph}_3\text{COH} \ (+\ \text{DMSO}_2) \quad (49)$$

$$\underset{(48)}{\text{Ph}_2\text{CH}_2} \xrightarrow[\text{DMSO, }t\text{-BuOH, 25}^\circ]{\text{O}_2\text{, }t\text{-BuOK}} \text{Ph}_2\text{CHOH} + \text{other products} \quad (50)$$

$$\underset{(49)}{\text{(Ph, Ph, Ph, Ph, CH}_2\text{)}} \xrightarrow[\text{HMPA}]{\text{O}_2\text{, NaOMe}} \text{(OH, Ph, Ph, Ph, Ph, CH}_2\text{OCH}_3\text{)} + \text{(Ph, Ph, Ph, Ph, O, CH}_2\text{)} \quad (51)$$

$$\underset{(50)}{\text{(H, Ph)}} \xrightarrow[\text{HMPA}]{\text{O}_2\text{, NaOMe}} \text{(OH, PhCHOCH}_3\text{)} + \text{(O, PhCH)} \quad (52)$$

a minimum value and have large kinetic isotope effects. In addition, agents capable of electron transfer such as aromatic nitro compounds which should catalyse radical chain oxidations are without effect[136]. On the other hand, the triarylmethane types which are stronger carbon acids[127, 128] e.g., tri-(*p*-nitrophenyl)methane or 9-phenylfluorene, show different behaviour. In these cases the rates of oxidation are much slower than ionization and the rate-determining stage is the reaction between carbanion and oxygen, although detailed kinetic analysis shows that the basicity of the reaction medium may also be a limiting feature[140]. Catalysis by aromatic nitro compounds has been observed. The comparative inactivity of these compounds is analogous to the inactivity of β-diketones, e.g., cyclohexane-1,3-dione, which form delocalized, stable carbanions which are not autoxidized[136]. Two mechanisms are currently suggested as reaction paths for carbanion autoxidation:

$$Ar_3CH + (CH_3)_3CO^- \rightleftharpoons Ar_3C^- + (CH_3)_3COH \quad (53)$$

$$Ar_3C^- + O_2 \longrightarrow Ar_3COO^- \quad (54a)$$

$$Ar_3C^- + O_2 \longrightarrow Ar_3C^{\cdot} + O_2^{\cdot -} \quad (54b)$$

$$Ar_3C^{\cdot} + O_2 \longrightarrow Ar_3COO^{\cdot} \quad (55b)$$

$$Ar_3COO^{\cdot} + Ar_3C^- \longrightarrow Ar_3COO^- + Ar_3C^{\cdot} \quad (56b)$$

Aromatic nitro compounds would catalyse a radical chain process by reactions such as[136]:

$$R^- + ArNO_2 \longrightarrow R^{\cdot} + ArNO_2^{\cdot -} \quad (57)$$

$$ROO^{\cdot} + ArNO_2^{\cdot -} \longrightarrow ROO^- + ArNO_2 \quad (58)$$

Path b was formerly preferred for many autoxidations and appears definitely to operate in the oxidation of 2-nitropropane. Many autoxidations do not show features characteristic of radical chain reactions and so path a may operate, e.g., in the oxidation of triphenylmethane and diphenylmethane, in spite of the objection to stage (54a) that it violates the spin conservation rule[139, 141]. The mechanism of autoxidation of carbanions derived from aliphatic ketones and esters is not certain[136, 141]; the oxidation of acetophenone is catalysed by aromatic nitro compounds[136]. The stereochemistry of the hydroperoxide products of suitably chosen hydrocarbons or carbonyl compounds does not appear to have been directly investigated. The autoxidation of steroidal 20-ketones[119, 120] mentioned earlier gives mainly the 17α-hydroperoxides indicating strong steric control. Further investigations of stereochemical aspects might yield information concerning mechanisms.

Volger and Brackman[142] have studied the copper(II)–pyridine complex catalysed autoxidation of α,β- and β,γ-unsaturated aldehydes and ketones which are oxidized specifically at the γ-carbon atom to yield subsequently dicarbonyl compounds, via base-catalysed decomposition of a hydroperoxide. The role of the copper(II) complex is to oxidize the dienolate anion to a dienoxy radical susceptible to oxygenation, and the peroxy radical resulting from this is reduced to hydroperoxy anion by copper(I) complex formed in a previous step. Adaptation of this type of system to other carbanion autoxidations may lead to useful methods in organic synthesis.

C. The Autoxidation of Metal Alkyls and Aryls

Continuing industrial interest in the autoxidation of aluminium and boron alkyls for the production of aliphatic alcohols is considerable if the incidence of patent registrations is any guide[143].

The processes are particularly suitable for preparing higher alcohols from the long chain alkyls of these metals which are themselves readily synthesized from the addition of lower aluminium alkyls or diborane to olefins. The alkyls are autoxidized in the liquid phase (this term will be used synonymously for reactions in solution) and the resulting alkoxides hydrolysed to the corresponding alcohols. Studies of these autoxidations have revealed much information as to mechanisms but the picture is not yet completely clear. Earlier work on autoxidation has been reviewed[144] and there are several reviews on the properties of organometal and metalloid peroxides[145, 146]. The gas and liquid phase autoxidations of boron and aluminium alkyls in particular probably exemplify the range of product-types and mechanisms encountered in autoxidations of metal alkyls generally. At present the mechanisms of the autoxidations are considered to differ principally according to the physical phase of the reaction mixture[147]. However, the dichotomy may be more apparent than real and perhaps only a reflexion of the comparative lack of investigation this area has received. It is more probable that a number of factors apart from phase such as temperature, nature of metal and organic group, together determine reaction paths. It is still convenient to subdivide the material according to reaction phase for the purposes of presentation.

The gas phase autoxidations of the alkyls of boron, zinc[147], aluminium[147–149] and indium [150] in the first instance are essentially similar in that they are free radical chain reactions involving peroxidic compounds or peroxy radicals. Peroxides have been isolated in the cases of boron (Me_2BO_2Me) and indium (Et_2InO_2Et), and are suspected in the case of zinc[151] because of observed catalysis. A peroxide was not detected in the slow gas phase autoxidation of trimethylaluminium[147]. The methyl derivatives are much the least active in all cases examined and for zinc alkyls the sequence of reactivity is: *n*–Pr > Et > Me[151, 152]. The methyl compounds also consume less oxygen even when this is present in excess and the stoichiometry approximates to, e.g., with zinc:

$$ZnMe_2 + \tfrac{1}{2}O_2 \longrightarrow ZnMe_2O \tag{59}$$

in comparison with homologues, e.g.,

$$ZnEt_2 + O_2 \longrightarrow ZnEt_2O_2 \tag{60}$$[151]

The eventual solid products in all cases are the metal alkoxides. The autoxidation of triethylindium produces considerable amounts of the hydrocarbons ethane and ethylene and the oxygenated products ethyl alcohol, diethyl ether and acetaldehyde[150]. In this latter respect it differs from other gas phase autoxidations reported in which only hydrogen and/or hydrocarbons are found, e.g., trimethylaluminium yields hydrogen (trace) and methane[147], dimethylzinc yields methane and ethane, and diethylzinc yields ethane and butane[151].

The complex kinetic features of gas phase metal alkyl autoxidation strongly suggest that the peroxy intermediates are formed by a free radical chain mechanism, and the dependence of the reaction rate on the nature and extent of the vessel surface also indicates the importance of heterogeneous steps[147, 150–152]. In addition, there is further support for homolytic mechanisms from two other sources. First, the autoxidation of metal alkyls induces the oxidation of hydrocarbons at low temperatures, and secondly, metal alkyl–oxygen mixtures induce the polymerization of olefins. In both cases, the reactions effected are considered to be initiated by radicals arising in the autoxidation of the metal alkyl[149, 153–155]. The polymerization reactions were carried out in the liquid phase. The mechanisms proposed to account for autoxidations differ in detail depending on the particular metal and organic group present. For example, the following scheme which has been suggested for the initiation and propagation stages of the slow autoxidation of trimethylaluminium involves a peroxy radical rather than a peroxidic compound[147]:

initiation: $$Al_2Me_6 + O_2 \longrightarrow Me_2AlOO^{\cdot} + Me_2Al^{\cdot} + 2Me^{\cdot} \qquad (61)$$

propagation: $$Me_2Al^{\cdot} + O_2 \longrightarrow Me_2AlOO^{\cdot} \qquad (62)$$

$$Me_2AlOO^{\cdot} + Al_2Me_6 \longrightarrow 2Me_2AlOMe + Me_2Al^{\cdot} \qquad (63)$$

The production of methyl radicals in the initiation stage is able to account for the quantity (ca 18%) of methane accompanying autoxidation. By contrast, the autoxidation of the less electron deficient triethylindium and possibly boron alkyls may proceed according to the scheme[150]:

initiation: $$Et_3In + O_2 \longrightarrow Et_2In^{\cdot} + EtOO^{\cdot} \qquad (64)$$

propagation: $$Et_2In^{\cdot} + O_2 \longrightarrow Et_2InOO^{\cdot} \qquad (65)$$

$$Et_2InOO^{\cdot} + Et_3In \longrightarrow Et_2InOOEt + Et_2In^{\cdot} \qquad (66)$$

An important feature here is the proposed formation of the intermediate peroxide, $Et_2InOOEt$. This may yield the predominant product of alkoxide by a nonsurface-catalysed, nucleophilic migration of alkyl or alkoxyl group either intra- (equation 67) or intermolecularly (equation 68)[150]:

$$RM(R)-O(OR) \longrightarrow RM(OR)_2 \quad (67)$$

$$R_2M(R) + OMR_2(OR) \longrightarrow 2R_2MOR \quad (68)$$

Some of the additional oxygenated products must arise from the homolytic decomposition of peroxides[150, 156, 157], thus diethyl ether from the process:

$$Et_2InOOEt \longrightarrow Et_2InO^{\cdot} + EtO^{\cdot} \quad (69)$$

$$EtO^{\cdot} + Et_3In \longrightarrow Et_2O + Et_2In^{\cdot} \quad (70)$$

and ethanol from:

$$EtO^{\cdot} + Et_3In \longrightarrow EtOH + Et_2In\dot{C}HCH_3 \quad (71)$$

The origin of the hydrocarbon products occurring in the gas phase autoxidation of other metal alkyls can also be rationalized as a consequence of the homolytic decomposition of peroxide or homolytic steps preceding this.

The chief apparent difference between the mechanisms of autoxidation of metal alkyls in the gas and liquid phases is in the universal formation of an intermediate peroxide in the case of the latter and in the manner of its formation. That this species is undoubtedly involved is attested by the numbers of different examples isolated and characterized (e.g., References 144, 157–160). The peroxides of organoboron compounds have been used in the synthesis of primary and secondary alkyl hydroperoxides. The procedure involves the autoxidation of trialkylboranes to the diperoxyboronates which are treated with a peracid to yield the alkyl hydroperoxide[161]. The formation of the peroxide is generally considered to involve coordination of the oxygen molecule with the metal followed or accompanied by a nucleophilic 1,3-migration of alkyl to electron-deficient oxygen, e.g., with boron[157]:

$$>B(R) + O_2 \longrightarrow >B^{-}(R)-O-O^{+} \longrightarrow >BO_2R \quad (72)$$

Recently, evidence has been reported in support of the reversible formation of a coordination-polymeric species of oxygen with boron trimethyl at 77 °K [162]. In common with the heterolytic autoxidation of carbanions is the requirement for oxygen to change its spin state and preliminary coordination of oxygen to metal may achieve this. The effect of the nature of the organic radical upon the reactivity of the organometal compound towards autoxidation has been investigated most completely in the case of boron where the following sequence of reactivity is observed: alkyl > aryl, vinyl; tertiary alkyl > secondary alkyl > primary alkyl > methyl[157]. This is not the order known for nucleophilic 1,2-migrations to electron-deficient carbon termini but resembles that for the Baeyer–Villiger oxidation[13] and the oxidative dealkylation of trialkylboranes with hydroperoxides[157]. The latter reaction is part of the useful synthetic route which effects anti-Markownikoff hydration of olefins via hydroboration. These oxidations have features in common with the mechanism (equation 72) suggested for the formation of a peroxide in the liquid phase. Both have nucleophilic addition as the first stage followed by a nucleophilic migration of alkyl to electron-deficient oxygen, e.g., the oxidation of a tributylborane[163]:

$$i\text{-}Bu_2BBu\text{-}t + HO_2H \longrightarrow \left[\begin{array}{c} t\text{-}Bu \\ | \\ i\text{-}Bu_2\bar{B}\text{—}O\text{—}OH \end{array} \right] \longrightarrow i\text{-}Bu_2BOBu\text{-}t + OH^- \quad (73)$$

At temperatures below about 80° the unimolecular homolysis of peroxides to alkoxy radicals is slow and under these conditions there is general agreement that alkoxides result from the intra- or intermolecular, nucleophilic rearrangements of alkyl and alkoxy groups in the manner of equations (67) and (68). The molecularity of this stage has been the object of some investigation and, e.g., in the case of trialkylboranes[164], is claimed to be intermolecular.

The overall picture, therefore, of the mechanism of autoxidation of metal alkyls in the liquid phase is that essentially heterolytic processes yield alkoxides. However there is evidence to suggest that this view is not representative of all liquid phase autoxidations under all conditions and is merely an extreme type or one of several reaction types occurring concurrently with homolytic routes. It has already been mentioned that metal alkyl–oxygen mixtures induce the liquid phase polymerization of olefins. The best known case is that of the polymerization of vinyl monomers with trialkylborane–oxygen mixtures[153–155, 165]. The presence of both oxygen and trialkylborane is essential and the derived peroxide initiates the polymerization at low temperatures only if the trialkylborane is also present. It is clear that the peroxide does not initiate polymerization via unimolecular homolysis to alkoxy radicals. The origin of the free radicals in these systems is not yet known but the following suggestions have been made:

(i) decomposition of peroxide by excess of boron alkyl[155]:

$$R_3B + R_2BOOR \longrightarrow R_3B + R_2BO^{\cdot} + RO^{\cdot} \quad (74)$$

(ii) decomposition of peroxide by a complex of monomer (M) and boron alkyl[165]:

$$R_3BM + R_2BOOR \longrightarrow R_3BMRO^{\cdot} + R_2BO^{\cdot} \quad (75)$$

(iii) a cage reaction following coordination of peroxidic oxygen to boron alkyl[166]:

$$\rangle BR + EtOOB\langle \longrightarrow \rangle BOEt + R^{\cdot} + \rangle BO^{\cdot} \quad (76)$$

It is also a possibility that radicals may arise from stages preceding and otherwise leading to the formation of the peroxide.

Other evidence for homolytic reactions comes from careful analysis of all reaction products formed in metal alkyl autoxidations. Tri-*n*-butylborane on autoxidation gives considerable quantities of the hydrocarbons butene, butane and *n*-octane, and lesser quantities of oxygenated products, e.g., *n*-butyraldehyde, di-*n*-butyl ether, and 4-octanone. These were all considered to be the consequence of radicals resulting from the homolytic decomposition of peroxide[156]:

$$\rangle BO_2R \longrightarrow \rangle BO_2^{\cdot} + R^{\cdot} \text{ and/or } \rangle BO^{\cdot} + RO^{\cdot} \quad (77)$$

The autoxidation of triethylaluminium in *n*-heptane gives as by-products hydrogen, ethylene and ethane[167, 168] and these are comparable with the by-products obtained from the gas phase autoxidation of trimethylaluminium[147]. The autoxidation of Grignard reagents appears to proceed along the lines of those of metal alkyls and it is of interest here that products formed from aromatic[160] and aliphatic[169] examples may be accounted for in terms of radical processes. An additional point of interest is that the autoxidation of aromatic Grignard reagents and of alkali metal aryls is accompanied by chemiluminescence but whether this originates in a mechanism of peroxide decomposition or electron transfer is not known[96, 170]. The latter compounds are autoxidized to several products including phenols and biphenyl but the yields are generally small[144, 171].

Other evidence suggesting the generation of free radicals derived from the organometal compound during liquid phase autoxidation comes from reactions carried out in solvents which are active towards radicals. Triphenylaluminium on autoxidation (80°) in ^{14}C-labelled benzene yields after hydrolysis of the reaction product mainly biphenyl and phenol both of which are partially ^{14}C-labelled. Phenyl radicals do not exchange between benzene and triphenylaluminium in the absence of oxygen at 80°, and the results may be interpreted in terms of radical processes[172]. Similar results were obtained from the autoxidation (room temperature) of diphenylzinc and diphenylmagnesium except that the phenol was formed entirely from the organometal compound. The use of D- or ^{14}C-labelled benzene showed that the solvent partly participated in the formation of biphenyl and part of the phenyl groups of the organometal compound were converted into benzene[173].

The autoxidation of cadmium dialkyls in a variety of solvents proceeds readily to the peroxides which may be isolated[174]. The autoxidation of diphenylcadmium possibly follows a similar route but, in solvents capable of chain transfer, products are formed which are best regarded as resulting from attack of phenyl radicals on those solvents. For example, when diphenylcadmium is autoxidized in carbon tetrachloride (room temperature) the hydrolysed products contain chlorobenzene, phenol, biphenyl, benzotrichloride and hexachloroethane[175]. The results of autoxidation carried out in chloroform, deuterochloroform and ^{14}C-labelled benzene substantiate the occurrence of homolytic processes. The oxidation of several mercury alkyls in organic solvents (50—60°) gives reaction products which contain mercury, alkoxyalkylmercury, alkylmercury hydroxide, products of the oxidation of the alkyl group of the original organomercury compound (the corresponding aldehydes and ketones), unsaturated and saturated hydrocarbons formed from the alkyl group of the mercury compound, and products of the oxidation of the solvent. When halohydrocarbons were used as solvent, products of the interaction of the solvent with the mercury compound were also formed. It was also observed that the ease of oxidation of the mercury alkyls increased with increasing nucleophilicity of the organic group[176]. The overall results were interpreted in terms of intermediate peroxide and homolytic processes; these will be mentioned later.

The majority of reported investigations of the autoxidation of metal alkyls and aryls have been directed towards establishing stoichiometries, the identities of intermediates and products, the influence of solvents and organic group, etc. Kinetic studies in the liquid phase are comparatively few in number and thus the reported investigation of the kinetics of the autoxidation of diisopropylmercury at 70° in *n*-nonane solution[177] is of interest. A variety of products are formed and these are mainly isopropylmercury isopropylate (greatest proportion), mercury, isopropylmercury hydroxide, acetone and isopropanol, and probably oxidation products of the solvent. The reaction is autocatalytic and displays the characteristics of a free radical chain process; e.g., it is almost completely suppressed by the addition of small quantities of the inhibitor *p*-hydroxydiphenylamine. The average chain length of the process is 160. The initiation of the oxidation in the early stages is probably due to reaction (78) and to a much lesser extent reaction (79):

$$R_2Hg + O_2 \longrightarrow [RHgOOR] \begin{cases} \xrightarrow{R_2Hg} 2\,RHgOR \\ \longrightarrow RHg^{\bullet} + RO_2^{\bullet} \end{cases} \quad (78)$$

$$R_2Hg \longrightarrow RHg^{\bullet} + R^{\bullet} \quad (79)$$

Later on, as the quantity of isopropylmercury isopropylate accumulates then this compound becomes on thermal decomposition the major source of radicals (equation 80) which can account for both the autocatalytic rate curve and the nature of most of the products:

$$RHgOR \longrightarrow RHg^{\bullet} + RO^{\bullet} \quad (80)$$

The mechanism of the autoxidation, therefore, involves a combination of concurrent and consecutive reactions of homolytic type with the formation and decomposition of isopropylmercury isopropylate playing a key role[177].

The picture which emerges with regard to the mechanisms operating in the thermal autoxidation of metal alkyls and aryls is unequivocal in only a few parts. Whereas the gas phase autoxidations generally on the one hand and the liquid phase oxidation of diisopropylmercury on the other appear to involve free radical chain reactions, evidence for homolytic processes in other liquid phase oxidations is absent, incomplete or circumstantial. For example, the series of investigations on the participation of the solvent during autoxidations of various metal aryls[168, 172, 173, 175] whilst strongly indicating homolytic processes is not able to identify the origin or mode of formation of the initiating species.

Of relevance to the subject of autoxidation of metal alkyls and aryls is the preliminary report of a homolytic mechanism in the autoxidation of optically active 1-phenylethaneboronic acid to the racemic peroxide, in benzene solution at room temperature[178]. The stereochemistry of this reaction is in contrast to the similar reaction with the oxidants hydrogen peroxide and trimethylamine oxide where retention of configuration is observed[179]. The rate of the reaction was affected only slightly by weak inhibitors but copper(II) *NN*-dibutyldithiocarbamate and galvinoxyl caused induction periods of about one and four half-lives respectively. The following propagation steps were suggested for a free radical chain process[178]:

$$R^{\bullet} + O_2 \longrightarrow RO_2^{\bullet}$$

$$RO_2^{\bullet} + {>}BR \longrightarrow \left[RO_2^{\bullet}\overset{|}{\underset{|}{B}}R \longleftrightarrow RO_2\overset{|}{\underset{|}{B}}{}^{\bullet}R \text{ etc.} \right] \longrightarrow RO_2B{<} + R^{\bullet} \quad (81)$$

Inhibition of autoxidation in other organoboron derivatives was also observed with the efficient radical scavenger, galvinoxyl[180].

It is perhaps to be expected that the homolytic mechanisms for the thermal autoxidation of metal alkyls resemble those suggested for photoinitiated autoxidation. There have been very few studies reported of the latter reactions but, e.g., the photoinitiated oxidation of tetraethyl-lead and tin involves the formation of a peroxide which decomposes by routes (82) and (83)[181]:

$$Et_3PbOOEt \xrightarrow{Et_4Pb} 2\,Et_3PbOEt \quad (82$$

$$Et_3PbOOEt \longrightarrow Et_3PbO^{\bullet} + EtO^{\bullet} \quad (83)$$

IV. INTRODUCTION: REDUCTION

Since certain of the processes and reagents which reduce oxygenated compounds to alcohols have been the subject of a previous volume in this series[182] and others have been reviewed elsewhere, these topics will be mentioned only briefly.

The photoreduction of ketones to alcohols involving inter- or intra-molecular transfer of hydrogen atom[183, 184] has not found much use in synthesis although the mechanistic aspects of the reactions have been well investigated. Many ketones are reduced to the corresponding pinacol in the presence of a hydrogen atom donor such as an alcohol, hydrocarbon[184] or amine[185, 186]. Even those ketones which have an excited state with the π,π^* configuration may reduce to pinacol in the presence of an efficient hydrogen atom source such as organometal hydrides. Photoreductions under basic conditions pursue a different course and in many cases lead to high yields of monomeric alcohol. These reductions appear to have been little investigated recently. The mechanisms and other aspects of photoreduction have been summarized elsewhere[184]; the characterization of an intermediate in the photoreduction of benzophenone is claimed[187].

The electroreduction of ketones was reviewed in an earlier volume[182] and comparatively few reports have appeared since. Diimide generated from potassium azodicarboxylate has been used to reduce aromatic aldehydes to the corresponding alcohols in good yields[188]. Chromium(II) acetate is able to reduce α,β-epoxyketones to β-hydroxyketones under controlled conditions[189].

V. REAGENTS AND PROCESSES WHICH REDUCE OXYGENATED COMPOUNDS TO ALCOHOLS

A. Metals

Metals effect the reduction or reductive fission of a wide variety of unsaturated and saturated, organic and organometallic compounds. The metals in common use for these purposes are lithium, sodium, potassium, calcium, magnesium, iron and tin. They are utilized in a variety of procedures including suspension in an inert solvent, or, particularly for the alkali metals and calcium, as a solution in liquid ammonia (the Birch reduction[182, 190–195]), or as a dissolving metal such as zinc amalgam in an aqueous solution of an acid (the Clemmensen reduction [182, 193, 196]) or sodium metal in an alcohol (the Bouveault–Blanc reduction). The first two methods require a proton source which may be incorporated into the reaction mixture or introduced during the recovery of products.

The Birch reduction is considered superior to the Bouveault–Blanc method and has found considerable synthetic use. Ketones, esters and epoxides are reduced to alcohols but the method has largely been superseded in these particular cases by reduction with metal hydrides. Aldehydes may undergo ammonolysis rather than reduction, aliphatic and alicyclic acetals and ketals are inert and this feature may be used to protect a particular carbonyl group from reduction[192]. Dialkyl ethers are also stable but aralkyl and diaryl ethers are cleaved to phenols[192].

The type of mechanism operating in the Birch reduction and associated reactions involving reduction by metals in various solvents, was first stated in the currently accepted form by Michaelis and Schubert[197]. The structure of the blue solutions which alkali and alkaline earth metals form when dissolved in ammonia and other solvents is not yet clear but it is possible that the active reducing species is the solvated electron[198, 199]. A nonenolizable ketone such as benzophenone (**51**) forms first a radical anion or ketyl (**52**). This species has several alternative reaction paths open to it, many of which are reversible[192], but the overall result for practical purposes is the eventual formation of the monomeric product benzhydrol (**53**) or the dimeric product benzopinacol (**54**):

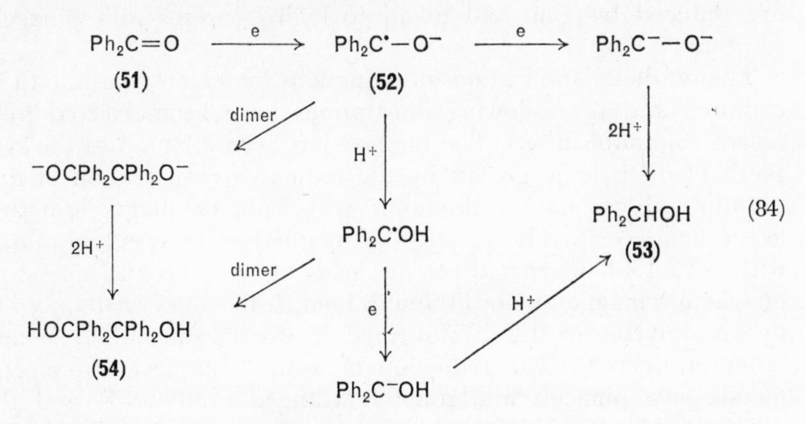

The choice of reaction conditions in the synthetic applications of the method is in part concerned with directing the reduction to one type of product or the other. The production of high concentrations of radical anions by using magnesium[200], zinc or aluminium in the absence of a proton source encourages dimerization as also do structural features in the ketyl which render it more stable by delocalization of the additional electron. A cosolvent is frequently used with reactions in liquid ammonia in order to dissolve more of the substrate. It now appears that the choice of cosolvent may partially control the yield of dimeric product[201]. The stereochemistry of the reduction of cyclanones is frequently but by no means exclusively such as to give the more stable equatorial alcohol[202]. The proportion of components in a product consisting of epimers depends both on the structure of the ketone and on the metal used for the reduction. α-Substituted ketones may suffer the loss of this substituent from the ketyl if it is a good leaving group and this is commonly observed with α-halo, α-acyloxy and α-hydroxy ketones. Enolizable ketones require the presence of a proton source, such as an alcohol, in the reaction mixture to carry the reaction forward which may otherwise remain at the stage of a stable enolate anion. Other details concerning the reduction of carbonyl compounds may be found in References 182 and 190—194.

There are few examples of the reduction of carbonyl compounds with metal–alkali combinations. It is noteworthy that a nickel–aluminium alloy added to an aqueous or aqueous-ethanolic solution of sodium hydroxide reduces aliphatic and aromatic aldehydes and ketones to carbinols[203], and that aromatic aldehydes and ketones are reduced by zinc and alkali to hydrobenzoins and pinacols, respectively[204].

The synthetic applications of reductions by metals continue to be explored and the following illustrations have been selected from recent communications. The method has been adapted to the synthesis of diethylenic glycols by the reductive dimerization of α,β-unsaturated carbonyl compounds with zinc or magnesium and acetic acid[205, 206]. The reaction of magnesium in tetrahydrofuran with γ- and δ-halogenated ketones leads to cyclanols[207]. Depending on whether magnesium or lithium is used, ε-diketones are reduced to linear ε-glycols or 1,2-disubstituted *cis*-cyclohexane-1,2-diols and other products[208]. The reduction of α-furyl ketones with several metals gave pinacols and some rearranged products[209], and the sodium reduction of hexamethylacetone does not lead to the pinacol but to a mixture of 3-ethyl-2,2,4,4-tetramethylpentan-3-ol and a 1,4-glycol[210]. Pinacols and unstable epimers have been prepared from A-norcholestanones by reduction with alkali metals in liquid ammonia[211].

Hexamethylphosphorotriamide dissolves alkali metals to form blue solutions of radical anions whose paramagnetic electron resonance spectra have been recorded[199, 212]. Such solutions containing also tetrahydrofuran effect the condensative reduction of a number of non-enolizable ketones to epoxides, e.g., benzophenone is converted into tetraphenylethylene oxide, benzaldehyde into *trans*-stilbene oxide, and acetophenone into *cis*- and *trans*-2,3-diphenyl-2-butene epoxide[212–214]. The active reducing agent is considered to be the species $(Me_2N)_2P^{-}O$ which attacks ketonic oxygen in a homolytic mechanism but the evidence gained so far is inconclusive. Similar results have been obtained in the reduction of benzophenone with the sodium derivative of the anion $(EtO)_2P^{-}O$ in a solution of HMPA and tetrahydrofuran[214].

The aromatic ketone fluorenone and several phenylpyridyl ketones have been reductively alkylated to 1,1-diarylalkanols by alkyl halides and sodium in liquid ammonia solution[215].

There are comparatively few reports of the reduction of epoxides to alcohols with metals. Under the Birch reduction conditions and in presence of a proton source the conversion of epoxide into alcohol might be expected to proceed:

$$>\overset{\overset{O}{\diagup\diagdown}}{C-C}< \xrightarrow{e} >\overset{O^-}{\overset{|}{C}}-\dot{C}< \xrightarrow{e} >\overset{O^-}{\overset{|}{C}}-\bar{C}< \qquad (85)$$

(55) (56)

$$\xrightarrow{H^+} >\overset{O^-}{\overset{|}{C}}-CH< \xrightarrow{H^+} >C(OH)-CH<$$

In the cases examined no pinacol-type products have been observed, suggesting that intermediate **55** is formed in low concentration, or is short-lived, or non-existent. The latter alternative necessarily implies a synchronous two-electron transfer to the epoxide giving intermediate **56**. The reductive cleavage of optically active 2-methoxy-2-phenylbutane by potassium in alcohols as solvents has suggested a stepwise addition of two electrons and the importance of solvation on the stereochemical course of the reaction[216].

Propylene oxide is reduced under Birch conditions to propan-2-ol, and indene oxide to indan-2-ol[192, 217]; styrene oxide is reduced by sodium and water to β-phenylethyl alcohol[217]. The structures of the dianions leading to these products must be the following:

$CH_3CHO^-CH_2^-$ [indan-2-oxide dianion, carbanion at C-1] $PhCH^-CH_2O^-$

It has been noted that of the two possible dianions, that containing the more stable carbanion moiety is observed. This principle has also been used to rationalize the orientation of reductive fission of aralkyl and substituted diaryl ethers under similar conditions[190, 192].

The reduction of epoxycyclohexanes of the steroid series is mainly similar to those effected by lithium aluminium hydride. The alcohols which result from the reduction of the following selection of epoxides[218–220] are shown: 5α,6α-epoxycholestane (**57**), 3β-acetoxy-7α,8α-epoxyergost-22-ene (**58**), 3β-acetoxy-9α, 11α-epoxyergostane

(**59**), 3β-acetoxy-9α,11α-epoxyergosta-7,22-diene (**60**), and 3α,5α-epoxycholestane (**61**):

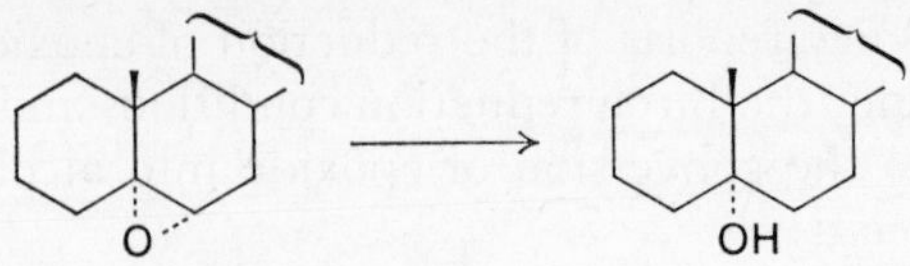

(57) → (86)

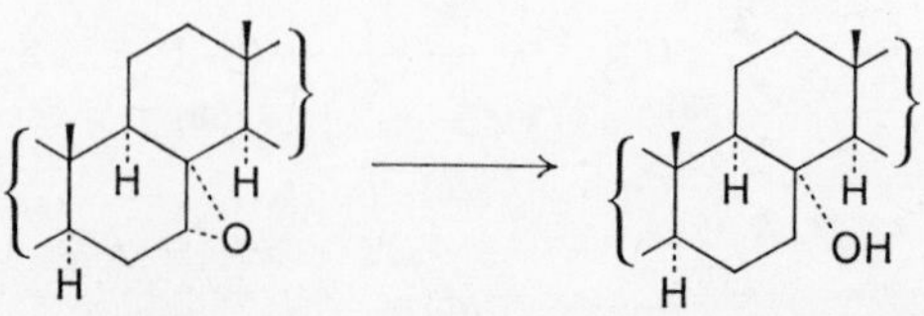

(58) → (87)

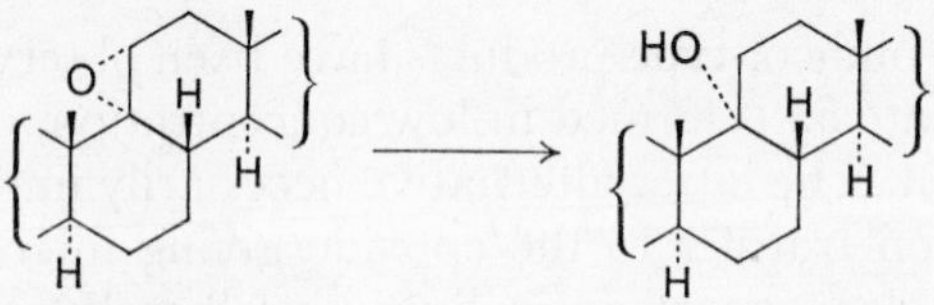

(59) → (88)

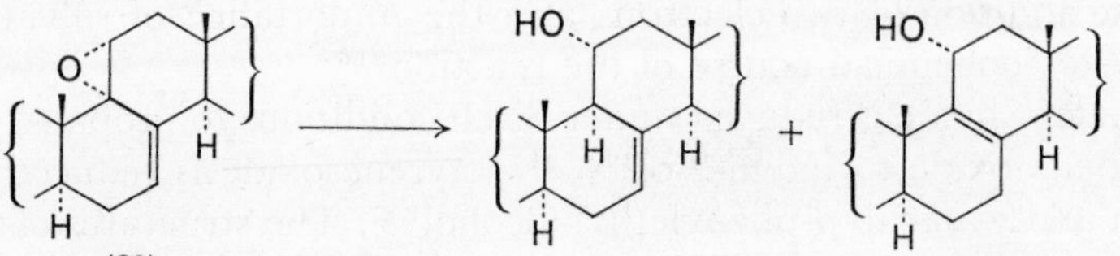

(60) → + (89)
ca. 1 : 1

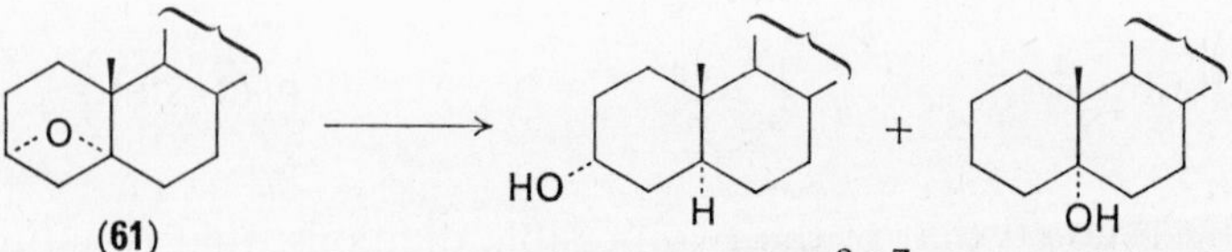

(61) → + (90)
3 : 7

In these illustrations the reductive fission can be interpreted as proceeding via the more stable carbanions, namely, a secondary rather than a tertiary carbanion, leading to tertiary alcohols. The secondary alcohols of equation (89) may be rationalized in terms of the intermediacy of an allylic carbanion[218]. Mainly axial alcohols result and this, like reductions with lithium aluminium hydride, is presumably a consequence of the preferred geometry for the transition state leading to nucleophilic, *trans*-diaxial opening of epoxide rings. The 7α,8α- and 9α,11α-epoxy steroids (**58** and **59** respectively) are not reduced by lithium aluminium hydride and this has led to the suggestion that solvated electrons of low steric requirement are responsible for Birch reductions[218]. Some other secondary–tertiary epoxides, e.g., 5β,6β-epoxycholestane[218], 2,3-epoxypinane[221] and α- and β-3,4-epoxycaranes[222], yield mainly or considerable proportions of secondary alcohols. This feature has also been noted with lithium aluminium hydride and ascribed to a conformational effect[223] which may also operate in the Birch reduction. The Birch reduction of α- and β-3,4-epoxycaranes gives different mixtures of alcohols from the reduction with lithium aluminium hydride[222].

B. Catalytic Hydrogenation

Catalytic hydrogenation and hydrogenolysis offer a general method for the reduction of oxygenated compounds to alcohols. Suitable combinations of metal and reaction conditions are described in several sources[224–226] for the conversion of aldehydes, ketones, epoxides, peroxides, acids and esters into alcohols, although the latter two are hydrogenated with some difficulty. Similar reagents are used to convert epoxides and peroxides into alcohols. The factors controlling the direction of ring opening accompanying hydrogenolysis of epoxides are numerous and have been noted elsewhere[227, 228]; there is little more recent work reported. Summaries of proposed mechanisms and stereochemical aspects of the hydrogenation of ketones may be found in recent publications[226, 229, 230]. The area continues to attract new investigations, e.g., the stereochemistry of the reduction of 2-acetoxy-l-tetralone and 3-acetoxy-4-chromanone by both catalytic hydrogenation and metal hydrides is reported[231]. In suitable cases the catalytic hydrogenation may be sterically controlled to effect asymmetric synthesis as, e.g., in the platinum and palladium catalysed hydrogenation of (−)-menthyl α-naphthylglyoxylate[232]. A relatively new development is the attempt to produce an asymmetric synthesis by the hydrogenation of ketones with modified Raney-nickel catalysts. The modification consists of treating the nickel with an aqueous solution of an optically active compound such as an amino acid[233], D-tartaric acid[234, 235], D-mandelic acid and (−)-ephedrine[235], which appear to be rapidly adsorbed on the nickel surface[233]. The reaction mainly studied has been the hydrogenation of ethyl acetoacetate to ethyl β-hydroxybutyrate.

The degree of conversion into optically active product depends both on the modifying agent and reaction conditions such as acidity. For example, the hydrogenation of ethyl acetoacetate with Raney-nickel modified by an aqueous solution of (+)-1,2-dihydroxycyclohexane-1,2-dicarboxylic acid has its greatest asymmetric activity in the range pH 5—7, and falls off sharply at pH < 5. The greatest activity coincides with the region in which the acid is half ionized and it is suggested that the carboxylate ion moiety is not adsorbed by nickel and is important in sterically controlling the approach of substrate to catalyst[236]. The relationship between the chemical and stereochemical properties of the modifying reagent and the asymmetric activity of the catalyst is not yet clear but preliminary accounts have appeared[233, 237]. In the hydrogenation of ethyl acetoacetate the stereochemistry of the product resembles that of the modifying agent, e.g., D-tartaric acid produces ethyl β-hydroxybutyrate with the D-configuration. The asymmetric activity of Raney-nickel modified with D- and L-tartaric acid is similar but for the sign of optical rotation; *meso* and racemic forms are inactive. An effective asymmetric catalyst must have a modifying reagent containing at least two functional groups such as hydroxy and carboxy with a strong ability to be adsorbed or chelated to Raney-nickel, although too high a stability of the complex may impair asymmetric catalysis[233, 237]. In addition, the modifying reagent should not contain a bulky substituent on the asymmetric carbon atom, e.g., L-α-methylmalic acid is less effective than L-malic acid. Chemical modification of functional groups by, e.g., acylation or esterification, may destroy the asymmetrical activity of the modified nickel.

C. Grignard Reagents and Organometal Compounds

Where the normal addition of a Grignard reagent to a ketone is sterically hindered by structural features present in either component, then reduction to carbinol often occurs. Limited synthetic use of the method has been made despite the degree of steric control of the product offered. Reductions which lead to asymmetric synthesis have been discussed recently[238], some examples of other various types follow. Ethanol-1-*d* of partial optical purity and other alcohols with H–D asymmetry have been prepared by the reduction of appropriate carbonyl compounds with isobornyloxymagnesium-2-*d* bromide[239]. The reduction of camphor to borneol and isoborneol with isopropylmagnesium chloride, bromide and iodide produced equal proportions of the products but the chloride afforded the higher overall yield; the latter effect was also noted with cyclohexylmagnesium chloride. 2-Butylmagnesium chloride gave more isoborneol than did the bromide and it was also observed that the isomer ratio was not affected by higher temperatures but the total yield increased[240]. An interesting variation with possible synthetic utility is the bimolecular reduction of aromatic ketones to high yields of glycols of the type $[ArAr'C(OH)]_2$ with ethylmagnesium bromide in the presence of cobalt(II) chloride[241]. The structure of Grignard reagents has been reviewed[242] and it appears that RMgX, R_2Mg and associated species (dimers and trimers) may coexist, the degree of association depending on the type of reagent, its concentration and the solvent. The mechanism of the addition of a Grignard reagent to a ketone is not settled, the main contending suggestions all involve two portions of magnesium compound to one of carbonyl compound in the transition state[242]. This may have a bearing on the structure of the transition state for the reduction of ketones but this reaction has received little kinetic investigation. Possibly the only kinetic study is the reduction of di-*t*-butyl ketone with *t*-butylmagnesium chloride and with di-*t*-butylmagnesium in tetrahydrofuran[243]. There were some similarities in the kinetics of reduction by the two reagents but the mechanistic conclusions are not clear. Organometal compounds continue to be used for reductions in synthesis, e.g., the pyridine-*n*-butyllithium-adduct reduces 4-*t*-butylcyclohexanone and 3,3,5-trimethylcyclohexanone to mainly equatorial alcohols[244]. Epoxides are also converted into alcohols by metal alkyls and aryls and Grignard reagents[33, 245]. A recent example is the reduction of styrene oxide with several Grignard reagents and alkyls of magnesium and cadmium to alcohols of the general formulae $PhCHRCH_2OH$ and $PhCH_2CHOHR$[246].

D. Meerwein–Ponndorf–Verley and Related Reductions

Grignard reagents reduce carbonyl compounds by serving as a source of hydride ion[247] and in this respect resemble the reduction of aldehydes in the Cannizzaro and Tishchenko reactions and the reduction of aldehydes and ketones in the Meerwein–Ponndorf–Verley (MPV) reaction[182]. The kinetics of the Tishchenko reaction have received little investigation and two mechanisms are proposed. In one, the preliminary stage is the addition of the aldehydic oxygen of one molecule to the carbonyl carbon atom of a second portion of aldehyde which is coordinated with metal alkoxide, in the other, the shift of an alkoxide group within a complex of aldehyde coordinated with metal alkoxide from the metal to the carbonyl carbon; in both suggestions this is followed by hydride transfer. A recent report concerning the kinetics of the Tishchenko conversion of acetaldehyde into ethyl acetate with aluminium isopropoxide in benzene lends support to the latter mechanism[248]. A substantial review in the Polish language of many aspects of the reaction is available[249].

The synthetic utility and other features of the MPV reduction continue to be explored and some illustrations of recent reports follow. The use of optically active MPV catalysts has not proved very successful hitherto in the synthesis of optically active alcohols[182]. However, it is now reported in preliminary form that optically active aluminium tri(2-methylbutoxide) in optically active amyl alcohol reduces methyl ethyl ketone, acetophenone and 3-methylcyclohexanone to good yields of alcohols with a high degree of optical purity[250]. Much less satisfactory results were obtained when an optically inactive reagent in an optically active solvent or an optically active reagent in an optically inactive solvent were used. Also reported are investigations into the asymmetric reductions of α-phthalimido-β-substituted propiophenones by aluminium isopropoxide in isopropanol[251] and of cyclohexanones, alkyl methyl and aralkyl ketones by isobornyloxyaluminium dichloride[252–254]. In the latter work, the optical purity of the aralkyl carbinols was much higher than that of the alkyl methyl carbinols. 1,2-Cyclohexanedione is reduced to the glycol by aluminium isopropoxide in toluene. The isomeric product is mainly *cis* but the proportion of this varies from 57 to 75% according to the relative concentrations of reactants and this is claimed to be an example of stoichiometric control of stereochemistry via several competing reaction paths[255, 256].

Formally similar reactions to the MPV reduction are effected by treating carbonyl compounds with an alcohol and an alkali[182, 257]. Potassium hydroxide in boiling ethylene glycol solution reduces benzophenone, norbornanone and certain of its derivatives to carbinols in moderate yields[258]. The reaction has received little investigation but probably involves a hydride transfer from alkoxide to the carbonyl carbon atom of the ketone. High yields of axial alcohols have been obtained by prolonged heating of an aqueous isopropanol solution of cyclohexanones containing chloroiridic acid and trimethyl phosphite. In this way, 3-*t*-butylcyclohexanone, 3,3,5-trimethylcyclohexanone and cholestanone have been reduced to axial alcohols in 92—99% yield[259]. Acetone is formed from the isopropanol solvent; dimethyl sulphoxide may be used in place of trimethyl phosphite but the proportion of axial alcohol is reduced. The presence of a tervalent iridium species with one or more phosphorus-containing ligands is suspected to be present and compounds containing dimethyl sulphoxide coordinated to iridium were isolated.

E. Metal and Organometal Hydrides

The number of different types available and selectivity in action make the metal hydrides and related reagents the most important single group of compounds currently available for reducing oxygenated compounds. The reduction of carbonyl compounds and epoxides has received attention in previous volumes of this series[182, 245] and elsewhere[260–264]. The use of organotin hydrides for reduction[265] and the chemistry of complex aluminohydrides[266] has been described.

The essentially nucleophilic character of sodium borohydride has been demonstrated in a Hammett study of the reduction of 2-, 3- and 4-substituted fluorenones to the corresponding alcohols in isopropanol solution[267]. The reduction of a large number of organic compounds including aldehydes, ketones, acids, esters, lactones and epoxides with lithium aluminium hydride, lithium trimethoxy- and tri-*t*-butoxy-aluminohydride and aluminium hydride under nearly similar conditions has been reported[268, 269]. The alkoxyaluminohydrides are predictably less reactive and more selective than lithium aluminium hydride, e.g., the latter reagent converts norcamphor into 90% of *endo*-norborneol whereas the trimethoxyaluminohydride produces 98%; aluminium hydride produces 93% of *endo*-norborneol. In another study, aluminium hydride was considered to react via a six-centred transition state which is less product-like than in the case of lithium aluminium hydride as the former reagent gave a relative enrichment of axial alcohol in the reduction of several steroidal ketones[270]. Aluminium hydride is more selective than lithium aluminium hydride in reducing α,β-unsaturated aldehydes and esters to unsaturated alcohols[271]. Diborane is another electrophilic reagent, its reduction of hindered monocyclic ketones is similar to that of lithium aluminium hydride in producing mainly the equatorial alcohol although there is increased stereospecificity observed in the reduction of the bicyclic ketone norcamphor to *endo*-norborneol[272]. The more hindered dialkylboranes, e.g., dicyclohexylborane, produce considerably more axial alcohol than lithium aluminium hydride in the reduction of 2-methylcyclanones[272]. Similarly, 3-methylcyclohexanone and *cis*- and *trans*-dimethylcyclohexanone-3,4-dicarboxylate yield on reduction with diborane mostly the equatorial alcohol, whereas dihydroisophorone gives mostly the axial isomer. Hindered alkoxyboranes give with dihydroisophorone more of the axial alcohol the bulkier the reducing agent[273]. The origin of the differing degrees of stereospecificity exhibited by nucleophilic and electrophilic reducing agents in their action on flexible monocyclic and rigid bicyclic ketones is not yet clear. The kinetics of the reduction of some substituted cyclohexanones by diborane have been measured and the rate is first-order in ketone and three-halves order in diborane; in the presence of high concentrations of boron trifluoride more axial isomer is produced and the rate tends to the first order in diborane[273].

Applications of hydride reagents to the synthesis of optically active alcohols is still limited but some progress is being made. The reagents (+)- and (−)-diisopinocampheylborane have been used in the asymmetric synthesis of (3*S*)(−)- and (3*R*)(+)-4-methylpentane-1-diols from carbonyl compounds[274]. Lithium aluminium hydride in the presence of (−)-quinine, (−)-menthol and others, effects some degree of asymmetric synthesis in the reduction of ketones[275, 276]. There is a correlation between the configuration of the optically active reducing agent and that of the product. Other reagents investigated are lithium aluminium hydride complexes with 1,2:3,4-di-*O*-isopropylidene-α-D-galactopyranose, 1,2:5,6-di-*O*-isopropylidene-α-D-glucofuranose[277], and 3-*O*-benzyl-1,2-*O*-cyclohexylidene-α-D-glucofuranose[278, 279]. It was shown in the case of the latter reagent that the configuration of the secondary alcohol product changed from (*S*) to *R*) with increasing quantities of added ethanol[278] and the reduction of, e.g., acetophenone yields optically active alcohols of up to 70% optical purity. This reagent shows no more stereospecificity in reducing 3,3,5-trimethylcyclohexanone than do lithium aluminium hydride and sodium borohydride, and this was interpreted by postulating the participation of the flexible form of the ketone[279].

Apart from the reduction of carbon–oxygen double bonds, metal hydrides are also able to effect the hydrogenolysis of carbon–oxygen single bonds in many compounds. The reduction of epoxides to alcohols has been particularly useful. A comparison of the direction of ring opening of styrene oxide by lithium trimethoxy- and tri-*t*-butoxy-aluminohydride, lithium aluminium hydride, aluminium hydride and mixed hydride shows that the proportions of primary alcohol product are 0, 1, 4, 24 and 95—98% respectively[269]. Similar results for styrene oxide and other unsymmetrical epoxides including those for reduction by hydridoaluminium halides have been reported[280–282]. The high proportion of primary alcohol produced by mixed hydride[283] may be explained in terms of the strong Lewis acidity of the hydridoaluminium halides present. The effect of substituents on the stereochemistry and direction of ring opening of cyclopentene[284] and cyclohexene[285] oxides by lithium aluminium hydride and mixed hydride has been reported. The mechanistic aspects of the anomalous reduction of cyclobutene oxides with lithium aluminium hydride to cyclobutanols and ring-cleaved primary alcohols have received attention[286]. The reduction of *cis*- and *trans*-2-alkoxy-3,4-epoxytetrahydropyrans by lithium aluminium hydride exclusively at the epoxide carbon atom remote from the alkoxy substituent is attributed to the polar influence of the two geminal oxygen atoms[287]. Skeletal rearrangements have been observed in the reductions with lithium aluminium hydride of terpene[288] and steroid[289] epoxides. Among other recent studies reported may be mentioned the reductive cleavage with hydrides of tetrahydropyranyl[290, 291] and furanyl[291] ethers, acetals and ketals[292], dioxolanes and dioxanes[293] and cyclic acylals[294].

VI. REFERENCES

1. T. A. Turney, *Oxidation Mechanisms*, Butterworths, London, 1965.
2. W. A. Waters, *Mechanisms of Oxidation of Organic Compounds*, Methuen, London, 1964.
3. R. Stewart, *Oxidation Mechanisms: Applications to Organic Chemistry*, W. A. Benjamin Inc., New York, 1964.
4. *Oxidation in Organic Chemistry* (*A*) (Ed. K. B. Wiberg), Academic Press, New York, 1965.
5. C. Walling, *Free Radicals in Solution*, John Wiley and Sons, New York, 1957, Chap. 9.
6. S. Fallab, *Angew. Chem. Intern. Ed. Engl.*, **6**, 496 (1967).
7. *Autoxidation and Antioxidants* (Ed. W. O. Lundberg), John Wiley and Sons, New York, 1961 (Vol. I), and 1962 (Vol. II).
8. K. A. Pecherskaya, M. F. Logus and A. Y. Tsybul'ko, *Geterogennye Reakts. Reakts. Sposobnost*, 234 (1964); *Chem. Abstr.*, **66**, 37248a (1967).
9. G. Zweifel and H. C. Brown in *Org. Reactions*, Vol. 13 (Ed. R. Adams), John Wiley and Sons, New York, 1963, Chap. 1.
10. H. C. Brown, *Hydroboration*, W. A. Benjamin Inc., New York, 1962.
11. A. Streitwieser, L. Verbit and R. Bittman, *J. Org. Chem.*, **32**, 1530 (1967).
12. H. C. Brown and C. D. Pfaffenberger, *J. Am. Chem. Soc.*, **89**, 5475 (1967).
13. H. O. House, *Modern Synthetic Reactions*, W. A. Benjamin Inc., New York, 1965, Chap. 5.
14. J. K. Kochi, *J. Am. Chem. Soc.*, **84**, 774 (1962).
15. J. K. Kochi, *Tetrahedron*, **18**, 483 (1962).

16. C. Walling and A. A. Zavitsas, *J. Am. Chem. Soc.*, **85,** 2084 (1963).
17. G. Sosnovsky and N. C. Yang, *J. Org. Chem.*, **25,** 899 (1960).
18. K. B. Wiberg in Reference 4, Chap. 2.
19. C. N. Rentea, *Stud. Cercet Chim.*, **14,** 627 (1966); *Chem. Abstr.*, **66,** 75422k (1967).
20. T. Matsuura and T. Suga, *Yuki Gosei Kagaku Kyoka Shi*, **25,** 214 (1967); *Chem. Abstr.*, **67,** 10852b (1967).
21. T. Sakao, T. Suga and T. Matsuura, *J. Sci Hiroshima Univ.*, Ser. A-II; *Phys. Chem.*, **31,** 51 (1967); *Chem. Abstr.*, **68,** 95458w (1968).
22. C. N. Rentea, I. Necsoiu, M. Rentea, A. Ghenciulescu and C. D. Nenitzescu, *Tetrahedron*, **22,** 3501 (1966).
23. N. A. Gibson and J. W. Hosking, *Australian J. Chem.*, **18,** 123 (1965).
24. R. Stewart in Reference 4, Chap. 1.
25. J. B. Aylward, *J. Chem. Soc.* (**B**), 1268 (1967).
26. R. Criegee in Reference 5, Chap. 5.
27. R. E. Partch, *J. Am. Chem. Soc.*, **89,** 3662 (1967).
28. H. O. House in Reference 13, p. 93.
29. J. F. Cairns and H. L. Roberts, *J. Chem. Soc.* (**C**), 640 (1968).
30. H. O. House in Reference 13, pp. 141–142.
31. H. O. House in Reference 13, Chap. 6.
32. R. O. C. Norman and C. B. Thomas, *J. Chem. Soc.* (**B**), 604 (1967).
33. R. E. Parker and N. S. Isaac, *Chem. Rev.*, **59,** 737 (1959).
34. J. Chatt, *Chem. Rev.*, **48,** 7 (1951).
35. N. S. Zefirov, *Russ. Chem. Rev.*, **34,** 527 (1965).
36. C. B. Anderson and S. Winstein, *J. Org. Chem.*, **28,** 605 (1963).
37. T. G. Traylor and A. W. Baker, *J. Am. Chem. Soc.*, **85,** 2746 (1963).
38. K. C. Pande and S. Winstein, *Tetrahedron Letters*, 3393 (1964).
39. J. K. Stille and S. C. Stinson, *Tetrahedron*, **20,** 1387 (1964).
40. F. R. Jensen and R. J. Ouellette, *J. Am. Chem. Soc.*, **83,** 4477 (1961).
41. M. M. Kreevoy and M. A. Turner, *J. Org. Chem.*, **30,** 373 (1965).
42. K. Ichikawa, K. Nishimura and S. Takayama, *J. Org. Chem.*, **30,** 1593 (1965).
43. E. R. Allen, J. Cartlidge, M. M. Taylor and C. F. H. Tipper, *J. Phys. Chem.*, **63,** 1437 and 1442 (1959).
44. A. P. Kreshkov and L. N. Balyatinskaya, *J. Gen. Chem. USSR*, **37,** 2099 (1967).
45. J. Oda, T. Nakagawa and Y. Inouye, *Bull. Chem. Soc. Japan*, **40,** 373 (1967).
46. J. Halpern and H. B. Tinker, *J. Am. Chem. Soc.*, **89,** 6427 (1967).
47. R. J. Ouellette, R. D. Robins and A. South, *J. Am. Chem. Soc.*, **90,** 1619 (1968).
48. W. L. Waters and E. F. Kiefer, *J. Am. Chem. Soc.*, **89,** 6261 (1967).
49. H. C. Brown and P. Geoghegan, *J. Am. Chem. Soc.*, **89,** 1522 (1967).
50. H. C. Brown and W. J. Hammar, *J. Am. Chem. Soc.*, **89,** 1524 (1967).
51. H. C. Brown, J. H. Kawakami and S. Ikegami, *J. Am. Chem. Soc.*, **89,** 1525 (1967).
52. T. G. Traylor, *J. Am. Chem. Soc.*, **86,** 244 (1964).
53. Sung Moon and B. H. Waxman, *Chem. Commun.*, 1283 (1967).
54. H. B. Henbest and B. Nicholls, *J. Chem. Soc.*, 227 (1959).
55. F. R. Jensen and J. J. Miller, *Tetrahedron Letters*, 4861 (1966).
56. M. Malaiyandi and G. Wright, *Can. J. Chem.*, **41,** 1493 (1963).

57. W. Treibs and M. Weissenfels, *Chem. Ber.*, **93**, 1374 (1960).
58. I. Alkonyi, *Chem. Ber.*, **95**, 279 (1962).
59. J. de P. Teresa and M. I. Bellida, *Annales Real Soc. Espan. Fis. Quim.* (*Madrid*), Ser. B, 62, 989 (1966); *Chem. Abstr.*, **67**, 43936k (1967).
60. R. Criegee in Reference 4, Chap. 5.
61. R. O. C. Norman and C. B. Thomas, *J. Chem. Soc.* (**B**), 771 (1967).
62. J. M. Davidson and C. Triggs, *Chem. Ind.*, 1361 (1967).
63. Z. Rappoport, P. D. Sleezer, S. Winstein and W. G. Young, *Tetrahedron Letters*, 3719 (1965).
64. P. D. Sleezer, S. Winstein and W. G. Young, *J. Am. Chem. Soc.*, **89**, 1890 (1963).
65. See 63 for references.
66. K. B. Wiberg and S. D. Nielsen, *J. Org. Chem.*, **29**, 3353 (1964), and references therein.
67. R. R. Grinstead, *J. Org. Chem.*, **26**, 238 (1961).
68. P. M. Henry, *J. Am. Chem. Soc.*, **87**, 990 and 4423 (1965).
69. J. B. Lee and M. J. Price, *Tetrahedron Letters*, 1155 (1962).
70. J. B. Lee and M. J. Price, *Tetrahedron Letters*, 936 (1963).
71. H. J. Kabbe, *Ann. Chem.*, **656**, 204 (1962).
72. F. A. L. Anet, *Tetrahedron Letters*, 3399 (1964).
73. P. M. Henry, *J. Am. Chem. Soc.*, **88**, 1597 (1966).
74. J. Smidt, *Chem. Ind.*, 54 (1962).
75. G. C. Bond, *Ann. Rep. Progr. Chem.*, **63**, 39 (1966), and references therein.
76. D. Clark and P. Hayden, Am. Chem. Soc., Div. Petrol. Chem. Preprints, 11, D5-D9 (1966).
77. B. C. Fielding and H. L. Roberts, *J. Chem. Soc.* (**A**), 1627 (1966).
78. J. C. Strini and J. Metzger, *Bull. Soc. Chim. France*, 3145 and 3150 (1966).
79. P. M. Henry, *J. Am. Chem. Soc.*, **88**, 1595 (1966).
80. W. Kitching, Z. Rappoport, S. Winstein and W. G. Young, *J. Am. Chem. Soc.*, **88**, 2054 (1966).
81. P. M. Henry, *J. Org. Chem.*, **32**, 2575 (1967).
82. S. Uemura and K. Ichikawa, *Bull. Chem. Soc. Japan*, **40**, 1016 (1967).
83. I. I. Moiseev and M. N. Vargaftik, *Izv. Akad. Nauk SSSR, Ser. Khim.*, 759 (1965).
84. P. M. Henry, *J. Am. Chem. Soc.*, **86**, 3246 (1964).
85. R. Jina, J. Sedlmeier and J. Smidt, *Ann. Chem.*, **693**, 99 (1966).
86. M. Green, R. N. Haszeldine and J. Lindley, *J. Organometal. Chem.*, **6**, 107 (1966).
87. W. C. Baird, *J. Org. Chem.*, **31**, 2411 (1966).
88. T. W. Campbell, H. G. Walker and G. M. Coppinger, *Chem. Rev.*, **50**, 279 (1952).
89. J. P. Schaefer and B. Horvath, *Tetrahedron Letters*, 2023 (1964).
90. G. O. Schenck, O. A. Neumueller, G. Ohloff and S. Schroeter, *Ann. Chem.*, **687**, 26 (1965).
91. C. S. Foote and S. Wexler, *J. Am. Chem. Soc.*, **86**, 3879 and 3880 (1964), and references therein.
92. R. W. Murray and M. L. Kaplan, *J. Am. Chem. Soc.*, **90**, 537 (1968).
93. H. H. Wasserman and J. R. Scheffer, *J. Am. Chem. Soc.*, **89**, 3073 (1967).
94. E. J. Corey and W. C. Taylor, *J. Am. Chem. Soc.*, **86**, 3881 (1964), and references therein.

95. A. M. Viner and K. D. Bayes, *J. Phys. Chem.*, **70,** 302 (1966).
96. F. McCapra, *Quart. Rev. (London)*, **20,** 485 (1966), and references therein.
97. P. Douzou, J. Capette and J. P. Gout, *Compt. Rend. Acad. Sci., Paris, Ser. C*, **266,** 993 (1968).
98. C. S. Foote, S. Wexler, W. Ando and R. Higgins, *J. Am. Chem. Soc.*, **90,** 975 (1968).
99. E. McKeown and W. A. Waters, *J. Chem. Soc.* (**B**), 1040 (1966).
100. J. A. Marshall and A. R. Hochstetler, *J. Org. Chem.*, **31,** 1020 (1966).
101. C. S. Foote, S. Wexler and W. Ando, *Tetrahedron Letters*, 4111 (1965).
102. K. R. Kopecky and H. J. Reich, *Can. J. Chem.*, **43,** 2265 (1965).
103. T. Wilson, *J. Am. Chem. Soc.*, **88,** 2898 (1966).
104. E. J. Forbes and J. Griffiths, *J. Chem. Soc.* (**C**), 601 (1967).
105. K. Gollnick and G. Schade, *Tetrahedron Letters*, 689 (1968).
106. K. H. Schulte-Elte and G. Ohloff, *Helv. Chim. Acta*, **51,** 494 (1968).
107. R. H. Young and H. Hart, *Chem. Commun.*, 827 (1967).
108. D. R. Kearns, R. A. Hollins, A. U. Khan, R. W. Chambers and P. Radlick, *J. Am. Chem. Soc.*, **89,** 5455 (1967).
109. D. R. Kearns, R. A. Hollins, A. U. Khan and P. Radlick, *J. Am. Chem. Soc.*, **89,** 5456 (1967).
110. D. B. Sharp, *Abs., 138th Natl. Meet.*, Am. Chem. Soc., New York, 1960, p. 79P.
111. A. Nickon and J. F. Bagli, *J. Am. Chem. Soc.*, **83,** 1498 (1961).
112. A. Nickon and W. L. Mendelson, *J. Am. Chem. Soc.*, **85,** 1894 (1963).
113. H. Gollnick, S. Schroeter, G. Ohloff, G. Schrade and G. O. Schenck, *Ann. Chem.*, **687,** 14 (1965).
114. E. Klein and W. Rojahn, *Tetrahedron*, **21,** 2173 (1965).
115. E. Klein and W. Rojahn, *Dragoco Rep.* (*German Edn.*), **14,** 95 (1967); *Chem. Abstr.*, **67,** 116955c (1967).
116. K. Gollnick and G. O. Schenck, *Pure Appl. Chem.*, **9,** 507 (1964).
117. K. Gollnick and G. O. Schenck in *1,4-Cycloaddition Reactions* (Ed. J. Hamer), Academic Press, New York, 1967, Chap. 10.
118. E. J. Forbes and J. Griffiths, *Chem. Commun.*, 427 (1967).
119. E. J. Bailey, D. H. R. Barton, J. Elks and J. F. Templeton, *J. Chem. Soc.*, 1578 (1962), and references therein.
120. E. J. Bailey, J. Elks and D. H. R. Barton, *Proc. Chem. Soc.*, 214 (1960).
121. J. B. Davis and B. C. L. Weedon, *Proc. Chem. Soc.*, 182 (1960).
122. D. H. R. Barton, S. K. Pradhan, S. Sternhell and J. F. Templeton, *J. Chem. Soc.*, 255 (1961).
123. H. R. Gersmann, H. T. W. Nieuwenhuis and A. F. Bickel, *Proc. Chem. Soc.*, 279 (1962).
124. M. Avramoff and Y. Sprinzak, *Proc. Chem. Soc.*, 150 (1962).
125. D. J. Cram, B. Rickborn, C. A. Kingsbury and P. Haberfield, *J. Am. Chem. Soc.*, **83,** 3678 (1961).
126. A. Schriesheim and C. A. Rowe, *J. Am. Chem. Soc.*, **84,** 3160 (1962).
127. D. J. Cram, *Chem. Eng. News*, **41,** 92 (1963).
128. D. J. Cram, *Fundamentals of Carbanion Chemistry*, Academic Press, New York, 1965, Chap. 1.
129. G. A. Russell, E. G. Janzen, H. D. Becker and F. J. Smentowski, *J. Am. Chem. Soc.*, **84,** 2652 (1962).

130. T. J. Wallace, A. Schriesheim and N. Jacobson, *J. Org. Chem.*, **29**, 2907 (1964).
131. J. E. Hofmann, A. Schriesheim and D. D. Rosenfeld, *J. Am. Chem. Soc.*, **87**, 2523 (1965).
132. Y. Sprinzak, *J. Am. Chem. Soc.*, **80**, 5449 (1958).
133. W. Bartok, D. D. Rosenfeld and A. Schriesheim, *J. Org. Chem.*, **28**, 410 (1963).
134. D. H. R. Barton and D. W. Jones, *J. Chem. Soc.*, 3563 (1965).
135. N. Kornblum and H. E. De La Mare, *J. Am. Chem. Soc.*, **73**, 880 (1951).
136. G. A. Russell, *Pure Appl. Chem.*, **15**, 185 (1967), and references therein.
137. G. A. Russell, *U.S. Patent* 3260570; *Chem. Abstr.*, **66**, 2378m (1967).
138. R. Lombard and B. Muckensturm, *Compt. Rend. Acad. Sci. Paris, Ser C*, **265**, 19 (1967).
139. G. A. Russell and A. G. Bemis, *J. Am. Chem. Soc.*, **88**, 5491 (1966).
140. D. Bethell and R. J. E. Talbot, *J. Chem. Soc.* (**B**), 638 (1968).
141. H. R. Gersmann, H. J. W. Nieuwenhuis and A. F. Bickel, *Tetrahedron Letters*, 1383 (1963).
142. H. C. Volger and W. Brackman, *Rec. Trav. Chim.*, **85**, 817 (1966), and references therein.
143. Examples are to be found in: *Chem. Abstr.*, **58**, P3317f (1963); **62**, P7796d (1965); **63**, P13315b (1965); **67**, 73136s (1967).
144. H. Hock, H. Kropf and F. Ernst, *Angew Chem.*, **71**, 541 (1959).
145. S. Sosnovsky and J. H. Brown, *Chem. Rev.*, **66**, 529 (1966).
146. A. G. Davies, *Organic Peroxides*, Butterworths, London, 1961.
147. C. F. Cullis, A. Fish and R. T. Pollard, *Proc. Roy. Soc.*, **288A**, 123 (1965), and references therein.
148. C. F. Cullis, A. Fish and R. T. Pollard, *Proc. Roy. Soc.*, **289A**, 413 (1966).
149. C. F. Cullis, A. Fish and R. T. Pollard, *Proc. Roy. Soc.*, **298A**, 64 (1967), and references therein.
150. C. F. Cullis, A. Fish and R. T. Pollard, *Trans. Faraday Soc.*, **60**, 2224 (1964).
151. C. H. Bamford and D. M. Newitt, *J. Chem. Soc.*, 688 (1946).
152. C. H. Bamford and D. M. Newitt, *J. Chem. Soc.*, 695 (1946).
153. F. S. Arimoto, *J. Polymer Sci.*, **4A**, 275 (1966).
154. R. L. Hansen, *J. Polymer Sci.*, **2A**, 4215 (1964).
155. F. J. Welch, *J. Polymer Sci.*, **61**, 243 (1962).
156. S. B. Mirviss, *J. Am. Chem. Soc.*, **83**, 3051 (1961).
157. A. G. Davies, *Progr. Boron Chem.*, **1**, 265 (1964), and references therein.
158. M. H. Abraham, *J. Chem. Soc.*, 4130 (1960).
159. A. G. Davies and C. D. Hall, *J. Chem. Soc.*, 1192 (1963).
160. C. Walling and S. A. Buckler, *J. Am. Chem. Soc.*, **77**, 6032 and 6039 (1955).
161. G. Wilke and P. Heimbach, *Ann. Chem.*, **652**, 7 (1962).
162. L. Parts and J. T. Miller, *U.S. Gov. Res. Develop. Rep.*, **68**, 56 (1968); *Chem. Abstr.*, **69**, 27469w (1968).
163. A. G. Davies, D. G. Hare and R. F. M. White, *J. Chem. Soc.*, 341 (1961).
164. S. B. Mirviss, *J. Org. Chem.*, **32**, 1713 (1967).
165. C. E. Bawn, H. D. Margerison and N. M. Richardson, *Proc. Chem. Soc.*, 397 (1959).
166. R. L. Hansen and R. R. Hamann, *J. Phys. Chem.*, **67**, 2868 (1963).
167. A. Grobler, A. Simon, T. Kada and L. Fazakas, *J. Organometal. Chem.*, **7**, P3 (1967).

168. G. A. Razuvaev, A. I. Graevskii, K. S. Minsker and M. D. Belova, *Proc. Acad. Sci. USSR, Chem. Sect.*, **152**, 696 (1963).
169. R. C. Lamb, P. W. Ayers, M. K. Toney and J. F. Garst, *J. Am. Chem. Soc.*, **88**, 4261 (1966).
170. H. A. Pacevitz and H. Gilman, *J. Am. Chem. Soc.*, **61**, 1603 (1939).
171. H. S. Chang and J. T. Edward, *Can. J. Chem.*, **41**, 1233 (1963), and references therein.
172. G. A. Razuvaev, E. V. Mitrofanova, G. G. Petukhov and R. V. Kaplina, *J. Gen. Chem. USSR*, **32**, 3390 (1962).
173. G. A. Razuvaev, R. F. Galiulina, G. G. Petukhov and N. V. Likhovidova, *J. Gen. Chem. USSR*, **33**, 3285 (1963).
174. A. G. Davies and J. E. Packer, *J. Chem. Soc.*, 3164 (1959).
175. V. N. Pankratova, V. N. Latyaeva and G. A. Razuvaev, *J. Gen. Chem. USSR*, **35**, 902 (1965).
176. Y. A. Aleksandrov, O. N. Druzhkov, S. F. Zhil'tsov and G. A. Razuvaev, *Dokl. Chem. USSR*, **157**, 798 (1964).
177. Y. A. Aleksandrov, O. N. Druzhkov, S. F. Zhil'tsov and G. A. Razuvaev, *J. Gen. Chem. USSR*, **35**, 1444 (1965).
178. E. C. J. Coffee and A. G. Davies, *J. Chem. Soc.* (**C**), 1493 (1966).
179. A. G. Davies and B. P. Roberts, *J. Chem. Soc.* (**C**), 1474 (1968), and references therein.
180. A. G. Davies and B. P. Roberts, *J. Chem. Soc.* (**B**), 17, (1967).
181. Y. A. Aleksandrov, B. A. Radbil and V. A. Shushunov, *J. Gen. Chem. USSR*, **37**, 190 (1967).
182. O. H. Wheeler in *The Chemistry of the Carbonyl Group* (Ed. S. Patai), John Wiley and Sons, New York, 1966, Chap. 11.
183. J. N. Pitts and J. K. S. Wan in *The Chemistry of the Carbonyl Group* (Ed. S. Patai), John Wiley and Sons, New York, 1966, pp. 851–5 and 885–6.
184. D. C. Neckers, *Mechanistic Organic Photochemistry*, Reinhold Publishing Co., New York, 1967, Chap. 7.
185. R. S. Davidson and P. F. Lambeth, *Chem. Commun.*, 1265 (1967).
186. S. G. Cohen and R. J. Baumgarten, *J. Am. Chem. Soc.*, **89**, 3471 (1967).
187. N. Filipescu and F. L. Minn, *J. Am. Chem. Soc.*, **90**, 1544 (1968).
188. D. C. Curry, B. C. Uff and N. D. Ward, *J. Chem. Soc.* (**C**), 1120 (1967).
189. J. R. Hanson and E. Premuzic, *Angew. Chem. Intern. Ed. Engl.*, **7**, 247 (1968), and references therein.
190. A. J. Birch, *Quart. Rev.* (*London*), **4**, 69 (1950).
191. A. J. Birch and H. Smith, *Quart. Rev.* (*London*), **12**, 17 (1958).
192. H. Smith, *Organic Reactions in Liquid Ammonia*, John Wiley and Sons, New York, 1963, Vol. 1, Part 2, Chap. H.
193. H. O. House in Reference 13, Chap. 3.
194. F. J. Kakis in *Steroid Reactions* (Ed. C. Djerassi), Holden-Day Inc., San Francisco, 1963, Chap. 6.
195. J. E. Starr in Reference 194, Chap. 7.
196. E. L. Martin in *Organic Reactions* (Ed. R. Adams), John Wiley and Sons, New York, 1942, Vol. 1, Chap. 7.
197. L. Michaelis and M. P. Schubert, *Chem. Rev.*, **22**, 439 (1938).
198. M. C. R. Symons, *Quart. Rev.* (*London*), **13**, 99 (1959).

199. G. Fraenkel, S. H. Ellis and D. T. Dix, *J. Am. Chem. Soc.*, **87**, 1406 (1965), and references therein.
200. M. D. Rausch, W. E. McEwen and J. Kleinberg, *Chem. Rev.*, **57**, 417 (1957).
201. J. Fried and N. A. Abraham, *Tetrahedron Letters*, 1879 (1964).
202. J. W. Huffman, D. M. Alabran and T. W. Bethea, *J. Org. Chem.*, **27**, 3381 (1962).
203. P. L. Cook, *J. Org. Chem.*, **27**, 3873 (1962).
204. G. E. Risinger, J. M. Garrett and J. A. Winkler, *Rec. Trav. Chim.*, **83**, 873 (1964).
205. J. Wiemann, M. R. Monot, G. Dana and J. Chuche, *Bull. Soc. Chim. France*, 3293 (1967).
206. N. Thoai, *Bull. Soc. Chim. France*, 1544 (1964).
207. Y. Leroux and H. Normant, *Compt. Rend. Acad. Sci. Paris, Ser. C*, **265**, 1472 (1967).
208. J. Wiemann and A. Jacquet, *Compt. Rend. Acad. Sci. Paris, Ser. C*, **263**, 313 (1966).
209. J. P. Morizur and J. Wiemann, *Bull. Soc. Chim. France*, 1619 (1964).
210. L. Eberson, *Acta Chem. Scand.*, **18**, 1255 (1964).
211. J. C. Espie, A. M. Giroud and A. Rassat, *Bull. Soc. Chim. France*, 809 (1967).
212. H. Normant, T. Cuvigny, J. Normant and B. Angelo, *Bull. Soc. Chim. France*, 3441 (1965).
213. H. Normant and M. Larcheveque, *Compt. Rend. Acad. Sci. Paris, Ser. C*, **260**, 5062 (1965).
214. J. F. Normant, *Bull. Soc. Chim. France*, 3601 (1966).
215. M. Miocque and C. Fauran, *Compt. Rend. Acad. Sci. Paris, Ser. C*, **259**, 408 (1964).
216. D. J. Cram in reference 128, pp. 165–168.
217. C. M. Suter and H. B. Milne, *J. Am. Chem. Soc.*, **65**, 582 (1943), and references therein.
218. A. S. Hallsworth and H. B. Henbest, *J. Chem. Soc.*, 4604 (1957).
219. A. S. Hallsworth and H. B. Henbest, *J. Chem. Soc.*, 3571 (1960).
220. J. G. Phillips and V. D. Parker in Reference 194, Chap. 14.
221. Z. Chabudzinski, D. Sedzik and Z. Rykowski, *Roczniki Chem.*, **41**, 1751 (1967); *Chem. Abstr.*, **68**, 78435n (1968).
222. Z. Chabudzinski, D. Sedzik and J. Szykula, *Roczniki Chem.*, **41**, 1923 (1967); *Chem. Abstr.*, **68**, 87403j (1968).
223. E. L. Eliel in *Steric Effects in Organic Chemistry* (Ed. M. S. Newman), John Wiley and Sons, New York, 1956, pp. 130–132.
224. R. L. Augustine, *Catalytic Hydrogenation*, Marcel Dekker Inc., New York, 1965.
225. P. N. Rylander, *Catalytic Hydrogenation over Platinum Metals*, Academic Press, New York, 1967.
226. H. O. House in Reference 13, Chap. 1.
227. R. L. Augustine in Reference 224, pp. 137–138.
228. P. N. Rylander in Reference 225, pp. 478–483.
229. O. H. Wheeler in Reference 182, pp. 510–514.
230. R. L. Augustine, *Ann. N.Y. Acad. Sci.*, **145**, 19 (1967).
231. K. Hanaya, *Bull. Chem. Soc. Japan*, **40**, 1884 (1967).
232. S. Mitsui and Y. Imai, *Nippon Kagaku Zasshi*, **88**, 86 (1967); *Chem. Abstr.*, **67**, 43934h (1967).

233. S. Tatsumi, *Bull. Chem. Soc. Japan*, **41**, 408 (1968), and references therein.
234. Y. Izumi, S. Tatsumi and M. Imaida, *Bull. Chem. Soc. Japan*, **39**, 2223 (1966).
235. Y. I. Petrov, E. I. Klabunovski and A. A. Balandin, *Kinetika i Kataliz*, **8**, 814 (1967).
236. Y. Izumi, S. Tatsumi and M. Imaida, *Bull. Chem Soc. Japan*, **39**, 1087 (1966).
237. E. I. Klabunovski and Y. I. Petrov, *Dokl. Akad. Nauk SSSR.*, **173**, 1125 (1967).
238. D. R. Boyd and M. A. McKervey, *Quart. Rev. (London)*, **22**, 95 (1968).
239. A. Streitwieser and M. R. Granger, *J. Org. Chem.*, **32**, 1528 (1967), and references therein.
240. P. J. Malkonen and J. Korvola, *Suomen Kemistilehti, B*, **39**, 267 (1966).
241. R. Pallaud and J. F. Treps, *Compt. Rend. Acad. Sci. Paris, Ser. C*, **260**, 1187 (1965).
242. E. C. Ashby, *Quart. Rev. (London)*, **21**, 259 (1967).
243. M. S. Singer, R. M. Salinger and H. S. Mosher, *J. Org. Chem.*, **32**, 3821 (1967).
244. R. A. Abramovitch, W. C. Marsh and J. G. Saha, *Can. J. Chem.*, **43**, 2631 (1965).
245. R. J. Gritter in *The Chemistry of the Ether Linkage* (Ed. S. Patai), John Wiley and Sons, New York, 1967, Chap. 9.
246. J. P. Denian, E. H. Basch and P. Freon, *Compt. Rend. Acad. Sci. Paris, Ser. C*, **264**, 1560 (1967).
247. N. C. Deno, H. J. Peterson and G. S. Saines, *Chem. Rev.*, **60**, 7 (1960).
248. Y. Ogata, A. Kawasaki and I. Kishi, *Tetrahedron*, **23**, 825 (1967).
249. L. Cichon, *Wiadomosci Chem.*, **20**, 641, 711 and 783 (1966); *Chem. Abstr.*, **66**, 94408b (1967).
250. S. Yamashita, *J. Organometal. Chem.*, **11**, 381 (1968).
251. D. Fles, B. Majhofer and M. Kovac, *Tetrahedron*, **24**, 3053 (1968).
252. E. L. Eliel and D. Nasipuri, *J. Org. Chem.*, **30**, 3809 (1965).
253. D. Nasipuri and G. Sarkar, *J. Indian Chem. Soc.*, **44**, 165 (1967).
254. D. Nasipuri, G. Sarkar and C. K. Ghosh, *Tetrahedron Letters*, 5189 (1967).
255. C. H. Snyder, *J. Org. Chem.*, **31**, 4220 (1966).
256. C. H. Snyder, *J. Org. Chem.*, **32**, 2904 (1967).
257. Y. Sprinzak, *J. Am. Chem. Soc.*, **78**, 466 (1956), and references therein.
258. D. C. Kleinfelter, *J. Org. Chem.*, **32**, 840 (1967).
259. Y. M. Y. Haddad, H. B. Henbest, J. Husbands and T. R. B. Mitchell, *Proc. Chem. Soc.*, 361 (1964).
260. H. O. House in Reference 13, Chap. 2.
261. H. Hormann in *Newer Methods of Preparative Organic Chemistry* (Ed. W. Foerst), Academic Press, New York, 1963, Vol. II, pp. 213–226.
262. W. G. Brown in *Organic Reactions* (Ed. R. Adams), John Wiley and Sons, New York, 1951, Vol. 6, Chap. 10.
263. P. P. Lynch, *Education in Chemistry*, **4**, 183 (1967); *Chem. Abstr.*, **67**, 116269g (1967).
264. N. A. Gaylord, *Reduction with Complex Metal Hydrides*, John Wiley and Sons, New York, 1956.
265. H. G. Kuivila in *Advances in Organometallic Chemistry* (Ed. F. G. A. Stone and R. West), Academic Press Inc., New York, 1964, pp. 47 ff.
266. E. C. Ashby, *Advan. Inorg. Chem. Radiochem.*, **8**, 283 (1966).

267. A. J. Harget, K. D. Warren and J. R. Yandle, *J. Chem. Soc.* (**B**), 214 (1968), and references therein.
268. H. C. Brown, P. M. Weissman and N. M. Yoon, *J. Am. Chem. Soc.*, **88**, 1458 (1966), and references therein.
269. H. C. Brown and N. M. Yoon, *J. Am. Chem. Soc.*, **88**, 1464 (1966), and references therein.
270. D. C. Ayres and R. Sawdaye, *J. Chem. Soc.* (**B**), 581 (1967).
271. M. J. Jorgenson, *Tetrahedron Letters*, 559 (1962).
272. H. C. Brown and V. Varma, *J. Am. Chem. Soc.*, **88**, 2871 (1966).
273. J. Klein and E. Dunkelblum, *Tetrahedron*, **23**, 205 (1967).
274. E. Caspi and K. R. Varma, *J. Org. Chem.*, **33**, 2181 (1968).
275. O. Cervinka and O. Belovsky, *Collection Czech. Chem. Commun.*, **32**, 3897 (1967), and references therein.
276. J. P. Fuette and G. Horean, *Bull. Soc. Chim. France*, 1747 (1967).
277. O. Cervinka and A. Fabryova, *Tetrahedron Letters*, 1179 (1967), and references therein.
278. S. R. Landor, B. J. Miller and A. R. Tatchell, *J. Chem. Soc.* (**C**), 197 (1967).
279. S. R. Landor and J. P. Regan, *J. Chem. Soc.* (**C**), 1159 (1967), and references therein.
280. E. C. Ashby and B. Cooke, *J. Am. Chem. Soc.*, **90**, 1625 (1968).
281. B. Cooke, E. C. Ashby and J. Lott, *J. Org. Chem.*, **33**, 1132 (1968).
282. L. I. Zakharkin and I. M. Khorlina, *Izv. Akad. Nauk SSSR, Ser. Khim.*, 862 (1965).
283. E. L. Eliel and M. N. Rerick, *J. Am. Chem. Soc.*, **82**, 1362 (1960).
284. P. T. Lansbury, D. J. Scharf and V. A. Pattison, *J. Org. Chem.*, **32**, 1748 (1967).
285. B. Rickborn and W. E. Lamke, *J. Org. Chem.*, **32**, 537 (1967).
286. L. A. Paquette, G. A. Youssef and M. L. Wise, *J. Am. Chem. Soc.*, **89**, 5246 (1967).
287. F. Sweet and R. K. Brown, *Can. J. Chem.*, **46**, 707 (1968).
288. J. P. Montheard and Y. C.-Bessiere, *Bull. Soc. Chim. France*, 336 (1968).
289. D. J. Collins, J. J. Hobbs and R. J. Rawson, *Chem. Commun.*, 135 (1967).
290. V. E. Diner and R. K. Brown, *Can. J. Chem.*, **45**, 2547 (1967).
291. E. L. Eliel, B. E. Novak, R. A. Daignault and V. G. Badding, *J. Org. Chem.*, **30**, 2441 (1965).
292. V. E. Diner, H. A. Davis and R. K. Brown, *Can. J. Chem.*, **45**, 207 (1967).
293. V. E. Diner and R. K. Brown, *Can. J. Chem.*, **45**, 1297 (1967).
294. A. Stephen and F. Wessely, *Monatsh. Chem.*, **98**, 184 (1967).

CHAPTER 5

Electrochemistry of the hydroxyl group

HENNING LUND

Department of Chemistry, University of Aarhus, 8000 Aarhus C, Denmark

I. INTRODUCTION

Electrolysis of an organic compound involves generally one or more steps in which electrons are transferred to or from the electrode—the electrochemical step(s)—and some chemical steps before and/or after the electrochemical steps. Electrode reactions may thus be divided into two main types (A and B) depending upon whether the electron transfer occurs directly between the electrode and the substrate (A) or the electron is transferred to (or from) another species which then reacts with the substrate (B).

A. Direct Electron Transfer Reactions

Reactions following the direct electron transfer mechanism may be classified according to whether the potential necessary for the electron transfer can be reached within the decomposition potentials of the medium or not. In the former case (A 1) a reaction at controlled electrode potential can occur with 100% current efficiency, in the latter (A 2) a certain part of the current is always consumed in the decomposition of the medium and the current yield depends on how well the substrate competes with the medium for the electrons.

I. The A I mechanism

Reactions of this type have been investigated in more detail than the others. On the basis of results obtained by some electroanalytical methods (e.g. polarography, cyclic voltammetry) meaningful predictions can be made about the optimum conditions for an electrolysis; e.g. the dependence of the electrode potential on experimental conditions and the number of electrons participating in the electrode reaction can be found. A short introduction to the use of polarography as a guide in controlled potential electrolysis together with a brief description of cells and apparatus used in such experiments was given in a previous volume of this series[1].

In the classical electrolytic reactions the current density, measured in A/dm^2, was the quantity which was controlled, possibly because it was the easiest factor to measure and keep constant. For a long time nearly all electrolytic reactions were performed with a control of the current density, although Haber[2, 3], as early as 1898, in his famous papers on the stepwise reduction of nitro compounds, realized that the potential of the working electrode was the proper quantity to control.

The differences in the two ways of controlling the electrolytic reaction are illustrated in Figure 1. In this figure curve I depicts the connexion between the current through the cell and the potential

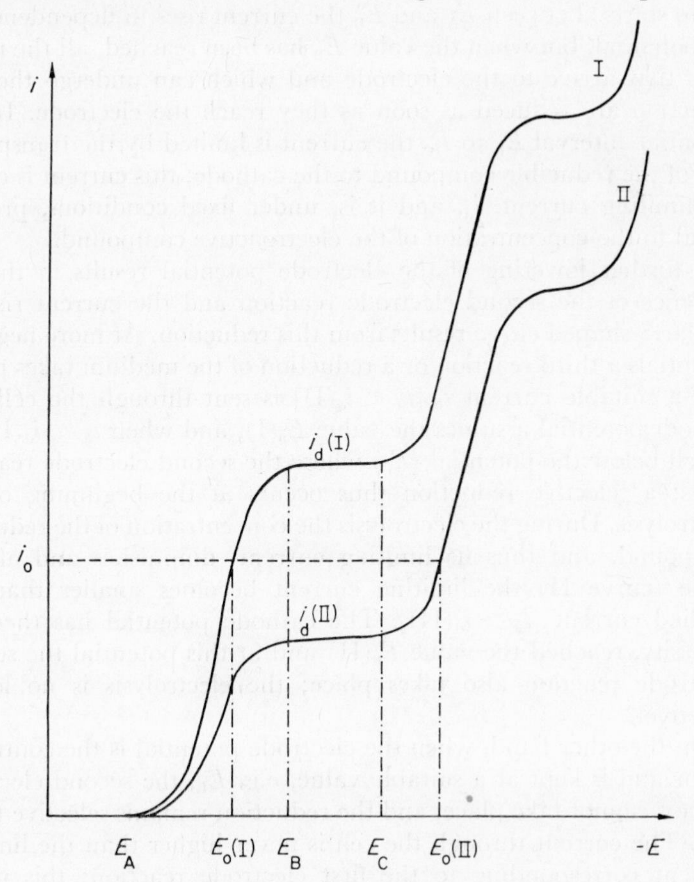

FIGURE 1. Schematic representation of the connexion between the current and the potential of the working electrode in a solution containing a compound with two groups reducible at different potentials. Curve I, before electrolysis; curve II, after the passage of some current; i_0 is an applied current, E_0(I) and E_0(II) the potentials corresponding to i_0; i_d is the limiting current, and E_A, E_B and E_C are applied potentials.

of the working electrode in the initial solution containing two reducible compounds or one compound with two groups reducible at different potentials. When the potential at the cathode is between 0 and E_A, no electron transfer across the electrical double layer can take place and thus no current runs through the cell. If the cathode potential is made more negative, the electron transfer becomes possible, that is, the reduction of the most easily reducible compound or group starts. Between E_A and E_B the current rises in dependence on the potential, but when the value E_B has been reached, all the molecules that arrive to the electrode and which can undergo the first reduction are reduced as soon as they reach the electrode. In the potential interval E_B to E_C the current is limited by the transportation of the reducible compound to the cathode; this current is called the limiting current, i_d, and it is, under fixed conditions, proportional to the concentration of the electroactive compound.

A further lowering of the electrode potential results in the occurrence of the second electrode reaction and the current rises; a similar S-shaped curve results from this reduction. At more negative potentials a third reaction or a reduction of the medium takes place.

If a suitable current i_0 [$i_0 < i_d(I)$] is sent through the cell, the cathode potential assumes the value $E_0(I)$, and when $i_0 < i_d(I)$ this is well below the potential (E_C) where the second electrode reaction starts; a selective reduction thus occurs at the beginning of the electrolysis. During the electrolysis the concentration of the reducible compound, and thus its limiting current, diminishes and after a while (curve II) the limiting current becomes smaller than the applied current [$i_0 > i_d(II)$]. The cathode potential has then, by necessity, reached the value $E_0(II)$ and at this potential the second electrode reaction also takes place; the electrolysis is no longer selective.

On the other hand, when the electrode potential is the controlled factor and is kept at a suitable value, e.g. E_B, the second electrode process cannot take place, and the reduction remains selective to the end. The current through the cell is never higher than the limiting current corresponding to the first electrode reaction; this means that the current decreases during the reduction and becomes very small towards the end of the reaction, as the limiting current is proportional to the concentration of the electroactive material.

2. The A 2 mechanism

Many of the 'classical' electrolytic reactions occur at a potential which is either more negative (reductions) or more positive (oxidations) than the decomposition potentials of the medium. The mechanism of such reactions must be investigated in each case, but can usually be classified as one of the following three cases. (i) A direct electron transfer from electrode to substrate (A 2), (ii) A formation of 'solvated electrons' which in turn reduce the substrate (B 1), or (iii). A formation of an active species in the electrochemical step (adsorbed hydrogen, active metals, hydrogen peroxide, hydroxyl radicals, etc.) which reacts chemically with the substrate (B 2).

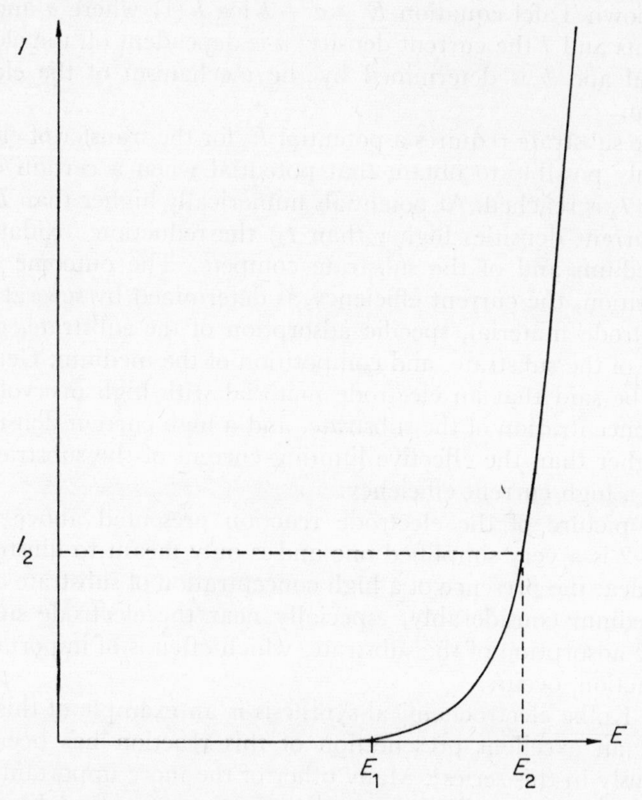

FIGURE 2. Schematic representation of the connexion between electrode potential E and the current density I in the decomposition reaction of the medium. Hydrogen (oxygen) evolution starts at E_1; if the substrate requires for the electrochemical step a potential (numerically) higher than E_2, transfer of electrons to the substrate only becomes appreciable at $I > I_2$.

In hydroxylic media the electrode reactions involve probably a direct electron transfer between electrode and substrate when the electrode material has a high overvoltage and a low catalytic effect; such reactions are not, in principle, different from those treated as A 1, only the potential necessary to bring about the electron transfer is (numerically) higher than that at which the medium is decomposed. This is illustrated in Figure 2.

In Figure 2 is shown the dependence of the current density I on the electrode potential E in an electrolysis involving the decomposition of the medium which may be evolution of hydrogen or oxygen, starting approximately at $E = E_1$. Curve I can be described by the well known Tafel equation $E = a + b \log I$ (1) where a and b are constants and I the current density; a is dependent on the electrode material and b is determined by the mechanism of the electrode reaction.

If the substrate requires a potential E_2 for the transfer of electrons it is only possible to obtain that potential when a certain current density I_2 is reached. At potentials numerically higher than E_2 (and thus current densities higher than I_2) the reduction (oxidation) of the medium and of the substrate compete. The outcome of this competition, the current efficiency, is determined by several factors as electrode material, specific adsorption of the substrate, concentration of the substrate, and composition of the medium. Generally, it can be said that an electrode material with high overvoltage, a high concentration of the substrate, and a high current density (but not higher than the effective limiting current of the substrate) will favour a high current efficiency.

The picture of the electrode reaction presented above and in Figure 2 is a very simplified one and is only meant to illustrate the basic idea; the presence of a high concentration of substrate changes the medium considerably, especially near the electrode surface if specific adsorption of the substrate, which often is of importance for the reaction, occurs.

The Kolbe electrochemical synthesis is an example of this mechanism; an excellent presentation of this reaction has been given previously in this series[4]. Many other of the more important 'classical' electrolytic reactions (e.g. the reduction of carboxylic acids to alcohols at lead cathodes) also follow this path, but many points in these processes are not clear; a rich field is here wide open for investigations by modern methods.

B. *Indirect Electrochemical Reactions*

I. Reduction by solvated electrons

Among the electrolytically produced reagents which have been considered to be operating is the solvated electron. It may be formed in reductions in nonaqueous media such as ammonia[5], ethylenediamine[6, 7], methylamine[8], polyethylene glycol dimethyl ether[9] and ethanol containing hexamethylphosphoramide[10], at electrodes with high hydrogen overvoltage and with tetraalkylammonium or lithium ions as supporting electrolyte. The reaction is formally

$$e^- + \text{solv} \rightleftarrows e^-_{\text{solv}}$$
$$e^-_{\text{solv}} + S \rightleftarrows S^- + \text{solv}$$

where S is the substrate.

The standard potential of the solvated electron is about $-2{\cdot}6$ *vs* Normal Hydrogen Electrode (NHE)[11] and solvated electrons can only be formed in the absence of more easily reducible substrates.

2. Other indirect electrochemical reactions

Whereas the reductions involving solvated electrons stand between the purely electrochemical and the indirect reductions, the reactions involving the formation of adsorbed hydrogen, amalgams, hydroxyl radicals, halogens, etc., are clearly indirect electrolytic reactions.

a. The electrocatalytic reduction may be important at electrodes with low hydrogen overvoltage and high catalytic activity, and its mechanism is closely related to the mechanism of the hydrogen evolution. The mechanism of this reaction at a platinum electrode in acid solution has been proposed[12] to be

$$H^+_{(\text{solv})} + e^- \rightleftarrows H_{(\text{ads})} \tag{2}$$
$$H^+_{(\text{solv})} + H_{(\text{ads})} + e^- \rightleftarrows H_{2(\text{ads})} \tag{3}$$
$$H_{2(\text{ads})} \rightleftarrows H_{2(\text{gas})} \tag{4}$$

In this reaction both equation (2) and equation (3) may be rate-controlling and the factor b in the Tafel equation (equation 1) acquires the value 0·116, or 0·038 when equation (2) or equation (3), respectively, is the rate-determining step.

An important point in the reaction mechanism for the electrocatalytic reactions may be illustrated by an investigation of the electrolytic reduction of acetone at a Raney-nickel cathode in alkaline solution[13]. It was found that the dependence of the electrode potential on log I had the form shown in Figure 3. In the absence of acetone the hydrogen evolution commenced at E_{II} and the slope of the straight line was 0·04 V cm^2/A. In the presence of acetone the current started to rise at a numerically lower potential (E_{I}), and between E_{I} and E_{II} the slope was 0·116; above E_{II} the slope again was 0·04. The current density, but not the potential at which the change in slope took place, was dependent on the concentration of acetone.

It has been suggested[14] that these results can be explained by assuming that the reactions (equations 2—4) occur also at Raney-nickel electrodes in alkaline solution, and that reaction (equation 5) is a fast reaction. Between E_I and E_{II} the following reactions are assumed to occur with a measurable rate.

$$H^+_{(solv)} + e^- \rightleftarrows H_{(ads)} \quad (2)$$

$$H_{(ads)} + \text{Substrate}_{(ads)} \longrightarrow \text{Product} \quad (5)$$

The Tafel slope 0·04 is in accordance with the assumption that equation (3) is the slow step in the hydrogen evolution in the absence

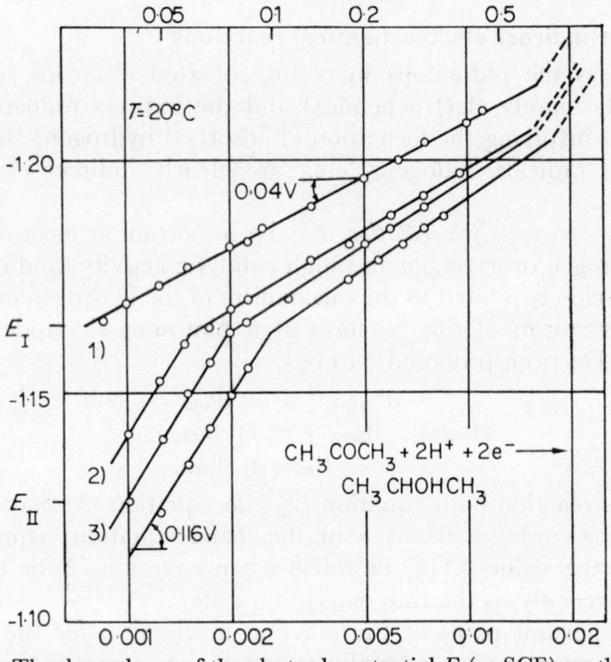

FIGURE 3. The dependence of the electrode potential E (vs SCE) on the current density I (on a semi-logarithmic scale) in the electrocatalytic reduction of acetone at a Raney-nickel electrode. Medium: aqueous solution of KOAc + KOH. Concentration of acetone: (1) 0; (2) 0·26; (3) 0·34; (4) 0·52 mole/l.

of acetone; reaction (5) is then able to compete successfully with equation (3) for the adsorbed hydrogen atoms; if reaction (5) also would be reasonably fast compared to reaction (2), the Tafel slope would change from 0·038 to 0·118, which compares well with that found experimentally (0·04, 0·116).

Recent investigations[15] have shown that two independent paths are accessible for the electrocatalytic reduction of acetone in 6N H_2SO_4 at a platinized platinum electrode; one leads to propane and the other to isopropyl alcohol. The rate of formation of these two products depends on the voltage and on the history of the electrode.

Two types of adsorption for hydrogen are found, one of them being interstitial. The 'interstitial' hydrogen is important in the electrocatalytic reduction of acetone to isopropyl alcohol, but not in the reaction leading to propane. At an anodic treated electrode no interstitial hydrogen is found, but the concentration of it is gradually built up. Besides being of importance in the reduction of substrate the interstitial hydrogen modifies the adsorption properties of the electrode, and the importance of the adsorption of the substrate to the electrode prior to the reduction is evidenced by the kinetics of the reaction.

The nature of the reaction changes with time; at a freshly anodized electrode acetone is preferentially reduced to propane whereas later isopropyl alcohol is the main product; eventually both electrocatalytic reactions are suppressed and hydrogen evolution becomes the main reaction.

In many cases the formation of amalgams, active metals, hydrogen peroxide, halogens, or hydroxyl radicals has been postulated as the electrochemical step, which is then followed by a purely chemical reaction. One of the usual arguments for these intermediates is that the reaction follows a route which may be duplicated by the chemical reagent, but this does not prove the presence of these intermediates in the electrolytic reaction. In some, but rather few cases, the occurrence of an indirect electrolytic reaction has been proved conclusively.

II. ELECTROLYTIC PREPARATION OF ALCOHOLS AND PHENOLS

A. By Reduction

Hydroxyl compounds may be formed in the electrolytic reduction of carboxylic acids, esters, amides, thioamides, acid chlorides, ketones and aldehydes, in some cases as main product, in other cases in minor amounts.

I. From acid derivatives

a. Carboxylic acids. The carboxyl group is rather difficult to reduce electrolytically, and its reduction potential is less negative than the decomposition potential of the medium only in the cases where the carboxyl group is activated by a suitable electron-attracting substituent. The substituent is also activated by the electron-attracting carboxyl group, and sometimes there is a competition between the reduction of the carboxyl group and of the substituent; in such cases the reduction of the carboxyl group will generally be favoured in acid solution and the reduction of the substituent in alkaline media.

Examples of this kind are found in the reduction of carboxy-derivatives of some π-electron deficient aromatic heterocyclic compounds. Thus, e.g., isonicotinic acid[16] (**1**), 2-carboxythiazole[17] and 2-carboxyimidazole[18] are reduced in acid solution, through the aldehyde, to the corresponding alcohols. The reduction of isonicotinic acid follows the scheme[16]:

$$\underset{(\mathbf{1})}{\overset{+}{\mathrm{H}}\mathrm{NC_5H_4COOH}} \xrightarrow{2e+2H^+} \underset{(\mathbf{2})}{\overset{+}{\mathrm{H}}\mathrm{NC_5H_4CHO}} \overset{H_2O}{\rightleftharpoons} \underset{(\mathbf{3})}{\overset{+}{\mathrm{H}}\mathrm{NC_5H_4CH(OH)_2}}$$

$$\underset{(\mathbf{2})}{\overset{+}{\mathrm{H}}\mathrm{NC_5H_4CHO}} \xrightarrow{2e+2H^+} \underset{(\mathbf{4})}{\overset{+}{\mathrm{H}}\mathrm{NC_5H_4CH_2OH}}$$

If the reduction of the acid is carried out at low temperature and at a suitable pH (1—4) and is stopped after the uptake of two electrons per molecule, the aldehyde **2** can be isolated in fair yield, as in this medium it is present predominantly in the non-reducible hydrated form, the *gem*-diol (**3**). When, however, the electrolysis is allowed to proceed to completion, the alcohol **4** is the isolated product, as the free aldehyde present in equilibrium with the hydrated form is reducible at a less negative potential (more easily reducible) than the acid. If the reduction is carried out at a slightly higher pH (e.g. 5—6), the aldehyde cannot be isolated as an intermediate and the carbinol is formed in a four-electron reduction. As only a few aldehydes are highly hydrated, the latter type of reduction is more typical for the reduction of carboxylic acids than the former in which the aldehyde is an isolable intermediate.

In alkaline solution the reduction of such heterocyclic acids takes place in the nucleus, and, using acids in which more than one nitrogen atom is present in a six-membered ring, reduction of the ring occurs even in acid solution.

Oxalic acid and its mono- and di-ester are polarographically reducible[19, 20] as the two carboxyl groups activate each other, and oxalic acid is reducible at a mercury or lead cathode through the aldehyde to glycolic acid. When the reduction is performed in dilute sulphuric acid at low temperatures, where the rate of the dehydration of the hydrated aldehyde is low, glyoxylic acid may be isolated; the yield, determined as glyoxylic acid phenylhydrazone, is reported to be 87·5%[21].

A similar activation by a carboxyl group is found in phthalic acid which may be reduced to phthalide[22] in weakly acid solution; in strongly acid solution, however, dihydro compounds were formed at a lead cathode from phthalic and terephthalic acid, whereas isophthalic acid produced the dialcohol under such conditions[23].

Generally, however, the carboxyl group is not reducible at a potential between the decomposition potentials of the medium. This means that a polarographic or voltammetric reduction wave cannot be obtained in such cases, and the valuable guidance with respect to, e.g., reduction potential, number of electrons in the electrode reaction, pH-dependence of reduction potential normally acquired from such curves, is not available.

The rather high negative reduction potential of the carboxyl group also means that a reduction of this group is not possible without a simultaneous reduction of the cations of the supporting electrolyte which most often in this kind of reduction are hydrogen ions; this electrode reaction is therefore an example of the kind illustrated in Figure 2, and, accordingly, in the competition between the reduction of hydrogen ions and carboxylic acid, the reduction of the latter is favoured by a high concentration of the substrate, an electrode with high overvoltage, and a high current density. An adsorption of the substrate to the electrode might be necessary for the reaction, but the importance of adsorption in such reactions has not been sufficiently investigated.

Aromatic acids are generally reducible at a lead cathode to the corresponding benzyl alcohol in a medium containing sulphuric acid[23–26]. A typical catholyte could consist of a mixture of 70 g alcohol and 30 g sulphuric acid; in this medium 20—40 g benzoic acid is reduced with a current density of 0·1 A/cm^2 [24]; yield of benzyl alcohol, 85%. Table 1 gives the yields of alcohols from the reduction of some carboxylic acids.

The yields of benzyl alcohols are generally good, except when the benzene ring carries two substituents *ortho* to the carboxyl group; the substituents may interfere with the coplanarity of the benzene ring and the (possibly protonated) carboxyl group. Lack of coplanarity might influence both the reduction potential and the adsorbability of the carboxylic acid. The potential at the cathode was generally not measured in the 'classical' electrolytic reactions, but it would be interesting to compare the electrode potential prevailing during the reduction of benzoic acid with that found in the reduction of phenylacetic acid. The latter is also reducible at a lead cathode in sulphuric acid, but the yield of alcohol is inferior to that obtained from benzoic acid[24]. If the interpretation of the reaction mechanism given above (and in section I.B.) is correct, the cathode potential found during the reduction of phenylacetic acid would be expected to be more negative than that prevailing during the reduction of benzoic acid. Conclusions drawn on the basis of structural considera-

TABLE 1. Yields of benzyl alcohols in the electrolytic reduction of substituted benzoic acids at a lead cathode in a medium containing sulphuric acid. Current density ca 0·1 A/cm².

Substituents	Yield (%) of subst. benzyl alcohol	Ref.
None	85	24
3-Bromo	75	24
3-Carboxy	60	23
3-Hydroxy	45	24
2-Amino	70	24
2-Amino-3-methyl	45	25
2-Amino-4-methyl	65	25
2-Amino-5-methyl	76	25
2-Amino-6-methyl	about 10	25
2-Amino-3,5-dimethyl	73	25
2-Amino-3,6-dimethyl	poor	25

tions give the same result, as the electron-withdrawing phenyl ring is further removed from the carboxyl group in phenylacetic acid than in benzoic acid.

In basic medium the carboxyl group is not reduced; instead the benzene ring is attacked, and from a reduction of benzoic acid under these conditions 1,2,3,4-tetrahydrobenzoic acid was isolated[23].

b. Esters. The reduction of esters is similar to the reduction of the acids; thus the carbethoxy pyridines are reduced through the aldehydes to the alcohols[16], ethyl phthalate yields phthalide in weakly acid solution, and methyl benzoate is reduced to benzyl alcohol in sulphuric acid at a lead cathode. In the latter case some methyl benzyl ether was obtained, and it was the main product when the solvent was methanolic sulphuric acid[21].

c. Amides and thioamides. On electrolytic reduction amides generally yield a mixture of alcohol and amine, but sometimes one of the products predominates. In acid solution the product distribution is determined by the relative rates of the loss of water or amine from the primarily formed reduction product, the *gem*-aminoalcohol (**5**), to either aldimine or aldehyde, according to the scheme.

$$RCONR'_2 \xrightarrow{2e,\ +2H^+} RCH(OH)NR'_2\ (\mathbf{5})$$

$$RCH(O\overset{+}{H}_2)NR'_2 \underset{}{\overset{H^+}{\rightleftharpoons}} RCH(OH)NR'_2\ (\mathbf{5}) \overset{H^+}{\rightleftharpoons} RCH(OH)\overset{+}{N}HR'_2$$

$$RCH(O\overset{+}{H}_2)NR'_2 \xrightarrow{-H_2O} R-CH=\overset{+}{N}R'_2 \xrightarrow{2e,\ +2H^+} RCH_2\overset{+}{N}HR'_2$$

$$RCH(OH)\overset{+}{N}HR'_2 \xrightarrow{-HNR_2} R-CH=\overset{+}{O}H \xrightarrow{2e,\ +H^+} RCH_2OH$$

An illustrative example is the reduction of different isonicotinic amides[27] at a mercury cathode; isonicotinic amide is in dilute hydrochloric acid reduced predominantly through the aldehyde, which it is possible to obtain in good yield, to the alcohol, whereas the anilide forms some 40% anilinomethylpyridine together with the alcohol; isonicotinic *N*-methylanilide yields predominantly the pyridylcarbinol. Similar results were obtained from other heterocyclic amides[17, 18].

Benzamide yields on reduction in sulphuric acid at a lead cathode a mixture of benzyl alcohol and benzylamine[28, 29], and other aromatic amides behave similarly. Table 2 gives the yields of amine and alcohol from the reduction of some ring substituted benzamides. From the reduction of *N*-substituted benzamides under these conditions no formation of alcohols has been reported.

Aliphatic amides[30] may be reduced electrolytically to alcohols in fair to excellent yield at a smooth platinum cathode in an amine-medium containing lithium chloride as supporting electrolyte. A typical catholyte would consist of 0·01 mole amide, 0·8 mole lithium chloride and 450 ml anhydrous methylamine; Table 3 lists the yields of alcohols from some primary, secondary and tertiary aliphatic amides.

In the presence of a hydrogen donor stronger than the amine, e.g., alcohol, a high yield of aldehyde or its reaction product with

TABLE 2. Yields of amines and alcohols in the electrolytic reduction of some substituted benzamides at a lead cathode[29] in alcoholic sulphuric acid; current density 0·2 A/cm^2.

Substituents	Yield of amine	Yield of alcohol
None	74	23
2-Methyl	83	11
3-Methyl	53	35
4-Methyl	79	18
4-Methoxy	73	22
3-Bromo	64	19
4-Bromo	67	—
4-Chloro	65	—

TABLE 3. Yields of alcohols in the electrolytic reduction of aliphatic amides at a smooth platinum cathode in methylamine containing lithium chloride[30].

Amide	(moles)	Methylamine (ml)	Yield of alcohol (%)
$CH_3(CH_2)_4CONHMe$	(0·05)	600	51
$CH_3(CH_2)_4CONH_2$	(0·05)	700	58
$CH_3(CH_2)_8CONH_2$	(0·05)	600	59
$CH_3(CH_2)_{10}CONH_2$	(0·02)	450	79
$CH_3(CH_2)_{10}CONHMe$	(0·02)	450	84
$CH_3(CH_2)_{12}CONH_2$	(0·02)	450	92
$CH_3(CH_2)_{12}CONMe_2$	(0·02)	450	97
$CH_3(CH_2)_{14}CONH_2$	(0·01)	450	86
$CH_3(CH_2)_{16}CONH_2$	(0·01)	450	79
$CH_3(CH_2)_{16}CONHCH_3$	(0·01)	450	72

methylamine is obtained; no reduction product from an *N*-methylimine has been isolated. From these observations the first scheme on the following page was proposed[30].

In the presence of a stronger proton donor than methylamine the *gem*-aminoalcohol (**6**) survives until the reaction mixture is worked up.

In some cases the *gem*-aminoalcohol primarily formed in the re-

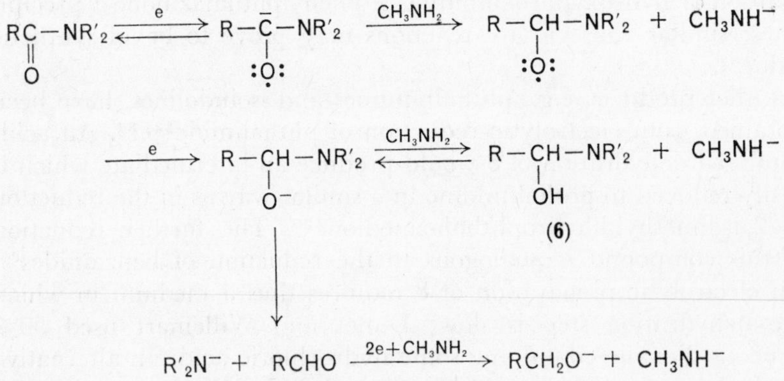

$$\underset{\displaystyle \overset{\|}{O}}{RC}{-}NR'_2 \underset{}{\overset{e}{\rightleftharpoons}} R{-}\underset{\displaystyle :\ddot{O}:}{\overset{-}{C}}{-}NR'_2 \xrightarrow{CH_3NH_2} R{-}\underset{\displaystyle :\ddot{O}:}{CH}{-}NR'_2 + CH_3NH^-$$

$$\xrightarrow{e} R{-}\underset{\displaystyle O^-}{CH}{-}NR'_2 \overset{CH_3NH_2}{\rightleftharpoons} R{-}\underset{\displaystyle OH}{CH}{-}NR'_2 + CH_3NH^-$$

(6)

$$\downarrow$$

$$R'_2N^- + RCHO \xrightarrow{2e+CH_3NH_2} RCH_2O^- + CH_3NH^-$$

duction of an amide or imide is sufficiently stable for its isolation. Derivatives of phthalimide (**7**) are reduced to hydroxyphthalimidines (**8**) in slightly acid solution[31]; in a medium of high alcohol content the isolated product is predominantly ethoxyphthalimidine. In weakly alkaline solution (**7**) is reduced in two one-electron reductions, and the result of the first one-electron reduction is a radical, which dimerizes.

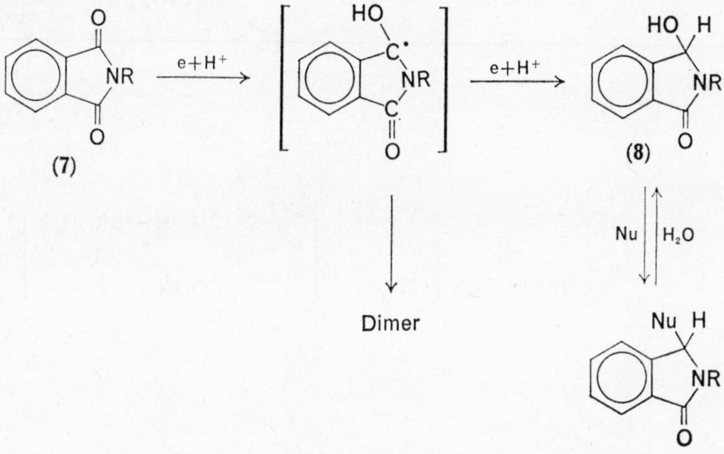

(7) $\xrightarrow{e+H^+}$ [radical: HO–C•, NR, C=O] $\xrightarrow{e+H^+}$ (8) (HO H, NR, O)

[radical] $\downarrow$ Dimer

(8) $\underset{H_2O}{\overset{Nu}{\rightleftharpoons}}$ (Nu H, NR, O)

In the presence of a nucleophile (Nu) the hydroxyphthalimidine **8**, which is a derivative of phthalaldehydic acid, reacts with the nucleophile; in the presence of ethyl alcohol ethoxyphthalimidine is formed and in the presence of isoindoline, 1-*N*-isoindolino-3-oxoisoindoline[31]. When phenylhydrazine is added to an aqueous solution of hydroxyphthalimidine, 2-phenylphthalazinone-1 precipitates; similar ring closure reactions may prove to be of synthetic value[32].

Other products, e.g., phthalimidines and isoindolines, have been obtained from electrolytic reduction of phthalimide[33, 34]. An acid-catalysed dehydration of **8** would produce an intermediate which is easily reduced to phthalimidine in a similar way as in the reduction of 2,3-dimethyldihydrophthalazinedione[35]. The further reduction of this compound is analogous to the reduction of benzamides[26]. An electrolytic preparation of **8** requires thus a medium in which the dehydration step is slow; Dunet and Willemart used 50% aqueous dioxane containing some hydrochloric acid; an alternative method is reduction in cold base at −1·6 V (SCE)[32].

In some derivatives of phthalimide, such as *N*-anilinophthalimide (**9**), another type of reaction may follow the initial reduction to a hydroxyphthalimidine, and from **9** is formed either 3-phenyl-*ψ*-phthalazinone (**10**) or 2-phenylphthalazinone (**11**). The following scheme has been suggested[36].

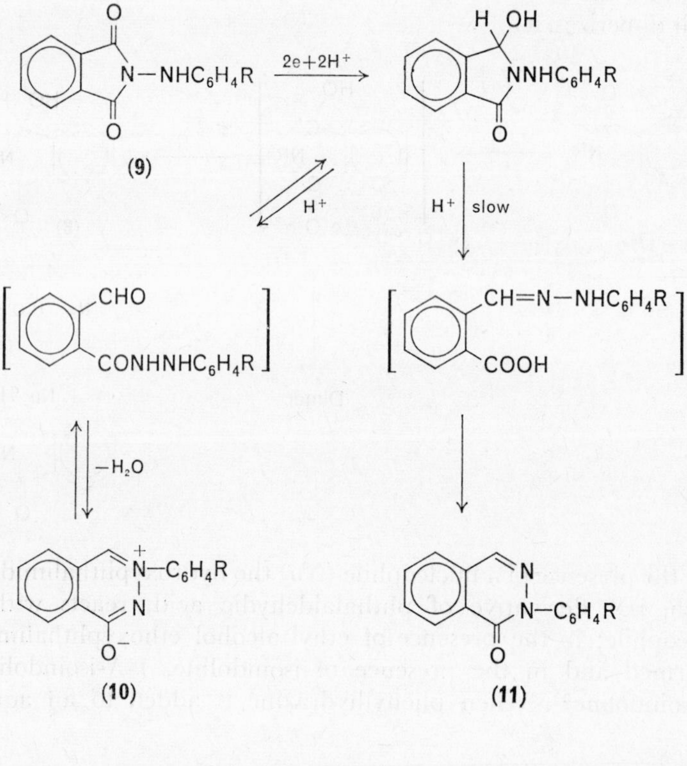

N−NHC$_6$H$_4$R (**9**) $\xrightarrow{2e+2H^+}$ H OH, NNHC$_6$H$_4$R

H$^+$; H$^+$ slow

[CHO, CONHNHC$_6$H$_4$R] ; [CH=N−NHC$_6$H$_4$R, COOH]

−H$_2$O

$\overset{+}{N}$−C$_6$H$_4$R, N, O$^-$ (**10**) ; N, N−C$_6$H$_4$R, O (**11**)

Thioamides are reduced more easily than the corresponding amides, and the primary reduction product, the *gem*-aminothiol, is generally more stable than the corresponding *gem*-aminoalcohol; the *gem*-aminothiol from the reduction of isonicotinic thioamide[27] is thus stable for hours in acid solution at 5—10°. The thioamides are generally not well suited for the preparation of alcohols.

d. Acid chlorides. The electrolytic reduction of an acid chloride must be performed in an indifferent medium as, e.g., acetonitrile; a polarogram of benzoyl chloride in this medium shows a reduction wave due to the hydrogenation of the carbon–chlorine bond followed by the reduction wave of the benzaldehyde thus formed[37].

When a preparative reduction is made in an aprotic medium such as acetonitrile the question of availability of protons arises. In a voltammetric experiment the amount of material reduced is so small that the residual water present in 'dry' acetonitrile can furnish the necessary protons, but when larger amounts are reduced, a suitable proton donor must be added if the reduction requires protons. The proton donor must be neither a very strong acid, as protons are then preferentially reduced, nor a very weak one. Phenol is in some cases an acceptable compromise.

Benzoyl chloride on reduction in dry acetonitrile yields benzil and its reduction products[37] whereas benzaldehyde and its reduction products are obtained in the presence of phenol as proton donor. The following reduction scheme may be proposed for the first step,

$$\mathrm{RCOCl} \xrightarrow{e} \mathrm{Cl^-} + \mathrm{R\dot{C}O} \xrightarrow{e+H^+} \mathrm{RCHO}$$

$$\mathrm{R\dot{C}O} \downarrow \quad \begin{array}{c} \mathrm{R{-}C{=}O} \\ | \\ \mathrm{R{-}C{=}O} \end{array}$$

The reductive cyclization of the dichlorides of dicarboxylic acids to the corresponding diketones may prove useful in the formation of rings as an alternative to the acyloin condensation.

2. From aldehydes and ketones

a. Reduction mechanism. Electrolytic reduction of aldehydes and ketones may produce pinacols, alcohols or hydrocarbons depending upon the experimental conditions, especially upon pH and electrode material. In the following the reduction to pinacols and alcohols is discussed.

Information concerning the reduction mechanism at a mercury electrode may be obtained from polarographic studies; references to such investigations may be found in a recent paper by Zuman[38]; the discussion below follows essentially the one given in that paper on the reduction of aromatic carbonyl compounds; aliphatic ketones are reduced at rather negative potentials which can only be reached in neutral and alkaline solution containing tetraalkylammonium ions as supporting electrolyte.

In acid solution a protonation of the carbonyl compound takes place prior to the electron transfer and the radical thus formed may either react chemically or be reduced further. Under these conditions the reduction may be described by the following steps:

$$H^+ + ArCOR \rightleftarrows Ar\overset{+}{C}(OH)R \tag{6}$$

$$Ar\overset{+}{C}(OH)R + e \underset{}{\overset{E_1}{\rightleftarrows}} Ar\dot{C}(OH)R \tag{7}$$

$$2\ Ar\dot{C}(OH)R \longrightarrow \text{dimer} \tag{8}$$

$$Ar\dot{C}(OH)R + Hg \longrightarrow \text{organometallic compound} \tag{9}$$

$$Ar\dot{C}(OH)R + \text{solvent} \longrightarrow \text{products} \tag{10}$$

$$Ar\dot{C}(OH)R + e \xrightarrow{E_2} Ar\overset{-}{C}(OH)R \tag{11}$$

$$Ar\overset{-}{C}(OH)R + H^+ \rightleftarrows ArCHOHR \tag{12}$$

The pH-dependent reduction potential E_1 of the first electron transfer is generally less negative than the pH-independent potential E_2 of the second electron transfer, which often at low pH is more negative than the reduction potential of the hydrogen ions, and only one reduction wave is then seen on a polarogram. A reduction of an aromatic carbonyl compound in acid solution at a mercury cathode with a cathode potential controlled at E_1 would thus be expected to give a high yield of pinacol, and this is also found in a preparative reduction. The stereochemistry of the products is discussed below.

As E_1, but not E_2, is dependent on pH, E_1 approaches E_2 at higher pH and they may merge at a certain pH above which only one two-electron wave is seen.

In the medium pH-range the preprotonation becomes unimportant as the equilibrium (equation 6) is shifted too far to the left. The electron transfer occurs then to the unprotonated carbonyl compound and the radical ion acquires a proton; this radical may react chemically or be reduced further.

$$\text{ArCOR} + e \xrightarrow{E_3} \text{Ar}\dot{\text{C}}(\text{O}^-)\text{R} \qquad (13)$$

$$\text{Ar}\dot{\text{C}}(\text{O}^-)\text{R} + \text{H}^+ \rightleftarrows \text{Ar}\dot{\text{C}}(\text{OH})\text{R} \qquad (14)$$

$$\text{Ar}\dot{\text{C}}(\text{OH})\text{R} \longrightarrow \text{dimer} \qquad (8)$$

$$\text{Ar}\dot{\text{C}}(\text{O}^-)\text{R} + \text{Ar}\dot{\text{C}}(\text{OH})\text{R} \longrightarrow \text{dimer} \qquad (15)$$

$$\text{Ar}\dot{\text{C}}(\text{OH})\text{R} + e \xrightarrow{E_2} \text{Ar}\bar{\text{C}}(\text{OH})\text{R} \qquad (11)$$

$$\text{Ar}\bar{\text{C}}(\text{OH})\text{R} + \text{H}^+ \rightleftarrows \text{ArCHOHR} \qquad (12)$$

Polarographically a two-electron wave is observed which means that under these conditions the second-order reactions (equations 8 and 15) cannot compete successfully with the further reduction; at higher concentration and current density varying yields of pinacols are formed.

In alkaline solution the polarographic behaviour depends—besides on pH—on the nature and concentration of the supporting electrolyte. The reduction of the primarily formed radical ion (equation 16) occurs at a more negative potential (E_4) than that of the first reduction (E_3); E_4 is sometimes more negative than the decomposition potential of the medium. The reduction potential of the coordinated radical (E_5) is most often between E_3 and E_4. (M^+ is a cation.)

$$\text{ArCOR} + e \xrightarrow{E_3} \text{Ar}\dot{\text{C}}(\text{O}^-)\text{R} \qquad (13)$$

$$\text{Ar}\dot{\text{C}}(\text{O}^-)\text{R} + e \xrightarrow{E_4} \text{ArC}^-(\text{O}^-)\text{R} \underset{}{\overset{2\text{H}^+}{\rightleftarrows}} \text{ArCHOHR} \qquad (16)$$

$$\updownarrow +\text{M}^+$$

$$\text{Ar}\dot{\text{C}}(\text{OM})\text{R} + e \xrightarrow{E_5} \text{ArC}^-(\text{OM})\text{R} \overset{2\text{H}^+}{\rightleftarrows} \text{ArCHOHR} + \text{M}^+ \qquad (17)$$

On electrostatic grounds the radical ion would not be expected to dimerize with another negatively charged radical ion; at least one of the partners must coordinate with a positively charged species, a proton or another cation, prior to the coupling as in equation (5); if the protonation took place near the electrode surface, the protonated species would be reduced immediately; this would indicate that if equation (5) were responsible for the formation of pinacols under basic conditions, the dimerization takes place at some distance from the electrode. A dimerization as equation (18) must also be taken into consideration.

$$2\ \text{Ar}\dot{\text{C}}(\text{OM})\text{R} \longrightarrow \text{dimer} \qquad (18)$$

The reaction mechanism must also take into account that other coupling products than pinacols may be formed, thus **12** and **13** are among the products obtained from reduction in alkaline solution of 1-acetonaphthone[39] and 2′-aminoacetophenone[40], respectively, at a mercury electrode.

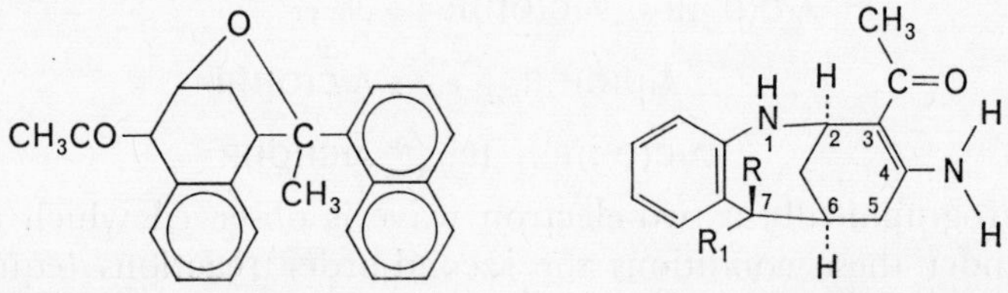

O
CH_3CO — C — CH_3

(12)

CH_3
H H C=O
N H
1 2 3
R 4 N
7 6 5
R_1 H
H

(13) R = OH, $R_1 = CH_3$
R = CH_3, R_1 = OH

13 is formed, as shown below, by coupling of a radical and a radical ion followed by a nucleophilic addition of the aromatic amino group to the α, β-unsaturated ketone:

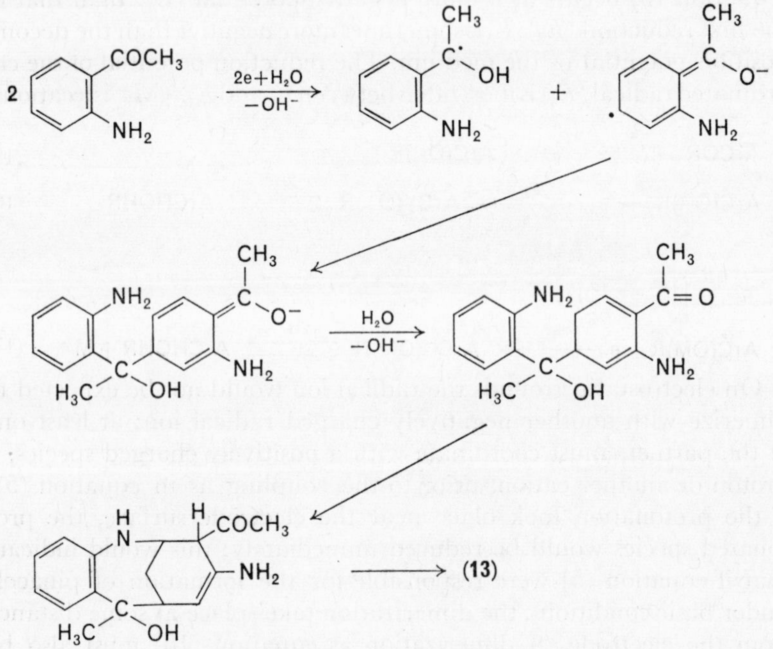

2 ($COCH_3$, NH_2) $\xrightarrow[-OH^-]{2e + H_2O}$ (CH_3, C·, OH, NH_2) + (CH_3, C, O^-, NH_2)

(NH_2, CH_3, C, O^-, NH_2, H_3C, C, OH) $\xrightarrow[-OH^-]{H_2O}$ (NH_2, CH_3, C=O, NH_2, H_3C, C, OH)

(H, N, H, $COCH_3$, NH_2, H_3C, C, OH) $\longrightarrow$ (13)

The effect of the cations on the polarographic behaviour of carbonyl compounds usually increases with size and charge of the ions, and the presence of a high concentration of tetraalkylammonium ions often results in the occurrence of a single rather than two polarographic waves. The effect of added metal cations on the stereochemistry of the product is not known.

The electrode reactions of ketones at other cathode materials have been much less thoroughly investigated, and the reductions have generally been performed without measurement and control of the electrode potential. At a platinum electrode acetophenone is reduced in alkaline solution to a pinacol[41]; unfortunately the stereochemistry (*d,l/meso*) was not reported; it would have been of interest to compare the results at an electrode with low hydrogen overvoltage and high catalytic activity with those obtained at a mercury cathode. The *d,l/meso* ratio of pinacols produced at a copper or tin electrode did not differ significantly from that found at a mercury cathode[42].

Acetone yields in acid solution at a lead cathode a mixture of pinacol, isopropyl alcohol and metalorganic compounds such as diisopropyllead and tetraisopropyllead[43]. The reduction of many other ketones follows the same pattern, and the results have been compiled in different monographs and reviews[44–46].

α,β-Unsaturated ketones, which are not conjugated to a phenyl ring, are usually reduced polarographically in a one-electron reduction. The primarily formed radical may form a mercury compound, dimerize at the β-carbon or at the carbonyl carbon, or form an unsymmetrical coupling product, so ε-diketones, pinacols, dihydrofurans, cyclopentenes and some other compounds may be formed[47, 48]. Often the hydrodimerization at the β-carbon to the saturated ε-diketone is the preferred reaction, but if a dimerization at this point is sterically unfavourable, α,β-unsaturated pinacols may be the major product; the latter reaction is important in the reduction of α,β-unsaturated steroid ketones[49]. In acid solution the reaction is:

$$\mathrm{RCH{=}CHCOR'} \xrightleftharpoons{\mathrm{H^+}} \mathrm{RCH{=}CHC^+(OH)R'}$$

$$\mathrm{RCH{=}CHC^+(OH)R'} \xrightarrow{e} \left\{ \begin{array}{c} \mathrm{RCH{=}CH\dot{C}(OH)R'} \\ \updownarrow \\ \mathrm{R\dot{C}HCH{=}C(OH)R'} \end{array} \right\} \longrightarrow \begin{array}{l} \text{dimers and} \\ \text{organometallic} \\ \text{compounds} \end{array}$$

In acid solution protonation takes place prior to the electron transfer; in alkaline solution a radical ion is formed primarily; this difference is reflected in the stereochemically different products found in acid and alkaline solution in the reduction of α,β-unsaturated steroid ketones.

The polarographic reduction of α,β-unsaturated aromatic carbonyl compounds is more complicated; often three waves are observed, and organomercury compounds, dimeric and saturated carbonyl compounds, pinacols and alcohols may be expected as products. A very thorough discussion of the polarographic behaviour of such compounds has recently been published[50].

Quinones are reduced to hydroquinones, and they form together a reversible redox system; sometimes semiquinones have considerable stability which can be judged from their polarographic behaviour. Of synthetic interest might be the fact that in a suitable medium a hydroquinone such as anthrahydroquinone may lose a hydroxyl group on reduction; in a medium consisting of 50 vol % ethanol and 50 vol % sulphuric acid anthraquinone is reduced at a suitable potential to anthrone[51] according to

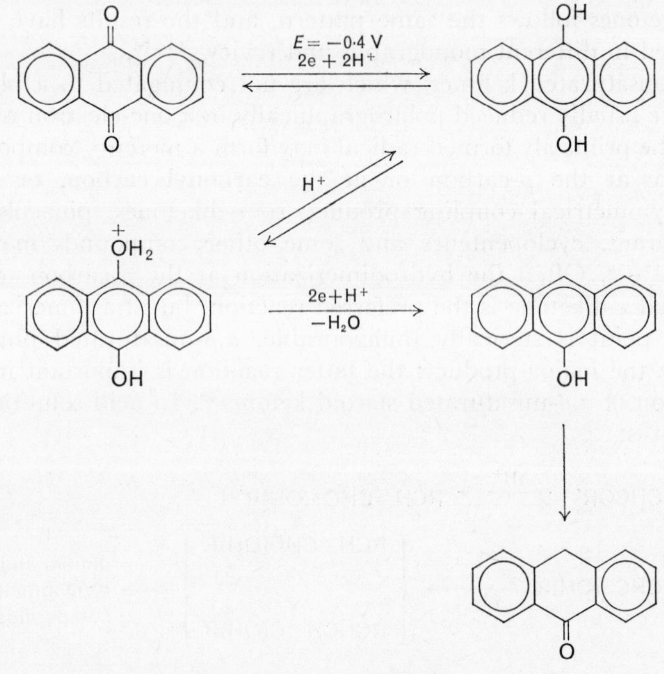

O
O
$E = -0{\cdot}4$ V
$2e + 2H^+$
OH
OH
H^+
$\overset{+}{O}H_2$
OH
$2e + H^+$
$-H_2O$
OH
O

The dehydration step (and the solubility) is favoured by higher temperatures and at 50—60° the reaction goes reasonably fast. By controlling the potential at $-0{\cdot}4$ V (SCE) further reduction of the anthrone is avoided.

b. Stereochemistry of the reduction of carbonyl compounds. Any reduction mechanism must take the stereochemistry of the products and its dependence on the medium into consideration. Several reports have appeared on the dependence on pH of the stereochemistry of the hydroxyl derivatives obtained by electrolytic reduction of carbonyl compounds. Thus different mixtures of pinacols were obtained in acid and alkaline solution from α,β-unsaturated steroid ketones[49], deoxybenzoin[42, 52], *p*-dimethylaminoacetophenone[53], and other acetophenones[42, 54, 55]. Also the stereochemistry of the reduction product, *erythro*-phenyl-α-phenylethyl-carbinol, from α-methyldeoxybenzoin has been reported[56]. The influence of added unsymmetrical constituents to the medium has recently received attention[57].

An important aspect of the pinacol formation is whether the dimerization occurs at the surface of the electrode or in the bulk of the solution. The electrical double layer is considered to consist of an inner layer of adsorbed molecules and a more diffuse outer layer; the electrical gradient is high (10^7 V/cm) in the inner layer, but falls rapidly to a negligible value. At a distance from the electrode of about 100 Å the influence of the electrode is no longer of importance.

Other questions are whether a radical, a radical anion, a radical anion coordinated with a cation, or a metalorganic compound is involved in the dimerization. A priori it cannot be excluded that a pinacol may be formed by reaction between a carbanion and an unreduced ketone.

One piece of evidence which may point to a fixation of the stereochemistry of the reaction product near the surface of the electrode is the reduction of benzil to stilbenediol[58]. The ratio of *cis*-stilbenediol to *trans*-stilbenediol was found to be dependent on the electrode potential in such a way that the *trans*-isomer predominates near the half-wave potential of benzil [$\sim$0·7 V (SCE) at pH 10] whereas *cis*-stilbenediol is the more abundant one about $-1{\cdot}0$ V (SCE); at more negative potentials the *trans*-isomer predominates again. It is well known that in the absence of strongly adsorbed species the mercury cathode is positively charged with respect to the solution at potentials more positive than $-0{\cdot}7$ V (SCE) and negatively charged at more negative values. The half-wave potential of benzil is thus near the point of zero charge; it has been suggested that the control of the stereochemistry in the reduction of benzil operates by the influence of the electric field in the electric double layer on the conformational equilibrium of benzil, as the dipoles in the carbonyl groups may be influenced by the field.

The analysis of the *cis–trans* ratio at the different potentials was made by anodic voltammetry at a hanging mercury electrode, at which the electrode reaction had proceeded for about 30 sec; as both of the stilbenediols tautomerize to benzoin (at different rates), an exhaustive electrolysis would not be applicable. The *trans*-isomer was taken to be the more easily reducible isomer, and this was substantiated by later work[59].

Another result which has been taken as evidence for a fixation of the stereochemistry near the surface of the electrode is the reduction of acetophenone to the alcohol in methanol in the presence of an optically active supporting electrolyte[57]. With 0·1M (−)-ephedrine hydrochloride as supporting electrolyte *R*-(+)-methylphenylcarbinol was obtained in 44% yield with an optical purity of 4·2%, whereas *S*-(−)-methylphenylcarbinol with an optical purity of 4·6% was obtained in 38% yield when (+)-ephedrine hydrochloride was used as electrolyte. The optical purity of the product was not raised when a 0·2M solution of ephedrine was used, but lowered to 3·1% when a 0·05M solution was employed.

These findings have been interpreted as evidence for the theory that it is the ephedrine adsorbed at the interface which predominantly influences the stereochemistry of the product and to a much lesser degree the electrolyte present in the bulk of the solution.

Another conclusion was reached in a work on the pinacol isomer distribution in the reduction of some acetophenones, benzaldehydes and related compounds[42, 53]. The results are given in Table 4. In acid solution the coupling of two neutral radicals is considered, whereas a neutral radical and a radical ion rather than two radical ions are believed to couple in alkaline solution. In order to explain the higher yield, especially in alkaline solution, of *d,l*-pinacols compared to the *meso*-form, which would be favoured on steric grounds, a combination of steric factors and hydrogen bonding interactions is considered to determine the stereochemistry of the pinacols. The conclusion is reached that the stereochemistry of the product is determined so far from the electrode that it does not play any significant role here, and it seems substantiated by the lack of dependence of the stereochemistry on the nature of the electrode material; approximately the same stereochemical results were

TABLE 4. Stereochemistry (*d,l/meso* ratio) of the pinacols obtained on reduction of some acetophenones and related compounds in 80% ethanol[42, 54, 55].

Electrode	Potential [−V vs (SCE)]	Medium	% Pinacol	Ratio (*d,l/meso*)
		Acetophenone		
Hg	1·1—1·2	1M LiCl/1·5M AcOH	48—74	1·0—1·2
Sn	1·1	1M LiCl/1·5M AcOH	58	1·0
Cu	1·1	1M LiCl/1·5M AcOH	68	1·0
Hg	1·7	2M KOAc	74	2·5
Sn	1·6	2M KOAc	56—70	2·9—3·0
Cu	1·5—1·6	2M KOAc	77—78	2·8—2·9
Hg	1·6—1·8	M KOH	66	2·7
		Propiophenone		
Hg	1·2	1M LiCl/1·5M AcOH	55	1·4
Hg	1·6	2M KOAc	41	2·7
Hg	1·6	0·1M KOH	52	2·8
Cu	1·7	2M KOAc	24	3·2
		p-*Chloroacetophenone*		
Hg	1·1	1M LiCl/1·5M AcOH	88	1·2
Hg	1·5	2M KOAc	95	3·1
		p-*Methoxyacetophenone*		
Hg	1·2	1M LiCl/1·5M AcOH	94	1·2
Hg	1·7	2M KOAc	96	3·0
		p-*Trifluoromethylacetophenone*		
Hg	1·2	1M LiCl/1·5M AcOH	87	1·0
		Deoxybenzoin		
Hg	1·2	1M LiCl/1·5M AcOH	44	1·3
Hg	1·4	1M KOH	98	3·2
		o-*Chloroacetophenone*		
Hg	1·6	2M KOAc	28	2·1
		o-*Methoxyacetophenone*		
Hg	1·7	2M KOAc	51	1·2
		2-*Acetopyridine*		
Hg	0·78	1M LiCl/1·5M AcOH	11	0·73
Hg	1·2	2M KOAc	98	0·28
Hg	1·22	1M KOH	68	0·46
Cu	1·6	1M KOH	55	0·53
		Benzaldehyde		
Hg	1·2	1M LiCl/1·5M AcOH	69	1·1
Hg	1·6	2M KOAc	85	1·2
Cu	1·6	2M KOAc	73	1·2

obtained at mercury, copper, and tin cathodes. Furthermore, the *d,l/meso* ratio found in the electrochemical reductions was essentially the same as that obtained in photopinacolization studies[60].

Further experimental evidence must be accumulated before a unified picture of the reduction mechanism can be presented, and especially stereochemical evidence as described above will be of great value in elucidation of the reaction path.

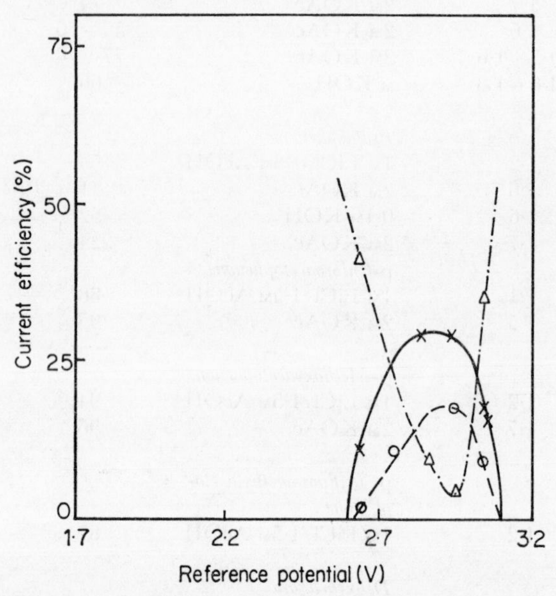

FIGURE 4. Plot of current efficiency against potential of the working electrode. Medium 1·0M sodium acetate, 0·4M sodium hydroxide. Products: —— methanol; ------ ethane; -·-·-·- oxygen; △ experimental points.

[Reproduced from *Trans. Faraday Soc.*, **63**, 1470 (1967), Fig. 2.[62]]

B. Preparation of Hydroxyl Compounds by Anodic Oxidation

I. Formation of alcohols

The well-known Kolbe electrochemical reaction which under 'normal' conditions produces predominantly hydrocarbons by anodic coupling of carboxylate ions may be directed by a suitable choice of reaction conditions towards a production of alcohols in the so-called Hofer–Moest reaction[61]. The reaction is schematically:

$$R—COO^- - 2e^- + OH^- \longrightarrow CO_2 + ROH$$

As illustrated in Figure 4, the anodic electrode reaction at a platinum electrode in an alkaline acetate solution is oxygen evolution below a certain potential; above that the Hofer–Moest reaction starts and, at a slightly higher potential, some of the Kolbe product is also obtained, while the oxygen evolution is suppressed to a high degree. At still higher potentials there is indication for oxygen evolution taking over again; this might be dependent on the acetate concentration.

The reaction mechanism is still under active discussion; the main question is whether the alcohol is formed by reaction between an adsorbed alkyl radical and an adsorbed hydroxyl radical or between an alkyl carbonium ion and a hydroxyl ion (or water molecule).

In the 'radical mechanism' the following steps are considered[62, 63],

$$CH_3COO^- \longrightarrow CH_3COO\cdot(M) + e \quad (19)$$

$$CH_3COO\cdot(M) \longrightarrow CH_3\cdot(M) + CO_2 \quad (20)$$

$$CH_3\cdot(M) + CH_3\cdot(M) \longrightarrow C_2H_6 \quad (21)$$

$$CH_3\cdot(M) + CH_3COO^- \longrightarrow C_2H_6 + CO_2 + e \quad (22)$$

$$OH^- \longrightarrow HO\cdot(M) + e \quad (23)$$

$$CH_3\cdot(M) + HO\cdot(M) \longrightarrow CH_3OH \quad (24)$$

$$CH_3\cdot(M) + \overline{O}H \longrightarrow CH_3OH + e \quad (25)$$

in which (M) indicates that the species is adsorbed to the electrode.

The first two steps are common for both the Kolbe and the Hofer–Moest reactions, and the branching occurs after the formation of methyl radicals which are considered to be stabilized by adsorption to the electrode. The methyl radicals may either form the Kolbe product (equations 21 and 22) or react with adsorbed hydroxyl radicals (equation 24) or hydroxyl ions in an electrochemical desorption step (equation 25) to the Hofer–Moest product. A loss of an electron from the alkyl radical thus forming a carbonium ion (equation 26) is considered for carboxylate ions having branched alkyl chains, but is thought less likely for methyl radicals.

$$R\cdot(M) \longrightarrow R^+ + e \quad (26)$$

$$R^+ + \overline{O}H \longrightarrow ROH \quad (27)$$

The formation of carbonium ions as general intermediates in the Hofer–Moest reaction has been assumed primarily from the carbonium ion-type rearrangements found in some products and from the stereochemical results. As a chapter in this series[4] covers this aspect of the Hofer–Moest reaction thoroughly, no further discussion of these points is necessary here.

Electrolysis of a sodium acetate solution at a smooth Pt-anode in a cell without diaphragm has been shown to give methanol in 93% yield (54% current yield)[64, 65], but in general the Hofer–Moest reaction has been investigated not as a preparative method in its own right, but rather as a side reaction to the Kolbe synthesis[66–68]. Besides the expected alcohol other alcohols and alkenes are often found in the reaction mixture, as would be reasonable from a rearrangement of or elimination from a primarily formed carbonium ion. The yields of alcohols are generally reasonably fair, but may possibly be raised by suitable choice of conditions. It would be expected that the use of a carbon anode would favour the alcohol formation; also an alkaline medium, e.g., made alkaline by addition of pyridine, containing a high concentration of difficultly oxidizable anions such as perchlorates, sulphates, carbonates or bicarbonates, would probably be favourable for the Hofer–Moest reaction.

It remains, however, to be seen how good yields may be obtained under such conditions; the further oxidation of alcohols must, of course, be avoided.

Some hydroxyl compounds have been prepared by anodic oxidation of different hydrocarbons to carbonium ions which react with solvent to alcohols. Thus among other products from the anodic oxidation of α-pinene at a platinum anode in sulphuric acid is found α-terpineol[69] and *cis*-terpine; also, *p*-nitrotoluene yields some *p*-nitrobenzyl alcohol at a platinum anode in glacial acetic acid containing sulphuric acid[70, 71]. The latter reaction is analogous to the anodic acyloxylation described in another chapter in this series[4].

III. ELECTROLYTIC REACTIONS OF HYDROXYL COMPOUNDS

Only few hydroxyl compounds can be reduced electrolytically, whereas many of them can be oxidized anodically.

A. Reductions

Certain activated hydroxyl groups may be reduced electrolytically, generally in acid solution. Thus in α-hydroxyketones, e.g., α-hydroxy steroid ketones[72] or derivatives of 2-hydroxyacetophenone[52] (**14**), and certain heterocyclic carbinols[73], as derivatives of 4-pyridylcarbinol, the hydroxyl group is reductively removed.

$$C_6H_5COC(CH_3)_2OH \xrightarrow[2e+H^+]{H^+} CH_3COCH(CH_3)_2 + H_2O$$

(14)

In these reactions there appears no indication of a primary loss of water; in some cases, such as the reduction of triaryl carbinols (**15**), in methanesulphonic acid[74] or of tropyl alcohol in aqueous buffer solution[75], an acid-catalysed dehydration precedes the reduction of the carbonium ion thus formed

$$Ar_3COH \xrightarrow{H^+} Ar_3C^+ + H_2O \xrightarrow{2e+H^+} Ar_3CH \quad (15)$$

It might also be mentioned that phenol can be reduced electrolytically to cyclohexanol in 2N sulphuric acid at a platinized platinum electrode[76].

B. Oxidations

1. Aliphatic alcohols

Primary aliphatic alcohols are generally oxidized[44, 45, 77] anodically through the aldehyde to the acid and secondary alcohols to the ketones or further; in some cases the aldehyde may be trapped. The reaction mechanism is not fully agreed upon; the oxidation of alcohols[78–80] at platinum has been subject to many investigations, especially in connexion with the development of fuel cells[81], and the following tentative scheme for the oxidation of methanol seems plausible in the light of recent results[82].

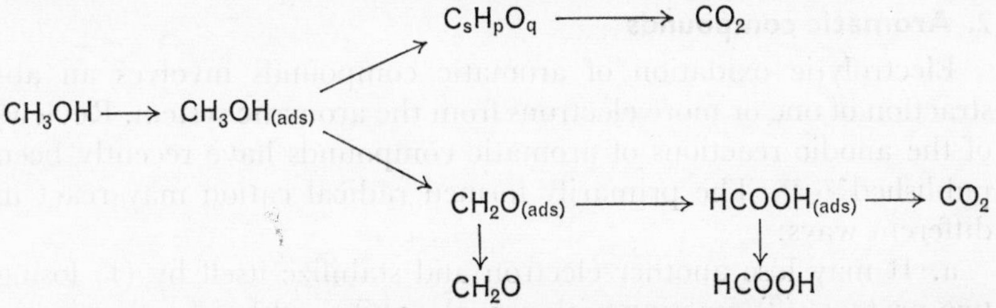

$$CH_3OH \longrightarrow CH_3OH_{(ads)} \nearrow C_sH_pO_q \longrightarrow CO_2$$

$$CH_3OH_{(ads)} \searrow CH_2O_{(ads)} \longrightarrow HCOOH_{(ads)} \longrightarrow CO_2$$

$$CH_2O_{(ads)} \downarrow CH_2O \qquad HCOOH_{(ads)} \downarrow HCOOH$$

The formation of a chemisorbed carbonaceous species, $C_sH_pO_q$, during the anodic oxidation of methanol on platinum is generally recognized, but the exact composition of this species is not known.

In an aqueous medium containing ammonium carbonate the anodic oxidation of ethanol at a platinum electrode yields acetamidine which was isolated as the nitrate. The following reaction path was suggested[83]:

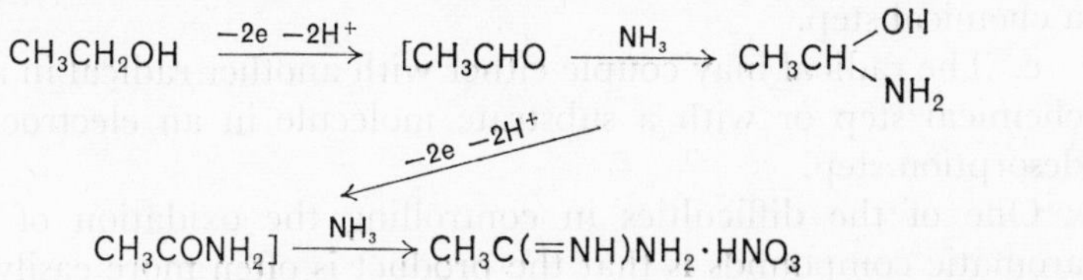

$$CH_3CH_2OH \xrightarrow{-2e\ -2H^+} [CH_3CHO \xrightarrow{NH_3} CH_3CH(OH)NH_2 \xrightarrow{-2e\ -2H^+} CH_3CONH_2] \xrightarrow{NH_3} CH_3C(=NH)NH_2 \cdot HNO_3$$

but the formation of acetamidine from acetamide under these conditions seems not attractive; perhaps an oxidation involving a carbonium ion rather than a dehydrogenation of the aldehyde-ammonia takes place.

In alkaline medium the anodic reactions of alcohols are more complex; the aldehyde formed may condense to resins before it is oxidized further. Other reactions lead to the formation of hydrocarbons and molecular hydrogen[84]. Some explanations of the latter reaction have been put forward[85, 86], but a reliable reaction path must await the results of investigations performed with modern techniques.

Ene-diols such as ascorbic acid are easily oxidizable[87] in a two-electron oxidation to the α-diketone

$$\mathrm{R{-}CO{-}CHOH{-}R'} \rightleftarrows \underset{\displaystyle \mathrm{HO}\ \ \ \mathrm{OH}}{\mathrm{R{-}C{=}C{-}R}} \xrightleftharpoons{-2e\ \ -2H^+} \mathrm{RCOCOR}$$

As mentioned in section II, the α-diketone is reducible through the ene-diol to the α-hydroxyketone.

2. Aromatic compounds

Electrolytic oxidation of aromatic compounds involves an abstraction of one or more electrons from the aromatic system. Reviews of the anodic reactions of aromatic compounds have recently been published[77, 88]. The primarily formed radical cation may react in different ways:

a. It may lose another electron and stabilize itself by (1) losing two protons, (2) reacting with a nucleophile and losing one proton, (3) reacting with two molecules of nucleophile. This reaction sequence may be represented by eecc (electron-transfer, electron-transfer, chemical step, chemical step).

b. It may lose a proton to a neutral radical which most likely will lose an electron and then react chemically, an ecec-sequence. Another ecec-reaction would be a reaction between the radical cation and a nucleophile followed by loss of an electron, followed by a chemical step.

c. The radical may couple either with another radical in a purely chemical step or with a substrate molecule in an electrochemical desorption step.

One of the difficulties in controlling the oxidation of organic aromatic compounds is that the product is often more easily oxidizable than the substrate. If, for example, the nucleophile is water or hydroxyl ion, the resulting phenol is more easily oxidizable than the parent hydrocarbon.

Another is that the oxidation potential of many aromatic compounds is rather positive, which restricts the choice of medium and electrode material. Acetonitrile, which is rather resistant towards oxidation, is often used as solvent; other useful solvents are dimethylformamide, methylene chloride, acetic acid or acetone; as supporting electrolytes perchlorates or tetrafluoroborates may be used.

3. Oxidation of aromatic alcohols

An investigation of the oxidation potentials of aromatic alcohols in acetonitrile at a platinum electrode[89] showed that the potential was mainly determined by the aromatic nucleus; substituents may influence the oxidation potential of an aromatic nucleus by their electron-donating or -attracting properties. Methoxy groups lower the oxidation potential; the $-CH_2OH$ group acts mostly as a weakly electron donating substituent, but in some difficultly oxidizable aromatic compounds the nonbonding electrons on the oxygen atom may be the most easily removable.

Controlled potential oxidations of aromatic alcohols have been made in a few cases; thus anisyl alcohol (**16**) was oxidized to anisaldehyde (**17**) in good yield at a platinum anode in acetonitrile containing sodium perchlorate as supporting electrolyte and pyridine as proton acceptor[89]. The reaction was formulated as

$$CH_3O-C_6H_4-CH_2OH \xrightarrow{-2e} CH_3\overset{+}{O}{=}C_6H_4{=}\overset{+}{}\!-CH_2OH$$

(16)

$$\xrightarrow[\text{Pyridine}]{-2H^+} CH_3OC_6H_4-CHO$$

(17)

without any attempt to determine whether or not a proton was lost before the second electron was removed from the system.

The product (**17**) is more difficult to oxidize under these conditions than **16** (anisyl alcohol: $E_{\frac{1}{2}} = 1{\cdot}22$ V (vs Ag^+/Ag), anisaldehyde: $E_{\frac{1}{2}} = 1{\cdot}63$); the reason is that the electron-attracting aldehyde group will raise the oxidation potential compared to anisole, whereas the slightly electron-donating $-CH_2OH$ group will lower it somewhat. This supports the view that the potential-determining step is the loss of an electron from the aromatic system; towards most chemical oxidants the aldehyde is more reactive than the alcohol.

When an aprotic solvent such as acetonitrile is used as a medium, it is necessary to add a proton acceptor to facilitate the removal of protons. Pyridine has been found to be a suitable base as it is oxidized at a rather positive potential. Aliphatic or aromatic amines are much more easily oxidized[76] and can thus not be used.

Very often side reactions take place when an attempt is made to oxidize an aromatic alcohol under these conditions. The reason is that a radical or radical ion intermediate starts a polymerization reaction at the electrode surface which is then fairly rapidly covered with a layer of tarrish product; this insulates the electrode from the solution and prohibits further transfer of electrons. This type of side reaction not only lowers the yields but in most cases prevents the use of the method for the preparation of aromatic aldehydes.

Certain aromatic alcohols, such as benzophenone pinacol and fluorenone pinacol[90], are oxidizable at controlled potential even in aqueous solution; the reaction involves a carbon–carbon cleavage with the formation of the parent ketone

$$R_2C(OH)C(OH)R_2 \longrightarrow 2e + 2H^+ + 2\,R_2CO$$

C. Oxidation of Phenols

The presence of a hydroxyl group on an aromatic nucleus lowers the oxidation potential of the system considerably; thus in acetonitrile the halfwave potentials, at a platinum anode, of benzene, phenol and hydroquinone are 2·00, 1·21 and 0·71 (vs Ag^+/Ag), respectively.

Electrolytic oxidation of phenols generally yields a mixture of compounds, but often the amount of high-molecular weight coupling products can be kept lower than in most chemical oxidations. The electrode reactions of phenols can be described by the general scheme for the electrochemical oxidation of aromatic compounds given above. The two-electron oxidations (a and b) often involve an attack by a nucleophile, whereas the one-electron oxidations result in coupling.

1. Two-electron oxidations

A simple overall two-electron oxidation is the oxidation of hydroquinones to quinones which is the common reaction of *o*- and *p*-dihydroxy benzenes; it might be mentioned, however, that on anodic oxidation[14, 91] of hydroquinone (**18**) in acetonitrile containing pyridine and sodium perchlorate *N*-(2,5-dihydroxyphenyl)-pyridinium perchlorate (**19**) was formed. This oxidation is analogous to the oxidation of anthracene to 9,10-dihydroanthranyl dipyridinium diperchlorate[92].

$$\text{(18)} \xrightarrow[+C_5H_5N]{-2e\ -H^+} \text{(19)}$$

In this case the electron-attracting properties of the pyridinium substituent make the system more difficultly oxidizable than the parent hydroquinone, and the product is not oxidized further at the potential used.

When the attacking nucleophile is a hydroxyl ion (or water) the product is more easily oxidizable than the starting material, and the primary product is oxidized further in preference to the starting material. The existence of the hydroxylated compound is then only indicated by its oxidation products. Thus phenol is oxidized to benzoquinone and maleic acid; when hydroquinone has been reported as a product, the oxidation has been performed in a non-divided cell and the hydroquinone is formed by reduction at the cathode of some of the quinone.

Under voltammetric conditions the anodic hydroxylation may be demonstrated; this is illustrated in Figure 5 which shows (curve A) the result of a cyclic voltammetric[93] investigation of 1,5-dihydroxynaphthalene (**20**) in 2M perchloric acid at a carbon paste electrode[94]. The initial anodic oxidation peak N [at +0·54 V vs (SCE)] is not followed by the expected cathodic reduction peak of the corresponding quinone. Instead, a rapid follow-up chemical reaction produces another product, and its reduction and subsequent reoxidation occurs at J and J′, respectively; as seen from Figure 5, they form a reversible redox system. This system was subsequently identified as 5-hydroxy-1,4-naphthoquinone (**21**) 1,4,5-trihydroxynaphthalene (**22**). Curve B depicts the cyclic voltammetric trace of this redox system, and it shows that the product obtained by anodic hydroxylation

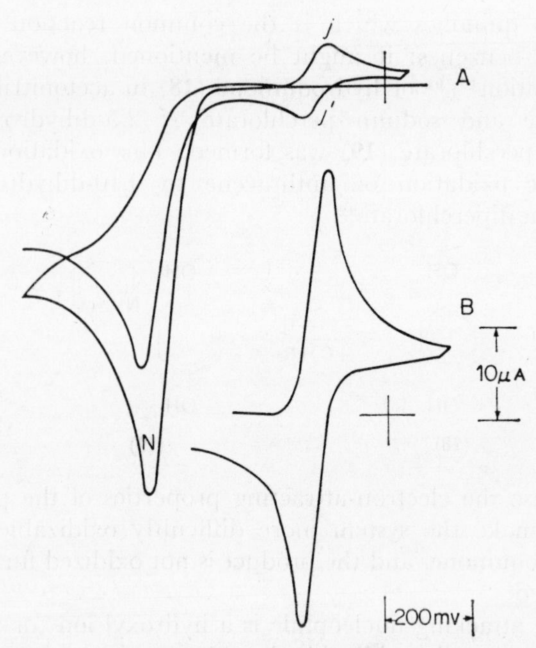

FIGURE 5. Cyclic voltammetry of anodic hydroxylation of 1,5-dihydroxynaphthalene: (A) cyclic polarogram of 1,5-dihydroxynaphthalene in 2M $HClO_4$; (B) cyclic polarogram for oxidation of 1,4,5-trihydroxynaphthalene in 2M $HClO_4$.
[Reprinted from *J. Am. Chem. Soc.*, **90**, 5620 (1968). Copyright 1968 by the American Chemical Society.]

of 1,5-dihydroxynaphthalene is oxidized at a potential about 0·35 V less positive than that of the starting material,

OH, OH **(20)** $\xrightarrow[+OH^-]{E = 0{\cdot}5\ V,\ -2e\ -H^+}$ OH, OH, OH **(22)** $\underset{}{\overset{E = 0{\cdot}17\ V,\ -2e\ -2H^+}{\rightleftharpoons}}$ O, OH, O **(21)**

Other nucleophiles such as halogens, thiocyanate ions, methoxide ions or methanol may attack during the anodic oxidation. In some cases it may be questionable whether an oxidation of the nucleophile rather than that of the phenol is the electrochemical step. An electrochemical oxidation of a phenol may be illustrated by the anodic reaction of 2,6-di-*t*-butyl-4-methylphenol (**23**) in acetonitrile containing some methanol[95]:

$$\text{(23)} \xrightarrow{-2e\ -H^+} \text{cation} \xrightarrow[-H^+]{CH_3OH} \text{(24)}$$

Besides the dienone (**24**) 2,6-di-*t*-butylbenzoquinone was isolated, formed by further oxidation of **24**.

The primary oxidation product from a phenol may also be stabilized by an intramolecular nucleophilic attack which results in a ring closure. The two following examples may illustrate this.

Electrolysis at a platinum anode of *p*-hydroxyphenylpropanoic acid (**25**) yielded the dienone lactone (**26**) in 20% yield. Using **25** labelled in the carboxyl group with ^{18}O it was demonstrated that the reaction proceeds intramolecularly[96].

$$\text{(25)} \xrightarrow{-2e\ -2H^+} \text{(26)} \xrightarrow{H_2SO_4} \text{6-hydroxychromanone}; \quad \text{(26)} \xrightarrow[H_2SO_4]{HOAc} \text{7-hydroxychromanone}$$

In a cyclic voltammetric investigation at the carbon paste electrode of various catecholamines such as adrenaline (**27**)[97], it was shown that in 1M H_2SO_4 the oxidation yielded the 1,2-benzoquinone (**28**). At pH 3, however, sufficient of the free amine of **28** was present to make an intramolecular nucleophilic addition to the *o*-quinone. As would be expected, the resulting catechol (**29**) is more easily oxidizable than adrenaline (**27**) and is converted into the quinone, adrenochrome (**30**) by chemical oxidation by adrenalinequinone (**28**).

A ring fission may also be induced anodically. Thus the chromane (**31**) is cleaved to the quinone (**32**)[98], as the initial oxidation product is attacked by water. A similar cleavage is often found when O-alkylated[99] or acylated[100] derivatives of hydroquinones are oxidized anodically to a quinone.

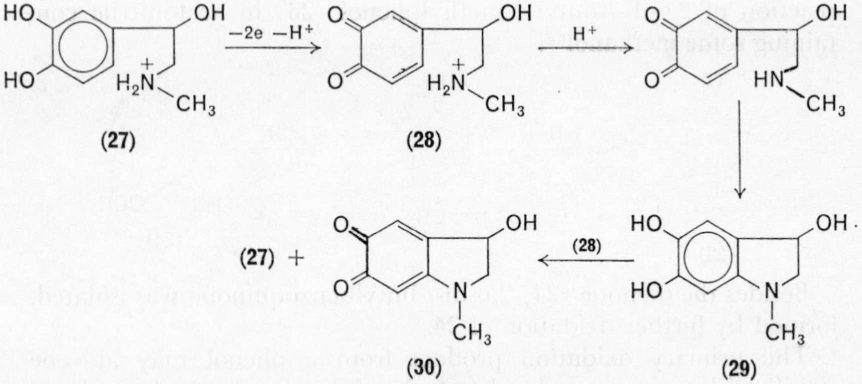

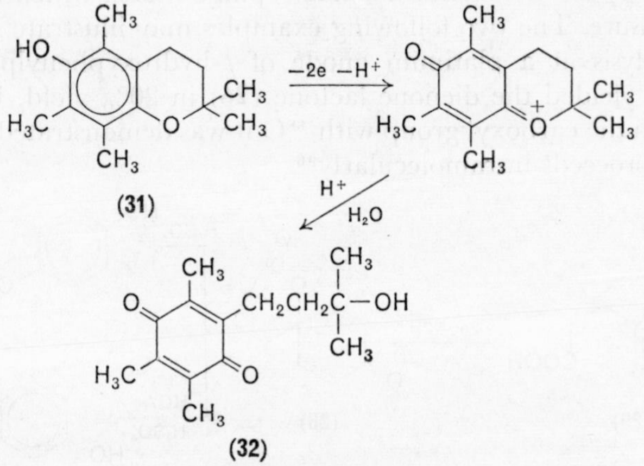

The oxidation of *p*-dimethylaminophenol (**33**)[101] shows some interesting features. In acid solution the oxidation is a two-electron reaction to yield *N,N*-dimethylquinoneimine (**34**) followed by hydrolysis to benzoquinone (**35**) with loss of dimethylamine (**36**). This reaction is similar to the anodic oxidation of *p*-aminophenol[102].

In alkaline solution, however, the reaction is complicated by an attack of **36**, formed by hydrolysis of the primarily obtained **34**, on unhydrolysed **34** to give 2,4-bis(dimethylamino)phenol (**37**), which then is oxidized by **34** to 3-dimethylamino-*p*-*N,N*-dimethylbenzoquinoneimine(**38**) (in the scheme below this is represented by an electrochemical oxidation in brackets). Hydrolysis of **38** yields dimethylamino-*p*-benzoquinone (**39**). **39** can also be obtained by electrochemical oxidation of 2-dimethylaminohydroquinone (**40**) which is formed by attack of **36** on **35**; this Michael addition is slower than the addition of **36** to **34**. The reactions are depicted above.

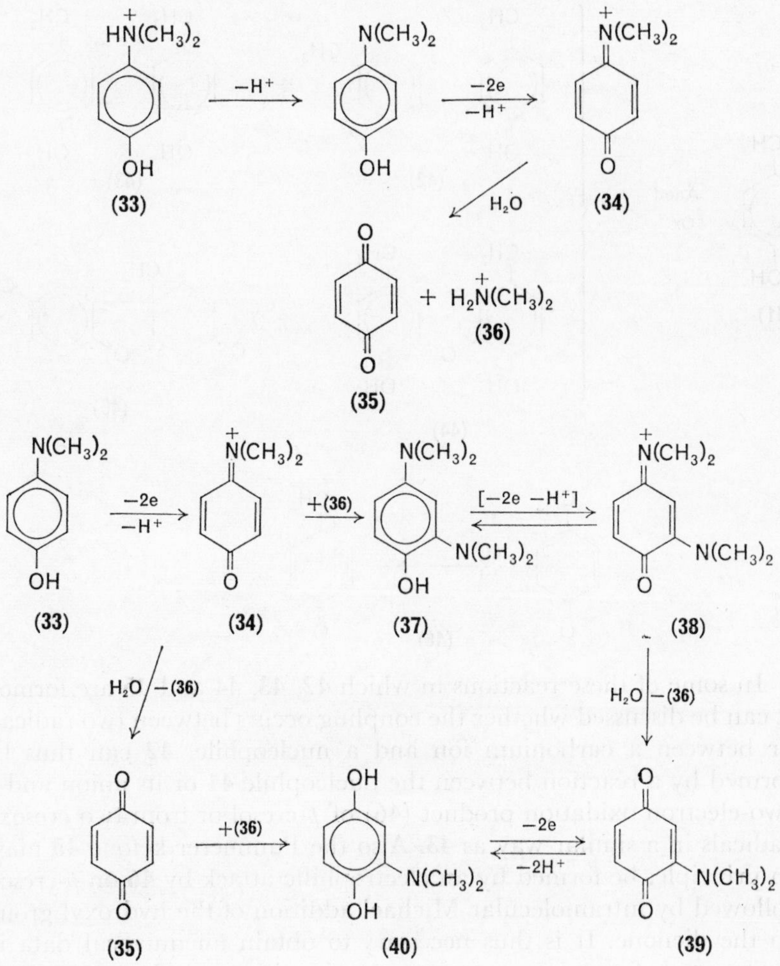

The overall reaction of **33** is thus a four-electron oxidation to **39**.

2. One-electron oxidations

When the primarily formed radical from the oxidation of a phenol reacts with another radical a C–C or a C–O coupling may occur[103–106]. The coupling products may then undergo further reactions. The oxidation of *p*-cresol (**41**) may be taken as a typical example[103, 104].

$$\text{(41)} \xrightarrow[\text{Ox}]{\text{Anod}} \text{(42)} + \text{(43)} + \text{(44)} + \text{(45)}$$

$$\text{(46)}$$

In some of these reactions in which **42**, **43**, **44** and **45** are formed it can be discussed whether the coupling occurs between two radicals or between a carbonium ion and a nucleophile. **42** can thus be formed by a reaction between the nucleophile **41** or its anion and a two-electron oxidation product (**46**) of *p*-cresol or from two cresoxy radicals in a similar way as **43**. Also the Pummerer ketone **45** may, in principle, be formed by an electrophilic attack by **46** on *p*-cresol, followed by intramolecular Michael addition of the hydroxyl group to the dienone. It is thus necessary to obtain unequivocal data in each case before any reaction mechanism is accepted.

Besides the simple coupling products described above further coupling may be induced anodically with the formation of polymerized material, but such tarrish material is often formed to a lesser degree than in many chemical oxidations.

IV. REFERENCES

1. H. Lund in *The Chemistry of the C═N Double Bond* (Ed. S. Patai), Interscience Publishers, London, 1970, p. 505.
2. F. Haber, *Z. Elektrochem.*, **4**, 506 (1898).
3. F. Haber, *Z. Physik. Chem.*, **32**, 193 (1900).

4. L. Eberson in *The Chemistry of Carboxylic Acids and Esters* (Ed. S. Patai), Interscience Publishers, London, 1969, p. 53.
5. A. J. Birch, *Nature*, **158**, 60 (1946).
6. H. W. Sternberg, E. M. Kaiser and R. F. Lambert, *J. Electrochem. Soc.*, **110**, 425 (1963).
7. H. W. Sternberg, R. E. Markby, J. Wender and D. M. Mohilner, *J. Electrochem. Soc.*, **111**, 1060 (1966).
8. R. A. Benkeser, E. M. Kaiser and R. F. Lambert, *J. Am. Chem. Soc.*, **86**, 5272 (1964).
9. T. Osa, T. Yamagishi, T. Kodama and A. Misono, *Preprints of Papers, Symposium, Durham, N.C.*, October 1968, p. 157.
10. H. W. Sternberg, R. E. Markby, J. Wender and D. M. Mohilner, *J. Am. Chem. Soc.*, **89**, 186 (1967); **91**, 4191 (1969).
11. E. J. Hart, S. Gordon and E. M. Frielden, *J. Phys. Chem.*, **70**, 150 (1966).
12. K. J. Vetter and D. Otto, *Z. Elektrochem.*, **60**, 1072 (1956).
13. X. de Hemptinne and J. C. Jungers, *Z. Physik. Chem.*, **15**, 137 (1958).
14. H. Lund, *Elektrodereaktioner i organisk polarografi og voltammetri*, Aarhuus Stiftsbogtrykkerie, Aarhus, 1961.
15. X. de Hemptinne and K. Schunk, *Ann. Soc. Sci. Bruxelles, Ser. I*, **80**, 289 (1966); *Chem. Abstr.*, **66**, 91139 (1967); *Trans. Faraday Soc.*, **65**, 591 (1969).
16. H. Lund, *Acta Chem. Scand.*, **17**, 972 (1963).
17. P. E. Iversen and H. Lund, *Acta Chem. Scand.*, **21**, 279 (1967).
18. P. E. Iversen and H. Lund, *Acta Chem. Scand.*, **21**, 389 (1967).
19. J. Kůta, *Collection Czech. Chem. Commun.*, **21**, 697 (1956).
20. J. Kůta, *Collection Czech. Chem. Commun.*, **22**, 1677 (1957).
21. J. Tafel and G. Friederichs, *Ber.*, **37**, 3187 (1904).
22. B. Sakurai, *Bull. Chem. Soc. Japan*, **7**, 127 (1932).
23. C. Mettler, *Ber.*, **39**, 2933 (1906).
24. C. Mettler, *Ber.*, **38**, 1745 (1905).
25. F. Mayer, W. Schäfer and J. Rosenbach, *Arch. Pharm.*, **267**, 571 (1929).
26. F. Fichter and I. Stein, *Helv. Chim. Acta.*, **12**, 821 (1929).
27. H. Lund, *Acta Chem. Scand.*, **17**, 2325 (1963).
28. Th. B. Baillie and J. Tafel, *Ber.*, **32**, 71 (1899).
29. K. Kindler, *Arch. Pharm.*, **265**, 389 (1927).
30. R. A. Benkeser, *Preprints of Papers, Symposium, Durham, N.C.*, October 1968, p. 189.
31. A. Dunet and A. Willemart, *Bull. Soc. Chim. France*, 887 (1948).
32. H. Lund, unpublished observations.
33. E. Späth and F. Brench, *Monatsh. Chem.*, **50**, 349 (1928).
34. B. Sakurai, *Bull. Chem. Soc. Japan*, **5**, 184 (1930).
35. H. Lund, *Collection Czech. Chem. Commun.*, **30**, 4237 (1965).
36. H. Lund, *Preprints of Papers, Symposium, Durham, N.C.*, October 1968, p. 197.
37. H. Lund, *Österr. Chem. Z.*, **68**, 43 (1967).
38. P. Zuman, *Collection Czech. Chem. Commun.*, **33**, 2548 (1968).
39. J. Grimshaw and E. J. F. Rea, *J. Chem. Soc.* (*C*), 2628 (1967).
40. H. Lund and A. D. Thomsen, *Acta Chem. Scand.*, **23**, 3567, 3582 (1969).
41. H. Kauffmann, *Z. Elektrochem.*, **4**, 461 (1898).
42. J. H. Stocker and R. M. Jenevein, *J. Org. Chem.*, **33**, 294 (1968).
43. J. Tafel, *Ber.*, **44**, 327 (1911).

44. F. Fichter, *Organische Elektrochemie*, Verlag von Th. Steinkoppf, Dresden und Leipzig, 1942.
45. M. J. Allen, *Organic Electrode Processes*, Chapman and Hall Ltd., London, 1958.
46. F. D. Popp and H. P. Schultz, *Chem. Rev.*, **62**, 19 (1962).
47. J. Wiemann and P. Maitte, *Bull. Soc. Chim. France*, 430 (1952).
48. J. Wiemann and M. Paget, *Bull. Soc. Chim. France*, 285 (1955).
49. H. Lund, *Acta Chem. Scand.*, **11**, 283 (1957).
50. P. Zuman, D. Barnes and A. Ryvolová-Kejharova, *Discussions Faraday Soc.*, **45**, 202 (1968).
51. H. Lund, unpublished results.
52. H. Lund, *Acta Chem. Scand.*, **14**, 1927 (1960).
53. M. Allen, *J. Chem. Soc.*, 1598 (1951).
54. J. H. Stocker and R. M. Jenevein, *J. Org. Chem.*, **33**, 294, 2145 (1968).
55. J. H. Stocker and R. M. Jenevein, *Preprints of Papers, Symposium, Durham, N.C.*, October 1968, p. 221.
56. L. Mandell, R. M. Powers and R. A. Day, *J. Am. Chem. Soc.*, **80**, 5284 (1958).
57. L. Horner and D. Degner, *Tetrahedron Letters*, 5889 (1968).
58. Z. R. Grabowski, B. Czochralska, A. Vincenz-Chodkowska and M. S. Balasiewicz, *Discussions Faraday Soc.*, **45**, 145 (1968).
59. H. E. Stapelfeldt and S. P. Perone, *Anal. Chem.*, **40**, 815 (1968).
60. J. H. Stocker and D. H. Kern, *J. Org. Chem.*, **31**, 3755 (1966); **33**, 291 (1968).
61. H. Hofer and M. Moest, *Ann.*, **323**, 284 (1902).
62. G. Atherton, M. Fleischmann and F. Goodridge, *Trans. Faraday Soc.*, **63**, 1468 (1967).
63. M. Fleischmann and F. Goodridge, *Discussions Faraday Soc.*, **45**, 254 (1968).
64. T. Kunugi, *J. Electrochem. Soc. Japan*, **20**, 111, 154 (1952); *Chem. Abstr.*, **47**, 421 (1953).
65. T. Kunugi, *J. Electrochem. Soc. Japan*, **20**, 69 (1952); *Chem. Abstr.*, **48**, 13485 (1954).
66. W. S. Koehl, *J. Am. Chem. Soc.*, **86**, 4686 (1964).
67. P. G. Gassmann and F. V. Zalar, *J. Am. Chem. Soc.*, **88**, 2252 (1966).
68. J. G. Traynham and J. S. Dehn, *J. Am. Chem. Soc.*, **89**, 2139 (1967).
69. F. Fichter and G. Schetty, *Helv. Chim. Acta*, **20**, 1304 (1937).
70. K. Elbs, *Z. Elektrochem.*, **2**, 522 (1896).
71. F. Fichter and G. Bonhôte, *Helv. Chim. Acta*, **3**, 39 (1920).
72. P. Kabasakalian and J. McGlotten, *Anal. Chem.*, **31**, 1091 (1959).
73. O. Manousek and P. Zuman, *Collection Czech. Chem. Commun.*, **29**, 1432 (1964).
74. S. Wawzonek, R. Berkey, E. W. Blaha and M. E. Runner, *J. Electrochem. Soc.*, **103**, 456 (1956).
75. P. Zuman, J. Chodkowski, H. Potesilova and F. Santavy, *Nature*, **182**, 1535 (1958).
76. F. Fichter and R. Stocker, *Ber.*, **47**, 2015 (1914).
77. N. L. Weinberg and H. R. Weinberg, *Chem. Rev.*, **68**, 449 (1968).
78. J. E. Oxley, G. K. Johnson and B. T. Buzalski, *Electrochim. Acta*, **9**, 897 (1964).
79. T. Takamura and K. Minamiyama, *J. Electrochem. Soc.*, **112**, 333 (1965).
80. M. Hollnagel and V. Lohse, *Z. Physik. Chem. (Leipzig)*, **232**, 237 (1966).
81. E. Gileadi and B. Piersma, 'The Mechanism of Oxidation of Organic Fuels'

in *Modern Aspects of Electrochemistry* (Ed. J. O'M. Bockris), Butterworth, London, 1966, pp. 47 ff.
82. M. W. Breiter, *Discussions Faraday Soc.*, **45,** 79 (1968).
83. F. Fichter, *Z. Elektrochem.*, **18,** 647 (1912).
84. E. Müller and F. Hochstetter, *Z. Elektrochem.*, **20,** 367 (1914).
85. E. Müller, *Z. Elektrochem.*, **28,** 101 (1922); **27,** 558, 563 (1921).
86. Ref. 44, p. 87.
87. J. Holubek and J. Volke, *Collection Czech. Chem. Commun.*, **25,** 3292 (1960).
88. K. Sasaki and W. J. Newby, *J. Electroanal. Chem.*, **20,** 137 (1969).
89. H. Lund, *Acta Chem. Scand.*, **11,** 491 (1957).
90. W. Kemula, Z. R. Grabowski and M. K. Kalinowski, *Collection Czech. Chem. Commun.*, **25,** 3306 (1960).
91. W. R. Turner and P. J. Elving, *J. Electrochem. Soc.*, **112,** 1215 (1965).
92. H. Lund, *Acta Chem. Scand.*, **11,** 1323 (1957).
93. J. R. Alden, J. Q. Chambers and R. N. Adams, *J. Electroanal. Chem.*, **5,** 152 (1963).
94. L. Papouchado, G. Petrie, J. H. Sharp and R. N. Adams, *J. Am. Chem. Soc.*, **90,** 5620 (1968).
95. F. J. Vermillion and I. A. Pearl, *J. Electrochem. Soc.*, **111,** 1392 (1964).
96. A. I. Scott, P. A. Dodson, F. McCapra and M. B. Meyers, *J. Am. Chem. Soc.*, **85,** 3702 (1963).
97. M. D. Hawley, S. V. Tatawawadi, S. Piekarski and R. N. Adams, *J. Am. Chem. Soc.*, **89,** 447 (1967).
98. L. I. Smith, I. M. Kolthoff, S. Wawzonek and P. M. Rouff, *J. Am. Chem. Soc.*, **63,** 1018 (1941).
99. M. D. Hawley and R. N. Adams, *J. Electroanal. Chem.*, **8,** 163 (1964).
100. C. A. Chambers and J. Q. Chambers, *J. Am. Chem. Soc.*, **88,** 2922 (1966).
101. M. F. Marcus and M. D. Hawley, *J. Electroanal. Chem.*, **18,** 175 (1968).
102. M. D. Hawley and R. N. Adams, *J. Electroanal. Chem.*, **10,** 376 (1965).
103. F. Fichter and F. Ackerman, *Helv. Chim. Acta*, **2,** 583 (1919).
104. T. Kametani, K. Ohtubo and S. Tatano, *Chem. Pharm. Bull.* (*Tokyo*), **16,** 1095 (1968).
105. J. M. Bobbitt, J. T. Stock, A. Marchand and K. H. Weisgoaber, *Chem. Ind.* (*London*), 2127 (1966).
106. G. F. Kirkbright, J. T. Stock, R. D. Pugliese and J. M. Bobbitt, *J. Electrochem. Soc.*, **116,** 219 (1969).

CHAPTER 6

Detection and determination of hydroxyl groups

SIDNEY SIGGIA

University of Massachusetts, Amherst, Massachusetts, U.S.A.

J. GORDON HANNA

Connecticut Agricultural Experiment Station, New Haven, Connecticut, U.S.A.

THOMAS R. STENGLE

University of Massachusetts, Amherst, Massachusetts, U.S.A.

I. DETECTION AND DETERMINATION OF HYDROXYL GROUPS

The discussion below includes alcohols, glycols, polyhydric alcohols and enols (including phenols). The hydroxyl group has some characteristic chemical reactions; these form the basis of *chemical methods* of detection and determination. In addition, the group has certain *optical absorption* properties which can also be used. Since there is a proton in the hydroxyl group, *nuclear magnetic resonance* is an excellent method for detection and determination.

Mass spectrometry is not discussed in this chapter since the fragmentation which occurs is not definitive for the hydroxyl group. The same is to be said for X-ray diffraction though it should be realized that both these approaches are very valuable in characterizing and proving structure. Powder X-ray patterns, however, make an excellent method for comparing unknown and known samples of hydroxyl compounds or their derivatives.

II. CHEMICAL METHODS

A. Qualitative

1. Identification of hydroxyl compounds with derivatization

References 1 and 2 yield boiling point or melting point data for a wide variety of liquid and solid hydroxyl compounds. Reference 3 gives boiling point, melting point, refraction index and density for a very wide range of alcohols and phenols.

X-ray diffraction parameters on solid hydroxyl compounds provide an excellent method of identification. This is discussed on page 298. The X-ray measurements on some hydroxyl compounds are listed in the ASTM compilation of X-ray data. Even if the X-ray data on the specific hydroxyl material are not listed in the ASTM compilation, X-rays can serve to compare the unknown with suspected known materials.

2. Dinitrobenzoate derivatives

These are the most common derivatives of hydroxyl compounds. They are generally solids with convenient melting points and characteristic X-ray diffraction patterns.

Preparation[4]: For alcohol samples containing less than 5% water, 1·5–2 millimoles are heated in a small test tube with 1 millimole of 3,5-dinitrobenzoyl chloride*. The mixture is heated gently at the lowest temperature at which it remains liquid. The lower alcohols are heated for 3–5 minutes and the higher alcohols for 10–15 minutes. The melt is cooled and the solidified mass is then broken up with a spatula. The resultant solid is shaken with 2% sodium carbonate at 50–60 °C for about 10–30 seconds and filtered. This removes the unreacted acid halide and any free acid from the derivative. Prolonged carbonate treatment should be avoided since it could cause hydrolysis of the ester. The product is crystallized from methanol or ethanol to which water is added to the first cloudiness before cooling. Usually only one crystallization is needed. However, a faulty carbonate wash could necessitate two or three crystallizations.

In cases of unreactive alcohols or polyhydric alcohols, the use of a solvent is recommended. Isopropyl or *n*-butyl ether are recommended (making sure that they are free of alcohol). The mixture of alcohol solution and reagent is refluxed for 0·5–1 hour. The mixture is then washed with sodium carbonate solution. The ether layer is separated and evaporated to dryness. The crude ester is recrystallized as above. With tertiary alcohols which are difficult to esterify, pyridine is used as a solvent. One millimole of reagent is mixed with 1·5 millimole of alcohol in a small test tube with 2 ml of pyridine. The mixture is refluxed for 0·4–1 hour, cooled and extracted with 4 ml of 1% sulphuric acid to remove the pyridine and precipitate the crude ester. Recrystallization is as described above.

Samples which contain more than 5% water might hydrolyse the acid chloride reagent. In this case, a sample containing 250–500 mg of alcohol is cooled to 0 °C and is shaken with a solution of 500 mg of the acid chloride in a mixture of 2 ml of specially purified hexane (ligroin or petroleum ether) and 3 ml of benzene. The hexane is

* 3,5-Dinitrobenzoyl chloride is susceptible to hydrolysis by atmospheric water on standing. It is best to check the melting point of the available reagent. If the melting point is 1–2° lower than the literature value, the acid halide should be recrystallized from carbon tetrachloride. On purchasing the acid halide, it is best to buy several small bottles rather than one large one to avoid atmospheric contamination of unused material.

purified by washing, first with concentrated sulphuric acid, next with water, drying over calcium chloride or sulphate and then distilling. The reaction mixture is kept below 5 °C with shaking for 15–30 minutes. Alcohol-free ether is then added, followed by vigorous shaking. The upper layer is separated, washed first with diluted sodium hydroxide, then with diluted hydrochloric acid, then with water. The solvent is evaporated and the crude ester is recrystallized as above.

Another reference to the above derivatization of hydroxyl compounds is given in Reference 5.

Tables of melting points of the 3,5-dinitrobenzoate esters of hydroxyl compounds are given in References 1–3. In addition, Garska, Doutkit and Yarborough[6] give the crystal data obtained by X-ray diffraction of 3,5-dinotrobenzoates. In addition, the identity of the components in mixtures of hydroxyl compounds can often be ascertained, and otherwise impure derivatives can be identified without complete purification being required. The authors of this chapter recommend the X-ray approach for comparing the derivative parameters of the unknown hydroxyl compound to those of the known.

3. Other ester derivatives

Less common ester derivatives for hydroxyl compounds include acetates, benzoates, *p*-nitrobenzoates, phthalates and nitrophthalates. Methods for preparation of these derivatives are given in References 7 and 8. The melting points of the derivatives are found in References 1–3.

4. Urethane derivatives

Among urethane derivatives of hydroxyl compounds the α-naphthyl urethanes are probably the most common since so many are solids. However, the phenyl urethanes are also used. Occasionally the *β*-naphthyl, *p*-bromophenyl, *o*-nitrophenyl, *m*-nitrophenyl and *p*-nitrophenyl urethanes are also used. The urethanes are prepared from the corresponding isocyanates:

$$R'OH + RNCO \longrightarrow RNHCOOR'$$

Preparation[9]: One millimole of hydroxyl compound is mixed with 1·25 millimoles of α-naphthyl isocyanate in a small test tube and heated in a water bath at 60–70 °C for 10–15 minutes. The crude product solidifies on cooling and is pulverized with a microspatula.

This powder is then extracted with a minimum amount of petroleum ether to remove the soluble impurities. The first extract contains a large amount of the urethane derivative along with the impurities. A second extraction is made; this extract usually contains rather purer urethane. The insoluble residue is usually di-α-naphthyl urea formed by reaction with water in the sample or reagents used. It is well to note that this method cannot be applied to samples which contain more than a trace of water.

In the case of rather unreactive hydroxyl compounds such as phenols, the isocyanate and the sample are mixed in a test tube and heated for 2–5 minutes. If no reaction is observed, 1 ml of pyridine is added along with one drop of 10% trimethylamine in hexane, and the mixture is heated for 20–30 minutes. If the urethane does not separate on cooling, 1 ml of 5% sulphuric acid is added. The crude urethane is crystallized from petroleum ether.

Another description of the preparation of urethane derivatives can be found in Reference 10.

Tables of melting points of the urethane derivatives of a wide range of hydroxyl compounds can be found in References 1–3.

5. Other derivatives

Xanthates are formed by reaction of primary and secondary alcohols with carbon disulphide and base can be used[10].

$$ROH + CS_2 + KOH \longrightarrow RO\overset{\overset{O}{\|}}{C}SK + H_2O$$

Tertiary alcohols are often difficult to derivatize by the above methods. However, these alcohols readily form the corresponding alkyl halides. The halide can then be derivatized further to the *S*-alkylthiuronium picrate[11].

Phenols react readily with bromine and these bromo derivatives can be used for identification[12]. However, the reaction is clear-cut only for a narrow range of phenols. Substituents on the phenol which are easily oxidized, such as –CHO, –SH, $-NH_2$ can cause problems. In alkyl phenols side-chain substitution may occur, giving a mixture of brominated products.

References 1–3 give melting points of some of these miscellaneous derivatives.

6. Handling of mixed derivatives

If the sample contains a mixture of alcohols, one usually obtains a mixture of derivatives. The mixed 3,5-dinitrobenzoates of alcohols can often be resolved by column or paper chromatography[13–15]. Mixtures of the 3,5-dinitrobenzoates of phenols can often be similarly resolved[13,14] observing additional conditions[16–18]. Separation of mixed xanthate derivatives is possible[19]. The use of *p*-phenylazobenzoate derivatives is advantageous because they are coloured and the chromatography is more easily carried out[20].

B. Quantitative

I. Esterification methods

Esterifications for quantitative analytical purposes include the three general reactions illustrated in equations (1–3).

$$RCOOH + R'OH \rightleftarrows RCOOR' + H_2O \qquad (1)$$

$$RCOCl + R'OH \longrightarrow RCOOR' + HCl \qquad (2)$$

$$(RCO)_2O + R'OH \longrightarrow RCOOR' + RCOOH \qquad (3)$$

Esterification with a carboxylic acid has the main disadvantage that it normally involves an unfavourable equilibrium. Bryant, Mitchell and Smith[21] succeeded in shifting the equilibrium almost completely to the right by using a large excess of acetic acid in the presence of boron trifluoride as catalyst and dioxane as solvent. After two hours at 67°C, pyridine was added to destroy the activity of the boron trifluoride. One mole of water is obtained per mole of alcoholic hydroxyl and is titrated with Karl Fischer reagent. This procedure is applicable to aliphatic, alicyclic and aralkyl alcohols and to hydroxy acids. Phenols react incompletely. Aldehydes and ketones interfere and amines decrease the activity of the catalyst. The accuracy reported was ±0·3%, but much care is required in the use of the sensitive Karl Fischer reagent.

Acid chlorides have not been widely used for hydroxyl group determinations. Although they react rapidly and completely their instability, due to great reactivity, discourages their use. However, procedures have been developed, based on the acid chloride reaction, and these have proved valuable, especially for some sterically hindered and tertiary hydroxyl groups which cannot be esterified by other methods.

Smith and Bryant[22] and Kaufmann and Funke[23] demonstrated quantitative esterifications with acetyl chloride in the presence of pyridine. One equivalent of titratable acid is produced for each mole of alcohol. Kappelmeier and Mostert[24] used a similar procedure to determine hydroxyl values of alkyd resins. Kepner and Webb[25] omitted the pyridine and used a toluene mixture and a semimicro technique. Bring and Kadleck[26] employed stearoyl chloride as the reagent to determine the hydroxyl groups in epoxy resins.

Probably the most practical of the acid chloride procedures is that of Robinson, Cundiff and Markunas[27] involving 3,5-dinitrobenzoyl chloride. The reaction is illustrated in equation (4), with Ar = 3,5-

$$ROH + ArCOCl + C_5H_5N \longrightarrow ArCOOR + C_5H_5N.HCl \qquad (4)$$

$(NO_2)_2C_6H_3$–. The excess dinitrobenzoyl chloride is hydrolysed as indicated by equation (5). The pyridinium hydrochloride and the

$$ArCOCl + H_2O + C_5H_5N \longrightarrow ArCOOH + C_5H_5N.HCl \qquad (5)$$

dinitrobenzoic acid titrate simultaneously as strong acids and give the first inflexion in a potentiometric curve when titrated with tetrabutylammonium hydroxide solution. The dinitrobenzoate titrates as a weak acid, represented by a second inflexion. The amount of dinitrobenzoate formed is a measure of the organic hydroxyl content. Another end-point indication which can be used is a colour change from yellow to red by a reaction product at the first equivalence point. Polyols, sugars, phenols, primary and secondary amines, and some oximes may be determined. Aldehydes, if present in amounts less than 40% of the alcohol being determined and ketones do not interfere.

Procedure[27]: A fresh 0·2M solution of 3,5-dinitrobenzoyl chloride (1·15 g in 25 ml of pyridine) is prepared for each series of analyses. The solution should not be unnecessarily exposed to moist air.

To determine liquid samples, approximately 4 meq of the hydroxyl compound is pipetted into a tared 10-ml volumetric flask containing 3 ml of pyridine. The flask and contents are reweighed and the contents are brought to volume with pyridine. Four millilitres of the dinitrobenzoyl chloride and 1 ml of the sample solution are pipetted into a 125-ml glass-stoppered flask. The flask is stoppered tightly and allowed to stand 5–15 minutes at room temperature.

To determine solid samples, 0·4 meq of the hydroxyl compound is weighed directly into a 125-ml glass-stoppered flask and 4·0 ml of the dinitrobenzoyl chloride solution are added. The flask is stoppered, swirled to dissolve the sample and allowed to stand 5–15 minutes at room temperature.

In either case, at the end of the reaction period, the stopper is removed and 7–10 drops of water are added.

To prepare a blank solution, 4·0 ml of the dinitrobenzoyl chloride solution are pipetted into a flask and 7–10 drops of water are added immediately.

For the visual titration, 40 ml of pyridine are added to the reaction mixture, the mixture is heated nearly to boiling, cooled, and then titrated with 0·2N tetrabutylammonium hydroxide to the first definite and permanent red colour. The titration is best performed with the titrant and solution protected from moisture and air and the tip of the burette immersed in the titrating solution. For the potentiometric titration, 25 ml of pyridine are added to the reaction mixture and heated nearly to boiling. The mixture is cooled and transferred to a 250-ml beaker. The flask is rinsed with two 10-ml portions of pyridine and the washings added to the beaker. The mixture is titrated under nitrogen with 0·2N tetrabutylammonium hydroxide. If a blank is determined, the titration need proceed only through the first inflexion point. The differences in volume between the end-points of the blank and samples are used to calculate the hydroxyl content. If no blank is determined, the titration is carried through both inflexions and the volume between the first and second end-points is used to calculate the hydroxyl content.

Quantitative esterifications are usually accomplished by anhydrides, with acetic anhydride, phthalic anhydride and pyromellitic dianhydride being the reagents of choice. A primary or secondary alcohol is acetylated conveniently by reaction with acetic anhydride in the presence of a catalyst, equation (6). Catalysts in-

$$RCH_2OH + (CH_3CO)_2O \xrightarrow{\text{Catalyst}} CH_3COOCH_2R + CH_3COOH \qquad (6)$$

clude bases, Lewis acids, mineral acids and other strong acids. Sodium acetate is a weak basic catalyst. Pyridine is a more effective basic catalyst and is the one which has been most extensively used in quantitative acetylations. Boron trifluoride catalyses the reaction but strong acids such as sulphuric, hydrochloric, perchloric, 2,4-dinitrobenzenesulphonic and *p*-toluenesulphonic acids are more effective.

Among the acetylation methods the most practical approach involves the calculation of the hydroxyl content based on the difference between the acid formed by the alcohol reaction and the acid formed when a blank is treated with water, equation (7). Early methods for

$$H_2O + (CH_3CO)_2O \rightarrow 2\ CH_3COOH \qquad (7)$$

the determination of hydroxyl values of fats, oils and glycerol depended on the acetylation with acetic anhydride, neutralization of the acetylation mixture and determination of the saponification value of the ester. This method is now rarely used, but the technique may still be useful for the determination of alcohol groups in the presence of compounds which react with the anhydride but do not form esters. For example, a method for the determination of hydroxyl groups in the presence of amine groups[28] involves the acetylation of both groups to form the corresponding esters and amides. Because esters in general hydrolyse much more rapidly than amides, a saponification reaction produces data for the calculation of the hydroxyl content of the original mixture.

Ogg, Porter and Willits[29] recommended a 1 : 3 mixture of acetic anhydride–pyridine and a 45 min reaction time on a steam bath followed by titration of the carboxylic acid formed. Primary and secondary amines and low molecular weight aldehydes interfere. Hydroxyl groups on tertiary carbon atoms and hydroxyls of 2,4,6-trisubstituted phenols react only very slightly[30]. Hydroxyl groups in less highly substituted phenols react readily with acetic anhydride.

Procedure[29]: A weighed sample containing from 0·010 to 0·016 equivalent of hydroxyl is placed in a 250-ml iodine flask and 10·00 ml of acetic anhydride–pyridine solution (1 : 3) are added. The flask stopper is moistened with pyridine and seated loosely. After the flask has been heated on a steam bath for 45 minutes, 5–6 ml of water are added to the cup of the flask and the stopper is loosened sufficiently to rinse it and the inside wall of the flask. The heating is continued for 2 minutes and the flask is then cooled under the tap with the stopper partly removed. The stopper and inside wall of the flask are rinsed with 10 ml of *n*-butanol. The contents are titrated with 0·5N alcoholic sodium hydroxide solution to a mixed indicator end-point (one part of 0·1% neutralized Cresol Red and three parts 0·1% neutralized Thymol Blue). A blank determination is made on 10 ml of the pyridine–acetic anhydride solution.

Triethylenediamine is claimed to be superior to pyridine for base-catalysed acetylations[31]. With this catalyst, the acetylation at reflux temperature of most primary and secondary alcohols is quantitative in 15–20 minutes. Salt catalysis with tetraethylammonium bromide was also found to produce quantitative results for the acetylation of cyclohexanol in 5 minutes and for tertiary alcohols in 45–65 minutes at reflux temperature[31].

Bring and Kadlecek[26] recommended acetic anhydride systems catalysed by sodium acetate, sulphuric acid, pyridinium chloride and pyridinium perchlorate. Fritz and Schenk[32] presented general methods based on the catalytic effect of perchlorate on the acetylation reaction. Ethyl acetate and pyridine were used as solvents. In most cases, the reaction was found to be complete in 5 minutes at room temperature. Even highly hindered phenols can be determined.

Procedure[32]: 2M acetic anhydride in ethyl acetate is prepared by dissolving 4 g (2·35 ml) of 72% perchloric acid in 150 ml of ethyl acetate in a 250-ml glass-stoppered flask. Eight millilitres of acetic anhydride are pipetted into the flask and allowed to stand at least 30 minutes at room temperature. The contents are cooled to 5°C and 42 ml of cold acetic anhydride are added. The flask is kept at 5°C for an hour and then allowed to come to room temperature. The reagent is stable for at least 2 weeks at room temperature.

To prepare 2M acetic anhydride in pyridine, 0·8 g (0·47 ml) of 75% perchloric acid is added dropwise to 30 ml of pyridine in a 50-ml flask with magnetic stirring.

The 0·55M sodium hydroxide titrant is prepared by adding 430 ml of water and 5400 ml of methyl Cellosolve or absolute methanol to 185 ml of saturated aqueous sodium hydroxide.

One part of 0·1% neutralized aqueous Cresol Red is mixed with 3 parts of 0·1% neutralized Thymol Blue to prepare the mixed indicator.

A weighed sample containing from 3 to 4 millimoles of hydroxyl is placed in a 125-ml glass-stoppered flask. Five millilitres of 2M acetic anhydride in ethyl acetate or pyridine are added. The mixture is stirred until solution is complete and the reaction is allowed to proceed for at least 5 minutes at room temperature (some alcohols require a somewhat longer reaction period if pyridine is used as the solvent). One or 2 ml of water are added, mixed, and then 10 ml of 3 : 1 pyridine–water are added. The flask is allowed to stand 5 minutes. The mixture is then titrated with 0·55M sodium hydroxide using the mixed indicator and the end-point is taken as the change from yellow to violet. A reagent blank is run by pipetting exactly 5 ml of acetylating reagent into a 125-ml flask containing 1–2 ml of water. Ten millilitres of 3 : 1 pyridine–water solution are added and the mixture is allowed to stand 5 minutes. The titration is performed as for the sample.

Caution! Solutions acetylated with perchloric acid present should not be heated and the sample and blank solutions should be disposed of promptly after the determination is completed.

The same workers[32] used *p*-toluenesulphonic acid in pyridine to catalyse the acetylation of sugars. This reaction was completed in 5–10 minutes at 50°C. Magnuson and Cerri[33] claimed that 1,2-dichloroethane is superior to ethyl acetate as the solvent for the perchloric acid-catalysed acetylation. The perchloric acid-catalysed system cannot be applied to polyethylene and polypropylene gycol ethers[30]. Erratic results are obtained, possibly due to oxidation of the chain by the perchloric acid. Pietrzyk and Belisle[34] suggested 2,4-dinitrobenzenesulphonic acid as a replacement for perchloric acid because it is highly acid and stable to heat and does not attack the polyglycol ethers.

A method has been reported for the analysis of mixtures of alcohols based on the difference in their esterification rates with acetic anhydride[35]. With this method the primary and secondary alcohol contents of mixtures can be determined, primary and secondary alcohol groups on the same molecule can be distinguished, and alcohols of a homologous series—even members different by only one carbon atom—can be distinguished. For the acetylation of mixtures of alcohols a conventional second-order rate plot shows a straight-line portion after the faster reacting hydroxyl has been consumed. The straight line is extrapolated to zero time and the amount of slower reacting component is calculated.

Procedure[35]: A sample containing 0·05 mole of hydroxyl is transferred to a 250-ml volumetric flask with pyridine and diluted almost to 240 ml with pyridine. Ten millilitres of acetic anhydride are pipetted into the flask. The mixture is rapidly diluted to volume with pyridine and the time noted. At intervals, 10 ml aliquots are pipetted into glass-stoppered flasks, 5 ml of water added and times again noted. Each is allowed to stand at least 10 minutes and is then titrated with 0·1N alcoholic potassium hydroxide to a mixed indicator end-point (a 2 : 1 mixture of 0·1% Nile Blue sulphate in 50% ethanol and 1% phenolphthalein in 95% ethanol). To determine the blank, 10 ml of acetic anhydride are pipetted into a 250-ml volumetric flask and diluted to volume with pyridine. A 10-ml aliquot of this is treated in the same manner as the sample.

Log $(b - x)/(a - x)$ is plotted against t, where x is the concentration of anhydride consumed in time t, a is the total initial hydroxyl group concentration and b is the initial concentration of anhydride. If the presence of two hydroxyl types is indicated by two slopes in the plot, a straight line is drawn representing the less reactive hydroxyl group (the second slope) and extrapolated to zero time. If the concentration of the more reactive hydroxyl group is designated a_1, then $x = a_1$ at the point (y) of intersection of the extrapolated line and the zero time coordinate. Substitution of a_1 for x gives the expression, $\log (b - a_1)/(a - a_1) = y$. The value of y can be obtained from the plot and the equation solved for a_1. Subtraction of the concentration of the more reactive hydroxyl group from the initial total hydroxyl concentration gives the concentration of the less reactive hydroxyl.

The phthalation reaction is less rapid and is less widely applicable for hydroxyl analysis but does not suffer from interference by aldehydes. Phenols fail to react with phthalic anhydride; therefore, alcohols can be determined in their presence with this reagent. Elving and Warshowsky[36] carried out the reaction either under reflux or in pressure bottles in the presence of pyridine. The titrations of the phthalic acid formed were made with aqueous sodium hydroxide solution to a phenolphthalein end-point.

Procedure[36]: For samples containing a high percentage of ethanol, a sample weighing 1·0 to 1·5 g is pipetted into a weighed 50-ml volumetric flask containing 30–40 ml of pyridine. For higher alcohols and dilute solutions of ethanol, larger samples should be taken. After reweighing, the solution is made to volume with pyridine. Into a pressure bottle are pipetted 25 ml of a solution of 20 g of phthalic anhydride in 200 ml of pyridine, and 10 ml of the sample solution are added. The sealed bottle is placed in an air oven at 100°C and is heated for 1 hour. After cooling to room temperature, the pressure is carefully released and 50 ml of water are added. The mixture is cooled under a cold water tap and titrated with 0·35N sodium hydroxide, with phenolphthalein as the indicator. A blank determination is made in the same manner and the reagents employed.

Ethanol could be accurately determined in samples containing as much as 85% water. However, water adversely affects the phthalation of alcohols in general[30]. It appears that the observed esterification of ethanol was actually the reaction of the alcohol and phthalic acid.

Pyromellitic dianhydride combines the advantages of acetic anhydride and phthalic anhydride—it can be used in the presence of aldehydes, it is not volatile, it can be used to determine alcohols in the presence of phenols and its rate of reaction compares favourably with that of acetic anhydride[37]. The reaction is best carried out in dimethyl sulphoxide in the presence of pyridine[38]. All four acid groups are neutralized at the phenolphthalein end-point (equation 8). Pyromellitic dianhydride does not attack ether linkages of polyglycol ethers and is useful therefore for the determination of the hydroxyl contents of these compounds.

Procedure[38]: Fifty millilitres of 0·5M pyromellitic dianhydride (109 g dissolved in 525 ml of dimethyl sulphoxide and 425 ml of pyridine added) are pipetted into a glass-stoppered 250-ml flask.

The sample containing 0·010 to 0·015 equivalents of alcohol is weighed and added to the reagent. The flask is placed on a steam bath and the stopper is moistened with pyridine and loosely seated. The contents are heated for 15–20 minutes (30 minutes for polyglycols). 20 ml of water are added and the heating continued for 2 minutes. The mixture is then cooled to room temperature and titrated with 1N sodium hydroxide solution to the phenolphthalein end-point. A blank in which only the sample is omitted is treated in the same manner.

$$C_6H_2(COOH)_4 + 4\,NaOH \longrightarrow C_6H_2(COONa)_4 + 4\,H_2O \qquad (8)$$

Other anhydrides have been recommended for hydroxyl determinations and may offer advantages in specific situations. These include stearic anhydride[39], *o*-sulphobenzoic anhydride[40], propionic anhydride[41], succinic anhydride[42] and 3-nitrophthalic anhydride with triethylamine as the basic catalyst[43]. The last three listed were specifically recommended for the determinations of polyglycol ethers.

2. Acid–base methods (enols, phenols and nitroalcohols)

Most alcohols are slightly acidic but not sufficiently to be titrated. However, enols, phenols and nitroalcohols are acidic enough to be titrated in nonaqueous media though not in water. Nitro- and polynitrophenols can also be titrated in aqueous media.

Various nonaqueous media and titrants are used. Moss, Elliot and Hall[44] used ethylenediamine as solvent and the sodium salt of ethanolamine as titrant. Fritz[45] extended the method to include dimethylformamide as solvent as well and used sodium ethoxide dissolved in benzene–methanol as titrant. Deal and Wyld[46] and Harlow, Noble and Wyld[47] summarize the earlier methods for titrating weak acids in various solvents. The currently most popular method is described on the following page.

Procedure[48]:

a. Reagents and apparatus

Beckman general purpose glass electrode, No. 4990–80.

Beckmann sleeve type calomel electrode, No. 1170–71, modified by replacing the saturated aqueous KCl solution in the outer jacket with a saturated solution of KCl in methanol (designated hereafter as methanol modified calomel electrode).

Tetrabutylammonium hydroxide, 0·1N in 10 : 1 benzene–methanol prepared as described below.

Technical grade. Acetonitrile, pyridine and dimethylformamide.

Benzene–isopropanol, 10 : 1.

Thymol Blue indicator solution: 0·3 gm in 100 ml isopropanol.

Azo Violet indicator, a saturated solution of *p*-nitrobenzeneazoresorcinol in benzene.

b. Preparation of titrant. Forty grammes of tetrabutylammonium iodide are dissolved in 90 ml of absolute methanol. Twenty grammes of finely powdered silver oxide are added, followed by vigorous agitation of the mixture for 1 hour. A few millilitres of the solution are centrifuged and the supernatant is tested for iodide with aqueous silver nitrate. If the test is positive, 2 grammes of additional silver oxide are added and agitation is continued for an additional 30 minutes. When the iodide test is negative, the mixture is filtered through a sintered glass filter funnel of fine porosity. The reaction flask and funnel are rinsed with three 50-ml portions of dry benzene which are added to the filtrate. The filtrate is diluted to one litre with dry benzene. This solution is flushed for 5 minutes with dry, prepurified nitrogen and then stored in a reservoir protected from carbon dioxide and moisture. The titrant remains stable on extended storage. It is standardized against pure benzoic acid using either the visual or potentiometric methods described below.

c. Potentiometric titrations. An accurately weighed sample, sufficient to consume 2–10 ml of titrant, is placed in a 250-ml beaker and dissolved in 50 ml of solvent (see discussion below for choice of solvent). Insert the glass and methanol-modified calomel electrodes. The burette is covered with an Ascarite tube. Best results are obtained if titrations are carried out under a nitrogen blanket when dimethylformamide or pyridine are used as solvents. Titrant can be added in 0·05 ml increments in the region of the end-point. A blank should be run on solvents to account for any acidic impurities. A curve is plotted of millivolts against millilitres of titrants, the inflexion point or points are then determined.

Pyridine is reported to be the best solvent and to give the best end-points. However, dimethylformamide is also usable. Acetonitrile is reported as the best of the neutral solvents.

Good titrations can be obtained with the glass and the normal saturated calomel electrode. However, the sharpness of the inflexion points was markedly increased when the methanol-modified calomel electrode was used.

Hydroxyl compounds titrated potentiometrically in the original work include phenol, resorcinol, hydroquinone, dimethyl dihydroresorcinol, thymol, pyrogallic acid, catechol, *o*-, *m*- and *p*-hydroxybenzoic acids, cresol.

Tertiary butyl alcohol, isopropanol and acetone, as well as pyridine, were used as solvents, with a modified electrode system[49]. However, the general utility of the method remains as shown.

d. Visual titrations. An accurately weighed sample, sufficient to consume 5–10 ml of titrant is added to a 125-ml Erlenmeyer flask. Twenty-five ml of solvent is added with four drops of indicator (Thymol Blue for the weak acids or Azo Violet for the very weak acids). The titration should be as rapid as possible to the blue end-point in the case of Thymol Blue indicator to a violet (sometimes blue) end-point with the Azo Violet indicator. Blanks should be run on the solvents.

o- and *p*-nitrophenols were successfully titrated using Thymol Blue as an indicator and pyridine as sample solvent. Phenol, *p*-benzylphenol, *o*-phenylphenol, 1- and 2-naphthols, 2,5-dimethylphenol, *p*-bromophenol, catechol, pyrogallic acid and dimethyldihydroresorcinol were successfully titrated in pyridine as solvent with Azo Violet as indicator. Acetonitrile as solvent was used to titrate 1-naphthol. Phenolic compounds which were successfully titrated potentiometrically but which could not be visually titrated are thymol, hydroquinone, *p*-toluhydroquinone, *m*- and *p*-cresols.

3. Determination of 1,2 dihydroxy compounds (glycols)

The most general method for specifically measuring 1,2 dihydroxy compounds is the oxidation with periodic acid.

$$\underset{\substack{| \\ OH}}{R-CH}-\underset{\substack{| \\ OH}}{CH}-R' + HIO_4 \longrightarrow \underset{\substack{\| \\ O}}{RCH} + \underset{\substack{\| \\ O}}{HCR'} + H_2O + HIO_3$$

This reaction is specific and not generally influenced by the presence of monohydric alcohols or by polyhydric alcohols where the hydroxyl groups are not on adjacent carbon atoms. Occasionally interference will be noted but this is rare; for example: 2-butyne-1,4-diol was found to be significantly oxidized by periodic acid under conditions of the analysis.

Procedure[50]: A sample containing 0·0005–0·001 moles of dihydroxy compound is weighed into a 50-ml glass-stoppered iodine flask. To this is added 100 ml of the reagent (5 g of HIO_4 dissolved in a mixture of 800 ml of acetic acid and 200 ml of distilled water). A blank is run on the reagent alone. The reaction mixture is allowed to stand for half an hour at room temperature; a few samples may require one hour but this can only be determined by trial. At the end of the reaction period 20 ml of 20% potassium iodide solution are added and the liberated iodine is titrated with 0·1N sodium thiosulphate. The titration on the sample should be more than 80% of the blank since the iodate formed in the reaction also liberates iodine, and, if all the periodic acid reacts, the back titration equals 75% of the blank.

$$\%\ \text{glycol compound} = \frac{A \times N \times MW \times 100}{g \times 2000}$$

A = ml blank − ml sample
N = normality thiosulphate
MW = mole weight glycol compound
g = grammes sample

Long reaction periods should be avoided in cases where the reaction products include formaldehyde or formic acid, which are subject to slow oxidation.

1,2,3-Trihydroxy compounds consume two moles of periodic acid and generally liberate one mole of formic acid from the central carbinol moiety. The method of Bradford et al[51] enables the determination of these compounds (i.e. glycerol) in the presence of 1,2 dihydroxy compounds via the titration of the formic acid formed from the trihydroxy material.

4. Determination of trace quantities of hydroxyl compounds

a. Primary and secondary alcohols are determined through formation of benzoate esters, extraction of the ester and spectrometric determination of the ester in the extract. Johnson and Critchfield[52] esterified with 3,5-dinitrobenzoyl chloride. The ester was extracted with hexane. A colour was developed by addition of 2N sodium hydroxide and acetone to the extract, and measured at 575 mμ.

Scoggins[53] esterified with *p*-nitrobenzoyl chloride, extracted with cyclohexane and measured the u.v. absorption of the ester in the extract at 253 mμ.

b. Secondary alcohols are oxidized to the ketone[54], the 2,4-dinitrophenylhydrazone of which is then prepared and measured colorimetrically. Primary alcohols do not interfere in this method since they are oxidized to carboxylic acids.

c. Tertiary alcohols can be determined by reaction with hydriodic acid to form the corresponding alkyl halide[55], which is extracted with cyclohexane and measured in the u.v. at the wavelength of maximum absorption (267–269 mμ).

Esterification[52] has been tried, but the reaction is very slow.

d. Phenols. The esterification methods[52, 53] may well be usable for measuring traces of phenols although apparently they had not been tried. However, an excellent method is the Azo dye formation when the phenol is coupled with a diazonium compound[56].

$$C_6H_5OH + [ArN{\equiv}N]^+ Cl^- \longrightarrow ArN{=}N{-}C_6H_4{-}OH + HCl \quad (9)$$

By using diazotized sulphanilic acid, the acid group lends water solubility to the resultant dyes. The method is fast, specific and sensitive to lower than 1 ppm of most phenols or naphthols.

III. PHYSICAL METHODS

A. Infrared Spectroscopy

1. Qualitative

In the infrared region the hydroxyl groups exhibit strong absorptions which are highly sensitive to structure. These absorptions provide an important means of characterizing compounds containing this function. Characteristic absorption positions are given in Table 1.

The hydroxyl group is highly polar and subject to strong association; therefore, the intensity of the hydroxyl stretching vibration is dependent on the degree of association in the system. Essentially, complete dissociation is observed only in the vapour state or, in some cases, when the hydroxyl-containing compound is diluted extensively with nonpolar solvents. In general, the presence of a band in the 2·7- to 3·0-μ region is a reliable indication of the presence of hydroxyl groups. Water, N–H groups and carbonyl groups cause interfering absorptions in this area.

TABLE 1. Infrared vibrations of hydroxyl compounds.

		Wavelength, μ	Ref.
Free hydroxyl stretching vibrations			
Alcohols			
Fundamental vibration—	in alcohols in general	2·75–2·77	57
	in primary alcohols	2·746–2·753	58
	in secondary alcohols	2·754–2·762	58
	in tertiary alcohols	2·764–2·769	58
First overtone—	in alcohols in general	1·40–1·46	59
	in primary alcohols	1·405–1·410	60
	in secondary alcohols	1·413–1·415	60
	in tertiary alcohols	1·418–1·420	60
Second overtone—	in alcohols in general	0·877–0·980	59
Third overtone—	in alcohols in general	0·738–0·744	61, 62
Combination band—	in alcohols in general	1·95–2·15	58
Phenols			
Fundamental vibration		2·77–2·78	63
First overtone		1·404–1·418	64
Second overtone (doublet)		1·00 and 0·971	62,65
Third overtone (doublet)		0·7466 and 0·7698	66
Bonded hydroxyl stretching vibrations			
Intermolecular hydrogen bonding—	dimeric	2·82–2·90	67
	polymeric	2·94–3·09	67
Intramolecular hydrogen bonding—	hydroxyl–organic group interaction	2·79–2·92	67
	hydroxyl–metal interaction (chelation)	3·12–4·00	68
Hydroxyl bending vibration			
Hydroxyls in general		7·14–7·70	69
Carbon–oxygen stretching vibrations			
Alcohols			
Primary—	straight chain	9·22–9·52	70
	α-branched and/or α-unsaturated	beyond 9·52	70
Secondary—	saturated aliphatic	8·90–9·20	70
	highly symmetrical	8·30–8·90	70
	branched at one α-carbon	9·10–9·20	70
	α-unsaturated	9·22–9·52	70
	alicyclic (5- or 6-membered ring)	9·22–9·52	70
	di-α-unsaturated	beyond 9·52	70
	α-branched and α-unsaturated	beyond 9·52	70
	alicyclic (7- and 8-membered ring)	beyond 9·52	70
Tertiary—	saturated aliphatic	8·30–8·90	70
	α-unsaturated	8·90–9·22	70
	cyclic	8·90–9·22	70
	highly unsaturated (e.g. triphenylcarbinol)	beyond 9·52	70
Phenols			
Phenols in general		about 8·3	70

Both intermolecular and intramolecular associations of the hydroxyl group are possible and displace the absorption to longer wavelengths. The type and degree of intermolecular bonding are dependent on temperature, molecular structure and environment. High temperature favours dissociation. Normally, most alcohols are found as polymeric species in the absence of steric effects. Indications are that cyclic dimers occur at relatively low concentrations as a result of nonlinear hydrogen bonding[71]. Solute–solvent hydrogen bonding can also influence the infrared spectra.

In contrast to intermolecular hydrogen bonding, which is highly concentration-dependent, intramolecular bonding is not affected by dilution with a nonpolar solvent such as carbon tetrachloride. Any electron-rich system, such as a double bond, a cyclopropyl ring, an aromatic ring, a halogen atom, and carbonyl, amino, nitro, ether and ester groups, will interact with the hydroxyl group if close enough in space and result in a shift and splitting of the free hydroxyl bond[72]. Diols show both intermolecular and intramolecular bonding[72–74].

The molar absorptivities of the hydroxyl groups of phenols are 3–4 times larger than those of alcohols which vary between 30–100 and remain within $\pm 10\%$ for alcohols of similar structure[58].

Hydrogen bonding also affects the hydroxyl bending vibrations, shifting the spectra to shorter wavelengths[75]. The C–O stretching vibrations are of great value in the differentiation of primary, secondary and tertiary alcohols for qualitative and structural work[70].

2. Quantitative

Many applications of infrared spectroscopy to the quantitative determination of hydroxyl-containing compounds have been made. However, because of the tendency of hydroxyl groups to form bonds with other polar groups, there are serious limitations of the technique for general use. With proper control of the conditions, infrared methods are successful in many specific situations. For example, samples have been diluted to the point where the concentration of the associated species is negligible and, in other cases, samples are run under conditions of complete association. Other methods depend on strict control of conditions so that the ratio of associated to unassociated species is constant and can be related to calibration curves prepared under the same conditions. Also, methods involving the preparation of derivatives and their examination at appropriate wavelengths have been used.

Ahlers and McTaggert[76] devised infrared methods for the determination of hydroxyl groups in autoxidized or copolymerized fatty esters and related compounds. Measurements were confined to dilute solutions in carbon tetrachloride where no absorption of the associated species was observed. The free hydroxyl determination was then based on the intensity of the absorption at 2·76 μ. Crisler and Burrill[64] determined aliphatic primary alcohols, using the hydroxyl stretching overtone band at 1·4 μ and in 0·04–0·06M solutions of the samples in carbon tetrachloride or tetrachloroethylene. It was found that a single calibration curve cannot be used for all hydroxyl-containing compounds because the band positions and intensities depend on the structures of the compounds. Hilton[77] used the 2·0–3·2-μ region to determine hydroxyl contents of polyesters and polyethers in 1 : 10 chloroform–carbon tetrachloride as the solvent. Temperature control of ±0·1 °C was found to be necessary and each new compound required calibration. Burns and Muraca[78] demonstrated that the hydroxyl absorptions at 2·84 μ of 14 different polypropylene glycols follow Beer's law with a standard deviation of 2·2%. The polypropylene glycols appear to exhibit intramolecular bonding through the hydroxyl group and an ether oxygen forming a five-membered ring. The hydroxyl band position was unaltered by dilution with benzene or carbon tetrachloride indicating the absence of intermolecular bonding. Murphy[79] determined the hydroxyl contents of alkyd resins in dichloromethane solutions at 2·85 μ. Corrections were required for the water and organic acid present in the resin. Adams[80] determined the hydroxyl content of epoxy resins at 3·08 μ. Pyridine was used as a solvent to produce associated hydroxyl bands exclusively. In the determination of the hydroxyl equivalents of steroids[81], the absorbance in pyridine was found to be linear with concentration for the associated band which appears near 3·05 μ and is essentially independent of the type of hydroxyl group, excepting those with phenolic hydroxyl groups. Dvoryantseva and Sheinker[82] used the molecular extinction coefficient of the band at approximately 3 μ to determine the number of hydroxyl groups in steroids that do not contain phenolic hydroxyl groups. It was demonstrated that the extinction coefficient does not depend on the position of the hydroxyl group in the molecule but is an additive value depending on the number of these groups.

Partridge and Kirby[83] used the absorption band at 2·80 μ to determine residual 2-ethylhexyl alcohol in di-2-ethylhexylphthalate. Interference due to water was eliminated by subtraction of the absorption at 2·74 μ. Mitchell, Bockman and Lee[84] determined the acetyl content of cellulose acetate by measuring the absorbance due to residual hydroxyl groups at 1·445 μ. Shauenstein and Puchner[85] also used the 1·4-μ band to determine unbranched aliphatic primary alcohols in chloroform solution. The relative content of primary and secondary fatty alcohols (C_9 to C_{18}) in various industrial products of the oxidation of paraffins was based on the relation between the molar extinction coefficient and wave number in the region 960–1200 cm^{-1} [86]. Oba[87] used the 9·84-μ band to determine 0·2 to 5·0% methanol in ethanol by a differential technique. Gronau, Broadlick and Hamilton[88] determined the ethanol content of Thimerol Tincture NF at 11·37 μ after extraction of the alcohol from the sample with carbon tetrachloride.

Using a very thin absorption cell with barium fluoride windows, Potts and Wright[89] determined 5% solutions of ethylene glycol and diethylene glycol in water. The absorption at about 8·8 μ determines uniquely the amount of diethylene glycol present; ethylene glycol can be determined by the absorption at 9·2 μ after correcting for diethylene glycol present. It was also shown that 5% phenol and 10% ethanol in water could be determined.

Friedel[90] studied the absorbance at 2·89–3·01 μ of 22 phenols, mostly alkyl derivatives. Compounds containing methyl groups in both *ortho* positions showed no associated hydroxyl band or shoulder but ethyl or larger groups *ortho* to the hydroxyl showed weak associated hydroxyl bands at 2·89–2·93 μ. The use of these bands for quantitative analysis was suggested. Goddu[91] used the 2·7- –3·0-μ range for the qualitative and quantitative analysis of phenols. Samples containing as little as 25 ppm phenolic hydroxyl were analysed. The intramolecular bonding shifts in the hydroxyl band, which differ in degree depending on the type of phenol, were used to analyse mixtures which contain several phenolic species. Phenolic hydroxyl end groups were determined in solutions of aromatic polycarbonates using the band at 2·79 μ. Average molecular weights obtained by this method were in good agreement with data from osmometry, fractionation methods and ultracentrifugation.

Lippmaa[92] measured the absorption of solutions of phenols in anisole at 2·82 μ and 2·95 μ in lithium fluoride cells. Beer's law was found applicable for concentrations of 0·1 to 0·4 g equivalents of phenolic hydroxyl per litre. The ratio of the molar extinction coefficients for the two maxima was consistent for the phenols studied.

Kyriacou[93] determined hydroxyl groups in polypropylene glycol by a procedure which basically involved a spectrophotometric titration with acetyl chloride. Different and definite amounts of acetyl chloride were added to equal weight portions of the sample. The amounts of acetylating reagent were chosen to obtain absorbances between 0·2 and 0·8. The absorbance was determined for each portion at 2·87 to 2·88 μ and plotted vs the respective acetyl chloride concentration. The amount of acetyl chloride required to react with the sample was taken as the point at which there was no further decrease in the hydroxyl absorption. Mamiya[94] determined hydroxyl groups in polyethylene and polypropylene glycols by titration of their toluene solutions containing zinc powder with toluene solutions of acetyl chloride. The absorption at 1·45 μ was used to determine the end-point.

Jaffe and Pinchas[95] determined dipentaerythritol in the presence of pentaerythritol by measuring the absorption of the corresponding acetates in carbon tetrachloride. The hydroxyl content of oxidized polyethylene was determined by infrared analysis after quantitative acetylation with acetic anhydride[96].

Hendrickson[97] determined primary hydroxyl groups in polyglycols by following the rate of disappearance of the hydroxyl band at 3·05 μ using triphenylchloromethane as the reactant. A rate plot was made and extrapolated to zero time. The rate of disappearance of the secondary hydroxyl band was measured and the quantity of primary alcohol determined by difference.

B. Nuclear Magnetic Resonance

Nuclear magnetic resonance is a powerful tool for functional group analysis. Each distinct type of atom is characterized by a specific chemical shift which does not change greatly from one molecule to another. In a manner analogous to vibrational frequencies, tables of chemical shifts have been compiled for the common functional groups containing hydrogen[98]. Utilizing these data together with the rules governing multiplet splitting, it is generally not difficult to assign the lines in an experimental spectrum. Consequently, n.m.r. has become one of the most commonly used instrumental methods in qualitative organic analysis.

Protons bound to an oxygen atom have two properties which complicate the interpretation of their spectra. First, the –OH groups have a variable chemical shift, depending on the extent of hydrogen bonding, if present. Secondly, although the O–H bond is thermodynamically stable, it is kinetically labile. The –OH proton is capable of exchanging with labile protons on other sites. If this exchange is rapid enough, several distinct types of protons may be observed as a single n.m.r. signal, thus confusing the assignment of the spectrum. Furthermore, this exchange will also lead to the collapse of the multiplet structure of the –OH signal, causing the loss of valuable analytical information.

Hydroxyl groups in liquid organic samples are often involved in intermolecular hydrogen bonding with solvents containing basic groups. If the solvent is inert, dimers and higher polymers may be formed, or intramolecular hydrogen bonds can be formed if the steric factor is favourable. The complete absence of hydrogen bonding is observed only when no intramolecular hydrogen bonding is possible and the sample is a very dilute solution in an inert solvent. Then the chemical shift of the –OH proton of saturated aliphatic alcohols is approximately 0·5 ppm downfield from tetramethylsilane (δ scale). However, the experimental conditions most often encountered in n.m.r. work involve a moderately concentrated solution of the sample in an inert solvent. Under these conditions, a chemical shift between 3·0 and 5·2 is seen for hydroxyl protons of saturated alcohols. A higher degree of hydrogen bonding is reflected by a larger downfield shift.

A finite time is required for a collection of protons to come to equilibrium with the radio frequency field generated by the spectrometer. This depends on the relaxation times of the sample; it is usually several seconds for ordinary samples. Chemical processes taking place within this time span will affect the observed n.m.r. spectrum. Hydroxyl hydrogens may exchange with labile protons in mineral acids, carboxylic acids, water and other –OH containing compounds; in the absence of these the –OH protons on a given compound will exchange with each other. The n.m.r. spectrum reflects the exchange by an averaging process. When protons are exchanging rapidly between sites that ordinarily would have different chemical shifts, only one line is observed. The shift of this line is the average shift of the two sites occupied by labile protons weighted for the relative numbers of each site present. It is given by the relation[99]:

$$\sigma_{obs} = p_{I}\sigma_{I} + p_{II}\sigma_{II}$$

where σ_{obs} is the experimental chemical shift, σ_{I} is the chemical shift at site I, and p_{I} is the probability that a given proton be found at site I. The quantities p_{II} and σ_{II} refer to site II. In cases where exchange takes place among equivalent sites, as in a pure liquid alcohol, the process has no effect on the chemical shift; but it does eliminate the spin–spin coupling between the –OH proton and neighbouring protons causing collapse of the multiplet structure of the line. Multiplet collapse also results in exchange between non-equivalent sites.

I. Direct analysis

Since –OH chemical shifts can vary over a large range, the presence of hydroxyl cannot be established unequivocally from a simple n.m.r. spectrum. If an –OH group is suspected, and a single line is observed in the proper region, it can be tentatively assigned as an –OH signal, but additional information will be required to confirm the assignment. The chemical shifts for several types of groups are given in Table 2.

TABLE 2. Chemical shift[a].

Group	Ordinary conditions	Infinite dilution
Saturated alcohol	3·0–5·2	0·5
Phenol	4·5–7·7	4·0–5·0
Enol	15·0–16·0	15·0–16·0

[a] Shifts are quoted in the δ scale, ppm downfield from tetramethylsilane.

a. Alcohols. If the sample contains no acidic impurities, the –OH signal usually falls in the predicted region. It is not uncommon for the line to be broadened slightly due to incipient spin coupling with neighbouring protons. This often shows up as a lack of ringing on the –OH peak. This effect cannot be used to make reliable assignments, however.

The common methods of identifying –OH signals are based on the lability of the hydroxyl proton. A small quantity of acid added to the sample will cause a pronounced downfield shift of the –OH signal. This is due to the rapid exchange of acid protons with –OH groups which causes an averaging of the signals from the two sites. Since the shift difference between an alcohol and an acid may be as large as 10 ppm, a relatively small amount of acid will cause a noticeable shift in the –OH resonance. Trifluoroacetic acid is a favourite for this test since it is soluble in most n.m.r. solvents, and the molecule contains no protons other than the acid proton itself.

Another experiment that is diagnostic of hydroxyl protons (or labile protons in general) depends on hydrogen exchange with water. If the sample is shaken with a quantity of deuterium oxide, the labile hydrogens will be replaced by deuterium. Since the latter give no signal, a drastic reduction in the intensity of the labile proton signal results.

Recently a degree of success has been achieved in hydroxyl detection through the artifice of tying up the proton in a strong hydrogen bond, using a strongly basic solvent such as acetone or dimethyl sulphoxide (DMSO). In such solvents the lability of the –OH proton is reduced and spin–spin coupling can be observed. On the basis of the splitting pattern, the –OH can be classified as primary, secondary or tertiary. Furthermore, since all of the hydrogen bonding is between alcohol and solvent, there is no alcohol association–dissociation equilibrium. Hence the chemical shift of the –OH group will be concentration independent, and useful shift measurements can be made at convenient concentrations.

Such measurements were first made in acetone solution[100, 101] where it was noted that the proton exchange of benzyl alcohol is strongly inhibited. Later, Chapman and King[102] introduced the use of DMSO which is the favoured solvent today. In DMSO the –OH resonance is shifted downfield to $\delta = 4$ or lower, and the multiplet structure of the hydroxyl signal can be resolved. An early application of this technique was made by Casu et al[103] in their study of reducing sugars. DMSO is an excellent solvent for sugars, and it has the effect of slowing the mutarotation to the point where the spectrum of individual anomers can be seen, which is not the case in water. Since signals of individual –OH groups can be distinguished, this is a powerful tool for the investigation of such molecules. This solvent has also been used in the study of the conformation of cyclohexanols[104]. The coupling constant between the –OH and the carbinol protons has been used to estimate the HO–CH dihedral angle in a manner analogous to the work of Karplus[105] on HC–CH couplings.

Unfortunately, the DMSO technique is not applicable to all alcohols. Traynham and Knesel[106] discovered that the –OH splitting does not always appear in alcohols which contain an electron-withdrawing group close to the –OH. Groups such as alkoxide, vinyl and halide are effective in this respect. In some cases the splitting can be developed by shaking the sample with solid K_2CO_3, but some alcohols are intractable even with this treatment. Later, Moniz et al[107] were able to observe –OH splitting in difficult cases by using dilute solutions in DMSO or in CCl_4 that had been rigorously dried. Replacement of CCl_4 by DMSO decreases the values of $J_{HC\text{-}OH}$ and has a levelling effect on their variation. Recently, Takino et al[108] were unable to observe splittings in certain alcohols. They suggested that acetone is more effective in producing splittings, although its effectiveness as a solvent is limited.

There have been several studies of chemical shifts of alcohols in DMSO solution. For example, Brook and Pannell[109] recorded hydroxyl shifts for a large number of substituted triphenylsilyl phenyl carbinols. They correlated the shifts with empirical parameters such as Hammett's σ function. In general, however, –OH analysis has been based on splitting patterns rather than chemical shifts. A typical example is shown in the work of Rothweiler and Tamm[110] on the structure of the antibiotic phomin. They showed that both –OH groups on the molecule were secondary since they appeared as doublets which disappeared on treatment with D_2O.

When recording spectra in DMSO, it is very advantageous to use the deuterated solvent to avoid interference from solvent protons. If common DMSO must be used, one can anticipate difficulties due to overlap of the sample spectrum with the DMSO signal. All factors considered, the higher expense of DMSO-d_6 is usually well justified.

b. Phenols and enols. These compounds have not been studied as extensively as alcohols. Since there are no protons neighbouring the phenolic group, the –OH shows no splitting and the chemical shift is the only parameter obtainable. In inert solvents the chemical shift is concentration dependent and ranges from $\delta = 4{\cdot}5$–8. Some *o*-substituted phenols are intramolecularly hydrogen bonded; their chemical shifts are nearly independent of concentration. The intermolecular association of phenols does not occur in strong hydrogen bonding solvents such as DMSO, and useful shift data can be obtained. Traynham and Knesel[111] studied a large number of phenols in DMSO, and observed values ranging from 8·5–11·0 ppm downfield from TMS. Dietrich et al[112] investigated a number of basic solvents and found that in hexamethylphosphoramide proton exchange is sufficiently slowed to allow observation of a distinct –OH signal from each individual phenol in a mixture. Hexamethylphosphoramide (HMPA) is potentially as useful as DMSO.

Stable enols owe their existence to strong intramolecular hydrogen bonds. The β-diketones are an especially favourable case and their enols have been studied in the greatest detail. The –OH chemical shift is far downfield ($\delta = 15–16$) and shows little concentration dependence. Since the keto–enol conversion is slow in these cases, the equilibrium can be studied by observing the relative areas under the peaks from the –CH= and the $-CH_2-$ groups. Many studies of keto–enol equilibria have been made, utilizing this approach. For example, Reeves[113] studied the effect of the solvent on the acetylacetone equilibrium. An interesting, but not generally applicable, method was developed by Gorodetsky et al[114]. They observed the oxygen resonance in ^{17}O enriched β-diketones. The spectra showed separate peaks for keto and enol forms, and it was possible to derive equilibrium constants for the conversion.

2. Derivativization methods

A different approach depends on making derivatives of the alcohol function, and then studying the derivatives. An early suggestion of this method was made by Jackman[115] who noted that esterification of an alcohol resulted in a downfield shift of the α hydrogens. Useful structural information can be obtained from the multiplicity and the area of the shifted peaks. Mathias[116] has made a systematic study of this effect. One frequently finds that α proton signals from primary and secondary alcohols overlap. When the alcohols are acetylated, the resonances separate. In some cases this effect can be used to obtain primary/secondary ratios.

Manatt[117] proposed studying the resonance of the protons of the acetyl group. However, these protons are insensitive to their environment, so the trifluoroacetate esters were prepared, and the ^{19}F signal utilized as the analytical probe. Fluorine chemical shifts are more sensitive than proton shifts in most cases. In the fluorine-containing esters, separate signals from primary and secondary groups are often observed. This approach requires the use of a ^{19}F spectrometer, but sometimes this problem can be circumvented by using the α protons. For example, Ludwig[118] showed that the α proton signal could be used to determine primary/secondary ratios in alcohol mixtures.

Babiec et al[119] prepared dichloroacetate esters by addition of dichloroacetic anhydride. The shielding of the lone proton on the dichloroacetyl group is sensitive to its environment, the order being tertiary > secondary > primary. The chemical shift will readily distinguish between tertiary and secondary, but there is sometimes overlap between secondary and primary. The latter shift can be enhanced by the use of DMSO-d_6 as a solvent. In common with other esterification techniques, the α hydrogens can also be used as the analytical probe, and they often serve as a useful check.

Goodlett[120] investigated the use of ketones and isocyanates as derivatizing agents. He concluded that trichloroacetyl isocyanate, TAI, (CCl_3—CO—N═C═O), was superior. This compound contains no protons, has a long shelf life and reacts smoothly in situ, even with hindered –OH groups. The α protons are the analytical probe. Primary protons are shifted downfield by 0·5 to 0·9 ppm while secondary protons are shifted by 1·0 to 1·5 ppm. The use of TAI is further discussed by Trehan et al[121] and by Butler and Mueller[122]. The latter group has extended its use to –SH groups.

Seikel et al[123] recently studied the structure of thomasic acid. Here the –OH spectrum is complicated by the fact that the carbinol carbon is an asymmetric centre. The TAI derivative was used to identify the peaks due to the carbinol hydrogens, while the splitting pattern of the –OH was obtained from a solution of the alcohol itself in DMSO. With the combined results of these experiments, the authors were able to locate the hydroxyl group within the molecule.

3. Hydrogen bonding

A voluminous literature has developed on the use of n.m.r. for the study of association equilibria in alcohols and phenols. This subject is too vast to be reviewed here, but several representative references will be cited. The work usually involves the determination of the –OH chemical shift as a function of concentration in some suitable solvent. The relative amounts of the various polymeric species can be calculated.

Saunders and Hyne[124] investigated the behaviour of *t*-butanol in CCl_4 solution. Their data indicate a simple equilibrium between the unassociated alcohol and a trimeric species. In reviewing the earlier literature, Littlewood and Willmott[125] were able to show that most alcohols associated into linear polymers of varying length. Phenols usually associate into trimers in solution[126].

IV. REFERENCES

1. N. D. Cheronis, J. B. Entrikin and E. M. Hodnett, *Semimicro Qualitative Organic Analysis*, Wiley-Interscience, New York, 1965, pp. 714–725, 924–945.

2. R. L. Shriner, R. C. Fuson and C. Y. Curtin, *A Systematic Identification of Organic Compounds*, 5th ed., Wiley, New York, 1964, pp. 316–319, 374–376.
3. 'Tables of Identification of Organic Compounds', in *Supplement to Handbook of Chemistry and Physics*, Chemical Rubber Publishing Company, Ohio, pp. 28–57.
4. Ref. 1, pp. 468–470.
5. Ref. 2, pp. 247–248.
6. K. J. Garska, R. C. Doutkit and V. A. Yarborough, *Anal. Chem.*, **33,** 392–395 (1961).
7. Ref. 1, pp. 473–474.
8. Ref. 2, pp. 246–248.
9. Ref. 1, p. 475.
10. Ref. 2, p. 246.
11. Ref. 1, pp. 480, 550–551.
12. Ref. 1, p. 490.
13. A. C. Rice et al, *Anal. Chem.*, **23,** 195 (1951).
14. E. O. Woolfolk, F. E. Beach and S. P. McPherson, *J. Org. Chem.*, **20,** 391 (1951).
15. E. Sundt and M. Winter, *Anal. Chem.*, **29,** 851–852 (1957).
16. E. Lederer, *Australian. J. Sci.*, **11,** 208 (1949).
17. R. A. Evans et al, *Nature*, **164,** 674 (1949); **170,** 249 (1952).
18. S. Rydel and M. Macheboeuf, *Bull. Soc. Chim. Biol.*, **31,** 1265 (1949).
19. E. O. Woolfolk and J. M. Taylor, *J. Org. Chem.*, **22,** 827–829 (1957).
20. J. W. Spanyer and J. P. Phillips, *Anal. Chem.*, **28,** 253 (1956).
21. W. M. D. Bryant, J. Mitchell Jr. and D. M. Smith, *J. Am. Chem. Soc.*, **62,** 1 (1940).
22. D. M. Smith and W. M. D. Bryant, *J. Am. Chem. Soc.*, **57,** 61 (1935).
23. H. P. Kaufmann and S. Funke, *Ber.*, **70B,** 2549 (1937).
24. C. P. A. Kappelmeir and J. Mostert, *Verfkroniek*, **31,** 61 (1958); *Chem. Abstr.*, **55,** 15955d (1961).
25. R. E. Kepner and A. W. Webb, *Anal. Chem.*, **26,** 925 (1954).
26. A. Bring and F. Kadlecek, *Plaste u. Kautschuk.*, **5,** 43 (1958); *Chem. Abstr.*, **52,** 12450h (1958).
27. W. T. Robinson Jr., R. H. Cundiff and P. C. Markunas, *Anal. Chem.*, **33,** 1030 (1961).
28. S. Siggia and I. R. Kervenski, *Anal. Chem.*, **23,** 117 (1951).
29. C. I. Ogg, W. L. Porter and C. O. Willits, *Ind. Eng. Chem.*, *Anal. Edition*, **17,** 394 (1945).
30. S. Siggia, *Quantitative Organic Analysis via Functional Groups*, 3rd ed., Wiley, New York, 1963.
31. G. H. Schenk, P. W. Wines and C. Mojzis, *Anal. Chem.*, **36,** 914 (1964).
32. J. S. Fritz and G. H. Schenk, *Anal. Chem.*, **31,** 1808 (1959); G. H. Schenk and J. S. Fritz, *Anal. Chem.*, **33,** 896 (1961).
33. J. A. Magnusen and R. J. Cerri, *Anal. Chem.*, **38,** 1088 (1966).
34. D. J. Pietrzyk and J. Belisle, *Anal. Chem.*, **38,** 1508 (1966).
35. S. Siggia and J. G. Hanna, *Anal. Chem.*, **33,** 896 (1961).
36. P. J. Elving and B. Warshowsky, *Anal. Chem.*, **19,** 1006 (1947).
37. S. Siggia, J. G. Hanna and R. Culme, *Anal. Chem.*, **33,** 900 (1961).
38. R. Harper, S. Siggia and J. G. Hanna, *Anal. Chem.*, **37,** 600 (1965).

39. B. D. Sully, *Analyst*, **87**, 940 (1962).
40. V. Iyler and N. K. Mathur, *Anal. Chim. Acta*, **33**, 554 (1965).
41. E. H. Vegelenzang and D. J. Stëver, *Pharm. Weekblad*, **93**, 550 (1958).
42. C. K. Narang and N. K. Mathur, *Indian J. Chem.*, **4**, 263 (1966).
43. J. A. Floria, I. Dobratz and J. H. McClure, *Anal. Chem.*, **36**, 2053 (1964).
44. M. Moss, J. Elliot and R. Hall, *Anal. Chem.*, **20**, 784 (1948).
45. J. S. Fritz, *Anal. Chem.*, **24**, 674–675 (1952).
46. V. Z. Deal and G. E. A. Wyld, *Anal. Chem.*, **27**, 47–55 (1955).
47. G. A. Harlow, C. M. Noble and G. E. A. Wyld, *Anal. Chem.*, **28**, 787–791 (1956).
48. R. H. Cundiff and P. C. Markunas, *Anal. Chem.*, **28**, 792–797 (1956).
49. L. W. Marple and J. S. Fritz, *Anal. Chem.*, **34**, 796–800 (1962).
50. W. D. Pohle, V. C. Mehlenbacher and J. H. Cook, *Oil Soap* (*Egypt*), **22**, 115–119 (1945).
51. P. Bradford, W. D. Pohle, J. K. Gunther and V. C. Mehlenbacher, *Oil Soap* (*Egypt*), **19**, 189–193 (1942).
52. D. P. Johnson and F. E. Critchfield, as described in *Quantitative Organic Analysis via Functional Groups*, by S. Siggia, 3rd ed., Wiley, New York, 1963, pp. 60–63.
53. M. W. Scoggins, *Anal. Chem.*, **36**, 1152–1154 (1964).
54. F. E. Critchfield and J. A. Hutchinson, *Anal. Chem.*, **32**, 862–865 (1965).
55. M. W. Scoggins and J. W. Miller, *Anal. Chem.*, **38**, 612–614 (1966).
56. J. J. Fox and J. H. Grange, *J. Chem. Ind.*, **39**, 206T (1920–1921): cf. S. Siggia, *Quantitative Organic Analysis via Functional Groups*, 3rd ed., Wiley, New York, 1963, pp. 71–72.
57. L. A. Smith and E. C. Creitz, *J. Res. Nat. Bur. Std.*, **46**, 145 (1951).
58. R. F. Goddu, *Advances in Analytical Chemistry and Instrumentation*, Vol. 1 (Ed. C. N. Reilley), Interscience, New York, 1960.
59. W. Kaye, *Spectrochim. Acta*, **6**, 257 (1954); **7**, 181 (1955).
60. G. Habermahl, *Angew. Chem., Intern. Ed. Eng.*, **3**, 309 (1964).
61. A. Naherniac, *Ann. Phys.*, **7**, 528 (1937).
62. O. R. Wulf, E. J. Jones and L. S. Deming, *J. Chem. Phys.*, **8**, 753 (1940).
63. R. G. White, *Handbook of Industrial Infrared Analysis*, Plenum Press, New York, 1964.
64. R. O. Crisler and A. M. Burrill, *Anal. Chem.*, **31**, 2055 (1959).
65. O. R. Wulf and E. J. Jones, *J. Chem. Phys.*, **8**, 745 (1940).
66. P. Barchewitz, *Compt. Rend.*, **203**, 1245 (1936).
67. A. D. Cross, *An Introduction to Practical Infrared Spectroscopy*, Butterworth, Washington D.C., 1960.
68. L. J. Bellamy, *Infrared Spectra of Complex Molecules*, Wiley, New York, 1958.
69. A. W. Stewart and G. B. B. M. Sutherland, *J. Chem. Phys.*, **24**, 559 (1956); S. Krimm, C. Y. Liang and G. B. B. M. Sutherland, *J. Chem. Phys.*, **25**, 778 (1956).
70. H. H. Zeiss and M. Tsutsui, *J. Am. Chem. Soc.*, **75**, 897 (1953).
71. U. Liddel and E. D. Becker, *Spectrochim. Acta*, **10**, 170 (1957).
72. M. St. C. Flett, *Spectrochim. Acta*, **10**, 21 (1957).
73. L. P. Kuhn, *J. Am. Chem. Soc.*, **74**, 2492 (1952).
74. A. R. H. Cole and P. R. Jeffries, *J. Chem. Soc.*, 4391 (1956).
75. E. K. Plyler, *J. Res. Nat. Bur. Std.*, **48**, 281 (1952).

76. N. H. E. Ahlers and N. G. McTaggert, *Analyst*, **79**, 70 (1954).
77. C. L. Hilton, *Anal. Chem.*, **31**, 1610 (1959).
78. E. A. Burns and R. F. Muraca, *Anal. Chem.*, **31**, 397 (1959).
79. J. F. Murphy, *Appl. Spectry.*, **16**, 139 (1962).
80. M. R. Adams, *Anal. Chem.*, **36**, 1688 (1964).
81. P. Kabasakalian, E. R. Townley and M. D. Yudes, *Anal. Chem.*, **31**, 375 (1959).
82. G. S. Dvoryantseva and Y. N. Sheinker, *Zh. Analit. Khim.*, **17**, 883 (1962).
83. B. R. Partridge and J. I. Kirby, *J. Chem. Ind. (London)*, 1495 (1965).
84. J. A. Mitchell, C. D. Bockman Jr. and A. V. Lee, *Anal. Chem.*, **29**, 499 (1957).
85. E. Shauenstein and H. Puchner, *Monatsh.*, **93**, 243 (1962).
86. G. B. Meluzova, B. P. Kotel'nikov and Z. A. Prokhorova, *Zh. Analit. Khim.*, **17**, 362 (1962).
87. T. Oba, *Eisei Shikensho Hokoku*, **76**, 53 (1958); *Chem. Abstr.*, **53**, 16465d (1959).
88. F. A. Gronau, D. E. Broadlick and J. E. Hamilton, *J. Pharm. Sci.*, **51**, 242 (1962).
89. W. J Potts Jr. and N. Wright, *Anal. Chem.*, **28**, 1255 (1956).
90. R. A. Friedel, *J. Am. Chem. Soc.*, **73**, 2881 (1951).
91. R. F.. Goddu, *Anal. Chem.*, **30**, 2009 (1958).
92. E. T. Lippmaa, *Tr. Tallinsk. Politekh. Inst. Ser. A*, 35 (1962).
93. D. Kyriacou, *Anal. Chem.*, **33**, 153 (1961).
94. M. Mamiya, *Japan Analyst*, **11**, 739 (1962); *Anal. Abstr.*, **11**, 2687 (1964).
95. J. H. Jaffe and S. Pinchas, *Anal. Chem.*, **23**, 1164 (1951).
96. D. E. Kramm, J. N. Lamonte and J. D. Mayer, *Anal. Chem.*, **36**, 2170 (1964).
97. J. G. Hendrickson, *Anal. Chem.*, **36**, 126 (1964).
98. J. R. Dyer, *Applications of Absorption Spectroscopy of Organic Compounds*, Prentice-Hall, Englewood Cliffs, New Jersey, 1965, p. 84.
99. J. A. Pople, W. G. Schneider and H. J. Bernstein, *High-resolution Nuclear Magnetic Resonance*, McGraw-Hill, New York, 1959, pp. 218–226.
100. P. L. Corio, R. L. Rutledge and J. R. Zimmerman, *J. Am. Chem. Soc.*, **80**, 3163 (1958).
101. P. L. Corio, R. L. Rutledge and J. R. Zimmerman, *J. Mol. Spectr.*, **3**, 592 (1959).
102. O. L. Chapman and R. W. King, *J. Am. Chem. Soc.*, **86**, 1256 (1964).
103. B. Casu, M. Reggiani, G. G. Gallo and V. Vigevani, *Tetrahedron Letters*, 2839 (1964); *Tetrahedron Letters*, 2253 (1965).
104. R. J. Ouellett, *J. Am. Chem. Soc.*, **86**, 4378 (1964); C. P. Rader, *J. Am. Chem. Soc.*, **88**, 1713 (1966); J. J. Uebel and H. W. Goodwin, *J. Org. Chem.*, **31**, 2040 (1966).
105. M. Karplus, *J. Chem. Phys.*, **30**, 11 (1959).
106. J. G. Traynham and G. A. Knesel, *J. Am. Chem. Soc.*, **87**, 4220 (1965).
107. W. B. Moniz, C. F. Poranski and T. N. Hall, *J. Am. Chem. Soc.*, **88**, 190 (1966).
108. Y. Takino, A. Ferritti, V. Flanagan, M. A. Gianturco and M. Vogel, *Can. J. Chem.*, **45**, 1949 (1967).
109. A. G. Brook and K. H. Pannell, *J. Organometal. Chem.*, **8**, 179 (1967).
110. W. Rothweiler and Ch. Tamm, *Experimentia*, **22**, 750 (1966).
111. J. G. Traynham and G. A. Knesel, *J. Org. Chem.*, **31**, 3350 (1966).
112. M. W. Dietrich, J. S. Nash and R. E. Keller, *Anal. Chem.*, **38**, 1479 (1966).

113. L. W. Reeves, *Can. J. Chem.*, **35,** 1351 (1957).
114. M. Gorodetsky, Z. Luz and Y. Mazur, *J. Am. Chem. Soc.*, **89,** 1183 (1967).
115. M. Jackman, *Applications of Nuclear Magnetic Resonance Spectroscopy in Organic Chemistry*, Pergamon, London, 1959, p. 55.
116. A. Mathias, *Anal. Chem. Acta*, **31,** 598 (1964).
117. S. L. Manatt, *J. Am. Chem. Soc.*, **88,** 1323 (1966); S. L. Manatt, D. D. Lawson, J. D. Ingham, N. S. Rapp and J. P. Hardy, *Anal. Chem.*, **38,** 1063 (1966).
118. F. J. Ludwig, *Anal. Chem.*, **40,** 1620 (1968).
119. J. S. Babiec, J. R. Barrante and G. D. Vickers, *Anal. Chem.*, **40,** 610 (1968).
120. V. W. Goodlett, *Anal. Chem.*, **37,** 431 (1965).
121. I. R. Trehan, C. Monder and A. K. Bose, *Tetrahedron Letters*, 67 (1968).
122. P. E. Butler and W. H. Mueller, *Anal. Chem.*, **38,** 1407 (1966).
123. M. K. Seikel, F. D. Hostettler and D. B. Johnson, *Tetrahedron*, **24,** 1475 (1968).
124. M. Saunders and J. B. Hyne, *J. Chem. Phys.*, **29,** 1319 (1958).
125. A. B. Littlewood and F. W. Willmott, *Trans. Faraday Soc.*, **62,** 3287 (1966).
126. V. S. Griffiths and G. Socrates, *J. Mol. Spectr.*, **21,** 302 (1966).

CHAPTER 7

Acidity and inter- and intra-molecular H-bonds

C. H. ROCHESTER

Nottingham University, England

I. INTRODUCTION

The percentage of chemical literature which is in some way concerned with hydrogen bonding has increased steadily over the years since the concept of the H-bond was first introduced in 1920[1, 2]. Furthermore, quantitative studies of the H-bonding of OH groups to a wide variety of acceptors (bases) have made a decisive contribution to our knowledge of the characteristic properties of H-bonds and their influence on chemical reactions and molecular structure. A comprehensive consideration of the appropriate data would be impossible here and is therefore not attempted. The authoritative text by Pimentel and McClellan[2] on H-bonding was published in 1960. We have therefore concentrated in this article on contributions which have appeared since that date with adequate reference to the older work where appropriate. In particular, results for phenols and alcohols are emphasized.

The theoretical interpretations of hydrogen bonding have been discussed by Coulson[371, 372] and Murrell[373]. The relative magnitudes of the contributions due to electrostatic, exchange, induction, dispersion and charge-transfer energies to the H-bond interaction are considered. A complete discussion would be inappropriate here. However, for weak hydrogen bonds the electrostatic energy is the dominant attractive contribution. Delocalization effects are only important for strong hydrogen bonds. The predominantly electrostatic nature of hydrogen bonds favours a linear bond. However, bent bonds are possible. The variation of interaction energy with angle will be of the form $E = E^{\circ} \cos \theta$ which reduces to zero when the bond angle is $\pi/2$. In general, molecular orbitals with low *s*-character are better H-bond acceptors than orbitals with high *s*-character.

The acidity and basicity of OH groups are considered in sections V and VI.

II. INTERMOLECULAR H-BONDING

A. Self Association

1. Alcohols

Studies of the association of alcohols in the vapour phase have the advantage that they are free from the effects of solute–solvent interactions. It is well established that alcohol vapours exist as monomer, dimer and tetramer units[3, 4]. Thermodynamic data for the formation of the dimers and tetramers of several alcohols are given in Table 1. The tetramers are believed to be cyclic[3-5] (structure **1**)

TABLE 1. Thermodynamic parameters for the formation of the dimers and tetramers of alcohols in the gas phase[4].

	$-\Delta H_2$ (kcal/mole)	$-\Delta S_2$ (e.u.)	$-\Delta H_4$ (kcal/mole)	$-\Delta S_4$ (e.u.)	Ref.
Methanol	3·22	16·5	24·2	81·3	3
	4·0	11·0	22·1	94	6
	4·1	17·5			374
	3·5		18·4		5, 7
Ethanol	4·0	11·0	20·1	88	8
	3·4	16·6	24·8	81·5	9
n-Propanol	3·4	15·4	25·2	75·4	10
2-Propanol	4·0	11	22·6	95	6
	5·3	22·4	22·3	74·2	10
	4·5	19·5	22·9	75·3	11
2-Butanol	5·3	21·4	23·1	74·7	12
2-Methyl-2-propanol	4·6	19·0	25·1	82·2	13
Methanol-*d* (CH_3OD)	5		14		7

whereas both linear (**2**) and cyclic (**3**) dimers probably exist. The results for methanol-*d* were deduced from infrared spectra (2400–2800 cm^{-1}) of deuterated methanol at 305 and 335°K[7]. It is

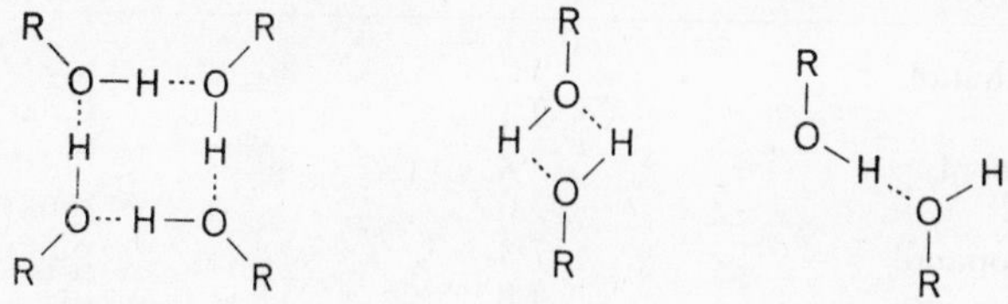

(1) (2) (3)

surprising that the enthalpy of dissociation of the CH_3OD tetramer is less than that for the CH_3OH tetramer. Further evidence for this result is however given by consideration of the Badger–Bauer relationship[14, 15]. This states that $\Delta\nu/\nu$ should be a linear function of ΔH where ν is the infrared OH stretch frequency of the free OH group and $\Delta\nu$ is the frequency change on forming a hydrogen-bonded complex with heat of formation ΔH. For the CH_3OD and CH_3OH tetramers $\Delta\nu/\nu$ was 0·070 and 0·082 respectively. Thus

$$\Delta H_4(CH_3OD) < \Delta H_4(CH_3OH).$$

The self association of alcohols in solution in nonpolar solvents has been widely studied in particular by measurement of infrared OH stretching frequencies[16–24]. The method is based on the principle that if an OH group undergoes H-bonding a shift to lower frequencies occurs. Furthermore in general ν(monomer) > ν(dimer) > ν(trimer) > ν(tetramer) for the association of a particular monomer[25, 26, 36]. This is exemplified by data for the OH stretch frequencies of methanol at 20°K[25]. Frequencies were assigned as follows: for the monomer 3660 cm^{-1}, cyclic dimer 3490 cm^{-1}, cyclic trimer 3445 cm^{-1}, tetramer 3290 cm^{-1}, and for higher (linear) polymers 3250 cm^{-1}. Explanations of why absorption bands in the region 2500–3400 cm^{-1} are obtained for H-bonded systems have recently been discussed and summarized[26–28]. Proton magnetic resonance studies of alcohol association similarly allow association constants (from chemical shifts as a function of concentration) and heats of association (from chemical shifts as a function of temperature and concentration) to be evaluated[24, 29–31]. Agreement between results from infrared and p.m.r. measurements is emphasized by the data in Table 2. Vapour pressures[17] and viscosities[32–34] have also

TABLE 2. Thermodynamic parameters for the dimerization of some alcohols in carbon tetrachloride[20, 24, 31].

Alcohol	$-\Delta H$ (kcal/mole)	$-\Delta S$ (e.u.)	Method of determination
Methanol	9·2	28	i.r.
	9·4		p.m.r.
Ethanol	7·2	20	i.r.
	7·6		p.m.r.
i-Propanol	7·3		p.m.r.
t-Butanol	4·8	11	i.r.
	4·4		p.m.r.
Di-*t*-butylcarbinol	4·12	13·1	i.r.
	4·21	13·9	p.m.r.

been used to study the intermolecular H-bonding of alcohols. Calorimetric measurements of heats of H-bonding of alcohols have been reported[35, 36]. The dimers of methanol and ethanol are probably cyclic[20, 24]. However, for higher alcohols in which steric factors hinder the formation of the cyclic dimer, the linear dimer becomes predominant[24]. The thermodynamic parameters for dimerization also reflect the effects of steric hindrance (Table 2)[16, 18, 19]. Cyclic trimers of methanol are supposed to exist[17, 29] but cyclic tetramers appear more likely[3, 19, 30]. The association of 1-octanol is considered to give both linear and cyclic tetramers for which free energies, enthalpies and entropies of formation have been deduced[23]. The dimerization constant and heat of dimerization of cholesterol in CCl_4 at 23° are 4·5 l/mole and −1·8 kcal/mole respectively[66]. A trimer has been reported for cholesterol in CCl_4 [66].

It does not follow that information about the H-bonding of alcohols in inert solvents may necessarily be equally applicable for the pure liquid alcohols. The heats of vaporization[37, 38], dielectric constants[39], viscosities[33, 34] and n.m.r. spectra[40] of aliphatic liquid alcohols have all shown that the geometry of the alcohol molecules through steric factors greatly influences their H-bonding ability. Both linear and cyclic dimers may exist[4, 39]. However, whether higher polymers are cyclic[4, 17] or linear[39, 40] appears to be uncertain. Radial distribution functions[41] for liquid methanol have shown that the O · · · O distance (2·7 Å) for two H-bonded molecules is about the same as for the solid[42, 43]. The structure of liquid alcohols and alcohol–water mixtures has been reviewed by Franks and Ives[44]. The effect of pressures up to 25,000 atm on the H-bonds of polyvinyl alcohol has been investigated[45].

Structures of carbohydrate crystals show that in general all the hydroxyl groups are hydrogen-bonded[46]. X-ray studies of β-D-glucopyranose[47] and methyl β-D-xylopyranoside[48] suggest that some of the OH groups are not H-bonded. However, the infrared spectra for these compounds are not compatible with this conclusion[49].

2. Phenols

The OH stretching frequencies and intensities of phenols[50, 51] both follow a Hammett $\rho\sigma$ relationship[52]. For *ortho*-substituted phenols where intramolecular H-bonding can occur it is the frequency of the *trans*-hydroxyl group which correlates with the substituent constant[53]. On H-bonding the OH stretch vibrations move to lower frequencies[26]. Hall and Wood[26] have listed the following four specific frequencies for different types of interaction:

type α: monomer, 3611 cm^{-1},

type β: acceptor end group (Ph—O(—H)· · ·), 3599 cm^{-1},

type γ: donor end group (Ph—O—H · · ·), 3481 cm^{-1},

type δ: OH acting as both acceptor and donor, 3393 cm^{-1}.

The frequencies quoted refer to phenol itself in carbon tetrachloride solution[21, 22]. The bands are shifted slightly for other sterically unhindered phenols.

The frequency shifts of phenols on dimerization are greater for the phenols with the greatest H-bonding ability[54, 55]. In particular a pronounced reduction in $\Delta\nu$ is produced by bulky substituents in the 2-position of phenols[59, 61]. A comparison of the frequency shifts with the dimerization constants K for some phenols is given in Table 3.

TABLE 3. Infrared frequency and p.m.r. signal shifts, association constants and heats of association for the dimerization of some phenols in carbon tetrachloride (room temperature).

Phenol	K (l/mole)	$-\Delta H$ (kcal/mole)	$\Delta\nu$ (cm^{-1})	$\Delta\tau$ (ppm)
Phenol	13 ± 7[a]	5·12[d]	262[e]	3·05[g]
4-Chlorophenol	7·77[d]	3·78[d]		
4-Cresol	10[d]	6·09[d]	276[e]	
2-Cresol	8 ± 4[a]		172[e]	2·04[g]
2-*i*-Propylphenol	1·7[b]		172[e]	1·78[g]
2-*t*-Butyl-4-methylphenol	1·37[c]		114[e]	
2-*t*-Butylphenol	1·0[b]		70[f]	0·22[g]
2,6-Di-*t*-butylphenol	⩽·05[b]		ca 3[e]	0·0[g]

[a] Ref. 56; [b] ref. 57; [c] ref. 18; [d] ref. 58; [e] ref. 59; [f] ref. 60; [g] ref. 62.

Also tabulated are the shifts $\Delta\tau$ in the p.m.r. signals caused by hydrogen bonding. In general both $\Delta\tau$[57, 62] and $\Delta\nu$[54, 59, 61] decrease as K decreases. The p.m.r. chemical shift or the infrared frequency change on hydrogen bonding for a particular phenol gives an indirect measure of the ability of that phenol to form H-bonded dimers. Association constants may be deduced from $\Delta\tau$ or $\Delta\nu$ measurements as a function of phenol concentration[18, 56, 57]. Varying the temperature allows the enthalpy of dimerization to be calculated[58, 63]. The three enthalpies given in Table 3 show the same trend as the free energies of ionization of phenol (ΔG°_{298} = 13·57 kcal/mole)[64], 4-chlorophenol (12·86 kcal/mole)[64] and 4-cresol (14·02 kcal/mole)[65]. When deducing ΔH from infrared frequency shifts it is unwise to assume that the temperature dependencies of the free and associated band absorptions are equal[63].

For *ortho*-substituted phenols in which intramolecular H-bonds are possible intermolecular H-bonding may be reduced[67–70]. For a series of substituted 2-bromophenols as liquid films or nujol mulls a sharp infrared band at ca 3500 cm^{-1} was assigned to intramolecular H-bonds whereas a broad (3450–3250 cm^{-1}) less intense absorption was attributed to intermolecular H-bonding[68]. Many of the phenols showed both forms of bonding. Allan and Reeves[69] have pointed out that the *trans* form of 2-halophenols is available in CS_2 solution as a hydrogen bond donor whereas the *cis* form can only act as a H-bond acceptor. Dimerization is considered to occur via hydrogen bonding from a *trans* to a *cis-* isomer as follows.

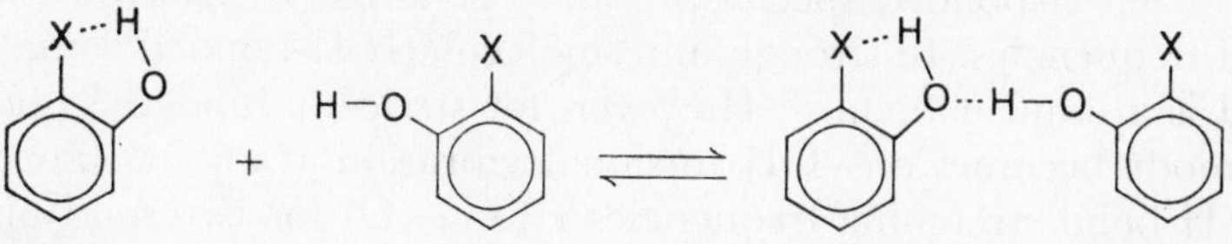

(*cis*) + (*trans*) ⇌ (*cis–trans* dimer)

Equilibrium constants (mole fraction units, 300°K) are 23·48, 11·22 and 5·24 for 2-chlorophenol, 2-bromophenol and 2-iodophenol respectively. The self association of phenols is also profoundly influenced by solvent[21, 57, 61, 68, 70]. A solvent which is itself a strong H-bond acceptor such as dioxan[57], ether[68, 70] or pyridine[61] reduces the extent of phenol dimerization considerably.

Evidence for the formation of phenol trimers in CCl_4 solution has been discussed[71–74]. Trimerization constants of 4·1 $(mole/l)^{-2}$ at 25° [71] and 4·78 $(mole/l)^{-2}$ at 21° [74] have been evaluated for phenol in carbon tetrachloride. Trimerization is prevented for phenols such as 2-nitrophenol, 2-hydroxyacetophenone and 2-fluorophenol in which intramolecular H-bonding occurs[72]. Unlike the dimers which probably have an open-chain structure[21] the phenol trimers are cyclic[73]. A trimolecular species containing two phenol molecules and one water molecule has been assigned the structure 4 [75] with the benzene nuclei of the phenol rings parallel and facing

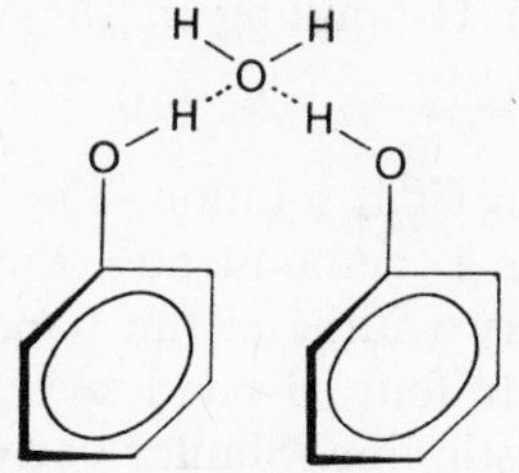

(4) 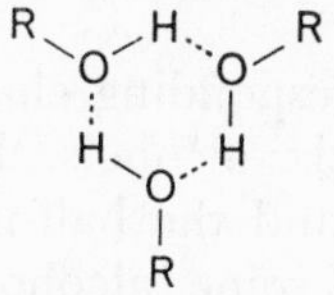 **(5)**

each other. However phenol (P) and water (W) in benzene, 1,2,2-1, tetrachloroethane and 1,2-dichloroethane form a series of aggregates of stoichiometry P_3, P_2W, PW_2 and W_3 [73]. These are considered to have the cyclic structure **5** with $R = H$ or C_6H_5.

The –O–H · · · O– vibrations in the far infrared of H-bonded forms of several phenols have been determined[76]. The stretching vibrations are in the range 98–187 cm^{-1}. Phenols for which different H-bonded associates are known to exist showed more than one –OH out of plane deformation mode (γOH) in the range 280–690 cm^{-1}. Thus each H-bonded species probably gives rise to its own distinct γOH frequencies. In strongly intermolecularly H-bonded complexes γOH is around 600 cm^{-1}. However, for sterically hindered phenols the mode becomes an –OH torsional vibration at about 300 cm^{-1}. The H-bond stretching frequencies $\nu(O \cdots O)$ for two solid phases of phenol are 175 cm^{-1} for a strongly H-bonded phase and 135 cm^{-1} for a weakly H-bonded phase[77]. The frequency $\nu(O \cdots O)$ of the H-bond vibration is therefore, as expected, smaller for a weaker H-bond. However, this result does not appear to be general[27].

Phenols are strongly hydrogen bonded in the solid state. Phenol forms H-bonded chains in the form of a threefold spiral[78, 79, 82]. The 2,3-, 2,5- and 2,6-xylenols[80, 81] and resorcinol[83] are similar. Bois[84] has proposed that 4-cresol exists as H-bonded discrete tetramers and not as H-bonded chains. A second more weakly H-bonded phase of phenol appears at pressure greater than 5 kbar[77].

B. Alcohols as H-bonding Acids

Infrared spectra of the complexes formed between six alcohols and several H-bond acceptors in the gas phase have been studied by Reece and Werner[85]. The OH stretch frequency shifts on H-bonding were found to obey a product rule $\Delta\nu = A.D$ cm^{-1} where A and D are suitably defined acceptor and donor capacities. Similar results were obtained for the same complexes in CCl_4 solution. Equation (1) relates the frequency change on H-bonding in the vapour phase

$$\Delta\nu_{\text{vapour}} = -1{\cdot}7 + 0{\cdot}816\Delta\nu_{CCl_4} + 1{\cdot}2 \times 10^{-4}(\Delta\nu_{CCl_4})^2 \qquad (1)$$

with the corresponding change in CCl_4 solution. The equation was obeyed for six alcohols. Figure 1 demonstrates the relationship between $\Delta\nu$ and the half-intensity widths of the bands due to the complexes of some alcohols with four H-bond acceptors[85]. Each acceptor gives its own characteristic line. Similar correlations of $\Delta\nu$ and $\nu_{\frac{1}{2}}$ have been described[86, 87] and the temperature dependence of $\Delta\nu$ and $\nu_{\frac{1}{2}}$ has been investigated[88]. Whereas $\Delta\nu$ varies significantly, $\nu_{\frac{1}{2}}$ is insensitive to temperature. The increase in intensity $\Delta B°$ of the OH stretching absorption on H-bonding at 25° is related to $\Delta\nu$ by equation (2) for the complexes of methanol, ethanol and *t*-butanol

$$\Delta B° = 225\Delta\nu \tag{2}$$

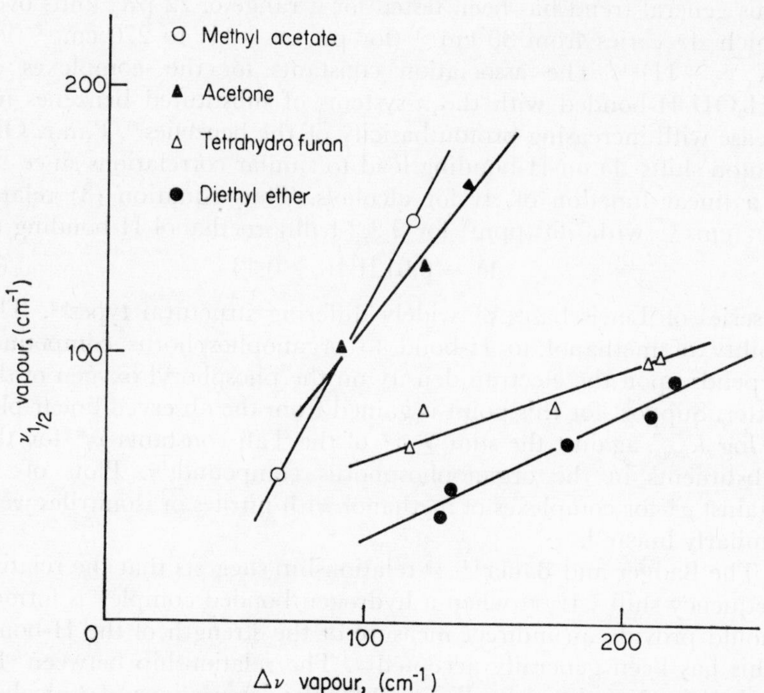

FIGURE 1. Correlation between $\nu_{\frac{1}{2}}$ for the H-bonded complex and $\Delta\nu$ on H-bonding for the complexes of some aliphatic alcohols with four H-bond acceptors[85]. Reproduced by permission from I. H. Reece and R. L. Werner, *Spectrochim. Acta*, **24A,** 1271 (1968).

with acetone, ethyl acetate, dioxan, benzophenone, dimethylformamide and pyridine[88]. The intensities of the OH bands of the complexes of methanol and *t*-butanol with three substituted benzene derivatives are linearly related to $\sum(\sigma_I + \sigma_R)$, the sum of the inductive and resonance σ parameters for the substituents in the benzene rings[89]. A linear relationship between band intensity and the nuclear quadrupole coupling constants of 5 acceptors has been demonstrated[89]. The coupling constants and $\sum(\sigma_I + \sigma_R)$ were also linear functions of $(\Delta\nu/\nu)$ the relative frequency shifts on hydrogen bonding.

Gordon has discussed correlations between the equilibrium constants K_{assoc} and the infrared frequency shifts $\Delta\nu$ for H-bond complex formation, pK_a for the H-bond donors and pK_b for the H-bond acceptors[90]. For example, the shift $\Delta\nu$ for methanol-*d* (CH_3OD) increases as the proton basicity of the acceptor molecules increases. This general trend has been tested for a range of 22 pK_a units over which $\Delta\nu$ varies from 30 cm^{-1} (for $pK_a \simeq -11$) to 270 cm^{-1} (for $pK_a \simeq +11$)[91]. The association constants for the complexes of CH_3OD H-bonded with the π-systems of substituted benzenes increase with increasing proton basicity of the benzenes[90]. P.m.r. OH proton shifts $\Delta\delta$ on H-bonding lead to similar correlations since $\Delta\delta$ is a linear function of $\Delta\nu$ for alcohols. Thus equation (3) relates $\Delta\nu$ (cm^{-1}) with $\Delta\delta$ (ppm) for 2,2,2-trifluoroethanol H-bonding to

$$\Delta\delta = 0{\cdot}0121\Delta\nu + 0{\cdot}43 \quad (3)$$

a series of Lewis bases of widely differing structural types[92]. The ability of methanol to H-bond to organophosphorus compounds depends upon the electron density on the phosphoryl oxygen of the latter. Support for this point is gained from the observed linear plot of log K_{assoc} against the sum $\sum \sigma^*$ of the Taft constants σ^* for the substituents in the organophosphorus compounds[93]. Plots of $\Delta\nu$ against σ^* for complexes of methanol with nitriles or isonitriles were similarly linear[94].

The Badger and Bauer[14, 15] relationship suggests that the relative frequency shift ($\Delta\nu/\nu$) when a hydrogen bonded complex is formed should provide an indirect measure of the strength of the H-bond. This has been generally accepted[86]. The relationship between ΔH and $\Delta\nu$ is exemplified by Figure 2 for the interactions of 3 alcohols with 9 Lewis bases[95]. However, the linear relationship is not always obeyed[86, 88, 96, 97]. The H-bonding of alcohols to ketones shifts the C=O stretching frequency of the ketones to lower wavenumbers[98]. A linear relationship between ($\Delta\nu_{CO}/\nu_{CO}$) and the H-bond energy has been recorded for methanol–ketone complexes[99]. The antisymmetric and symmetric stretching frequencies of the NH_2 group are shifted to lower frequencies when alcohols (or phenols) H-bond to aniline. The magnitude of the shifts correlates both with the OH stretch shift $\Delta\nu_{OH}$ and with pK_a for the donor alcohol[101]. Thus for a more acidic alcohol a stronger H-bond is obtained as indicated by the larger changes in ν_{OH} and ν_{NH_2} on H-bonding. An approximate correlation exists between ΔH for H-bond formation and the intensity of the infrared OH stretching bands of the H-bonded complexes of methanol, ethanol and *t*-butanol, each with the acceptors acetone, benzophenone, ethyl acetate, dioxan, dimethylformamide and pyridine[88].

Heats of H-bond formation give somewhat scattered but approximately linear graphs when plotted against the corresponding entropies or free energies of ionization[88, 96, 97, 100, 103]. Some selected

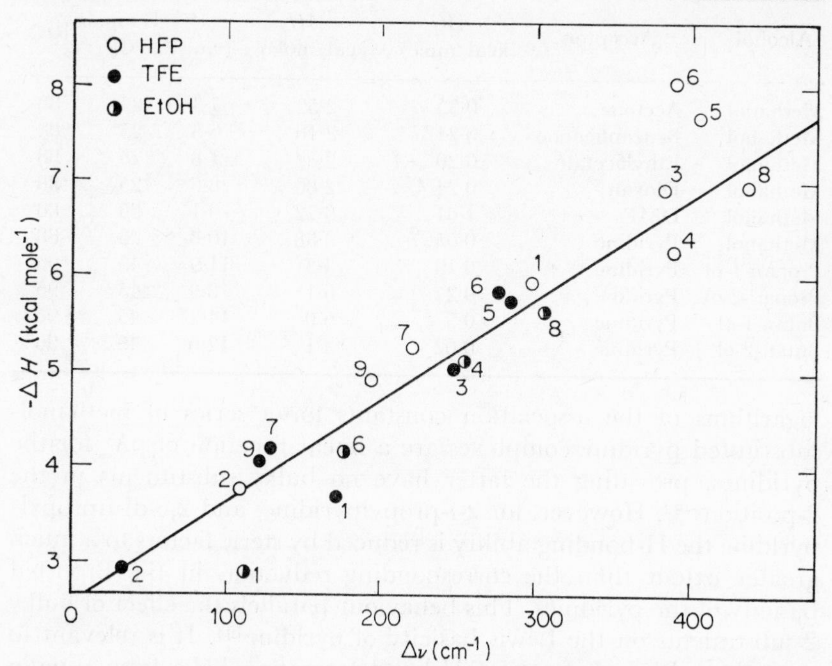

FIGURE 2. Relationship between ΔH and $\Delta\nu$ for 1,1,1,3,3,3-hexafluoro-2-propanol (HFP), ethanol, and 2,2,2-trifluoroethanol (TFE) as H-bond donors[95]. Acceptors were (1) acetone, (2) 1,1,1-trifluoroacetone, (3) di-*i*-propylether, (4) tetrahydrofuran, (5) *N,N*-dimethylacetamide, (6) tetramethylurea, (7) *N,N*-dimethyltrifluoroacetamide, (8) dimethyl sulphoxide, (9) sulpholane. Reproduced by permission from A. Kivinen, J. Murto and L. Kilpi, *Suomen Kemistilehti*, **40B**, 301 (1967).

values of ΔG, ΔH and ΔS are given in Table 4. In general, as ΔH becomes more negative ΔS becomes more negative. A more negative ΔH implies a stronger H-bond and therefore a more restricted configuration in the H-bond complex[96, 100]. Therefore ΔS should also be more negative. It follows that the differences between ΔH and ΔS for series of structurally related H-bond donors or acceptors contribute to ΔG in a compensating manner and therefore ΔG is comparatively insensitive to change. Increasing the acidity of the donor[101] or the basicity of the acceptor[91, 102] produces a stronger H-bond. These correlations may be upset if there is any steric hindrance to the formation of the H-bonded complex[91]. Thus the

TABLE 4. Free energy, enthalpy and entropy changes on forming H-bonded complexes of alcohols and H-bond acceptors in carbon tetrachloride solvent.

Alcohol	Acceptor	$-\Delta G$ (kcal/mole)	$-\Delta H$ (kcal/mole)	$-\Delta S$ (e.u.)	Temp. (°C)	Ref.
Methanol	Acetone	0·35	2·52	7·3	25	88
Methanol	Benzophenone	0·24	2·16	6·5	25	88
Methanol	Ethylacetate	0·20	2·52	7·8	25	88
Methanol	Dioxan	0·24	2·80	8·6	25	88
Methanol	DMF	1·01	3·72	9·1	25	88
Methanol	Pyridine	0·65	3·88	10·8	25	88
Propan-1-ol	Pyridine	0·19	4·3	11·9	45	96
Propan-2-ol	Pyridine	0·2	6·1	18·9	45	96
Butan-1-ol	Pyridine	0·5	5·0	14·2	45	96
Butan-2-ol	Pyridine	0·02	4·1	12·6	45	96

logarithms of the association constants for a series of methanol-substituted pyridine complexes are a linear function of pK_a for the pyridines, providing the latter have no bulky substituents in the 2-positions[102]. However, for 2-*i*-propylpyridine, and 2,6-di-*i*-propylpyridine the H-bonding ability is reduced by steric factors to a much greater extent than the corresponding reduction in the Brønsted basicity of the pyridines. This behaviour parallels the effect of bulky 2-substituents on the Lewis basicity of pyridine[104]. It is relevant to note that although heats of H-bond formation have been usually deduced indirectly from infrared[63] or p.m.r.[105] measurements direct calorimetry gives results close to the spectroscopic values[103].

The infrared ν_{OH} stretching frequencies and the H-bond enthalpies of alcohol-acceptor complexes are sensitive to changes in solvent composition[22, 109, 110]. Typical frequency shifts are given in Table 5. Allerhand and Schleyer[109] found that an empirical relationship, equation (4), fitted the observed solvent shifts of the H-bond ν_{OH}

$$(\nu^0 - \nu^S)/\nu^S = aG \tag{4}$$

absorption bands. Here ν^0 is the infrared frequency for the H-bonded complex for which ν^S is the corresponding frequency in a particular solvent, a is a function of the particular vibration being studied and G is a function only of the solvent. Equation (4) represents an empirical form of the Kirkwood–Bauer–Magat relation-

TABLE 5[a]. Infrared OH stretching frequency of the methanol (0·05 mole/l)–ether complex in ether–chloroform mixtures[22].

% $CHCl_3$	0	10	20	30	35	40	45
ν (cm^{-1})	3508	3504	3498	3491	3482	3473	3465

% $CHCl_3$	50	60	70	80	90	98
ν (cm^{-1})	3463	3459	3454	3452	3449	3444

[a] Reproduced by permission from L. J. Bellamy, K. J. Morgan and R. J. Pace, *Spectro chim. Acta*, **22**, 535 (1966).

ν = 3558 cm^{-1} for MeOH–Et_2O complex in gas phase[111].

ship[112]. The latter only fits the H-bond frequency shift data for the phenol–acetonitrile complex in CCl_4 solutions; equation (4) appears much more generally applicable. However, this approach has been criticized by Bellamy and co-workers[22] who suggest that specific interactions between solvent molecules and H-bonded complexes are largely responsible for the observed results. Thus, for example, for the methanol–ether complex (Table 5) in chloroform the interaction between the proton donor solvent chloroform and the complex may be represented as

$$Cl_3C-H\cdots O(CH_3)-H\cdots O(CH_2CH_3)_2$$

(6)

and would be expected to lead to a decrease in frequency as the chloroform concentration is increased. Proton acceptor solvents give a similar effect. The magnitude of the shifts is therefore dependent on the extent to which the solvent molecules can act as H-bond donors or acceptors.

Alcohols can hydrogen bond to the π-electron systems of olefins or aromatic hydrocarbons. H-bonding of this type is favoured by increasing the number of conjugated double bonds or the number of condensed rings in the π-system[107]. It has been suggested that this type of interaction is responsible for the affinity between many dyes and cellulosic substrates[108]. Calorimetric measurements of heats of mixing of benzyl alcohol, ethylbenzene and cyclohexane have provided evidence for an OH$\rightarrow\pi$ H-bond interaction with $\Delta H = -1{\cdot}48$ kcal/mole[106]. For the *m*-cresol–*m*-xylene–cyclohexane system $\Delta H = -3{\cdot}14$ kcal/mole: a stronger H-bond for a more acidic proton donor. A weak H-bond interaction exists between alcohols and nitro groups[113, 114]. Thus, comparison of the OH stretching frequency of methanol in carbon tetrachloride, nitromethane, nitrobutane and nitrobenzene shows that ν_{OH} is displaced to lower frequencies in the latter three solvents[114]. Also the band intensity and half-width increase, both effects being characteristic of H-bond interactions[86]. When a nitro compound is added to methanol in CCl_4 a reduction in intensity (but no change in position) of the free OH absorption occurs and a new broad maximum at a lower frequency appears[113]. The nitro compounds must therefore be acting as H-bond proton acceptors. Phenol gives spectral shifts 2–3 times greater than those for methanol in accord with its greater strength as an acid.

A study of the effect of H-bonding on the ring–chain tautomerism **7** $\rightleftarrows$ **8** of oxazolidines demonstrates how H-bonding solvents can in-

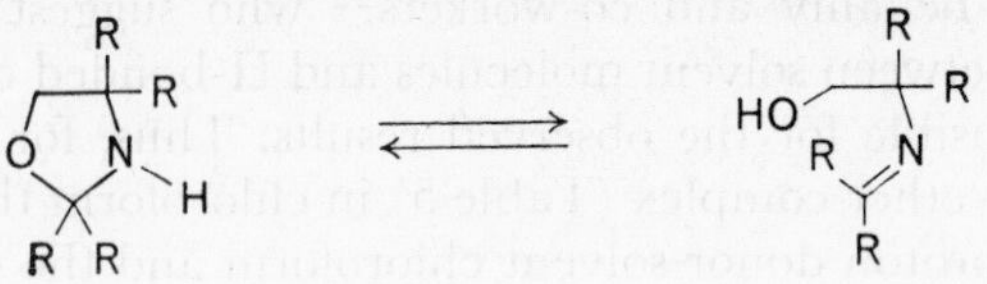

(7) $\rightleftarrows$ (8)

fluence tautomeric equilibria of solute species[115]. Structure **8** was favoured relative to **7** by using solvents which can H-bond strongly with the alcoholic group in **8**. For a series of 8 solvents the H-bonding ability was assessed by measuring $\Delta\nu_{OH}$ for di-*t*-butylcarbinol (DTBC) when mixed with each solvent in CCl_4 solution. The enthalpy changes $\Delta H°$ for equilibrium **7** $\rightleftarrows$ **8** in the 8 solvents were a linear function of $\Delta\nu_{OH}$(DTBC) in the sense that as $\Delta\nu_{OH}$ became greater so $\Delta H°$ became less positive or more negative. The effect of solvent on the equilibrium therefore primarily arises through the influence of H-bonding.

C. Phenols as H-bonding Acids

I. Phenol

General studies of H-bonding of phenol to a variety of H-bond acceptors of different structural types have been made by Gramstad[116] and by Drago and co-workers[117–119]. Acceptors included ethers, amines, aldehydes, ketones, esters, amides, sulphoxides, phosphoryl compounds, acid fluorides and alkyl halides. The p.m.r.[118] and infrared[116, 117] methods give results which are consistent with each other[120]. Thus the p.m.r. chemical shifts of the phenol proton in the H-bonded complex and the infrared ν_{OH} frequency shift on forming the complex were both linear functions of the enthalpy of H-bond formation. Equations (5)[118] and (6) fit the experimental

$$\delta_{obs} = 0{\cdot}748\Delta H - 4{\cdot}68 \quad (5)$$

$$-\Delta H = 0{\cdot}016\Delta\nu_{OH} + 0{\cdot}63 \quad (6)$$

results with δ in ppm, $\Delta\nu_{OH}$ in cm^{-1} and ΔH in kcal/mole. A theoretical justification for equation (6) has been deduced and dis-

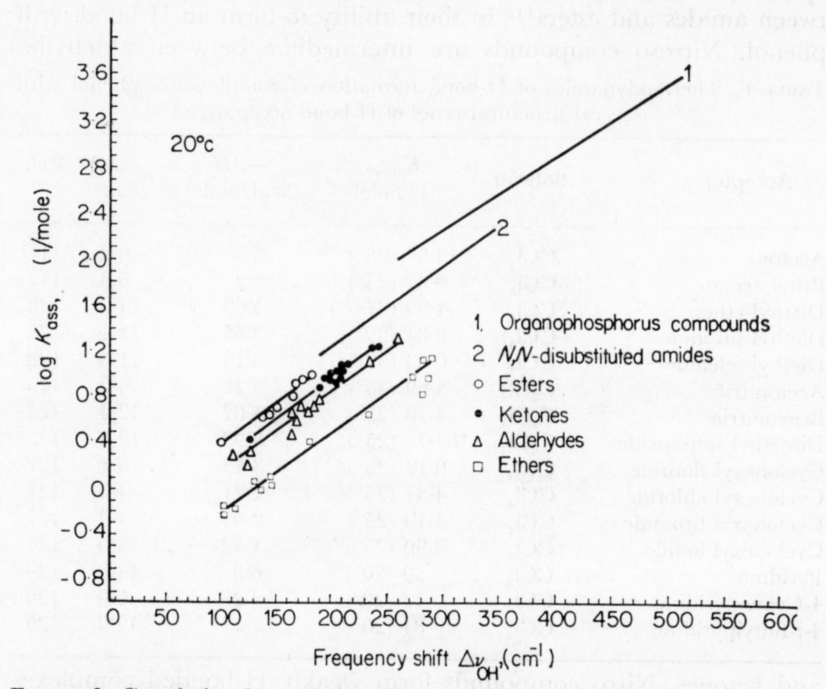

FIGURE 3. Correlation between log K_{assoc} and the ν_{OH} frequency shifts for the association of phenol with H-bond bases of different structural types[116]. Reproduced by permission from T. Gramstad, *Spectrochim. Acta*, **19**, 497 (1963).

cussed[119]. The correlation between the association constant and the infrared frequency shift for H-bond formation is not as general as the corresponding enthalpy correlations[117]. However, Gramstad[116] has shown that a plot of $\Delta\nu_{OH}$ against log K_{assoc} is linear for H-bond bases of similar structural class. Figure 3 demonstrates these results. The changes in ν_{OH} frequency or δ_{OH} chemical shift produced with phenolic H-bonds are greater the more strongly basic the acceptor molecule[120]. The magnitudes of the shifts also vary inversely as the overall length (R_{xy} for X–H···Y) of the H-bond at equilibrium[27, 86, 120]. The shifts in carbonyl stretch frequency when phenol H-bonds to ketones is a linear function of the enthalpy of formation of the H-bonds[99].

H-bonding of phenol with ten sulphoxides and three nitroso compounds has been studied by Gramstad[121]. Plots of log K_{assoc} against $\Delta\nu_{OH}$, ΔH against $\Delta\nu_{OH}$, ΔG against $\Delta\nu_{OH}$, ΔS against $\Delta\nu_{OH}$ and $\nu_{\frac{1}{2}}$ against $\Delta\nu_{OH}$ were all linear. Sulphoxides are intermediate between amides and esters[116] in their ability to form an H-bond with phenol. Nitroso compounds are intermediate between aldehydes

TABLE 6. Thermodynamics of H-bond formation of complexes of phenol with several structural types of H-bond acceptor.

Acceptor	Solvent	K_{assoc} (l/mole)	$-\Delta H$ (kcal/mole)	$-\Delta S$ (e.u.)	Ref.
Acetone	CCl_4	13·5 (25°)	3·3	6·2	117
Ethyl acetate	CCl_4	9·3 (25°)	3·2	6·3	117
Diethyl ether	CCl_4	4·99 (37·5°)	5·63	14·9	128
Diethyl sulphide	CCl_4	0·84 (39°)	3·35	11·1	128
Diethyl selenide	CCl_4	0·63 (39·5°)	3·15	11·0	128
Acetonitrile	C_2Cl_4	5·69 (25°)	5·22	14·0	125
Benzonitrile	C_2Cl_4	4·70 (25°)	4·62	12·4	125
Dimethyl sulphoxide	C_2Cl_4	187·7 (25°)	6·93	12·8	125
Cyclohexyl fluoride	CCl_4	9·12 (25°)	3·13	6·1	127
Cyclohexyl chloride	CCl_4	4·34 (25°)	2·21	4·5	127
Cyclohexyl bromide	CCl_4	4·19 (25°)	2·05	4·0	127
Cyclohexyl iodide	CCl_4	3·99 (25°)	1·72	3·0	127
Pyridine	CCl_4	59 (20°)	6·5	14·0	129
4-Cyanopyridine	CCl_4	12 (20°)	3·2	6·0	129
4-*t*-Butylpyridine	CCl_4	84 (20°)	7·1	12·0	129

and ketones. Nitro compounds form weakly H-bonded complexes with phenol[113, 136]. Phenol can H-bond to sulphones and sulphonates[122]. The frequency shifts $\Delta\nu_{OH}$ correlate linearly with the symmetric and asymmetric stretching frequencies of the sulphuryl ($-SO_2-$) group. A higher basicity gives a lower stretching frequency ν_{SO_2}. $\log_{10} K_{assoc}$ and the Taft σ^* constants[52] for the substituents in a series of sulphones were both linear functions of the observed OH stretch frequency shift for phenol on H-bonding. Similar results to those for organosulphur compounds have also been obtained for organophosphorus compounds[116, 123, 124].

The association constants of phenol with eight pyridines show a smooth dependence on the σ-constants for the pyridine substituents[129]. Electron-withdrawing substituents give less negative enthalpies and entropies of H-bond formation in accord with the decrease in both the Lewis and Brønsted basicity of the pyridine nitrogen atom[104]. Linear correlations between $\Delta\nu_{OH}$ and σ^* were also observed for the H-bonding of phenol to nitriles and isonitriles[94]. For the latter, which probably exist as $R{-}N^+{\equiv}C^-$, phenol hydrogen bonds to the carbon and not to the nitrogen atom. The H-bond of phenol to isocyanides is stronger than that to cyanides. For cyanides typical values for the enthalpy of H-bond formation are $-3{\cdot}54 > \Delta H > -5{\cdot}72$ kcal/mole[125]. The more negative enthalpy changes are accompanied also by more negative entropies of H-bond formation. The generality of this ΔH–ΔS correlation is exemplified by the data in Table 6 and also by the plot in Figure 4 which compares results for several different H-bond donors as well as acceptors.

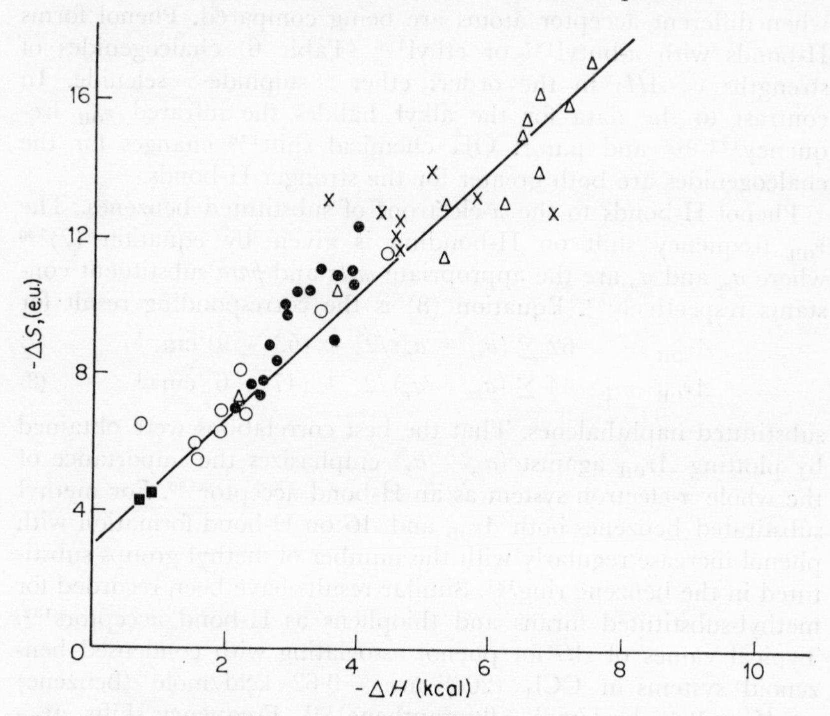

FIGURE 4. Relationship between ΔH and ΔS for the formation of H-bonded complexes (25°)[125]. X—phenol with nitriles; ○—*t*-butanol with nitriles; ■—2,6-di-*t*-butylphenol with nitriles; ●—alcohols with ethers, ketones etc.[88]; △—thiocyanic acid with ethers. Reproduced by permission from M. C. Sousa Lopes and H. W. Thompson, *Spectrochim. Acta*, **24A**, 1367 (1968).

Enthalpy changes of $-2{\cdot}13$, $-1{\cdot}65$, $-1{\cdot}57$ and $-1{\cdot}25$ kcal/mole have been observed when phenol in tetrachloroethylene H-bonds to *n*-heptyl fluoride, *n*-heptyl chloride, *n*-heptyl bromide and *n*-heptyl iodide respectively[126]. The corresponding ν_{OH} frequency shifts were 39·7, 58·7, 69·2 and 71 cm^{-1}. The enthalpy and frequency shift results therefore apparently lead to conflicting conclusions about the order of H-bond strength for the four interactions. The order of enthalpies is $I < Br < Cl < F$, that is, the correct order if the H-bond accepting ability of the halide is predominantly influenced by the electronegativity of the halogen atom. The Badger–Bauer relationship[14, 15] is not applicable in this case. Similar results were obtained for the association of phenol with the four cyclohexyl halides in CCl_4 (Table 6)[127]. The relative frequency shifts ν_{OH} for several acceptors are not a reliable measure of H-bond strength when different acceptor atoms are being compared. Phenol forms H-bonds with *n*-butyl[127] or ethyl[128] (Table 6) chalcogenides of strengths ($-\Delta H$) in the order: ether > sulphide > selenide. In contrast to the data for the alkyl halides the infrared ν_{OH} frequency[127, 128] and p.m.r. OH chemical shift[128] changes for the chalcogenides are both greater for the stronger H-bonds.

Phenol H-bonds to the π-electrons of substituted benzenes. The ν_{OH} frequency shift on H-bonding is given by equation (7)[130] where σ_m and σ_p are the appropriate *meta* and *para* substituent constants respectively[52]. Equation (8) is the corresponding result for

$$\Delta\nu_{OH} = [-62 \sum (\sigma_m + \sigma_p)/2] + (53 \pm 9)\ cm^{-1} \quad (7)$$

$$\Delta\nu_{OH} = [-44 \sum (\sigma_m + \sigma_p)/2] + (47 \pm 6)\ cm^{-1} \quad (8)$$

substituted naphthalenes. That the best correlations were obtained by plotting $\Delta\nu_{OH}$ against $(\sigma_m + \sigma_p)$ emphasizes the importance of the whole π-electron system as an H-bond acceptor[130]. For methyl substituted benzenes both $\Delta\nu_{OH}$ and ΔG on H-bond formation with phenol increase regularly with the number of methyl groups substituted in the benzene ring[131]. Similar results have been recorded for methyl-substituted furans and thiophens as H-bond acceptors[132]. Typical values of ΔG for phenol associating with condensed benzenoid systems in CCl_4 (20°) are $-0{\cdot}62$ kcal/mole (benzene) $< \Delta G < 0{\cdot}46$ kcal/mole (fluoranthene)[131]. Frequency shifts $\Delta\nu_{OH}$ are generally below 100 cm^{-1}, although hexamethylbenzene ($\Delta\nu_{OH} = 127\ cm^{-1}$) is an exception. 6,6-Dialkylfulvenes are stronger proton acceptors than methylbenzenes[132]. With arylfulvenes and azulenes as acceptors two infrared absorption maxima both characteristic of H-bonded complexes were observed[132]. These may be assigned to the H-bond complex of phenol to the benzene **9** and to the fulvene **10** ring for the arylfulvenes. It is logical tentatively to

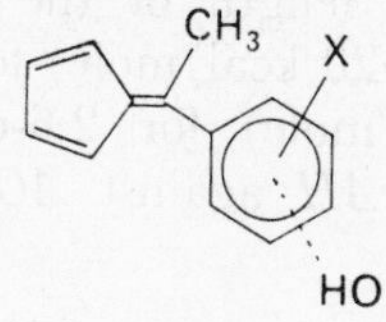

(9) 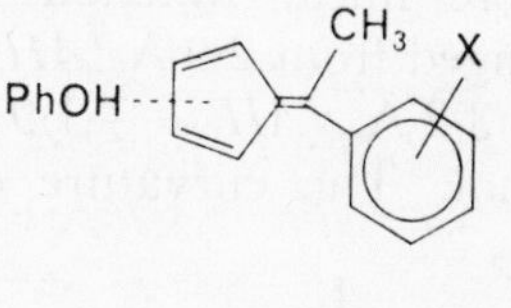 (10)

assign the two bands for the azulenes to phenol H-bonding to the seven **11** and to the five **12** membered ring. Alkyl groups increase

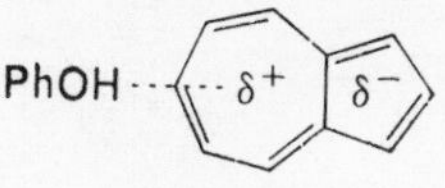

(11) 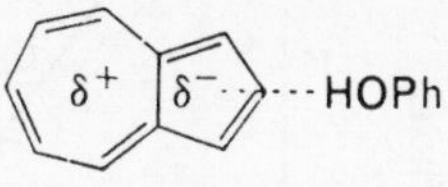 (12)

the H-bonding basicity of olefins as measured by K_{assoc} or $\Delta\nu_{OH}$ on complexing with phenol[133, 134]. 1,3-Butadienes give smaller frequency shifts but larger association constants than the monolefins.

The effect of deuteration on the H-bond strength of phenol to several structurally dissimilar bases has been investigated by Singh and Rao[135]. The enthalpy of formation of the H-bond was always greater than that for the deuterium bond. Also $\Delta\nu_{OH} > \Delta\nu_{OD}$. The ratio of association constants K_H/K_D was dependent on the structure of the acceptor molecule and varied from 0·2 for self-dimerization to 4·3 for association with tetra-*n*-heptylammonium iodide.

2. Comparison of phenols

a. Unhindered phenols. The ability of a phenol to form a strong H-bond increases with increasing Brønsted acidity of the phenol. The ν_{OH} frequency shift, $\log_{10} K_{assoc}$, and ΔH for H-bond formation of a series of phenols with a given acceptor are all linear functions of pK_a for the phenols. Acceptors for which one or other of these correlations have been established include methyl acetate[90], triethylamine[90], tri-*n*-butylamine[137], carbon disulphide[138], *n*-heptyl fluoride[126], several nitro compounds[136], and acetone[97]. The typical relationship between acidity and H-bond strength is exemplified by the results of Huyskens and co-workers[137] for tri-*n*-butylamine acceptor (Figure 5).

The usual linear correlations of ΔH with ΔG and $\Delta\nu_{OH}$ for H-bond formation were not observed when results for several phenols complexing with acetone were considered[97]. However, the enthalpy changes were linear functions of the length of the H-bonds[97]. Lengths ranged from 2·9 Å ($\Delta H = -5{\cdot}20$ kcal/mole) for 3-bromophenol to 3·2 Å ($\Delta H = -1{\cdot}91$ kcal/mole) for 2,6-di-*t*-butyl-4-methylphenol. The curvature of the ΔH against ΔG and $\Delta\nu_{OH}$

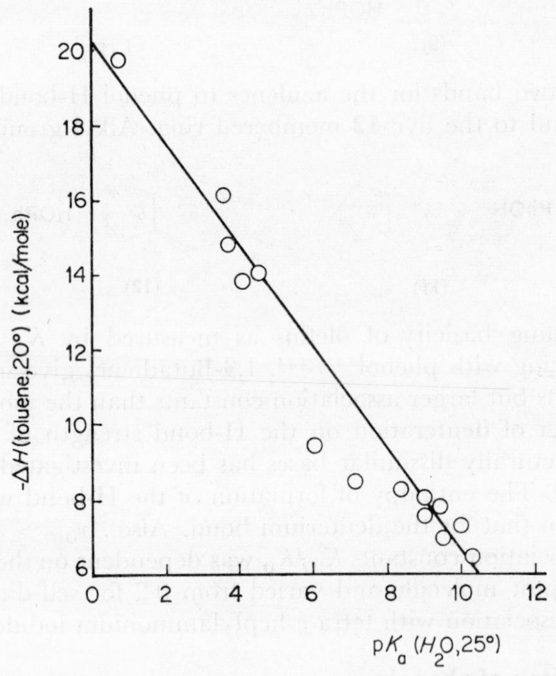

FIGURE 5. Correlation between acidity of phenols and ΔH for the formation of H-bonds between the phenols and tri-*n*-butylamine. ΔH data from reference 137.

graphs probably arises because some of the phenols studied had bulky *ortho* substituents which introduce a severe steric restriction to H-bonding. When *meta* and *para* substituted phenols only are considered good linear correlations are obtained[126, 139]. Thus equations (9) and (10) (ΔH kcal/mole; $\Delta\nu_{OH}$ cm^{-1}) describe the lines for a series of sterically unhindered phenols complexing with *n*-heptyl fluoride[126] and diphenyl sulphoxide[139] respectively, both in C_2Cl_4

$$-\Delta H = 0{\cdot}053\Delta\nu_{OH} \text{ (}n\text{-heptyl fluoride)} \quad (9)$$

$$-\Delta H = 0{\cdot}0134\Delta\nu_{OH} + 3{\cdot}378 \text{ (diphenyl sulphoxide)} \quad (10)$$

solvent. From the significant difference between the two equations Ghersetti and Lusa[139] concluded that $\Delta\nu_{OH}$ can only be used to estimate H-bond strengths providing a given acceptor and a series of structurally related donors are being compared.

For a given phenol bonding to a series structurally related acceptors both $\Delta\nu_{OH}$ and $\log_{10} K_{assoc}$ are linear functions of the σ-constants for the substituents in the acceptors[116, 123, 136]. However, for a given acceptor both $\Delta\nu_{OH}$ and $\log_{10} K_{assoc}$ are larger for a more acidic phenol. The shifts in OH stretching frequencies for a series of *p*-substituted phenols on H-bonding to dioxan in cyclohexane increased linearly with the increment in OH bond moment calculated to explain the measured dipole moments of the phenols in dioxan[140]. Equation (11) is applicable and emphasizes the simple

$$y = 0{\cdot}00814\Delta\nu_{OH} - 1{\cdot}43 \qquad (11)$$

relationship between bond moment and H-bonding strength. Pentafluorophenol and 2,3,5,6-tetrafluorophenol form 2 : 1 crystalline complexes of type **13** (Y = H or F) with dioxan[141]. Pentafluoro-

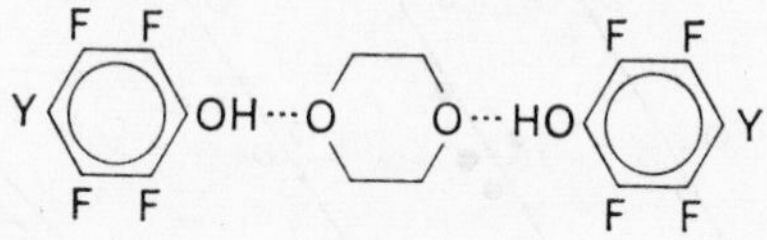

(13)

phenol also forms complexes with triphenylphosphine oxide of 1 : 1, 1 : 2 and 2 : 1 stoichiometry[142]. Heats and enthalpies of formation of the complexes in CCl_4 are for $(C_6H_5)_3PO \cdots HOCF_{65}$ $-5{\cdot}5$ kcal/mole and $-1{\cdot}1$ e.u., for $(C_6H_5)_3PO \cdots 2\,HOC_6F_5$ $-4{\cdot}8$ kcal/mole and $-9{\cdot}5$ e.u., and for $C_6F_5OH \cdots 2\,OP(C_6H_5)_3$ $-1{\cdot}9$ kcal/mole and $-3{\cdot}6$ e.u.

The H-bonding of phenols to aniline has been studied by Zeegers-Huyskens[101]. The OH stretch frequencies in the H-bonded complexes showed a steady drift to lower frequencies as the acidity of the phenols became greater. Thus $\nu_{OH} = 3350$ cm^{-1} for 2,4,6-trimethylphenol (pK_a = 10·88) and $\nu_{OH} = 3160$ cm^{-1} for 4-nitrophenol (pK_a = 7·15) represent the extremes for the series. The asymmetric and symmetric stretching frequencies of the aniline NH_2 group showed corresponding shifts to lower frequency. 2,4-Dinitrophenol, 2,6-dinitrophenol, pentachlorophenol, 2,3,5-trichlorophenol, and 2,6-dichloro-4-nitrophenol form complexes with aniline in the solid phase which have a broad infrared absorption at 2100–2900 cm^{-1}. This arises because proton transfer has occurred and the

H-bond becomes $NH^+\cdots O^-$. The fluorescence spectrum of the α-naphthol–triethylamine H-bonded complex is identical with the fluorescence spectrum of α-naphthol in alkaline solution. Thus proton transfer from α-naphthol to triethylamine occurs when the H-bonded complex absorbs radiation. The excited state of the complex is in the ion-pair form[143].

b. Steric effects. Bulky *ortho* substituents in phenols hinder the formation of H-bonds which involve the OH group[144]. The effect of *ortho* substituents on the relative frequency shifts ($\Delta\nu_{OH}/\nu_{OH}$) for several phenols H-bonding to many acceptors of different structural types has been investigated by Bellamy and Williams[61]. Plots (Figure 6) of

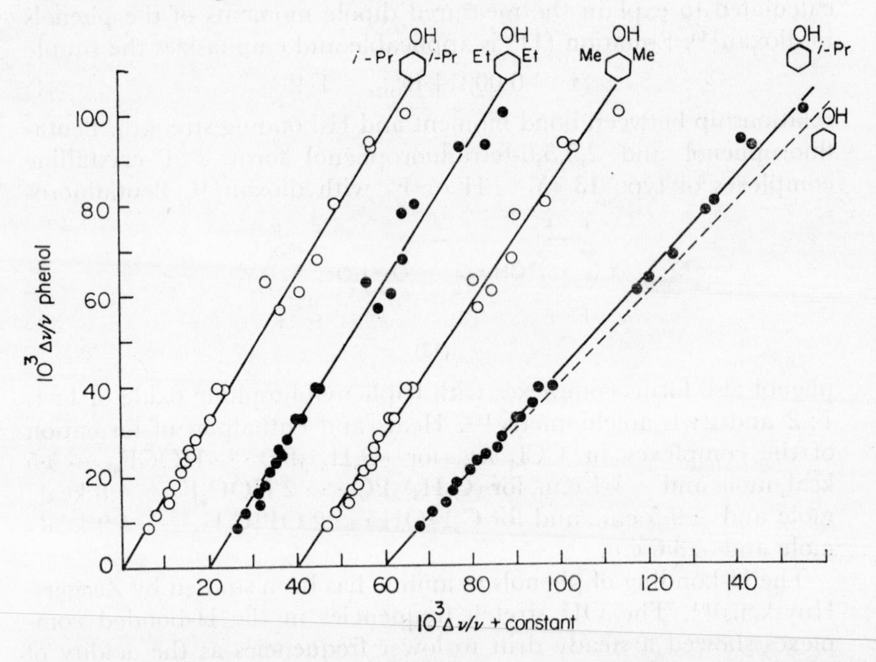

FIGURE 6. Relative frequency shifts for four phenols compared with those for phenol itself on H-bonding to 23 acceptor solvents[61]. The intercepts on the $10^3\Delta\nu/\nu$ axis are staggered by 20 units for clarity. Reproduced by permission from L. J. Bellamy and R. L. Williams, *Proc. Roy. Soc.*, **A254**, 119 (1960).

($\Delta\nu_{OH}/\nu_{OH}$) for 2-*i*-propyl, 2,6-dimethyl, 2,6-diethyl and 2,6-di-*i*-propyl phenols against ($\Delta\nu_{OH}/\nu_{OH}$) for phenol were linear, suggesting the absence of steric effects. For 2,6-di-*t*-butylphenol the geometry of the acceptor molecules has a decisive influence on whether H-bonding can occur. There is no association with pyridine, triethylamine, dimethylamine or ether except at high concentrations. However, 2,6-di-*t*-butylphenol does H-bond to dioxan. It appears uncertain whether H-bonding to π-electron systems can occur[61] or not[130].

Although the frequency shift data only give a clear indication of steric hindrance for 2,6-di-*t*-butylphenol, the association constants of several phenols with ethers show that smaller *ortho* substituents than *t*-butyl also have an effect[145]. This is exemplified by the figures in Table 7. One *ortho* alkyl group is not sufficient to cause an appreciable

TABLE 7. Equilibrium constants (l/mole) for the association of some substituted phenols with ethers in CCl_4 at 29° [145].

Ether	Et_2O	*i*-Pr_2O	*n*-Bu_2O	*t*-Bu_2O
Phenol substituents				
2-Methyl	3·4	3·2	2·6	1·7
2-*i*-Propyl	3·9	3·9	3·0	2·0
2-*t*-Butyl	3·7	4·1	2·9	2·3
2,6-Dimethyl	0·67	0·62	0·63	0·22
2,6-Di-*i*-propyl	0·51	0·48	0·23	0·19
2,6-Di-*t*-butyl	~0	~0	~0	~0

steric effect. However 2,6-dialkyl substitution produces a steric inhibition to H-bond formation which is more severe either for larger alkyl groups or for a more sterically hindered acceptor site. Association constants for H-bonding of *N*-methylacetamide or *N,N*-dimethylacetamide with three phenols were in the order 2,6-di-*t*-butylphenol < 2-methyl-6-*t*-butylphenol < 2-*t*-butylphenol, which is consistent with the expected relative steric requirements of the phenols[146]. The steric hindrance to H-bonding of 2,4,6-tri-*t*-butylphenol with acetone[147] is paralleled by a corresponding hindrance to the H-bonding of phenol with the oxygen atom attached to C_1 of 2,6-di-*t*-butyl-1,4-benzoquinone[148]. In general, steric factors have a greater influence on H-bonding association than on the proton acidity of phenols[97]. This difference parallels the relative effects of bulky 2,6-substituents on the Lewis and Brønsted basicity of pyridines[104].

The conclusion[61, 145] that the reduction in K_{assoc} caused by steric interference is an entropy rather than an enthalpy effect does not seem to be wholly justified[97, 125, 146]. Thus for 2,6-di-*t*-butylphenol H-bonding to acetonitrile (C_2Cl_4, 25°) $K_{assoc} = 0{\cdot}41$ l/mole,

$\Delta H = -0.8$ kcal/mole and $\Delta S = -4.4$ e.u., and H-bonding to benzonitrile $K_{assoc} = 0.43$ l/mole, $\Delta H = -0.9$ kcal/mole and $\Delta S = -4.6$ e.u.[125]. Comparison with the corresponding results for phenol (Table 6) shows that both the enthalpy and entropy changes are affected. In general steric hindrance leads to a weaker ($-\Delta H$ smaller) and longer H-bond[97] than would be expected from electronic considerations alone.

3. Solvent effects

The frequency shifts observed for H-bonded complexes on changing the solvent may be approximately fitted to the Kirkwood–Bauer–Magat equation (12)[112] provided no specific interactions occur between the solvent molecules and the complex[149]. C is a constant

$$\nu = \nu_0 - C\frac{\varepsilon - 1}{2\varepsilon + 1} \tag{12}$$

for a particular complex, ε is the dielectric constant of the solvent and ν_0 is the OH stretching frequency for the complex in the vapour phase. Thus ν_{OH} is a linear function of $(\varepsilon - 1)/(2\varepsilon + 1)$ for the H-bonded complexes phenol–acetonitrile, phenol–propionitrile, phenol–methyl iodide and phenol–dimethyl sulphoxide in various solvents[100, 149]. The values of C for the complexes of five phenols with acetonitrile in acetonitrile–tetrachloroethene mixtures were a linear function of the σ-constants[52] for the substituents in the phenol rings[149]. Equation (12) is not consistent with the frequency shifts for the phenol–ether complex in ether–chloroform mixtures[109]. In this context the empirical equation (4) is found to have a wider generality than equation (12)[109]. However, a more rigorous theory of solvent effects should also take variations of refractive index n of the solvent into account[150]. Thus the Buckingham[150] equation (13) often gives an impressive agreement between experimental and cal-

$$\nu = \nu_0 - C_1\frac{\varepsilon - 1}{2\varepsilon + 1} - C_2\frac{n^2 - 1}{2n^2 + 1} \tag{13}$$

culated frequency shifts (Table 8)[149]. Extensions of the Buckingham equation have also been tested[89, 149].

The effect of specific solute–solvent interactions on the behaviour of H-bonded complexes in different solvents must also be considered[22, 110, 151, 152]. Thus for the phenol–triphenylphosphine oxide complex in CBr_4/CCl_4 mixtures ΔH, ΔG and ΔS for H-bond formation all become less negative as the mole fraction of CBr_4 in the solvent is increased[152]. This probably arises because a specific interaction between triphenylphosphine oxide and CBr_4 occurs and is much stronger than the corresponding interaction with CCl_4. Specific solvent effects are most likely to influence the H-bond in a solute complex when the solvent molecules themselves can act as H-bond donors or acceptors[22]. An example has been given in structure **6** above. The OH stretch frequencies for the H-bonded complexes of phenol with di-*n*-butyl ether or mesitylene in *n*-hexane, carbon tetrachloride or chloroform are all dependent on temperature. It has been argued[151] that these results indicate that specific solvent–solute interactions are the predominant cause of solvent effects on H-bond equilibria. The 'inductive association' concept of solvation[109, 149] is considered unsatisfactory.

TABLE 8. Comparison of the experimental OH stretch frequencies for the phenol–tetrahydrofuran H-bonded complex in CCl_4 with the values calculated from equation (13)[149].

Volume % tetrahydrofuran	ν_{OH} (experimental) (cm^{-1})	ν_{OH} (calculated) (cm^{-1})
1	3323	3323·5
10	3319	3318·5
30	3313	3312·0
50	3307	3307·8
70	3305	3305·3
100	3304	3303·7

o-t-Alkylphenols exist as *cis* and *trans* isomers[153]. The *cis–trans* ratio at equilibrium is increased by solvents which can act as acceptors for an H-bond from the OH of the *cis* isomer[154]. H-bonding from the *trans* isomer is sterically hindered. The solvation by water of the conjugate acid of 2,4,6-trimethoxybenzene is increased by progressively substituting OH groups for OCH_3 groups[155]. In the tri-substituted compounds solvation by H-bonding decreases in the order $(OH)_3 > (OH)_2(OCH_3) > (OH)(OCH_3)_2 > (OCH_3)_3$. In a similar way, substituting hydroxyl groups in aromatic nuclei should increase their solvation through H-bonding and therefore their solubility in water. The reverse is often true, particularly for heterocyclic nuclei[156]. The hydroxypteridines are a good example: the solubilities are in general in the order tetrahydroxypteridine < trihydroxypteridines < dihydroxypteridines < hydroxypteridines < pteridine. The increase in H-bonding ability in the solid state clearly outweighs the increased solvation of the molecules on adding an OH group.

4. Spectroscopy and H-bonding

H-bonding alters the p.m.r. chemical shifts and infrared stretching frequencies of the OH group. Shifts in the stretching frequencies associated with the acceptor site in molecules to which phenols are H-bonding have also been noted[98, 99, 101]. H-bonding by 2- or 4-substituted phenols produces appreciable shifts in p.m.r. spectra of the ring protons of the phenols[120]. The perturbation to the OH group caused by the formation of an H-bond is spread over the whole phenol molecule. The π–π^* transition in the electronic spectrum of phenols is generally shifted to lower frequencies (longer wavelength) when a phenol becomes an H-bond donor[138, 157–160, 162]. The stabilization of the appropriate ground ($^1A_1$) and excited ($^1B_2$–) states on H-bonding is greater for the latter than for the former and therefore a smaller excitation energy is necessary to induce the electronic transition in the H-bonded molecule[160]. The conclusion[159] that 4-nitrophenol is a stronger H-bond donor in the excited state than in the ground state is analogous to the increase in acidity of some phenols in going from the ground to the excited states[161].

Far infrared[76, 77, 163–168] studies have given values for the vibrational frequencies of the H-bonds between phenols and various electron acceptors. For phenol the H-bond stretching frequency ν_σ is 175 cm^{-1} for the solid[76, 77, 163], 162 cm^{-1} for the liquid[76], and 130–150 cm^{-1} for solutions in CCl_4 [163, 165]. The frequencies for substituted phenols are shifted[76, 163, 167] but there is no direct correlation between the values of ν_σ and the corresponding changes in $\Delta\nu_{OH}$ for the OH stretching frequencies[76, 167]. If both the H-bond frequency ν_σ and the OH frequency shift $\Delta\nu_{OH}$ were a reliable measure of H-bond strength a smooth relationship between the two would be expected. For the phenol–triethylamine, phenol–trimethylamine and phenol–pyridine complexes the ν_σ frequencies are 123 cm^{-1}, 143 cm^{-1} and 134 cm^{-1} respectively[165]. The values are 120 cm^{-1}, 141 cm^{-1} and 130 cm^{-1} for the corresponding C_6H_5OD complexes. The ν_σ frequency for a particular complex varies with solvent composition. Thus, for the phenol–pyridine complex in pyridine–carbon tetrachloride mixtures ν_σ (cm^{-1}) varies linearly with the mole fraction X_{py} of pyridine according to equation (14)[166].

$$X_{py} = 0{\cdot}126\nu_\sigma - 16{\cdot}455 \qquad (14)$$

III. INTRAMOLECULAR H-BONDING*

A. Alcohols

The easiest test for intramolecular H-bonding is to measure the variation of infrared or Raman spectra at low concentrations or pressures[169]. The spectral characteristics of intermolecular H-bonding disappear at low concentrations when H-bonding intermolecular association becomes absent. Intramolecular H-bonding does not disappear. A general comparison of the properties of intra- and intermolecular H-bonds has been given by Pimentel and McClellan[169].

Intramolecular H-bonding in dihydroxy compounds was studied by Kuhn[170]. In general, only those compounds for which the calculated length of the H-bond was less than 3·3 Å formed intramolecular H-bonds. Thus, cyclohexane-1,4-diol does not H-bond intramolecularly. However, cyclohexane-1,3-diol forms an H-bond in its *cis* form but not in its *trans* form. The strongest intramolecular bond (measured as that giving the largest shift $\Delta\nu_{OH}$) was for 1,2-dimethylolcyclohexane in which the H-bond leads to the formation of a seven-membered ring. In the series $HO(CH_2)_nOH$ the $\Delta\nu_{OH}$ differences between the frequencies for the free and the intramolecularly H-bonded OH groups were (CCl_4 solvent) 32 cm^{-1} for $n = 2$, 78 cm^{-1} for $n = 3$, and 156 cm^{-1} for $n = 4$. The H-bond is apparently stronger when the geometry of the molecule allows a close approach of the two OH groups. For $n = 6$ however no intramolecular H-bond is formed. Such bonding is favoured when it leads to the formation of five-, six- or seven-membered ring structures[174].

Although cyclohexane-1,4-diol does not form an intramolecular H-bond[170] its *cis, cis, cis*-2,5-dialkyl derivatives (**14**) in CCl_4 (25°) can exist in intramolecularly H-bonded twist conformations (**15**)[171, 172]. The population of the twist form at equilibria depends upon the size of the alkyl groups R_1 and R_2. It is about 5% for

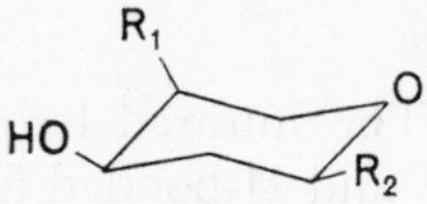

R1 HO OH R2 **(14)**   O H···O H R1 R2 **(15)**

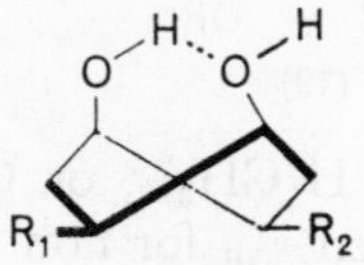

* Intramolecular H-bonding leads to cyclic structures. The number of atoms in an H-bonded ring is counted in this section to include the hydrogen atom involved in the H-bond.

$R_1 = R_2 = CH_3$, >98% for $R_1 = R_2 = t$-alkyl, 14% for $R_1 = CH_3$, $R_2 = CH(CH_3)_2$ or $R_1 = CH_3$, $R_2 = C(CH_3)_3$, and about 80% or $R_1 = R_2 =$ *sec*-alkyl. Increasing the size of R_1 or R_2 increases the proportion of the H-bonded twist structure. Structure **16** which resembles the shape of *trans* decalin has been suggested for the H-bonded form of heptane-1,4,7-triol-1-methyl ether[173]. The H-bonded structures of the isomeric triols 1,1,1-trimethylolethane and pentane-1,3,5-triol are probably **17** and **18** respectively.

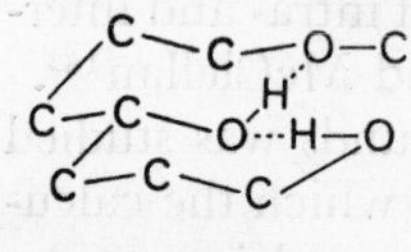

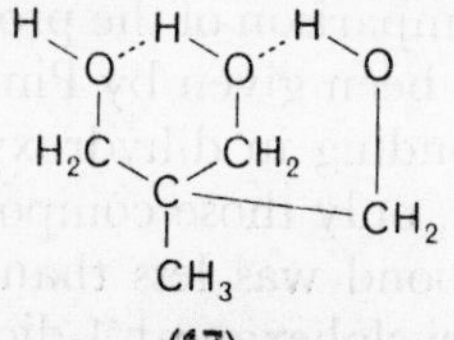

Intramolecular H-bonding from OH to the ether linkage in the series $MeO(CH_2)_nOH$ is favoured when a five-, six- or seven-membered ring results[174]. For $n = 2,3$ and 4 the ν_{OH} frequency shifts in going from the nonbonded to the H-bonded conformations were 30, 86 and 180 cm^{-1} respectively[175]. The corresponding heats of H-bond formation (CCl_4 solvent) were 2200, 2100 and 2700 cal/mole respectively. The Badger–Bauer relationship[14, 15] is not applicable for these intramolecular H-bonds.

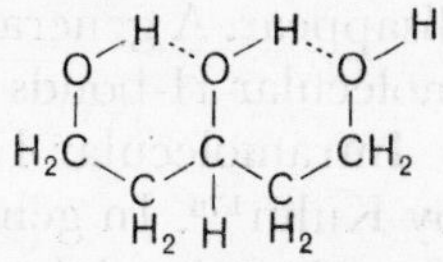

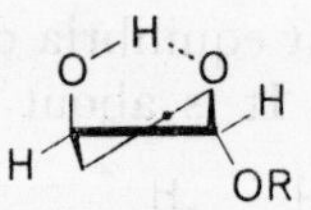

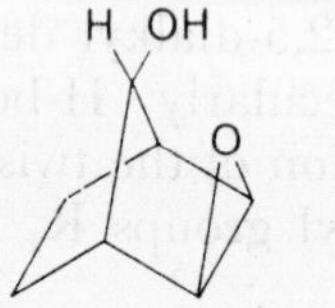

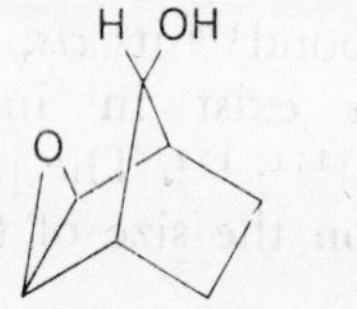

Intramolecular H-bonding occurs in both the *cis* and *trans*-2-alkoxy-3-hydroxytetrahydrofurans[176]. For the latter, because the alkoxy and hydroxy groups are *trans*, H-bonding between OH and the heterocyclic oxygen must occur. A possible structure for the H-bonded form of the *trans* isomer is **19** in which $R = CH_3$, CH_3CH_2, $CH(CH_3)_2$, or $C(CH_3)_3$. The infrared frequency difference between ν_{OH} for non H-bonded and H-bonded forms is about 24 cm^{-1} for all four alkyl substituents. Proton magnetic resonance hydroxy proton chemical shifts and rates of change of chemical shift with concentration at infinite dilution sometimes enable intra- and inter-molecular H-bonding to be distinguished. Thus for fifteen isomeric epoxyalcohols with the bicyclo[2,2,1] heptane skeleton the chemical shifts (relative to tetramethylsilane) were characteristically large and the limiting slopes of the chemical shift against concentration plots were small in cases where intramolecular H-bonding occurred[177]. Compounds **20** and **21** serve as examples. For **20** in which H-bonding to an epoxide linkage occurs, the limiting chemical shift (CCl_4 solvent) was 192·0 cps and the limiting slope 85 cps/mole fraction. The corresponding figures for **21** which can only H-bond intermolecularly were 44·0 cps and 3580 cps/mole fraction. The large concentration dependence of the chemical shift for intermolecular H-bonding compared to that for intramolecular H-bonding is a general result[169]. Infrared evidence has been recorded for the intramolecular H-bonding by OH groups to the oxirane rings of some epoxyalcohols[178]. Thus for glycidol (in CCl_4) three ν_{OH} absorptions exist at 3638, 3611·8 and 3590 cm^{-1}. The first is due to free OH and the last to OH H-bonded to the epoxide oxygen atom. The middle band has been ascribed to H-bonding of OH to the electrons of the CO bond of the oxirane ring. This interaction is favoured when the OH is orientated in the plane of the ring.

The strength of the intramolecular H-bond between the alcoholic OH and the ethylenic double bond π-electrons in ethylenic alcohols depends upon the relative positions of the OH group and the double bond in the molecules[179, 180]. Hydroxy stretching frequency shifts on H-bonding are 1–2 cm^{-1} for α,β ethylenic alcohols, 25–45 cm^{-1} for a β,γ double bond, and 60 cm^{-1} for a γ,δ bond. The corresponding enthalpies of H-bond formation ΔH are $-0{\cdot}8$ kcal/mole (α,β), $-1{\cdot}0$ kcal/mole (β,γ) and $-2{\cdot}3$ kcal/mole (γ,δ). The heat of formation of an intramolecular H-bond between an alcohol group and a triple bond in β,γ acetylenic alcohols is about $-1{\cdot}1$ kcal/mole[181]. Frequency shifts $\Delta\nu_{OH}$ on H-bonding for the intramolecular interactions between the hydroxyl groups and the π-electrons of unsaturated bonds in the methyl esters of unsaturated monohydroxy acids may be divided into three groups depending upon the geometry of the molecules[182]. It has been concluded that studies of the strength of H-bonds in novel compounds might prove a useful aid in the determination of molecular structure[182]. A further example involving OH$\rightarrow\pi$ H-bonding is analogous to the difference (discussed above) in H-bonding properties of structures **20** and **21**. Thus for **22** and **23** the limiting chemical shifts were 60·5 ppm and 121·0 ppm and the limiting chemical shift against concentration slopes were 2380 and 320 ppm/mole fraction respectively[177]. Clearly

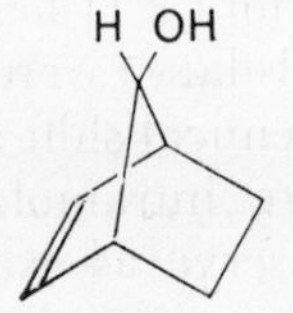

(22) 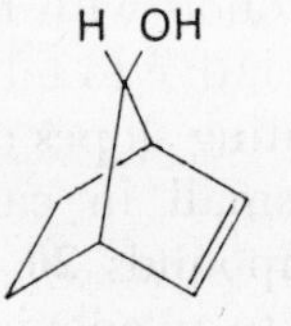 (23)

the geometry of **23** is such that OH→π intramolecular bonding can occur whereas in **22** it becomes impossible.

In many compounds competition between different types of intramolecular H-bonding can occur. Thus for the phenyl-α,ω-alkanediols $PhCHOH(CH_2)_nOH$ the H-bonding of the hydroxyl groups with each other and with the phenyl π-electron system has been studied[183]. The primary hydroxyl group H-bonds to both the secondary OH and the phenyl ring if $n = 1$. However for $n = 2$ or 3 only bonding to the secondary OH groups occurs. The primary OH is not intramolecularly H-bonded when $n = 4$. The secondary OH H-bonds to the primary OH and the phenyl ring when $1 \leqslant n \leqslant 3$ but only to the π-system when $n > 3$. For example, for 1-phenylpropane-1,3-diol ($n = 2$) there are four infrared OH stretching absorptions which have been assigned in accord with structures **24**

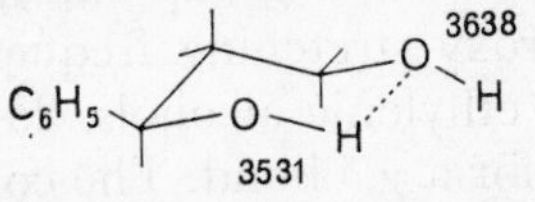

(24) 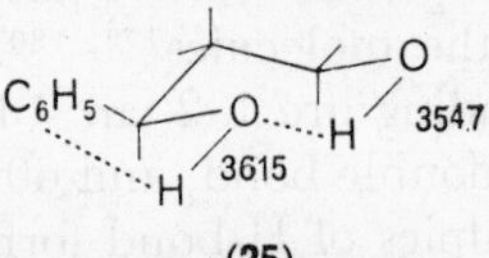 (25)

and **25**. Competition between different intramolecular H-bonding possibilities is further exemplified by a study of substituted benzyl alcohols with structure **26** in which Y = H or OH and X is C≡C, CH=CH (*cis* or *trans*) or CH_2CH_2 [184]. Four H-bonded structures (**27–30**) are possible. The infrared ν_{OH} frequencies were around 3640 cm^{-1} for **26**, 3618 cm^{-1} for **27**, 3585 cm^{-1} for **28** and 3535 cm^{-1}

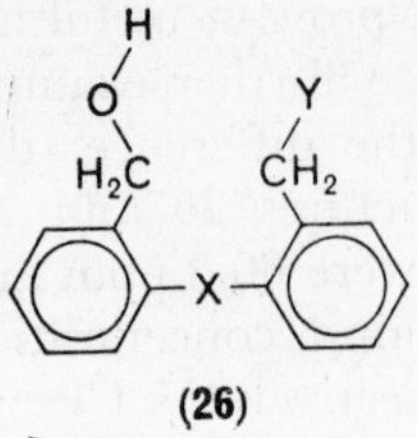

(26) 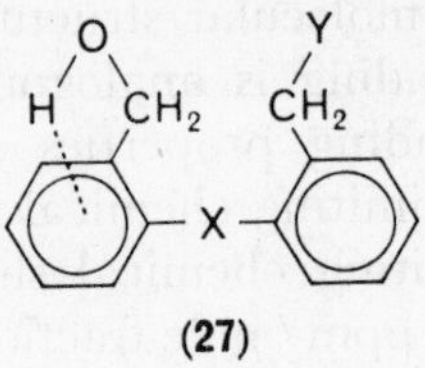 (27) 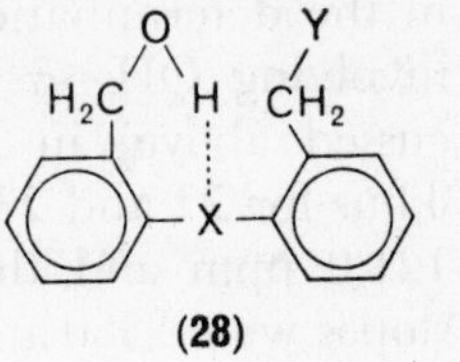 (28)

(29) (30)

for **29**. Conformer **30** was only observed for two of the alcohols for which ν_{OH} (**30**) were 3507 and 3575 cm^{-1}. From the relative intensities of the absorptions the proportions of each conformer at equilibrium in CCl_4 were evaluated. The results are given in Table 9. For the dihydroxy compounds (Y = OH) the sum of the conformations for both OH groups contributes to the infrared intensities.

TABLE 9[a]. Approximate percentages of each conformer of the substituted benzyl alcohols (**26**) present in CCl_4 at about 30°C [184].

Substituents		Conformers				
X	Y	**26**	**27**	**28**	**29**	**30**
C≡C	H	35	40	25		
C≡C	OH	25	40	15		20
CH=CH (*cis*)	H	25	50	15	10	
CH=CH (*cis*)	OH	25	50	15	10	Not obsd.
CH=CH (*trans*)	H	20	55	15	10	
CH=CH (*trans*)	OH	20	55	15	10	Not obsd.
CH_2–CH_2	H	10	75		15	
CH_2–CH_2	OH	20	60		10	10

[a] Reproduced by permission from I. D. Campbell, G. Eglington and R. A. Raphael, *J. Chem. Soc.* (**B**), 338 (1968).

Intramolecular H-bonding in hydroxyketones and hydroxyesters leads to a reduction in the carbonyl stretching frequency. The p.m.r. chemical shifts of the OH protons for a series of such compounds were a linear function of the carbonyl stretching frequencies in the H-bonded molecules[185]. The stronger the H-bond the lower is $\nu_{C=O}$. Intramolecular H-bonding of OH to a carbonyl group also produces the usual shift to lower frequencies of the OH stretching vibration. For a series of ketoalcohols the shift $\Delta\nu_{OH}$ may be correlated with the relative geometrical orientation of the hydroxyl group and the lone pair electrons on the oxygen atom of the carbonyl group[186]. The frequency shift is biggest for shorter H-bonds and for bonds in which the maximum overlap occurs between the orbitals of the hydroxyl hydrogen atom and the lone pair molecular orbitals of the carbonyl oxygen atoms. There is evidence that the latter are predominantly 2*p* in character. In some cases where this interaction is unlikely from steric considerations there is still a small $\Delta\nu_{OH}$ shift (8–34 cm^{-1}) indicative of a weak intramolecular H-bond. This has been attributed to an interaction between the hydroxyl groups and the π-electrons of the carbonyl group. An example is 5α-cholestan-5α-ol-4-one (**31**) for which $\Delta\nu_{OH} = 11$ cm^{-1}. For 5α-hydroxyergosta-7,22-dien-3-one (**32**) competition occurs between H-bonding to a carbonyl group and an ethylenic double bond. The relevant frequency shifts are 10 and 24 cm^{-1} respectively.

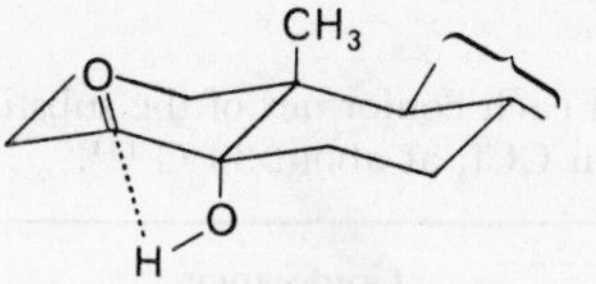

CH_3 O H O (**31**) H_3C O OH (**32**)

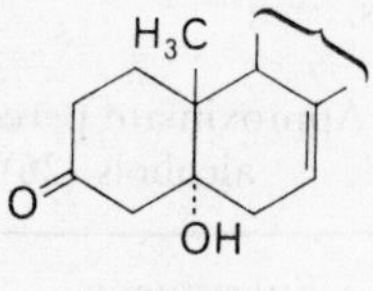

H-bonding intramolecular interactions between the alcoholic group and the π-electrons of the cyano group occur in α-cyanoalkanols[94, 187]. Thus, for example, the OH stretch absorption for cyclohexanone cyanohydrin is symmetric (ν_{max} 3591 cm^{-1}) indicating that the molecule exists predominantly in one of the interacting conformations **33** or **34** and not as the non-interacting forms

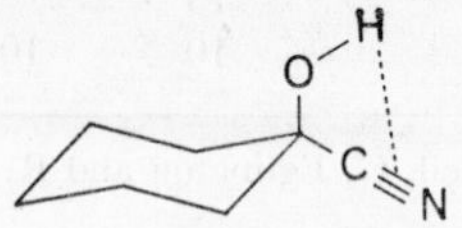

(**33**) 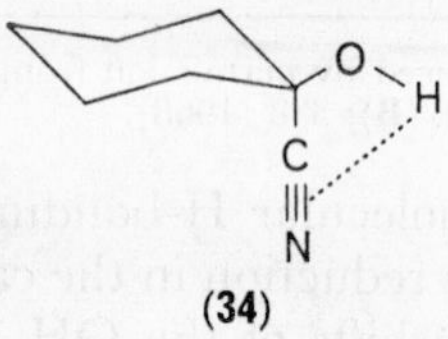 (**34**) 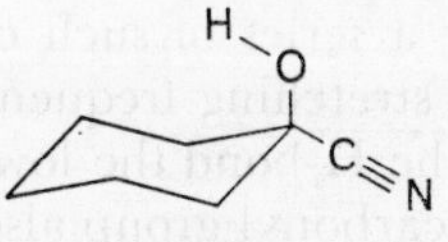 (**35**) 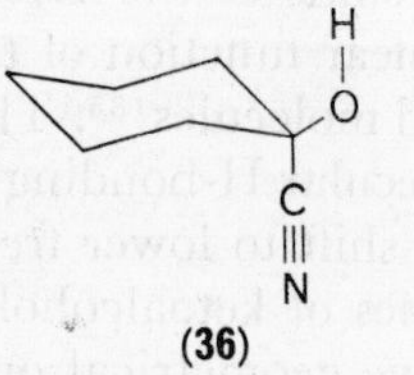 (**36**)

35 or **36**. For the cyanoalkanols $NC(CH_2)_nOH$ an H-bond interaction with the CN π-electrons occurs only if $n = 1$ or 2.

The frequency shifts $\Delta\nu_{OH}$ for intramolecular H-bonding in the 2-haloethanols were 12 cm^{-1} for F, 32 cm^{-1} for Cl, 38 cm^{-1} for Br and 46 cm^{-1} for I [188]. However, the strengths ($-\Delta H$) of the H-bonds are in the order $F > Cl \simeq Br > I$ [189]. The Badger–Bauer relationship[14, 15] is therefore not applicable. In general the H-bond strength increases with increasing electronegativity of the halogen atom. The frequency shifts $\Delta\nu_{OH}$ cannot be used to compare H-bond strengths when the interactions involve different acceptor atoms or groups. None of the 3-halopropanols can form an intramolecular H-bond[190]. The enthalpy of H-bond formation is too small compared with the entropy loss on forming a cyclic H-bonded structure when a ring of six or more atoms would result.

Aminoalkanols can form either NH · · · O or OH · · · N intramolecular H-bonds. For example, the enthalpies of interconversion of the three predominant conformers of *N*-methylethanolamine are as follows[191] (C_2Cl_4 solvent).

CH_2CH_2 / H—O···H—N—CH_3 **(37)** $\xrightarrow{-1{\cdot}0\ kcal/mole}$ HO—CH_2CH_2—$NHCH_3$ **(38)** $\xleftarrow{-0{\cdot}7\ kcal/mole}$ CH_2CH_2 / O···H—N(H)—CH_3 **(39)**

The OH · · · N interaction provides the most stable conformer. For the diethylaminoalkanols $(C_2H_5)_2N(CH_2)_nOH$ the $\Delta\nu_{OH}$ values are much larger than those for the corresponding halo, cyano or alkoxyalkanols[188, 192]. Furthermore, cyclic H-bonded conformers could be detected for $n \leqslant 5$ although the equilibrium constants for formation of the conformers from the non H-bonded forms decreased in the sequence 14, 4·7, 3·5, 0·11 for $n = 2$, 3, 4 and 5. For $n = 6$ intramolecular H-bonding could not be detected.

Evidence for intramolecular H-bonding between hydroxyl groups and nitro groups in 2-nitroalcohols includes the detection of ν_{OH} absorption bands for the alcohols which are at 10–28 cm^{-1} lower frequencies than the bands for the non H-bonded alcohols[113]. Spectra of two 2-alkyl-2-nitropropane-1,3-diols showed three absorptions: a free OH band at 3632 cm^{-1}, an OH · · · O_2N bonded band at 3604 cm^{-1} and an OH · · · OH absorption at 3550 cm^{-1}. The nitro group is a weaker acceptor than the second hydroxyl group but is suitably orientated for an interaction to occur.

A concentration-independent sideband at lower frequencies than he main absorption appears on the hydroxyl stretch band contour for aliphatic alcohols in carbon tetrachloride[193]. This has been attributed to an interaction between the CH group in the γ position and the lone pair electrons on the oxygen atom of the hydroxyl group. Interaction enthalpies of $-0{\cdot}44$ kcal/mole for *n*-heptanol and $-1{\cdot}06$ kcal/mole for 4-heptanol have been deduced.

B. Phenols

2-Substituted phenols exist as *cis* **40** or *trans* **41** isomers[51, 53, 61, 154, 155]. The hydroxyl group is in the plane of the benzene ring even for 2,6-di-*t*-butyl phenols[51, 61]. For bulky 2-substituents the *trans* isomer becomes preferred to the *cis* isomer with increasing size of the group[53, 154, 155]. However, if the group is capable of acting as an H-

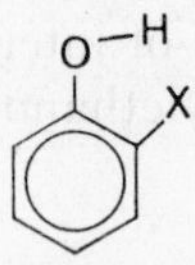

(40) 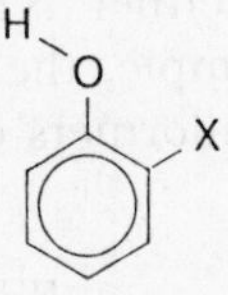 (41) 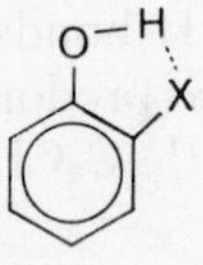 (42)

bond acceptor intramolecular H-bonding **42** occurs and leads to an increase in the thermodynamic stability of the *cis* isomer[53, 67, 69, 194]. The experimental differences (Table 10) in ν_{OH} frequencies for **40** and **42** are characteristic of an H-bonding interaction in the latter. The corresponding shifts of the hydroxyl proton magnetic resonance signal δ_{OH} are approximately a linear function of $\Delta\nu_{OH}$ [67, 194]. This is an analogous result to the one, exemplified by equation (3), for intermolecular H-bonds. Typical values of $\Delta\delta_{OH}$ and $\Delta\nu_{OH}$ for the intramolecular H-bonding of phenols are given in Table 10. The 1,2-diol catechol shows two ν_{OH} absorptions of about equal intensity, one at 3618 cm^{-1} due to an OH group acting as an H-bond acceptor and one at 3570 cm^{-1} due to the other OH group which is the H-bond donor[173]. The 1,2,3-triol pyrogallol exists as a conformer in which two of the OH groups act as H-bond donors.

The out of plane deformation vibration γ_{OH} of the OH group in intramolecularly H-bonded 2-substituted phenols occurs in the 300–860 cm^{-1} frequency region[202]. A curved plot of γ_{OH} against ν_{OH} has been obtained for about 50 phenols with γ_{OH} increasing as ν_{OH} decreases. Comparing H-bonds of quite different strengths, the strongest bonds give rise to the highest γ_{OH} frequencies. The far infrared ν_σ H-bond stretching frequency for 2-chlorophenol is at 84 cm^{-1} which is about 40 cm^{-1} lower than ν_σ for the cresols,

TABLE 10. Typical changes in the position of the OH p.m.r. signal $\Delta\delta_{OH}$ and infrared shifts $\Delta\nu_{OH}$ caused by intramolecular H-bonding of phenols in carbon tetrachloride[67, 194, 195, 203].

Phenol	Acceptor group	$\Delta\delta_{OH}$ (ppm)[a]	$\Delta\nu_{OH}$ (cm^{-1})
Salicylaldehyde	CHO	6·71	471
2-Nitrophenol	NO_2	6·34	364
Methyl salicylate	$COOCH_3$	6·32	395
2-Iodophenol	I	1·33	105
2-Methoxypheno	OCH_3	1·18	60
2-Chlorophenol	Cl	1·17	61
2-Bromophenol	Br	1·15	92
2-Allylphenol	$CH_2CH{=}CH_2$	1·07	66
2-Fluorophenol	F	0·80	18
2-Cresol	CH_3	0·32	−8

[a] Measured with respect to a cyclohexane internal standard and corrected for ring currents.

3-chlorophenol, and 4-chlorophenol which cannot form intramolecular H-bonds[167].

For 2-halophenols the *trans–cis* ratios and the infrared frequency shifts $\Delta\nu_{OH}$ (Table 10) both increase in the order F < Cl < Br < I [67, 194–197]. However, measurements of the enthalpies of formation of the H-bonds (Table 11) have shown that this is not the same

TABLE 11. Strengths of the intramolecular H-bond of 2-halophenols as a function of solvent.

Phenol	$-\Delta H$ (kcal/mole)[198] in vapour	$-\Delta H$ (kcal/mole)[67] in CS_2	$-\Delta H$ (kcal/mole)[197] in CCl_2CCl_2
2-Chlorophenol	3·41	2·36	1·28
2-Bromophenol	3·13	2·14	1·86
2-Iodophenol	2·75	1·65	0·99
2-Chlorophenol-*d*	2·81		
2-Bromophenol-*d*	2·65		
2-Iodophenol-*d*	2·65		

as the order of H-bond strengths[67, 197, 198]. In the vapour and in CCl_4 solution the order of strength is I < Br < Cl whereas in ethylene tetrachloride solvent the OH · · · Br H-bond becomes the strongest. From a study of the infrared spectra of unsymmetrical 2,6-dihalophenols in CCl_4 it has been deduced that the OH · · · F intramolecular H-bond is intermediate in strength between the bonds to Br and to I [199]. The Badger–Bauer relationship[14, 15] is clearly not obeyed for these H-bonds. The infrared frequency shifts are anomalous because of an orbital–orbital repulsion between the donated lone pair orbital of the halogens and the O–H bonding orbital[199]. The greatly varying size of the halogen atoms also has an effect[197, 199]. Anomalies in the correlation between p.m.r. chemical shift and H-bond strength have been attributed to the varying diamagnetic anisotropies of the *ortho* C–X group for the four halogens[67]. Comparison of the enthalpies in Table 11 for phenol vapours and in CCl_2CCl_2 or CS_2 solutions emphasizes the effect of solvent on the strength of intramolecular H-bonds. The reduction in H-bond strength is due to the stabilization of the *trans* isomers **41** of the phenols through intermolecular interaction with the solvent[68, 200]. In the vapour phase the OD · · · X intramolecular bonds are weaker than the corresponding OH · · · X interactions (Table 11)[198]. However, in general a D-bond can be weaker or stronger than the corresponding H-bond depending on the shape of the potential function and the geometry of the particular interaction being considered. A quantum mechanical tunnel effect may also be significant[201].

The ν_{OH} frequency shifts and enthalpies of intramolecular H-bond formation for some interactions between OH groups and π-electrons are given in Table 12. The frequency shifts are the same order of

TABLE 12. Frequency shifts and enthalpies of intramolecular H-bonding of phenol OH group to π-electrons (CCl_4 solvent)[203, 204].

Phenol	$\Delta\nu_{OH}$ (a)[a] (cm^{-1})	$\Delta\nu_{OH}$ (b)[a] (cm^{-1})	$-\Delta H$ (kcal/mole)
6-Methyl-2-(β-methylallyl)phenol		99·2	0·98
2-(β-Methylallyl)phenol		91·8	1·04
2-Isopropenylphenol		82·0	0·76
2-Isobutenylphenol	37·0	70·5	
2-Allylphenol		65·6	0·46
2-(*cis*-Propenyl)phenol	24·2	63·6	
2-(*trans*-Propenyl)phenol		59·7	−0·60
2-Benzylphenol		50·8	0·33
2-Phenylphenol		42·0	1·45

[a] For explanation see text.

magnitude as those for the 2-halophenols (Table 10) although the enthalpies (Table 11) show that the OH $\cdots \pi$ interactions are somewhat weaker. That an interaction occurs for 2-phenylphenol implies that the two benzene rings are not coplanar[195, 202]. For 2-isobutenylphenol and 2-(*cis*-propenyl)phenol three distinct OH stretching absorptions have been observed[203]. Because of steric requirements in these compounds the coplanarity of the 2-substituents and the benzene ring is not possible. However, because a minimum resonance interaction occurs when the substituents are at 90° to the plane of the ring this orientation is also not energetically favoured. It follows that there are two stable conformers, one in which the angle of twist θ of the C=C bond is somewhere between 0° and 90° out of the plane of the benzene ring, and the other for which $90° < \theta < 180°$. The actual value of θ depends on the balance between the steric repulsions which increase as $\theta \rightarrow 0°$ or $\theta \rightarrow 180°$ and the stabilizing resonance interaction which decreases as $\theta \rightarrow 90°$. H-bonding interaction between OH and the π-electrons can occur in both conformers and gives rise to the two ν_{OH} frequencies (Table 12) due to H-bonded OH. For 2-(*trans*-propenyl)phenol the propenyl group can exist coplanar with the phenyl ring but orientated away from the OH group. A very weak OH $\cdots \pi$ interaction occurs, making the enthalpy of the H-bond appear to be smaller than the resonance energy involved.

Large shifts of the OH p.m.r. signal and infrared stretch frequency occur when an OH group intramolecularly H-bonds to the lone pair electrons on a carbonyl oxygen atom[67, 185, 194, 195, 202, 205]. The ν_{OH} absorption is moved about 350–500 cm^{-1} (Table 10) and becomes very broad because of the contributions of several resonance forms exemplified by **43**, **44** and **45**[98, 202]. A six-membered H-

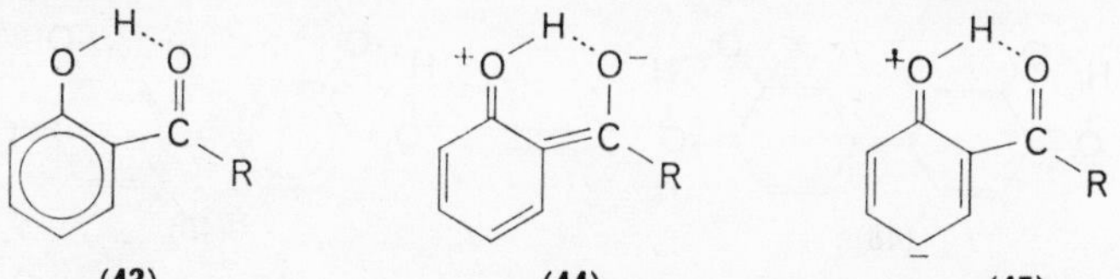

(43) **(44)** **(45)**

bonded ring is formed. The spectroscopic results suggest a strong H-bond and this is confirmed by the enthalpies of H-bond formation which were ($-\Delta H$ kcal/mole) 5·7, 6·9, 9·0 and 8·4 for the methyl esters of 3-hydroxy-2-naphthoic acid, salicylic acid, 2-hydroxy-1-naphthoic acid and 1-hydroxy-2-naphthoic acid respectively, and

6·8 for 2-hydroxy-1-naphthaldehyde[99]. The shifts in the carbonyl stretching frequency for these H-bonded phenols correlate both with the p.m.r. chemical shifts[185] and with the strengths ($-\Delta H$) of the H-bonds[99]. The strengths of the H-bonds are increased by substituents in the 3-position which force the carbonyl group nearer to the hydroxyl group thus increasing the interaction[206]. A different steric effect is observed for 2-hydroxy-4,6-di-*t*-butylbenzophenone (**46**) which shows two ν_{OH} bands at 3609 cm^{-1} and 3588 cm^{-1} [98]. The latter is attributed to a weak interaction **47** between OH and the

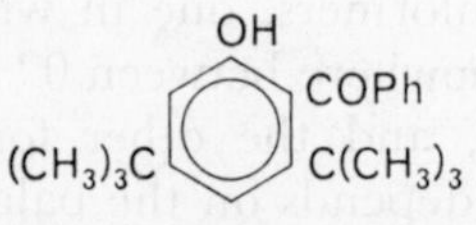

OH, COPh, $(CH_3)_3C$, $C(CH_3)_3$ **(46)** $(CH_3)_3C$, $C(CH_3)_3$ **(47)**

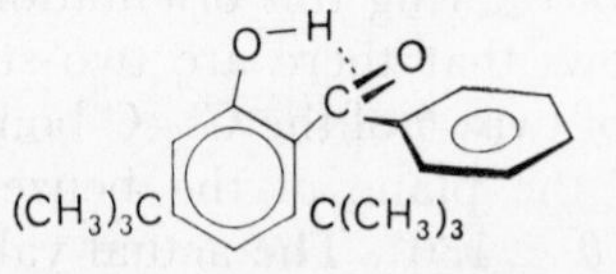

π-electrons of the carbonyl bond which is twisted out of the plane of the phenol ring. The twisting occurs because of the steric repulsion between a *t*-butyl group and the second benzene ring. A similar OH $\cdots\pi$ (C=O) interaction has been proposed for aliphatic systems[186].

The electronic spectra of 2-hydroxyacetophenone and of 2-hydroxybenzaldehyde are nearly identical in ether and in cyclohexane[157]. The intermolecular H-bond in these compounds is sufficiently strong to prevent intermolecular H-bonding to solvent ether. However, the tautomeric equilibria of some dihydroxydiphenoquinones is influenced by solvent. Thus 3,3′-dihydroxy-4,4′-diphenoquinone (**48**) (**49**) exists predominantly as its diphenoquinone form (**48**) in dioxan and as its 2-benzoquinone form (**49**) in

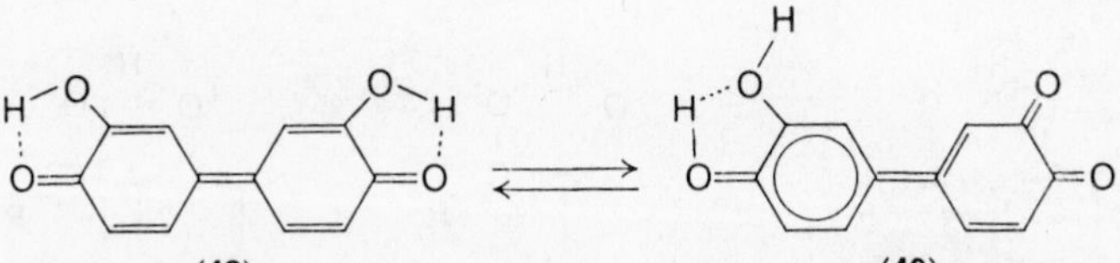

(48) **(49)**

methanol[207]. In the H-bond donor solvent intermolecular interactions become more significant than intramolecular H-bonding.

The high first ionization constant of salicylic and related acids may be attributed to the strong intramolecular H-bond between the *ortho* phenol and carboxylate groups in the acid anions[208-212]. The second ionization constant of substituted salicylic acids is unusually small[213]. A kinetic study[214] has shown that the rate of abstraction of the salicylate OH proton by hydroxide ions is slow since the intramolecular H-bond must be broken if the reaction is to take place.

A strong intramolecular H-bond **50** from OH to a nitro group exists in 2-nitrophenols[216]. The canonical formulation **51** requires

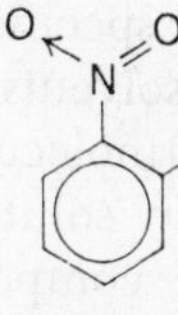

(50) 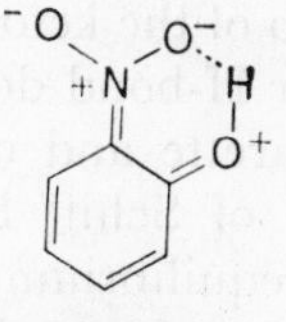 **(51)**

that the nitro group should be coplanar with the benzene ring; a requirement for the formation of a strong bond[113]. Small 3- or 6-substituents strengthen the H-bond interaction probably because the nitro or hydroxyl groups are pushed slightly closer together[202]. Thus 3- and 6-methyl-2-nitrophenol have stronger H-bond interaction than 2-nitrophenol[215]. For 3,6-dimethyl-2-nitrophenol the effect is further magnified. However, more bulky 3-substituents [e.g., Cl, C(CH)$_3$, CF_3] twist the nitro group out of the plane of the aromatic ring and weaken the H-bond[215, 217, 218]. Similar effects have been observed for the nitro coumarins[215]. Changes in solvent can also alter the extent and strength of intramolecular H-bonds. Thus 3-trifluoromethyl-2-nitrophenol has a weak intramolecular H-bond interaction in cyclohexane solvent but no such interaction in ether, methanol or water[218]. Intermolecular solute–solvent H-bonding becomes predominant in these three solvents. For 3-hydroxy-2-nitrophenol which has two intramolecular interactions even ethanol solvent fails to disrupt both H-bonds completely. The ability of a polar solvent to disrupt an intramolecular H-bond depends not only on the overall strength of the H-bond but also on the individual acidity of the donor and the basicity of the acceptor in the H-bond interaction[219]. Thus, comparing the effect of an H-bond accepting solvent (dioxan, acetone) on two intramolecular H-bonds of equal strength the H-bond which owes its strength to the high acidity of the H-bond donor rather than to the basicity of the acceptor will be ruptured more easily.

Intramolecular H-bonding in Schiff bases has been investigated by p.m.r.[220–223], infrared[224] and electronic[225, 226] spectroscopy. Schiff bases undergo a keto (**52**)–enol (**53**) equilibrium the position of which is influenced by solvent[220-222]. Specific H-bonding interactions of the solvent are more important than variations in dielectric constant. Thus in chloroform (dielectric constant $\varepsilon = 4{\cdot}81\ \mu$) there is a greater proportion of **52** (R = C_6H_5) at equilibrium than there is in acetonitrile ($\varepsilon = 37{\cdot}5\mu$). This is unexpected but explicable in terms of a specific solute–solvent interaction between $CHCl_3$ and the carbonyl group of the keto form **52** [222]. Other specific solute–solvent interactions for H-bond donor and acceptor solvents have been discussed by Charette and co-workers[225]. Intermolecular H-bonding (dimerization) of Schiff bases also occurs in solution to a small extent[225]. At equilibrium there is therefore competition between intramolecular, solute–solute intermolecular, and solute–solvent intermolecular H-bonding. H-bonding in aromatic azo compounds

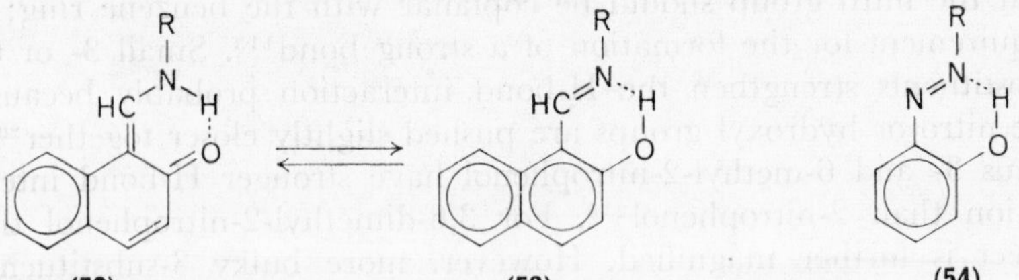

(52) (53) (54)

54 is analogous to that in Schiff bases. The intramolecular H-bond in substituted α-benzeneazo-β-naphthol compounds is stronger than that in the corresponding 2-benzeneazophenol compounds[227]. Intramolecular H-bonding from OH to the π-electrons of a –CH=N– double bond occurs in benzylidine-2-aminophenol (**55**) and salicylidene-2-aminophenol (**56**) [228]. The infrared ν_{OH} stretching frequency for the latter may be compared with those for salycilidene-2-

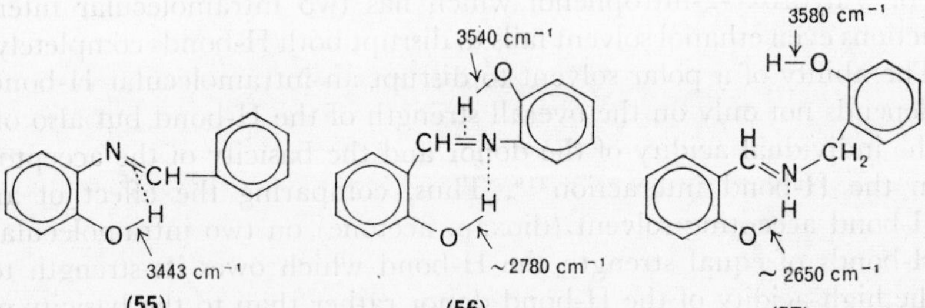

3443 cm⁻¹ (55) 3540 cm⁻¹ ~2780 cm⁻¹ (56) 3580 cm⁻¹ ~2650 cm⁻¹ (57)

hydroxybenzylamine (**57**). The presence of an OH · · π (CH=N) interaction in **56** results in the OH · · · N (lone pair) interaction being weaker in **56** than in **57**. In general for intramolecular H-bond interactions OH · · · N bonds are stronger than OH · · · O bonds[228, 229].

IV. H-BONDING AND KINETICS

The influence of solvent on the kinetics of reactions in solution embraces many different effects. Specific H-bonding solute–solvent interactions play an important role in determining the reaction rates. Thus there are significant differences between the variations with increasing acid concentration of the rates of aromatic hydrogen exchange for 1,3,5-trihydroxybenzene and 1,3,5-trimethoxybenzene. For the HCl catalysed reactions plots of $\log_{10} k_{exp}$ against the Hammett acidity function[230–232] had slopes of 0·80 for 1,3,5-trihydroxybenzene, and 1·14 for 1,3,5-trimethoxybenzene[233]. These results are in part due to specific H-bonding interactions between solvent water and the phenolic OH groups in the reactant 1,3,5-trihydroxybenzene and 1-methoxy-3,5-dihydroxybenzene[234]. The correlation between acidity function dependencies and the solvation of reacting molecules and of transition states has been discussed for both acid-catalysed[232, 235] and base-catalysed reactions[232, 236].

The second-order rate constants for the alcoholysis of acetic anhydride in carbon tetrachloride or cyclohexane decrease as the concentration of the alcohol is increased[237]. Similar decreases occur if typical H-bond acceptors such as dioxan or acetone are added. Hydrogen bonding of the reacting alcohols either with themselves or with other molecules results in the need to break the H-bonds before the transition state in the reaction can be formed. The activation energy for the reaction is increased by an amount corresponding to the energy required to break the H-bonds. The relative rates of ethanolysis in benzene and cyclohexane are 0·264 : 1 which is consistent with an estimate that a fraction 0·24 of ethanol in benzene is monomeric whereas the other 76% is H-bonded in an ROH · · · $\pi(C_6H_6)$ interaction[237].

The activation parameters for the inversion of the methyl ether of 1-hydroxy-5,7-dihydrodibenz[*c*,*e*]oxepin are insensitive to change of solvent from $CDCl_3$ to dimethyl sulphoxide[238].

The inversion may be represented, looking end on along the two benzene rings, as shown at the top of the next page.

For the parent phenol the activation energy is about 1 kcal/mole less in $(CH_3)_2SO$ than in $CDCl_3$. In the latter solvent the phenol is stabilized with respect to the transition state by an intramolecular

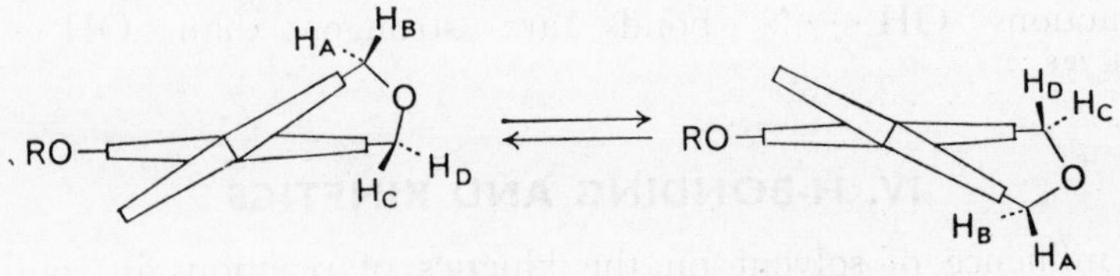

$OH \cdots \pi$ interaction **58** which leads to a non-planar configuration for the molecule. In dimethyl sulphoxide a strong intermolecular H-bond is formed which stabilizes, through canonical structures such as **59**, the planar configuration which is required for the

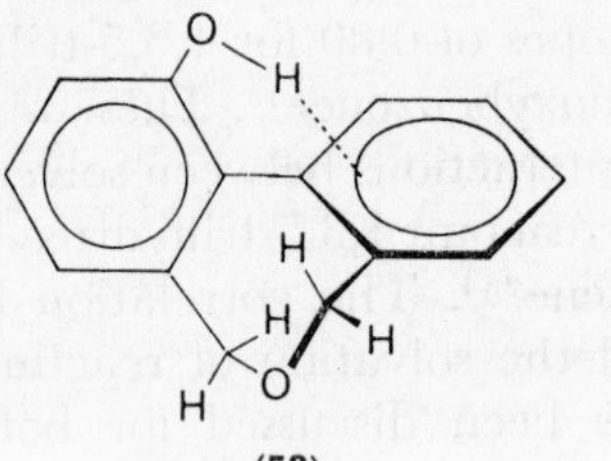

(58)

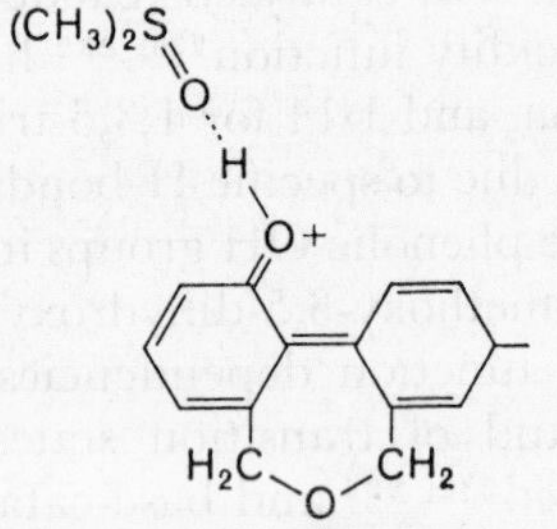

(59)

activated complex in the inversion. The $OH \cdots \pi$ intramolecular H-bond does not exist in $(CH_3)_2SO$. The stabilization of the transition state in $(CH_3)_2SO$ through intermolecular H-bonding and the ground state in $CDCl_3$ through intramolecular H-bonding may both contribute to the observed difference between the activation energies in the two solvents.

The influence of solute–solvent interactions on the rates of acid–base reactions of some phenols and their anions in methanol has been studied[239]. The rate constants and deuterium isotope effects for the ionization of some phenolic azobenzene derivatives are influenced by the intramolecular $OH \cdots NH$-bonds in the reacting molecules[229]. The removal of the second phenolic proton from the salicylate anion is retarded by intramolecular H-bonding with the neighbouring carboxylate group[214]. The relative rates of removal of tritium from OH-labelled 2- and 4-nitrophenol by methyl radicals show that for this reaction tritium participating in an intramolecular H-bond is more reactive than tritium in an intermolecular H-bond[240]. The rates of intermolecular proton transfer reactions are in accord with the order of H-bond strengths[241]. For example, in the absence of charge or steric effects, proton transfers from OH to O are faster than those from OH to N.

V. THE ACIDITY OF HYDROXYL GROUPS

A. Alcohols

The most reliable measurements of the acidity of alcohols in water were made by Long and Ballinger[242–244] using a conductivity method (Table 13). Data for several carbinols[259] and fluorinated alcohols[246, 247, 258] are also available. The high acidity of perfluoropinacol has been attributed to OH · · · O⁻ intramolecular H-bonding in the anion formed when an OH proton is removed from the

TABLE 13. The acidity of some alcohols in water at 25°C[a].

Alcohol	pK_a	Reference
Water	15·74	245
Methanol	15·5	243
Ethanol	15·9	243
2,2,2-Trichloroethano	12·24	243
2,2,2-Trifluoroethanol	12·3; 12·37; 12·42	246, 242, 247
2,2,3,3-Tetrafluoropropan-1-ol	12·74	243
2,2-Dichloroethanol	12·89	243
Allyl alcohol	15·5	243
Prop-2-yn-1-ol	13·55	243
2-Methoxyethanol	14·8	243
2-Chloroethanol	14·31	252
1,1,1-Trifluoro-3-aminopropan-2-ol	12·29	247
1,1,1-Trifluoro-3-diethylaminopropan-2-ol	12·56	247
Ethylene glycol	15·4[b]	243
Glycerol	14·4[b]	243
Pentaerythritol	14·1[b]	243
2,2,3,3-Tetrafluoro-1,4-butanediol	12·1[b]	246
	13·7[c]	246
2,2,3,3,4,4,5,5-Octafluoro-1,6-hexanediol	12·1[b]	246
	12·8[c]	246
Perfluoro-*t*-butanol	9·52	258
Perfluoropinacol	5·95[b]	258
Hexafluoro-2-propanol	9·30	258
1-Phenyl-2,2,2-trifluoromethylethanol	11·90	259
1-(4-Methoxyphenyl)-2,2,2-trifluoroethanol	12·24	259
1-(4-Methylphenyl)-2,2,2-trifluoroethanol	12·04	259
1-(3-Bromophenyl)-2,2,2-trifluoroethanol	11·50	259
1-(3-Nitrophenyl)-2,2,2-trifluoroethanol	11·23	259

[a] See also J. Murto, this volume, Chap. 20, Table 3.
[b] First dissociation; [c] second dissociation.

diol[258]. The pK_a values for the substituted methanols (RCH_2OH)[243] are a linear function of the Taft σ^* constants for the R substituents[248, 249]. Equation (15) is applicable, the observed pK_a values

$$pK_a = 15{\cdot}9 - 1{\cdot}42\sigma^* \tag{15}$$

being within $\pm 0{\cdot}2$ unit of the predicted values[244, 250]. The figure of 1·42 for ρ^* is close to that of 1·36 deduced[249] from measurement of the relative acidities of some alcohols in *i*-propanol solution[251]. Ionization constants of 2-chloroethanol[252] and 2,2,2-trifluoroethanol[242] in D_2O are given by $pK_a = 14{\cdot}99$ and $pK_a = 13{\cdot}02$ respectively. The magnitudes of the isotope effects are consistent with the corresponding magnitudes for certain phenols and carboxylic acids[243, 244]. The pK_a of several fluorinated alcohols in 50% aqueous ethanol have been reported[253]. However, potentiometric measurement[246, 247, 253] of such high pK_a values is probably not very accurate[244]. The acidity of some hydroxyalkylpyridines (and their conjugate acids) in water and in *i*-propanol are influenced by the electronic and steric effects of the substituent groups[254]. The intramolecular OH $\cdots$ π and OH $\cdots$ NH-bonds which exist in these compounds in CCl_4 are absent in polar hydroxylic solvents where solute–solvent intermolecular H-bonding becomes predominant.

The thermodynamics of ionization of several carbohydrates and their derivatives have been studied by thermometric titrimetry (Table 14)[255, 256, 257]. The ionization occurs from the 1-position in the monosaccharides. Changing from a pentose to a hexose or altering the stereochemistry of the hydroxyl groups only has a small effect on the observed thermodynamic quantities. Replacement of a hydroxyl group by a hydrogen atom (cf. ribose and 2-deoxyribose; glucose and 2-deoxyglucose) produces only small changes in pK_a, ΔH and ΔS. Both 2′ and 3′ hydroxy groups are necessary if adenosine (**60**) and its derivatives are to show their acidity. Thus, the substitution of CH_3 for H on the 2′ hydroxyl or the substitution of H for either the 2′ or the 3′ hydroxyl in adenosine leads to a considerable reduction in acidity for the molecule as a whole[255]. The enhanced

TABLE 14. Thermodynamics of ionization of some carbohydrates and their derivatives in water (25°)[255, 256, 257].

Compound	pK_a	$\Delta H°$ (kcal/mole)	$-\Delta S°$ (e.u.)
Fructose	12·27	8·2	28·6
Glucose	12·46	7·7	31·3
2-Deoxyglucose	12·52	8·2	29·7
Mannose	12·08	7·9	28·9
Galactose	12·48	9·0	26·9
Arabinose	12·54	8·3	29·6
Xylose	12·29	8·2	28·7
Ribose	12·22	8·1	28·7
2-Deoxyribose	12·67	7·7	32·1
Lyxose	12·11	8·0	28·6
Adenosine[a]	12·35	9·7	24·0
9-β-D-Xylofuranosyladenine[a]	12·34	8·4	28·3
Ribose 5-phosphate[a]	13·05	6·1	39·4
Glucose 6-phosphate[a]	11·71	8·4	25·0
Adenosine 5′-monophosphate[a]	13·06	10·9	23·3

[a] Ionization $HA^{2-} + H_2O \rightleftharpoons A^{3-} + H_3O^+$.

acidity given by vicinal OH-groups is independent of whether the OH groups are *cis* or *trans* because the thermodynamic quantities for adenosine and 9-β-D-xylofuranosyladenine (**61**) are very similar[257]. The small differences in ΔS for the two compounds probably arise because of different interactions with solvent caused by the change in stereochemistry.

Several measurements of the acidity of *gem* diols have been reported and are summarized in Table 15. The *gem* diols are carbonyl

TABLE 15. First ionization constants of some *gem* diols in water (25°).

gem Diol	pK_a	Ref.	*gem* Diol	pK_a	Ref.
$C_6H_5C(OH)_2CF_3$	10·00	259	$CF_3C(OH)_2CF_3$	6·58	258
$4\text{-}CH_3OC_6H_4C(OH)_2CF_3$	10·18	259	$CF_2ClC(OH)_2CF_2Cl$	6·67	258
$4\text{-}CH_3C_6H_4C(OH)_2CF_3$	10·15	259	$CF_2ClC(OH)_2CFCl_2$	6·48	258
$3\text{-}BrC_6H_4C(OH)_2CF_3$	9·51	259	$CFCl_2C(OH)_2CFCl_2$	6·42	258
$3\text{-}NO_2C_6H_4C(OH)_2CF_3$	9·18	259	$CF_2HC(OH)_2CF_2Cl$	7·90	258
$CH_2(OH)_2$	13·27	260	$CF_2HC(OH)_2CF_2H$	8·79	258
$CH_3CH(OH)_2$	13·57	260	$CF_3C(OH)_2CHBr_2$	7·69	258
$CCl_3CH(OH)_2$	10·04	260	$(CH_3)_2CHCH(OH)_2$	13·77	261

hydrates which are formed by the addition of water to an aldehyde or ketone according to the equilibrium

$$R_1R_2CO + H_2O \rightleftharpoons R_1R_2C(OH)_2$$

Equilibrium constants for the hydration equilibria have been summarized by Bell[262]. The anion of the *gem* diol can be assumed to be formed either by the loss of a proton from the diol or by the addition of an hydroxide ion to the unhydrated carbonyl compound. Aldehydes and ketones can also ionize to enolate anions in solution. However, for those *gem* diols for which pK_a has been determined the amount of enolate anion formed at equilibrium is negligibly small. Correlations between the pK_a values and Hammett (σ) or Taft (σ^*) constants have been noted[250, 259, 260, 262]. Evaluation of pK_a values from studies of the ionization of the enolic tautomers of ketones requires knowledge of the equilibrium constants for the keto–enol tautomerism of the neutral molecules in solution[263]. Thus for acetylacetone the ratio of enolate anion to neutral keto + enol species is in accord with $pK_a = 8{\cdot}9$. However, the equilibrium constant for the tautomerism is $[C(\text{enol})/C(\text{diketo})] = 0{\cdot}25$ and therefore 8·2 is the true pK_a of the enol form of the diketone. Several combined studies of keto–enol equilibria and acidity have been made[263–268]. The enols of some substituted cyclohexane-1,3-diones (e.g., dimedone $pK_a = 5{\cdot}23$ for the enol form)[264] have $pK_a \sim 5$ in water[263]. The strongest acid of this type appears to be 2,4-dimethylcyclobutane-1,3-dione with $pK_a = 2{\cdot}8$. In general in 1,3-diketones or β-ketoesters enolization leads to a conjugated system and ionization leads to resonance stability and delocalization of charge in the ion. These compounds are more acidic than monoketones (e.g., acetone[269]) in which these effects cannot occur. The enols of ring 1,3-diketones or 2-acetyl or 2-formyl cycloalkan-1-ones are more acidic than analogous non-cyclic compounds[264]. The effect of substituent R on the acidity of the enols $CH_3COCR{=}C(OH)CH_3$ of acetylacetones correlates with the Taft σ^* constants for R[250].

The ability to undergo self-ionization is an important property of hydroxylic solvents. The ionic products (Table 16) of four alcohols have been determined by e.m.f. measurements[270, 271]. Values for methanol from 0° to 45° have also been determined and give $\Delta H = 11{\cdot}0$ kcal/mole for the self-ionization at 25°C [272]. Conductivity results are probably less reliable[273].

TABLE 16. The ionic product of alcohols at 25°C.

$$K_s = \frac{a_{RO^-} \cdot a_{ROH_2^+}}{a^2_{ROH}} \quad (a_{ROH} = 1 \text{ in pure liquid})$$

Alcohol	$pK_s{}^a$	$pK_s{}^b$	Ref.
Methanol	16·707	16·916	270
Ethanol	18·67	18·88	271
n-Propanol	19·24	19·43	271
i-Propanol	20·58	20·80	271

[a] K_s in units of (molality)^2.
[b] K in units of (molarity)^2.
($pK_s{}^a = pK_s{}^b + 2 \log_{10} d$, where d is the density of the particular alcohol at 25°.)

B. Phenols

I. Ionization of phenols in water

a. General correlations. There have been many studies of the acidity of phenols in aqueous solution[64, 65, 274–322]. These include not only measurement of pK_a (compilations[250, 301, 321, 322]) but also of enthalpies, entropies, heat capacities (Table 17) and volumes[320] of ionization. Results for many substituted naphthols are also available[250, 321–323]. Biggs and Robinson[293] showed that equation (16)

$$pK_a = 9{\cdot}919 - 2{\cdot}229\sigma \tag{16}$$

was applicable for the ionization of fourteen 3- and 4-substituted phenols. The substituent σ-constants in the Hammett equation[52] were deduced by taking $\rho = 1$ for the corresponding benzoic acids. 4-Substituents which can undergo mesomeric interaction with the aromatic ring may not fit the correlation[324]. For 3-substituted phenols the correlation between pK_a and σ is excellent at any temperature in the range 10°C $\leqslant$ to $\leqslant$ 55°C [277]. Substituted 2-nitrophenols[302, 303] and 2,4-dinitrophenols[301] and 2,4,6-trinitrophenols[301] also exhibit linear relationships between pK_a and σ-constants although substituents which twist nitro groups out of coplanarity with the ring[325] give anomalous results. In general equation (17) is applicable and has been tested for a large number of sterically unhindered phenols[301]. Here $\sum \sigma$ is the sum of the σ-constants for the

$$pK_a = 9{\cdot}94 - 2{\cdot}26 \sum \sigma \tag{17}$$

TABLE 17. Thermodynamic quantities for the ionization of some phenols in water at 25°.

Phenol	pK_a	$\Delta G°$ (kcal/mole)	ΔH (kcal/mole)	$-\Delta S°$ (e.u.)	$-\Delta C_p$ (cal/deg)	References
Phenol	9·99	13·63	5·55	27·1	53 (32)	64, 65, 274–6, 285
2-Cresol	10·33	14·10	5·73	28·1	36·0	64, 65
3-Cresol	10·10	13·78	5·52	27·7	38·6	64, 65, 274
4-Cresol	10·28	14·02	5·50	28·6	52·9	64, 65
2,3-Xylenol	10·54	14·39	5·70	29·1	38·0	64, 65
2,4-Xylenol	10·60	14·46	5·76	29·2	28·0	64, 65
2,5-Xylenol	10·40	14·19	5·58	28·9	24·2	64, 65
2,6-Xylenol	10·62	14·49	5·46	30·3	36·7	64, 65
3,4-Xylenol	10·36	14·13	5·37	29·4	31·8	64, 65
3,5-Xylenol	10·20	13·92	5·34	28·8	27·2	64, 65, 274
2-Fluorophenol	8·73	11·91	4·66	24·3		286
3-Fluorophenol	9·29	12·67	5·52	24·0		286
4-Fluorophenol	9·89	13·49	5·93	25·4		286
2-Chlorophenol	8·555	11·67	4·26	24·9	89 (52)	64, 65, 276, 283
3-Chlorophenol	9·119	12·44	5·35	23·8	23·5	278, 279
4-Chlorophenol	9·42	12·86	5·73	23·9	34·7	64, 276
2-Bromophenol	8·452	11·53	4·40	23·9	39	283
3-Bromophenol	9·031	12·32	5·35	23·4	35·7	278
4-Bromophenol	9·34	12·74	5·74	23·5	25·5	64
2-Iodophenol	8·513	11·61	4·27	24·6	49	283
3-Iodophenol	9·033	12·31	5·53	22·7	40·6	278
4-Iodophenol	9·33	12·72	5·38	24·6	43·8	64
3-Cyanophenol	8·56	11·69	5·20	21·8		282
4-Cyanophenol	7·95	10·87	4·92	20·0		282
2-Methoxyphenol	9·99	13·63	5·74	26·5		280

3-Methoxyphenol	9·652	13·17	5·02	27·3	35·1	277, 280
4-Methoxyphenol	10·20	13·92	5·70	27·6		280
2-Ethoxyphenol	10·109	13·79	6·18	25·5	23	283
3-Ethoxyphenol	9·655	13·17	5·08	27·1	56·5	274
3-Ethylphenol	10·069	13·74	5·29	28·3	47·5	277
2-Propylphenol	10·50	14·33	6·00	28·0		288
2-Allylphenol	10·28	14·02	6·01	26·9		288
2-Formylphenol	8·38	11·42	5·15	21·0		281
3-Formylphenol	8·983	12·26	4·99	24·4	33·3 (11·4)	277, 281, 287
4-Formylphenol	7·62	10·39	4·26	20·6	55·6	281, 287
2-Nitrophenol	7·230	9·88	4·60	17·7	38·1	64, 276, 290
3-Nitrophenol	8·360	11·41	4·79	22·2	20·9	276, 277, 291
4-Nitrophenol	7·156	9·78	4·72	16·8	34·6	64, 276, 284, 290
3,5-Dichlorophenol	8·179	11·16	4·88	21·1	24·5	274
3,5-Dibromophenol	8·056	10·99	5·25	19·3	79·5	274
3,5-Diiodophenol	8·103	11·06	5·45	18·8	79·4	274
3,5-Dinitrophenol	6·732	9·19	3·76	18·2	16·8	274
3,5-Diethoxyphenol	9·370	12·79	4·65	27·3	46·7	274
2,4,6-Trimethylphenol	10·89	14·85	5·44	31·6		282
3,4,5-Trimethylphenol	10·24	13·98	5·68	27·9		282
2,4,5-Trimethylphenol	10·57	14·41	6·40	26·9		282
2,3,5-Trimethylphenol	10·59	14·45	6·60	28·7		282
2-Hydroxymethylphenol	9·84	13·6	5·4	27·4		292
4-Hydroxymethylphenol	9·73	13·4	4·5	30·0		292
2,4-Dihydroxymethylphenol	9·69	13·4	4·9	28·4		292
2,4,6-Trihydroxymethylphenol	9·45	13·1	4·5	28·6		292
3-Methyl-4-nitrophenol	7·409	10·12	4·58	18·7	34·3	64, 290
2-Methoxy-4-formylphenol	7·396	10·09	3·75	21·3		281, 289
2-Methoxy-5-formylphenol	8·889	12·13	4·62	25·2		281, 289
2-Methoxy-6-formylphenol	7·912	10·79	4·13	22·3		281, 289

phenol substituents. These correlations have been recently reviewed[250]. The effect of substituents on the acidity of 1-naphthol and 2-naphthol has been correlated with σ-constants[250].

Intramolecular H-bonding accounts for the high first ionization constant[208-212] and the low second ionization constant[213, 214] of salicylic acid. The intramolecular $OH \cdots \pi$ interaction in 2-allylphenol has been invoked to explain the observed entropy of ionization of this phenol in water[288]. Comparison of entropies of ionization of nitrophenols suggests there is negligible intramolecular H-bonding for 2-nitrophenol in aqueous solution[299, 330].

Linear correlations between free energies, enthalpies, and entropies of ionization (Table 17) have been analysed in detail for several classes of phenols[275, 277]. In general 2- and 4-substituted phenols show excellent linear correlations but the plots for 3-substituted phenols are not so good. The entropies of ionization of 4-substituted phenols are a linear function of the σ^n substituent constants[275, 324]. Providing there are no appreciable steric effects the free energies of ionization (or pK_a) of phenols are additive, a particular substituent in a particular ring position always producing a similar increment to $\Delta G°$ or pK_a [65, 274]. For 3,5-disubstituted phenols the entropies of ionization are also additive[274]. Equation (18) is applicable. This

$$\Delta S_{3,5} = 1{\cdot}91\ (\pm 0{\cdot}08)\ \Delta S_3 - 24{\cdot}5\ (\pm 0{\cdot}4) \tag{18}$$

result is demonstrated in Figure 7 (data are in Table 17).

The volume of ionization of phenol in water is $-17{\cdot}0$ cm^3/mole[320].

The acidity of many phenols in D_2O has been measured[330-333]. Deuteration results in increases in pK_a in the range $0{\cdot}44 \leqslant \Delta pK_a \leqslant 0{\cdot}62$. ΔpK_a is a linear function of $pK_a(H_2O)$ with positive slope 0·018[332, 333]. The following equation is applicable[333]. A few results

$$pK_a(D_2O) - pK_a(H_2O) = \Delta pK_a = 0{\cdot}41 + 0{\cdot}018 pK_a(H_2O) \tag{19}$$

are included in Table 18.

b. Hepler's theory. Hepler and co-workers[279, 280, 282, 326, 327] have considered the effect of a substituent on the acidity of phenol in terms of equilibrium (20) in which HA_s is a substituted phenol and

$$HA_s(aq) + A_u^-(aq) \rightleftharpoons A_s^-(aq) + HA_u(aq) \tag{20}$$

HA_u is phenol itself. The increment $\delta\Delta G°$ in the free energy of ionization in going from HA_u to HA_S ($\delta\Delta G° = \Delta G_s° - \Delta G_u°$) is equal to the free energy change for reaction (20). The corresponding

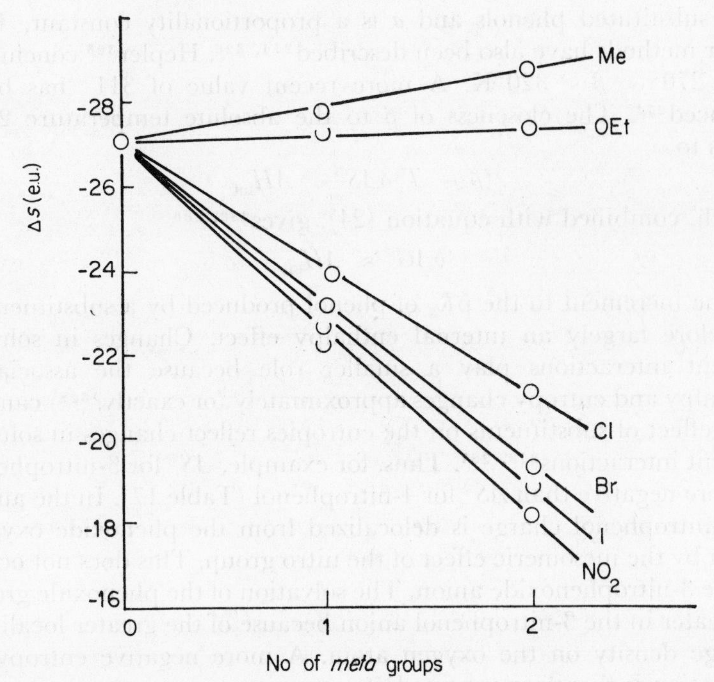

FIGURE 7. Effect of 3-substituents on the entropy of ionization of phenol in water[274]. Reproduced by permission from P. D. Bolton, F. M. Hall and J. Kudrynski, *Australian J. Chem.*, **21**, 1541 (1968).

increments in enthalpy and entropy change may each be written in terms of internal and external contributions as follows

$$\delta \Delta H^\circ = \Delta H_{\text{int}} + \Delta H_{\text{ext}} \quad \text{and} \quad \delta \Delta S^\circ = \Delta S_{\text{int}} + \Delta S_{\text{ext}} \tag{21}$$

Taking $\Delta S_{\text{int}} \sim 0$ [326, 327] gives $\delta \Delta S^\circ = \Delta S_{\text{ext}}$. Also equation (22) is

$$\delta \Delta H^\circ = \Delta H_{\text{int}} + \beta \Delta S_{\text{ext}} \tag{22}$$

applicable providing equation (23) is assumed valid[326, 327]. Combination of these equations leads to equation (24) for the increment in ΔG° produced by a substituent.

$$\Delta H_{\text{ext}} = \beta \Delta S_{\text{ext}} = \beta \delta \Delta S^\circ \tag{23}$$

$$\delta \Delta G^\circ = \Delta H_{\text{int}} + (\beta - T)\delta \Delta S^\circ \tag{24}$$

Estimation of β is difficult. A method dependent on the assumption in equation (25) was originally suggested[327].

$$\Delta H_{\text{int}} = a(\nu_s{}^2 - \nu_u{}^2) \tag{25}$$

Here ν_u and ν_s are the OH stretching frequencies of the unsubstituted and substituted phenols and a is a proportionality constant. Two other methods have also been described[274, 326]. Hepler[326] concluded that $270° < \beta < 320°K$. A more recent value of 311° has been deduced[274]. The closeness of β to the absolute temperature 298° leads to

$$(\beta - T)\delta\Delta S° \ll \Delta H_{\text{int}} \tag{26}$$

which, combined with equation (24), gives[282, 286]

$$\delta\Delta G° \approx \Delta H_{\text{int}} \tag{27}$$

The increment to the pK_a of phenol produced by a substituent is therefore largely an internal enthalpy effect. Changes in solute–solvent interactions play a smaller role because the associated enthalpy and entropy changes approximately (or exactly?[328]) cancel. The effect of substituents on the entropies reflect changes in solute–solvent interactions[276, 287]. Thus, for example, $\Delta S°$ for 3-nitrophenol is more negative than $\Delta S°$ for 4-nitrophenol (Table 17). In the anion of 4-nitrophenol charge is delocalized from the phenoxide oxygen atom by the mesomeric effect of the nitro group. This does not occur in the 3-nitrophenoxide anion. The solvation of the phenoxide group is greater in the 3-nitrophenol anion because of the greater localized charge density on the oxygen atom. A more negative entropy of ionization is therefore expected[276].

c. Steric effects. Bulky substituents in the 2-position of phenols lead to decreases in acidity because of steric effects[311]. Thus, for example, pK_a = 10·23 for 4-*t*-butylphenol[300, 312] and pK_a = 11·34 for 2-*t*-butylphenol[300]. Equation (17) is not obeyed[283]. Linear Hammett plots[52] are obtained for series of substituted phenols all with the same 2- or 2,6-substituent groups. However, the slopes ρ differ from that (2·26) for the unhindered phenols. Thus $\rho = 2{\cdot}610$ for 4-substituted 2,6-dichlorophenols[315], $\rho = 2{\cdot}700$ for 4-substituted 2,6-dimethylphenols[314] and $\rho = 3{\cdot}50$ for 4-substituted 2,6-di-*t*-butylphenols[312]. The increment in pK_a produced by a given 4-substituent is increased when bulky substituents are present in the 2,6-positions. This has been attributed to steric inhibition of solvation of the phenol anions[104, 312, 314, 315]. The anomalous acidities of 4-nitroso-2,6-dimethylphenol[314] and 4-nitroso-2,6-di-*t*-butylphenol[312] have been ascribed to the predominant existence of the quinone monoxime tautomers of these molecules in solution.

The additive nature of pK_a [65, 274] is often absent when steric factors are operating[300, 313, 329]. For the cresols and xylenols the pK_a increments are additive and so steric effects are not obvious from consideration of $\Delta G°$ alone[65]. However, comparison of the pK_a values for these phenols with the corresponding values for the picolines and lutidines suggests that one 2-methyl substituent in a phenol introduces an increment to pK_a of +0·14 units through steric hindrance to solvation of the phenol anions[319].

A further steric effect arises when substituents which can undergo mesomeric interaction with the benzene nucleus are twisted out of the plane of the ring[325]. Thus the increment in pK_a on introducing 3,5-dimethyl groups into 4-nitrophenol is 1·06 units[309] compared with +0·21 units[65] for the same substitution in phenol itself.

TABLE 18. Effect of deuterium substitution and electronic excitation on the acidity of phenols[161, 333]. Values for the excited states were deduced via the Förster cycle.

Phenol	$pK_a(H_2O)$	$pK_a^*(H_2O)$	$pK_a(D_2O)$	$pK_a^*(D_2O)$
4-Methoxyphenol	10·20	5·7	10·85	6·2
Phenol	9·99	4·1	10·62	4·6
3-Methoxyphenol	9·65	4·6	10·20	5·1
2-Naphthol	9·46	2·5, 3·0	10·06	
4-Bromophenol	9·34	2·9	9·94	3·4
1-Naphthol	9·23	2·0		
2-Naphthol-5-sulphonate	9·18	0·53		
2-Naphthol-6-sulphonate	9·10	1·65		
4-Phenolsulphonate	9·03	2·3	9·52	2·7
4-Hydroxyphenyl trimethylammonium chloride	8·34	1·6	8·90	2·0
3-Hydroxypyrene-5,8,10-trisulphonate	7·30	1·0		

d. Acidity of phenols in excited electronic states. Phenols in their electronically excited states are more acidic than the ground state molecules. The pK_a of an excited state may be deduced from the absorption spectrum of the ground state and the fluorescence spectrum of the excited state using equation (28) in which pK_a refers to the ground state and $\Delta\bar{\nu}_h$ (cm^{-1}) is the arithmetic mean of the spectral shifts in absorption and fluorescence in going from the neutral acid to its anion.

$$pK_a^* = pK_a - \frac{0{\cdot}625}{T}\Delta\bar{\nu}_h \qquad (28)$$

The determination of pK_a of excited states has been discussed in detail by Weller[161].

Results for the first excited singlet state of some phenols in water and in D_2O are given in Table 18. The deuterium isotope effect

ΔpK_a^* for the excited molecules is consistent with equation (19) and therefore the isotope effects for the ground and excited states conform to the same equation[333]. The values of pK_a^* deduced from the Forster cycle are in reasonable agreement with values evaluated from kinetic measurements[161]. For a series of 4-substituted phenols ($pK_a - pK_a^*$) correlates with the substituent σ-constants and is greater for more electron-withdrawing groups[334]. A value $pK_a(293°) = 8{\cdot}1$ has been reported for the triplet state of β-naphthol in water[335]. pK_a^* values for thirteen phenols in their lowest singlet excited states have been measured by Avigal, Feitelson and Ottolenghi[375].

2. Non-aqueous and mixed aqueous solvents

The effect of solvent on acid–base equilibria has been discussed in detail by Bell[336]. He concluded that 'the relative strengths of acids of the same charge and chemical type are independent of the solvent'. This conclusion is justified for the ionization of phenols[337–340]. For phenol itself at 25°C $pK_a = 14{\cdot}46$ in methanol[340] and 15·58 in ethanol[341, 271] (both with K_a in molarity units). Ionization constants for the cresols and xylenols in methanol have been measured[339]. An increment of +0·33 units due to steric effects has been estimated to contribute to the pK_a of 2-methyl phenols[342]. For pentamethylphenol $pK_a(25°) = 16{\cdot}35$ in methanol[343]. The acidity of several nitrophenols in methanol[337, 338, 344], ethanol[345], acetone[346], acetonitrile[347] and alcohol–water mixtures[351–354] has been determined. Among the factors affecting relative acidities of a given phenol in different solvents[336] must be included comparison of solvent–solute interactions and particularly solvation effects around the phenol anions[337]. Hammett $\rho\sigma$ correlations have been tested for 4-substituted phenols, 2,6-dimethylphenols and 2,6-dichlorophenols in hydroxylic and aprotic solvents[316]. For each solvent ρ is higher for 2,6-dimethyl or 2,6-dichloro phenols than for unhindered phenols. This is attributed to steric inhibition to solvation in all solvents.

The effect of steric hindrance on acidity is particularly marked when 2,6-di-*t*-butyl groups are introduced into phenol, 4-cresol or 4-*t*-butylphenol[310, 312, 329]. For methanol solvent the change in pK_a is about +2·7 units[329]. However, 2,6-*t*-butyl-4-nitrophenol is a *stronger* acid than 4-nitrophenol[312, 340, 348]. In general, for a series of 4-substituted 2,6-di-*t*-butylphenols in methanol a plot of pK_a against σ gave $\rho = 4{\cdot}76$ compared with $\rho = 2{\cdot}24$ for the corresponding 4-substituted phenols[349]. Similar results are obtained for aqueous and 50–50 volume % ethanol–water solutions[312]. Rochester and Rossall have measured the free energies[340], enthalpies, entropies[349] and volumes[350] of ionization of 4-*t*-butylphenol, phenol, 4-bromophenol, 4-formylphenol, 4-nitrophenol and their 2,6-di-*t*-butyl analogues in methanol. The effects of 2,6-di-*t*-butyl groups on the acidity of phenols are largely caused by the influence of steric factors on solute–solvent interactions particularly for the phenol anions.

C. Heteroaromatic Hydroxyl Compounds

The study of the ionization of hydroxy substituted heteroaromatic molecules is often complicated by the possibility of several tautomeric structures for the parent molecule. In fact, measurement of the acidity of these compounds provides information about the extent of the tautomeric equilibria[356, 357]. The method is exemplified by the work of Mason[355]. Thus for 2-hydroxypyridine (**62**) a tautomer with either zwitterionic (**63**) or amide (**64**) character exists in equilibrium with **62**. Mason deduced that $pK_a(20°) = 8{\cdot}66$ for the ionization of the OH proton in **62** and $pK_a = 0{\cdot}75$ for the corresponding OH dissociation of the conjugate acid **65** of **62**. Care must

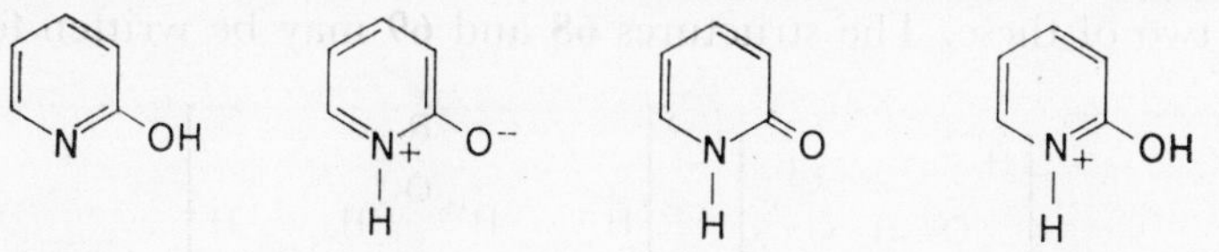

(**62**) (**63**) (**64**) (**65**)

be taken to distinguish between the loss of NH and OH protons in these molecules. This can only be done if a complete knowledge of the equilibrium concentrations of all the possible tautomers of each acid and base species is obtained. Infrared, ultraviolet and p.m.r. spectroscopy are often useful in this respect[370].

The ionization constants of the hydroxyl groups in eighteen heterocyclic hydroxyl compounds have been correlated with the π-electron energies of the species present in the equilibria[355]. The effect of substituents on the acidity of heterocyclic hydroxyl compounds is consistent with the Hammett $\rho\sigma$ relationship[250].

The true OH dissociation constants of the ground states of 3-hydroxyquinoline (**66**) and its conjugate acid **67** are given by $pK_a(20°) = 8{\cdot}03$ and $5{\cdot}52$ respectively[355]. In their lowest electronically excited states these acidities are increased to $pK_a(18°) = 3{\cdot}6$

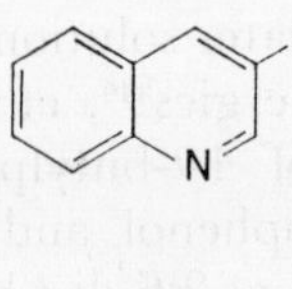

(66) 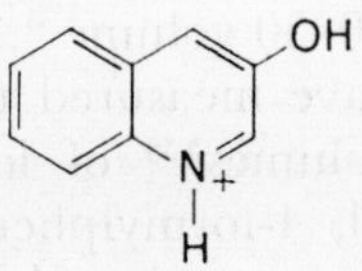 (67)

and $-0{\cdot}3$ respectively[358]. This effect is similar to that observed for phenols[161, 333, 334].

VI. THE PROTONATION OF HYDROXYL GROUPS

Aliphatic alcohols are protonated in concentrated acid solutions[359, 360]. The basicities of eight alcohols have been tabulated by Arnett[91]. They range from $pK_a(CH_3OH_2^+) = -2{\cdot}2$ to $pK_a = -7{\cdot}0$ for the conjugate acid of β-phenyl-β-hydroxypropionic acid. For the conjugate acids of a series of glycols $4{\cdot}4 \leqslant pK_a \leqslant 1{\cdot}5$ [361] and for phenol $pK_a = -6{\cdot}74$ at 0°C [362]. The protonation of three alcohols in acetonitrile has been studied by Kolthoff and Chantooni[363]. Methyl alcohol and *n*-butyl alcohol form the species $R\overset{+}{O}H_2$, $(ROH)_2H^+$ and $(ROH)_3H^+$ whereas *t*-butyl alcohol only forms the first two of these. The structures **68** and **69** may be written for the

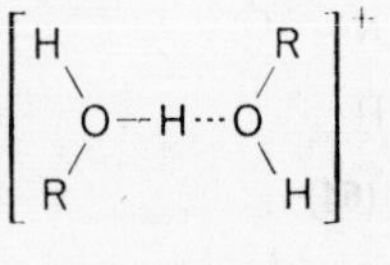

(68) 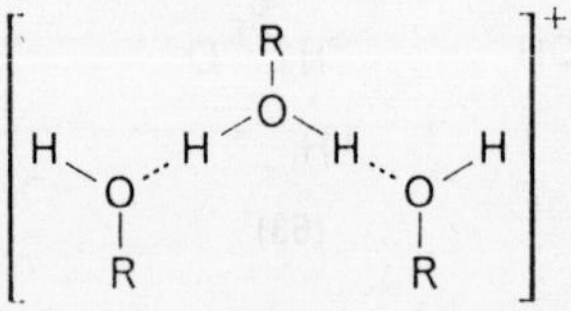 (69)

protonated alcohol dimer and trimer respectively. An appreciable concentration of **69** is only present at alcohol concentrations about or >1M.

In 100% sulphuric acid triarylcarbinols give a cryoscopic van't Hoff factor $i = 4$ [364–366]. The relevant ionization equilibrium is given by equation (29). Equilibrium constants K for equilibrium

$$Ar_3COH + 2\,H_2SO_4 \rightleftharpoons Ar_3C^+ + H_3O^+ + 2\,HSO_4^- \qquad (29)$$

$$Ar_3COH + H^+ \rightleftharpoons Ar_3C^+ + H_2O \qquad (30)$$

(30) have been deduced from measurements using concentrated sulphuric[367], hydrochloric[368], perchloric[369], nitric[369] and phosphoric acids[368]. Thus pK ranges from $-0{\cdot}82$ for 4,4′,4″-trimethoxytri-

phenylcarbinol to 16·27 for 4,4′,4″-trinitrotriphenylcarbinol in water at 25°C [367].

The author wishes to thank Drs. P. Jackson, R. B. Cundall, J. C. Roberts and B. Rossall for discussion and advice during the preparation of this article.

VII. REFERENCES

1. W. M. Latimer and W. H. Rodebush, *J. Am. Chem. Soc.*, **42**, 1419 (1920).
2. G. C. Pimentel and A. L. McClellan, *The Hydrogen Bond*, W. H. Freeman San Francisco, 1960, Chapter 1.
3. W. Weltner and K. S. Pitzer, *J. Am. Chem. Soc.*, **73**, 2606 (1951).
4. N. S. Berman, *Am. Inst. Chem. Engrs. J.*, **14**, 497 (1968).
5. R. G. Inskeep, J. M. Kelliher, P. E. McMahon and B. G. Somers, *J. Chem. Phys.*, **28**, 1033 (1958).
6. C. B. Kretschmer and R. Wieke, *J. Am. Chem. Soc.*, **76**, 2579 (1954).
7. R. G. Inskeep, F. E. Dickson and H. M. Olsen, *J. Mol. Spectr.*, **5**, 284 (1960).
8. G. M. Barrow, *J. Chem. Phys.*, **20**, 1739 (1952).
9. J. F. Mathews and J. J. McKetta, *J. Phys. Chem.*, **65**, 753 (1961).
10. N. S. Berman, C. W. Larkam and J. J. McKetta, *J. Chem. Eng. Data*, **9**, 218 (1964).
11. J. L. Hales, J. D. Cox and E. B. Lees, *Trans. Faraday Soc.*, **59**, 1544 (1963).
12. N. S. Berman and J. J. McKetta, *J. Phys. Chem.*, **66**, 1444 (1962).
13. E. T. Beynon and J. J. McKetta, *J. Phys. Chem.*, **67**, 2761 (1963).
14. R. M. Badger and S. H. Bauer, *J. Chem. Phys.*, **5**, 859 (1937).
15. R. M. Badger, *J. Chem. Phys.*, **8**, 288 (1940).
16. S. C. Stanford and W. Gordy, *J. Am. Chem. Soc.*, **62**, 1247 (1940).
17. R. Mecke, *Discussions Faraday Soc.*, **9**, 161 (1950).
18. N. D. Coggeshall and E. L. Saier, *J. Am. Chem. Soc.*, **73**, 5414 (1951).
19. A. Ens and F. E. Murray, *Can. J. Chem.*, **35**, 170 (1957).
20. U. Liddel and E. D. Becker, *Spectrochim. Acta*, **10**, 70 (1957).
21. L. J. Bellamy and R. J. Pace, *Spectrochim. Acta*, **22**, 525 (1966).
22. L. J. Bellamy, K. J. Morgan and R. J. Pace, *Spectrochim. Acta*, **22**, 535 (1966).
23. A. N. Fletcher and C. A. Heller, *J. Phys. Chem.*, **71**, 3742 (1967).
24. L. K. Patterson and R. M. Hammaker, *Spectrochim. Acta*, **23A**, 2333 (1967).
25. M. Van Thiel, E. D. Becker and G. C. Pimentel, *J. Chem. Phys.*, **27**, 95, 486 (1957).
26. A. Hall and J. L. Wood, *Spectrochim. Acta*, **23A**, 2657 (1967).
27. H. Ratajczak and W. J. Orville-Thomas, *J. Mol. Structure*, **1**, 449 (1968).
28. P. Sohar and Gy. Varsanyi, *Spectrochim. Acta*, **23A**, 1947 (1967).
29. A. D. Cohen and C. Reid, *J. Chem. Phys.*, **25**, 790 (1956).
30. M. Saunders and J. B. Hyne, *J. Chem. Phys.*, **29**, 1319 (1958).
31. J. C. Davis, K. S. Pitzer and C. N. R. Rao, *J. Phys. Chem.*, **64**, 1744 (1960).
32. H. Elmgren, *J. Chim. Phys.*, **65**, 206 (1968).
33. L. H. Thomas and R. Meatyard, *J. Chem. Soc.*, 1986 (1963).
34. L. H. Thomas, *J. Chem. Soc.*, 1995 (1963).
35. R. H. Stokes, *Australian J. Chem.*, **21**, 1343 (1968).

36. H. C. Van Ness, J. Van Winkle, H. H. Richtol and H. B. Hollinger, *J. Phys. Chem.*, **71**, 1483 (1967).
37. A. Bondi and J. Simkin, *J. Chem. Phys.*, **25**, 1073 (1956).
38. I. A. Wiche and E. B. Bagley, *Am. Inst. Chem. Engrs. J.*, **13**, 836 (1967).
39. W. Dannhauser, *J. Chem. Phys.*, **48**, 1911 (1968).
40. J. Feeney and S. M. Walker, *J. Chem. Soc.* (**A**), 1148 (1966).
41. D. L. Wertz and R. K. Kruh, *J. Chem. Phys.*, **47**, 388 (1967).
42. W. H. Zachariasen, *J. Chem. Phys.*, **3**, 158 (1935).
43. K. J. Tauer and W. N. Lipscomb, *Acta Cryst.*, **5**, 606 (1952).
44. F. Franks and D. J. G. Ives, *Quart. Rev.* (*London*), **20**, 1 (1966).
45. J. Reynolds and S. S. Sternstein, *J. Chem. Phys.*, **41**, 47 (1964).
46. G. A. Jeffrey and R. D. Rosenstein, *Advan. Carbohydrate Chem.*, **19**, 7 (1964).
47. W. G. Ferrier, *Acta Cryst.*, **16**, 1023 (1963).
48. C. J. Brown, G. Cox and F. J. Llewellyn, *J. Chem. Soc.* (**A**), 922 (1966).
49. A. J. Mitchell, *Carbohydrate Res.*, **5**, 229 (1967).
50. A. W. Baker, *J. Phys. Chem.*, **62**, 744 (1958).
51. K. U. Ingold, *Can. J. Chem.*, **38**, 1092 (1960).
52. H. H. Jaffe, *Chem. Rev.*, **53**, 191 (1953).
53. N. A. Puttnam, *J. Chem. Soc.*, 5100 (1960).
54. N. D. Coggeshall, *J. Am. Chem. Soc.*, **69**, 1620 (1947).
55. W. C. Sears and L. J. Kitchen, *J. Am. Chem. Soc.*, **71**, 4110 (1949).
56. C. M. Huggins, G. C. Pimentel and J. N. Shoolery, *J. Phys. Chem.*, **60**, 1311 (1956).
57. B. G. Somers and H. S. Gutowsky, *J. Am. Chem. Soc.*, **85**, 3065 (1963).
58. M. M. Maguire and R. West, *Spectrochim. Acta*, **17**, 369 (1961).
59. N. A. Puttnam, *J. Chem. Soc.*, 486 (1960).
60. R. S. Bowman, D. R. Stevens and W. E. Baldwin, *J. Am. Chem. Soc.*, **79**, 87 (1957).
61. L. J. Bellamy and R. L. Williams, *Proc. Roy. Soc.*, **A254**, 119 (1960).
62. V. F. Bystrov and V. P. Lezina, *Opt. Spectr.* (*USSR*), **16**, 542 (1964).
63. A. W. Baker, H. O. Kerlinger and A. T. Shulgin, *Spectrochim. Acta*, **20**, 1467 (1964).
64. P. D. Bolton, F. M. Hall and I. H. Reece, *Spectrochim. Acta*, **22**, 1149 (1966).
65. D. T. Y. Chen and K. J. Laidler, *Trans. Faraday Soc.*, **58**, 480 (1962).
66. F. S. Palker and K. R. Bhaskar, *Biochemistry*, **7**, 1286 (1968).
67. E. A. Allan and L. W. Reeves, *J. Phys. Chem.*, **66**, 613 (1962).
68. I. Brown, G. Eglington and M. Martin-Smith, *Spectrochim. Acta*, **19**, 463 (1963).
69. E. A. Allan and L. W. Reeves, *J. Phys. Chem.*, **67**, 591 (1963).
70. J. C. Dearden and W. F. Forbes, *Can. J. Chem.*, **38**, 896 (1960).
71. J. R. Johnson, S. D. Christian and H. E. Affsprung, *J. Chem. Soc.*, 1 (1965).
72. V. S. Griffiths and G. Socrates, *J. Mol. Spectr.*, **21**, 302 (1966).
73. J. R. Johnson, S. D. Christian and H. E. Affsprung, *J. Chem. Soc.* (**A**), 764 (1967).
74. M. Saunders and J. B. Hyne, *J. Chem. Phys.*, **29**, 1319 (1958).
75. R. M. Badger and R. C. Greenough, *J. Phys. Chem.*, **65**, 2088 (1961).
76. R. J. Jakobsen and J. W. Brasch, *Spectrochim. Acta*, **21**, 1753 (1965).
77. J. W. Brasch, R. J. Jackobsen, W. G. Fately and N. T. McDevitt, *Spectrochim. Acta*, **24A**, 203 (1968).

78. C. Scheringer, O. J. Wehrahn and M. v. Stackelberg, *Z. Elektrochem.*, **64,** 381 (1960).
79. C. Scheringer, *Z. Krist.*, **119,** 273 (1963).
80. H. Gillier-Pandraud, *Compt. Rend.*, **262C,** 1860 (1966).
81. H. Brusset, H. Gillier-Pandraud and Ch. Viossat, *Compt. Rend.*, **263C,** 53 (1966).
82. H. Gillier-Pandraud, *Bull. Soc. Chim. France*, 1988 (1967).
83. G. E. Bacon and N. A. Curry, *Proc. Roy. Soc.*, **A235,** 552 (1956).
84. C. Bois, *Bull. Soc. Chim. France*, 4016 (1966).
85. I. H. Reece and R. L. Werner, *Spectrochim. Acta*, **24A,** 1271 (1968).
86. Reference 2, Chapter 3.
87. C. M. Huggins and G. C. Pimentel, *J. Phys. Chem.*, **60,** 1615 (1956).
88. E. D. Becker, *Spectrochim. Acta*, **17,** 436 (1961).
89. A. R. H. Cole, L. H. Little and A. J. Mitchell, *Spectrochim. Acta*, **21,** 1169 (1965).
90. J. E. Gordon, *J. Org. Chem.*, **26,** 738 (1961).
91. E. M. Arnett, *Progr. Phys. Org. Chem.*, **1,** 223 (1963).
92. K. F. Purcell and S. T. Wilson, *J. Mol. Spectr.*, **24,** 468 (1967).
93. T. Gramstad, *Spectrochim. Acta*, **20,** 729 (1964).
94. A. Allerhand and P. von R. Schleyer, *J. Am. Chem. Soc.*, **85,** 866 (1963).
95. A. Kivinen, J. Murto and A. Viitala, *Suomen Kemistilehti*, **B40,** 301 (1967).
96. T. J. V. Findlay and A. D. Kidman, *Australian J. Chem.*, **18,** 521 (1965).
97. L. Lamberts, *J. Chim. Phys.*, **62,** 1404 (1965).
98. A. T. Shulgin and H. O. Kerlinger, *Chem. Commun.*, 249 (1966).
99. B. T. Zadorozhnyi and I. K. Ishchenko, *Opt. Spectr. (USSR)*, **19,** 306 (1965).
100. Reference 2, Chapter 7.
101. Th. Zeegers-Huyskens, *Spectrochim. Acta*, **23A,** 855 (1967).
102. T. Kitao and C. H. Jarboe, *J. Org. Chem.*, **32,** 407 (1967).
103. A. Kolbe, *Z. Physik. Chem.*, **58,** 75 (1968).
104. V. Gold, *Progr. Stereochem.*, **3,** 169 (1962).
105. L. W. Reeves, *Adv. Phys. Org. Chem.*, **3,** 187 (1963).
106. S. Murakami and R. Fujishiro, *Bull. Chem. Soc. Japan*, **40,** 1784 (1967).
107. Z. Yoshida, E. Osawa and R. Oda, *J. Phys. Chem.*, **68,** 2895 (1964).
108. C. H. Bamford, *Discussions Faraday Soc.*, **16,** 229 (1954).
109. A. Allerhand and P. von R. Schleyer, *J. Am. Chem. Soc.*, **85,** 371 (1963).
110. A. R. H. Cole and A. J. Mitchell, *Australian J. Chem.*, **18,** 102 (1965).
111. R. G. Inskeep, F. E. Dickson and J. M. Kelliher, *J. Mol. Spectr.*, **4,** 477 (1960).
112. E. Bauer and M. Magat, *J. Phys. Radium*, **9,** 319 (1938).
113. W. F. Baitinger, P. von R. Schleyer, T. S. S. R. Murty and L. Robinson, *Tetrahedron*, **29,** 1635 (1964).
114. H. E. Ungnade, E. M. Roberts and L. W. Kissinger, *J. Phys. Chem.*, **68,** 3225 (1964).
115. J. V. Paukstelis and R. M. Hammaker, *Tetrahedron Letters*, 3557 (1968).
116. T. Gramstad, *Spectrochim. Acta*, **19,** 497 (1963).
117. M. D. Joesten and R. S. Drago, *J. Am. Chem. Soc.*, **84,** 3817 (1962).
118. D. P. Syman and R. S. Drago, *J. Am. Chem. Soc.*, **88,** 1617 (1966).
119. K. F. Purcell and R. S. Drago, *J. Am. Chem. Soc.*, **89,** 2874 (1967).
120. I. Gränacher, *Helv. Phys. Acta*, **34,** 272 (1961).

121. T. Gramstad, *Spectrochim. Acta*, **19**, 829 (1963).
122. P. Biscarini, G. Galloni and S. Ghersetti, *Spectrochim. Acta*, **20**, 267 (1964).
123. T. Gramstad, *Spectrochim. Acta*, **20**, 729 (1964).
124. U. Blindheim and T. Gramstad, *Spectrochim. Acta*, **21**, 1073 (1965).
125. M. C. Sousa Lopes and H. W. Thompson, *Spectrochim. Acta*, **24A**, 1367 (1968).
126. D. A. K. Jones and J. G. Watkinson, *J. Chem. Soc.*, 2366 (1964).
127. R. West, D. L. Powell, L. S. Whatley, M. K. T. Lee and P. von R. Schleyer, *J. Am. Chem. Soc.*, **84**, 3221 (1962).
128. J. Chojnowski and W. N. Brandt, *J. Am. Chem. Soc.*, **90**, 1384 (1968).
129. J. Rubin and G. S. Panson, *J. Phys. Chem.*, **69**, 3089 (1965).
130. E. Osawa, T. Kato and Z. Yoshida, *J. Org. Chem.*, **32**, 2803 (1967).
131. Z. Yoshida and E. Osawa, *J. Am. Chem. Soc.*, **87**, 1467 (1965).
132. Z. Yoshida and E. Osawa, *J. Am. Chem. Soc.*, **88**, 4019 (1966).
133. R. West, *J. Am. Chem. Soc.*, **81**, 1614 (1959).
134. L. P. Kuhn and R. E. Bowman, *Spectrochim. Acta*, **23A**, 189 (1967).
135. S. Singh and C. N. R. Rao, *Can. J. Chem.*, **44**, 2611 (1966).
136. Y. S. Su and H-K Hong, *Spectrochim. Acta*, **24A**, 1461 (1968).
137. D. Neerinck, A. Van Audenhaege, L. Lamberts and P. Huyskens, *Nature*, **218**, 461 (1968).
138. A. B. Sannigrahi and A. K. Chandra, *Bull. Chem. Soc. Japan*, **40**, 1344 (1967).
139. S. Ghersetti and A. Lusa, *Spectrochim. Acta*, **21**, 1067 (1965).
140. D. A. Ibbitson and J. P. B. Sandall, *J. Chem. Soc.*, 4547 (1964).
141. D. G. Holland, N. T. McDevitt, J. V. Pustinger and J. E. Strobel, *J. Org. Chem.*, **32**, 3671 (1967).
142. T. Gramstad and G. Van Binst, *Spectrochim. Acta*, **22**, 1681 (1966).
143. S. Suzuki and H. Baba, *Bull. Soc. Chem. Japan*, **40**, 2199 (1967).
144. N. D. Coggeshall and E. M. Lang, *J. Am. Chem. Soc.*, **70**, 3283 (1948).
145. L. J. Bellamy, G. Eglington and J. F. Morman, *J. Chem. Soc.* 4762 (1961).
146. F. Takahashi and N. C. Li, *J. Phys. Chem.*, **69**, 1622 (1965).
147. N. Shishka and I. V. Berezin, *Russian J. Phys. Chem.*, **40**, 1555 (1966).
148. H. Fritzsche, *Z. Physik. Chem.*, **43**, 154 (1964).
149. M. Horak, J. Polakova, M. Jakoubkova, J. Moravec and J. Pliva, *Collection Czech Chem. Commun.*, **31**, 622 (1966).
150. A. D. Buckingham, *Proc. Roy. Soc.*, **A248**, 169 (1958).
151. E. Osawa and Z. Yoshida, *Spectrochim. Acta*, **23A**, 2029 (1967).
152. T. Gramstad, *Spectrochim. Acta*, **19**, 1363 (1963).
153. K. U. Ingold and D. R. Taylor, *Can. J. Chem.*, **39**, 471 (1961).
154. K. U. Ingold and D. R. Taylor, *Can. J. Chem.*, **39**, 481 (1961).
155. W. M. Schubert and R. H. Quacchia, *J. Am. Chem. Soc.*, **85**, 1278 (1963).
156. A. Albert, J. H. Lister and C. Pedersen, *J. Chem. Soc.*, 4621 (1956).
157. J. C. Dearden and W. F. Forbes, *Can. J. Chem.*, **38**, 896 (1960).
158. W. A. Lees and A. Burawoy, *Tetrahedron*, **19**, 419 (1963).
159. A. J. Parker and D. Brody, *J. Chem. Soc.*, 4061 (1963).
160. F. Takahashi, W. J. Karoly, J. B. Greenshields and N. C. Li, *Can. J. Chem.*, **45**, 2033 (1967).
161. A. Weller, *Progr. Reaction Kinetics*, **1**, 187 (1961).
162. A. Burawoy in *Hydrogen Bonding* (Ed. D. Hadzi), Pergamon, London, 1959, p. 259.

163. A. E. Stanevich, *Opt. Spectr.* (*USSR*), **16,** 425, 539 (1964).
164. A. Hall and J. L. Wood, *Spectrochim. Acta*, **23A,** 1257 (1967).
165. S. G. W. Ginn and J. L. Wood, *Spectrochim. Acta*, **23A,** 611 (1967).
166. A. Hall and J. L. Wood, *Spectrochim. Acta*, **24A,** 1109 (1968).
167. W. J. Hurley, I. D. Kuntz and G. E. Leroi, *J. Am. Chem. Soc.*, **88,** 3199 (1966).
168. K. Fukushima, *Bull. Chem. Soc. Japan*, **38,** 1694 (1965).
169. Reference 2, Chapter 5.
170. L. P. Kuhn, *J. Am. Chem. Soc.*, **74,** 2492 (1952).
171. R. D. Stolow, P. M. McDonagh and M. M. Bonaventura, *J. Am. Chem. Soc.*, **86,** 2165 (1964).
172. R. D. Stolow, *J. Am. Chem. Soc.*, **86,** 2170 (1964).
173. L. P. Kuhn and R. E. Bowman, *Spectrochim. Acta*, **17,** 650 (1961).
174. A. B. Foster, A. H. Haines and M. Stacey, *Tetrahedron*, **16,** 177 (1961).
175. L. P. Kuhn and R. A. Wires, *J. Am. Chem. Soc.*, **86,** 2161 (1964).
176. W. W. Zajac, F. Sweet and R. K. Brown, *Can. J. Chem.*, **46,** 21 (1968).
177. R. J. Ouellette, K. Liptak and G. E. Booth, *J. Org. Chem.*, **32,** 2394 (1967).
178. M. Oki and T. Murayama, *Bull. Chem. Soc. Japan*, **40,** 1997 (1967).
179. P. Arnaud and Y. Armand, *Compt. Rend.*, **255C,** 1718 (1962).
180. P. Arnaud and Y. Armand, *Compt. Rend.*, **256C,** 4450 (1963).
181. M. M. Dominique Audo, Y. Armand and P. Arnaud, *Compt. Rend.*, **266C,** 1129 (1968).
182. J. Grundy and L. J. Morris, *Spectrochim. Acta*, **20,** 695 (1964).
183. N. Mori, S. Omura and Y. Tsuzuki, *Bull. Chem. Soc. Japan*, **38,** 1631 (1965).
184. I. D. Campbell, G. Eglington and R. A. Raphael, *J. Chem. Soc.* (**B**), 338 (1968).
185. R. W. Hay and P. P. Williams, *J. Chem. Soc.*, 2270 (1964).
186. M. Oki, H. Iwamura, J. Aikara and H. Iida, *Bull. Soc. Chem. Japan*, **41,** 176 (1968).
187. N. Mori, S. Omura, H. Yamakawa and Y. Tsuzuki, *Bull. Soc. Chem. Japan*, **38,** 1627 (1965).
188. P. von R. Schleyer and R. West, *J. Am. Chem. Soc.*, **81,** 3164 (1959).
189. P. J. Krueger and H. D. Mettee, *Can. J. Chem.*, **42,** 326 (1964).
190. N. Mori, E. Nakamura and T. Tsuzuki, *Bull. Soc. Chem. Japan*, **40,** 2189 (1967).
191. P. J. Krueger and H. D. Mettee, *Can. J. Chem.*, **43,** 2970 (1965).
192. N. Mori, E. Nakamura and Y. Tsuzuki, *Bull. Chem. Soc. Japan*, **40,** 2191 (1967).
193. E. L. Saier, L. R. Cousins and M. R. Basila, *J. Chem. Phys.*, **41,** 40 (1964).
194. L. W. Reeves, E. A. Allan and K. O. Stromme, *Can. J. Chem.*, **38,** 1249 (1960).
195. A. W. Baker and A. T. Shulgin, *J. Am. Chem. Soc.*, **80,** 5358 (1958).
196. A. W. Baker, *J. Am. Chem. Soc.*, **80,** 3598 (1958).
197. D. A. K. Jones and J. G. Watkinson, *J. Chem. Soc.*, 2371 (1964).
198. Tien-Sung Lin and E. Fishman, *Spectrochim. Acta*, **23A,** 491 (1967).
199. A. W. Baker and W. W. Kaeding, *J. Am. Chem. Soc.*, **81,** 5904 (1959).
200. A. W. Baker and A. T. Shulgin, *Spectrochim. Acta*, **22,** 95 (1966).
201. R. E. Rundle, *J. Phys.* (*Paris*), **25,** 487 (1964).
202. R. A. Nyquist, *Spectrochim. Acta*, **19,** 1655 (1965).

203. A. W. Baker and A. T. Shulgin, *Spectrochim. Acta*, **20**, 153 (1964).
204. M. Oki and H. Iwamura, *Bull. Chem. Soc. Japan*, **33**, 717 (1960).
205. G. Pala, *Nature*, **204**, 1190 (1964).
206. I. M. Hunsberger, H. S. Gutowsky, W. Powell, L. Morin and V. Bandurco, in *Hydrogen Bonding* (Ed. D. Hadzi), Pergamon, London, 1959, p. 461.
207. H. Musso and H. Pietsch, *Chem. Ber.*, **100**, 2854 (1967).
208. G. E. K. Branch and D. L. Yabroff, *J. Am. Chem. Soc.*, **56**, 2568 (1934).
209. D. Chapman, D. R. Lloyd and R. H. Prince, *J. Chem. Soc.*, 550 (1964).
210. G. E. Dunn and F. L. Kung, *Can. J. Chem.*, **44**, 1261 (1966).
211. G. E. Dunn and T. L. Penner, *Can. J. Chem.*, **45**, 1699 (1967).
212. A. O. McDougall and F. A. Long, *J. Phys. Chem.*, **66**, 429 (1962).
213. Z. L. Ernst and J. Menashi, *Trans. Faraday Soc.*, **59**, 230, 1803 (1963).
214. M. Eigen, W. Kruse, G. Maass and L. De Maeyer, *Progr. Reaction Kinetics*, **2**, 285 (1964).
215. R. H. Laby and T. C. Morton, *Australian J. Chem.*, **20**, 2279 (1967).
216. U. Dabrowska and T. Urbanski, *Roczniki Chem.*, **37**, 805 (1963).
217. J. C. Deardon, *Nature*, **206**, 1147 (1965).
218. A. Balasubramanian, W. F. Forbes and J. C. Dearden, *Can. J. Chem.*, **44**, 961 (1966).
219. U. Dabrowska and T. Urbanski, *Spectrochim. Acta*, **21**, 1765 (1965).
220. G. O. Dudek and E. P. Dudek, *J. Am. Chem. Soc.*, **86**, 4283 (1964).
221. G. O. Dudek and E. P. Dudek, *Chem. Commun.*, 464 (1965).
222. G. O. Dudek and E. P. Dudek, *J. Am. Chem. Soc.*, **88**, 2407 (1966).
223. J. J. Charette, *Spectrochim. Acta*, **19**, 1275 (1963).
224. P. Teyssie and J. J. Charette, *Spectrochim. Acta*, **19**, 1407 (1963).
225. J. Charette, G. Faltlhansl and P. Teyssie, *Spectrochim. Acta*, **20**, 597 (1964).
226. M. D. Cohen, Y. Hirshberg and G. M. J. Schmidt, in *Hydrogen Bonding* (Ed. D. Hadzi), Pergamon, London, 1959, p. 293.
227. L. W. Reeves, *Can. J. Chem.*, **38**, 748 (1960).
228. H. H. Freedman, *J. Am. Chem. Soc.*, **83**, 2900 (1961).
229. J. L. Haslam and E. M. Eyring, *J. Phys. Chem.*, **71**, 4470 (1967).
230. L. P. Hammett and A. J. Deyrup, *J. Am. Chem. Soc.*, **54**, 2721 (1932).
231. M. A. Paul and F. A. Long, *Chem. Rev.*, **57**, 1 (1957).
232. C. H. Rochester, *Acidity Functions*, Academic Press, to be published.
233. A. J. Kresge, R. A. More O'Ferrall, L. E. Hakka and V. P. Vitullo, *Chem. Commun.*, 46 (1965).
234. W. M. Schubert and R. H. Quacchia, *J. Am. Chem. Soc.*, **85**, 1284 (1963).
235. J. F. Bunnett, *J. Am. Chem. Soc.*, **83**, 4956, 4968, 4973, 4978 (1961).
236. C. H. Rochester, *J. Chem. Soc.* (**B**), 1076 (1967).
237. J. Koskikallio and K. Koivula, *Suomen Kemistilehti*, **B40**, 138 (1967).
238. M. Oki, H. Iwamura and T. Nishida, *Bull. Chem. Soc. Japan*, **41**, 656 (1968).
239. E. Grunwald, C. F. Jumper and M. S. Puar, *J. Phys. Chem.*, **71**, 492 (1967).
240. W. Köhler, N. F. Kasanskaya, L. G. Nagler and I. W. Beresin, *Ber. Bunsengesellschaft Phys. Chem.*, **71**, 736 (1967).
241. M. Eigen, *Discussions Faraday Soc.*, **39**, 7 (1965).
242. P. Ballinger and F. A. Long, *J. Am. Chem. Soc.*, **81**, 1050 (1959).
243. P. Ballinger and F. A. Long, *J. Am. Chem. Soc.*, **82**, 795 (1960).
244. F. A. Long and P. Ballinger, in *Electrolytes* (Ed. B. Pesce), Pergamon, Oxford, 1962, p. 152.

245. H. S. Harned and R. A. Robinson, *Trans. Faraday Soc.*, **36,** 973 (1940).
246. E. T. McBee, W. F. Marzluff and O. R. Pierce, *J. Am. Chem. Soc.*, **74,** 444 (1952).
247. C. W. Roberts, E. T. McBee and C. E. Hathaway, *J. Org. Chem.*, **21,** 1369 (1956).
248. R. W. Taft, *J. Am. Chem. Soc.*, **74,** 3120 (1952).
249. R. W. Taft, *J. Am. Chem. Soc.*, **75,** 4231 (1953).
250. G. B. Barlin and D. D. Perrin, *Quart. Rev. (London)*, **20,** 75 (1966).
251. J. Hine and M. Hine, *J. Am. Chem. Soc.*, **74,** 5266 (1952).
252. P. Ballinger and F. A. Long, *J. Am. Chem. Soc.*, **81,** 2347 (1959).
253. R. N. Haszeldine, *J. Chem. Soc.*, 1757 (1953).
254. M. Tissier and C. Tissier, *Bull. Soc. Chim. France*, 3155 (1967).
255. R. M. Izatt, L. D. Hansen, J. H. Rytting and J. J. Christensen, *J. Am. Chem. Soc.*, **87,** 2760 (1965).
256. R. M. Izatt, J. H. Rytting, L. D. Hansen and J. J. Christensen, *J. Am. Chem. Soc.*, **88,** 2641 (1966).
257. J. J. Christensen, J. H. Rytting and R. M. Izatt, *J. Am. Chem. Soc.*, **88,** 5105 (1966).
258. W. J. Middleton and R. V. Lindsey, *J. Am. Chem. Soc.*, **86,** 4948 (1964).
259. R. Stewart and R. Van der Linden, *Can. J. Chem.*, **38,** 399 (1960).
260. R. P. Bell and D. P. Onwood, *Trans. Faraday Soc.*, **58,** 1557 (1962).
261. J. Hine, J. G. Houston and J. H. Jensen, *J. Org. Chem.*, **30,** 1184 (1965).
262. R. P. Bell, *Advan. Phys. Org. Chem.*, **4,** 1 (1966).
263. B. Eistert, E. Merkel and W. Reiss, *Chem. Ber.*, **87,** 1513 (1954).
264. G. Schwarzenbach and E. Felder, *Helv. Chim. Acta*, **27,** 1701 (1944).
265. G. Schwarzenbach and K. Lutz, *Helv. Chim. Acta*, **23,** 1147, 1162 (1940).
266. G. Schwarzenbach, H. Suter and K. Lutz, *Helv. Chim. Acta*, **23,** 1191 (1940).
267. P. Rumpf and R. La Riviere, *Compt. Rend.*, **244,** 902 (1957).
268. I. Eidinoff, *J. Am. Chem. Soc.*, **67,** 2072, 2073 (1945).
269. R. P. Bell, *Trans. Faraday Soc.*, **39,** 253 (1943).
270. J. Koskikallio, *Suomen Kemistilehti*, **B30,** 111 (1957).
271. R. Schaal and A. Teze, *Bull. Soc. Chim. France*, 1783 (1961).
272. J. Koskikallio, *Suomen Kemistilehti*, **B30,** 155 (1957).
273. G. Briere, B. Crochow and N. Felici, *Compt. Rend.*, **254,** 4458 (1962).
274. P. D. Bolton, F. M. Hall and J. Kudrynski, *Australian J. Chem.*, **21,** 1541 (1968).
275. P. D. Bolton, F. M. Hall and I. H. Reece, *Spectrochim. Acta*, **22,** 1825 (1966).
276. L. P. Fernandez and L. G. Hepler, *J. Am. Chem. Soc.*, **81,** 1783 (1959).
277. P. D. Bolton, F. M. Hall and I. H. Reece, *J. Chem. Soc.* (**B**), 709 (1967).
278. P. D. Bolton, F. M. Hall and I. H. Reece, *Spectrochim. Acta*, **22,** 1825 (1966).
279. W. F. O'Hara and L. G. Hepler, *J. Phys. Chem.*, **65,** 2107 (1961).
280. F. J. Millero, J. C. Ahluwalia and L. G. Hepler, *J. Chem. Eng. Data*, **9,** 192 (1964).
281. F. J. Millero, J. C. Ahluwalia and L. G. Hepler, *J. Chem. Eng. Data*, **9,** 319 (1964).
282. H. C. Ko, W. F. O'Hara, T. Hu and L. G. Hepler, *J. Am. Chem. Soc.*, **86,** 1003 (1964).
283. P. D. Bolton, F. M. Hall and I. H. Reece, *J. Chem. Soc.* (**B**), 717 (1966).
284. G. F. Allen, R. A. Robinson and V. E. Bower, *J. Phys. Chem.*, **66,** 171 (1962).

285. A. I. Biggs, *J. Chem. Soc.*, 2572 (1961).
286. F. T. Crimmins, C. Dymek, M. Flood and W. F. O'Hara, *J. Phys. Chem.*, **70**, 931 (1966).
287. C. L. Liotta, K. H. Leavell and D. F. Smith, *J. Phys. Chem.*, **71**, 3091 (1967).
288. W. F. O'Hara, T. Hu and L. G. Hepler, *J. Phys. Chem.*, **67**, 1933 (1963).
289. R. A. Robinson and A. K. Kiang, *Trans. Faraday Soc.*, **51**, 1398 (1955).
290. R. A. Robinson and A. Peiperl, *J. Phys. Chem.*, **67**, 1723 (1963).
291. R. A. Robinson and A. Peiperl, *J. Phys. Chem.*, **67**, 2860 (1963).
292. A. A. Zavitsas, *J. Chem. Eng. Data*, **12**, 94 (1967).
293. A. I. Biggs and R. A. Robinson, *J. Chem. Soc.* 388, (1961).
294. E. H. Binns, *Trans. Faraday Soc.*, **55**, 1900 (1959).
295. L. Canonica, *Gazz. Chim. Ital.*, **77**, 92 (1947).
296. D. J. G. Ives and P. G. N. Moseley, *J. Chem. Soc.* (**B**), 757 (1966).
297. R. Y. Kirdani and M. J. Burgell, *Arch. Biochem. Biophys.*, **118**, 33 (1967).
298. C. L. Liotta and D. F. Smith, *Chem. Commun.*, 416 (1968).
299. L. B. Magnusson, C. A. Craig and C. Postmus, *J. Am. Chem. Soc.*, **86**, 3958 (1964).
300. C. H. Rochester, *J. Chem. Soc.*, 4603 (1965).
301. P. J. Pearce and R. J. J. Simkins, *Can. J. Chem.*, **46**, 241 (1968).
302. M. Rapoport, C. K. Hancock and E. A. Meyers, *J. Am. Chem. Soc.*, **83**, 3489 (1961).
303. R. A. Robinson, *J. Res. Nat. Bur. Std.*, **71A**, 385 (1967).
304. R. A. Robinson, *J. Res. Nat. Bur. Std.*, **71A**, 213 (1967).
305. R. A. Robinson and A. I. Biggs, *Trans. Faraday Soc.*, **51**, 901 (1955).
306. R. A. Robinson, *J. Res. Nat. Bur. Std.*, **68A**, 159 (1964).
307. R. A. Robinson, M. M. Davis, M. Paabo and V. E. Bower, *J. Res. Nat. Bur. Std.*, **64A**, 347 (1960).
308. B. S. Smolyakov, *Izv. Sibirsk. Otd. Akad. Nauk Ser. Khim. Nauk*, 8 (1967).
309. G. W. Wheland, R. M. Brownell and E. C. Mayo, *J. Am. Chem. Soc.*, **70**, 2492 (1948).
310. N. D. Coggeshall and A. S. Glessner, *J. Am. Chem. Soc.*, **71**, 3150 (1949).
311. D. R. Boyd, *J. Chem. Soc.*, 1538 (1915).
312. L. A. Cohen and W. M. Jones, *J. Am. Chem. Soc.*, **85**, 3397 (1963).
313. P. Demerseman, J. P. Lechartier, R. Reynaud, A. Cheutin, R. Royer and P. Rumpf, *Bull. Soc. Chim. France*, 2559 (1963).
314. A. Fischer, G. J. Leary, R. D. Topsom and J. Vaughan, *J. Chem. Soc.* (**B**), 782 (1966).
315. A. Fischer, G. J. Leary, R. D. Topsom and J. Vaughan, *J. Chem. Soc.* (**B**), 686 (1967).
316. A. Fischer, G. J. Leary, R. D. Topsom and J. Vaughan, *J. Chem. Soc.* (**B**), 846 (1967).
317. C. M. Judson and M. Kilpatrick, *J. Am. Chem. Soc.*, **71**, 3110 (1949).
318. E. F. G. Herington and W. Kynaston, *Trans. Faraday Soc.*, **53**, 138 (1957).
319. C. L. de Ligny, H. J. H. Kreutzer and G. F. Visserman, *Rec. Trav. Chim.*, **85**, 5 (1966).
320. S. D. Hamann and S. C. Lim, *Australian J. Chem.*, **7**, 329 (1954).
321. G. Kortüm, W. Vogel and K. Andrussov, *Pure Appl. Chem.*, **1**, 187 (1960).
322. R. A. Robinson and R. H. Stokes, *Electrolyte Solutions*, Butterworths, London, 1959, appendix 12.1.

323. A. Bryson and R. W. Mathews, *Australian J. Chem.*, **16,** 401 (1963).
324. H. van Bekkum, P. E. Verkade and B. M. Wepster, *Rec. Trav. Chim.*, **78,** 815 (1959).
325. B. M. Wepster, *Progr. Stereochem.*, **2,** 99 (1958).
326. L. G. Hepler, *J. Am. Chem. Soc.*, **85,** 3089 (1963).
327. L. G. Hepler and W. F. O'Hara, *J. Phys. Chem.*, **65,** 811 (1961).
328. D. J. G. Ives and P. D. Marsden, *J. Chem. Soc.*, 649 (1965).
329. C. H. Rochester, *J. Chem. Soc.*, 676 (1965), see note in Ref. 340.
330. A. O. McDougall and F. A. Long, *J. Phys. Chem.*, **66,** 429 (1962).
331. D. C. Martin and J. A. V. Butler, *J. Chem. Soc.*, 1366 (1939).
332. R. P. Bell and A. T. Kuhn, *Trans. Faraday Soc.*, **59,** 1789 (1963).
333. E. L. Wehry and L. B. Rogers, *J. Am. Chem. Soc.*, **88,** 351 (1966).
334. R. Cetina, S. Meza and J. L. Mateos, *Bol. Inst. Quim. Univ. Na. Auton. Mexico*, **19,** 41 (1967).
335. J.-P. Grivet and M. Ptak, *Compt. Rend.*, **266B,** 848 (1968).
336. R. P. Bell, *The Proton in Chemistry*, Methuen, London, 1959, Chapter 4.
337. B. W. Clare, D. Cook, E. C. F. Ko, Y. C. Mac and A. J. Parker, *J. Am. Chem. Soc.*, **88,** 1911 (1966).
338. J. Juillard, *Bull. Soc. Chim. France*, 1727 (1966).
339. C. H. Rochester, *Trans. Faraday Soc.*, **62,** 355 (1966), see note in Ref. 340.
340. C. H. Rochester and B. Rossall, *J. Chem. Soc.* (**B**), 743 (1967).
341. B. D. England and D. A. House, *J. Chem. Soc.*, 4421 (1962).
342. C. L. de Ligny, *Rec. Trav. Chim.*, **85,** 1114 (1966).
343. C. H. Rochester, *J. Chem. Soc.* (**B**), 121 (1966), see note in Ref. 340.
344. J. Juillard and M.-L. Dondon, *Bull. Soc. Chim. France*, 2535 (1966).
345. W. D. Treadwell and G. Schwarzenbach, *Helv. Chim. Acta*, **11,** 386 (1928).
346. F. Aufauvre, *Bull. Soc. Chim. France*, 2802 (1967).
347. I. M. Kolthoff, M. K. Chantooni and S. Bhowmik, *J. Am. Chem. Soc.*, **88,** 5430 (1966).
348. W. R. Vaughan and G. K. Finch, *J. Org. Chem.*, **21,** 1201 (1956).
349. C. H. Rochester and B. Rossall, *Trans. Faraday Soc.*, **65,** 992 (1969).
350. C. H. Rochester and B. Rossall, *Trans. Faraday Soc.*, **65,** 1004 (1969).
351. B. J. Steel, R. A. Robinson and R. G. Bates, *J. Res. Nat. Bur. Std.*, **A71,** 9 (1967).
352. P. Vetesnik, R. M. Hanikainem, J. Lakomy and M. Vecera, *Collection Czech. Chem. Commun.*, **32,** 1027 (1967).
353. G. Kortüm and K.-W. Koch, *Ber. Bunsengesellschaft Phys. Chem.*, **69,** 677 (1965).
354. D. Jannakoudakis and J. Moumtzis, *Chim. Chronika (Athens)*, **33A,** 7 (1968).
355. S. F. Mason, *J. Chem. Soc.*, 674 (1958).
356. G. F. Tucker and J. L. Irvin, *J. Am. Chem. Soc.*, **73,** 1923 (1951).
357. S. J. Angyal and C. L. Angyal, *J. Chem. Soc.*, 1461 (1952).
358. J. C. Haylock, S. F. Mason and B. E. Smith, *J. Chem. Soc.*, 4897 (1963).
359. C. F. Wells, *Trans. Faraday Soc.*, **62,** 2815 (1966).
360. C. F. Wells in *Hydrogen-Bonded Solvent Systems*, Taylor and Francis, London, 1968, p. 323.
361. S. Solway and P. Rosen, *Science*, **121,** 832 (1955).
362. E. M. Arnett and C. Y. Wu, *J. Am. Chem. Soc.*, **82,** 5660 (1960).
363. I. M. Kolthoff and M. K. Chantooni, *J. Am. Chem. Soc.*, **90,** 3320 (1968).

364. A. Hantzsch, *Z. Physik. Chem.*, **65**, 41 (1909).
365. L. P. Hammett and A. J. Deyrup, *J. Am. Chem. Soc.*, **55**, 1901 (1933).
366. M. S. Newman and N. C. Deno, *J. Am. Chem. Soc.*, **73**, 3644 (1951).
367. N. C. Deno, J. J. Jaruzelski and A. Schriesheim, *J. Am. Chem. Soc.*, **77**, 3044 (1955).
368. E. M. Arnett and G. W. Mach, *J. Am. Chem. Soc.*, **88**, 1177 (1966).
369. N. C. Deno, H. E. Berkheimer, W. L. Evans and H. J. Peterson, *J. Am. Chem. Soc.*, **81**, 2344 (1959).
370. A. R. Katritzky and J. M. Lagowski, *Advan. Heterocyclic Chem.*, **1**, 311 (1963).
371. C. A. Coulson, *Research*, **10**, 149 (1957).
372. C. A. Coulson in *Hydrogen Bonding* (Ed. D. Hadzi), Pergamon, London, 1959, p. 339.
373. J. N. Murrell, *Chemistry in Britain*, **5**, 107 (1969).
374. A. D. H. Claque, G. Govil and H. J. Bernstein, *Can. J. Chem.*, **47**, 625 (1969).
375. I. Avigal, J. Feitelson and M. Ottolenghi, *J. Chem. Phys.*, **50**, 2614 (1969).

CHAPTER 8

Directing and activating effects

D. A. R. Happer and J. Vaughan
University of Canterbury, Christchurch, New Zealand

I. INTRODUCTION

In this chapter we are concerned only with the hydroxyl group as a substituent and with its effects on chemical reactions occurring at some other site in the molecule. No consideration is given to reactions which lead to chemical modification of the hydroxyl group, unless this is temporary and leads to a product with the hydroxyl group apparently unchanged.

The hydroxyl group is highly polar; it displays strong directing and activating effects and it is often a convenient scapegoat when reaction rates or products are unexpected. There has had to be much selection from the large body of information; we have chosen the data which appeared to be most generally useful and we have preferred the quantitative study to the qualitative observation.

In discussing polar effects, our use of symbols is similar to that of Chuchani[1]. For example, electron attraction by an inductive mechanism is a $-I$ effect, and electron release by the conjugative (resonance) mechanism is a $+R$ effect. Work on linear free-energy relationships has given rise to several scales for substituent constants. Of the symbols to which reference is made in the following sections, σ is the Jaffé–Hammett substituent constant[2]; $\bar{\sigma}_m$ (or $\bar{\sigma}_p$) is the constant derived for the *meta-* (or *para-*) position from a particular reaction series; σ^o and σ^n are, respectively, Taft's[3] and Wepster's[4] versions of substituent constants free from resonance interaction; σ_I and σ_R are the polar and resonance contributions of a given substituent as resolved by Taft[5] ($\sigma = \sigma_I + \sigma_R$); σ^+ is the constant applicable to a $+R$ substituent when conjugated to an electron-deficient reaction site[6], and assumed to apply to normal electrophilic substitution at the benzene ring; σ^- is the constant for a $-R$ substituent when conjugated to a negatively charged site.

II. REACTIONS OF ALIPHATIC SYSTEMS

A. Electronic Effects

Attempts to apply Hammett-type structure–reactivity relationships to aliphatic systems have been limited in number and in scope by difficulties in the assessment and control of steric effects.

The first attempt to eliminate steric influences was that of Roberts and Moreland[7] who measured the relative reactivities of a number of 4-substituted bicyclo[2,2,2]octane-1-carboxylic acids, in which the carbon cage has a rigidity comparable with that of the benzene ring. They demonstrated the existence of a linear free-energy relationship between the acidity and the rates of both the reaction with diphenyldiazomethane and ester hydrolysis.

The derived substituent constants (σ' values) were indicative of polar (inductive) effects and may be considered as equivalent to σ_I values. The results showed that, in the absence of steric and resonance effects, the hydroxyl group behaves as a moderately strong electron-withdrawing substituent roughly comparable in strength with the carbethoxy substituent.

$$\mathrm{H} < \mathrm{OH} \sim \mathrm{CO_2Et} < \mathrm{Br} < \mathrm{CN}$$

Later work by other groups[8, 9, 10] extended this series. Other relevant studies involving relatively rigid systems are those by Siegel and Komarmy[11] on the 1,4-disubstituted cyclohexane series and by Stetter and Mayer[12] on 1,3-derivatives of adamantane.

A more extensive treatment of aliphatic systems became possible when Taft derived substituent constants (σ^* values) from the measurement of base-catalysed ester hydrolysis rates[13]. He did this through a comparison with corresponding rates of acid hydrolysis, in which similar steric effects are expected, but in which polar effects are unimportant. Most of Taft's polar substituent constants referred to CH_2X groups and he used these constants to correlate other aliphatic reaction data. The results showed that the experimental approaches of Taft and of Roberts were compatible and reinforced one another.

Taft's substituent constant for the hydroxymethyl group (0·555) was based on eleven reaction series. It was found that the mean deviation for this substituent was considerably greater than for most others. However, the deviation was not large enough to be unacceptable and, since a wide variety of solvents and reaction conditions were used, the variation could possibly be attributed to changes in solvation of the hydroxyl group. For this group, the differences between individual $\bar{\sigma}$ values (both for Taft's reactions and for other, more recent studies) cannot be systematically linked to changes in either solvent or reaction type but there are in fact no spectacular departures from the mean. In all of the reactions the hydroxyl group appears to act as a moderately consistent and quite strongly electron-withdrawing substituent.

Several groups of workers[14–21] have been able to correlate aliphatic reactivities by means of Hammett aromatic substituent constants in unsaturated structures which allow conjugation between substituent and reaction site. In these cases the system would be expected to bear some resemblance to that of a *para*-substituted benzene derivative. These reactions cannot be satisfactorily correlated by means of inductive substituent constants.

It is therefore assumed that the resonance component of a normal aromatic substituent constant (σ_R) is making an effective contribution. In later work[5], Taft changed from σ values to the more convenient σ_I scale by using the relationship $\sigma_I^X = 0{\cdot}45\ \sigma^{CH_2X}$. This σ_I scale, derived from aliphatic reactivities, is but one of three available scales which differ from each other in their experimental origins but which should give, within experimental limits, identical values for a given substituent. A second scale is based on the separation of normal aromatic σ values into inductive and resonance components[22], and the third is based on ^{19}F shielding parameters of *m*-substituted fluorobenzenes[23]. Table 1, compiled from figures in the review by Ritchie and Sager[24], shows the differences between the scales that exist for representative substituents. As an example of

TABLE 1. Aliphatic substituent constants.

X	σ_I (aliphatic)	σ_I (aromatic)	σ_I (n.m.r.)
$-N(CH_3)_3^+$	0·92	0·90	0·93
$-NO_2$	0·63	0·68	0·60
–CN	0·56	0·52	0·53
–F	0·52	0·45	0·52
–Cl	0·47	0·42	—
–Br	0·45	0·45	0·44
–I	0·38	0·42	—
$-CO_2R$	0·30	0·34	0·21
$-COCH_3$	0·28	0·32	0·23
$-OCH_3$	0·25	0·28	0·29
–OH	0·25	0·32	0·16
$-NH_2$	0·10	0·04	0·05
–H	0·00	0·00	0·00
$-CH_3$	0·00	−0·03	−0·08
$-O^-$	—	−0·12	−0·16

the general success of the treatment, Figure 1 shows the good correlation obtained between the pK_a's for aliphatic acids of the type XCH_2CO_2H in water and σ_I (n.m.r.), the scale for which the greatest number of σ_I values have been measured.

B. *Proximity Effects*

1. Nucleophilic substitution

Compounds of the type R–CH(OH)(R)–X, where X is a good leaving group (as X^-), are often unstable, breaking down to give carbonyl compounds and HX.

With the hydroxyl substituent on a *β*-carbon or on a more distant atom, the compounds are stable, and several mechanistic studies of nucleophilic displacement are available. In most cases the leaving group X has been Cl, Br or I but the same principles should apply to other cases in which X is any good leaving group. A general prediction based on electronic factors is that the presence of the electron-withdrawing hydroxyl group should result in a decrease in the rate of halide displacement, because the positive nature of the attached carbon atom is increased in the transition state. This is particularly the case for tertiary halides which solvolyse by a S_N1 mechanism. The effect of the corresponding ionized substituent, (O^-) which should be inductively electron-releasing, is more difficult to assess. Through its $+I$ effect, it should stabilize the transition state but, because of its negative charge, approach of the attacking nucleophile may be hindered. However, under the basic conditions producing the alkoxide ion, the normal reaction of $HO-(CH_2)_n-X$ is intra-

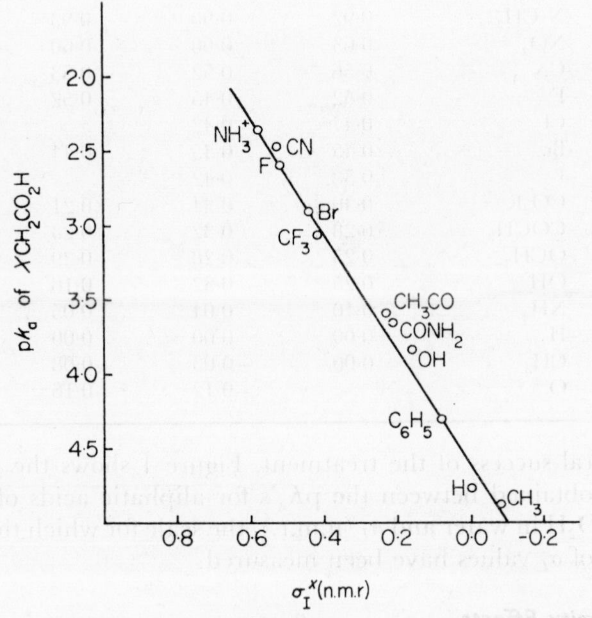

FIGURE 1. Correlation of the acid dissociation constants of XCH_2CO_2H in water at 25°[25] with σ_I (n.m.r.).

molecular and results in the formation of cyclic ethers[26], even in the case where $n = 2$. Such reactions, leading to the conversion of OH into some other functional group, do not come within the scope of this review, unless the cyclic ether is a transient intermediate in a reaction regenerating the hydroxyl substituent. Of the cyclic ethers formed, only the ethylene oxides (oxiranes) show any tendency to cleave in the presence of alkoxide or hydroxide ions, and the conditions are frequently more vigorous than those required for oxirane formation from the halohydrin.

There are at least two cases in which such breakdown is unavoidable and in which the cyclic intermediate is not isolated. Myszkowski and co-workers[27] examined the kinetics of alkaline hydrolysis of 1,3-dichloro-2-propanol and observed that the rate of liberation of chloride ion was greater than the rate of formation of glycerol. They suggested that glycidol was an intermediate in the reaction and proposed the following mechanism.

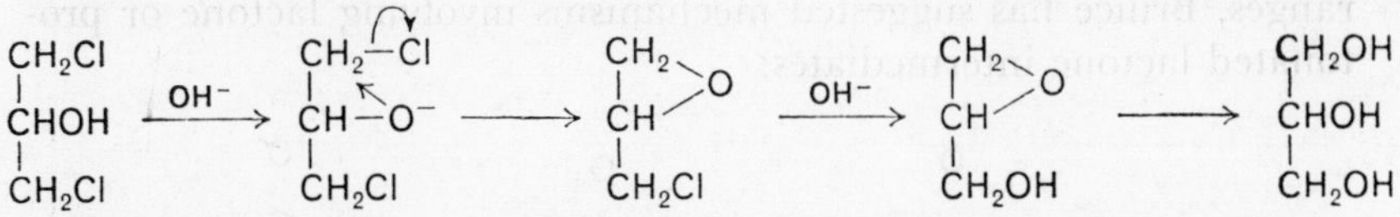

$$\begin{matrix} CH_2Cl \\ | \\ CHOH \\ | \\ CH_2Cl \end{matrix} \xrightarrow{OH^-} \begin{matrix} CH_2{-}Cl \\ | \\ CH{-}O^- \\ | \\ CH_2Cl \end{matrix} \longrightarrow \begin{matrix} CH_2\diagdown \\ | \quad O \\ CH\diagup \\ | \\ CH_2Cl \end{matrix} \xrightarrow{OH^-} \begin{matrix} CH_2\diagdown \\ | \quad O \\ CH\diagup \\ | \\ CH_2OH \end{matrix} \longrightarrow \begin{matrix} CH_2OH \\ | \\ CHOH \\ | \\ CH_2OH \end{matrix}$$

Buchanan and Oakes[28] reported examples of intramolecular nucleophilic catalysis of the opening of an oxetane ring. They found, for instance, that the 3,5-oxide ring of 3,5-anhydro-1,2-*O*-isopropylidene-α-D-glucofuranose is opened more rapidly than that of the corresponding xylose derivative. The ring-opening reaction is believed to involve the reversible formation of a 5,6-epoxide intermediate:

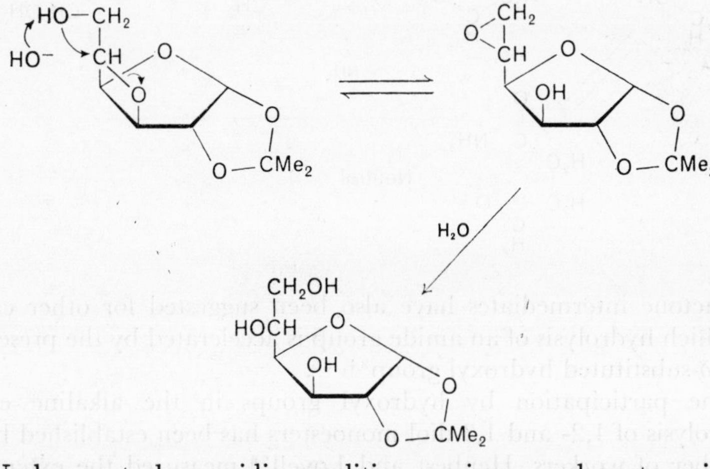

HO−CH₂ / HO⁻ / CH / O / O / O—CMe₂ ⇌ CH₂ / O / CH / O / OH / O / O—CMe₂ —(H₂O)→ CH₂OH / HOCH / O / OH / O / O—CMe₂

Under neutral or acidic conditions, oxirane and oxetane intermediates are not formed[29]. Normal solvolysis products are obtained and the reaction rates are compatible with those predicted from simple electronic effects.

2. Hydrolysis of carboxyl derivatives

Bruice and Benkovic[30] have given an authoritative account of hydroxyl group participation in some ester and amide reactions;

sections of Capon's comprehensive review[31] are also relevant. In the hydrolysis of simple hydroxy-amides, the combined work of Bruice[32] and of Zürn[33] has established neighbouring group participation for the γ- and δ-hydroxyl substituents, which leads to much faster hydrolysis than that shown by the corresponding unsubstituted amides. For γ-hydroxybutyramide hydrolysis over different pH ranges, Bruice has suggested mechanisms involving lactone or protonated lactone intermediates:

Lactone intermediates have also been suggested for other cases in which hydrolysis of an amide group is accelerated by the presence of a δ-substituted hydroxyl group[34, 35].

The participation by hydroxyl groups in the alkaline ester hydrolysis of 1,2- and 1,3-diol monoesters has been established by a number of workers. Henbest and Lovell[36] measured the extent of hydrolysis under standard conditions for cholestane-3,5-diol monoacetates. The results shown on p. 401 indicate that *cis*-1,3-diaxial compounds (**2**) and (**4**) are hydrolysed more rapidly than *trans*-1,3-compounds, (**1**) and (**3**)*.

Other significant increases in the rates of alkaline methanolysis

* This is in spite of the general observation for alicyclic compounds that equatorial acetates are more susceptible to alkaline hydrolysis than axial ones[40].

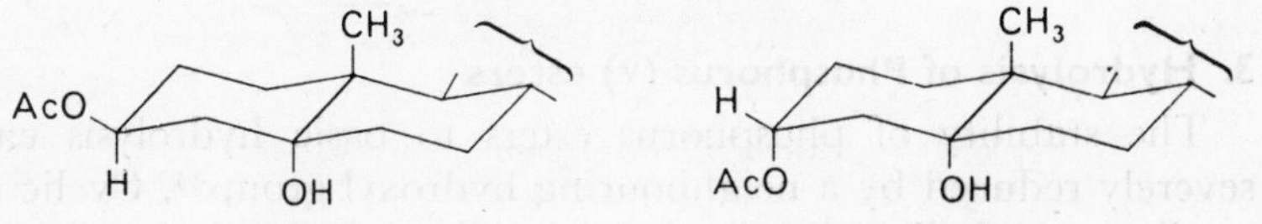

(1) 18 (2) 70

% reaction

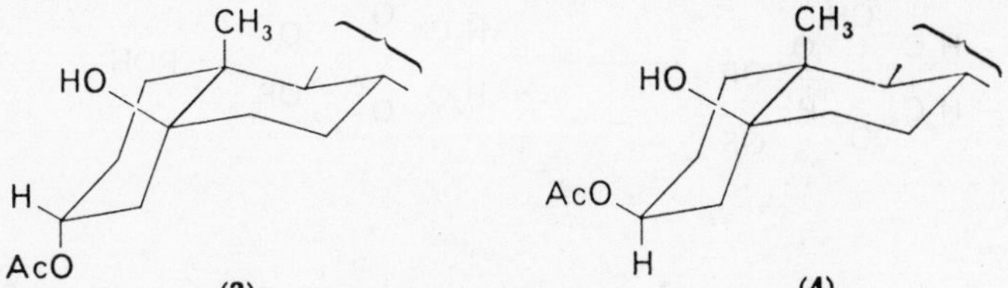

(3) 13 (4) 78

of *cis*-1,3-diaxial diol monoesters have been noted for derivatives of the steroidal alkaloids, germine and cervine[37–39]. Similar accelerations have been observed in the alkaline hydrolysis of 1,2-diol monoesters in cholestane-3β,4β-diol and its derivatives[41].

For 1,2-diol monoesters in five-membered rings, in both substituted cyclopentanes[42] and tetrahydrofurans[43], there are marked rate accelerations, the effect of a *cis*-hydroxyl group being greater than a *trans*-group.

For all these observations a hydrogen-bonding explanation is generally accepted, but there is difficulty in deciding whether the effective hydrogen bond is to the ether oxygen or to the carbonyl oxygen atom:

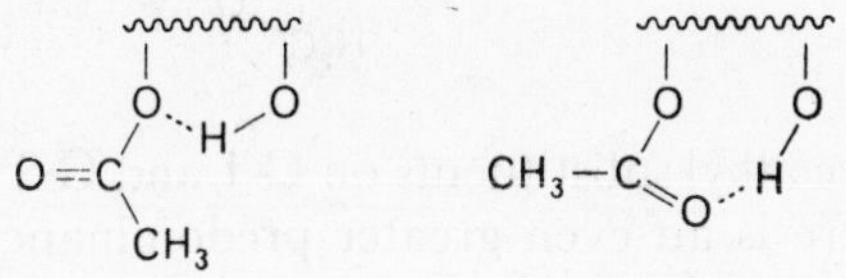

Henbest and Kupchan have both favoured ether–oxygen participation in the cyclohexane derivatives examined by them. Bruice and Fife[42], after kinetic and infrared studies, firmly concluded that, for the hydrolysis of 1,2- and 1,3-cyclopentanediol monoacetates, the transition state is one in which the activating hydrogen bond is to the carbonyl oxygen atom. However, generalizations are not readily made and, as has been emphasized in more than one report, the geometry may be all-important. Capon[31] has also pointed out that the rate enhancement is not usually very great and that this also makes it difficult to generalize.

3. Hydrolysis of Phosphorus (v) esters

The stability of phosphorus esters to basic hydrolysis can be severely reduced by a neighbouring hydroxyl group[44]. Cyclic intermediates are believed to be formed and subsequently hydrolysed:

$$\text{HOCH}_2\text{CH}_2\text{OP(O)(OR)}_2 \longrightarrow \text{cyclic } (\text{CH}_2\text{O})_2\text{P(O)OR} + \text{ROH}$$

$$\longrightarrow \text{HOCH}_2\text{CH}_2\text{OP(O)(OR)OH} \longrightarrow \text{cyclic } (\text{CH}_2\text{O})_2\text{P(O)OH} + \text{ROH} \longrightarrow \text{CH}_2\text{OH}\text{–}\text{CH}_2\text{OPO}_3\text{H}_2$$

In the case of the alkaline hydrolysis of *O*-cyclohexyl *O*-2-hydroxy ethyl phosphate, an alternative hydrolysis path, involving an epoxide as the cyclic intermediate, has also been observed:

$$\text{HOH}_2\text{CCH}_2\text{OP(O)(OC}_6\text{H}_{11})\text{O}^- \xrightarrow{33\%} \text{cyclic } (\text{CH}_2\text{O})_2\text{P(O)O}^- + \text{C}_6\text{H}_{11}\text{OH}$$

$$\text{HOH}_2\text{CCH}_2\text{OP(O)(OC}_6\text{H}_{11})\text{O}^- \xrightarrow{67\%} \text{H}_2\text{C(O)CH}_2 \text{ (epoxide)} + \text{C}_6\text{H}_{11}\text{OP}\bar{\text{O}}_3\text{H}$$

When there are alkyl substituents on C-1 and C-2 of the 2-hydroxy ethyl group, there is an even greater predominance of the epoxide route. In these cases the products are mainly cyclohexyl phosphate and the glycol.

The accelerating effect of similarly placed, β-hydroxyl, substituents has also been observed by Larsson and Wallerberg[45] for the alkaline hydrolysis of alkoxy-diethyl-phosphine oxides, ROP(O)(Et)$_2$. Rate constants are as follows:

R	$k \times 10^4$ (1/mole sec)
CH_3CH_2–	0·573
$CH_3OCH_2CH_2$–	2·55
$HOCH_2CH_2$–	4·42

The authors attribute the rate enhancement to acid catalysis via hydrogen-bonding:

$$Et_2P(=O)(OH)\text{-}O\text{-}CH_2\text{-}CH_2 \xrightarrow{OH^-} \left[HO\cdots P(Et)_2(\cdots O\cdots HO\text{-}CH_2)\cdots O\text{-}CH_2 \right] \longrightarrow Et_2P\bar{O}_2 + \begin{matrix} CH_2OH \\ | \\ CH_2OH \end{matrix}$$

They believe that ester hydrolysis in compounds of this type is S_N2 (on phosphorus) and, if so, this would seem to be the only reported case in which an un-ionized hydroxyl group acts as an acid catalyst for a nucleophilic displacement.

4. Aliphatic acid–base equilibria

In the series glycollic acid ($K_a = 1{\cdot}5 \times 10^{-4}$), hydracrylic acid ($HOCH_2CH_2CO_2H$; $0{\cdot}311 \times 10^{-4}$) and γ-hydroxybutyric acid ($0{\cdot}193 \times 10^{-4}$)[46], the changes in acidity constant are close to those predicted from application of the Taft fall-off factor[13] of 1/2·8. There is therefore no participation by the hydroxyl substituent.

For 1,2-disubstituted-4-*t*-butylcyclohexane and *trans*-decalin derivatives, Sicher and co-workers[47] examined the influence of a neighbouring hydroxyl group on the pK_a values of equatorial and axial carboxylic acids and ammonium ions. Their results indicated that the introduction of an axial hydroxyl group adjacent to an axial carboxyl or ammonium group lowers the pK_a by about 0·45 unit (for acids) or 0·90 unit (for amines). The arrangement of the two groups is *trans* and the effect can be regarded as a simple polar one. The assumption was made that any marked variation from these figures for other orientations is caused by interaction between the hydroxyl group and the functional group. Hydrogen-bonding was proposed but no detailed mechanism was offered. Sicher's results are summarized below; ΔpK is the difference between the *trans* diaxial value and the pK actually found:

Conformation			ΔpK
–OH	$-CO_2H$	$-NH_3^+$	
e	e		+0·12
a	e		+0·36
e	a		+0·94
e		e	−0·23
a		e	−0·49
e		a	−0·24
trans[a]		boat	−0·16
cis[a]		boat	−0·72

[a] From bicyclo[2,2,2]octane derivatives.

No allowance was made for any steric effect caused by the presence of the hydroxyl group but, if this is accepted as minor, the results indicate that hydroxyl group participation increases acid strength for carboxylic acids and decreases base strength for amines. These effects are predictable on the basis of stabilizing hydrogen bonds in the carboxylate ion and the free amine:

O–H···O=C–O   O–H···NH₂

Kilpatrick and Morse[48] have measured the dissociation constants of the various hydroxycyclohexanecarboxylic acids in water, glycol, methanol and ethanol. The results are given in Table 2.

TABLE 2. pK's of *cis*- and *trans*-hydroxycyclohexanecarboxylic acids.

Position of OH group	Conformation relationship	$K_{a(H_2O)} \times 10^5$	$K_{c(glycol)} \times 10^9$	$K_{a(MeOH)} \times 10^{10}$	$K_{a(EtOH)} \times 10^{11}$
cis 2-OH	ea or ae	1·60	30·6	4·42	17·0
trans 2-OH	ee or aa	2·08	21·0	2·33	7·79
cis 3-OH	ee or aa	2·50	18·2	1·70	3·09
trans 3-OH	ea or ae	1·53	9·06	0·817	1·56
cis 4-OH	ea or ae	1·46	8·62	0·761	1·48
trans 4-OH	ee or aa	2·10	14·8	1·50	2·45
Cyclohexanecarboxylic acid		1·25	9·50	0·923	1·70

The situation is much more complex than that in Sicher's system; for each case there are two possible conformations for the molecule in the ionized and in the un-ionized states and the pK_a's of axial and equatorial carboxyl groups differ considerably. In spite of this, it is apparent from the Table that as the solvent becomes less polar, and consequently less capable of effectively solvating carboxylate anions, the strength of each 2-hydroxy acid relative to that of cyclohexanecarboxylic acid in the same solvent is increased markedly. For example, although the *cis*-2-hydroxy acid and the unsubstituted acid have roughly comparable strengths in water, the former is ten times stronger in ethanol. The acid-strengthening effect is presumably due to hydrogen-bonding between the hydroxyl group and the carboxylate ion and seems to be greater for ea conformational relationships than ee (the most probable conformation for the *trans*-2-hydroxy compound). This is supported by Sicher's observations.

Since the acid-strengthening effect (relative to cyclohexanecarboxylic acid) associated with solvent change is only significant for the *cis*-2-hydroxyl derivatives, participation by the hydroxyl group in the 3- and 4-positions may be presumed to be relatively unimportant.

C. Miscellaneous Aliphatic Reactions

Howard[49] has reported that 2-cyclopentylidene-cyclopentanol (**5**), on hydrogenation over Raney-nickel gives at least 96% *trans*-2-cyclopentylcyclopentanol (**6**) and only 1–2% of the *cis*-isomer. The

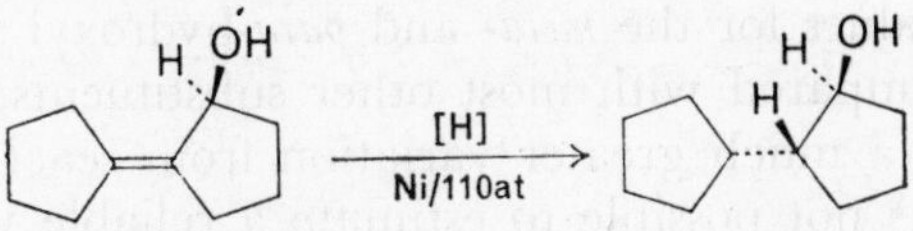

H OH [H] Ni/110at H OH H

(5) (6)

overwhelming preference for the *trans*-isomer in spite of the essentially planar nature of **5** is attributed to some form of bonding between the hydroxyl group and the catalyst surface, resulting in *cis*-addition of hydrogen from the 'hydroxyl' side of the molecule. A similar effect has been reported by Henbest[50] and by Nishimura[51] in cholestene derivatives.

However, in the course of other work on steroid derivatives in which the hydroxyl group is more remote from the double bond, it was noted that *cis*-hydrogenation appeared to be the result of a main attack from the side of the ring remote from the hydroxyl group[52], which is simply exerting a steric effect. The difference in findings may arise from differences in catalyst and in reaction conditions.

III. REACTIONS OF AROMATIC SIDE-CHAINS

A. Electronic Effects

In this section we consider the effects of *meta*- and *para*-hydroxyl substituents on side-chain reactivities. The proximity effects normally associated with *ortho*-substituents are particularly prominent with hydroxyl substituents; because of their high polarity and their tendency to form hydrogen-bonds, these substituents often complicate the transition state through direct interaction with groups at the reaction site*.

* In spite of this, Tribble and Traynham[53] have recently determined σ^- values for a large number of *ortho*-substituents from n.m.r. measurements. The measured value for *o*-hydroxyl ($-0{\cdot}40$) proved to be within the range of reported *p*-hydroxyl values.

A large amount of general information is available on the effect of *m*- and *p*-hydroxyl groups on side-chain reactivity. Much of this is qualitative, and a useful picture of the electronic effects of the hydroxyl substituent is only obtained by restricting discussion to those cases in which there are quantitative data on reactivity effects relative to other substituents. This restriction results in a marked reduction in the number of reactions to be considered.

Most of these reaction series were looked at by Jaffé[2], who correlated each set of data by means of the Hammett equation and calculated σ values for the *meta*- and *para*-hydroxyl substituent. He found that, compared with most other substituents, the derived σ values showed a much greater variation from reaction to reaction and that it was not possible to estimate a reliable value for either σ_m or σ_p.

There seem to be about sixty reactions (see Tables 3–8) from which we can derive, with varying reliability, about forty σ_m values and about fifty σ_p values for the hydroxyl group. The observed variations in these substituent constants may be caused by insufficient or inaccurate data (1), or they may reflect the variation of substituent constants with solvent (2), or with the nature of the side-chain (3).

The first of these causes is the most serious and can hinder attempts to systematize variations of types (2) and (3). Some early, imprecise work is readily recognized and may be ignored. In other studies, inaccurate experimental information on known 'well-behaved' substituents (e.g., H, *m*-CH_3, *m*-Cl, *m*-NO_2) can lead to quite serious errors in σ_p for the hydroxyl substituent because this adds to the uncertainties in extrapolation. For $\sigma_{m\text{-OH}}$ the problem is not serious because the value of this constant is normally within the range covered by the reliable substituents. In fact, for the *meta*-hydroxyl substituent, one can clearly see the variation of σ with solvent. With $\sigma_{p\text{-OH}}$ random scatter is often increased by the inclusion of results derived from studies which cover too few 'well-behaved' substituents; this is especially true of some earlier investigations. In addition, in compiling tables for this section, we have excluded reaction series which have to be built up from the studies of separate working groups; any exceptions are specified.

I. Solvent effects

In Table 3 are listed the data on acidity constants of *meta*- and *para*-substituted benzoic acids in various solvents. It will be seen

TABLE 3. Ionization of $ArCO_2H$ in various solvents.

Reaction	System	Probable range of $\bar{\sigma}_m$ and $\bar{\sigma}_p$[a]	Ref.
(1)	H_2O, 25°	p m	54
(2)	H_2O, 25°	p m	54
(3)	H_2O, 25°	p m	55
(4)	H_2O, 25°	p m	56
(5)	26·5% Dioxan–H_2O, 25°	p m	57
(6)	43·5% Dioxan–H_2O, 25°	p m	57
(7)	73·5% Dioxan–H_2O, 25°	p m	57
(8)	20% EtOH–H_2O[b], 25°	p m	58
(9)	40% EtOH–H_2O, 25°	p m	58
(10)	50% EtOH–H_2O, 25°	p m	58
(11)	50% EtOH–H_2O, 25°	m	59
(12)	70% EtOH–H_2O, 25°	p m	58
(13)	80% EtOH–H_2O, 25°	p m	58
(14)	90% EtOH–H_2O, 25°	p m	58
(15)	95% EtOH–H_2O, 25°	p m	58
(16)	EtOH, 25°	p m	60
(17)	MeOH, 25°	p m	61
(18)	*n*-PrOH, 25°	p m	62
(19)	*n*-BuOH, 25°	p m	63
(20)	Glycol, 25°	p m	64
(21)	Benzene[c], 25°	p m	65
		−0·6 −0·4 −0·2 0·0 0·2 $\bar{\sigma}$	

[a] The range of σ values indicated in the table was determined by basing Hammett plots on only a small number of 'well-behaved substituents' (*m*-NO_2, *m*-CN, *m*-Cl, *m*-Br, *m*-I, H, *m*-CH_3). We would consider any sigma value within the ranges tabulated to be acceptable. The spread is not intended as an indication of the experimental error in any of the individual points.

[b] Insufficient points for a reliable rho value. An estimate of $\rho = 1·2$ was used.

[c] Based on the equilibrium constant for the reaction between the benzoic acid and diphenylguanidine.

that in only two cases has the benzoic acid series been systematically studied by more than one group of investigators—in water, and in 50% ethanol. Of the four sets of data on aqueous solution, those from studies (3) and (4) must be considered the most reliable by a considerable margin.

Much of Table 3 is based on the work of two groups. Most results support a generalization that, in moving from water to solutions containing substantial proportions of low molecular-weight alcohols, the value of $\bar{\sigma}_m$ shifts from a value of about $+0{\cdot}1$ to a figure of approximately $-0{\cdot}1$. This is an appreciable change. In the move from water to dioxan-rich solvents, the shift in sigma values is very much smaller. There is one report on benzene as solvent, and the sigma value is intermediate between those in water and alcohols; on the whole, this conclusion is supported by other isolated studies on non-polar solvents.

The $\bar{\sigma}_p$ values from Bright's data[58] on ethanol–water mixtures perhaps imply changes similar to those observed for σ_m—a shift to more negative values as the proportion of alcohol is increased. But the work of Kilpatrick[60–64] does not indicate that the values in pure alcoholic solvents are substantially different from the figure in water.

2. Dependence on side-chain

Two key reviews of σ values are relevant: that by Taft and Lewis[3] in which they compiled a list of σ^o values, and that by Wepster[4], who derived σ^n values. Taft and Lewis do not quote a value for $\sigma^o{}_m$ or $\sigma^o{}_p$ for the hydroxyl group in hydroxylic solvents, but merely state that the value is strongly dependent on the nature of the solvent. They give figures for $\sigma_m{}^o$ and $\sigma_p{}^o$, relating to nonaqueous solvents, of $+0{\cdot}04$ and $-0{\cdot}13$ respectively, and these may be based on Reaction 21 in Table 3. In his paper, Wepster reports $\sigma^n{}_{m\text{-OH}}$ as $+0{\cdot}095 \pm 0{\cdot}025$ and concludes that $\sigma_{p\text{-OH}}$ is variable.

There are no Hammett data covering the effect of either a *meta*- or *para*-hydroxyl substituent on the reactivity of a functional group which is insulated from the benzene ring (e.g., by a methylene group) and there is therefore no standard of reference for any discussion of variations in interaction between such a substituent and the side-chain. In Tables 4–7 data on a selection of different reactions are given. If a reaction has been investigated by more than one group under similar conditions, we have made the choice of what appears to be the most reliable figure. The reactions are classified on the basis of the electronic nature of the side-chain.

There are only four reactions in the Tables in which the values for σ_m differ significantly from those expected from the nature of the reaction solvent. These are reactions (33), (34), (38) and (43). For reactions (33) and (34) the differences may be caused by hydrogen-bonding between solvent and phenolic oxygen; in each

TABLE 4. Reactions of weak $-R$ side-chains.

	Reaction	Probable range of $\bar{\sigma}_m$ and $\bar{\sigma}_p$ [a]	Ref.
(1–21)	Ionization of $ArCO_2H$	see Table 3	
(22)	Rate, $ArCO_2H + HN_3$, trichloroethylene, 40°	m	66
(23)	Rate, $ArCO_2H + Ph_2CN_2$, EtOH, 30°	p m	59
(24)	pK_a, $ArP(O)(OH)_2$, H_2O, 25°	p m	67
(25)	pK_a, $ArP(O)(OH)_2$, 50% $EtOH–H_2O$, 25°	p m	67
(26)	pK_a, $ArPO_2H^-$, H_2O, 25°	p m	67
(27)	pK_a, $ArPO_2H^-$, $EtOH–H_2O$, 25°	p m	67
(28)	pK_a, $ArAs(O)(OH)_2$, H_2O [b], 22°	p	68
(29)	pK_a, $ArAsO_2H^-$, H_2O [c], 22°	p	68
(30)	pK_a, $ArCH{=}CHCO_2H$, H_2O, 25°	m	69
(31)	pK_a, $ArC_6H_4CO_2H$, 50% Butyl cellosolve–H_2O, 25°	p	70
(32)	Rate, $ArCOOCH_3 + PhNH_2$, $PhNO_2$, 100° [d,e]	m p	71

[a] See footnote to Table 3.
[b] Other data have also been reported for this reaction at 18° [72,73].
[c] The authors assign pK_2 to the OH group and pK_3 to the AsO_2H^-. This would lead to a very high value for the substituent constant for the AsO_2H^- group. We have assigned pK_3 to the OH ionization which leads to a $\sigma_p^{AsO_2H^-}$ of approximately zero.
[d] The conditions for this reaction are rather vigorous and the stability of phenols in this system is unknown.
[e] The authors also report data at other temperatures that lead to similar $\bar{\sigma}$ values.

TABLE 5. Reactions of strong *–R* side-chains.

	Reaction	Probable range of $\bar{\sigma}_m$ and $\bar{\sigma}_p$[a]	Ref.
(33)	pK_a, $ArCO_2H_2^+$, H_2SO_4–H_2O, 25°	p m	74
(34)	pK_a, ArC^+OHCH_3, H_2SO_4–H_2O, 25°	p m	75
(35)	pK_a, Ar_3C^+, H_2O, 25°	p	76
(36)	pK_a, Ar—$C_7H_6^+$, 50% CH_3CN–H_2O, 25°	p	77
(37)	pK_a, Ar=NNHPh$^+$, 20% EtOH–H_2SO_4[b], 25°	p	78
(38)	pK_a, ArC(=NH_2^+)Ph[c, d], H_2O	m p	79
(39)	pK_a, ArCH=NH_2^+Ph, MeCN[e], 25°	p m	80
(40)	Rate, ArC(=NH_2^+)Ph + H_2O[e], H_2O[f], 25°	p m	79
(41)	Rate, $ArNO_2$ + $SnCl_2$, H_2O, 90°	p	81

-0·8 -0·6 -0·4 -0·2 0·0 0·2

$\bar{\sigma}$

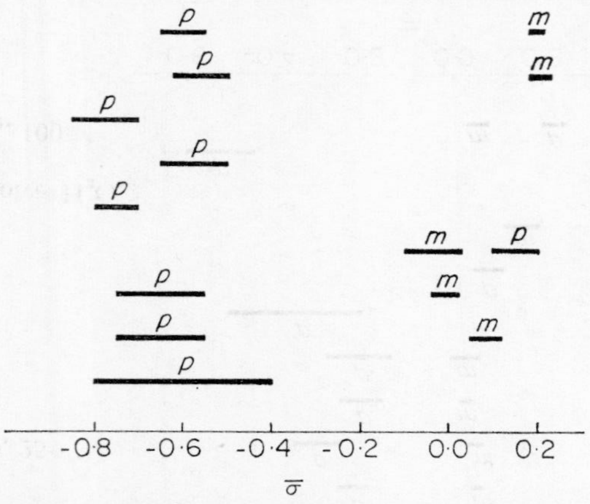

[a] See Table 3.

[b] The author assumes proton sharing by the two nitrogens to account for the single Hammett plot for both $+T$ and $-T$ substituents.

[c] The 3,5-dimethyl substituent was used in constructing the Hammett plot for this reaction.

[d] The measured dissociation constant for the *p*-OH product may be that of phenolic group.

[e] The same series was also studied using compounds with substituents in the other aromatic ring. Differences in $\bar{\sigma}$ values were not considered significant and correlation was poorer than in the series listed here.

[f] An additional value for $\bar{\sigma}_m$ based on the rate at 0° was available from the same source but this was not significantly different from the one here.

instance similar positive deviations in $\bar{\sigma}$ have been noted for the *meta*-alkoxy group[74, 75].

The $\bar{\sigma}_p$ values cover a wide range but, as a rough guide, they can be placed in one of two groups—the values between −0·2 and −0·4, which may arbitrarily be called 'normal' substituent constants, and the 'exalted' values lying between −0·5 and −0·9. The first range is generally linked with the reactions of weak $-R$ and with $+R$ side-chains; the second range is applicable to strong $-R$ side-chains. Clear exceptions to the generalization are reactions (31), (42), (43) and (50). The high value for $\bar{\sigma}_p$ in reaction (31) is matched by the correspondingly high negative σ value required for the *para*-amino group in this reaction (but not for the *p*-methoxyl group). However, the rho value for the reaction is rather low (about 0·45) and the experimental error may be greater than usual.

The high sigma value in reaction (50) is interesting. Although the reaction is tabulated as that of a $+R$ (amino) side-chain, the suggested mechanism[85] involves an electron-deficient nitrogen intermediate:

$$\mathrm{Ar}-\overset{+}{\mathrm{N}}-\underset{\mathrm{NH_2}}{\underset{|}{\mathrm{C}}}=\mathrm{N}-\mathrm{N}=\overset{-}{\mathrm{N}}$$

We should therefore treat the reaction as that of a strong $-R$ side-chain, and it yields a $\bar{\sigma}_p$ value within the expected range for such a reaction.

The value for $\bar{\sigma}_p$ for phenol dissociation appears to be out of line, but this trend towards less negative $\bar{\sigma}_p$ values for phenolic ionizations is also discernible for the *p*-NH_2 group—and is possibly true for other $+R$ substituents.

The exalted sigma values obtained for reactions either involving 'strong' $-R$ side-chains and/or attack at aromatic centres can, of course, be attributed to large changes in resonance interaction in the system in going from the ground state to the transition state. Within the given range, the values appear to either be around −0·6 ($-R$ side-chains) or around −0·8 (aromatic centres), but once again the data are not numerous.

3. Comparison of substituent constants for methoxyl and hydroxyl groups

The substituent constants for the methoxyl group, like those for the hydroxyl group, appear to vary from reaction to reaction. The

TABLE 6. Reactions of $+R$ side-chains.

	Reaction	Probable range of $\bar{\sigma}_m$ and $\bar{\sigma}_p$[a]	Ref.
(42)	pK_a, ArOH, H_2O, 25°[b]	p m	[b]
(43)	pK_a, ArOH, 8% Dioxan, 38°	p m	82
(44)	pK_a, ArOH 48·9% EtOH–H_2O, 20–2°	p	83
(45)	pK_a, ArOH 95% EtOH–H_2O, 20–2°	p	83
(46)	pK_a, ArSH 48·9% EtOH–H_2O, 20–2°	p	83
(47)	pK_a, ArSH 95% EtOH–H_2O, 20–2°	p	83
(48)	pK_a, $ArNH_3^+$, H_2O[c], 25°	p m	54
(49)	pK_a, Pyridine N-oxides, H_2O[d], 23–5°	p	84
(50)	K_{eq}, H_2N—C=N—N=N—N(Ar) (ring) ⇌ HN—N=N—N=C(NHAr) (ring), glycol, 193–4°	p	85
(51)	pK_a, ArN=CHPh MeCN[e], 25°	p	86
(52)	pK_a, $ArNH_3^+$, MeCN, 25°	p	86

Axis: $\bar{\sigma}$: -0·6 -0·4 -0·2 0·0 0·2 0·4

[a] See Table 3.
[b] A composite graph based on pK_a, H_2O, 25° taken from Kortum[25]; pK_a for m-OH is an estimate based on the value at 18°, pK_a p-OH is that of Bishop and co-workers[87].
[c] Additional data on this reaction at 25° and at 20° are available[54, 88]. The range given is consistent with all three.
[d] Additional data (also unreliable) are available[89].
[e] The same series was also studied using compounds with substituents in the other aromatic ring. Differences in $\bar{\sigma}$ values were not considered significant, and correlation was poorer than in the series listed here.

TABLE 7. Reactions of aromatic nuclei.

	Reaction	Probable range of $\bar{\sigma}_m$ and $\bar{\sigma}_p$ [a]	Ref.
(52)	pK_a, Pyridines, H_2O, 23–5°	m	84
(53)	Rate, $ArGe(Et)_3 + H_3O^+$, 71·4% $MeOH–H_2O$, 50°	p	90
(54)	Rate, $ArSi(Me)_3 + H_3O^+$, 71·4% $MeOH–H_2O$, 51·2°	p	91
(55)	Rate, $ArH + Br_2$, CH_3COOH, 25°	p	92

$\bar{\sigma}$: -1·0 -0·8 -0·6 -0·4 -0·2 0·0

[a] See Table 3.

TABLE 8. Reactions involving the O^- substituent.

	Reaction	Probable range of $\bar{\sigma}_m$ and $\bar{\sigma}_p$ [a]	Ref.
(56)	pK_a, $ArC(=N^+H)NBu_2{}^n.HCl$, 50% $MeOH–H_2O$, 25°	p	94
(57)	Rate, $ArCO_2Et + OH^-$, H_2O [b], 25°	p m	95
(58)	Rate, $ArCO_2Et + OH^-$, 56% $Me_2CO–H_2O$, 25°	p m	96
(59)	Rate, $ArCONH_2 + OH^-$, H_2O, 100°	p	97
(60)	pK_a, ArOH, H_2O [c], 25°	m and p	[c]
(61)	Rate, X–$C_6H_3(NO_2)$–Br + piperidine, piperidine[d], 25°	p	98

$\bar{\sigma}$: -1·2 -1·0 -0·8 -0·6 -0·4

[a] See Table 3.
[b] The data for substituents other than *m*-OH and *p*-OH are obtained from reactions in mixed aqueous organic solvents by extrapolation.
[c] See Footnote [b], Table 6. Data on *p*-O^- are from Bishop[87], data on *m*-O^- are unreliable.

para-methoxyl group appears to parallel the *para*-hydroxyl group in reactivity but to be slightly less reactive.

For the *meta*-methoxyl substituent Taft[3] suggests a value of +0·13 for pure aqueous solutions and +0·06 for nonhydroxylic media and most mixed aqueous organic solvents. Only a minority of the listed reaction studies (Tables 3–7) yield data on both the *m*-OH and *m*-OMe substituents and most of these refer to aqueous solution. In his review, Wepster[4] has tabulated $\bar{\sigma}$ values for *m*-OMe for forty reactions and although there is a tendency for $\bar{\sigma}_{OMe}$ to fall to about +0·05 in alcoholic solvents it rarely becomes negative, and never as negative as *m*-OH in similar solvents. This implies that negative shifts of *m*-OH in alcoholic solvents may be caused by hydrogen-bonding of the phenolic hydroxyl group, perhaps in the manner proposed by de la Mare[93].

4. The oxide ion substituent as an activating group

In contrast to the –OH group which is inductively electron-withdrawing ($-I$) and conjugatively electron-releasing ($+R$), the negatively charged oxygen of a phenoxide ion is strongly electron-releasing by both mechanisms ($+I$, $+R$). There are few data available, however, to give a quantitative estimate of its reactivity. For only six reaction series can the effect of an O^- substituent in the *meta*- (or *para*-) position be estimated as a $\bar{\sigma}$ value. These are given in Table 8. It is clear from the Table that few useful conclusions can be drawn, and practically all that can be said about $\bar{\sigma}_m$ is that, as would be predicted, it is negative.

Reactions (58), (59) and (60) involve attack by the negatively charged hydroxide ion on a species already bearing a negative charge. Such attack would be comparatively slow and would lead, in these cases, to $\bar{\sigma}_m$ and $\bar{\sigma}_p$ values more negative than otherwise expected. It is noteworthy that $\bar{\sigma}_p$ for reaction (56) is less negative than for other reactions.

The one apparently abnormal value in Table 8 is the value for $\bar{\sigma}_m$ in reaction (61). There is only one reported measurement of pK_2 for resorcinol[99] and until this figure receives further support, it should be treated with caution.

5. The substituent constant for the hydroxyl group

The generally accepted substituent constants for the *meta*- and *para*-hydroxyl substituents are given in Table 9.

TABLE 9. Substituent constants for the hydroxyl group.

Author	Symbol	σ_m	σ_p
Jaffé[2]	σ	-0.002 ± 0.106	-0.357 ± 0.104
McDaniel and Brown[100]	σ	$+0.121 \pm 0.02$	-0.357 ± 0.04
Wepster[4]	σ^n	$+0.095 \pm 0.025$	-0.178 ± 0.036
Taft[3]	σ^o	$(+0.04)^a$	(-0.13)
Brown and Okamoto[6]	σ^+	—	-0.92
Deno[76]	σ^+	—	-0.82
Hine[101]	σ	$+0.165$	-0.21
Yukawa–Tsuno[102]	σ^o	—	-0.16

[a]Nonhydroxylic solvent only. The value in hydroxylic solvents is variable.
In addition, for the $-O^-$ substituent, Jaffé gives σ_m as -0.708 and σ_p as -0.52; Hine suggests -0.47 and -0.81 respectively.

The available data lead us to suggest that the following hydroxyl sigma values can be expected under the given conditions:

$\sigma_{m\text{-OH}}$ $+0.10 \pm 0.05$ for aqueous and water-rich aqueous mixtures.
-0.10 ± 0.03 for alcohols and alcohol-rich solvents.
0.0 ± 0.05 for nonhydroxylic solvents.
$+0.20 \pm 0.05$ for very strongly acidic solutions (e.g., H_2SO_4)

These values are considered to be independent of the nature of the side-chain.

$\sigma_{p\text{-OH}}$ -0.20 to -0.40 for reactions of weak $-R$ and $+R$ side-chains, e.g., CO_2H, NH_2
-0.50 to -0.80 for reactions of strong $+R$ side-chains, e.g., $CO_2H_2^+$, $-C(R){=}NH_2^+$
-0.70 to -0.90 for electrophilic aromatic substitution.

For $\sigma_{p\text{-OH}}$ solvent dependence is considered unlikely to shift the values outside the given ranges.

6. The substituent constant for the oxide ion

There are two influential features of this substituent: the electronic effect of the group, and the overall charge on the molecule in the transition state. With only three reactions as a guide, the rough estimate of σ_m is -0.35 ± 0.1. For σ_p, the best that can be said is that the value is expected to lie between -0.5 and -1.0, higher values being associated with nucleophilic attacks on negatively charged substrates.

B. *Proximity Effects*

I. Aromatic acid–base equilibria

The best known and most widely quoted example of the proximity effect of a hydroxyl group on an acid–base reaction is the abnormally high acidity of salicylic acid[103]. This can be seen most readily by comparison with the other monohydroxybenzoic acids and methoxybenzoic acids:

$10^5\ K_a$ in H_2O, 25°

	o	*m*	*p*
hydroxy (K_1)	105	8·3	2·9
methoxy	8·06	8·17	3·38

This high acidity is attributed to abnormal stabilization of the salicylic acid mono-anion by internal hydrogen-bonding:

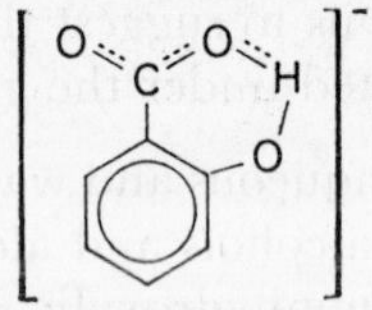

The negative charge is shared by all three oxygen atoms. The importance of the proton in stabilizing the system is illustrated by the fact that K_2 for salicylic acid is more than 2500 times smaller than that for either the *meta*- or the *para*-isomer. However, the issue is complicated by the effect of high charge repulsion in the *ortho*-dianion.

Dippy and co-workers[104] have reported that 6-nitrosalicylic acid (**7**) is a weaker acid than *o*-nitrobenzoic acid (**8**) indicating that the acid-strengthening factor in salicylic acid is no longer present in **7**.

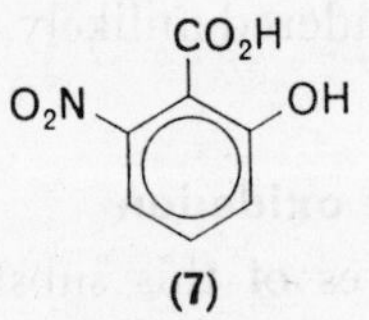

(**7**) $K_a = 5{\cdot}81 \times 10^{-3}$ 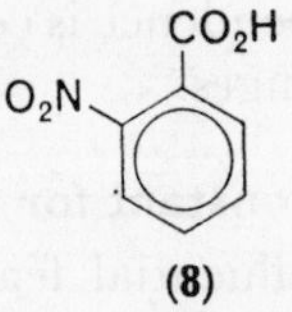 (**8**) $K_a = 6{\cdot}71 \times 10^{-3}$

Dippy considers that the carboxyl group is forced out of plane. In these circumstances hydrogen-bonding is less effective and stabilization of the mono-anion accordingly lower.

In the case of 2,6-dihydroxybenzoic acid ($pK_a = 1{\cdot}3$) the proximity effect in the primary dissociation is even greater[105]; the pK_a of 2,6-dimethoxybenzoic acid is 3·44[106]. A similar but much smaller exaltation of pK_a is found for catechol.

pK_a of phenols[25], H_2O, 25°

	o	*m*	*p*
hydroxy	9·48	9·44	9·96
methoxy	9·98	9·65	10·21
hydroxymethyl	9·92	9·83	9·82

It will be seen that for the hydroxymethyl phenols, differences are too small to be significant.

For *o*-aminophenols, there is no evidence for hydroxyl participation. Furthermore, no pK_a value is available for the *o*-hydroxybenzylammonium ion, and no conclusion can be drawn from the closely similar figures for 2-hydroxy-3-methoxybenzylamine (8·70) and 3-hydroxy-2-methoxybenzylamine (8·89)[107].

2. Hydrolysis of carboxyl derivatives

Bender, Kezdy and Zerner[108] found that the alkaline hydrolysis of *p*-nitrophenyl-5-nitrosalicylate was unexpectedly fast. The data were consistent with either **9**, reaction of hydroxide ion with the un-ionized ester (intramolecular general acid-nucleophilic catalysis) or **10**, reaction of the ionized ester with water (intramolecular general-base catalysis). The accelerations observed for the reaction

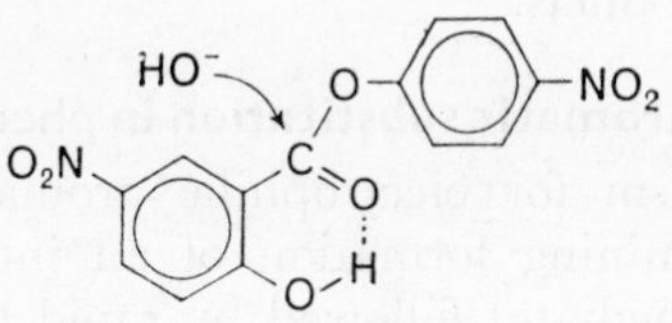

(9) 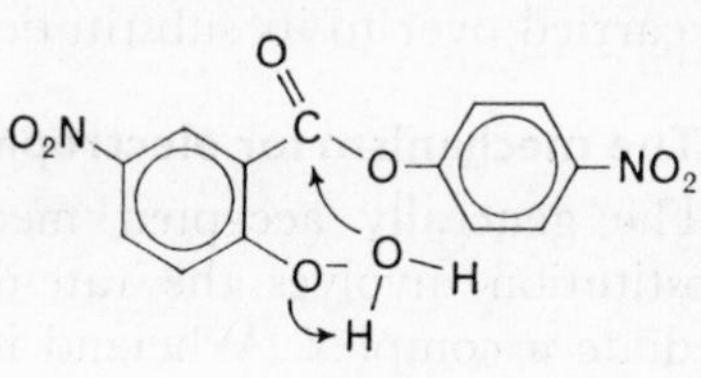 (10)

were much greater than those for corresponding reactions involving nucleophiles in which acidic hydrogens are absent (e.g. azide ion) and mechanism **10** was considered to be the more likely.

Other studies of esterification and ester hydrolysis have led to the suggestion of lactone intermediates. Kupchan and Saettone[109] have proposed that the esterification of *o*-hydroxyphenoxyacetic acid proceeds through the lactone since the rate is at least ten times greater than that of the *o*-OCH_3, *o*-Cl or *p*-OH compounds. The considerable accelerating effect of a 2-hydroxy substituent on the rate of hydrolysis of methyl triptoate has also been accounted for in terms of the intermediacy of a lactone[110].

The occurrence of intramolecular general-base catalysis in amide hydrolysis has been established by Bruice and Tanner[111], who investigated the mechanism of hydrolysis of salicylamides. If general-base catalysis were operative, the introduction of a nitro group in the 5-position should increase the reactivity of the amide side-chain and decrease the ability of the oxide ion substituent to act as a general base. If intramolecular general-acid catalysis occurred, on both grounds the introduction of the nitro group should increase the reaction rate. The rate of hydrolysis for salicylamide proved to be lower than for the 5-nitro compound, a result that supported the general-base mechanism.

IV. AROMATIC SUBSTITUTION REACTIONS

A. Electrophilic Substitutions—General Considerations

The very large amount of information available in this field enforced selection. Most of the subsequent sections deal with this type of aromatic substitution. We have decided that the more profitable results are those related to phenol itself. The extremely powerful directing and activating effect of the hydroxyl group gives it a strong tendency to swamp the effects of most other substituents in phenol derivatives*, and it is often true that trends observed for phenol can be carried over to its substitution products.

1. The mechanism for electrophilic aromatic substitution in phenol

The generally accepted mechanism for electrophilic aromatic substitution involves the rate-determining formation of an intermediate σ complex (Wheland intermediate) followed by rapid loss of proton[112], e.g., for benzene,

There is no evidence against a Wheland intermediate in the case of substitution in phenols, but there is considerable support for the opinion that the actual scheme is much more complex than that

* A notable exception is that 2,4-dimethoxyphenol undergoes bromination in the 5-position; Fries rearrangement of the acetate also leads to the 5-ketone[113].

pictured here. For instance, it has been established that, in certain halogenations of phenol, dienone intermediates may be formed and that weakening or breaking of the O–H bond may be involved in the rate-determining step[93]. Whether this is caused by hydroxyl 'hyperconjugation' or solvent interactions is still a matter for argument. However, the observations that phenol is more activated than anisole in the *para*-position and that phenol is brominated in acetic acid nearly twice as rapidly as in *O*-deuteroacetic acid demonstrate the inadequacy of the simple mechanism, at least for halogenations*.

Electrophilic attack at the *ortho*-position of phenol may be complicated by interaction with the hydroxyl group. In some cases the electrophile attacks oxygen first and later migrates; in others some form of weak bonding between the attacking electrophile and the phenolic oxygen results in abnormally high *ortho–para* ratios. In the case of phenoxides, participation in *ortho* attack may be even more pronounced. One case has been reported where the phenoxide ion oxygen acts as a general base in removing the proton from the Wheland intermediate. In other cases, *ortho* attack may be preceded by the formation of some sort of preliminary complex involving the O^- substituent, the associated metal ion and the electrophile. It is probable, indeed, that electrophilic substitution at the *ortho*-position of phenols and phenoxides is rarely a 'normal' reaction.

2. The *ortho–para* ratio

The usual site for electrophilic attack on the aromatic ring of phenol is *ortho* or *para* to the hydroxyl group. There is some activation of the *meta*-position. Illuminati[114] estimates a value for $\sigma_m{}^+$ of $-0{\cdot}133$ for bromination in acetic acid but in view of the very high rho value (>10) for this system and the relatively high negative value expected for σ_o and σ_p ($\sigma_p \approx -0{\cdot}8$), one would not predict appreciable amounts of *meta*-substitution products. The occurrence of *meta*-substitution products is most common in cases such as sulphonation or Friedel–Crafts reactions, where substitution is reversible and the *meta*-isomer is thermodynamically the most stable.

In a monosubstituted benzene derivative, the steric effect of the substituent normally decreases the accessibility of the *ortho*-positions. In addition, the $-I$ and $+R$ electronic characteristics of the hydroxyl group combine to favour *para*- substitution. Overall then, an *ortho–para* ratio of less than the statistical 2 : 1 should be found. Some

* The mechanism of bromination of phenols is discussed thoroughly by de la Mare[93].

experimental observations support this prediction but in the majority of cases the *ortho–para* ratio is high, sometimes greater than 2 : 1. An accounting in terms of participation by the OH or O^- group has been referred to in the previous section.

3. Replacement of substituents other than hydrogen

Substitutions of this type have been observed quite frequently in aromatic systems[115]. They can be divided conveniently into two groups: those in which the replacement group is hydrogen, and those in which the electrophile is some other group or atom. The replacement of substituents by hydrogen is the most widespread, probably because the hydrogen ion is an extremely powerful electrophile and can readily be generated in high concentrations. The ease of reaction varies considerably. Some groups such as MgX are displaced extremely readily while others require strong acids and/or activation by other substituents in the ring, e.g., protodehalogenation. One or two of the reactions are mentioned elsewhere in this review—dealkylation, desulphonation, desilylation and degermylation for instance. The pronounced activating effect of the hydroxyl substituent *ortho* or *para* to the leaving electrophile is most apparent in the more difficult displacements, e.g., dehalogenation and decarboxylation[115].

The replacement of substituents by groups other than hydrogen are not as common, because these reactions are reversible and restricted to powerful attacking electrophiles. A number of such displacements have been observed in nitration and halogenation. For example, *para*-substituents are often replaced during nitration or halogenation of phenols[116]. However, in these cases the mechanism may not be simple. Addition products of phenols have been isolated during the course of halogenations[93], and the normal product of bromination of phenol in aqueous solution is the ketone, 2,4,4,6-tetrabromocyclohexadienone[117].

B. *ortho–para* Ratios in Electrophilic Substitution

1. Nitration and nitrosation

The nitration of aromatic compounds has been extensively studied, and some impressive and comprehensive reviews are available[118–120].

a. Nitration in organic solvents. On the nitration of phenol, perhaps the most surprising feature of the data is the constancy of the *ortho–para* ratio for nitration in organic solvents. Table 10 is from the work of Arnall[121].

TABLE 10. Nitration of phenol with nitric acid in various solvents.

Solvent	Nitrophenol (%)		
	ortho	*para*	*meta*
Acetic anhydride	59·6	37·8	2·6
Acetic acid	59·2	38·1	2·7
Acetone	57·4	39·6	3·0
Ether	57·8	39·2	3·0
Ethanol–acetic acid (2 : 1)	57·6	39·3	3·1
Ethanol	57·7	39·2	3·1

The change in orientation with temperature is not great at room temperatures and above. Presumably this is because nitration, unlike sulphonation and Friedel–Crafts alkylation and acylation, is not reversible under the reaction conditions. However, at low temperatures a change in the *ortho–para* ratio has been observed by Spryskov[122] for nitration with nitric acid–acetic anhydride mixtures.

TABLE 11. Nitration of phenol with 99·6% HNO_3 in acetic anhydride.

<table>
<tr><th rowspan="2">T(°C)</th><th colspan="3">Nitrophenol (%)</th></tr>
<tr><th>ortho</th><th>para</th><th>meta</th></tr>
<tr><td>+20</td><td>54·9</td><td colspan="2">—— 45 ——</td></tr>
<tr><td>0</td><td>62·5</td><td>32·3</td><td>5·2</td></tr>
<tr><td>−15</td><td>68·8</td><td colspan="2">—— 31·2 ——</td></tr>
<tr><td>−30</td><td>72·0</td><td colspan="2">—— 28·0 ——</td></tr>
<tr><td>−50</td><td>79·5</td><td>16·2</td><td>4·3</td></tr>
<tr><td>−56</td><td>80·0</td><td>16·3</td><td>3·7</td></tr>
<tr><td>−70</td><td>79·0</td><td>16·0</td><td>5·0</td></tr>
</table>

A limiting $o:p:m$ ratio of 80: 16: 4 appears to be obtained for temperatures below −50°. The increased amount of *ortho*-product may be attributable to increased participation by the hydroxyl group at low temperatures. Participation by the lone pairs on the oxygen has been proposed[123] to explain the anomalously large amount of *ortho*-product formed during the nitration of anisole by acetyl nitrate, e.g.,

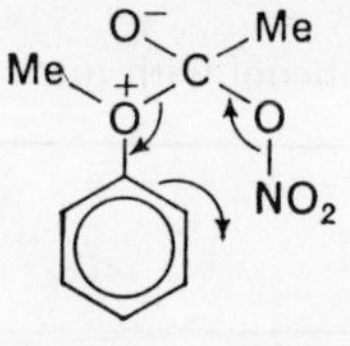

An alternative explanation based on a change in the nitrating agent is also possible. Bordwell and Garbisch[124] have shown that whereas acetyl nitrate is rapidly formed by the reaction of nitric acid and acetic anhydride at room temperature, at $-10°$ most of the nitric acid does not react, and nitration is a much slower process. Spryskov's nitration technique involved the addition of the nitric acid last; accordingly the nitrating agent under his conditions is more likely to be nitric acid or $H_2NO_3^+$ than acetyl nitrate except for his runs at 0° and above. It may be significant that the runs with the highest *ortho–para* ratios are also those which would be expected to involve reaction of nitric acid rather than the larger acetyl nitrate species. However, since the *ortho–para* ratio has in many cases proved to be almost independent of the nitrating agent, this explanation is not as attractive as the first.

b. Nitration in aqueous solution. Ingold and co-workers[125] have investigated the nitration of phenol in aqueous solution and found that the reaction is catalysed by nitrous acid which, if not initially present, is formed as a by-product. In addition to the effect of the catalyst on reaction rates, the *ortho–para* ratio is drastically changed. The same workers found that in acetic acid solvent similar trends were observed, the *ortho–para* ratio changing from about 45: 55 (cf. Arnall's figures) to about 74: 26 when considerable quantities

TABLE 12. The effect of nitrous acid on the *ortho–para* ratio for nitration by nitric acid in strongly acid aqueous solutions[125].

T = 20° [PhOH] = 0·45M, [HNO_3] = 0·50M, [H_2SO_4] = 1·75M

[HNO_2]	*ortho* : *para*
0·00	73 : 27
0·25	55 : 45
1·00	9 : 91

of N_2O_4 were added. These changes are probably caused by changes in the nature of the nitrating species, but at present it is uncertain which species are involved.

c. meta-*Nitration.* Arnall suggested a figure of 2–3% for *meta*-nitration under all conditions[121]. Spryskov[122] gave a somewhat higher value of 4·5% but it is unlikely that the difference is significant.

d. Nitrosation. Phenol reacts with acidified solutions of sodium nitrite to form the *o*- and *p*-nitroso derivatives. The reaction has been studied in detail by Veibel[126]. He found that the main point of attack was *para* to the hydroxyl group. The proportion of *ortho*-compound in the product was about 6% at 0°, and rose to 10% at 40°. The kinetics of nitrosation have been studied by Suzawa and co-workers[127] but they were unable to identify the nitrosating agent.

2. Halogenation

In the halogenation of phenol, the tendency to high *ortho–para*-substitution ratios is much less marked than in nitration. Data on these particular ratios are not easily found in spite of the existence of a considerable body of published work on phenol halogenations[118]. Most mechanistic studies on phenol have been concerned with reaction rates determined by following the rate of disappearance of halogen. Isomer ratios have seldom been determined. However, from preparative halogenations, the following information can be drawn.

TABLE 13. Halogenation of phenol.

Halogen	*ortho*	*para*
Cl_2, CCl_4 [128]	74	26
Br_2, CCl_4 [129]	11·4	88·6
I_2, benzene[a] [130]	23	77

[a] In the iodine case, the reaction may be with the phenoxide ion[131], and the $+I$ effect of the O^- group should to some extent offset the depressing effect of the large halogen on *ortho*-substitution.

It would appear that some participation by the hydroxyl group is likely because the corresponding data for anisole ($o:p = 21:79$ for chlorination and $2:98$ for bromination)[132] show a much stronger tendency for anisole to be attacked in the *para*-position. In addition, chlorination of phenol in methanol at 0–5° has been stated to give *o*-chlorophenol[133]. Both the higher overall reactivity of phenol as compared with anisole, and the greater proportion of *ortho*-isomer formed, would be consistent with a greater concentration of negative charge on the phenolic oxygen. Norman and Harvey[128] have reported on the chlorination of phenol using Bu^tOCl. They conclude that in acidic or neutral solutions the halogenating agent is Cl^+ and that in base the reactive species is HOCl. Comparative figures are given in Table 14.

TABLE 14. Chlorination of phenol in different systems[128].

	System	*ortho*	*para*
(1)	*t*-BuOCl, CCl_4	51	49
(2)	Cl_2, CCl_4	74	26
(3)	*t*-BuOCl, H_2SO_4	50·8	49·2
(4)	Cl_2, PhOH (molten)[a], 60°	39·5	60·5
(5)	Cl^+, H_2O	51·4	48·6

[a] Bing and co-workers[134] report an *ortho–para* ratio of 37 : 63 for this system.

Systems (1) and (3) are those that appear to involve attack on the phenol by Cl^+. Presumably in the case of (1) the phenol is sufficiently acidic to generate this species. To account for the difference in *ortho–para* ratio between (2) and (4), Norman and Harvey assumed that in (2) hydrogen–bonding involved molecular chlorine, while in (4) the hydrogen bonds were between phenol molecules.

The same authors have also measured *ortho–para* ratios for the chlorination of phenoxide ion under various conditions[128]. The attacking reagent is probably HOCl in every case, and the experimental results are as follows.

TABLE 15. Chlorination of phenoxide ion in different systems[128].

	System	*ortho*	*para*
(6)	*t*-BuOCl, 4N NaOH, 25°	78·9	21·1
(7)	HOCl, H_2O, 25°	80·7	19·3
(8)	*t*-BuOCl, 15N NaOH, 50°	78·9	21·1
(9)	*t*-BuOCl, 15N KOH, 50°	81·3	18·7
(10)	*t*-BuOCl, 3·5N $HNMe_3^+OH^-$, 25°	63	37

There are three possible causes of the high percentage of *ortho* attack: (a) the high electron density in the vicinity of the O^- group,

(b) interaction between HOCl and the O^- group, (c) coordination involving the HOCl, O^- and M^+ functions (cf Lederer–Manasse reaction). Suggestion (c) would seem to be the most probable.

Chlorination of phenol using sulphuryl chloride has been studied[135]. In nitrobenzene, nitromethane, ether, and in the absence of solvent the reaction has been found to lead to *ortho–para* ratios of around 70 : 30.

3. Sulphonation[136]

Phenol is sulphonated in the *ortho-* and *para*-positions. In all investigations there have been analytical difficulties but it is clear that the site of attack is temperature-dependent, the *ortho*-product being favoured at low temperatures, and the *para*-product at high temperatures. Obermiller[137] noted that the proportion of *ortho*-product was less than 40% even at low temperatures (0–5°), while Muramoto[138] estimated that the amount of *ortho*-product changed from 39% at 20° to 4% at 100°. Olsen and Goldstein[139] have also studied the reaction. Table 16 is based on the work of Chase and McKeown[140].

TABLE 16. Temperature dependence in the sulphonation of phenol.

T (°C)	*ortho* : *para* (20% oleum)	*ortho* : *para* (98% H_2SO_4)
20	42 : 58	49 : 51
30	—	41 : 59
40	37·5 : 62·5	35 : 65
50	35·5 : 64·5	—
60	34 : 66	33 : 67
70	27 : 73	25 : 75
80	18 : 82	14·5 : 85·5
100	12 : 88	9·5 : 90·5
120	9 : 91	11 : 89
140	—	10·5 : 89·5

Changes in the *ortho–para* ratio with temperature for aromatic sulphonation have been attributed to (i) desulphonation of the sulphonic acid products, the rate of this process being dependent on the position of substitution, (ii) direct isomerization of the products—this has been demonstrated in some cases, (iii) variation in selectivity due to variations in the reactivity of the attacking species. In the case of phenol (iii) can be excluded because it should result in a limiting *ortho* figure of 40% at high temperatures.

Spryskov has investigated the sulphonation of phenol, in dichloroethane between 0° and −40°, using chlorosulphonic acid[141]. He found that the variation in product composition was not large, and that *ortho*- and *para*-isomers were formed in approximately equal quantities, accompanied by about 2% of the 2,4-disulphonic acid. Typical *ortho–para* product ratios were 14·2 : 9·9 at −40° and 42·4 : 48·5 at 0°, the amount of unchanged phenol being very much larger at the lower temperature. When the solvent was changed to carbon disulphide, the product composition at −15° was 45·5% *ortho*, 48% *para* and 6·5% 2,4-disulphonic acid. By the use of labelled phenol he established the mechanism below for the reaction in dichloroethane:

$$C_6H_5OH + HOSO_2Cl \underset{}{\overset{1}{\rightleftharpoons}} C_6H_5OSO_3H + HCl$$

$$\downarrow\uparrow \; 3 \;|\; 2\ (HOSO_2Cl)$$

$$HOSO_2Cl + HOC_6H_4SO_3H \overset{4}{\rightleftharpoons} HO_3SOC_6H_4SO_3H + HCl$$

The sulphate ester of the *ortho*-sulphonic acid cleaves much faster in the presence of hydrogen chloride (via path 3) and there is a tendency for the *ortho–para* ratio to decrease as the temperature is raised. In view of the present belief that it is the phenylsulphuric acid that is being sulphonated this reaction is not strictly comparable with the normal sulphonation of phenol. The reaction should be considered more akin to neighbouring group participation rather than illustrative of electronic effect of the hydroxyl group.

Since sulphonation is a reversible reaction and *m*-hydroxybenzenesulphonic acid is the most resistant to hydrolysis, equilibrium control of sulphonation should lead to the formation of the *meta*-product. This had never been isolated from low-temperature sulphonations but Spryskov was able to obtain, after prolonged sulphonation at higher temperatures, yields of the *meta*-isomer approaching 40%[142].

TABLE 17. Sulphonation of phenol with sulphuric acid.

T (°C)	Time (hours)	*ortho* (%)	*meta* (%)	*para* (%)
120	30	low	3·7	96
160	150	low	20·8	79
180	50	low	23·2	76
209	20	low	38·1	61

4. Diazonium coupling[143]

Arene diazonium salts will attack the aromatic nucleus of highly activated aromatic compounds such as phenols and amines, but only in basic solution. A kinetic investigation of the reaction with phenols showed that the reaction is second-order and that the rate-determining step involves either the diazonium cation and a phenoxide ion or a diazotate ion and a neutral phenol molecule[144]:

$$\frac{\mathrm{d(product)}}{\mathrm{d}t} = k(\mathrm{ArN_2}^+)(\mathrm{ArO}^-) \text{ or } k(\mathrm{ArN_2O}^-)(\mathrm{ArOH})$$

For aqueous solutions, the two expressions are kinetically indistinguishable. However, Pütter[145] made a careful study of the pH dependence of the rate of coupling in naphtholsulphonic acids. Using substrates which differed greatly from each other in the $\mathrm{p}K_a$ of the naphtholic hydroxyl group, he was able to show that for this case at least, the two reacting species are the diazonium cation and the naphthoxide ion. He excluded not only attack of the diazotate on the naphthol, but also attack of the diazonium cation on the naphthol. The result is not too surprising since diazonium cations, unless highly activated, do not attack aromatic ethers. Subsequently, semiquantitative experiments in 60–80% sulphuric acid have shown that under these conditions phenols are attacked but that such attack on the neutral molecule is slower (by a factor of 10^{10}) than reaction with the phenoxide ion[146].

Bamberger[147] determined the *ortho–para* product distribution for the reaction with phenol itself. He found that the product contained about 98% *para*-compound, with 1% of *ortho* and 1% of 2,4-disubstituted product. Most studies of *ortho–para* ratios have involved 1-naphthol derivatives, for which the *ortho–para* ratio is much nearer unity. In these cases the ratio has been shown to depend on the nature of the diazo component[148] and the reaction conditions.

Stamm and Zollinger's[149] results on the coupling reaction between the *o*-nitrobenzene diazonium ion and 1-hydroxynaphthalene-3-sulphonic acid are:

Buffer system	pH	$k_o \times 10^5$	$k_p \times 10^5$
0·05M NaOAc/0·05M AcOH	4·59	2·9	0·21
0·17M NaOAc/0·17M AcOH	4·61	3·6	0·37
0·50M NaOAc/0·50M AcOH	4·64	4·4	1·04
0·50M NaOAc/0·05M AcOH	5·60	4·2	0·88

k values are in l/mole sec.

The reaction was general base-catalysed at both *ortho*- and *para*-positions by external bases but extrapolation to zero buffer concentration revealed that the water-catalysed reaction at the *ortho*-position was much faster. This was attributed to the naphthoxide ion acting as a general base in removing a proton from the sigma-complex for *ortho* attack.

5. Friedel–Crafts alkylation*

In Friedel–Crafts alkylation an alkyl group is introduced into an aromatic substrate by means of a combination of an alkylating agent and a Lewis acid catalyst. The most commonly encountered alkylating agents are alkyl halides, alkenes and alcohols, although various other reagents (aldehydes, ketones, alkynes, inorganic esters, ethers, alkanes, mercaptans, sulphides, thiocyanates) have been used. Lewis acid catalysts used include $AlCl_3$, $FeCl_3$, BF_3, $SbCl_5$, $ZnCl_2$, $TiCl_4$. In addition, Brønsted–Lowry acids have been used—HF, H_2SO_4, H_3PO_4, etc. Acidic-oxide catalysts of the silica-oxide type, and cation exchange resins have also been used. Except for the case of alkylation with alcohols, only catalytic quantities of acids are needed. The order of catalytic activity for metal halides in the acetylation of toluene is[151]:

$$AlCl_3 > SbCl_5 > FeCl_3 > TeCl_2 > SnCl_4 > TiCl_4 > BiCl_3 > ZnCl_2$$

and a similar order would be expected for alkylation. For proton acids, the order appears to be[152]:

$$HF > H_2SO_4 > H_3PO_4$$

The order may change with the conditions. The choice of a catalyst

* Exhaustive reviews of this and other reactions of the Friedel–Crafts type are given in the treatise edited by G. A. Olah[150].

for a particular Friedel–Crafts alkylation is governed by the activities of substrate and alkylating agent, the solvent, reaction temperature, etc.

In the alkylation of phenols, the substrates are strongly nucleophilic. They can be alkylated under mild conditions but then ethers are often formed, i.e., *O*-alkylation occurs rather than *C*-alkylation. The major complication is reaction with the catalyst. With some catalysts, e.g., H_2SO_4, the nucleus is attacked but in most cases initial reaction is with the hydroxyl group. Consequently it is not surprising that *ortho–para* ratios vary widely with catalyst type and reaction conditions.

The measurement of *ortho–para* ratios for alkylation of phenols is often easy but the meaning of the results obtained is seldom clear. The major complicating factors are:

(1) Alkylation is reversible under normal reaction conditions and the kinetic and thermodynamic *ortho–para* ratios are invariably different.
(2) The introduction of an alkyl group activates the ring towards further substitution. Polyalkylation is difficult to prevent and, when it occurs, the *ortho–para* ratio loses its significance.
(3) A combination of (1) and (2) can lead to disproportionation of monoalkylated products.
(4) Ring alkylation can occur through intramolecular rearrangement of an intermediate ether. (Intermolecular rearrangement leading to the *para*-product is a dealkylation–realkylation reaction.)
(5) Ether formation can lead to abnormal results if substitution takes place in the ether and the product is subsequently *O*-dealkylated.

To all these factors must be added the modification of the electronic nature of the hydroxyl substituents by reaction with the catalyst (particularly Lewis acids). Interpretation of particular findings is, therefore, often difficult and unwise. Some practical generalizations can be made, of which the widest and the most obvious is that alkylation of phenol almost invariably occurs *ortho* or *para* to the hydroxyl group. There have been instances involving apparent *meta*-alkylation but they are very rare[153, 154].

a. Thermal alkylation using alkenes (no catalyst). Phenol can be alkylated by heating with alkenes to 320° under pressure[155]. Under these conditions the principal product is *ortho*. Direct alkylation of the phenol via a cyclic transition state rather than a thermal Fries rearrangement has been suggested:

b. Catalytic alkylation using alkenes[156]. Although systematic investigations are rare the amount of available information is quite large. Conditions and catalysts may be altered to favour either *ortho-* or *para-*products. Briefly, results show that the *ortho-* isomer is the kinetically favoured product, with the *para-*isomer being thermodynamically the more stable[157]. Low concentrations of catalyst lead to the *ortho-*product, as do certain catalysts, particularly aluminium phenoxide[158]. With this catalyst, alkylation appears to involve the species $HAl(OPh)_4$ and a six-membered transition state:

Aluminium chloride reacts with phenol, and most reactions involving the use of $AlCl_3$ probably involve $Al(OPh)_3$ as the true catalyst. An interesting example of the effect of time on product composition has been reported by Buls and Miller[159] (Table 18).

TABLE 18. Reaction of isobutylene (4 moles) with phenol (1 mole) in the presence of aluminium chloride.

Time (hours)	Phenol (moles) recovered	Substituted phenols (moles)				
		2-	4-	2,4-	2,6-	2,4,6-
0	1·0	0·00	0·00	0·00	0·00	0·00
0·16	0·35	0·4	0·02	0·03	0·05	0·06
0·23	0·24	0·61	—	0·05	0·03	0·03
0·36	0·18	0·54	0·02	0·03	0·19	0·02
0·50	0·05	0·32	0·01	0·03	0·45	0·11
1·14	—	0·05	0·03	0·16	0·38	0·37
7·10	—	0·02	0·01	0·25	0·04	0·66

c. Catalytic alkylation using alkyl halides and alcohols. Apart from $AlCl_3$, the most commonly used catalyst for the alkylation of phenols is $ZnCl_2$[160]. This has the advantage of being strong enough to catalyse *C*-alkylation but weak enough to complex only loosely with the hydroxyl group. Strong Lewis acids such as $AlCl_3$ decrease the reactivity of the phenol by complexing with the hydroxyl group. In all cases substitution occurs mainly in the *para*-position.

6. Alkylation of phenoxides (no catalyst)

The phenoxide ion behaves as an ambident nucleophile towards reactive alkyl halides of the benzylic or allylic type and it undergoes both *O*- and *C*-alkylation. Kornblum[161] found that the relative importance of the two pathways depended mainly on the solvent, and, in protic solvents at least, selective solvation of the phenoxide ion appeared to favour *C*-alkylation. In homogeneous solution, the *C*-alkylation occurs at both the *ortho*- and *para*-positions. Kornblum obtained the results in Table 19. The attacking electrophile in these

TABLE 19. Alkylation of sodium phenoxide with allylic and benzylic halides[161].

Reaction conditions	*ortho* : *para* ratio
Allyl bromide/NaOPh/H_2O, 27°	35 : 65
Benzyl chloride/NaOPh/H_2O, 27°	38 : 62
Allyl bromide/NaOPh/PhOH, 43°	48 : 52
Benzyl chloride/NaOPh/PhOH, 43°	49 : 51
Benzyl chloride/NaOPh/CF_3CH_2OH, 27°	52 : 48

cases was considered to be the alkyl halide, and the possibility of a main reaction between the phenoxide ion and the derived carbonium ion was ruled out.

Under heterogeneous conditions, where reaction is believed to occur at the crystal surface, the *ortho*-isomer is the only *C*-alkylation product isolated[162].

7. Hydroxyalkylation: reaction with aldehydes and ketones

a. Baeyer reaction[163]. This refers to the reaction between phenol and ketones or aldehydes other than formaldehyde. Hydroxyalkyl derivatives are initially formed and sometimes react with other phenol molecules to form bis-arylalkanes under the reaction conditions. The general reaction requires the use of Friedel–Crafts catalysts, usually either $AlCl_3$ or anhydrous HCl/HOAc. Aldehydes other than formaldehyde usually react with phenol to form polymeric products which decompose on heating to give alkylphenols[164, 165]. Ketones usually give bis-hydroxyarylalkanes directly[166]. In all the reported cases attack on phenol is at the *para*-position; if this is blocked, then *ortho* attack will occur. Yields are usually less than 50%.

b. Lederer–Manasse reaction[167]. This is the reaction of phenols with formaldehyde to form hydroxymethyl derivatives. The reaction takes place under acidic or alkaline conditions. Further reaction of the hydroxymethyl derivative often occurs to form dihydroxydiarylalkanes. The reaction goes with great ease even in the absence of catalysts. Resin formation, through polysubstitution and cross-condensation, is common. Under controlled conditions it is possible to examine the monohydroxyalkylation reaction. As expected, the *ortho*- and *para*-positions are the most susceptible to attack. The *ortho–para* ratio depends upon the reaction conditions and experimental findings indicate that attack on the *para*-position is favoured by polar solvents and acidic conditions, while attack on the *ortho*-position is favoured by nonpolar solvents, alkaline conditions and Group II metal oxide, hydroxide or acetate catalysts[167]. These conclusions may be rationalized by assuming that acid catalysis involves reaction of the hydroxymethylcarbonium ion ($^{+}CH_2OH$) with phenol and that nonpolar solvents encourage participation of the phenolic hydroxyl group in stabilizing the transition state for *ortho* attack.

Alkaline reaction involves attack by formaldehyde on the phenolate anion which is more *ortho*-directing than phenol. Highest yields of *ortho*-product occur in those cases where the greatest amount of ion-pairing between the metal ion and the phenoxide is to be expected. A mechanism involving some form of chelate is therefore likely (cf. aluminium phenoxide-catalysed alkylation). The solvent has a major effect (Table 20).

TABLE 20. Alkaline phenol–formaldehyde condensation[167].

Catalyst	Solvent	*ortho* : *para* ratio
NaOH	H_2O	65 : 35
$Mg(OH)_2$	H_2O	69 : 31
Et_3N	H_2O	55 : 45
Et_3N	C_2Cl_4	77 : 23
Et_3N	toluene	87 : 13

8. Haloalkylation[168]

Haloalkyl groups can be introduced directly into aromatic nuclei using methods similar to those for alkylation. By far the most common haloalkylation reaction is chloromethylation[168, 169]. This reaction is usually carried out with a mixture of formaldehyde (or one of its polymers) and hydrogen chloride in the presence of a Friedel–Crafts catalyst such as zinc chloride. Other haloalkylating agents include chloromethyl ether[170] or chloromethyl sulphide[171].

Phenols normally react so readily with HCHO/HCl that a catalyst is unnecessary, and even then monomeric products are difficult to obtain unless deactivating groups are present[172]. The usual product with phenol is a bicyclic methylene ether[173].

OH (phenol) $\xrightarrow[HCl]{CH_2O}$ (bicyclic methylene ether: H_2C, O, $\overset{H_2}{C}$, O; CH_2Cl; CH_2Cl)

The mechanism of chloromethylation is not known with certainty, but in the case of the HCHO/HCl/phenol system there is evidence that the reaction involves initial hydroxymethylation (see Lederer–Manasse reaction) followed by replacement of hydroxyl by halogen[174].

9. Friedel–Crafts acylation[175]

Most of the general remarks on Friedel–Crafts alkylation also apply to acylation. The major difference is that acylation requires at least 1 mole of catalyst per mole of reagent because Lewis acids form complexes by coordination with the carbonyl oxygen of the acyl derivative. The most common acylating agents are acid halides and anhydrides, but ketones and carboxylic acids have also been used. The same problems of reversibility and kinetic versus thermodynamic control are encountered, and with phenols competing *O*-acylation also occurs. Acylation, like alkylation, occurs mainly *ortho* and *para* to the hydroxyl group. Since an acyl group is deactivating, acylation differs from alkylation in leading to far less polysubstitution and disproportionation. The *ortho–para* ratio is usually less than unity and depends mainly on the catalyst employed. The largest amount of *ortho*-product is obtained using aluminium chloride or polyphosphoric acid as catalyst. The *para–ortho* ratios for acetylation of phenol are PPA 3·2, $ZnCl_2$ 3·5, $AlCl_3$

4·6, $TiCl_4$ 6·3, BF_3 30 [176]. The ratio can also be influenced by solvent. For example, the octanoylation of phenol with $AlCl_3$ catalyst gives a *para–ortho* ratio of 0·24 in tetrachlorethane and 2·7 in nitrobenzene[176]. The use of phenoxyaluminium chloride as a reagent cum catalyst leads to approximately equal amounts of *ortho*- and *para*-product[177] (cf. aluminium phenoxide alkylation). An extensive series of tables covering acylation of phenols has been drawn up by Gore[178].

a. The Fries rearrangement[179]. Phenolic esters rearrange on heating with Friedel–Crafts catalysts to give *o*- and *p*-acylphenols. The mechanism has not been elucidated. The *para*-product is formed by an intermolecular deacylation–reacylation[180], but in the cases so far investigated, the *ortho*-product appears to form by an intramolecular reaction[180]. These are therefore independent reactions and competitive processes. This should be reflected in a changing *ortho–para* ratio with change in reaction conditions. In general, yields of *ortho*-product are greater at higher temperatures. *Para*-substitution is facilitated by the use of polar solvents and by the presence of hydrochloric acid[181]. It is possible that the *para*-Fries rearrangement is a simple Friedel–Crafts acylation process, while the *ortho*-Fries is a true rearrangement.

10. Formylation

a. Gattermann–Koch reaction[182]. Many aromatic compounds can be successfully formylated by a mixture of carbon monoxide and hydrogen chloride in the presence of aluminium chloride. Normally high pressures are required but with added cuprous chloride, which appears to act as a carrier, the reaction proceeds at atmospheric pressure. Generally, phenols and phenol ethers cannot be formylated by this latter and more common method. This was attributed by Gattermann to insolubility of cuprous chloride in the system, but this explanation has since been challenged[183]. The only hydroxyl compound successfully formylated under Gattermann–Koch conditions is 3-hydroxyretene, which is said to give a mixture of aldehydes in good yield[184]. Formyl derivatives of phenols are usually prepared by a closely related reaction, the Gattermann aldehyde synthesis.

b. Gattermann aldehyde synthesis[185, 186]. The Gattermann aldehyde synthesis in its original form was the reaction of a mixture of HCN and HCl with an aromatic substrate in the presence of $AlCl_3$ or $ZnCl_2$. A later modification[187] removed the necessity for working with large amounts of HCN by passing HCl into $Zn(CN)_2$ to generate the required mixture. The reaction is most satisfactory for strongly activated aromatic compounds such as phenols and ethers.

Monohydric phenols usually require $AlCl_3$ as the catalyst. The yields vary with the structure of the phenol. Phenol itself is formylated in 30% yield, entirely in the *para*-position[188]. In general, only one formyl group is introduced and it always enters *para* to the phenolic hydroxyl group if that position is unoccupied. If the *para*-position is blocked, the reaction may not proceed at all or it may give a poor yield of the *ortho*-product. An exception is 2-naphthol, which gives good yield of 2-hydroxyl-1-naphthaldehyde[189], but there are cases in which even replacement of a *para*-substituent is preferred to *ortho* attack[190].

The nature of the attacking electrophile is not known, but an isolable nitrogen-containing intermediate is formed and may be hydrolysed to the aldehyde. Possibly this is the imino-hydrochloride $ArCH{=}NH_2^+Cl^-$. The Gattermann reaction may be regarded as a special case of the Houben–Hoesch reaction (see later).

c. Vilsmeier reaction[191, 192]. This is currently the most common method for the formylation of aromatic rings. The formylating agent is a mixture of a substituted amide (usually dimethylformamide or *N*-methylformanilide) and phosphorus oxychloride. It is only applicable to substrates such as amines, phenols and certain aromatic hydrocarbons and heterocycles that are highly activated. It is closely related to both the Gattermann synthesis and Friedel–Crafts acylations. The attacking species (for the case of *N,N*-dimethylformamide) is believed to be[193]

$$\left[\mathrm{Me{-}N(Me){-}\overset{+}{C}H{-}Cl}\right] \quad [\mathrm{OPOCl_2}]^-$$

and the assumed mechanism (in the case of phenol) is

$$\mathrm{C_6H_5OH} \xrightarrow{\mathrm{Me_2N\overset{+}{C}HCl}} \mathrm{Me_2N{-}CH(Cl){-}C_6H_5{=}\overset{+}{O}H}\ (\text{H at ring carbon}) \longrightarrow \mathrm{Me_2N{-}CH(Cl){-}C_6H_4{-}OH} \longrightarrow \mathrm{OHC{-}C_6H_4{-}OH}$$

Like the Gattermann synthesis, reaction with phenols occurs at the *para*-position if this is free. If it is blocked, then *ortho* attack will occur[194]. Phenol reacts with DMF/$POCl_3$ to give an 85% yield of the *p*-hydroxybenzaldehyde[195]. No *ortho*-product has been reported but, as was the case with the Gattermann synthesis, no serious attempt has apparently been made to obtain it.

d. Reaction with dichloromethyl ether. Dichloromethyl ethers, e.g., CH_3OCHCl_2, in the presence of Lewis acids, react with aromatic compounds to form α-alkoxybenzyl chlorides which decompose, either on heating or on the addition of water, to form aldehydes[196].

$$\mathrm{ArH} \xrightarrow[\text{F.C. catalyst}]{\mathrm{HCCl_2OR}} \mathrm{Ar{-}CH(Cl)(OR)} \xrightarrow[\text{or } \mathrm{H_2O}]{\text{heat}} \mathrm{Ar{-}CHO}$$

The catalysts most frequently used are $TiCl_4$, $SnCl_4$, $SnBr_4$ and $AlCl_3$. The reaction is closely related to Friedel–Crafts alkylation and acylation and also to formylation by trialkyl orthoformates (see below). No data are reported for formylation of phenol, but a number of substituted phenols have been formylated, often in good yields[197] (Table 21). The tabulated figures indicate that *ortho*-substitution is more important than in most other formylation reactions involving electrophilic substitution (see, however, the Reimer–Tiemann reaction).

e. Reaction with trialkyl orthoformates. Gross and co-workers have reported a direct aldehyde synthesis using trialkyl orthoformates in the presence of aluminium chloride[197]. Using this method, phenol aldehydes were obtained in good yield. This reaction, although similar in many ways to formylation using dichloromethyl methyl ether, gives far less *ortho*-product, which is in keeping with a much greater steric requirement for the attacking group.

f. The Reimer–Tiemann reaction. Phenols (as phenoxide ion) and certain highly reactive heterocyclic compounds may be formylated by reaction with chloroform in the presence of alkali[198]. The reactive electrophile is believed to be dichlorocarbene. Although this species is neutral its carbon is highly electron-deficient. The proposed mechanism is as shown on p. 438. Both *ortho*- and *para*-products are formed. Yields of the mixed aldehydes are usually less than 50%. The main product is usually the *ortho*-isomer but the *ortho–para* ratio varies quite considerably with the reaction conditions. Experimental results all indicate, however, that *ortho* attack is favoured over *para* (this is usual for electrophilic attack on the phenoxide ion) but that

TABLE 21. Alkylation with dichloromethyl ether : $TiCl_4$.

Phenol	Yield (total %)	Products		*ortho* : *para* ratio
OH; Me	52%	OH; OCH—; Me	OH; Me; CHO	77 : 23
OH; Me; Me	60%	OH; OCH—; Me; Me	OH; Me; Me; CHO	33 : 67
OH; Me; Me	84%	OH; CHO; Me; Me	OH; Me; Me; CHO	82 : 18
OH; Me; Me	61%	OH; OCH—; Me; Me		—
OH; *i*Pr; Me	77%	OH; OCH—; *i*Pr; Me	OH; *i*Pr; Me; CHO	53 : 47
OH	78%	OH; CHO	OH; CHO	13 : 87
OH	82%	CHO; OH		—
OH; OH	68%	OH; OH; CHO		—
OH; Me; OH	48%	OH; CHO; Me; OH	OH; OCH—; Me; OH	40 : 60
OH; HO; OH	52%	OH; HO; OH; CHO		—
OH; HO; OH	36%	OH; CHO; HO; OH		—

$$\text{C}_6\text{H}_5\text{O}^- \xrightarrow{:CCl_2} \text{(cyclohexadienone with H and } \bar{C}Cl_2\text{)} \xrightarrow{H_2O} \text{(cyclohexadienone with H and } CHCl_2\text{)} \rightleftharpoons o\text{-}CHCl_2C_6H_4O^- \xrightarrow{OH^-} o\text{-}OHCC_6H_4O^-$$

para attack can be made more favourable by hindering the *ortho*-position by introducing large cations which complex with the O⁻ group[199].

g. The Duff reaction. Hexamethylenetetramine will condense with phenols to form intermediate products which, on acid hydrolysis, give hydroxy aldehydes[200]:

$$C_6H_5OH \xrightarrow{C_6H_{12}N_4} o\text{-}HOC_6H_4-CH_2-NH-CH_2-C_6H_4OH\text{-}o \xrightarrow[H_2O]{HCl} o\text{-}HOC_6H_4CHO + o\text{-}HOC_6H_4CH_2NH_2$$

Attack on phenol occurs at the *ortho*-position, but the yields are even lower than those of the Reimer–Tiemann reaction.

11. Houben–Hoesch ketone synthesis[201]

This ketone synthesis is a variation of the Gattermann synthesis of aldehydes. The reaction is one between a phenol and a nitrile in the presence of hydrogen chloride (and sometimes a Lewis acid such as $ZnCl_2$) to form a ketimine hydrochloride, which readily hydrolyses to give a ketone.

$$\text{ArH} \xrightarrow[ZnCl_2]{RCN/HCl} \text{Ar}-\underset{\substack{\| \\ {}^+NH_2}}{C}-\text{R} \;\; Cl^- \xrightarrow{H_2O} \text{Ar}-\underset{\substack{\| \\ O}}{C}-\text{R} + NH_4Cl$$

The precise nature of the attacking electrophile is not known but it is believed to be a complex of $R{-}C^+{=}NH$ with HCl. Phenol itself does not undergo nuclear acylation with most nitriles; it forms instead the hydrochloride of an iminophenyl ester[202]. An exception is the reaction of phenol with trichloroacetonitrile in the presence of aluminium chloride to give a 95% yield of ω-trichloro-4-hydroxy-acetophenone[203]. The formation of the *para*-product is consistent with the *para*-directing effect of the hydroxyl substituent in the Gattermann aldehyde synthesis. It is possible that this product is formed by a Fries rearrangement of the imino ester but independent rearrangement of such esters has not been achieved.

12. Carboxylation

a. Kolbe–Schmitt reaction[204]. From the middle of the nineteenth century this reaction has been used for the preparation of aromatic hydroxy acids. The reaction is between the dry metal phenoxide and carbon dioxide and normally is carried out at 120–130° under pressure.

It is usually assumed that reaction occurs through attack on the electron-rich ring of the phenoxide ion by the electrophilic carbon of the CO_2 molecule. In general, substitution occurs *ortho* to the hydroxyl group but cases of *para*-substitution are known. The *ortho–para* ratio for phenol is strongly dependent on the nature of the metal ion in the phenoxide as well as on reaction conditions such as temperature, reaction time, etc. With different alkali metal phenoxides, the proportion of *ortho*-product diminishes in the order Na > K > Cs[205]. The substitution is apparently reversible and *ortho–para* migration occurs at higher temperatures[206]. It would seem therefore that there is probably association between the phenoxide oxygen and the metal ion and that *ortho*-substitution may go through a cyclic transition state involving CO_2, the phenoxide oxygen and (perhaps) the metal ion (cf. aluminium phenoxide alkylations).

13. Other electrophilic substitutions

a. Hydrogen exchange in aromatic systems under acidic conditions is an electrophilic substitution. It has been established that three nuclear hydrogens are readily exchanged in phenol under these conditions and that these are at the *ortho*- and *para*-positions[207]. The *ortho*- and *para*-hydrogens in phenol are also rapidly exchanged in alkaline solution; in these circumstances the attacking species is thought to be a Brønsted acid, and the substrate the highly activated phenoxide ion[208]. The relative reactivities of the *ortho*- and *para*-positions have not been measured but for anisole an *ortho–para* ratio of 29·5 : 70·5 is indicated[209]. It is unlikely that the figure for phenol would be lower than this.

b. Mercuration of aromatic compounds has been shown to occur through either a radical or electrophilic mechanism. The electrophilic mechanism is favoured by the presence of acids, while in nonpolar solvents the homolytic path is often followed[210]. For highly reactive aromatics such as phenol, however, the electrophilic path is likely to be the more important under all conditions.

An accurate *ortho–para* ratio for the mercuration of phenol is not available. A preparative procedure for *o*-chloromercuri-phenol via the acetate[211] leads to an isolated yield of the *ortho*-isomer of up to 44% with an amount of the *para*-product which is substantially lower. In contrast, mercuration of anisole at 25° in acetic acid leads to an *o–p* ratio of only 14:86 [212]. The considerable difference between phenol and anisole can be attributed to participation of the hydroxyl group and/or preferential reaction of the mercurating species with phenoxide ion—the system would not be strongly acidic under the reaction conditions.

c. Hydroxylation of aromatic systems may be either a homolytic or electrophilic substitution, but, even in homolytic substitution, the attacking species is likely to be sufficiently electrophilic to prefer the normal substitution pattern, i.e. for phenol, *ortho* and *para* to the hydroxyl group. The most common hydroxylating agents are acidified hydrogen peroxide or solutions of peracids. Hydroxylation of phenols is usually accompanied by oxidation of the product to *ortho*- and *para*-quinones[213]. Phenol on treatment with peracetic acid gives a mixture of benzoquinone (40%) and a muconic acid (35%), indicating an *ortho–para* ratio around unity for initial attack[213]. Diacyl peroxides are also effective hydroxylating (or acyloxylating) agents and have been reported to attack phenols mainly in an *ortho*-position if this is free[214]. There is a large negative entropy of activation and a cyclic mechanism rather similar to that proposed for the Claisen rearrangement has been proposed:

H O O C Ph O—OCOPh slow O O C Ph O fast OH OCOPh O OCOPh H

In a similar reaction, *N*-benzoyloxypiperidine was found to react with phenol in the presence of boron trifluoride to give *N*-(*o*-hydroxyphenyl)piperidine. A cyclic mechanism was also proposed[215], the key intermediate structure being one of the following:

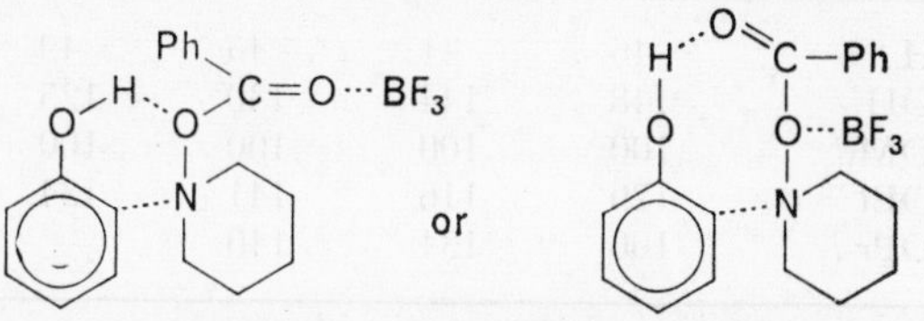

or

C. Comparative Directing Power of the Hydroxyl Group

If the σ^+ values of Brown and Okamoto[6] apply reasonably well to electrophilic substitution reactions then the order of relative *ortho–para* directing powers are:

NMe_2, −1·7; NH_2, −1·3; OH, −0·92;
OMe, −0·78; CH_3, −0·31.

In view of the high rho values (>5) normally encountered in electrophilic aromatic substitutions the expected rate differences found between the hydroxyl group and amino groups should be at least 100-fold and the hydroxyl substituent should in turn be at least five times as activating as methoxyl. It is to be expected, then, that in systems containing the hydroxyl group and any of the other substituents, electrophilic substitution would be dominated by the more reactive of them. This is the usual observation. As examples, the bromination of *p*-aminophenol ($Br_2/CHCl_3$) gives 3,5-dibromo-4-aminophenol[216], and the nitration of *p*-hydroxyacetanilide gives 4-hydroxy-3-nitroacetanilide[216]. However, electrophilic substitutions are often carried out under strongly acidic conditions, with accompanying protonation of amino groups, which then become deactivating. Thus, sulphonation of *o*-aminophenol results in attack *para* to the hydroxyl group[217] and sulphonation of *p*-aminophenol occurs *ortho* to the hydroxyl group[218].

Chuchani and co-workers have observed an interesting case where the directing power of the phenolic hydroxyl group is, at first sight, more influential than its activating power[219]. They examined the kinetics of tritylation of a series of *ortho*-substituted phenols and obtained the results below (Table 22). Although the *o*-isopropoxyphenol (and in some cases the *o*-ethoxy compound also) can react more rapidly than catechol the only product of the reaction is that resulting from attack of the trityl cation *para* to the hydroxyl group.

TABLE 22. Relative rates of tritylation of o-$C_6H_4(OH)(X)$ with $Ph_3C^+ClO_4^-$ in nitromethane.

X	$T =$ 30°	40°	50°	60°
H	46	41	45	49
OH	148	144	130	125
OMe	100	100	100	100
OEt	129	116	141	134
OPr-*i*	166	134	140	—

Chuchani attributed this to hydrogen-bonding between the hydroxyl group and the alkoxyl-oxygen. The activating power of the hydroxyl group would thereby be increased, and the strength of the hydrogen bond would presumably increase with the electron-donating power of the alkyl group. In monosubstituted benzene derivatives, such internal hydrogen-bonding is not possible, but the observed reactivity order is still not simple.

$$\text{OPr-}i > \text{OH} > \text{OEt} > \text{OMe}$$

Here, it was assumed, the hydroxyl group is hydrogen-bonded to the solvent. There is support for this explanation in the work of Campbell and co-workers[135] who noted that chlorination of *o*-cresol in carbon tetrachloride gives 50% of the 6-isomer, while in ether only 15% is formed. However, de la Mare has offered much evidence for hyperconjugation of the hydroxyl group in phenol and enhanced reactivity can be accounted for on this basis[93].

In spite of the demonstrated, very high *o,p*-directing power of the phenolic hydroxyl group, there are cases in which *meta* reaction products are isolable[220, 221], e.g.,

(i) OH, CH_3, CH_3 —$\xrightarrow[AlCl_3]{AcCl}$→ OH, Ac, CH_3, CH_3 (80%) + OH, CH_3, Ac, CH_3 (9%)

(ii) OH, OCH_3 —$\xrightarrow[ZnCl_2]{Ac_2O}$→ OH, OCH_3, Ac (1–2%) + normal products

Each of these reactions was being carried out under conditions which could lead to *O*-acetylation, and it is possible that some *C*-acylation of the *O*-acetyl derivatives is followed by hydrolysis of the acetate side-chain.

D. Rates of Electrophilic Aromatic Substitution

The rate of an electrophilic aromatic substitution reaction can be expressed either as the rate of introduction of the attacking electrophile into the aromatic nucleus, regardless of the position taken, or else as the rate of attack at a specific site in the molecule. Most recorded rate measurements for electrophilic substitution in phenol are overall rates, and the most commonly studied reactions are nitration and halogenation. Direct experimental comparison of phenol and benzene rates, to give the relative activating power of the phenolic hydroxyl group, is impracticable because the difference in reactivity is always great and may be as high as 10^{10}–10^{12}.

Comparisons with other activating groups have, however, been made[222, 223]. Most of these have been based on competition experiments, involving either two aromatic substrates or else one aromatic substrate with two functional groups[219, 224] (see preceding section). All of these studies are in agreement and suggest that the order of activating power of the common strongly activating groups is

$$NMe_2 > NHMe > NH_2 > OH > OMe > Me$$

Alkyl-substituted amino groups, it will be noted, are more activating than the amino group itself, whereas the methoxyl group is less activating than the hydroxyl group. This finding may well be related to de la Mare's observation[93] that in the halogenation of phenol O–H bond breaking is involved in the rate-determining step. If so, the N–H bond is presumably too strong in amines for a similar reaction path to be involved.

Direct comparisons of phenoxide ion with other substrates have not been made but it is generally assumed that the oxide ion group is a very strongly activating substituent, stronger even than NR_2. Bell and Ramsden have estimated that *N,N*-dimethylaniline are about 10^{16}–10^{17} times as reactive as benzene in aqueous solution[225]. For the dimethyl compound at 25° they found a second-order rate constant of almost 10^9 l/mole sec. In a subsequent paper Bell and Rawlinson[226] reported that bromination of both the *p*-bromophenoxide and *m*-nitrophenoxide ions gave second-order rate constants around 10^9 l/mole sec, and that for the more active, unsubstituted phenoxide ion, the rate could not be measured. Halogenation of aryloxides has, in fact, been stated as often occurring 'at about the encounter rate' in aqueous solution[227].

The most comprehensive description of reaction rate for aromatic substitution is in terms of partial rate factors. These represent the rate of substitution at a given position in the ring relative to that at one of the six positions in benzene. That is, for phenol.

$$f_{ortho} = \frac{k_{ortho}}{2} \Big/ \frac{k_{benzene}}{6}$$

$$f_{meta} = \frac{k_{meta}}{2} \Big/ \frac{k_{benzene}}{6}$$

$$f_{para} = k_{para} \Big/ \frac{k_{benzene}}{6}$$

A very large number of partial rate factors covering a wide variety of reactions have been calculated[228]. However, for phenol only three reactions have been examined in this way and even then, partial rate factors were obtained only for the *para*-position. The three reactions are (54), (55) and (56) in Table 7 (section III.A.2). Only one of these reactions (56), bromination in acetic acid at 25°, is a 'normal' electrophilic substitution; the other two involve replacement of groups other than hydrogen. The calculated partial rate factors are given below.

Reaction	f_{para}
Protodetriethylgermylation, $HClO_4$, $CH_3OH–H_2O$, 50°	$2{\cdot}73 \times 10^3$
Protodetrimethylsilylation, $HClO_4$, $CH_3OH–H_2O$, 51°	$1{\cdot}07 \times 10^4$
Bromination, CH_3CO_2H, 25°	$3{\cdot}7 \times 10^{12}$

The value of f_p for the bromination reaction was based on an assumed 100% attack at the *para*-position. On the basis of the measured *ortho–para* ratio for this system[93] the value of f_o should be about 2·5–3% of this figure. Illuminati[114] has estimated $\sigma_m{}^+$ for the hydroxyl group to be $-0{\cdot}133$ for bromination in acetic acid and on this basis, f_m should be about 50. However, Illuminati's figure comes from work on mesitylene derivatives, in which a steric factor could well operate.

E. Nucleophilic Aromatic Substitution

Nucleophilic aromatic substitution does not take place very readily unless the substrate is activated by strong $-R$ or $-I$ groups in the aromatic nucleus. Most substitutions of this type actually occur under basic conditions (cf. nucleophilic aliphatic substitution) and if an aromatic hydroxyl is present it is normally there as the O^- group. This group is strongly $+I$, $+R$, and therefore has a marked tendency to deactivate the system. This is apparent from the study of Berliner[98] on the reaction of 4-*X*-2-nitrobromobenzenes (section III.A.4) where the O^- group was found to have a $\bar{\sigma}_p$ value of about $-0{\cdot}8$. In this study the O^- group proved to be less deactivating than $-NH_2$. A similar order has been found for the reaction of *m*-substituted chlorobenzenes with sodium methoxide at 150° [229]. Here the activation order for substituents was $NO_2 > H > O^- > NH_2$. However, for the reaction of *p*-substituted chlorobenzenes under the same conditions, differences between the deactivating groups were less significant[229]. In this study, the activation order proved to be $NO_2 > Cl > H > NH_2 > CH_3 > m\text{-}O^-$. Other work on related dichloro compounds[230] confirms the CH_3 group as less deactivating than the O^- group. Clark and co-workers[231] give the following figures (Table 23) for the extent of halide ion liberated from

TABLE 23. Extent of reaction of ArCl with sodium methoxide after 50–60 hours at 155°.

X	$[OMe^-]$	$[Cl^-]$ (% yield)
H	2·0	4
H	2·5	7
2-O^-	2·0	0·0
3-O^-	2·0	12
4-O^-	2·0	0·0

chlorobenzene and the three monochlorophenols on heating with NaOMe. From these results, the *m*-O^- substituent would appear to act as an activating substituent and this surprising result requires corroborative evidence.

Miller and co-workers[232] have measured rates for the reaction of a number of 5-substituted 1-chloro-2,4-dinitrobenzenes, and have derived from this a $\bar{\sigma}_m$ for the O^- group of $-1{\cdot}358$. The very high value obtained here, compared with that of Berliner for $\bar{\sigma}_p$ value in reaction (61) $(\sim -0{\cdot}8)$[98], may result from the combination of a rather low $+I$ effect on the reaction site, and a very high $+R$ effect operating on the two nitro groups and thereby reducing their activating effect.

Polyfluoro compounds of the type C_6F_5X are attacked to varying degrees (depending on the nature of X) by strong nucleophiles, with the displacement of fluoride ion[233]. If X has no powerful electronic effect, the fluorine *para* to the substituent is displaced. Strongly electron-releasing substituents react much more slowly and there is an increasing tendency for a fluorine *meta* to the substituent to be displaced. Reaction with potassium hydroxide in *t*-butyl alcohol at 170° converts pentafluorophenol into tetrafluororesorcinol. Pentafluorophenol, in which the activating group is the oxide ion, is the least reactive of all compounds investigated.

V. REFERENCES

1. G. Chuchani in *The Chemistry of the Amino Group* (Ed. S. Patai), Interscience New York, 1968, pp. 205–277.
2. H. H. Jaffé, *Chem. Rev.*, **53,** 191 (1953).
3. R. W. Taft Jr., *J. Phys. Chem.*, **64,** 1805 (1960).
4. H. van Bekkum, P. E. Verkade and B. M. Wepster, *Rec. Trav. Chim.*, **78,** 815 (1959).
5. R. W. Taft Jr. and I. C. Lewis, *J. Am. Chem. Soc.*, **80,** 2436 (1958).
6. H. C. Brown and Y. Okamoto, *J. Am. Chem. Soc.*, **80,** 4979 (1958).
7. J. D. Roberts and W. T. Moreland, *J. Am. Chem. Soc.*, **75,** 2167 (1953).
8. C. D. Ritchie and E. S. Lewis, *J. Am. Chem. Soc.*, **84,** 591 (1962).
9. H. D. Holtz and L. M. Stock, *J. Am. Chem. Soc.*, **86,** 5188 (1964).
10. C. F. Wilcox and J. S. McIntyre, *J. Org. Chem.*, **30,** 777 (1965).
11. S. Siegel and J. M. Komarmy, *J. Am. Chem. Soc.*, **82,** 2547 (1960).
12. H. Stetter and J. Mayer, *Chem. Ber.*, **95,** 667 (1962).
13. R. W. Taft Jr. in *Steric Effects in Organic Chemistry* (Ed. M. S. Newman), John Wiley & Sons, New York, 1956, pp. 556–675.
14. M. Charton and H. Meislich, *J. Am. Chem. Soc.*, **80,** 5940 (1958).
15. M. Charton and H. Meislich, *Can. J. Chem.*, **38,** 2493 (1960).
16. J. Hine and W. C. Bailey, *J. Am. Chem. Soc.*, **81,** 2075 (1959).
17. L. Herk, A. Stefani and M. Szwarch, *J. Am. Chem. Soc.*, **83,** 3008 (1961).
18. M. Charton, *J. Org. Chem.*, **26,** 735 (1961).
19. P. B. D. de la Mare, *J. Chem. Soc.*, 3823 (1960).
20. O. Exner and J. Jonas, *Collection Czech. Chem. Commun.*, **27,** 2296 (1962).
21. J. R. Knowles and R. O. C. Norman, *J. Chem. Soc.*, 2938 (1961).
22. R. W. Taft Jr. and I. C. Lewis, *J. Am. Chem. Soc.*, **81,** 5343 (1959).
23. R. W. Taft Jr., E. Price, I. R. Fox, I. C. Lewis, K. K. Andersen and G. T. Davis, *J. Am. Chem. Soc.*, **85,** 709 (1963).
24. C. D. Ritchie and W. F. Sager in *Progress in Physical Organic Chemistry*, Vol. II (Eds. S. G. Cohen, A. Streitwieser Jr. and R. W. Taft), Interscience, New York, 1964, pp. 323–400.
25. G. Kortum, W. Vogel and K. Andrusson in *Dissociation Constants of Organic Acids in Aqueous Solution*, Butterworths, London, 1961.
26. A. Streitwieser Jr. in *Solvolytic Displacement Reactions*, McGraw-Hill, New York, 1962, pp. 110–112.

27. J. Myszkowski, A. Z. Zielinski and E. Laskowska, *Przemysl Chem.*, **44**, 565 (1965); *Chem. Abstr.*, **64**, 6427 (1966).
28. J. G. Buchanan and E. M. Oakes, *Carbohydrate Res.*, **1**, 242 (1965).
29. Ref. 26, pp. 112–113.
30. T. C. Bruice and S. Benkovic in *Bioorganic Mechanisms*, Vol. I, W. A. Benjamin Inc., New York, 1966, pp. 146–166.
31. B. Capon, *Quart. Rev. (London)*, **18**, 45 (1964).
32. T. C. Bruice and F.-H. Marquardt, *J. Am. Chem. Soc.*, **84**, 365 (1962).
33. L. Zürn, *Ann. Chem.*, **631**, 56 (1960).
34. H. Zahn and L. Zürn, *Ann. Chem.*, **613**, 76 (1958).
35. M. L. Wolfrom, R.B. Bennett and J. D. Crum, *J. Am. Chem. Soc.*, **80**, 944 (1958).
36. H. B. Henbest and B. J. Lovell, *J. Chem. Soc.*, 1965 (1957).
37. S. M. Kupchan, W. S. Johnson and S. Rajagopalan, *Tetrahedron*, **7**, 47 (1958).
38. S. M. Kupchan and W. S. Johnson, *J. Am. Chem. Soc.*, **78**, 3864 (1956).
39. S. M. Kupchan and C. R. Narayanan, *J. Am. Chem. Soc.*, **81**, 1913 (1959).
40. N. B. Chapman, R. E. Parker and P. J. A. Smith, *J. Chem. Soc.*, 3634 (1960).
41. S. M. Kupchan, P. Slade and R. J. Young, *Tetrahedron Letters*, **24**, 22 (1960).
42. T. C. Bruice and T. H. Fife, *J. Am. Chem. Soc.*, **84**, 1973 (1962).
43. H. G. Zachan and W. Karau, *Chem. Ber.*, **93**, 1830 (1960).
44. D. M. Brown and D. A. Usher, *Proc. Chem. Soc.*, 309 (1963).
45. L. Larsson and G. Wallerberg, *Acta Chem. Scand.*, **20**, 1247 (1966).
46. T. Karneko and H. Katsura in *Handbook of Organic Structural Analysis* (Ed. Y. Yukawa), W. A. Benjamin Inc., New York, 1963, p. 627.
47. J. Sicher, M. Tichy, F. Sipos, M. Svoboda and J. Jonas, *Collection Czech. Chem. Commun.*, **29**, 1561 (1964).
48. M. Kilpatrick and J. G. Morse, *J. Am. Chem. Soc.*, **75**, 1846 (1953).
49. T. J. Howard, *Chemistry and Industry*, 1899 (1963).
50. M. C. Dart and H. B. Henbest, *J. Chem. Soc.*, 3563 (1960).
51. S. Nishimura and K. Mori, *Bull. Chem. Soc. Japan*, **36**, 318 (1963).
52. J. R. Lewis and C. W. Shoppee, *J. Chem. Soc.*, 1365 (1955) and references therein cited.
53. M. T. Tribble and J. G. Traynham, *J. Am. Chem. Soc.*, **91**, 379 (1969).
54. J. M. Vandenbelt, C. Henrich and S. G. Vandenberg, *Anal. Chem.*, **25**, 726 (1954).
55. G. Briegleb and A. Bieber, *Z. Elektrochem.*, **55**, 250 (1951).
56. J. F. J. Dippy and F. R. Williams, *J. Chem. Soc.*, 1888 (1934); J. F. J. Dippy and R. H. Lewis, *J. Chem. Soc.*, 644 (1936); J. F. J. Dippy and J. E. Page, *J. Chem. Soc.*, 357 (1938); L. G. Bray, J. F. J. Dippy and S. R. C. Hughes, *J. Chem. Soc.*, 265 (1957); L. G. Bray, J. F. J. Dippy, S. R. C. Hughes and L. W. Laxton, *J. Chem. Soc.*, 2405 (1957).
57. J. H. Elliot and M. Kilpatrick, *J. Phys. Chem.*, **45**, 485 (1941).
58. W. L. Bright and H. T. Briscoe, *J. Phys. Chem.*, **37**, 787 (1933).
59. J. D. Roberts, E. A. McElhill and R. Armstrong, *J. Am. Chem. Soc.*, **71**, 2923 (1949); J. D. Roberts and W. T. Moreland, **75**, 2267 (1953).
60. J. H. Elliot and M. Kilpatrick, *J. Phys. Chem.*, **45**, 466 (1941); M. Kilpatrick and R. D. Eanes, *J. Am. Chem. Soc.*, **65**, 589 (1943).
61. J. H. Elliot and M. Kilpatrick, *J. Phys. Chem.*, **45**, 454 (1941); M. Kilpatrick and R. D. Eanes, *J. Am. Chem. Soc.*, **65**, 589 (1943).

62. J. H. Elliot, *J. Phys. Chem.*, **46,** 221 (1942).
63. J. H. Elliot and M. Kilpatrick, *J. Phys. Chem.*, **45,** 472 (1941).
64. M. Kilpatrick and R. D. Eanes, *J. Am. Chem. Soc.*, **65,** 589 (1943), and Ref. 63.
65. M. M. Davis and H. B. Hetzer, *J. Res. Nat. Bur. Std.*, **60,** 569 (1958).
66. L. H. Briggs and J. W. Lyttleton, *J. Chem. Soc.*, 421 (1943).
67. H. H. Jaffé, L. D. Freedman and G. O. Doak, *J. Am. Chem. Soc.*, **75,** 2209 (1953).
68. D. Pressman and D. H. Brown, *J. Am. Chem. Soc.*, **65,** 540 (1943).
69. J. F. J. Dippy and J. E. Page, *J. Chem. Soc.*, 357 (1938).
70. E. Berliner and E. A. Blomers, *J. Am. Chem. Soc.*, **73,** 2479 (1951).
71. N. T. Vartak, N. L. Phalnikar and B. V. Bhide, *J. Indian Chem. Soc.*, **24,** 131A (1947).
72. B. Breyer, *Ber.*, **71B,** 163 (1938).
73. A. I. Portnov, *Zhur. Obshch. Khim.*, **18,** 594 (1948); *Chem. Abstr.*, **43,** 57 (1949).
74. R. Stewart and K. Yates, *J. Am. Chem. Soc.*, **82,** 4059 (1960).
75. R. Stewart and K. Yates, *J. Am. Chem. Soc.*, **80,** 6355 (1958).
76. N. C. Deno, J. J. Jaruzelski and A. Schriesheim, *J. Am. Chem. Soc.*, **77,** 3044 (1955); N. C. Deno and A. Schriesheim, *J. Am. Chem. Soc.*, **77,** 3051 (1955); N. C. Deno and W. L. Evans, *J. Am. Chem. Soc.*, **79,** 5804 (1957).
77. C. Jutz and F. Voithenleitner, *Chem. Ber.*, **97,** 29 (1964).
78. H. H. Jaffé and R. W. Gardner, *J. Am. Chem. Soc.*, **80,** 319 (1958); Si-Jung Yeh and H. H. Jaffé, *J. Am. Chem. Soc.*, **81,** 3274, 3279 (1959).
79. J. B. Culbertson, *J. Am. Chem. Soc.*, **73,** 4818 (1951).
80. V. I. Minkin and V. A. Bren, *Reakts. Sposobnost Org. Soedin., Tartu Gos. Univ.*, **4,** 112 (1967).
81. Y. Ogata and I. Sugiyama, *Science (Japan)*, **19,** 185 (1949); *Chem. Abstr.*, **45,** 5116 (1951).
82. O. Gawron, M. Duggan and C. J. Grelecki, *Anal. Chem.*, **24,** 969 (1952).
83. G. Schwarzenbach and H. Egli, *Helv. Chim. Acta.*, **17,** 1176, 1183 (1934); G. Schwarzenbach and E. Rudin, *Helv. Chim. Acta*, **22,** 360 (1939).
84. H. H. Jaffé and G. O. Doak, *J. Am. Chem. Soc.*, **77,** 4441 (1955).
85. R. A. Henry, W. G. Finnegan and E. Lieber, *J. Am. Chem. Soc.*, **76,** 88 (1954).
86. V. A. Bren, E. N. Malysheva and V. I. Minkin, *Reakts. Sposobnost Org. Soedin., Tartu Gos. Univ.*, **4,** 523 (1967); *Chem. Abstr.*, **69,** 43279 (1968).
87. C. A. Bishop and L. K. W. Tong, *J. Am. Chem., Soc.*, **87,** 501 (1965).
88. A. V. Willi and W. Meier, *Helv. Chim. Acta*, **39,** 318 (1956).
89. J. N. Gardner and A. R. Katritzky, *J. Chem. Soc.*, 4375 (1957).
90. C. Eaborn and K. C. Pande, *J. Chem. Soc.*, 297 (1961).
91. C. Eaborn, *J. Chem. Soc.*, 4858 (1956).
92. P. W. Robertson, P. B. D. de la Mare and B. E. Swedlund, *J. Chem. Soc.*, 782 (1953); P. B. D. de la Mare, *J. Chem. Soc.*, 4450 (1954).
93. P. B. D. de la Mare, O. M. H. El Dusouqui, J. G. Tillet and M. Zeltner, *J. Chem. Soc.*, 5306 (1964); P. B. D. de la Mare and O. M. H. El Dusouqui, *J. Chem. Soc.* (**B**), 251 (1967).
94. E. Lorz and R. Baltzly, *J. Am. Chem. Soc.*, **71,** 3992 (1949).
95. L. Pekkarinen and E. Tommila, *Acta Chem. Scand.*, **13,** 1019 (1959).

96. E. Tommila, *Suomen Kemistilehti*, **A17**, 1 (1944).
97. E. E. Reid, *Am. Chem. J.*, **24**, 397 (1900).
98. E. Berliner and L. C. Monack, *J. Am. Chem. Soc.*, **74**, 1574 (1952).
99. C. T. Abichandani and S. K. K. Jatkar, *J. Indian Inst. Sci.*, **A21**, 417 (1938).
100. D. H. McDaniel and H. C. Brown, *J. Org. Chem.*, **23**, 420 (1958).
101. J. Hine, *J. Am. Chem. Soc.*, **82**, 4877 (1960).
102. Y. Yukawa and Y. Tsuno, *Bull. Chem. Soc. Japan*, **32**, 971 (1959).
103. G. E. K. Branch and D. L. Yabroff, *J. Am. Chem. Soc.*, **56**, 2568 (1934).
104. L. G. Bray, J. F. J. Dippy, S. R. C. Hughes and L. W. Laxton, *J. Chem. Soc.*, 2405 (1957).
105. W. Baker, *Nature*, **137**, 236 (1936).
106. M. M. Davis and H. B. Hetzer, *J. Phys. Chem.*, **61**, 125 (1957).
107. R. A. Robinson and A. K. Kiang, *Trans. Faraday Soc.*, **52**, 327 (1956).
108. M. L. Bender, F. J. Kezdy and B. Zerner, *J. Am. Chem. Soc.*, **85**, 3017 (1963).
109. S. M. Kupchan and M. F. Saettone, *Tetrahedron*, **18**, 1403 (1962).
110. P. D. Bartlett and F. D. Greene, *J. Am. Chem. Soc.*, **76**, 1088 (1954).
111. T. C. Bruice and D. W. Tanner, *J. Org. Chem.*, **30**, 1668 (1965).
112. E. Berliner in *Progress in Physical Organic Chemistry*, Vol. 2 (Eds. S. G. Cohen, A. Streitwieser Jr. and R. W. Taft), Interscience, New York, 1964, pp. 253–321.
113. A. Ballio, *Gazz. Chim. Ital.*, **79**, 924 (1949); **81**, 782 (1951).
114. G. Illuminati and G. Marino, *J. Am. Chem. Soc.*, **78**, 4975 (1956); G. Illuminati, *J. Am. Chem. Soc.*, **80**, 4945 (1958).
115. Ref. 120, pp. 225–279.
116. Ref. 118, pp. 212–214, 217–218.
117. R. Benedikt, *Ann. Chem.*, **199**, 129 (1879).
118. P. B. D. de la Mare and J. H. Ridd, *Aromatic Substitution—Nitration and Halogenation*, Butterworths, London, 1959.
119. A. V. Topchiev, *Nitration of Hydrocarbons and other Organic Compounds*, Pergamon Press, London, 1959.
120. R. O. C. Norman and R. Taylor in *Electrophilic Substitution in Benzenoid Compounds*, Elsevier, Amsterdam, 1965, pp. 61–91.
121. F. Arnall, *J. Chem. Soc.*, 811 (1924).
122. A. A. Spryskov and I. K. Barbinskaya, *J. Org. Chem.* (*USSR*), (Engl. Trans.) **1**, 1978 (1965).
123. K. Halvarson and L. Melander, *Arkiv Kemi*, **11**, 77 (1957).
124. F. G. Bordwell and E. W. Garbisch Jr., *J. Am. Chem. Soc.*, **82**, 3588 (1960).
125. C. A. Bunton, E. D. Hughes, C. K. Ingold, D. I. H. Jacobs, M. H. Jones, G. J. Minkoff and R. I. Reed, *J. Chem. Soc.*, 2628 (1950).
126. S. Veibel, *Ber.*, **63**, 1577 (1930).
127. T. Suzawa, Z. Yasuoka, O. Manabe and H. Hiyama, *Sci. Ind.* (*Japan*), **29**, 7 (1955); *Chem. Abstr.*, **49**, 13749 (1955).
128. D. R. Harvey and R. O. C. Norman, *J. Chem. Soc.*, 3604 (1961).
129. A. F. Holleman and I. J. Rinkes, *Rec. Trav. Chim.*, **30**, 48 (1911).
130. B. V. Tronov and S. F. Kolesnikova, *Soobshch. o Nauchn.-Issled. Rabot. Vses. Khim. Obshchestva im. Mendeleeva* **No, 1**, 46 (1953); *Chem. Abstr.*, **49**, 8173 (1955).
131. B. G. Painter and F. G. Soper, *J. Chem. Soc.*, 342 (1947). E. Berliner, *J. Am. Chem. Soc.*, **72**, 4003 (1950).

132. G. Kohnstam and D. L. H. Williams in *The Chemistry of the Ether Linkage* (Ed. S. Patai), Interscience, New York, 1967, p. 137.
133. E. Plazek, *Roczniki Chem.*, **10,** 761 (1930).
134. G. H. Bing, W. W. Kennard and D. N. Matthews, *Australian J. Chem.*, **13,** 317 (1960).
135. A. Campbell and D. J. Shields, *Tetrahedron*, **21**, 211 (1965).
136. H. Cerfontain in *Mechanistic Aspects of Aromatic Sulphonation and Desulphonation*, Interscience, New York, 1968, pp. 95–100.
137. J. Obermiller, *Ber.*, **40**, 3623 (1907).
138. Y. Muramoto, *Sci. Ind.* (*Japan*), **29**, 315 (1955); *Chem. Abstr.*, **50,** 9946 (1956).
139. F. Olsen and J. C. Goldstein, *Ind. Eng. Chem.*, **16,** 66 (1924).
140. B. H. Chase and E. McKeown, *J. Chem. Soc.*, 50 (1963).
141. A. A. Spryskov and B. G. Gnedin, *J. Org. Chem.* (*USSR*), (Engl. Trans.) **1,** 1983 (1965).
142. B. I. Karavaer and A. A. Spryskov, *J. Gen. Chem.* (*USSR*), (Engl. Trans.) **33,** 1840 (1963).
143. H. Zollinger in *Azo and Diazo Chemistry—Aliphatic and Aromatic Compounds* (Engl. Trans. by H. E. Nursten), Interscience, New York, 1961, pp. 221–243, 253–257.
144. J. B. Conant and W. D. Peterson, *J. Am. Chem. Soc.*, **52,** 1220 (1930); R. Wistar and P. D. Bartlett, *J. Am. Chem. Soc.*, **63,** 413 (1941).
145. R. Pütter, *Angew. Chem.*, **63,** 188 (1951).
146. Z. J. Allan, *Collection Czech. Chem. Commun.*, **16–17,** 620 (1952); H. Zollinger, *Helv. Chim. Acta*, **36,** 1070 (1953).
147. E. Bamberger, *Ber.*, **33,** 3188 (1900).
148. L. N. Ogoleva and B. I. Stepanov, *J. Org. Chem.* (*USSR*), (Engl. Trans.) **1,** 2126 (1965) and references therein cited.
149. O. A. Stamm and H. Zollinger, *Helv. Chim. Acta*, **40,** 1105, 1955 (1957).
150. *Friedel–Crafts and Related Reactions* (Ed. G. A. Olah), Vols I–IV, Interscience, New York, 1963–5.
151. O. C. Dermer, D. M. Wilson, F. M. Johnson and W. H. Dermer, *J. Am. Chem. Soc.*, **63,** 2881 (1941).
152. C. C. Price in *Organic Reactions*, Vol. III (Eds. R. Adams, W. E. Bachmann, L. F. Fieser, J. R. Johnson and H. R. Snyder), John Wiley & Sons, New York, 1946, p. 3.
153. J. F. Olin, *U.S. Pat. 3,014,079* (Dec. 19, 1961).
154. Coalite and Chemical Products Ltd., *Belg. Pat. 609,029* (open Feb. 1, 1962).
155. E. A. Goldsmith, M. J. Schlatter and W. G. Toland, *J. Org. Chem.*, **23,** 1871 (1958).
156. S. H. Patinkin and B. S. Friedman in Ref. 150, Vol. II, pp. 75–97.
157. E. Weingaertner, *Brennstoff-Chem. Abstract*, **42,** 361 (1961).
158. Ref. 150, Vol. II, pp. 94–97.
159. V. W. Buls and R. S. Miller, *U.S. Pat. 2,923,745* (Feb. 2, 1960).
160. D. Bethell and V. Gold, *J. Chem. Soc.*, 1930 (1958).
161. N. Kornblum, P. J. Berrigan and W. J. le Noble, *J. Am. Chem. Soc.*, **85,** 1141 (1963).
162. N. Kornblum and A. P. Lurie, *J. Am. Chem. Soc.*, **81,** 2705 (1959).
163. J. E. Hofmann and A. Shriesheim in Ref. 150, Vol. II, pp. 597–640.
164. J. von Braun, *Ann. Chem.*, **507,** 15 (1933).

165. J. B. Niederl, *J. Am. Chem. Soc.*, **59,** 1113 (1937).
166. M. E. McGreal, V. Niederl and J. B. Niederl, *J. Am. Chem. Soc.*, **61,** 345 (1939); I. P. Tsukervanik and Z. N. Nazarova, *J. Gen. Chem. (USSR)*, **9,** 33 (1939).
167. A. M. Partansky in *A.C.S. Div. Org. Coatings Plast. Chem. Preprints*, **27,** 115 (1967).
168. G. A. Olah and W. S. Tolgyesi in Ref. 150, Vol. II, pp. 659–784.
169. R. C. Fuson and C. H. McKeever in *Organic Reactions*, Vol. I (Ed. R. Adams), John Wiley & Sons, New York, 1942, p. 63.
170. H. Stephen, W. F. Short and G. Gladding, *J. Chem. Soc.*, 510 (1920).
171. D. I. Randall and E. E. Renfrew, *U.S. Pat. 2,642,444* (1953); *Chem. Abstr.*, **48,** 7640 (1954).
172. C. A. Buehler, F. K. Kirchner and G. F. Deebel, *Org. Syn.*, Coll. Vol. III, 468 (1955).
173. H. Arnold, *Chem. Ind. (London)*, **76,** 777 (1943).
174. E. Ziegler and H. Ludde, *Monatsh.*, **79,** 55 (1948); E. Ziegler, *Monatsh.*, **79,** 142 (1948).
175. P. H. Gore in Ref. 150, Vol. III, pp. 1–381.
176. Ref. 150, Vol. III, p. 46.
177. G. Sandulesco and A. Girard, *Bull. Soc. Chim. France, Mem.*, [4] **47,** 1300 (1930).
178. Ref. 150, Vol. III, pp. 168–169, 170–179, 256–257.
179. A. Gerecs in Ref. 150, Vol. III, pp. 499–533.
180. R. Baltzly, W. S. Ide and A. P. Phillips, *J. Am. Chem. Soc.*, **77,** 2522 (1955).
181. Ref. 150, Vol. III, pp. 507–511.
182. G. A. Olah and S. J. Kuhn in Ref. 150, Vol. III, pp. 1154–1179.
183. Ref. 150, Vol. III, p. 1157.
184. K. J. Karmann, *Svensk. Kem. Tidskr.*, **58,** 293 (1946); *Chem. Abstr.*, **41,** 2721 (1947).
185. Ref. 150, Vol. III, pp. 1191–1210.
186. W. E. Truce in *Organic Reactions* (Ed. E. Adams), Vol. IX, John Wiley & Sons, New York, 1957, p. 37.
187. R. Adams and I. Levine, *J. Am. Chem. Soc.*, **45,** 2375 (1923); R. Adams and E. Montgomery, *J. Am. Chem. Soc.*, **46,** 1518 (1924).
188. L. Gattermann, *Ann. Chem.*, **357,** 313 (1907); L. Gattermann and W. Beschelmann, *Ber.*, **31,** 1765 (1898).
189. L. Gattermann, *Ann. Chem.*, **357,** 313 (1907); L. Gattermann and T. von Horlacher, *Ber.*, **32,** 284 (1899); R. Adams and I. Levine, *J. Am. Chem. Soc.*, **45,** 2373 (1923).
190. K. von Auwers and W. Mauss, *Ber.*, **61,** 1495 (1928).
191. Ref. 150, Vol. III, pp. 1211–1240.
192. V. I. Minkin and G. N. Dorofeenko, *Russ. Chem. Rev.*, (Engl. Trans.) **29,** 599 (1960).
193. Z. Arnold and A. Holy, *Collection Czech. Chem. Commun.*, **27,** 2886 (1962); G. Martin and M. Martin, *Bull. Soc. Chim. France*, 1637 (1963).
194. N. P. Buu-Hoi, G. Lejeune and M. Sy, *Compt. Rend.*, **240,** 2241 (1955).
195. N. P. Buu-Hoi, N. D. Xuong, M. Sy, G. Lejeune and N. B. Tien, *Bull. Soc. Chim. France*, 1594 (1955).
196. A. Rieche, H. Gross and E. Hoft, *Chem. Ber.*, **93,** 88 (1960).

197. H. Gross, A. Rieche and G. Matthey, *Chem. Ber.*, **96**, 308 (1963).
198. H. Wynberg, *Chem. Rev.*, **60**, 169 (1960).
199. O. L. Brady and J. Jakobovits, *J. Chem. Soc.*, 767 (1950).
200. J. C. Duff, *J. Chem. Soc.*, 547 (1941).
201. W. Ruske in Ref. 150, Vol. III, pp. 383–497.
202. J. Houben, *Ber.*, **59**, 2878 (1926).
203. J. Houben and W. Fischer, *J. Prakt. Chem.*, [2] **123**, 262 (1929).
204. A. S. Lindsey and H. Jeskey, *Chem. Rev.*, **57**, 583 (1957).
205. O. Baine, G. F. Adamson, J. W. Barton, J. L. Fitch, D. R. Swayampati and H. Jeskey, *J. Org. Chem.*, **19**, 510 (1954), and Ref. 203.
206. F. Wessely, K. Benedikt, H. Benger, G. Friedrich and F. Prillinger, *Monatsh.* **81**, 1071 (1950).
207. H. Hart, *J. Am. Chem. Soc.*, **72**, 2900 (1950).
208. A. P. Best and C. L. Wilson, *J. Chem. Soc.*, 28 (1938); C. K. Ingold, C. G. Raisin and C. L. Wilson, *J. Chem. Soc.*, 1637 (1936); M. Koizumi and T. Titani, *Bull. Chem. Soc. Japan*, **13**, 681 (1938); P. A. Small and J. H. Wolfenden, *J. Chem. Soc.*, 1811 (1936).
209. D. P. N. Satchell, *J. Chem. Soc.*, 3911 (1956).
210. Ref. 120, p. 194.
211. F. C. Whitmore and E. R. Hanson, *Org. Syn.*, Coll. Vol. I, 161 (1941).
212. H. C. Brown and M. Dubek, *J. Am. Chem. Soc.*, **82**, 1939 (1960).
213. J. Boeseken and M. L. von Konigsfelt, *Rec. Trav. Chim.*, **54**, 313 (1935); J. Boeseken, C. F. Metz and J. Pluim, *Rec. Trav. Chim.*, **54**, 345 (1935).
214. C. Walling and R. B. Hodgdon, *J. Am. Chem. Soc.*, **80**, 228 (1958).
215. P. Kovacic, R. P. Bennett and J. L. Foote, *J. Org. Chem.*, **26**, 3013 (1961).
216. W. Fuchs, *Monatsh.*, **38**, 331 (1917).
217. J. Post, *Ann. Chem.*, **205**, 51 (1880).
218. G. Cohn, *Ann. Chem.*, **309**, 236 (1899).
219. G. Chuchani, H. Diaz and J. Zabicky, *J. Org. Chem.*, **31**, 1573 (1966); N. Barroeta, G. Chuchani and J. Zabicky, *J. Org. Chem.*, **31**, 2330 (1966).
220. N. M. Cullinane and B. F. R. Edwards, *J. Appl. Chem.*, **9**, 133 (1959).
221. T. Reichstein, *Helv. Chim. Acta*, **10**, 392 (1927).
222. V. Kese and G. Chuchani, *J. Org. Chem.*, **27**, 2032 (1962).
223. G. Chuchani and V. Rodriguez-Uzcanga, *Tetrahedron*, **22**, 2665 (1966).
224. G. Chuchani, *Acta Cient. Venezolana*, Supl. 1, 200 (1963); G. Chuchani, *J. Chem. Soc.*, 1753 (1959); 325 (1960); G. Chuchani and J. Zabicky, *J. Chem. Soc.*, (**C**), 297 (1966).
225. R. P. Bell and E. N. Ramsden, *J. Chem. Soc.*, 161 (1958).
226. R. P. Bell and D. J. Rawlinson, *J. Chem. Soc.*, 63 (1961).
227. J. H. Ridd in *Ann. Rep. Chem. Soc.* (*London*), 163 (1961).
228. L. M. Stock and H. C. Brown, in *Advances in Physical Organic Chemistry*, Vol. 1 (Ed. V. Gold), Academic Press, New York, 1963, pp. 35–154.
229. E. A. Kryuger and M. S. Bednova, *J. Gen. Chem.* (*USSR*), **3**, 67 (1933).
230. G. M. Kraay, *Rec. Trav. Chim.*, **49**, 1082 (1930).
231. R. H. Clark and R. H. Hall, *Trans. Roy. Soc. Can.* [3] **21**, Sect. 3, 311 (1922).
232. M. Liveris, P. G. Lutz and J. Miller, *J. Am. Chem. Soc.*, **78**, 3375 (1956).
233. J. Burdon, W. B. Hollyhead and J. C. Tatlow, *J. Chem. Soc.*, 5152 (1965); J. Burdon, *Tetrahedron*, **21**, 3373 (1965).

CHAPTER 9

Electrophilic attacks on the hydroxyl group

PENTTI SALOMAA, ALPO KANKAANPERÄ and KALEVI PIHLAJA

Department of Chemistry, University of Turku, Turku, Finland

I. INTRODUCTION

The problems involved in electrophilic attacks on the hydroxyl group are, in the last analysis, one aspect of the more fundamental problem about the underlying pattern of structural factors which determine the mutual reactivities of different electron-donor and electron-acceptor molecules. It is, therefore, quite natural that the reactions like, e.g., acylations, in which the hydroxyl group in an organic molecule becomes the primary site of the attack, have several counterparts in the chemistry of other functional groups.

Some characteristics of the hydroxyl group, made evident from its reactions with electrophiles, require especial attention to be drawn to them. In the first place, the hydroxyl group possesses only relatively moderate nucleophilic properties, that is, as a functional group of a Lewis base it is of an intermediate strength (section II.A). In terms of the concept of hard and soft acids and bases, this implies that the hydroxyl group is moderately reactive with Lewis acids of widely varying strengths. In fact, the spectrum of reactions involving the hydroxyl group as the electron donor seems to be extraordinarily wide in respect of the nature of the electrophile. In the second place, the influence of the structure of the alcohol upon the facility of the attack is normally very material. Apart from the steric factors involved, this can be attributed to the fact that the transmission of electronic effects from the remnant of the molecule through the C–O bond is facilitated by the similar sizes of the valence orbitals in the carbon and oxygen atoms, both being second-row elements.

The experimental material on various kinds of electrophilic attacks on the hydroxyl group is both diffuse and rich, and many extensive review articles and monographs on special topics have been published. Therefore, a rather drastic limitation in the scope of the present chapter seemed unavoidable. While some representative older results were included, the main part of the material dealt with elsewhere was simply omitted, making, wherever possible, citations to the pertinent literature. Primary attention was focused on relatively recent work covering mechanistically and practically important aspects of the function of the hydroxyl group as the nucleophilic partner in various reactions.

II. ALKYLATION. CARBON ATOM AS THE ELECTROPHILIC CENTRE

A. Nucleophilic Reactivity of the Hydroxyl Group

The nucleophilic reactivity of the hydroxyl group towards carbon can be illustrated by comparing the formation of ethers from alkyl halides and alcohols (equation 1) with the reactions of alkyl halides

$$ROH + RX \longrightarrow ROR + HX \qquad (1)$$

with other nucleophiles. The relative rates of methyl iodide with methanol and some other nucleophiles are shown in Table 1. The data confirm conclusively that the proton basicity of the nucleophile is not the sole factor influencing its attacking capability; all the nucleophiles listed in Table 1 react more rapidly than methanol, notwithstanding the fact that many of them are less basic than methanol.

TABLE 1. Relative rates for the electrophilic attack of methyl iodide on the hydroxyl group and on various other nucleophiles[1, 2]. Temperature 25°C, methanol solution.

Nucleophile N	$\log (k_N/k_{MeOH})$	pK_a (in water)
MeOH	0·0	−1·7[a]
Cl^-	4·4	−7
Me_2S	5·3	−5·3
PhO^-	5·8	9·9
Br^-	5·8	−9
MeO^-	6·3	16
CN^-	6·7	9·1
Et_3As	7·1	2
I^-	7·4	−9·5
Et_3P	8·7	8·9
PhS^-	9·9	6·5

[a] Deno and Turner[3] have obtained a pK_a value of −2·5.

Another important factor affecting the reactivity of an alcohol towards electrophiles is the polarizability of the hydroxyl group. Normally, the oxygen atoms in hydroxyl groups are of moderately low polarizability. In terms of the concept of hard and soft acids and bases, advocated by Pearson[1, 4, 5], this implies that alcohols are Lewis bases of an intermediate hardness. In the reactions between Lewis acids and bases hard bases have the strongest tendency to react with hard acids, and, correspondingly, soft acids with soft bases.

An existence of 'symbiotic effects' in nucleophilic displacement reactions has been suggested[2, 6]; a grouping of either several hard bases or soft bases around the central carbon atom will stabilize the transition state and thus increase the reaction rate. The relative rate coefficients shown in Table 2 illustrate this point. In these reactions the tosylate ion is a hard Lewis base and the iodide ion a soft base.

TABLE 2. Attack of MeOTs and MeI on the hydroxyl group and on various other nucleophiles[2, 6]. Relative rate coefficients are in methanol solution at 25°C.

Nucleophile and its classification		k_{OTs}/k_I
MeOH	hard	210
MeO^-	hard	4·6
Cl^-	hard	2·8
PhS^-	soft	0·13
$SeCN^-$	soft	0·23
I^-	soft	0·13
Ph_3P	soft	0·18

The decrease of the ratio k_{OTs}/k_I with increasing softness of the attacking nucleophile can be rationalized in this way.

The factors influencing the reactivity of alcohols have also been represented by equation (2), proposed by Edwards[7, 8]. Here a and b are constants, E is the susceptibility of the alcohol to an electrophilic attack, P is the polarizability factor of the alcohol in question and H is a factor depending on the basicity of the oxygen atom. In short, according to equation (2), the sole factors determining the

$$E = \mathrm{a}\,P + \mathrm{b}\,H \tag{2}$$

reactivities of alcohols would be their proton basicities and polarizabilities. This, of course, must be a very crude simplification.

One obvious reason for the inadequacy of simplifications like equation (2) is that, in a series of formally similar compounds, the reaction mechanism may change when going from one compound to another. The data in Table 3 illustrate this point. The ratio of the alcoholysis rates of ROTs and RBr increase greatly in the sequence Me < Et < *i*-Pr < *t*-Bu. It has been assumed that this indicates

TABLE 3. Attacks of ROTs and RBr on the hydroxyl group. Relative rate coefficients are in ethanol at 25°C.

Nucleophile	R	k_{OTs}/k_{Br}	Mechanism	Reference
EtOH	Me	16	S_N2	9, 10
EtOH	Et	15	S_N2	9, 10
EtOH	*i*-Pr	73		9, 10
EtOH	*t*-Bu	>4000	S_N1	11, 12

an increasing charge separation between the alkyl group and the leaving group in the transition state[11]. The deuterium isotope effect on the solvolysis of *t*-butyl-d_9 chloride has been studied in various solvents[13-15]; the independence of the isotope effect of the solvent shows that there is only a weak covalent interaction between the solvent and the *t*-butyl group at the transition state. So the study of the deuterium isotope effect confirms the accepted view that the alcoholysis of *t*-butyl chloride occurs by the S_N1 mechanism; thus, differing from primary halides, the attacking electrophile is in this case a carbonium ion. The data in Table 3 point out that the alcoholyses of isopropyl tosylate and isopropyl bromide may also have some S_N1 character.

It is interesting to note in this context that bicyclic bridgehead halides are solvolysed slower than the corresponding *t*-butyl halides by several powers of ten, owing to the steric strain associated with the carbonium ion geometry. In contrast, Wiberg and Williams[16] have recently found that 1-chlorobicyclo(1,1,1)pentane (see equation 3) is about three times more reactive than *t*-butyl chloride and

$$\text{(bicyclo[1.1.1]pentyl)}-Cl \xrightarrow[\text{EtOH}]{80\%} CH_2=\text{(cyclobutyl)}-OEt + CH_2=\text{(cyclobutyl)}-OH \qquad (3)$$

10^{14} times more reactive than 1-chloronorbornane. This exceptional reactivity is probably due to the fact that much more energy will be gained when releasing the strain of the bicyclic system than will be lost when approaching the carbonium ion geometry.

Detailed information on reactions of various alcohols with a saturated carbon atom as the reaction centre is available in articles on nucleophilic substitution reactions[17-23].

B. Alkylation with α-Halo Ethers, Vinyl Ethers, Acetals and Carbonyl Compounds

1. Kinetics and mechanism

Experimental evidence derived from different independent sources indicates that oxonium–carbonium ions play an important role as intermediates in the reactions of α-halo ethers, vinyl ethers and acetals. The oxonium–carbonium ions are strong electrophiles and they are very fast to react with alcohols (equation 4). In a relatively

$$\mathrm{RO\overset{+}{=\!=}CR^1_2 + ROH \longrightarrow ROCR^1_2OR + H^+} \quad (4)$$

recent article Schmitz and Eichhorn[24] have discussed the formation of acetals. Most of the reactions involve an electrophilic attack on the hydroxyl group. Some additional observations, in particular those concerning the mechanistic and thermodynamic aspects of the reaction, may be mentioned in this context.

In their reactions in alcohol solution, α-halo ethers, vinyl ethers and acetals form unstable monoalkoxy carbonium ions as intermediates (equation 5). Although the intermediate is the same, the mechanism of its formation differs in three cases: the reaction (A) is

$$\left.\begin{array}{l} \mathrm{ROCH(CH_3)Cl} \xrightarrow{(A)} \\ \mathrm{ROCH{=}CH_2} \xrightarrow{(B)} \\ \mathrm{ROCH(CH_3)OR} \xrightarrow{(C)} \end{array}\right\} \mathrm{RO\overset{+}{=\!=}CHCH_3} \quad (5)$$

solvolytic[23], reaction (B) exhibits general acid catalysis[25, 26], and reaction (C) is specifically hydronium ion-catalysed[27].

The alcoholysis reaction of α-halo ethers proceeds by the S_N1 mechanism[23]. The enhanced electrophilic reactivity of halo ethers as compared with that of alkyl halides (section II.A) has been explained in terms of the resonance stabilization present in oxonium-carbonium ions.

The reactions of α-halo esters with alcohols have also been used for the preparation of acetals[28–29]. Baldwin and Walker[29] synthesized in this way a number of acetals of α-bromophenyl acetaldehyde. These acetals were subsequently used for the preparation of keten e acetals by dehydrobromination. It is very likely that the attacking electrophile is the oxonium–carbonium ion (equation 6).

$$\begin{array}{l} \mathrm{RCHBrCHBrOCOCH_3 \longrightarrow RCHBrCH\overset{+}{=\!=}OCOCH_3 \xrightarrow{MeOH}} \\ \mathrm{RCHBrCH(OCH_3)OCOCH_3 \longrightarrow RCHBrCH\overset{+}{=\!=}OCH_3 \xrightarrow{MeOH}} \\ \mathrm{RCHBrCH(OCH_3)_2} \end{array} \quad (6)$$

The thermodynamics and kinetics of reactions (5) have been recently studied[30]. Although ethyl vinyl ether is initially on a higher free energy level than diethyl acetal (about 3·3 kcal/mole), the two compounds produce oxonium–carbonium ions at rates which are virtually the same, because the transition state of the vinyl ether hydrolysis is on a higher energy level than that of the acetal hydrolysis.

On the basis of the above results it is possible to estimate the rate of an electrophilic attack on the hydroxyl group by the oxonium-carbonium ion. In the acetal solvolysis, the transition state is only by about 4 kcal/mole less stable than the intermediate ion. The smallness of this free energy difference shows that the energy well relating to the intermediate is not deep because the intermediate must further react at a rate which corresponds to a free energy of activation that is less than this difference. In this way the rate coefficient for the attack of the intermediate ion on water was estimated to be about 10^{10} s^{-1}. It is very probable that the rate of the attack in an alcohol solution is smaller by only one or two powers of ten.

Great variations in the stabilities of alkoxy carbonium ions become evident when the substituents are varied. The rate of an attack on alcohol depends largely on the stability of this ion. The following observations made on the hydrolysis of acetals provide further information on this point, which is of interest in the synthesis of structurally different acetals (see section II.B.3).

The overall hydrolysis rate of acetals can be dissected to partial rate factors relating to the leaving group and to the remnant of the molecule[31]. For reaction (7), in which R is varied and R^1 is held unchanged, the following partial rate factors have been calculated:

$$ROCH_2OR^1 \longrightarrow RO\overset{+}{\text{---}}CH_2 \longrightarrow ROCH_2OH \tag{7}$$

R	FCH_2CH_2	$ClCH_2CH_2$	$CH_3OCH_2CH_2$	CH_3	CH_3CH_2	$(CH_3)_2CH$
k_{rel}	0·0801	0·0480	0·201	1	4·48	22·1

The corresponding partial rate factors for various leaving groups (R^1 is varied) differ only slightly and exhibit a minimum:

R^1	FCH_2CH_2	$ClCH_2CH_2$	$CH_3OCH_2CH_2$	CH_3	CH_3CH_2	$(CH_3)_2CH$
k_{rel}	2·02	1·96	1·53	1	1·21	2·27

The minimum is caused by the circumstance that the structural effects on the pre-equilibrium protonation and on the rate-determining heterolysis act in opposite directions. Further information on the leaving group effects is available from a study of the basicities of symmetrical and unsymmetrical acetals[32].

2. Equilibrium studies

Kubler and co-workers[33, 34] have measured equilibrium constants for the reactions of methanol with various aldehydes and ketones (equations 8 and 9). The equilibrium constant for the hemiacetal

$$R^1COR^2 + CH_3OH \rightleftharpoons R^1R^2C(OH)OCH_3 \quad (8)$$

$$R^1R^2C(OH)OCH_3 + CH_3OH \rightleftharpoons R^1R^2C(OCH_3)_2 + H_2O \quad (9)$$

$$K = K_8K_9 = \frac{a_{\text{acetal}}\, a_{H_2O}}{a_{\text{ald.}}\, a^2_{CH_3OH}} \quad (10)$$

formation, K_8, was measured spectrophotometrically in neutral or basic solutions, in which no acetal was formed. Only a few of the aldehydes or ketones studied formed significant amounts of hemiacetals. If the carbonyl compound contained strongly electronegative substituents the extent of hemiacetal formation was substantial. For instance, *p*-nitrobenzaldehyde was predominantly in the form of a hemiacetal in neutral or basic methanol, but in acidic media the corresponding acetal was formed.

In acidic solutions the acetalization of saturated aliphatic aldehydes with primary alcohols is almost complete; in cases of aromatic aldehydes and α,β-unsaturated aldehydes the equilibria are less favourable. The conversion of ketones to acetals (ketals) is ordinarily so unfavourable that the latter cannot be isolated from the equilibrium mixture; however, the preparation of acetal can be successfully accomplished if one of the reaction products is removed from the mixture.

Kinetic studies of acetal hydrolysis provide additional information about the equilibria between carbonyl compounds and alcohols. The hydrolysis of 1,3-dioxan and its alkyl derivatives differs from the hydrolysis reactions of acyclic acetals[35] and alkyl-1,3-dioxolanes[36] in that the equilibrium is not completely on the side of the hydrolysis products[37] (see Table 4). Similar results have been reported by Aftalion and co-workers[38–40].

TABLE 4. Hydrolysis equilibria for six-membered cyclic acetals in dilute solutions (<0·1M) in water at 25°C [37].

Acetal	% Unhydrolysed at equil.
1,3-Dioxan	24
2-Methyl-1,3-dioxan	33
4-Methyl-1,3-dioxan	47
4,4,6-Trimethyl-1,3-dioxan	65
2,4,4,5,6-Pentamethyl-1,3-dioxan	71

3. Applications to synthesis

The facts discussed above make it possible to rationalize a number of observations made in connection with the synthesis of acetals. Thus, e.g., attempts to prepare the trichloroethyl acetal of formaldehyde from other formaldehyde acetals via *trans*acetalization (equation 11) were unsuccessful[41], the unsymmetrical acetal being the sole

$$(EtO)_2CH_2 \longrightarrow EtO\overset{+}{\cdots}CH_2 \xrightarrow[Cl_3CCH_2OH]{} EtO(Cl_3CCH_2O)CH_2 \longrightarrow$$

$$Cl_3CCH_2O\overset{+}{\cdots}CH_2 \xrightarrow[Cl_3CCH_2OH]{} (Cl_3CCH_2O)_2CH_2 \qquad (11)$$

product that could be isolated after the ethanol liberated had been distilled off. This is easily understood by the low stability of the chlorinated oxonium–carbonium ion; its energy level has been estimated from the partial rate factors of the corresponding acetals[42].

From the above it follows that the acetalization of alcohols having strongly electron-attracting substituents can be effected only under drastic conditions. Shipp and Hill[43] have used concentrated sulphuric acid for a number of negatively substituted alcohols. The yields were reportedly at least 70%.

The reaction of chloral with ethylene glycol (equation 12) provides a further illustration. The hemiacetal formed is exceptionally

$$\begin{matrix} CH_2OH \\ | \\ CH_2OH \end{matrix} + O{=}C\begin{matrix} H \\ CCl_3 \end{matrix} \longrightarrow \begin{matrix} CH_2OH \\ | \\ CH_2OCHCCl_3 \\ | \\ OH \end{matrix} \xrightarrow[H_2SO_4]{conc.} \begin{matrix} CH_2O \\ | \\ CH_2O \end{matrix}{>}C\begin{matrix} H \\ CCl_3 \end{matrix} + H_2O \qquad (12)$$

stable in this case, drastic conditions (e.g., concentrated sulphuric acid) are therefore required to produce the corresponding acetal[44, 45]. This is quite understandable because here also the intermediate oxonium–carbonium ion is unstabilized by strongly electronegative substituents.

As discussed above (section II.B.2) the acetalization of aromatic aldehydes is usually facile. It is to be noted that here the oxonium–carbonium ion is stabilized by the resonance effect of the aromatic ring, the latter becoming conjugated with the partial double bond of the oxonium–carbonium ion.

C. Alkylation with Ketene Acetals and Orthoesters

The stabilities of oxonium–carbonium ions are greatly increased when additional alkoxy groups are attached to the central carbon atom. The following stabilization energies, relative to the methyl cation, have been reported[46]:

Ion	$CH_3OCH_2^+$	$(CH_3O)_2CH^+$	$(CH_3O)_3C^+$
Stabilization energy (kcal/mole)	66	85	90

Although the above values refer to the gaseous state and are not directly applicable to liquid solutions, it may be expected that reactions of alcohols with such substrates, which are capable of generating dialkoxy carbonium ions, take place much more readily than the reactions which involve monoalkoxy carbonium ions (section II.B).

In the presence of an acid catalyst, ketene acetals and orthoesters are readily transformed to dialkoxy carbonium ions (equation 13), which subsequently react rapidly with the nucleophiles present, e.g.,

$$CH_2{=}C(OR)_2 \text{ or } CH_3C(OR)_3 \longrightarrow CH_3C\overset{OR}{\underset{OR}{\langle +}} \tag{13}$$

with an alcohol. Some data on the kinetics of ketene acetals in water solution are available[47, 48]. The reaction of orthoesters has been the subject of a number of studies[35, 47c]. The electrophilic attack of the intermediate ion on a hydroxyl group of an alcohol yields orthoesters (equation 14). When alcohol R^1OH is present in great excess all of

$$CH_3C\overset{OR}{\underset{OR}{\langle +}} + R^1OH \longrightarrow CH_3C(OR)_2OR^1 \tag{14}$$

the original alkoxy groups will be replaced. However, the preparation of the symmetrical orthoester from methyl orthoformate and 2,2,2-trichloroethanol was not possible except under relatively drastic reaction conditions[47b]. In this case, the stability of the intermediary alkoxy methyl cation was strongly reduced by the electronegative substituents.

Kinetic data on the hydrolysis of unsubstituted ketene acetals[47] suggest that the dialkoxy carbonium ion involved is only about 2 kcal/mole less stable than the reactant in the initial state. Because of the stability of the intermediate ion, the electrophilic attack with this ion may be slower than that with the corresponding monoalkoxy carbonium ion. However, it is evident from the kinetic data that this attack is anyway much faster than the formation of the ion.

Kuryla and co-workers[49, 50] have recently described a new route to ketene acetals and orthoesters from dichloro olefins (equation 15).

$$2\ ROCH_2CH_2ONa + CH_2{=}CCl_2 \longrightarrow (ROCH_2CH_2O)_2C{=}CH_2 \qquad (15)$$

This reaction is characteristic of β-alkoxy alcoholates; unsubstituted alcohols do not give ketene acetals. When alcohol is used as the solvent the ketene acetal formed is then rapidly transformed to the corresponding orthoester. The proposed scheme for reaction (15) involved an electrophilic attack on alcohol by an acetylenic intermediate (equation 16)[49, 50].

$$RONa + CH_2{=}CCl_2 \longrightarrow CH{\equiv}CCl + ROH + NaCl \longrightarrow CH_2{=}C(Cl)OR \longrightarrow$$
$$CH{\equiv}COR + ROH + NaCl \longrightarrow CH_2{=}C(OR)_2 \qquad (16)$$

D. Alkylation with Ethylene and Acetylene Derivatives

The general features of alkylation reactions with alkenes have been extensively discussed quite recently[51]. Only a few additional observations will be mentioned here.

It is evident from a great number of reactions that electron-withdrawing substituents attached directly to a sp^2-hybridized carbon atom enhance the electrophilic character of the double bond. Addition of alcohols to activated vinyl compounds like $R_2C{=}CRX$, where X is an electronegative substituent, has been extensively studied recently[52]. The activating effect of X was found to increase in the sequence: $CONHR < CONH_2 < CONR_2 < COOR < SO_2NR_2 < COR < {}^+PR_3$. The rate-enhancing effect of the $CONR_2$ group was rather surprising. No satisfactory explanation for this could be offered. In all cases the alkoxy group of the alcohol became attached to the neighbouring carbon atom having no electronegative substituents (equation 17).

$$R_2C{=}CRX + R^1OH \longrightarrow R_2(R^1O)C{-}CRXH \qquad (17)$$

In contrast to its electrophilic properties, the nucleophilic strength of the carbon–carbon double bond is increased by electron-releasing substituents attached to the trigonally hybridized carbon atoms. Thus it is possible to prepare ethers from alkyl-substituted olefins, for instance, from $Me_2C{=}CH_2$, in alcohol solution containing added acid. In these reactions it is a carbonium ion that attacks the alcohols.

The addition of mercuric salt to a simple olefin involves a formation of a bridged or π-complexed mercurinium ion, followed by a product-determining *trans* attack on the solvent, e.g., on an alcohol (equation 18)[53, 54]. The structure of the intermediate, which subse-

$$>C{=}C< \underset{}{\overset{^{+}HgX}{\rightleftharpoons}} >C\overset{HgX}{\overset{+}{\cdots}}C< \xrightarrow{ROH} -\underset{|}{\overset{HgX}{\overset{|}{C}}}-\underset{OR}{\underset{|}{C}}- \qquad (18)$$

quently attacks on the hydroxyl group, has been studied in the case of allene and its methyl derivatives[54]. These compounds reacted with methanol faster than simple olefins by factors of several powers of ten. The results were best interpreted in terms of a stable σ-bonded cyclic mercurinium ion **1**. The π-complex **2** was less probable as the reaction intermediate.

$$\underset{\textbf{(1)}}{R_2C\overset{\overset{OAc}{|}}{\overset{Hg}{\overset{+}{\cdots}}}C{=}CR_2} \qquad \underset{\textbf{(2)}}{R_2C{=}\overset{\overset{OAc}{|}}{\overset{Hg^{+}}{\vdots}}C{=}CR_2}$$

Several technical methods for the alkylation of alcohols with acetylenes are available[55].

An intramolecular reaction of a hydroxyl group with the acetylenic linkage has been shown to occur in the transformation of acetylenic epoxides to the derivatives of furan[56].

E. Alkylation with Divalent Carbon

Carbenes may react as electrophiles with alcohols, leading to ethers (equation 19). Alternatively, the conjugate acid of the carbene

$$R^1OH + R{:}\ddot{C}{:}R \longrightarrow R^1OCR_2H \qquad (19)$$

may function as the attacking reagent. Kirmse[57, 58] has drawn the conclusion that in the case of phenyl-substituted carbenes there exists an acid–base equilibrium with the carbene and the corresponding carbonium ion as the conjugate acid–base pair. The carbonium ion then attacks on the hydroxyl group (equation 20).

$$Ph_2C{:} \rightleftharpoons Ph_2CH^{+} \xrightarrow{ROH} Ph_2CHOR \qquad (20)$$

Carbenes are easily produced, for instance, from diazo compounds (e.g., equation 21). Methods of preparation of ethers with carbenes

$$\begin{aligned} CH_2N_2 &\rightarrow :CH_2 + N_2 \\ N_2CHCOOC_2H_5 &\rightarrow :CHCOOC_2H_5 + N_2 \end{aligned} \tag{21}$$

as the intermediates have been reviewed recently[22].

The cupric chloride-catalysed reactions of diazo acetates with butyl, benzyl and allyl alcohols are not selective[59]. In addition to the desired ether many other compounds have been found in the reaction product. The same is the case with the carbene produced from α-diazo-*p*-methoxyacetophenone in a light-induced reaction[60].

The carbene reactions of hindered phenols have been studied[61]. A normal diazo reaction was found in the case of 2-*t*-butyl-4-nitro

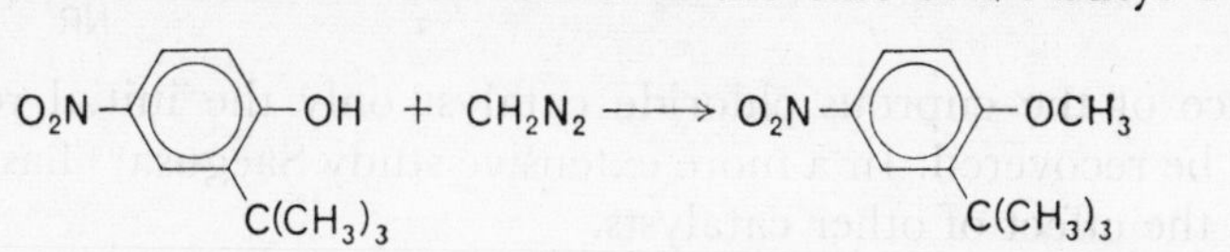

$$O_2N\text{—}C_6H_3(C(CH_3)_3)\text{—}OH + CH_2N_2 \rightarrow O_2N\text{—}C_6H_3(C(CH_3)_3)\text{—}OCH_3 \tag{22}$$

phenol (equation 22). In the case of the di-*t*-butyl derivative **3** the above reaction was sterically hindered and the nitronic ester **4** was the sole product. In the case of the 2,6-diisopropyl analogue of **3**

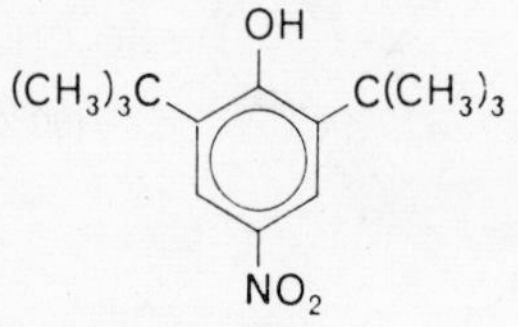

OH; $(CH_3)_3C$, $C(CH_3)_3$; NO_2

(3)

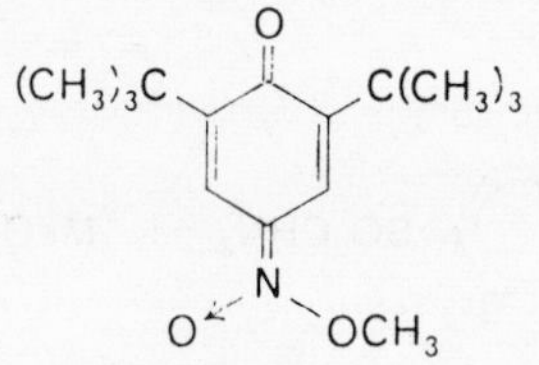

O; $(CH_3)_3C$, $C(CH_3)_3$; N; O^-, OCH_3

(4)

the steric hindrance can be assumed to be smaller. In fact, Meek and co-workers[61] have found that with this compound both the anisole derivative and the nitronic ester were formed. On the basis of these results it is surprising that the corresponding 2,6-diiodo analogue of **3** produces only anisoles, notwithstanding the bulkiness of the iodine atoms. It is very likely that, although *t*-butyl groups and iodine atoms are approximately of the same sizes, the alkylation of the hydroxyl group is much more hindered sterically by the *t*-butyl groups than by the iodine atoms. This is because the attacking reagent approaches the hydroxyl group from a direction which is almost perpendicular to the plane of the ring, and the hydroxyl group is much better shielded from these attacks by the *t*-butyl groups than by the iodine atoms. In addition, inductive effects of electronegative substituents like iodine atoms at 2- and 6-positions will reduce the tendency of the *p*-nitro group to react as a nucleophile, which makes the formation of the nitronic ester less favourable.

Schönberg, Junghans and Singer[62] have reported that attacks of diphenyldiazomethane on alcohols (methanol, benzyl alcohol) in the presence of ninhydrin give acetals (equation 23) instead of ethers.

$$Ph_2CN_2 + 2\ ROH \xrightarrow{\text{ninhydrin}} Ph_2C(OR)_2 \qquad (23)$$

Saegusa and co-workers[63] have reported that isocyanides react with allyl alcohol producing allyl formimidate (equation 24). In the

$$CH_2{=}CHCH_2OH \quad + \quad RN{\equiv}C \xrightarrow{\text{CuCl}} CH_2{=}CHCH_2OCH{=}NR \qquad (24)$$

absence of the cuprous chloride catalyst only the initial reactants could be recovered. In a more extensive study Saegusa[64] has investigated the effect of other catalysts.

The light-induced reaction of α-diazo sulphone with methanol gives aryl methoxymethyl sulphone and benzyl sulphonate (equation 25)[65]. The formation of benzyl sulphonate can be accounted for

$$X{-}C_6H_4{-}SO_2CHN_2 \quad + \quad MeOH \longrightarrow X{-}C_6H_4{-}SO_2CH_2OMe\ (90\%) \ + \ X{-}C_6H_4{-}CH_2SO_2OCH_3\ (10\%) \qquad (25)$$

$X = MeO, Me, H, NO_2$

by a rearrangement in which a minor part of the sulphonyl carbene is isomerized to sulphene (equation 26).

$$Ar{-}SO_2\ddot{C}H \longrightarrow Ar{-}CH{=}SO_2 \qquad (26)$$

III. ACYLATION. CARBON ATOM AS THE ELECTROPHILIC CENTRE

A. Introductory Remarks

The kinetics and mechanisms of the esterification of carboxylic acids and the reverse hydrolysis reaction (equation 27) have been

$$R^1COOH + R^2OH \rightleftharpoons R^1COOR^2 + H_2O \qquad (27)$$

the subject of considerable literature[66–70]. Other related reactions, the reactions of carbo diimides with alcohols, tioalcohols and phenols[71], and the synthesis and reactions of carbamate esters[72] have been reviewed recently.

The above-mentioned publications cover the main area of the subject; the following discussion is limited to studies published during the last few years and, in particular, to those dealing with stereochemical and mechanistic aspects.

B. *Structure and Mechanism*

1. Conformational aspects

The acetylation rates of cyclohexanol and its derivatives have been used for the evaluation of the conformational energy of the hydroxyl group. It has been generally accepted[73] that the rate constant of a pure axial conformation (k_a) and that of the pure equatorial conformation (k_e) may be represented by the rate constants of the respective *cis*- and *trans*-4-*t*-butyl-substituted compounds. Then, generally, the observed rate coefficient is given by $k = N_e k_e + N_a k_a$, in which N_e and N_a are the respective mole fractions of the two conformational isomers at equilibrium. The latter mole fractions can be readily calculated from the measured values of the rate coefficients, and a value derived for the conformational equilibrium constant $K = N_e/N_a$.

Thus Eliel and Biros[73] have derived the value $-\Delta G_{OH}{}^0 = 0{\cdot}56$ kcal/mole for the interaction free energy of an axial hydroxyl group in cyclohexanol (**5**), from the acetylation rates of cyclohexanol (k) and *cis* (k_e) and *trans* (k_a) forms of 4-*t*-butylcyclohexanol. This estimation agrees closely with other estimates[74] (equation 28).

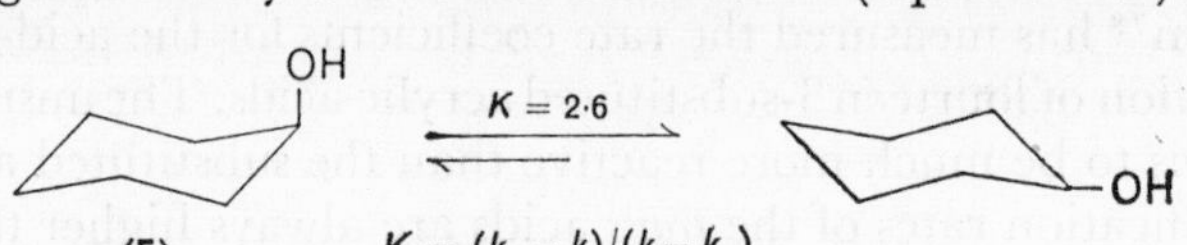

OH

$K = 2{\cdot}6$

—OH (28)

(**5**) $K = (k_a - k)/(k - k_e)$.

The *trans* form of 3-isopropylcyclohexanol is acetylated at a rate which corresponds to a conformational equilibrium (equation 29) with 93% of **6** and 7% of **7**. The value of the equilibrium constant

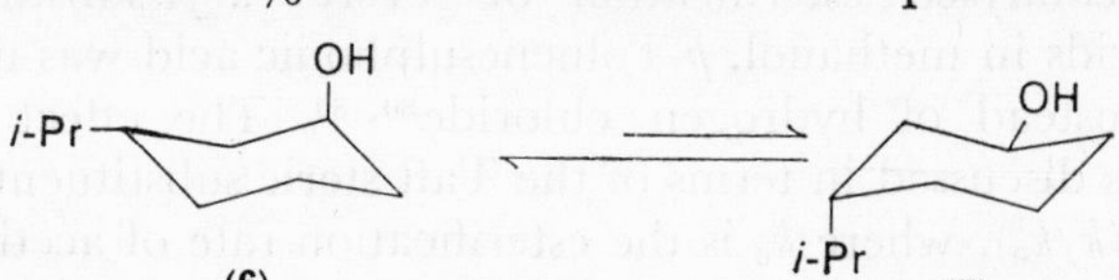

OH

i-Pr

OH

(29)

i-Pr

(**6**) (**7**)

gives a free energy difference of 1·6 kcal/mole, in agreement with the difference $\Delta G_{OH}{}^0 - \Delta G_{i\text{-}Pr}^0 = -0{\cdot}7 + 2{\cdot}3 = 1{\cdot}6$ kcal/mole[74, 75].

An estimation of a *syn*-axial Me–OH interaction energy from the acetylation rates of **8**, **9** and **10** led to the value 1·8 kcal/mole (the *syn*-axial Me–H interaction energy and the *syn*-axial OH–H interaction energy are 0·85 and 0·3 kcal/mole, respectively), which is in poor agreement with the thermodynamic value, 2·4 kcal/mole[74, 76].

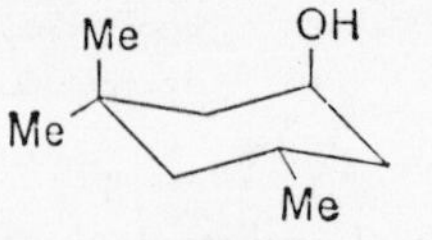

(8) 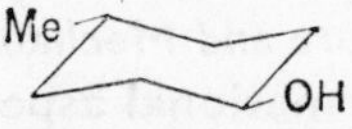 **(9)**

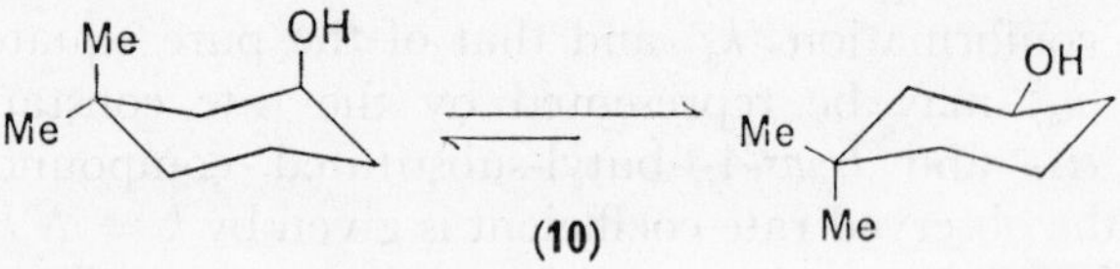

(10)

Eliel and Biros[73] have also given additional examples of instances in which estimations of conformational energies from the acetylation rates give erroneous values, proposing that only *cis* and *trans* *t*-butyl compounds[77] lead to reliable results.

2. Steric effects and esterification rates

Steric effects in the acid-catalysed esterification of 3-substituted acrylic acids[78] and several other acids[79] in methanol have been investigated recently.

Bowden[78] has measured the rate coefficients for the acid-catalysed esterification of fourteen 3-substituted acrylic acids. The unsubstituted acid seems to be much more reactive than the substituted acids, and the esterification rates of the *trans* acids are always higher than those of the *cis* acids. The latter effect indicates that the steric effects in the *trans* acids are not material.

Bowden, Chapman and Shorter[79] have studied the kinetics of the acid-catalysed esterification of several aryl-substituted carboxylic acids in methanol. *p*-Toluenesulphonic acid was used as the catalyst instead of hydrogen chloride[80, 81]. The effect of substitution was discussed in terms of the Taft steric substituent constant, $E_s = \log (k/k_0)$, where k_0 is the esterification rate of acetic acid[82].

The initial and transition states in the esterification of the acid

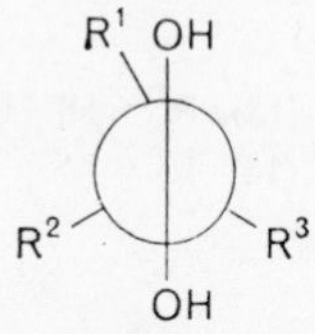

(11) 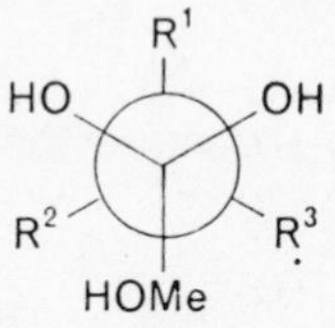 (12)

$R^1R^2R^3CCOOH$ were considered as **11** and **12**, respectively, both bearing a positive charge. The observed steric retardation was suggested to be a combination of a steric strain effect and of internal repulsive interactions. The increase in E_s caused by further substitution was found to depend on the bulk of all substituents R^1, R^2 and R^3. The following are examples of the results obtained by Bowden, Chapman and Shorter[79]:

R^1	R^2	R^3	E_s
Me	H	H	—
Ph	H	H	−0·37
Ph	Ph	H	−1·43
Ph	Ph	Ph	−4·68

Most of the observed effects could be interpreted in terms of conformational factors. As an alternative line of argumentation the build-up of the total steric effects was discussed[78, 82].

C. Applications to Synthesis

I. Esterification of hindered acids

The conventional methods of esterification of sterically hindered acids have been discussed by Newman[83]. A recent report of Parish and Stock[84] deals with the limitations involved in the reaction of methanol or other alcohols with the unsymmetrical anhydride of a hindered acid and trifluoroacetic acid[85, 86, 87]. The hindered acid was dissolved in trifluoroacetic anhydride and the alcohol was added, or, alternatively, a mixture of the acid and alcohol was treated with trifluoroacetic anhydride. It was found that **13** is easily formed from mesitoic acid and 2,6-dimethylphenol, whereas **15** is obtained from the same acid and 2,6-di-*t*-butylphenol in 83% yield. This alternative path, *C*-acylation, is facile in the latter case and therefore the esterification reaction may assume the minor role[87–89]. With a

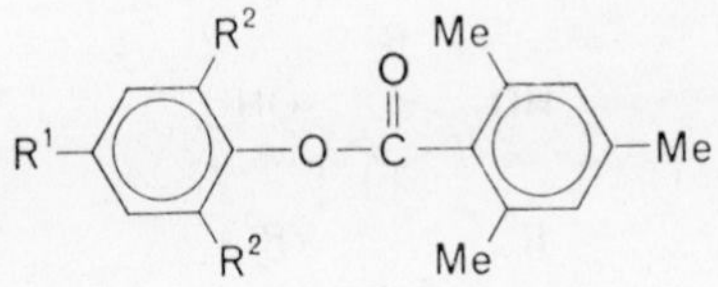

(13) $R^1 = H$, $R^2 = Me$
(14) $R^1 = Me$, $R^2 = t\text{-Bu}$

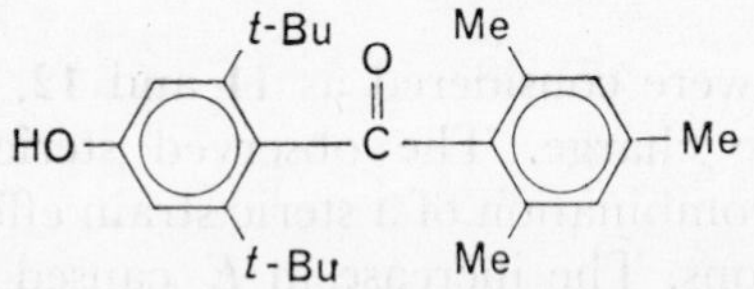

(13) (14)

(15)

blocked 4-position, like 2,6-di-*t*-butyl-4-methylphenol, the reaction with mesitoic acid takes place very slowly (23% conversion to **14** took 3 days, whereas the complete conversion of 2,6-dimethylphenol to **13** took 5 minutes). The fast solvolytic reactions of tertiary alcohols with trifluoroacetic acid also influenced the yields[86]. Thus *t*-butyl trifluoroacetate was formed competitively with *t*-butyl mesitoate. The purest products and most complete conversions were achieved when *t*-butyl alcohol was used in large excess[86].

The major component in the solution of a hindered acid in trifluoroacetic anhydride is the unsymmetrical anhydride **16**[90, 91]. Acylation is generally assumed to occur via an oxocarbonium ion **17** which is formed from the acid-catalysed ionization of the anhydride (equations 30 and 31)[86, 92]. Parish and Stock[84] listed a number of observations which supported the protonated anhydride **18** as the reactive intermediate (equation 32). Experiments with mesitoic acid and benzoic acid showed, however, that both the carbon and oxygen acylations were much more rapid with the former acid. The

$$\underset{(16)}{\text{ArC(=O)}-\text{O}-\text{C(=O)CF}_3} + \text{H}^+ \rightleftharpoons \text{ArC(=O}\cdots\text{H}\cdots\text{O}^+\text{=)}-\text{O}-\text{CCF}_3 \qquad (30)$$

$$\text{ArC(=O}\cdots\text{H}\cdots\text{O}^+\text{=)}-\text{O}-\text{CCF}_3 \rightleftharpoons \text{CF}_3\text{COOH} + \underset{(17)}{\text{ArCO}^+} \overset{\text{ROH}}{\rightleftharpoons} \text{ArCOOR} + \text{H}^+ \qquad (31)$$

$$\underset{(18)}{\text{ArC(=O}\cdots\text{H}\cdots\text{O}^+\text{=)}-\text{O}-\text{CCF}_3} + \text{ROH} \rightleftharpoons \text{ArCOOR} + \text{CF}_3\text{COOH} + \text{H}^+ \qquad (32)$$

great reactivity of mesitoic acid is also good evidence for the acylonium ion intermediate (equations 30 and 31). The relative esterification rates of *p*-toluic, benzoic and *m*-chlorobenzoic acids with phenol in trifluoroacetic anhydride at 25° were: *p*-Me 12, H 1·0 and *m*-Cl 0·062, respectively. These large substituent effects also indicate a relatively high electron deficiency in the transition state, and are consistent with the importance of the acylonium ion path.

2. Thermodynamic and kinetic control in esterification

In addition to some kinetic studies on acid-catalysed esterification in methanol[78, 79] (section III.B.2) Newman and Courduvelis[93] have investigated the esterification of several *o*-benzoylbenzoic acids in the same solvent. A result of the most general interest was the observation that the composition of the reaction product could be

(19) ⇌ (MeOH / A1) ⇌ (A2 / H_2O) (NE)

(19) ⇌ (MeOH, B1) ⇌ (B2) ⇌ (B3 / H_2O) (PE) ⇌ (MeOH / C3)

(NE) ⇌ (C1, MeOH) ⇌ (C2)

(33)

altered, depending on whether thermodynamic or kinetic control predominated under different conditions.

The reaction scheme (33) shows the possible routes to a normal (NE) or a pseudo ester (PE) in the acylation of *o*-benzoylbenzoic acids. The amounts of NE and PE were determined after 15 minutes (mainly kinetic control) and after 5 hours (thermodynamic control) in each experiment. For instance, *o*-benzoylbenzoic acid (**19**, R = H) forms at first mainly PE by route *B*, which is then rapidly converted into NE by route *C*. The conclusion that the first attack of methanol is on the carbonyl group has been confirmed experimentally[94, 95]. An additional, parallel route (equation 34) was proposed for the esterification of 2-benzoylbenzoic acid (**20**, R = H, R^1 = Me) and 2-(2,4-dimethylbenzoyl)benzoic acid (**20**, R = Me, R^1 = H) because of their enhanced esterification rates. The contribution of this route will depend on the equilibrium between the keto acid (**20**) and hydroxy lactone (**21**) forms.

(20) ⇌ (21) ⇌[MeOH] … ⇌[H₂O] (34)

3. Esterification with mixed anhydrides

Besides the esterification of hindered acids using mixed anhydrides[84] (section III.C.1) the preparation of formate and acetate esters in formic acid–acetic anhydride mixtures has been recently reinvestigated[96]. The method was originally introduced by Béhal[97], who did not report his experimental details. According to early reports[97–99] the esterification of alcohols with a 1:1 mixture of formic acid and acetic anhydride (equation 35) yielded only formates. Stevens and Van Es[96] utilized gas–liquid chromatography

$$\mathrm{MeC(=O)OC(=O)Me} + \mathrm{HCOOH} \rightleftharpoons \mathrm{MeC(=O)OC(=O)H} + \mathrm{MeCOOH} \quad (35)$$

and n.m.r. measurements for the determination of the product composition and found appreciable amounts of acetates from simple primary alcohols. Benzyl alcohol was an exception. Generally, the amount of acetate and the rate of esterification decreased in the sequence: primary > secondary > tertiary. In all cases detectable amounts of acetates were formed. It was concluded that the earlier reports[97, 100, 101] were erroneous because of the inadequacy of the available analytical methods.

Similar experiments with phenols[102] gave formates with less than 10 mole-% of acetates as the by-products. This is in contrast to the report of Béhal[97], but in substantial agreement with the preliminary experiments of Ducasse[103]. The yield of separated formates exceeded 80% in each case[96, 102].

4. Autocatalytic esterification

In general, esterification without added catalyst is a very slow process. Sometimes the acid itself is acidic enough to catalyse its own esterification. Perfluoro acids have been frequently found to be esterified by autocatalytic reactions with nonfluorinated alcohols[104–106]. In a study of the autocatalytic esterification of acetylenic acids and fluoro acids with alcohols[107], only acetylene dicarboxylic acid and perfluoro acids showed autocatalytic activity. The diester was the sole product formed from acetylene dicarboxylic acid, which was understood in view of the similar pK-values of both carboxyls. The triple bond enhances the acidity of the two carboxyl groups in a similar manner, and the lack of steric effects due to the linearity of the molecule also suggests that the two pK values will not grossly differ. Moreover, the resonance structures stabilize the dianion (**22**) more than the monoanion (**24**) or the free acid (**23**). From the experiments carried out in benzene[107] it was concluded that acids with p$K \leq 2{\cdot}6$ will be autocatalytic, whereas with p$K \geq 3{\cdot}4$ they will not exhibit autocatalysis.

D. Miscellaneous Acid Derivatives

Prajecus and co-workers have studied the kinetics and mechanism of esterification via an amine-catalysed addition of alcohols to substituted ketenes[108–110]. At 20 °C the reactions were formally of the first-order with respect to the alcohol, whereas the rate became

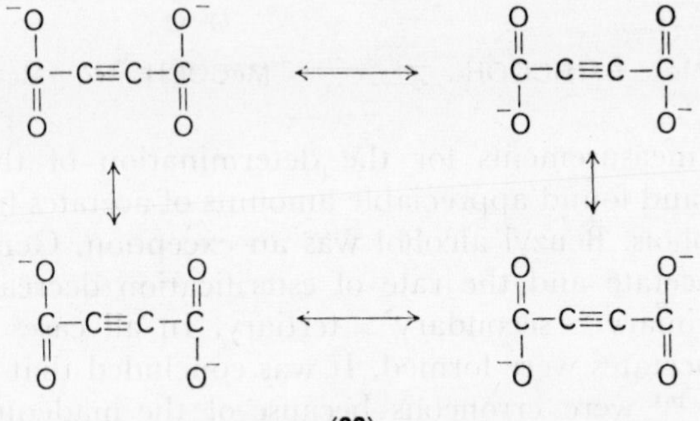

$$\begin{array}{c} ^{-}O-\overset{\displaystyle O}{\overset{\|}{C}}-C\equiv C-\overset{\displaystyle O}{\overset{\|}{C}}-O^{-} \end{array}$$

(22)

$$HO-\overset{-O}{C}=C=C=\overset{O^{+}}{C}-OH \longleftrightarrow HO-\overset{+O}{C}=C=C=\overset{O^{-}}{C}-OH \longleftrightarrow HO-\overset{\displaystyle O}{\overset{\|}{C}}-C\equiv C-\overset{\displaystyle O}{\overset{\|}{C}}-OH$$

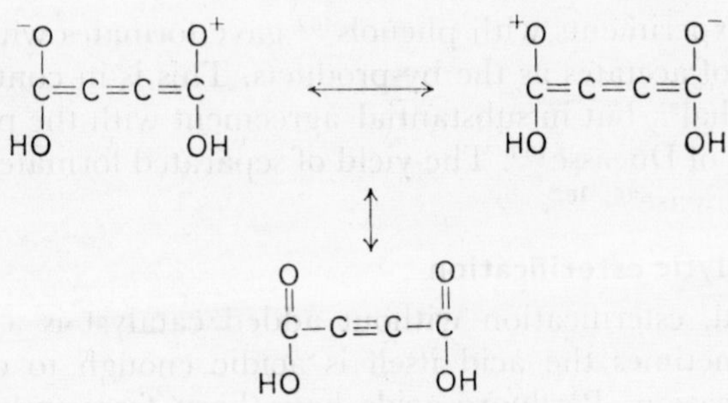

(23)

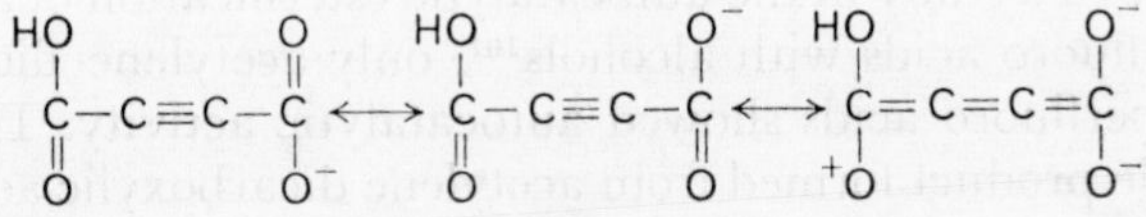

(24)

virtually independent of the alcohol concentration when the temperature was lowered to −95 °C. Although no unequivocal distinction between the different possibilities could be made, it was shown that, with certain additional assumptions, the observed formal kinetics was in accordance with a rate-determining reaction between the ketene and an alcohol-amine adduct (equation 36).

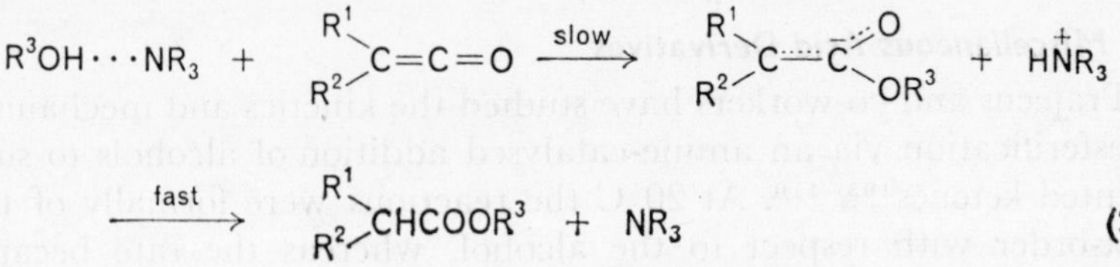

$$R^3OH\cdots NR_3 + \underset{R^2}{\overset{R^1}{>}}C=C=O \xrightarrow{\text{slow}} \underset{R^2}{\overset{R^1}{>}}C\!\!=\!\!C\overset{O^{-}}{\underset{OR^3}{<}} + H\overset{+}{N}R_3$$

$$\xrightarrow{\text{fast}} \underset{R^2}{\overset{R^1}{>}}CHCOOR^3 + NR_3 \qquad (36)$$

Brechbühler and co-workers[111] have prepared esters of several amino acids in methylene chloride solution in the presence of dimethylformamide dineopentylacetal (equation 37), the yields varying from 70 to 90%.

$$\begin{matrix} RCOOH \\ R^1OH \end{matrix} + \begin{matrix} Me \\ Me \end{matrix}{>}NCH{<}\begin{matrix} O\text{-neopentyl} \\ O\text{-neopentyl} \end{matrix} \longrightarrow RCOOR^1 + 2Me_3CCH_2OH + DMF \quad (37)$$

Satchell and co-workers[112–114] have investigated the kinetics and mechanism of some Schotten–Baumann type acylations. These include the acylation of β-naphthol by acetyl halides in the solvents nitromethane, acetic acid and acetonitrile, and the acylation of several phenols by acetyl, propionyl, butyryl, β-chloropropionyl and chloroacetyl chlorides in acetonitrile. The mechanism of the acylation was presumably similar for all these systems, the ionization (acylium ion) routes predominating[67, 112–114]. With chloroacetyl chloride and to a minor extent with β-chloropropionyl chloride, a concerted displacement of chlorine by the phenol predominates and the process is greatly facilitated by the presence of added salts[113].

McFarland and Howard[115] have investigated the reaction of sulphonyl isocyanates with hindered phenols and alcohols obtaining normal urethan products in high yields. In contrast to tertiary alkyl carbinols, tertiary aryl carbinols gave in most cases products other than urethans. When arylsulphonyl isocyanate (**25**) reacted with triphenylmethanol or with triaryl carbinols containing electron-releasing groups, *N*-(triarylmethyl) sulphonamides (**27**) and carbon dioxide were formed (equation 38).

$$\underset{\mathbf{(25)}}{ArSO_2N{=}C{=}O} + Ar_3COH \longrightarrow \underset{\mathbf{(26)}}{ArSO_2N(H)COOCAr_3}$$

$$\downarrow \quad (38)$$

$$CO_2 + \underset{\mathbf{(27)}}{ArSO_2NHCAr_3} \longleftarrow ArSO_2\overset{H}{\underset{..}{N}}{-}C{=}O \; (\text{with } Ar_3C{-}O)$$

Reaction (38) yielded no urethan (**26**) even at 0°C, whereas diphenylmethanol and **25** gave urethan in good yield[115, 116]. Two possible interpretations were proposed. Either the mechanism (38) was not applicable, or, alternatively, the intermediate **26** was extremely unstable. An alternative mechanism without the urethan intermediate was also presented (equation 39). In view of the stability of triarylmethyl carbonium ions the mechanism (39) seems very plausible.

$$\underset{(\mathbf{25})}{ArSO_2N{=}C{=}O} + Ar_3COH \longrightarrow \left[ArSO_2N{=}\overset{O^-}{\underset{OH}{C}} \quad Ar_3C^+ \right]$$

$$\downarrow$$

$$CO_2 + \left[ArSO_2NH^- \quad Ar_3C^+ \right] \longleftarrow \left[ArSO_2\overset{H}{N}{-}C(O^-){=}O \quad Ar_3C^+ \right]$$

$$\downarrow$$

$$\underset{(\mathbf{27})}{ArSO_2NHCAr_3} \qquad (39)$$

Ulrich and co-workers[117] have reported on the preparation of carbamates and 2,4-dialkyl allophanates from methyl isocyanate and phenols in the presence of potassium *t*-butoxide (equation 40).

$$R^1N{=}C{=}O + ROH \longrightarrow R^1NHCOOR \xrightarrow{+R^1N=C=O} R^1NH\overset{O}{\overset{\|}{C}}\underset{R^1}{N}COOR \qquad (40)$$

The structure of the products was demonstrated by pyrolysis and by an independent synthesis starting from the appropriate phenol and 2,4-dialkyl allophanoyl chloride. However, similar reactions with aliphatic alcohols gave a mixture of carbamate, allophanate and trimethyl isocyanate[117].

Taylor and McLay[118] have recently reported that thallium(I) ethoxide is an excellent reagent for the acylation of phenols and carboxylic acids (equations 41 and 42).

$$\begin{cases} ArOH + TlOEt \longrightarrow ArOTl + EtOH \\ ArOTl + RCOCl \longrightarrow RCOOAr + TlCl \end{cases} \qquad (41)$$

$$\begin{cases} RCOOH + TlOEt \longrightarrow RCOOTl + EtOH \\ RCOOTl + R^1COCl \longrightarrow RCOOCOR^1 + TlCl \end{cases} \qquad (42)$$

Because the thallium(I) salts of phenols and carboxylic acids are crystalline, sharp-melting compounds, they can be prepared and isolated relatively free from any impurities. Furthermore, the authors report that the yields of aryl esters and carboxylic acid anhydrides are virtually quantitative.

E. Intramolecular Catalysis by the Hydroxyl Group

A neighbouring hydroxyl group has been found to catalyse the hydrolysis and solvolysis of carboxylic esters (for earlier literature see Reference 66), the acylation of polyols, the hydrolysis of amines and amides, and the amination of esters.

Kupchan and co-workers[119-121] have shown that the alkaline hydrolysis of an alicyclic axial acetate is facilitated by a hydroxyl group bearing a 1,3-diaxial juxtaposition to the acetate. The solvolysis of 1,3-diaxial hydroxy acetates exhibits general base catalysis with simultaneous general acid catalysis by the neighbouring hydroxyl group[66, 122]. An introduction of a hydroxyl group to C-5 of coprostanol acetate, leading to the 1,3-diaxial hydroxy acetate, resulted in a 300-fold increase in the rate of solvolysis, the possible mechanism being shown in **28**. Correspondingly, an examination of the molecular models of strophanthidin 3-acetate 19-aldehyde and neogermitrine (**29** and **30**, respectively) showed that the hydrogen

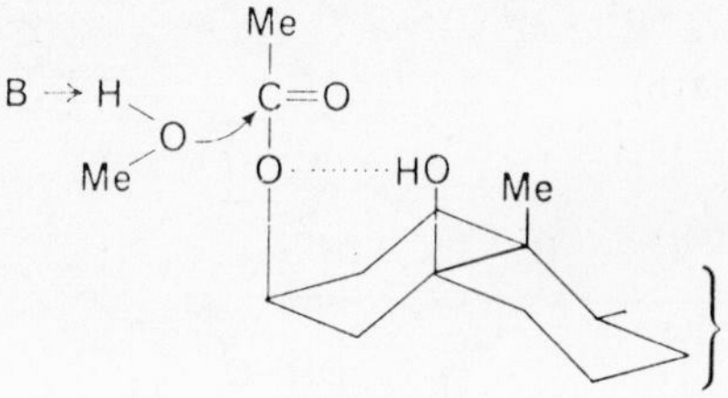

(28)

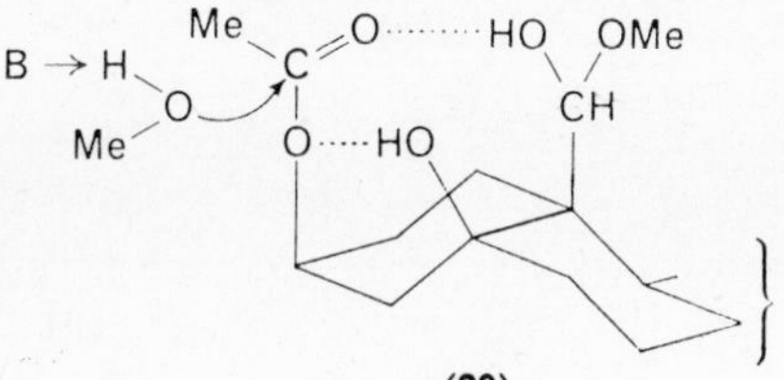

(29)

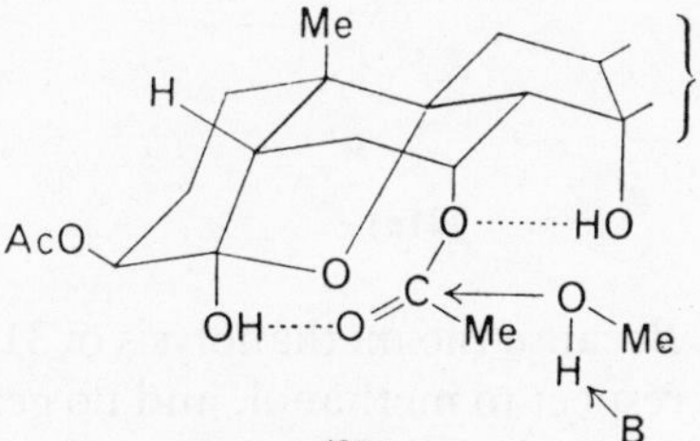

(30)

bonding via the acidic hemiacetal hydroxyl groups would polarize the carbonyl group and facilitate attack of the nucleophile.

Bruice and Fife[123] have suggested that a neighbouring hydroxyl group may assist ester solvolysis by changing the microscopic medium surrounding the ester group, for instance, by effecting a specific binding or orientation of water molecules in the critical complex. To investigate this possibility in the methanolysis of gallotannins, Biggins and Haslam[124] studied the effect of methanol on the i.r. spectra of compound **31** in carbon tetrachloride. When the concentration of methanol was increased the proportion of **31d** was greatly favoured over **32c**, which was probably due to the intramolecular hydrogen bond present in **31c.** In addition, the results showed that electron-attracting groups in the benzoyl residue favoured the formation of **31b** and electron-releasing groups the formation of **31d**; the ratio of the two species had a linear correlation to the

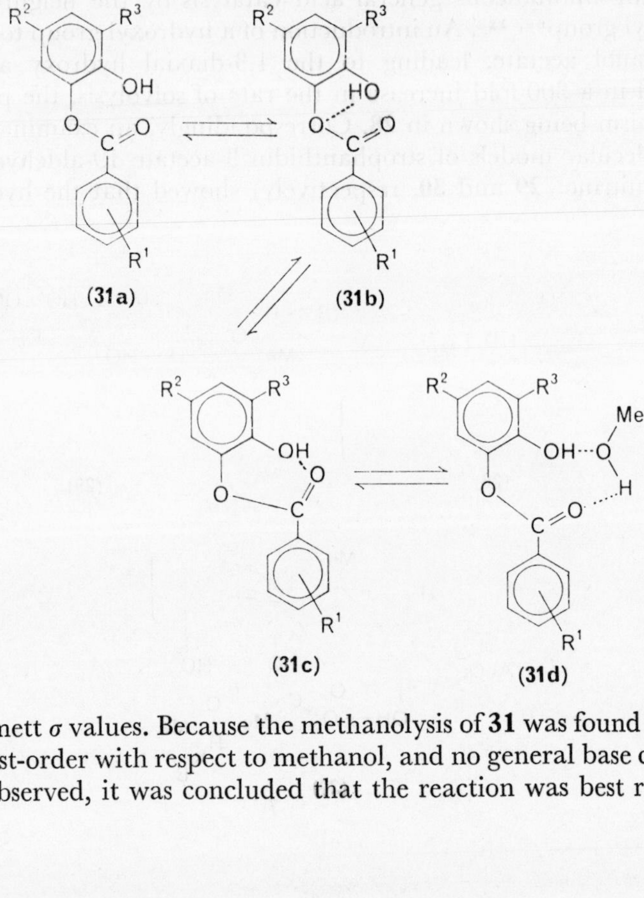

(31a) **(31b)**

(31c) **(31d)**

Hammett σ values. Because the methanolysis of **31** was found to be of the first-order with respect to methanol, and no general base catalysis was observed, it was concluded that the reaction was best rationalized in terms of the mechanism presented by Bender[66, 125], which involves an attack of a neutral methanol molecule on the ionized ester (**32**).

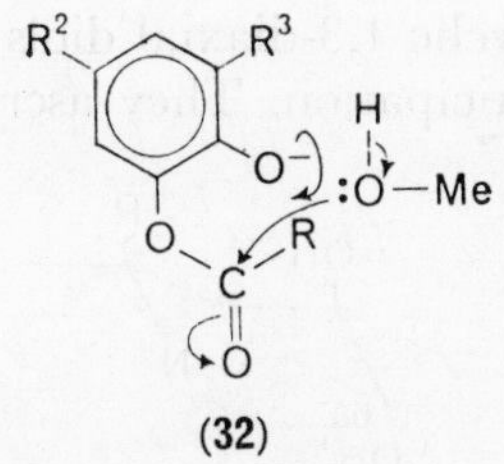

(**32**)

Capon and Ghosh[126] have proposed the mechanism (43) for the hydrolysis of phenyl salicylate, involving intramolecular general base catalysis by the ionized phenolic group. The solvent deuterium isotope effect ($k_H/k_D = 1{\cdot}7$) is also consistent with this mechanism[127]. Similarly, in contrast to the report of Hansen[128] for catechol monoacetate, Capon and Ghosh[126] suggested catechol monobenzoate to be hydrolysed by a mechanism with intramolecular general base catalysis ($k_H/k_D = 1{\cdot}8$).

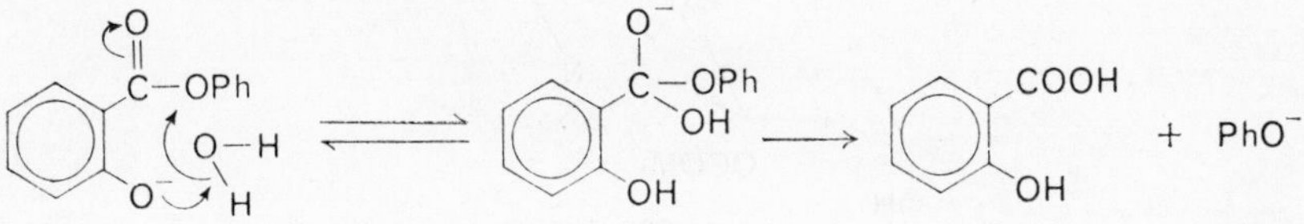

(43)

An interesting report has recently been made[129] about the hydroxyl group participation in amide hydrolysis. The mechanism (44) was presented for the hydrolysis of 4-hydroxybutyranilide in neutral and alkaline medium. The rate was found to be about ten-fold

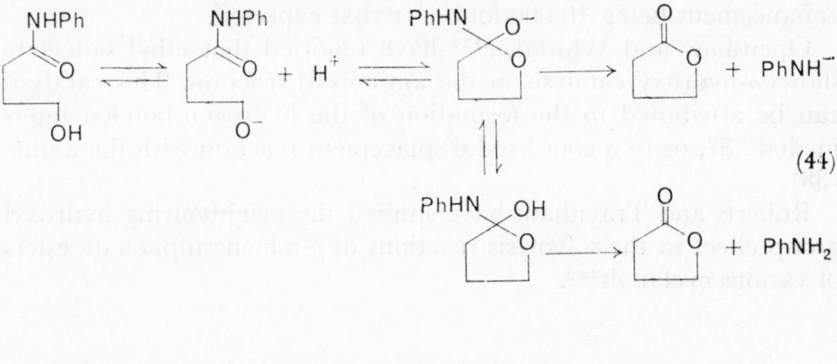

(44)

as compared with the hydrolysis rate of butyranilide. Similarly, the rate ratio of 4-hydroxybutryamide to butyramide was about twenty.

According to Kupchan and co-workers[130] the acylation of tertiary hydroxyl groups in alicyclic 1,3-diaxial diols may also be facilitated by hydroxyl group participation. They ascribed the ease of direct

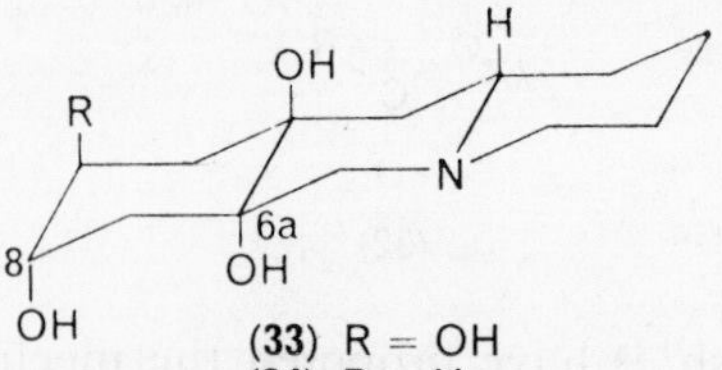

(33) R = OH
(34) R = H

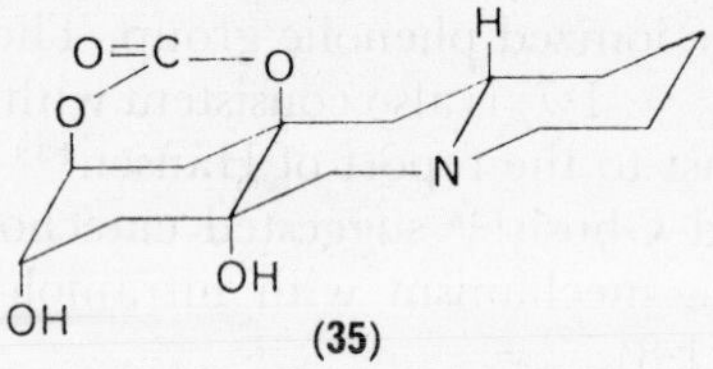

(35)

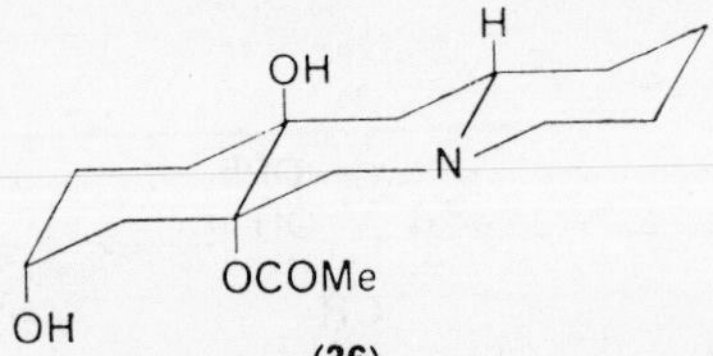

(36)

acylation of the tertiary hydroxyl groups in **33**, **34** and **35** to the intramolecular catalysis by the C-8 secondary hydroxyl group and the nearby tertiary nitrogen atom.

In addition to the acylation reaction, the solvolysis of the triol monoacetate **36** featured intramolecular base catalysis[130], the rate enhancement being 10,000-fold over that expected.

Openshaw and Whittaker[131] have reported that ethyl salicylate shows *o*-hydroxy catalysis in the aminolysis reaction. This catalysis can be attributed to the formation of the hydrogen-bonded intermediate **37**, or to a concerted displacement reaction with the amine (**38**).

Roberts and Traynham have studied the neighbouring hydroxyl group effect in the solvolysis reactions of *p*-toluenesulphonate esters of various cyclanols[132].

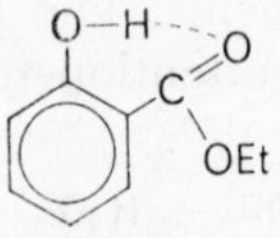

(37) 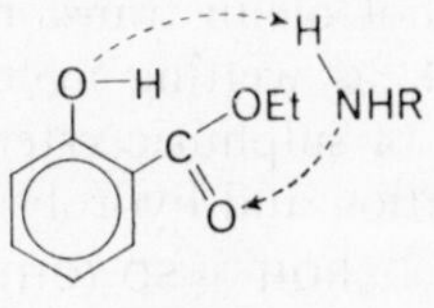 (38)

IV. HETEROATOMS AS THE ELECTROPHILIC CENTRES

A. *Esterification with Inorganic Oxyacids*

I. General

The reactions discussed below are formally represented as

$$ROH + HO\text{–}Y\text{–} = RO\text{–}Y\text{–} + H_2O \qquad (45)$$

where Y stands for a heteroatom, like sulphur, nitrogen or boron. Strictly, only in cases in which the reaction involves a cleavage of the O–Y bond is the alcoholic hydroxyl group actually attacked by the heteroatom; otherwise, when the R–O bond of the alcohol is cleaved, the reaction should be considered rather as a nucleophilic attack on the α-carbon atom of the alcohol.

Esterification of alcohols with inorganic oxyacids has been the subject of considerable study, and many exhaustive reviews on practical synthetic methods, both for laboratory and industry, may be found in the literature. The discussion below is therefore limited to a number of particularly chosen examples which serve to illustrate some of the general features of reaction (45).

In many instances the reversal of reaction (45), the hydrolysis of the ester, has been studied in more detail than the esterification itself. Examples of such cases are also given, bearing in mind that the principle of microscopic reversibility requires the forward and reverse reactions to pass through the same transition state, which makes the mechanistic information derived from both sources of study equally valuable.

2. Sulphuric acid

a. Primary and secondary alcohols. The reaction of ethanol with sulphuric acid leading to the formation of diethyl ether has probably been known for several centuries, although the intermediary formation of alkyl hydrogen sulphates (equation 46) remained unnoticed for a long time. Recent interest in these esters has been enhanced by, in addition to their importance as intermediates in alcohol–ether and in alcohol–olefin transformations, a variety of practical applications, e.g., as wetting agents and detergents. The physiological significance of sulphuric esters has drawn attention to their enzymatic formation and hydrolysis reactions[133, 134].

$$ROH + SO_2(OH)_2 = SO_2(OH)(OR) + H_2O \qquad (46)$$

Several studies on primary and secondary alcohols[135–139], dealing with both the sulphation reaction (46) and its reverse hydrolysis under acidic conditions, indicate that the R–O bond of the alcohol remains intact during the reaction. The evidence includes structural rate effects, retention of configuration in the case of optically active alcohols, and oxygen-18 tracer experiments. A formation of a carbonium ion R^+ as a reaction intermediate seems thus to have been excluded, at least for primary and secondary alkanols, and the reaction can be considered, in the strict sense, as one involving an electrophilic attack on the hydroxyl group of the alcohol.

The effect of the structure of the alcohol on the rate and equilibrium of esterification has been studied in detail by Deno and Newman[136]. Some of their results are shown in Table 5. Most of

TABLE 5. Equilibrium constants and relative rates[136] for the sulphation of alcohols (equation 46) at 25°C. The values are calculated to correspond to 70·4% aqueous sulphuric acid.

Alcohol	$K_{equil.}$	Relative rate
Methyl	2·3	1
Ethyl	1·7	0·43
2-Propyl	0·54	0·12
1-Butyl	1·9	0·29
2-Butyl	0·50	0·12
Isobutyl	2·2	0·25
2-Pentyl	0·64	0·11
3-Pentyl	—	0·10
Neopentyl	1·7	0·23
Pinacolyl	—	0·05
Cyclohexyl	0·70	0·16

the differences in the relative rates can be accounted for by steric hindrance at the site of the attack. Secondary alcohols react slower than primary alcohols, whereas the various primary alcohols show but minor differences. Deno and Newman considered various mechanistic possibilities, of which a bimolecular mechanism (47, 48

$$ROH_2^+ + H_2SO_4 = ROSO_3H + H_3O^+ \quad (47)$$

$$ROH + H_3SO_4^+ = ROSO_3H + H_3O^+ \quad (48)$$

$$ROH + H_2SO_4 = ROSO_3H + H_2O \quad (49)$$

or 49) seemed the most acceptable. Similar studies of the sulphation of 2,4-dinitrobenzyl alcohol, made by Williams and Clark[140], gave results which suggested an alternative, unimolecular rate-determining step (equation 50). The nature of the hypothetical reaction intermediate XH, the alcohol–sulphuric acid complex, was left open.

$$ROH + HSO_4^- \underset{}{\overset{\text{fast}}{\rightleftharpoons}} X^-;\ X^- + H^+ \underset{}{\overset{\text{ast}}{\rightleftharpoons}} XH \xrightarrow{\text{slow}} ROSO_3H + H_2O \quad (50)$$

The existence of intermediary alcohol–sulphuric acid complexes in reaction (46) has found some support from a number of recent studies of the hydrolysis of hydrogen sulphate esters[138, 139, 141]. Thus, Batts[139] observed that the hydrolysis rates of methyl and ethyl hydrogen sulphates were increased by factors of about 10^7 when going from water to moist dioxan. Similar accelerations by solvents of low polarity had been previously noticed in the case of steroid hydrogen sulphates[142, 143]. These results can be rationalized only if the reactions involve charged species as the reactants and relatively nonpolar transition states. Accordingly, Batts[139] proposed the mechanism (51) with a rate-limiting decomposition of a zwitterion. A unimolecular decomposition of the zwitterion, with a sulphur

$$R{-}O{-}SO_2{-}O^- + H^+ \underset{}{\overset{\text{fast}}{\rightleftharpoons}} R{-}\overset{H}{O^+}{-}SO_2{-}O^- \xrightarrow[H_2O]{\text{slow}} ROH + H_2SO_4 \quad (51)$$

trioxide-like transition state, was preferred to covalent involvement of a water molecule. This hypothesis, which is discussed in some detail below, is essentially that made by Williams and Clark[140] for the reverse sulphation reaction (50), if one considers the complex XH as the zwitterion and does not account in detail for the individual reaction steps. It can be also argued that the apparent contradiction to the sulphation mechanism of Deno and Newman[136] with a bimolecular reaction between ROH and H_2SO_4 as the rate-limiting step (equation 49) is only in the degree to which the sulphur trioxide present in the transition state resembles the acid itself or the anhydride.

The main arguments given by Batts[139] for essentially unimolecular rate-determining steps in reaction (51) were, first, the variation of the hydrolysis rate with the Hammett acidity function, and, second, the values of the activation entropies which were more positive than those generally associated with bimolecular *A*-2 reactions. One may further ask, whether the structural effects of the leaving groups on the hydrolysis rates, not discussed in detail by Batts, could be of value when making mechanistic conclusions. The relative hydrolysis rates in perchloric acid solutions were about 1, 1 and 3 for methyl, ethyl and *i*-propyl hydrogen sulphates, respectively. As these values are made up of the equilibrium constants of the protonation and of the rate coefficients for heterolysis of the alcohol molecules from the complex (equation 51), they may be compared directly with the effect of the leaving group on the rates of *A*-1 reactions involving the same alcohols. Thus, in the case of hydrolysis of formaldehyde acetals[31], the relative leaving rates (including the equilibrium protonation and the rate-determining heterolysis) are 1, 1·2 and 2·3, for methyl, ethyl and *i*-propyl alcohols, respectively. The effects of the leaving groups thus seem to follow the same structural pattern in both the reactions.

Some other recent studies also favour the unimolecular decomposition step with a sulphur trioxide-like transition state. In one of these investigations, the acid hydrolyses of phenyl and *p*-nitrophenyl sulphates were compared with that of methyl selenate[144]. Bunton and Hendy[145] had earlier presented strong evidence according to which the hydrolysis of the latter compound involved an *A*-2 type mechanism. The anhydride-like transition state **39** is relatively

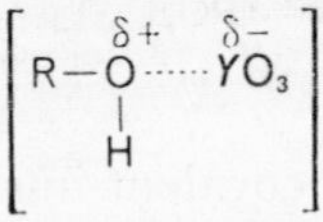

$$\left[\begin{array}{c} \overset{\delta+}{\mathrm{R{-}O}}\cdots\overset{\delta-}{Y\mathrm{O_3}} \\ | \\ \mathrm{H} \end{array}\right] \qquad \left[\begin{array}{c} \mathrm{R{-}\overset{+}{O}{-}}Y\mathrm{O_3^-} \\ \mathrm{H} \quad \vdots \\ \mathrm{H{-}O{-}H} \end{array}\right]$$

(39) Y = S, Se 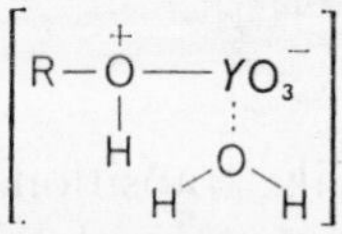 **(40)** Y = S, Se

nonpolar, whereas the acid-like transition state **40** (*A*-2 mechanism), in which the $^+$O–Y bond is cleaved to a smaller extent, still maintains much of its zwitterion character. One might therefore expect that in the former case the rate should increase much more markedly in solvents of low polarity than in the latter. This is actually the case as illustrated by the relative rates given in Table 6*.

* Alternatively, in terms of the theory advocated by Robertson[146], these solvent effects can be accounted for by the breakdown of the initial state solvation shells when the transition states **30** and **40** are approached, this breakdown being more complete in **39** than in **40**.

TABLE 6. Relative solvent effects on the acid hydrolyses of aryl sulphates and methyl selenate[144]. For both reactions, 40% dioxan–water is chosen as the reference solvent.

Solvent	Relative rate	
	Aryl sulphates	Methyl selenate
40% dioxan–water	1	1
60% dioxan–water	5·0	1·6
80% dioxan–water	67	6·3

Benkovic[147] has investigated carboxyl-substituted aryl sulphates **41** and **42**. The particular aim was to study the expected intra-

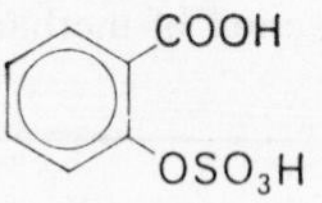

(41) 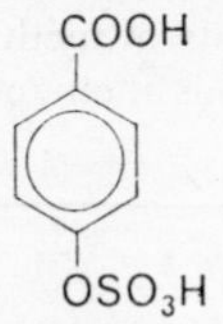 (42)

molecular carboxyl group catalysis in the case of salicyl sulphate (**41**). In fact, the pH–rate profile for **41** exhibited a plateau, as implied by involvement of the neighbouring carboxyl group, whereas in the case of **42**, with no possibility of such a neighbouring group participation, the rate decreased steadily with increasing pH. For the acid-catalysed reactions of both compounds the *A*-1 mechanism was suggested, first, because the activation entropies had positive values, and second, because the rate coefficients, along with earlier values, showed a marked increase with electron-attracting substituents. The relative rates in heavy and light water, k_D/k_H, were 1·30 and 2·34 for **41** and **42**, respectively. It was recognized that the former value, 1·30, is definitely within the range generally associated with an *A*-2 reaction. To overcome this controversy, the explanation was offered that the low value was due to the circumstance that the p*K* value of the protonation equilibrium (see equation 51) was lowered by intramolecular hydrogen bonding in the salicyl sulphate anion. Although this may be one of the factors involved, the magnitude of the isotope effect still seems to be surprisingly low, because in a series of similar acids the solvent deuterium isotope effect changes only slightly with the acid strength[148]. It is more likely that, if the mechanism really is *A*-1 for both compounds, **41** and **42**, the main factors responsible for the different isotope effects are the free energies of transfer of the reaction participants from light to heavy water; energies which are incorporated in the experimental values of k_D/k_H[148]. This pertains particularly to the anions of **41** and **42**, the activity coefficients of which are expected to vary quite differently when going from H_2O to D_2O because of the intramolecular hydrogen bond present in the former anion. Additional studies of the solvent deuterium isotope effects on this and similar systems would seem desirable.

In a recent study, Benkovic and Benkovic[149] have further advocated sulphur trioxide-like transition states in the case of aryl sulphates. Table 7 shows some results of experiments in which the

TABLE 7. Amounts of methyl sulphate formed from aryl sulphates and sulphur trioxide in aqueous methanol[149]. The mole fraction of methanol in the solvent was 0·303.

Substrate	Reaction	Mole fraction *l* $MeOSO_3H$ in the product
Salicyl sulphate	Intramolecular catalysis	0·58
Salicyl sulphate	Hydronium ion catalysis	0·36
p-Carboxyphenyl sulphate	Hydronium ion catalysis	0·54
Monomeric sulphur trioxide	Solvolysis	0·58

relative amounts of methyl hydrogen sulphate and sulphuric acid formed from aryl sulphates in aqueous methanol were compared with those formed from sulphur trioxide in the same solvent. Except the acid-catalysed reaction of **41**, which also behaves abnormally here, the similar product compositions suggest that the product-forming stages of the reactions are similar.

b. Tertiary alcohols. In comparison with the conjugate acids of primary and secondary alcohols, those of tertiary alcohols have an enhanced tendency to dissociate to carbonium ions, which, even under relatively mild conditions, lead to products other than sulphate esters (e.g., olefins). In fact, tertiary alkyl sulphates have never been isolated from the reaction product, although their presence as reaction intermediates has sometimes been postulated.

An idea about the rate of formation of carbonium ions from tertiary alcohols in aqueous sulphuric acid, compared to that of esterification of primary and secondary alcohols in the same medium, may be gained in the following way. Using the oxygen-18 technique, Taft and co-workers[150] measured the rate of reaction (52) in aqueous

$$(Me)_3COH + H^+ \underset{}{\overset{fast}{\rightleftharpoons}} (Me)_3COH_2^+ \xrightarrow{slow} (Me)_3C^+ + H_2O \quad (52)$$

sulphuric acid solutions at various acidities. When the rate coefficients are extrapolated to correspond to 70% sulphuric acid (see Table 5) it can be estimated that, in this solvent, *t*-butyl alcohol produces carbonium ions at a rate which is roughly 200,000 times that of the reaction of primary alcohols (forming alkyl hydrogen sulphates). Although the subsequent formation of isobutene is slower than the primary formation of the carbonium ion (by a factor of about thirty[151]), its rate still greatly exceeds that of the esterification.

c. Diesters of sulphuric acid. The second stage of the esterification of sulphuric acid (equation 53) has not been studied in detail, al-

$$ROSO_3H + ROH = (RO)_2SO_2 + H_2O \quad (53)$$

though the presence of these esters in minor quantities has been recognized when treating primary alcohols with sulphuric acid. They are best obtained from alcohols with other sulphating agents[152, 153]. The mechanistic aspects of hydrolysis of the diesters, both open-chain and cyclic, have been dealt with in numerous studies (for literature, see References 138, 154–157). In general, the hydrolyses of the diesters are much more facile than those of the corresponding monoesters and involve both C–O and S–O cleavages, depending on the structure of the ester and external conditions.

From the experimental values given by Breslow, Hough and Fairlough[158] one can calculate a value of about 0·4 for the equilibrium constant of the disproportionation reaction (54) at temperatures of 25–50°C. As the equilibrium constant of reaction (55) is

$$2\,EtOSO_3H = (EtO)_2SO_2 + H_2SO_4 \quad (54)$$

$$EtOH + H_2SO_4 = EtOSO_3H + H_2O \quad (55)$$

about 2 (Table 5), one gets an estimate for the relative amounts of the mono- and di-esters at equilibrium. Thus, for example, if one starts from one mole of ethanol and one mole of sulphuric acid, the equilibrium mixture is calculated to contain 0·40 mole of ethyl hydrogen sulphate and 0·14 mole of diethyl sulphate, respectively.

3. Phosphoric acid

Although other phosphorylating agents (section IV.B.2), rather than phosphoric acid itself or its mixtures with anhydrides, are most commonly used for the preparation of phosphate esters from alcohols, some structural effects involved in the formation and hydrolysis of these esters are discussed here because of their general interest. More extensive accounts have been published relatively recently[159-162].

Noting that esterifications with phosphoric acid and their reverse hydrolysis reactions are formally similar (equation 56), some rationalizations can be made. First, the nature of the phosphate species

$$O{=}P(OR^1)_3 + R^2OH \rightleftharpoons O{=}P(OR^1)_2(OR^2) + R^1OH \qquad (56)$$
$$R^1, R^2 = H, \text{alkyl, aryl}$$

under attack, whether uncharged, deprotonated (anion) or protonated (cation), will depend on the acidity of the medium as well as on the relative rates of attack on the various species present simultaneously, even in minor quantities. This circumstance becomes evident from the complicated nature of the pH–rate profiles frequently observed in reaction (56).

Second, many, though not all, of the structural effects observed can be fairly well explained in terms of the electronic and steric factors involved in (56). Electron-releasing substituents in the phosphate species, like alkyl groups, render the attack on the hydroxyl group of the alcohol (or water) more difficult, and vice versa. Thus, the replacement of the second and third hydroxyl groups of phosphoric acid is extremely difficult as compared to that of the first. Similarly, trialkyl phosphates are much more stable toward hydrolysis than di- or mono-alkyl phosphates.

One peculiar aspect of nucleophilic attacks on phosphorus, which cannot be rationalized in terms of classical structure–rate relations, is illustrated by the fact that the cyclic ethylene phosphate anion (**43**) is hydrolysed faster than its acyclic analogue **44** by a factor which is about eight powers of ten[163, 164]. In view of the observation

$$\begin{array}{l} CH_2O \\ | \qquad\quad PO_2^- \\ CH_2O \end{array} \qquad\qquad \begin{array}{l} CH_3O \\ \qquad\quad PO_2^- \\ CH_3O \end{array}$$

(43) **(44)**

that the hydrolysis of **44**, unlike that of **43**, also contains a contribution from simultaneous C–O fission, the actual rate enhancement in the nucleophilic attack on phosphorus is even greater than the above-mentioned ratio. As the six-membered analogue of **43**, trimethylene phosphate anion, reacts at a rate comparable to that of **44**[165], it was assumed earlier that the acceleration was due to release of the strain present in the five-membered ring. However, calorimetric measurement of the heats of hydrolysis of **43** and **44** showed that this strain is only about 5·5 kcal/mole[154], which would account for only 10^4-fold acceleration even in the extreme situation in which the strain would be wholly released at the transition state. It is more likely that open-chain phosphates like **44** are stabilized by π-type bonding involving the $3d$ orbitals on the central phosphorus atom[154, 166], whereas such a stabilization, on stereochemical grounds, is not possible for **43**. Similar argumentations may, of course, apply to esters derived from the oxyacids of sulphur, but not to esters with a second-row element as the central atom, like carbon, nitrogen or boron.

Molecular orbital calculations on the oxygen-$2p$–phosphorus-$3d$ interactions in phosphate esters[167] are in fair agreement with the observations discussed above. One of the most striking results of these calculations is that the electrophilic reactivity of the central phosphorus, as measured by its calculated charge density, is greatly dependent on the conformation of the groups about this atom. This factor may have an important role, e.g., in bioorganic systems involving formation and decomposition of phosphate esters.

Whereas extensive data exist on the hydrolytic cleavage of phosphates and related esters[154, 159–173], very few and fragmentary studies have been made on the phosphorylation of alcohols and phenols with phosphoric acid. Mesnard and Bertucat[174] phosphorylated primary, secondary and tertiary alcohols with phosphoric acid using pyridine as the solvent. Very low yields were obtained in the case of tertiary alcohols.

If stronger phosphorylating agents are used, such as phosphoric acid or polyphosphoric acid, tertiary alcohols are mainly dehydrated[175] without noticeable ester formation. However, tertiary alcohols in which one of the alkyl groups has been replaced by a –COOR, –CN or –$CONH_2$ group, are reportedly acetylated to an extent of about 15%[175].

Clarke and Lyons[176] have studied the various products formed when phosphorylating alkanols with polyphosphoric acids of different average chain lengths. Acid–base titrations and phosphorus-31 n.m.r. were used in the analyses. No dialkyl esters of orthophosphoric acid could be detected among the reaction products, which was explained in terms of the steric and inductive effects, rendering an alkoxy-substituted phosphorus less susceptible to a nucleophilic attack in comparison to that of phosphoric acid (see above).

TABLE 8. Sites of cleavage and relative rates for the hydrolytic decomposition of monoalkyl phosphates $ROPO_3H_2$ in 4M perchloric acid at 100°C [169, 171].

R	Relative rate	% P–O cleavage	Mechanism
Methyl	1	27	*A*-2
Ethyl	0·66	48	*A*-2
i-Propyl	70	1	*A*-1
t-Butyl	1400 [a]	0	*A*-1

[a] Extrapolated from data at lower temperatures.

Table 8 illustrates some of the structural effects observed in the hydrolytic cleavage of alkyl phosphates. Differing from the acid hydrolysis of the esters of carboxylic acids, the change to an *A*-1 type mechanism occurs here when going from ethyl to *i*-propyl ester.

The hydrolyses of monoaryl phosphates are subject to acid catalysis only if the aryl group possesses a strongly electron-withdrawing substituent, such as a nitro or an acetyl group in the *para* position[172].

Intramolecular catalysis in a number of *o*-carboxyaryl phosphates has been studied in detail[168].

An observation of considerable interest for the study of acid-catalysed ester hydrolysis and esterification has been made by Bunton and co-workers[173, 177]. They investigated the hydrolyses of triphenyl and *p*-nitrophenyl diphenyl phosphates in moderately concentrated mineral acids and found the rates to exhibit maxima at 1·5–6M acid. Similar maxima had been observed earlier in such acid-catalysed reactions in which the substrate was basic enough to become wholly protonated, but the point here was that the maxima were not caused by the protonation, as shown by basicity measurements conducted independently. The results, which were rationalized in terms of the activity coefficient behaviour of the reactants and transition states in solutions of strong acids, expose the Achilles heel in mechanistic conclusions drawn from the rate-acidity relations.

4. Nitric acid

Esterification of alcohols with nitric acid is most familiar from the preparation of common explosives, the nitrate esters of glycerol ('nitroglycerine') and cellulose ('nitrocellulose'). The usual procedure is to add the alcohol slowly to cold acid (100% acid, its solution in an inert solvent, or, most commonly, its mixture with sulphuric acid), and to separate the ester by pouring the mixture into cold water[178]. Small amounts of urea are added to the mixture to destroy the nitrous acid present, since otherwise violent explosions might occur.

Klein and Mentser[179], using the oxygen-18 tracer technique, were the first to prove that the oxygen atom of the alcohol remains intact through the process; the reaction thus involves an electrophilic attack on this oxygen atom. The reaction mechanism, shown in equations (57) to (59), has been subsequently clarified by Ingold and co-workers[180].

$$HNO_3 + HNO_3 = H_2NO_3^+ + NO_3^- \text{ (fast)} \tag{57}$$

$$H_2NO_3^+ = NO_2^+ + H_2O \tag{58}$$

$$ROH + NO_2^+ = R\overset{+}{\underset{\displaystyle\text{H}}{\underset{|}{O}}}NO_2 \tag{59}$$

Under conditions in which the medium does not contain substantial amounts of water, the reaction rate is of the zeroth order with respect to the alcohol and, moreover, different alcohols are esterified at equal rates which are the same as those for the *N*-nitration of amines and *C*-nitration of benzenoid hydrocarbons in the same media. When sufficient amounts of water are added to the solution the reaction becomes first-order with respect to the alcohol and exhibits individual differences which depend on the alcohol. An addition of sulphuric acid preserves the zeroth-order kinetics, but enhances the rate of esterification.

The facts enumerated above are readily understood in terms of reactions (57) to (59). In solutions of low water content, the reversal of the nitronium ion formation will be slow and it does not effectively compete with the nitronium ion–alcohol reaction (59). Put another way, the nitronium ions are captured by the alcohol molecules (or by other reactants, such as benzenoid hydrocarbons) as fast as they become available from the dehydration of the nitric acidium ion. In contrast, in solvents with appreciable amounts of water, the rate of the reaction between the nitronium ion and water becomes significant and, consequently, the electrophilic attack on the alcoholic hydroxyl by the nitronium ion (reaction 59) becomes rate-determining. The relative rates of attack by this ion on different nucleophiles were estimated as: water ~0·03; benzene 1; *N*-methyl-2,4,6-trinitroaniline 1·4; toluene 24; methanol 30.

Depending on the structure and external conditions the hydrolytic decomposition of nitrate esters may lead to three different products (equations 60–62). The experiments conducted by Baker

$$RONO_2 + H_2O = ROH + HNO_3 \tag{60}$$

$$RCH_2CH_2ONO_2 + H_2O = RCH{=}CH_2 + H_2O + HNO_3 \tag{61}$$

$$RCH_2ONO_2 + H_2O = RCHO + H_2O + HNO_3 \tag{62}$$

and co-workers[181–183] showed that for primary and secondary alkyl nitrates (and for aryl nitrates, having no available sites for elimination), the substitution reaction (60) predominated whereas concurrent elimination normally took place with tertiary alkyl nitrates. The structural and solvent effects observed followed the general pattern of substitution–elimination reactions.

Cryoscopic and spectrophotometric measurements on solutions of alkyl nitrates in 100% sulphuric acid[184] indicate the occurrence of reactions (63) to (65), which produce four ionic species from one molecule of nitrate ester. Because the alcohol formed in reaction (64) (reversal of reaction 59) is subsequently esterified by sulphuric acid (reaction 65), the nitrate ester is not regenerated when the reaction solution is poured into water.

$$RONO_2 + H_2SO_4 \longrightarrow R\overset{H}{O}{}^{+}{-}NO_2 + HSO_4^- \tag{63}$$

$$R\overset{H}{O}{}^{+}{-}NO_2 \longrightarrow ROH + NO_2^+ \tag{64}$$

$$ROH + 2\,H_2SO_4 \longrightarrow ROSO_3H + H_3O^+ + HSO_4^- \tag{65}$$

5. Boric acid

The esterification of boric acid with alcohols, in particular with polyhydroxylic compounds, is the subject of a considerable literature[185, 186]. Although many practically and mechanistically important features of the reaction are becoming understood, yet definite, quantitative equilibrium and kinetics data on relatively simple model systems are still virtually nonexistent.

The preparation of simple, symmetric alkyl or aryl triesters of boric acid usually takes place according to equation (66). The ester

$$B(OH)_3 + 3\,ROH = B(OR)_3 + 3\,H_2O \tag{66}$$

formation is monitored by removing the triester or water from the reaction zone, usually by means of azeotropic distillation. No mono- or di-esters have been isolated from the reaction mixture, other methods having been used for their preparation in particular cases[185]. For simple alkanols the equilibrium constants given in Table 9 have been reported[187]. It is seen that a lengthening of the alkyl chain

TABLE 9. Equilibrium constants for reaction (66) in acetone solution at 0°C [187].

Alcohol	Ester percentage at equil.	$K_{equil.}$
Methanol	33	0·062
n-Propanol	44	0·37
n-Butanol	45·5	0·48
n-Pentanol	46·5	0·55

increases the amount of ester at equilibrium; a comparable effect also becomes evident from the kinetic stability of these esters towards hydrolysis[188]. The general observation that the hydrolytic displacement of the first alkoxy group in boric triesters is much slower than those of the second and the third is in harmony with the polar and steric factors influencing the Lewis acidity of the central boron atom.

An interesting kinetic study, which clarifies many features of the formation and hydrolysis of aryl borates, has been recently published by Tanner and Bruice[189].

B. *Miscellaneous Heteroatomic Electrophiles*

I. Sulphur trioxide, halogen compounds of sulphur, etc.

The reaction of an alcohol with sulphur trioxide, oleum or a pyrosulphate differs from esterification with the acid itself in that the reaction is virtually irreversible. Ordinarily, these reagents bring about side reactions (oxidation, dehydration) and are therefore rarely used as such. A great number of sulphur trioxide complexes with various electron donors have been devised to prevent the side reactions, e.g., complexes with dimethylformamide[190, 191], dioxan[192] and tertiary amines[193]. In all cases the function of these electron donors is to moderate the reactivity of the anhydride, leading to more specific sulphating properties. They thus act in the same way as various bases in the Schotten–Baumann acylation of alcohols.

Preparation of sulphate esters with other sulphating agents, like chlorosulphonic acid, has been discussed elsewhere[152, 153].

A novel route to symmetric and unsymmetric tertiary amines from alcohols and sulphamoyl chlorides (equation 67) has been

$$R^1OH + ClSO_2NR^2_2 \longrightarrow R^1OSO_2NR^2_2 \xrightarrow{\Delta} R^1R^2_2N^+\text{–}SO_3^- \xrightarrow{H_2O} R^1R^2_2N \qquad (67)$$

reported by White and Ellinger[194]. The reaction was studied with R^2 = methyl, though it was considered probable that the reaction could also be applied to the preparation of primary and secondary amines (with two or one of the groups R^2, respectively, replaced by hydrogen atoms). The rearrangement step following the attack with sulphamoyl chloride was proposed to take place by the S_Ni mechanism, in accordance with the observed retention of configuration at R^1 and with the influence of the polar character of R^1.

Moffatt and co-workers[195–197] have recently studied the reaction with sulphoxide-carbodiimide adducts (equation 68). The reaction involves an electrophilic attack on the hydroxyl group, as shown by

$$R^1N{=}C(-NHR^1)-O-S^+Me_2 + HOCH_2R^2 \longrightarrow R^1NHCONHR^1 + R^2CH_2\overset{+}{O}SMe_2 \qquad (68)$$

experiments with oxygen-18[196]. As the alkoxy sulphonium ion is readily transformed (probably via an intramolecular rearrangement) to dimethyl sulphide and an aldehyde or a ketone (equation 69), the

$$R^2CH_2\overset{+}{O}SMe_2 \longrightarrow RCHO + Me_2S + H^+ \qquad (69)$$

reaction is useful for a facile oxidation of alcohols to the corresponding carbonyl compounds under extremely mild conditions. In the case of phenols substituted at the *ortho* position, a number of different products were obtained. Strongly acidic phenols gave phenol ethers (equation 70).

$$Me_2\overset{+}{S}O-C_6H_4-NO_2 \longrightarrow CH_3SCH_2O-C_6H_4-NO_2 + H^+ \qquad (70)$$

Several new, useful syntheses have been described which probably involve intramolecular electrophilic attacks on the hydroxyl group

$$R^1R^2C(OH)-CH_2SONHR \xrightarrow{\Delta} R^1R^2C{=}CH_2 + SO_2 + RNH_2 \qquad (71)$$

$$\begin{array}{c} \text{OH} \\ | \\ \text{O}-\text{S}-\text{N}< \\ | \qquad | \\ -\text{C}-\text{CH}_2 \\ | \end{array}$$

(45)

by a sulphur atom. As an example, the thermal decomposition of β-hydroxysylphinamides (themselves obtained from sulphinamides, BuLi and R^1R^2CO) to alkenes[198] may be mentioned (equation 71). The reaction proceeds most probably through a cyclic intermediate, **45**, which subsequently, via 1,2-cycloelimination, leads to the olefin. This mechanism found strong support from experiments which showed that the reaction took place by the *cis* elimination.

2. Electrophiles containing phosphorus

Various phosphorylation methods have been reviewed by Brown[159]. Whereas several convenient procedures are available for the preparation of di- and tri-substituted phosphate esters, relatively few methods have been designed for the synthesis of monoalkyl dihydrogen phosphates. One of the latter is that of Kirby[199], in which the solution of phosphorous acid in a large excess of alcohol is oxidized with iodine. The major disadvantage is that the method is not economic in case of alcohols which are available in only small amounts.

A novel synthesis of monoalkyl phosphates has been described by Obata and Mukaiyama[200]. A variety of monoalkyl phosphates could be obtained in good yields from reaction of alcohols with phosphorous acid and mercuric salts in the presence of tertiary amines (equation 72). When acetonitrile was used as the solvent

$$HPO(OH)_2 + ROH + HgX_2 \longrightarrow ROPO_3H_2 + Hg + 2\,HX \qquad (72)$$

only small excesses of alcohol were required. The investigators proposed an electrophilic attack of an intermediate metaphosphate anion (equation 73) on the hydroxyl group as the key step of the

$$O_2P{-}O^- + ROH \longrightarrow ROP(=O)(O^-)OH \qquad (73)$$

reaction, the anion being first formed by oxidation with mercuric salt. It can be seen that reaction (73) has its counterpart in the formation and hydrolysis of sulphate esters (section IV.A.2).

In addition to the halides and oxyhalides of phosphorus, some substituted halides of phosphorus and their reactions with alcohols have drawn the attention of several investigators. Thus, e.g., methyl chloromethyl phosphinate (**46**), which is easily obtained by reaction (74), is shown to have many synthetic uses[201].

Ramirez and co-workers[202, 203] have made extensive studies on the formation and reactions of pentaoxy phosphoranes, $P(OR)_5$.

$$ClCH_2PCl_2 + 2\ CH_3OH \longrightarrow ClCH_2P(=O)(H)-OCH_3 + CH_3Cl + HCl \qquad (74)$$

(46)

They showed that the earlier reports on the formation of pentaphenoxy phosphorane in reactions (75) and (76) were inconsistent.

$$PCl_5 + 3\ PhOH \xrightarrow{140°C} (PhO)_3PCl_2 + 3\ HCl \qquad (75)$$

$$(PhO)_3PCl_2 + 2\ PhOH \xrightarrow{25°C} (PhO)_5P + 2\ HCl \qquad (76)$$

However, when the reactions were carried out at low temperatures in hexane–benzene solution and in the presence of a tertiary amine, the pentaoxy phosphorane could be synthesized. The structure of the reaction product was ascertained by phosphorus-31 n.m.r. spectroscopy. A number of other cyclic pentaoxy phosphoranes with alkoxy groups attached to the central phosphorus atom were also described.

3. Electrophiles containing nitrogen, boron, etc.

In addition to the reaction with nitric acid (section IV.A.4), the *O*-nitration of alcohols can be effected by the use of other *O*-nitro or *N*-nitro compounds. The most commonly used reagents are benzoyl and acetyl nitrates[204, 205] (equation 77). The kinetics of the reaction have been recently studied by n.m.r. spectroscopy[205].

$$CH_3COONO_2 + ROH \longrightarrow RONO_2 + CH_3COOH \qquad (77)$$

Transesterification (e.g., equation 78) is in several instances a convenient means for the preparation of esters of inorganic acids. Ordinarily, an ester of lower molecular weight is converted to one of higher molecular weight. The transesterification is easily effected if the liberated alcohol is of lower boiling point than the other components of the reaction mixture and can be removed by distillation. Alternatively, the required ester may be the most volatile constituent. The practical and mechanistic aspects of these transesterification reactions are usually similar to those of the hydrolysis reactions of the esters in question.

$$B(OR^1)_3 + 3\ R^2OH \longrightarrow B(OR^2)_3 + 3\ R^1OH \qquad (78)$$

Electrophilic attacks on the hydroxyl group by halogens, yielding 'positive' halogen compounds, have several uses in organic synthesis. A familiar example is the formation of *t*-butyl hypochlorite from *t*-butanol and chlorine[206] in alkaline solution (equation 79). A

$$Me_3COH + Cl_2 + NaOH \longrightarrow Me_3COCl + NaCl + H_2O \qquad (79)$$

recent application of *t*-butyl hypochlorite is the radical chain halogenation of alkenes and alkynes[207]. Ring cleavage of cyclopropanols with various 'positive' halogen compounds, including *t*-butyl hypochlorite, has been recently studied by DePuy, Arney and Gibson[208].

Several oxidation reactions of alcohols, including the reaction with lead tetraacetate[209–214], are initiated by electrophilic attacks on the hydroxyl group. As the subsequent stages of these reactions, leading to a variety of products, are those of primary interest in their applications, a detailed discussion is beyond the scope of this chapter.

V. REFERENCES

1. R. G. Pearson and J. Songstad, *J. Am. Chem. Soc.*, **89,** 1827 (1967).
2. R. G. Pearson, H. Sobel and J. Songstad, *J. Am. Chem. Soc.*, **90,** 319 (1968).
3. N. C. Deno and J. O. Turner, *J. Org. Chem.*, **31,** 1969 (1966).
4. R. G. Pearson, *Science*, **151,** 172 (1966).
5. R. G. Pearson, *Chem. in Britain*, **3,** 103 (1967).
6. R. G. Pearson and J. Songstad, *J. Org. Chem.*, **32,** 2899 (1967).
7. J. O. Edwards, *J. Am. Chem. Soc.*, **76,** 1540 (1954).
8. J. O. Edwards, *J. Am. Chem. Soc.*, **78,** 1819 (1956).
9. R. E. Robertson, *Can. J. Chem.*, **31,** 589 (1953).
10. S. Winstein, E. Grundwald and H. W. Jones, *J. Am. Chem. Soc.*, **73,** 2700 (1951).
11. H. M. R. Hoffmann, *J. Chem. Soc.*, 6753, 6762 (1965).
12. A. H. Fainberg and S. Winstein, *J. Am. Chem. Soc.*, **79,** 1602 (1957).
13. V. J. Shiner Jr., B. L. Murr and G. Heinemann, *J. Am. Chem. Soc.*, **85,** 2413 (1963).
14. L. Hakka, A. Queen and R. E. Robertson, *J. Am. Chem. Soc.*, **87,** 161 (1965).
15. G. J. Frisone and E. R. Thornton, *J. Am. Chem. Soc.*, **90,** 1211 (1968).
16. K. B. Wiberg and V. Z. Williams Jr., *J. Am. Chem. Soc.*, **89,** 3373 (1967).
17. J. F. Bunnett, *Quart. Rev. (London)*, **12,** 1 (1958).
18. J. Sauer and R. Huisgen, *Angew. Chem.*, **72,** 294 (1960).
19. S. D. Ross in *Progress in Physical Organic Chemistry*, Vol. 1 (Ed. S. G. Cohen, A. Streitwieser Jr. and R. W. Taft), Interscience Publishers, New York, 1963, p. 1.
20. A. Streitwieser Jr., *Chem. Rev.*, **56,** 571 (1956).
21. C. A. Bunton, in *Nucleophilic Substitution at a Saturated Carbon Atom*, Vol. 1 (Ed. E. D. Hughes), Elsevier Publishing Co., Amsterdam, 1963.
22. H. Feuer and J. Hooz, in *The Chemistry of the Ether Linkage* (Ed. S. Patai), Interscience, London, 1967, Chap. 10.
23. P. Salomaa, in *The Chemistry of the Carbonyl Group* (Ed. S. Patai), Interscience, London, 1965, Chap. 3.
24. E. Schmitz and I. Eichhorn, in *The Chemistry of the Ether Linkage* (Ed. S. Patai), Interscience, London, 1967, Chap. 7.

25. P. Salomaa, A. Kankaanperä and M. Lajunen, *Acta Chem. Scand.*, **20**, 1790 (1966).
26. A. J. Kresge and Y. Chiang, *J. Chem. Soc.* (**B**), 53, 58 (1967).
27. C. K. Ingold, *Structure and Mechanism in Organic Chemistry*, G. Bell & Sons, London, 1953, p. 333.
28. P. Z. Bedoukian, *J. Am. Chem. Soc.*, **66**, 1325 (1944).
29. J. E. Baldwin and L. E. Walker, *J. Org. Chem.*, **31**, 3985 (1966).
30. P. Salomaa and A. Kankaanperä, *Acta Chem. Scand.*, **20**, 1802 (1966).
31. P. Salomaa, *Ann. Acad. Sci. Fennicae, Ser.* **A II**, No. 103 (1961).
32. A. Kankaanperä, *Acta Chem. Scand.*, **23**, 1728 (1969).
33. J. M. Bell, D. G. Kubler, P. Sartwell and R. G. Zepp, *J. Org. Chem.*, **30**, 4284 (1965).
34. R. Garrett and D. G. Kubler, *J. Org. Chem.*, **31**, 2665 (1966).
35. E. H. Cordes, in *Progress in Physical Organic Chemistry*, Vol. 4 (Ed. A. Streitwieser Jr. and R. W. Taft), Interscience Publishers, New York, 1967, pp. 1–44.
36. A. Kankaanperä, *Ann. Univ. Turku., Ser.* **A I**, No. 95 (1966).
37. K. Pihlaja, *Ann. Univ. Turku., Ser.* **A I**, No. 114 (1967).
38. F. Aftalion, D. Lumbroso, M. Hellin and F. Coussemant, *Bull. Soc. Chim. France*, 1950, 1958 (1965).
39. M. Garnier, F. Aftalion, D. Lumbroso, M. Hellin and F. Coussemant, *Bull. Soc. Chim. France*, 1512 (1965).
40. B. Fremaux, M. Davidson, M. Hellin and F. Coussemant, *Bull. Soc. Chim. France*, 4243, 4250 (1967).
41. P. Salomaa and R. Linnantie, *Acta Chem. Scand.*, **14**, 777 (1960).
42. A. Kankaanperä and M. Lahti, *Acta Chem. Scand.*, **23**, 2465 (1969).
43. K. G. Shipp and M. E. Hill, *J. Org. Chem.*, **31**, 853 (1966).
44. H. Hibbert, J. C. Morazain and A. Paquet, *Can. J. Research*, **2**, 131 (1930).
45. S. M. McElvain and M. J. Curry, *J. Am. Chem. Soc.*, **70**, 3781 (1948).
46. R. H. Martin, F. W. Lampe and R. W. Taft, *J. Am. Chem. Soc.*, **88**, 1353 (1966).
47a. A. Kankaanperä and H. Tuominen, *Suomen Kemistilehti*, **B40**, 271 (1967).
47b. A. Kankaanperä and M. Lahti, *Suomen Kemistilehti*, **B42**, 406 (1969).
47c. A. Kankaanperä and M. Lahti, *Suomen Kemistilehti*, **B43**, 75, 101, 105 (1970).
48. V. Gold and D. C. A. Waterman, *J. Chem. Soc.* **(B)**, 839, 849 (1968).
49. W. C. Kuryla and D. G. Leis, *J. Org. Chem.*, **29**, 2773 (1964).
50. W. C. Kuryla, *J. Org. Chem.*, **30**, 3926 (1965).
51. S. Patai and Z. Rappoport, in *The Chemistry of Alkenes* (Ed. S. Patai), Interscience, New York, 1964, Chap. 8.
52. R. N. Ring, G. C. Tesoro and D. R. Moore, *J. Org. Chem.*, **32**, 1091 (1967).
53. J. Chatt, *Chem. Rev.*, **48**, 7 (1951).
54. W. L. Waters and E. F. Kiefer, *J. Am. Chem. Soc.*, **89**, 6261 (1967).
55. M. F. Shostakovskii, A. V. Bogdanova and G. I. Plotnikova, *Usp. Khim.*, **33**, 129 (1964); *Chem. Abstr.*, **60**, 13132 (1964).
56. D. Miller, *J. Chem. Soc.* (**C**), 12 (1969).
57. W. Kirmse, L. Horner and H. Hoffmann, *Liebigs Ann. Chem.*, **614**, 19 (1958).
58. W. Kirmse, *Liebigs Ann. Chem.*, **666**, 9 (1963).
59. T. Saegusa, Y. Ito, S. Kobayashi, K. Hirota and T. Shimizu, *J. Org. Chem.*, **33**, 544 (1968).

60. N. R. Ghosh, C. R. Ghoshal and S. Shah, *Chem. Commun.*, 151 (1969).
61. J. S. Meek, J. S. Fowler, P. A. Monroe and T. J. Clark, *J. Org. Chem.*, **33,** 223 (1968).
62. A. Schönberg, K. Junghans and E. Singer, *Tetrahedron Letters*, 4667 (1966).
63. T. Saegusa, Y. Ito, S. Kobayashi and K. Hirota, *Tetrahedron Letters*, 521 (1967).
64. T. Saegusa, Y. Ito, S. Kobayashi, N. Takeda and K. Hirota, *Tetrahedron Letters*, 1273 (1967).
65. R. J. Mulder, A. M. van Leusen and J. Strating, *Tetrahedron Letters*, 3057 (1967).
66. E. K. Euranto, in *The Chemistry of Carboxylic Acids and Esters* (Ed. S. Patai), John Wiley & Sons, New York, 1969, Chap X.
67. M. L. Bender, *Chem. Rev.*, **60,** 53 (1960).
68. L. P. Hammett, *Physical Organic Chemistry*, McGraw-Hill, New York, 1940, Chaps. IV, VI, VII and IX.
69. J. Hine, *Physical Organic Chemistry*, 2nd ed., McGraw-Hill, New York, 1962, Chap. XII.
70. C. K. Ingold, *Structure and Mechanism in Organic Chemistry*, Cornell University Press, Ithaca, New York, 1953, pp. 751–782.
71. F. Kurzer and K. Douraghi-Zadeh, *Chem. Rev.*, **67,** 118 (1967).
72. P. Adams and F. A. Baron, *Chem. Rev.*, **65,** 567 (1965).
73. E. L. Eliel and F. J. Biros, *J. Am. Chem. Soc.*, **88,** 3334 (1966).
74. E. L. Eliel and S. Schroeter, *J. Am. Chem. Soc.*, **87,** 5031 (1965).
75. E. L. Eliel and T. J. Brett, *J. Am. Chem. Soc.*, **87,** 5039 (1965).
76. E. L. Eliel and H. Haubenstock, *J. Org. Chem.*, **26,** 3504 (1961).
77. E. L. Eliel, N. L. Allinger, S. J. Angyal and G. A. Morrison, *Conformational Analysis*, Interscience, New York, 1965, Chap. 2.2.*c*.
78. K. Bowden, *Can. J. Chem.*, **44,** 661 (1966).
79. K. Bowden, N. B. Chapman and J. Shorter, *J. Chem. Soc.*, 5239 (1963).
80. C. N. Hinshelwood and A. R. Legard, *J. Chem. Soc.*, 587 (1935).
81. H. A. Smith and J. Burn, *J. Am. Chem. Soc.*, **66,** 1494 (1944).
82. R. W. Taft Jr., in *Steric Effects in Organic Chemistry* (Ed. M. S. Newman), John Wiley & Sons, New York, 1956, Chap. XIII.
83. M. S. Newman, *Steric Effects in Organic Chemistry*, John Wiley & Sons, New York, 1956, pp. 204–217.
84. R. C. Parish and L. M. Stock, *J. Org. Chem.*, **30,** 927 (1965).
85. E. J. Bourne, M. Stacey, J. C. Tatlow and J. M. Tedder, *J. Chem. Soc.*, 2976 (1949).
86. E. J. Bourne, M. Stacey, J. C. Tatlow and R. Worrall, *J. Chem. Soc.*, 3268 (1958).
87. J. M. Tedder, *Chem. Rev.*, **55,** 787 (1955).
88. E. J. Bourne, M. Stacey, J. C. Tatlow and J. M. Tedder, *J. Chem. Soc.*, 718 (1951).
89. M. S. Newman, *J. Am. Chem. Soc.*, **67,** 345 (1945).
90. W. D. Emmons, K. S. McCallum and A. F. Ferris, *J. Am. Chem. Soc.*, **75,** 6047 (1953).
91. E. J. Bourne, M. Stacey, J. C. Tatlow and R. Worrall, *J. Chem. Soc.*, 2006 (1954).
92. E. J. Bourne, J. E. B. Randles, M. Stacey, J. C. Tatlow and J. M. Tedder, *J. Am. Chem. Soc.*, **76,** 3206 (1954).

93. M. S. Newman and C. Courduvelis, *J. Org. Chem.*, **30,** 1795 (1965).
94. M. L. Bender and M. S. Silver, *J. Am. Chem. Soc.*, **84,** 4589 (1962).
95. F. Ramirez, B. Hansen and N. B. Desai, *J. Am. Chem. Soc.*, **84,** 4588 (1962).
96. W. Stevens and A. Van Es, *Rec. Trav. Chim.*, **83,** 1287 (1964).
97. A. Béhal, *Compt. Rend.*, **128,** 1460 (1900); *Ann. Chim.* (*17*), **20,** 411 (1900).
98. A. Verley, *Bull. Soc. Chim. France* (*4*), **41,** 803 (1927).
99. E. R. Schierz, *J. Am. Chem. Soc.*, **45,** 455 (1923).
100. V. Gold and E. G. Jefferson, *J. Chem. Soc.*, 1416 (1953).
101. C. D. Hurd, S. S. Drake and O. Fancher, *J. Am. Chem. Soc.*, **68,** 789 (1946).
102. W. Stevens and A. Van Es, *Rec. Trav. Chim.*, **83,** 1294 (1964).
103. J. Ducasse, *Bull. Soc. Chim. France* (*5*), **12,** 918 (1945).
104. M. Hudlicky, *Chemistry of Organic Fluorine Compounds*, The MacMillan Co., New York, 1962, p. 197.
105. J. Radell and J. W. Connolly, *Chem. Eng. Data*, **6,** 282 (1961).
106. E. E. Burgoyne and F. E. Condon, *J. Am. Chem. Soc.*, **72,** 3276 (1950).
107. J. Radell, B. W. Brodman, A. Hirshfeld and E. D. Bergmann, *J. Phys. Chem.*, **69,** 928 (1965).
108. H. Prajecus and U. Kellner, *Z. Chem.*, **4,** 226 (1964).
109. H. Prajecus and J. Leška, *Z. Naturforsch.*, **21b,** 30 (1966).
110. H. Prajecus and A. Tille, *Chem. Ber.*, **100,** 196 (1967).
111. H. Brechbühler, H. Büchi, E. Hatz, J. Schreiber and A. Eschenmoser, *Helv. Chim. Acta*, **48,** 1746 (1965).
112. D. P. N. Satchell, *J. Chem. Soc.*, 558, 564 (1963).
113. J. M. Briody and D. P. N. Satchell, *J. Chem. Soc.*, 3724 (1964); 168 (1965).
114. D. P. N. Satchell, *Quart. Rev.*, **17,** 160 (1963).
115. J. W. McFarland and J. B. Howard, *J. Org. Chem.*, **30,** 957 (1965).
116. J. W. McFarland, D. E. Lenz and D. J. Grosse, *J. Org. Chem.*, **31,** 3798 (1966).
117. H. Ulrich, B. Tucker and A. A. R. Sayigh, *J. Org. Chem.*, **32,** 3938 (1967).
118. E. C. Taylor and G. W. McLay, *J. Am. Chem. Soc.*, **90,** 2422 (1968).
119. S. M. Kupchan and W. S. Johnson, *J. Am. Chem. Soc.*, **78,** 3864 (1956).
120. S. M. Kupchan, W. S. Johnson and S. Rajagopalan, *Tetrahedron*, **7,** 47 (1959).
121. S. M. Kupchan and C. R. Narayanan, *J. Am. Chem. Soc.*, **81,** 1913 (1959).
122. S. M. Kupchan, S. T. Eriksen and M. Friedman, *J. Am. Chem. Soc.*, **88,** 343 (1966).
123. T. C. Bruice and T. H. Fife, *J. Am. Chem. Soc.*, **84,** 1977 (1962).
124. R. Biggins and E. Haslam, *J. Chem. Soc.*, 6883 (1965).
125. M. L. Bender, F. J. Kezdy and B. Zerner, *J. Am. Chem. Soc.*, **85,** 3017 (1963).
126. B. Capon and B. Ch. Ghosh, *J. Chem. Soc.* (**B**), 472 (1966).
127. F. A. Long, *Ann. New York Acad. Sci.*, **84,** 596 (1960).
128. B. Hansen, *Acta Chem. Scand.*, **17,** 1375 (1963).
129. B. A. Cunningham and G. L. Schmir, *J. Am. Chem. Soc.*, **89,** 917 (1967).
130. S. M. Kupchan, J. H. Block and A. C. Isenberg, *J. Am. Chem. Soc.*, **89,** 1189 (1967).
131. H. T. Openshaw and N. Whittaker, *J. Chem. Soc.* (**C**), 89 (1969).
132. J. D. Roberts and J. G. Traynham, *J. Org. Chem.*, **32,** 3177 (1967); J. D. Roberts, *J. Org. Chem.*, **33,** 118 (1968).
133. K. S. Dodgson, *Proc. Intern. Congr. Biochem.*, **13,** 23 (1960).

134. A. B. Roy, *Advan. Enzymol.*, **22,** 205 (1960).
135. F. C. Whitmore and H. S. Rothrock, *J. Am. Chem. Soc.*, **54,** 3431 (1932).
136. N. C. Deno and M. S. Newman, *J. Am. Chem. Soc.*, **72,** 3852 (1950).
137. R. L. Burwell Jr., *J. Am. Chem. Soc.*, **74,** 1462 (1952).
138. J. S. Brimacombe, A. B. Foster, E. B. Hancock, W. G. Overend and M. Stacey, *J. Chem. Soc.*, 201 (1960).
139. B. D. Batts, *J. Chem. Soc.* (**B**), 547, 551 (1966).
140. G. Williams and D. J. Clark, *J. Chem. Soc.*, 1304 (1956).
141. S. J. Benkovic, *J. Am. Chem. Soc.*, **88,** 5511 (1966).
142. S. Burstein and S. Lieberman, *J. Am. Chem. Soc.*, **80,** 5235 (1958).
143. J. McKenna and J. K. Norymberski, *J. Chem. Soc.*, 3889 (1957).
144. J. L. Kice and J. M. Anderson, *J. Am. Chem. Soc.*, **88,** 5242 (1966).
145. C. A. Bunton and B. N. Hendy, *J. Chem. Soc.*, 3130 (1963).
146. R. E. Robertson, *Progr. Phys. Org. Chem.*, **4,** 213 (1967).
147. S. J. Benkovic, *J. Am. Chem. Soc.*, **88,** 5511 (1966).
148. P. Salomaa, R. Hakala, S. Vesala and T. Aalto, *Acta Chem. Scand.* **23,** 2116 (1969).
149. S. J. Benkovic and P. A. Benkovic, *J. Am. Chem. Soc.*, **90,** 2646 (1968).
150. R. H. Boyd, R. W. Taft Jr., A. P. Wolf and D. R. Christman, *J. Am. Chem. Soc.*, **82,** 4729 (1960).
151. I. Dostrovsky and F. S. Klein, *J. Chem. Soc.*, 791 (1955).
152. C. M. Suter, *Organic Chemistry of Sulphur*, John Wiley & Sons, New York, 1944.
153. N. Kharash (Ed.), *Organic Chemistry of Sulphur Compounds*, Vol. 1, Pergamon Press, New York, 1961.
154. E. T. Kaiser, M. Panar and F. H. Westheimer, *J. Am. Chem. Soc.*, **85,** 602 (1963).
155. R. E. Robertson and S. E. Sugamori, *Can. J. Chem.*, **44,** 1728 (1966).
156. F. P. Boer, J. J. Flynn, E. T. Kaiser, O. R. Zaborsky, D. A. Tomalia, E. A. Young and Y. C. Tong, *J. Am. Chem. Soc.*, **90,** 2970 (1968).
157. E. T. Kaiser and O. R. Zaborsky, *J. Am. Chem. Soc.*, **90,** 4626 (1968).
158. D. S. Breslow, R. R. Hough and J. T. Fairlough, *J. Am. Chem. Soc.*, **76,** 5361 (1954).
159. D. M. Brown in *Advances in Organic Chemistry*, Vol. 3 (Ed. R. A. Raphael, E. C. Taylor and H. Wynberg), Interscience, New York, 1963, pp. 75–158.
160. B. Capon, M. J. Perkins and C. W. Rees, *Organic Reaction Mechanisms 1967*, Interscience, New York, 1968, pp. 358–365.
161. A. J. Kirby and S. G. Warren, *Organic Chemistry of Phosphorus*, Elsevier, Amsterdam, 1967.
162. T. C. Bruice and S. J. Benkovic, *Bioorganic Mechanisms*, Vol. 2, Benjamin, New York, 1966, Chap. 2.
163. C. A. Bunton, M. M. Mhala, K. G. Oldham and C. A. Vernon, *J. Chem. Soc.*, 3293 (1960).
164. P. C. Haake and F. H. Westheimer, *J. Am. Chem. Soc.*, **83,** 1102 (1961).
165. H. G. Khorana, G. M. Tener, R. S. Wright and J. G. Moffatt, *J. Am. Chem. Soc.*, **79,** 430 (1957).
166. D. A. Usher, E. A. Dennis and F. H. Westheimer, *J. Am. Chem. Soc.*, **87,** 2320 (1965).
167. R. L. Collin, *J. Am. Chem. Soc.*, **88,** 3281 (1966).

168. M. L. Bender and J. M. Lawlor, *J. Am. Chem. Soc.*, **85**, 3010 (1963).
169. A. Lapidot, D. Samuel and M. Weiss-Broday, *J. Chem. Soc.*, 637 (1964).
170. T. Higuchi, G. L. Flynn and A. C. Shah, *J. Am. Chem. Soc.*, **89**, 616 (1967).
171. L. Kugel and M. Halman, *J. Org. Chem.*, **32**, 642 (1967).
172. C. A. Bunton, E. J. Fendler, E. Humeres and Kui-Un Yang, *J. Org. Chem.*, **32**, 2806 (1967).
173. C. A. Bunton, S. J. Farber and E. J. Fendler, *J. Org. Chem.*, **33**, 29 (1968).
174. P. Mesnard and M. Bertucat, *Bull. Soc. Chim. France.*, 307 (1959).
175. E. Cherbuliez, C. Gandillon, A. de Picciotto and J. Rabinowitz, *Helv. Chim. Acta*, **42**, 2277 (1959).
176. F. B. Clarke and J. W. Lyons, *J. Am. Chem. Soc.*, **88**, 4401 (1966).
177. P. W. C. Barnard, C. A. Bunton, D. Kellerman, M. M. Mhala, B. Silver, C. A. Vernon and V. A. Welch, *J. Chem. Soc.* (**B**), 227 (1966).
178. R. Boschan, R. T. Merrow and R. W. Van Dolah, *Chem. Rev.*, **55**, 485 (1955).
179. R. Klein and M. Mentser, *J. Am. Chem. Soc.*, **73**, 5888 (1951).
180. E. L. Blackall, E. D. Hughes, Sir Christopher Ingold and R. B. Pearson, *J. Chem. Soc.*, 4366 (1958).
181. J. W. Baker and D. M. Easty, *J. Chem. Soc.*, 1193 (1952).
182. J. W. Baker and E. J. Neale, *J. Chem. Soc.*, 608 (1955).
183. J. W. Baker and T. G. Heggs, *J. Chem. Soc.*, 616 (1955).
184. L. P. Kuhn, *J. Am. Chem. Soc.*, **69**, 1974 (1947).
185. H. Steinberg, *Organoboron Chemistry*, Vol. 1, John Wiley & Sons, New York, 1964, Chaps. 4–7.
186. W. Gerrard, *Organic Chemistry of Boron*, Academic Press, London, 1961, pp. 5–21.
187. J. A. Bradley and P. M. Christopher, *129th Meeting of the American Chemical Society*, Dallas, 1956, Abstracts of Papers, p. 39-N.
188. H. Steinberg and D. L. Hunter, *Ind. Eng. Chem.*, **49**, 174 (1957).
189. D. W. Tanner and T. C. Bruice, *J. Am. Chem. Soc.*, **89**, 6954 (1967).
190. D. W. Clayton, J. A. Farrington, G. W. Kenner and J. M. Turner, *J. Chem. Soc.*, 1398 (1957).
191. R. G. Schweiger, *Chem. Ind.*, 900 (1966).
192. E. E. Gilbert, B. Veldhuis, E. J. Carlson and S. L. Giolito, *Ind. Eng. Chem.*, **45**, 2065 (1953).
193. A. B. Burg, *J. Am. Chem. Soc.*, **65**, 1629 (1943).
194. E. H. White and C. A. Ellinger, *J. Am. Chem. Soc.*, **87**, 5261 (1965).
195. K. E. Pfitzner and J. G. Moffatt, *J. Am. Chem. Soc.*, **87**, 5661, 5670 (1965).
196. A. H. Fenselau and J. G. Moffatt, *J. Am. Chem. Soc.*, **88**, 1762 (1966).
197. M. G. Burdon and J. G. Moffatt, *J. Am. Chem. Soc.*, **88**, 5855 (1966); **89**, 4725 (1967).
198. E. J. Corey and T. Durst, *J. Am. Chem. Soc.*, **90**, 5553 (1968).
199. G. W. Kirby, *Chem. Ind.*, 1877 (1963).
200. T. Obata and T. Mukaiyama, *J. Org. Chem.*, **32**, 1063 (1967).
201. H. Goldwhite and D. G. Roswell, *J. Am. Chem. Soc.*, **88**, 3572 (1966).
202. F. Ramirez, *Acc. Chem. Research*, **1**, 168 (1968); this article contains references to the pertinent literature.
203. F. Ramirez, K. Tasaka, N. B. Desai and C. P. Smith, *J. Am. Chem. Soc.*, **90**, 751 (1968).

204. F. E. Francis, *J. Chem. Soc.*, 1 (1906).
205. B. Östman, *Acta Chem. Scand.*, **21**, 1257 (1967).
206. H. M. Teeter and E. W. Bell in *Organic Syntheses*, Vol. 32 (Ed. R. T. Arnold), John Wiley & Sons, New York, 1952, pp. 20–22.
207. C. Walling, L. Heaton and D. D. Tanner, *J. Am. Chem. Soc.*, **87**, 1715 (1965).
208. C. H. DePuy, N. C. Arney Jr. and D. H. Gibson, *J. Am. Chem. Soc.*, **90**, 1830 (1968).
209. R. E. Partch, *J. Org. Chem.*, **30**, 2498 (1965).
210. A. C. Cope, M. Gordon, Sung Moon and Chung Ho Park, *J. Am. Chem. Soc.*, **87**, 3119 (1965).
211. M. Lj. Mihailović, Ž. Čeković, Ž. Maksimović, D. Jeremić, Lj. Lorenc and R. I. Mamuzić, *Tetrahedron*, **21**, 2799 (1965).
212. M. Lj. Mihailović, Ž. Čeković and D. Jeremić, *Tetrahedron*, **21**, 2813 (1965).
213. Sung Moon and P. R. Clifford, *J. Org. Chem.*, **32**, 4017 (1967).
214. W. H. Starnes Jr., *J. Org. Chem.*, **33**, 2767 (1968).

CHAPTER **10**

Oxidation and reduction of phenols

MIHAILO LJ. MIHAILOVIĆ and ŽIVORAD ČEKOVIĆ

Department of Chemistry, Faculty of Sciences, University of Belgrade and Institute for Chemistry, Technology and Metallurgy, Belgrade, Yugoslavia

I. INTRODUCTION

One of the characteristic chemical properties of phenols is their facile oxidative conversion to compounds of different structural types. The diversity of phenol oxidation products offers interesting synthetic possibilities for the preparation of simple and polymeric molecules containing phenolic and/or quinonoid structural elements, particularly of those resulting from oxidative coupling of both like and unlike intermediate radical species[1–8]. In this way various natural products have been successfully synthesized from phenols[1, 4–6, 8]. In addition, mechanistic studies of oxidation reactions of phenols[1–8] have lent strong support to the interpretation that a number of biosynthetic processes actually proceed by biogenetic pathways involving the oxidative utilization of phenolic substrates[1, 4–6, 8].

The long-known fact that substituted phenols and the corresponding phenoxy radicals are efficient inhibitors in autoxidation processes of organic substances[9–11] has stimulated in recent years the development of chemical and physical methods, particularly of electron spin resonance (e.s.r.) spectroscopy, for the study of the detailed structure of phenoxy (and other) radicals and of their role as intermediates in free radical reactions[12, 13].

In this chapter the principles and the scope of oxidation and reduction of monohydric and polyhydric phenols are reviewed.

II. OXIDATION OF PHENOLS

A. *Mechanism of One-Electron Oxidations*

1. Phenoxy radicals and quinone methides

The first step in the oxidation of monohydric phenols, by oxidizing agents capable of one-electron abstraction[6, 8, 12, 14], such as lead dioxide, silver oxide, manganese dioxide, ferric and ceric ions, electrochemical methods[15], alkaline potassium ferricyanide[2] and others, consists in the generation of free phenoxy radicals (**3**) (equation 1), either by homolytic cleavage of the O–H bond in the phenol (**1**) with loss of the hydrogen atom or by loss of one electron from the corresponding phenoxide anion (**2**). These phenoxy (or aryloxy)

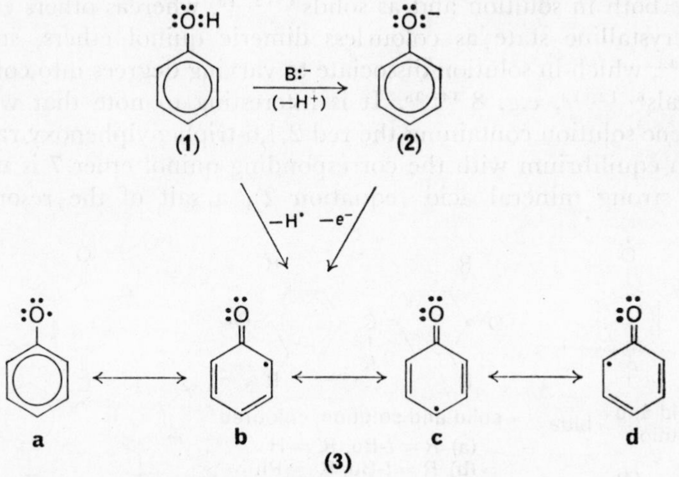

:Ö:H (1) —B:⁻ / (−H⁺)→ :Ö:⁻ (2)

−H˙ −e⁻

(1)

a ⟷ b ⟷ c ⟷ d

(**3**)

radicals (**3**), which may be formally defined as monovalent oxygen radical species, are resonance-stabilized by delocalization of the unpaired electron over the aromatic ring, as shown by structures **3a–d**[1–8, 12, 13].

Although phenoxy radicals with *ortho* and/or *para* unsubstituted positions usually undergo further reactions very rapidly[1–8, 12, 13], because of resonance stabilization they have a longer lifetime than alkyl or aryl radicals and they do not attack the solvent or initiate polymerization (e.g., by removal of hydrogen from C–H bonds) as readily[6, 7]. Their transient existence, which was postulated mainly on the basis of chemical reactivity, has recently been confirmed by e.s.r. spectroscopy using a flow system technique[7, 16, 17]. The mean lifetime of the simplest phenoxy radical (**3**), derived from phenol itself, has been estimated to be about 10^{-3} second[7, 16].

For the production of 'stable' phenoxy radicals, which may survive for a sufficient time to be used as substrates for other experiments or which may exist in solution (and in certain cases in the solid state) over periods of hours or even days (in the presence of limited amounts of air), it is necessary that the reactive *ortho-* and *para*-positions be blocked by suitable groups which give increased resonance stabilization or steric protection, since 2,4,6-trisubstitution usually retards or prevents further reaction (i.e. addition or substitution)[1–8] and the absence of α-CH groups in the substituents prohibits formation of quinone methides[18–21]. Some of these free aryloxy mono- and diradicals, e.g., **4** [22–25], **5** [26–32] and **6** [33], are highly coloured and stable both in solution and as solids[6, 12, 13], whereas others exist in the crystalline state as colourless dimeric quinol ethers, such as **7** [34–38], which in solution dissociate to varying degrees into coloured radicals[6, 12, 13], e.g. **8** [34–38]. It is interesting to note that when a benzene solution containing the red 2,4,6-triphenylphenoxy radicals (**8**) in equilibrium with the corresponding quinol ether **7** is treated with strong mineral acid (equation 2), a salt of the resonance-

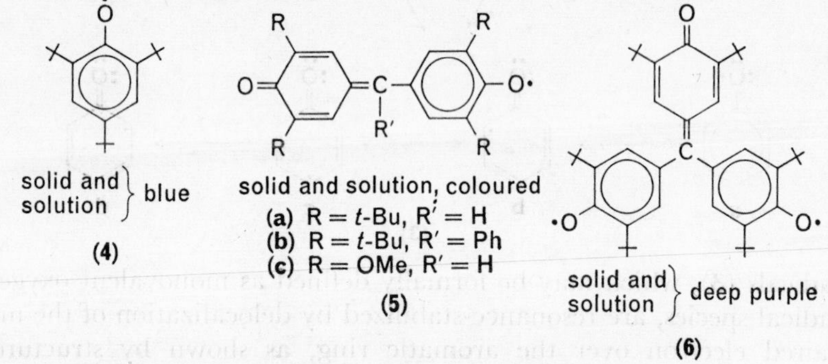

solid and solution } blue

(4)

solid and solution, coloured

(a) R = *t*-Bu, R′ = H
(b) R = *t*-Bu, R′ = Ph
(c) R = OMe, R′ = H

(5)

solid and solution } deep purple

(6)

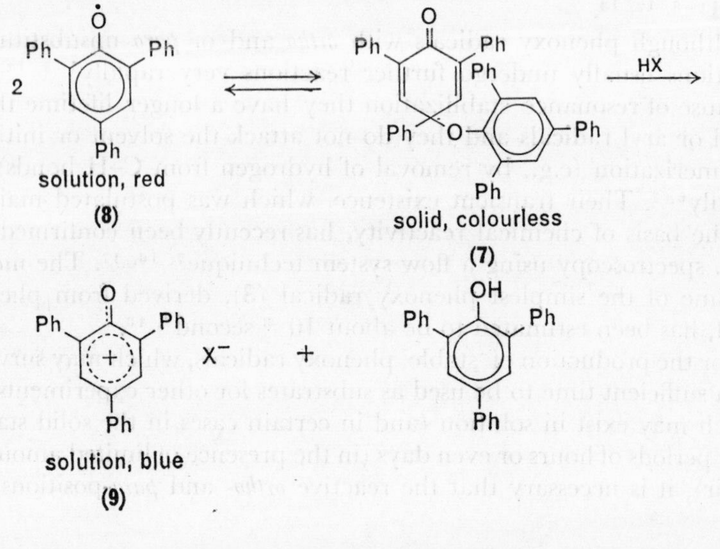

solution, red

(8)

solid, colourless

(7)

HX

(2)

solution, blue

(9)

stabilized phenoxonium ion (**9**) is formed and the reaction mixture turns deep blue[39].

Aryloxy radicals with heteroatoms can also be prepared, as illustrated by examples of the nitrogen-containing radicals **10** [40] and **11** [41], which are very stable to oxygen and exist as such in the solid state, and of the moderately stable to unstable radicals **12** [42] and **13** [43] containing phosphorus and sulphur, which have been generated only in solution.

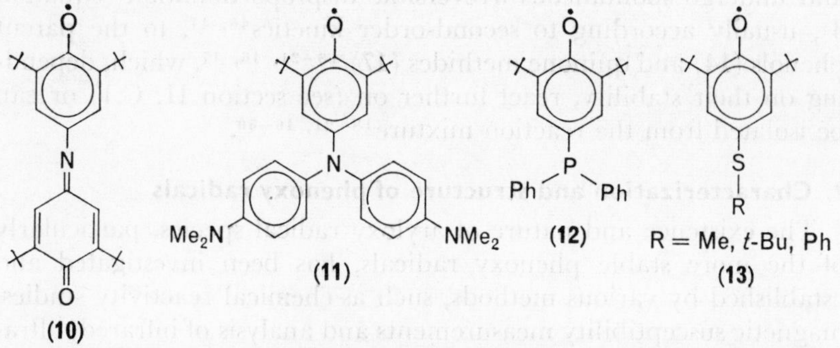

Me_2N NMe_2 Ph Ph R = Me, *t*-Bu, Ph

(**10**) (**11**) (**12**) (**13**)

Stable free aryloxy radicals, such as galvinoxyl (**5a**) and 2,6,3′,5′-tetra-*t*-butylindo-phenoxyl [BIP] (**10**), which remain unchanged in

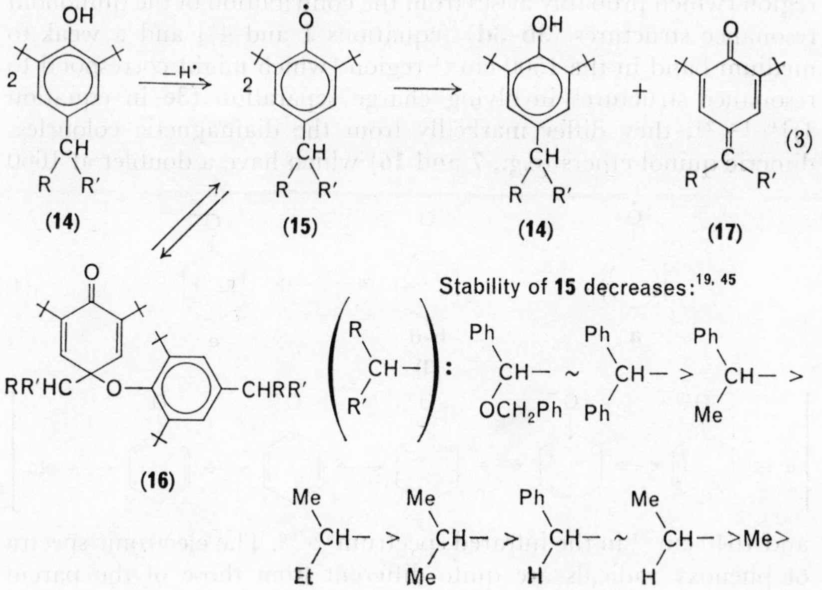

(3)

(**14**) (**15**) (**14**) (**17**)

(**16**) RR′HC CHRR′

Stability of **15** decreases:[19, 45]

(R, R′ CH—): CH(Ph)(OCH$_2$Ph)— ~ CH(Ph)(Ph)— > CH(Ph)(Me)— > CH(Me)(Et)— > CH(Me)(Me)— > CH(Ph)(H)— ~ CH(Me)(H)— >Me>

the solid state for months or years, have been used as efficient scavengers of alkyl and alkoxy or phenoxy radicals[28, 44, 45].

Hindered phenols (e.g., phenols with 2,6-di-*t*-butyl groups) containing an α-CH in the *para*-substituent can also be oxidized by one-electron abstracting agents (alkaline potassium ferricyanide, lead dioxide, silver oxide, 2,3-dichloro-5,6-dicyano-1,4-benzoquinone, etc.) to the corresponding phenoxy radicals (**15**) [12, 13]. These are only of moderate stability (a few hours) or of transient existence and undergo spontaneous irreversible disproportionation (equation 3), usually according to second-order kinetics[46, 47], to the parent phenols (**14**) and quinone methides (**17**) [18–21, 46, 47], which, depending on their stability, react further on (see section II. C.1) or can be isolated from the reaction mixture[18–21, 46–50].

2. Characterization and structure of phenoxy radicals

The existence and nature of aryloxy radical species, particularly of the more stable phenoxy radicals, has been investigated and established by various methods, such as chemical reactivity studies, magnetic susceptibility measurements and analysis of infrared, ultraviolet-visible, n.m.r. and e.s.r. spectral data.

Phenoxy radicals are paramagnetic[24, 30, 51–53], and usually show in their infrared spectra a broad strong band in the 1560–1600 cm^{-1} region [which probably arises from the contribution of the quinonoid resonance structures (**3b**–**3d**) (equations 1 and 4)] and a weak to medium band in the 1500 cm^{-1} region [which might correspond to resonance structures involving charge separation (**3e** in equation 4)][1, 12, 24]; they differ markedly from the diamagnetic colourless dimeric quinol ethers (e.g., **7** and **16**) which have a doublet at 1660

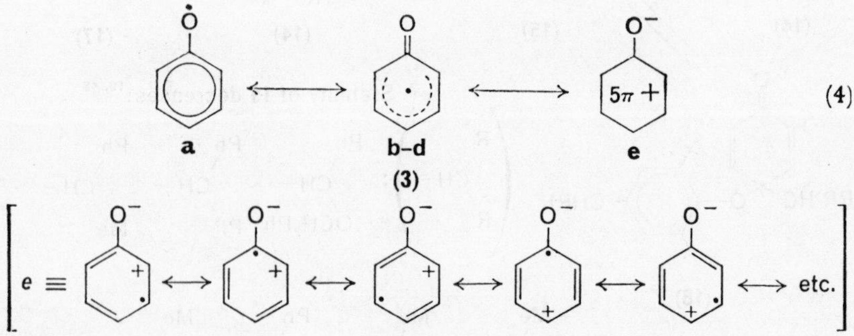

a ⟷ **b–d** ⟷ **e** (**3**) (4)

[e ≡ ⟷ ⟷ ⟷ ⟷ ⟷ etc.]

and 1640 cm^{-1} in the infrared spectrum[35, 52]. The electronic spectra of phenoxy radicals are quite different from those of the parent phenols and show several maxima in the ultraviolet and visible regions[12]. These bands and the characteristic brilliant colours of most phenoxy radicals represent further evidence for the contribution of quinonoid and dipolar resonance structures **3b–3d** and **3e** (equations 1 and 4), respectively.

Particularly useful information regarding the detailed structures of phenoxy radicals, the mechanisms of phenol oxidation which involve radical species as intermediates and the characterization of new radicals has been obtained by the application of electron spin resonance spectroscopy. The determination of the *g*-values shows that stable aryloxy radicals, such as 2,4,6-tri-*t*-butylphenoxy radicals (**4**), have a very high radical content, which is close to the value ($g = 2{\cdot}0023$) for the completely free electron[24, 53, 54]. Analysis of the hyperfine splitting constants of the e.s.r. spectra of phenoxy radicals, particularly of the stable 2,4,6-tri-*t*-butylphenoxy (**4**) [53, 55, 56], 2,4,6-tris(diphenylmethyl)phenoxy (**18**) [55, 56], various 2,6-di-*t*-butyl-4-(substituted phenyl)phenoxy (**19**)[57], and 2,4,6-triphenylphenoxy radicals (**8**)[58], including the corresponding ^{17}O- and in the ring ^{2}H- and ^{13}C-labelled radicals[55–62], indicates[12, 13, 16, 17, 19, 31, 42, 52, 53, 55–60] (a) that all six carbon atoms in the central ring as well as the oxygen show a spin density for the unpaired electron; (b) that

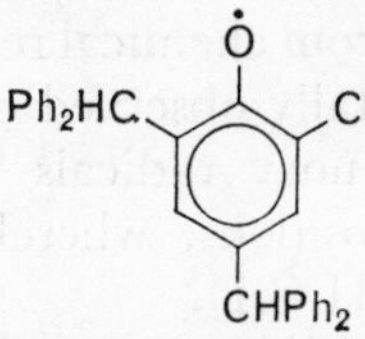

Ph$_2$HC CHPh$_2$ CHPh$_2$ **(18)**

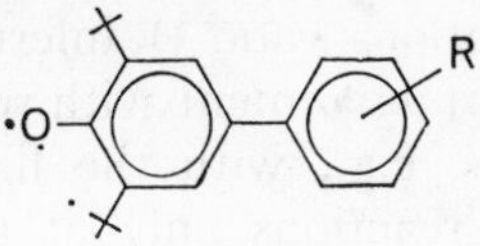

R •O **(19)**

a relatively high spin density resides on the oxygen; (c) that of the central ring carbon atoms the *para*-carbon (C-4) shows a considerable spin density which is higher than that on the *ortho*-carbons (C-2 and C-6), and that the densities at C-1 and the *meta*-carbons (C-3 and C-5) are low but not zero; (d) that the unpaired electron can also distribute itself over alkyl substituents and heteroatoms; (e) that in phenoxy radicals carrying phenyl groups in the *para*- and/or *ortho*-positions (such as **8**), the spin density of the odd electron is distributed over all phenyl substituents present in the radical, but that it is higher in the *p*-phenyl group than in the *o*-phenyl residue, probably because the latter is forced out of the molecular plane by the phenolic oxygen. Spin densities of the unpaired electron in some short-lived and stable phenoxy radicals, found experimentally or calculated from available spectral data, are given in Table 1.

TABLE 1. Spin density distribution for some aryloxy radicals.

Radical	Spin density ρ					Refs.
	O	C-1	C-2, C-6	C-3, C-5	C-4	
Phenoxy[a]	[c]	—	0·28	−0·075	0·42	7, 16, 17, 63
4-Methylphenoxy	[c]	—	0·25	−0·06	0·44	16, 17
2,4-Dimethylphenoxy	[c]	—	0·24	−0·07	0·40	16, 17
2,4,6-Trimethylphenoxy	[c]	—	0·22	−0·06	0·44	16, 17
2,4,6-Tris(diphenyl-methyl)phenoxy[b]	0·5	−0·06	0·176	−0·071	0·352	55, 56
2,4,6-Triphenylphenoxy	0·26	0·03	0·211	−0·055 (0·065)	0·218	58

[a] The recalculated values are as follows[58]: C-2 and C-6, 0·255 (0·246); C-3 and C-5, 0·070 (0·066); C-4, 0·374 (0·374).

[b] It was stated[55, 56] that a similar distribution should be generally valid for 4-R-2,6-di-*t*-butylphenoxy radicals[57, 60].

[c] Not calculated.

These and other data on the distribution of the odd electron spin density substantiate the resonance-hybrid structure of aryloxy radicals (equations 1 and 4), inferred previously from chemical reactivity, and are in agreement with some experimentally observed chemical properties, e.g., with the finding that phenoxy radicals undergo coupling reactions only at the *o*- and *p*-positions, whereby *para*-coupling usually predominates (see section II.C.1).

Although e.s.r. spectroscopy has already given valuable information about the ground state structures of phenoxy radicals and therefore offers the best approach to relative radical stabilities, further and more detailed studies are necessary in order to obtain additional and reliable knowledge on the influence of electronic and steric effects of substituents on the stability (as measured by intensity and duration of e.s.r. signals) and reactivity (as measured, for example, by rate of reaction with oxygen or of coupling dimerization) of aryloxy radicals[12].

Catechol (**20**) and hydroquinone (**22**), and their derivatives, are converted by most oxidizing agents to the corresponding *o*- (**21**) and *p*-quinones (**26**), respectively (equations 5 and 6) which, depending on reaction conditions, can either be isolated or react further, particularly with nucleophilic reactants[4, 6]. Complex polymers of the humic acid-type are often the main products. Here too, the initial step (equation 6) consists of a one-electron transfer with formation

(20) —$-2H^+, -e^-$→ (21) (5)

(22) —$-2H^+$ (A)→ (23) —$-e^-$→ (24) —$-e^-$→ (26)

(22) —$-e^-$ (B)→ (25) —$-e^-, -2H^+$→ (26) (6)

of a semiquinone radical which has usually a sufficiently long life-time—in alkaline solution (pathway A) in the form of the symmetrically resonance-stabilized anion **24** (which arises from the intermediate dianion **23**) and in acidic solution as the cation **25**—that it can be detected and characterized by titration[64], spectroscopically[65, 66] or by e.s.r. measurements[67–70]. In the case of 2,4-di-*t*-butylcatechol, the alkali-metal salts of the semiquinone radical anion have been isolated in the form of their crystalline etherates[71]. The oxidation of resorcinol (**27**) and its derivatives[5, 6], which proceeds at a slower rate than that of catechols and hydroquinones[72], cannot give rise to products of the quinone type on removal of two electrons. Hence resorcinols (**27**) behave as monophenols, whose oxidation potentials are lowered by the second hydroxyl group, and afford as primary product, upon one-electron oxidation (equation 7), an unstable *m*-benzosemiquinone, either in the anion form **29** (e.g., when the oxidation of **27** is performed with an alkaline solution of potassium ferricyanide; pathway A)[5, 6, 73] or in the cationic (**30**)

(28) (29) (27) (30) (31) (7)

or neutral form (**31**) (e.g., when the oxidation is carried out with an acid solution of ceric sulphate; pathway B)[74]. These short-lived radicals have been detected and analysed by e.s.r. spectroscopy[73], using a rapid-flow technique[16, 17].

B. Oxidation Potentials

The relative ease of oxidation of phenols to phenoxy radicals can be estimated from relative oxidation potentials corresponding to the half-wave potentials determined by polarographic methods[6, 15, 72, 74, 75]. Actually, thermodynamic redox potentials should be used for comparing oxidation rates, but these can only be measured for reversible systems, such as catechol/*o*-quinone (equation 5)[6, 72], hydroquinone/*p*-quinone* (equation 6)[6, 72] and phenol/stable phenoxy radical (equation 1) equilibria[76]. Most of the monohydric phenol (and resorcinol) oxidations are practically irreversible on account of further fast reactions of the initially formed aryloxy radicals[6, 12, 72], and therefore thermodynamic redox potentials for such systems cannot be determined experimentally†. However, since it was shown[72, 76, 81] that polarographic half-wave potentials for reversible systems are comparable, at least in their sequence, with the actual redox potentials, it is possible to use relative oxidation (e.g., half-wave) potentials of irreversible phenol oxidations, obtained by polarographic analysis, as a measure of phenol oxidizability.

* The equilibrium (reaction 6A) quinone (**26**) + hydroquinone (**22**) ⇌ semiquinone (**24**) is established very rapidly, with $k_1 = 2{\cdot}6 \times 10^8$ mole^{-1} sec^{-1} [65].

† The ease of phenol oxidation can be approximately gauged in terms of Fieser's 'critical' or 'apparent redox' potentials[77], which refer essentially to the oxidizing powers of suitable inorganic oxidants[6, 78, 79]. A rough but significant correlation exists between Hammett's σ-values and Fieser's critical potentials[80].

The values given in Table 2 [12, 15, 74] show on general lines that

TABLE 2. Relative oxidation potentials of phenols (E in millivolts).

Phenol	$E_{\frac{1}{2}}$	Ref.	Phenol	$E_{\frac{1}{2}}$	Ref.
Phenol	+1004	82[a]	4-Phenyl-2,6-dicyano	>+800	
4-Methyl	543		4,6-Diphenyl-2-cyano	593	
3-Methyl	607		2,6-Diphenyl-4-cyano	549	
2-Methyl	556		Pentaphenyl	366	
4-Methoxy	406		2,4,6-Triphenyl-3-cyano	433	
3-Methoxy	619		3-Chloro-2,4,6-triphenyl	347	
2-Methoxy	456		2,3,4,6-Tetraphenyl	238	
4-Nitro	924		2,4,6-Triphenyl	211	
3-Nitro	855	83[b]	4-Fluoro-2,6-diphenyl	164	75[d]
2-Nitro	846		2,4,6-Tri-*p*-phenoxyphenyl	179	
4-Chloro	653		2,4,6-Tri-*p*-methoxyphenyl	124	
3-Chloro	734		4-*t*-Butyl-2,6-diphenyl	120	
2-Chloro	625		6-*t*-Butyl-2,4-diphenyl	112	
4-*t*-Butyl	578		4,6-Di-*t*-butyl-2-phenyl	76	
2-*t*-Butyl	552		2,6-Di-*t*-butyl-4-phenyl	−14	
4-Phenyl	534		2,4,6-Tri-*t*-butyl	−59	
4-Phenyl-2-chloro	560		1-Naphthol	+740	
2,4-Dichloro	660	84[c]	2-Naphthol	820	82[a]
4-Carboxy-2-chloro	920		2-Hydroxybiphenyl	970	
			4-Hydroxybiphenyl	890	

Dihydroxybenzene	E_0[e] (redox)[72]	$E_{\frac{1}{2}}$[f] (pH 5·6)[85]	$E_{\frac{1}{2}}$[g] (pH 0)[72]
Hydroquinone	699	234	560
Methylhydroquinone	644	—	505
2,6-Dimethylhydroquinone	593	—	454
Tetramethylhydroquinone	463	—	324
2,5-Di-*t*-butylhydroquinone	522	—	383
Catechol	792	349	600
Hydroxyhydroquinone	600	—	460
Resorcinol	—[h]	613	800
5-Methylresorcinol (Orcinol)	—[h]	—	750
2,5-Dimethylresorcinol	—[h]	—	700

[a] Pt anode, Ag/0·1N Ag^+ ref. electrode, solvent system: acetonitrile + perchlorate salt.
[b] Graphite anode, saturated calomel ref. electrode, aqueous buffered solution, pH 5·6.
[c] Pt anode, saturated calomel ref. electrode, solvent system: 0·14M aqueous LiCl.
[d] Graphite anode, Ag/AgCl ref. electrode, solvent system: acetonitrile–H_2O + $(CH_3)_4NOH$.
[e] Hydrogen ref. electrode.
[f] Graphite anode, Ag/AgCl ref. electrode, aqueous buffered solution.
[g] Saturated calomel ref. electrode.
[h] Irreversible system.

the oxidation potential of phenols, i.e., the oxidizing power of phenoxy radicals, decreases and therefore the ease of oxidation of phenols to the corresponding radicals increases with increasing steric crowding in the 2- and 6-positions, and with decreasing electron-withdrawing, i.e., with increasing electron-releasing properties of substituents in these and the 4-position. For example, in the case of the 2,4,6-triphenylphenoxy radical (**8**) and the 2,4,6-tri-*t*-butylphenoxy radical (**4**), the equilibrium (equation 8) is shifted to the right, since **8** is a stronger oxidizing agent than **4** (Table 2). The influence of steric and polar effects of 3- and 5-substituents on the relative

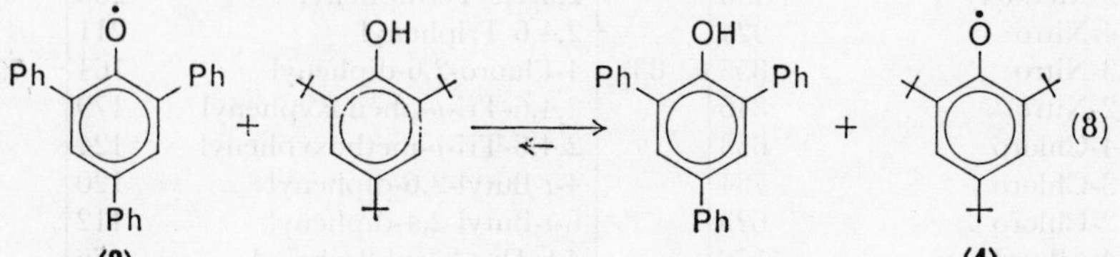

(**8**) + (**4**) ⇌ (8)

oxidation potentials of phenols is usually less pronounced. Since oxidation potentials refer to the oxidizing power of phenoxy radicals, they do reflect to a certain extent the relative stabilities and reactivities of aryloxy radicals, particularly of those which carry substituent groups in the positions 2, 4 and/or 6 [12]. Therefore, attempts have been made to correlate redox potentials (for reversible systems) and relative oxidation potentials (for irreversible systems) with the effectiveness of phenols (i.e., phenoxy radicals) as autoxidation inhibitors[9, 74, 86].

Phenoxy radicals, formed in the first step of the oxidation of phenols by one-electron transfer oxidants, may undergo a variety of reactions, depending on the reactivity and substitution pattern of the radicals, on experimental conditions, on the amount of oxidizing agent and on the presence of other compounds in the reaction mixture[12, 13]. The most important reactions of aryloxy radicals, which are related to the oxidation of the parent phenols, will be discussed in the following sections.

C. Oxidation by Ferric Salts

1. Potassium ferricyanide

One of the most widely used oxidizing agents for the generation of phenoxy radicals from phenols is potassium ferricyanide (KFC), $K_3[Fe(CN)_6]$, in alkaline solution[2, 6, 8]. Phenol oxidation with this reagent appears to be a process of considerable complexity, the mechanistic details of which have not yet been fully elucidated. Kinetic studies and the requirement of a large excess of oxidizing agent to approach quantitative yields of aryloxy radicals are in agreement with a reversible one-electron transfer reaction (9) involving phenoxide anions as oxidizable substrate, the rate of oxidation being dependent on the basicity of the solution and on the ferricyanide/ferrocyanide ratio[78].

$$ArO^- + [Fe(CN)_6]^{3-} \rightleftharpoons ArO^{\cdot} + [Fe(CN)_6]^{4-} \tag{9}$$

Most of the stable aryloxy radicals mentioned in the preceding section were obtained by the ferricyanide oxidation of the corresponding phenols, e.g., **4** [22–25, 55], **5** [27, 29, 30], **6** [33], **7** [34–39], **18** [55], **19** [57] and others [18, 51, 52, 59–62, 87]. Unstable and moderately stable phenoxy radicals, i.e., radicals with free *ortho*- and/or *para*-positions and 2,4,6-trisubstituted phenoxy radicals containing α-hydrogens in the 4-substituent, can also be generated by the same method, but once formed, and in the absence of other reactive molecules, they usually undergo radical dimerization (and further polymerization) by way of self-coupling to furnish dimeric and trimeric products, and often ill-defined polymeric materials as well.

Phenoxy radical dimerizations, some of which are reversible, can be divided into six classes.

a. Carbon–carbon dimerization involving nuclear C–C coupling can take place at *ortho–ortho*, *ortho–para* or *para–para* positions, and it has been reported for a large number of phenols[2]. Thus, when *p*-cresol and other 4-alkylphenols (**32**) are oxidized with potassium ferricyanide in aqueous alkaline solution (equation 10), in addition to polymeric products, the *ortho*-linked dimer **34** and trimer **35**, and Pummerer's ketone (**37**) are obtained[78, 88–91]. The correct structure of Pummerer's ketone (**37**) (R = CH_3) was established by Barton[89], who also explained its formation (reaction 10), on the basis of initial *o,p'*-coupling of two 4-methylphenoxy radicals (**33**), followed by intramolecular β-addition of the *o*-hydroxyl group on to the enone system of the resulting cyclohexadienone(**36**). This reaction scheme was confirmed by the elegant synthesis of usnic acid, involving potassium ferricyanide oxidation of methylphloroacetophenone[89]. The dual course of the oxidation (equation 10) of *p*-cresol (**32**) lends further support to the resonance-hybrid structure of phenoxy radicals (**33**).

If in the starting phenol (**38**) the *p*-position or two of the three *o*- and *p*-positions are occupied by substituents, *ortho–ortho* and *para–para* C–C bonded dimeric phenols, e.g. **41**, are usually produced

$$\text{(32)} \xrightarrow[\text{HO}^-]{\text{KFC}} \text{(33)}\ \mathbf{a} \longleftrightarrow \mathbf{b} \longleftrightarrow \mathbf{c} \longleftrightarrow \mathbf{d}$$

(34), (35), (37), (36) (10)

(equation 11) in high yields[2, 78, 91–102], although in the case of 2,6-di-*t*-butylphenol (**38**, R = R′ = Me_3C) the major product isolated was the keto-tautomer **40** (R = R′ = Me_3C)[92, 93]. Depending on the nature and basicity of the reaction medium, on the

$$2\ \text{(38)} \xrightarrow{\text{KFC}} 2\ \text{(39)} \longrightarrow \text{(40)} \longrightarrow \text{(41)} \xrightarrow{\text{KFC}} \text{(42)} \qquad (11)$$

amount of the oxidizing agent and on its oxidation potential (relative to that of the bisphenol **41**), oxidation can proceed further to furnish the corresponding diphenoquinones, such as **42** [2, 78, 92, 93, 97, 102–105]. The 4,4′-diphenoquinones (**42**) are stable compounds and, according to n.m.r. measurements, may show *cis–trans* isomerism (when substituents R and R′ are different)[97], whereas the formation of 2,2′-diphenoquinones (such as **43** and **44**) by further oxidation of the *o–o′* coupled bisphenols has been rarely observed[101, 106], probably because most of these products are relatively unstable (e.g., **44**), and more favourable routes are offered for the stabilization of their radical precursors (equation 11).

3,5-Di-*t*-butyl-4-hydroxybenzoic acid (**45**) undergoes oxidative decarboxylation when treated with alkaline ferricyanide in the absence of oxygen (equation 12) and affords 3,3′,5,5′-tetra-*t*-butyl-4,4′-diphenoquinone (**46**) in nearly quantitative yield[103]. The same

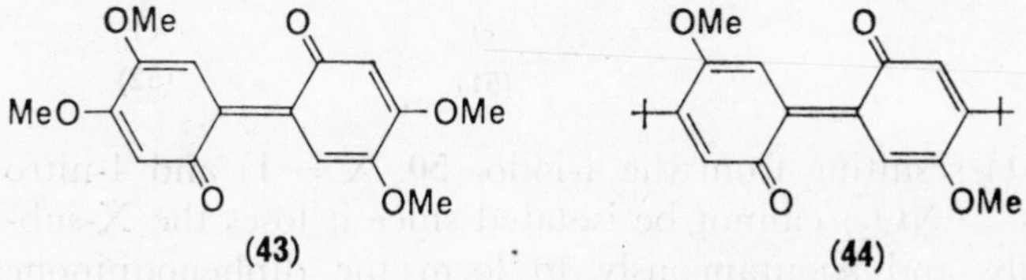

(43) (44)

product is obtained in high yield by oxidation of the benzaldehyde **47** [103], and the benzyl alcohols **48** (R = H or C_6H_5)[19]; reaction schemes have been suggested involving intermediate anions and radicals[12, 103] or only radicals[19].

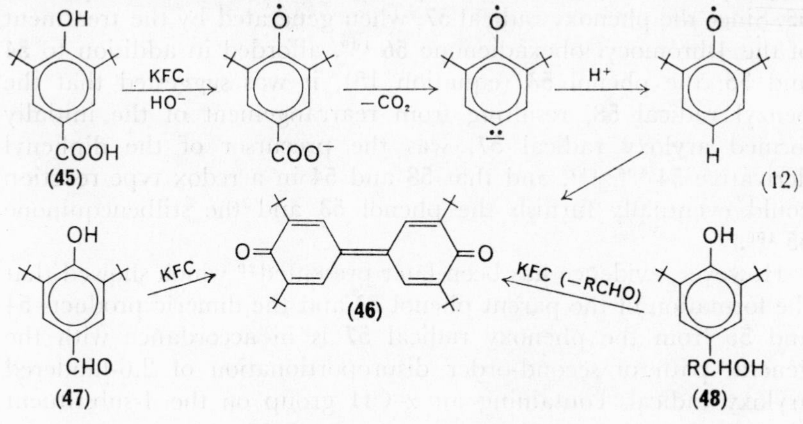

(12)

(45) (46) (47) (48)

When oxidized with alkaline ferricyanide, the 2,6-di-*t*-butylphenols(**49**) carrying in position 4 a chlorine (X = Cl) or bromine atom (X = Br) are rapidly converted (equation 13), through their phenoxy radicals **50**, to the *p–p'* C–C coupled dimers **51**, which readily oxidize further to the corresponding diphenoquinones **52** [107, 108]. For the chloro compound (**51**, X = Cl) this conversion (to **52**) requires silver or mercury, whereas for the bromo compound (**51**, X = Br) bromine is already slowly lost at room temperature.

$$2\,(\mathbf{49}) \xrightarrow{\text{KFC}} 2\,(\mathbf{50}) \longrightarrow (\mathbf{51}) \xrightarrow[\text{(Ag or Hg)}]{-2X^{\cdot}} (\mathbf{52}) \qquad (13)$$

X = Cl, Br, I, NO_2

The dimer (**51**) resulting from the 4-iodo (**50**, X = I) and 4-nitro radical (**50**, X = NO_2) cannot be isolated since it loses the X-substituent rapidly and spontaneously to form the diphenoquinone **52** [107, 108].

b. Carbon-carbon dimerization involving coupling of substituent α-carbon atoms has been particularly studied in the case of 2,6-di-*t*-butyl-4-methylphenol (**53**). Upon oxidation with alkaline ferricyanide[109] and other oxidizing agents[110–116], **53** affords (equation 14) two major products, the dimeric bisphenol **54** and the stilbenequinone **55**. Since the phenoxy radical **57**, when generated by the treatment of the 4-bromocyclohexadienone **56** [109], afforded in addition to **54** and **55**, the phenol **53** (equation 15), it was suggested that the benzyl radical **58**, resulting from rearrangement of the initially formed aryloxy radical **57**, was the precursor of the diphenyl derivative **54** [109–111], and that **58** and **54** in a redox type reaction could eventually furnish the phenol **53** and the stilbenequinone **55** [109].

However, evidence has been later presented[20] which showed that the formation of the parent phenol **53** and the dimeric products **54** and **55** from the phenoxy radical **57** is in accordance with the general path of second-order disproportionation of 2,6-hindered aryloxy radicals containing an α–CH group on the 4-substituent

2 (53) $\xrightarrow{K_3Fe(CN)_6}$ (54) + (55) (14)

2 (56) $\xrightarrow{Hg}$ 2 [(57)] $\longrightarrow$ 2 (58) $\longrightarrow$ 54 (15)

54 + 4 (58) $\longrightarrow$ 4 (53) + 55

8 [(57)] $\longrightarrow$ 4 (53) + 4 (59) (16)

4 (59) $\longrightarrow$ 2 [(60)] $\longrightarrow$ 54 + 55

(**15**→**14** + **17**, equation 3)[18–21, 46–50, 108, 117] and that therefore the unstable quinone methide **59** (reaction 16), rather than the benzyl radical **58**, is an intermediate in the oxidative coupling through 4-methyl groups of 2,6-di-*t*-butyl-4-methylphenol (**53**) (reaction 14)[20*]. The formation of **54** and **55** from the intermediate quinone methide **59** (equation 16) involves free radical species of uncertain structure[20, 48, 49], one of which might well be the biradical **60** [49].

c. Carbon–carbon dimerization involving coupling of substituent β-carbon atoms has been observed upon alkaline ferricyanide oxidation (equation 17) of 4-hydroxyphenylethylene derivatives (**61**) related to coniferyl alcohol[118]. Depending on the nature of the R group, the dimeric products were either the bisquinone methides **62** (for R = COOR′) or the tetra-*t*-butyl homologues of isoxanthocillin (**63**) (for R = CN and CHO), which could be further oxidized to the ethylenic bisquinone methides **64**.

2 (**61**) —KFC→ 2 [•CH radical] → (**62**) —O→ (17)

(**63**) —$-H^{\bullet}$→ —$-H^{\bullet}$→ (**64**)

R = COOR′, CN, CHO

d. Carbon–oxygen dimerization involving nuclear o- *or* p-*carbon–phenoxyl oxygen coupling* (equation 18) is often observed when 2,4,6-trisubstituted and other hindered phenols (**65**) are oxidized by potassium ferricyanide to the corresponding phenoxy radicals **66**, and these radicals then form by intermolecular reaction simple dimeric quinol

* Spectral and other data reported in connection with the mechanism of reaction (14) have been discussed in a previous review[12].

ethers, such as **67**, which are in equilibrium with the aryloxy radicals **66** (e.g., **7**⇌**8** in reaction (2), and **16**⇌**15** in reaction (3); section II.A.1)[6, 8, 12, 13, 21, 34–38, 102, 105, 119–121]. When the 4-position is free (R'' = H), the quinol ethers **67** of certain 2-mono- and 2,6-disubstituted phenols rearrange rapidly by 4-hydrogen shift to the corresponding hydroxy-ethers **68**, which undergo further coupling reactions to polyphenylene ethers **69** (for the mechanism of this polymerization see section II.G). Thus, upon alkaline ferricyanide oxidation of guaiacol (**65**, R = OCH_3, R′ = R″ = H) and 4-(2′,6′-dimethylphenoxy)-2,6-dimethylphenol (**68**, R = R′ = CH_3, R″ = H) the major products obtained were polymeric ethers of type **69** [122, 122a, 123].

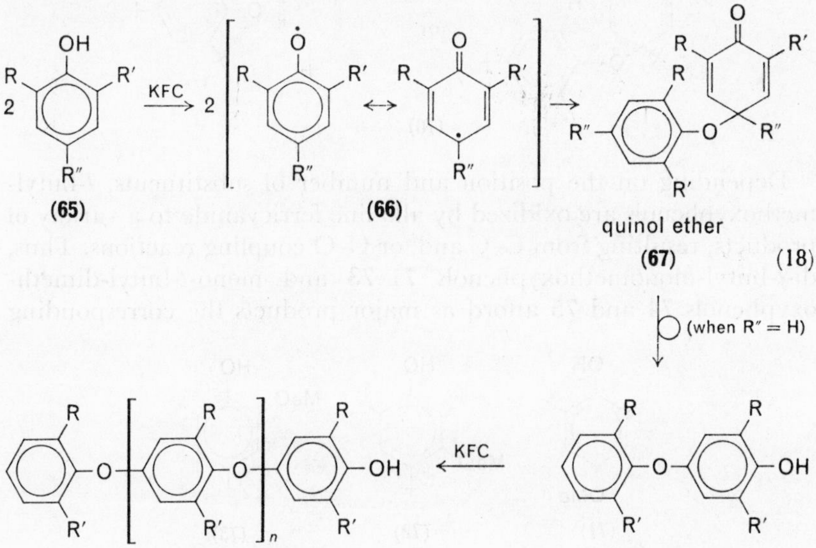

(65) —KFC→ **(66)** → quinol ether **(67)** (18)

↓ (when R″ = H)

polyphenylene ether **(69)** ←KFC— *p*-phenoxyphenol **(68)**

Hindered phenoxy radicals, such as the 2,4,6-tri-*t*-butylphenoxy radical (**4**), are quite stable in solutions of hydrocarbon solvents, but upon longer standing a slow decrease of radical content is observed, even when air and moisture are rigorously excluded[25]. This fact and the variation in yield of phenoxy radicals when prepared by oxidation of phenols in different solvents suggest that radical decomposition due to disproportionation may involve the solvent. A study of the rates of reaction of 2,4,6-tri-*t*-butylphenoxy radicals (4) with various hydrocarbons (*n*-decane, isodecane, toluene, ethylbenzene and cumene) over a temperature range of 70–150° has shown that the reaction was first-order in phenoxy radical, and that the rate was fastest in ethylbenzene and slowest in isodecane[124]. Moreover, it was found that 2,4,6-tri-*t*-butylphenoxy radicals (4) decompose upon heating to give the parent phenol, isobutylene and two higher molecular weight products[87, 125]; the C–O coupled dimeric structure **70**, corresponding to a quinol ether which has lost an *ortho*-*t*-butyl group, was tentatively assigned to one of these products[87, 125].

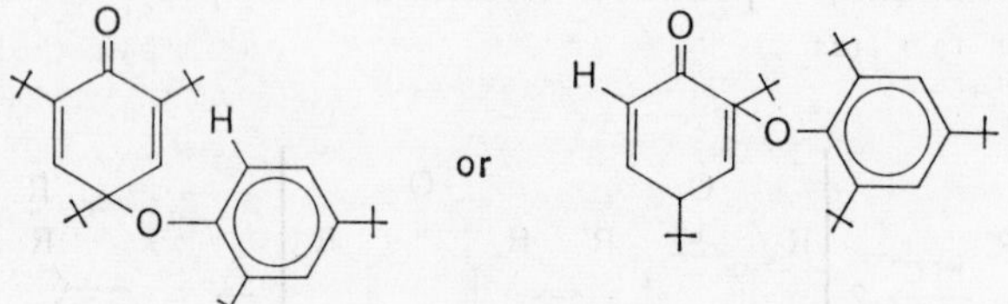

(70)

Depending on the position and number of substituents, *t*-butylmethoxyphenols are oxidized by alkaline ferricyanide to a variety of products, resulting from C–C and/or C–O coupling reactions. Thus, di-*t*-butyl-monomethoxyphenols **71**–**73** and mono-*t*-butyl-dimethoxyphenols **74** and **75** afford as major products the corresponding

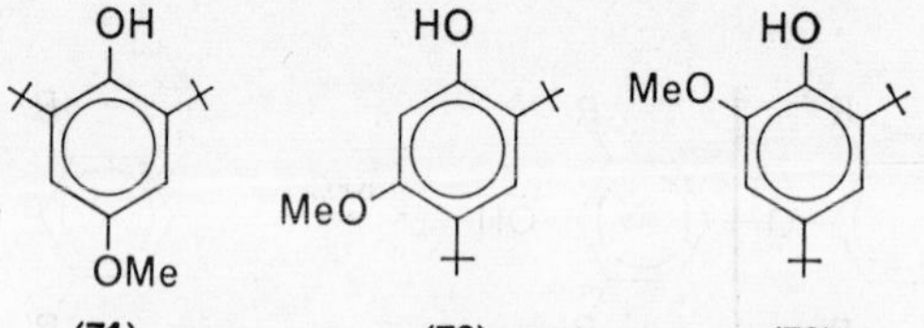

(71) **(72)** **(73)**

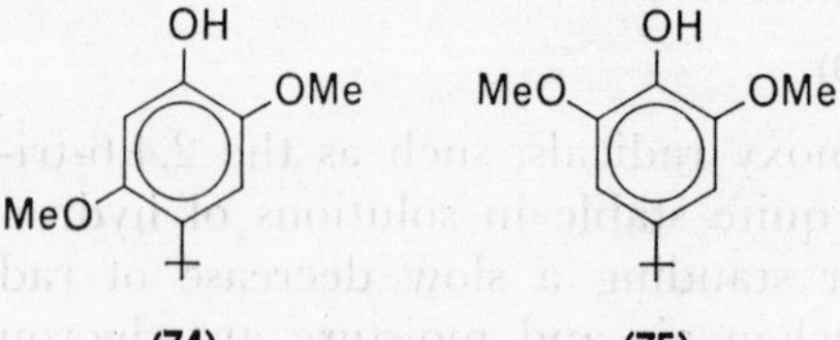

(74) **(75)**

dimeric simple quinol ethers (of type **67**, equation 18) and, when an *o*-position is free (as in **72**), 2,2′-dihydroxy-diphenyl derivatives[102, 105, 126, 127]. On the other hand, oxidation of 2,5-di-*t*-butyl-

4-methoxyphenol (**76**) in methanol or light petroleum with alkaline potassium ferricyanide yields a C–O coupled trimer **81**, according to reaction scheme (19)[127, 128]. The same oxidation of **76** in benzene

(76) KFC → [(77)] → (78) → (79) → (80) +77 → (81) (19)

occurs with demethylation and affords only 2,5-di-*t*-butyl-1,4-benzoquinone[127].

Less hindered dimethoxyphenols **82–84** and mono-*t*-butyl-mono-methoxyphenols **85** and **86** are oxidized to the corresponding dimeric dihydroxydiphenyls or diphenoquinones[78, 97, 101, 102], whereas the mono-*t*-butyl-monomethoxyphenols **87–89** furnish only polymeric materials[97, 102, 105]. The ferricyanide oxidation of 3-alkyl-4-methoxyphenols (**90**) and 4-*t*-butyl-2-methoxyphenol (**91**) proceeds, via the corresponding 2,2′-dihydroxydiphenyls (of type **94**), to trimeric spiroketals (of type **95**) as one of the reaction products (equation

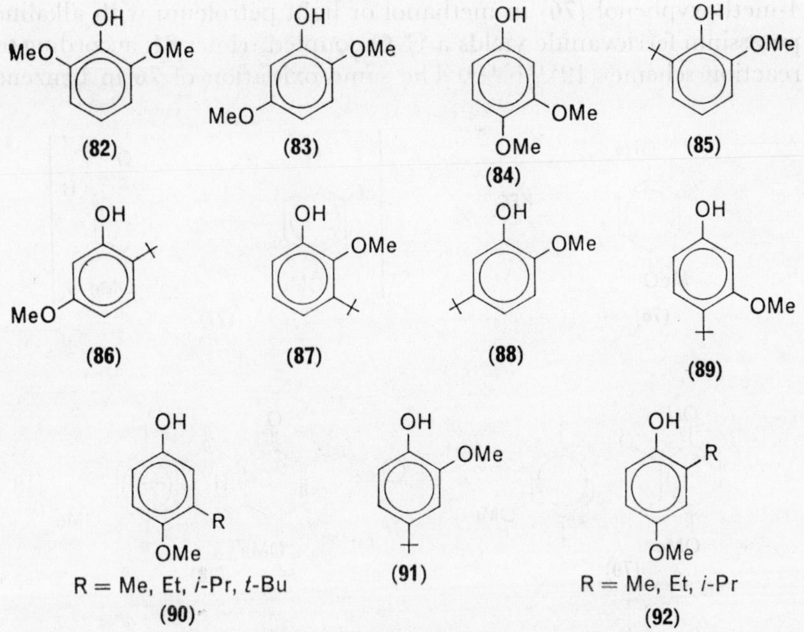

(82) (83) (84) (85)

(86) (87) (88) (89)

R = Me, Et, *i*-Pr, *t*-Bu (90) (91) R = Me, Et, *i*-Pr (92)

20)[91, 96–99, 129]. Whereas 4-methoxyphenols with relatively small alkyl substituents in position 2 (**92**) may be also oxidized to spiroketals (of type **95**)[99], 2-*t*-butyl-4-methoxyphenol and 2,4-di-*t*-butyl-

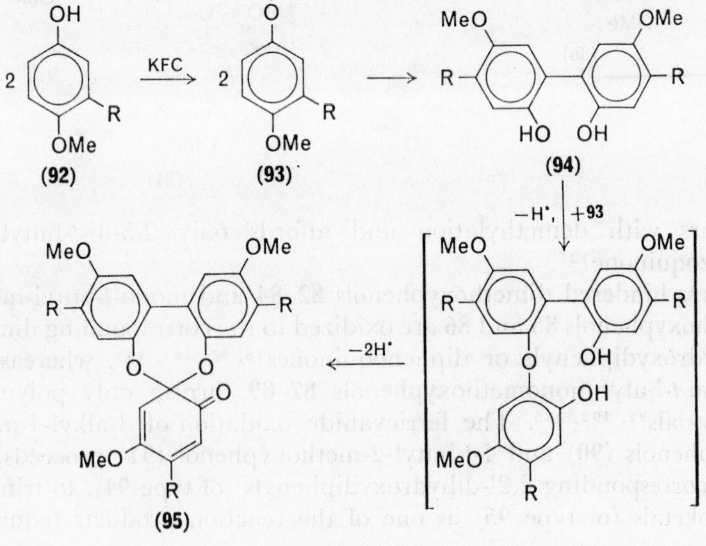

(92) —KFC→ 2 (93) → (94) —−H•, +93→ [ ] —−2H•→ (95) (20)

phenol (**96**) are converted (equation 21), via the corresponding 2,2′-bisphenols **97** (which have been isolated), to dimeric internal quinol ethers containing a four-membered oxetane ring

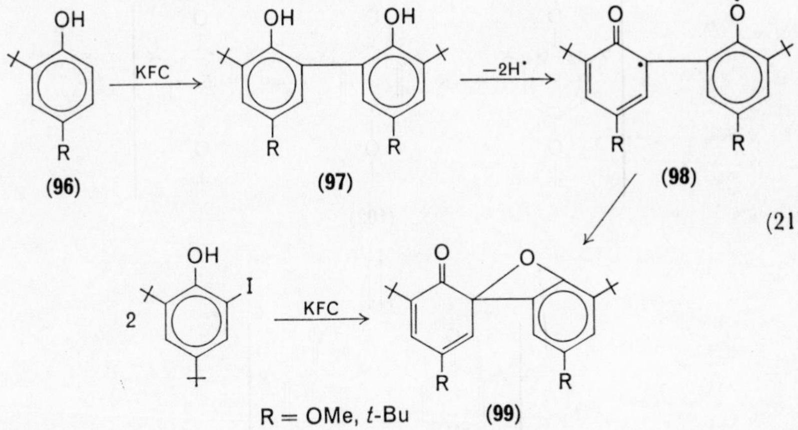

(21)

R = OMe, *t*-Bu

(**99**)[94, 95, 100, 130]. Similar intramolecular quinol ethers (**101**), but with larger heterocyclic rings, are formed (equation 22) upon oxidation of substituted bisphenolmethanes (**100**, $n = 0$) and 1,2-bisphenolethanes (**100**, $n = 1$)[94, 131], as well as of analogous β-naphthol derivatives[119, 132, 133].

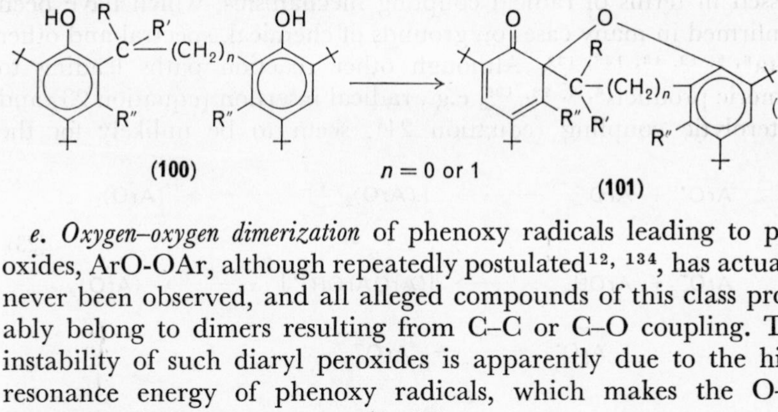

(22)

$n = 0$ or 1

e. Oxygen–oxygen dimerization of phenoxy radicals leading to peroxides, ArO-OAr, although repeatedly postulated[12, 134], has actually never been observed, and all alleged compounds of this class probably belong to dimers resulting from C–C or C–O coupling. The instability of such diaryl peroxides is apparently due to the high resonance energy of phenoxy radicals, which makes the O–O bond dissociation energy very low[135].

f. Charge-transfer complexes have been proposed for certain phenoxy radical dimers, on the basis of chemical, steric or spectral evidence[31, 60, 126, 136, 137]. For example, for the diamagnetic, colourless solid dimer of 2,6-di-*t*-butyl-4-*t*-butoxyphenoxy radicals (**102**), which even in nonpolar solvents dissociates nearly quantitatively to the highly coloured paramagnetic monomeric radical species, structure **103**, involving complete electron transfer, was suggested.

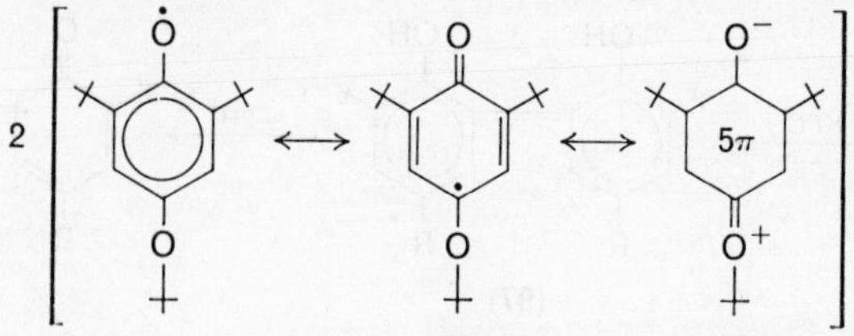

(102)

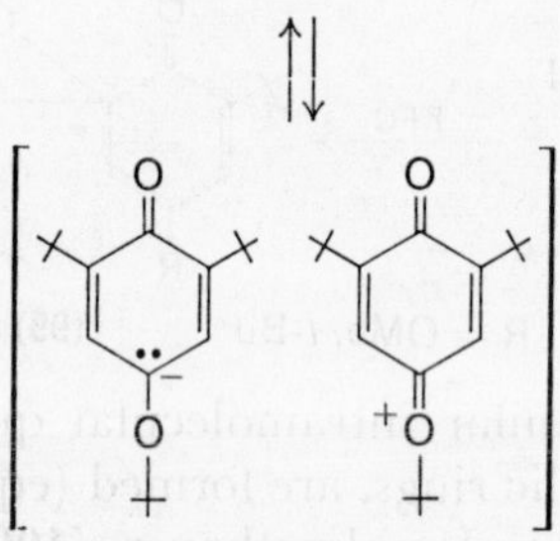

(103)

All the above described oxidative dimerizations have been discussed in terms of radical coupling mechanisms, which have been confirmed in many cases on grounds of chemical, spectral and other data[6, 8, 12, 13, 138, 139]. Although other reaction paths leading to dimeric products[6, 8, 78, 139], e.g., radical insertion (equation 23) and heterolytic coupling (equation 24), seem to be unlikely for the

$$ArO^{\bullet} + ArO^{-} \longrightarrow [(ArO)_2^{-\bullet}] \xrightarrow{-e^-} (ArO)_2$$

or

$$ArO^{\bullet} + ArOH \longrightarrow [(ArOArOH)^{\bullet}] \xrightarrow{-H^{\bullet}} (ArO)_2 \qquad (23)$$

$$ArO^{\bullet} \xrightarrow{-e^-} ArO^{+}$$

$$ArO^{+} + ArO^{-} \longrightarrow (ArO)_2 \qquad (24)$$

or

$$ArO^{+} + ArOH \longrightarrow [(ArOArOH)^{+}] \xrightarrow{-H^{+}} (ArO)_2$$

majority of phenol oxidations performed under usual conditions[6, 8, 138, 139], it is not impossible that these and similar mechanisms may operate in special circumstances and in different structural conditions[7, 21, 139–141].

Controlled oxidation of catechols (**20**, equation 5) and hydroquinones (**22**, equation 6) with alkaline ferricyanide affords the corresponding benzoquinones (**21** and **26**, respectively)[6], which can then be converted, by nucleophilic attack of hydroxide ions, to hydroxybenzoquinones, e.g., **104** (equation 25)[69].

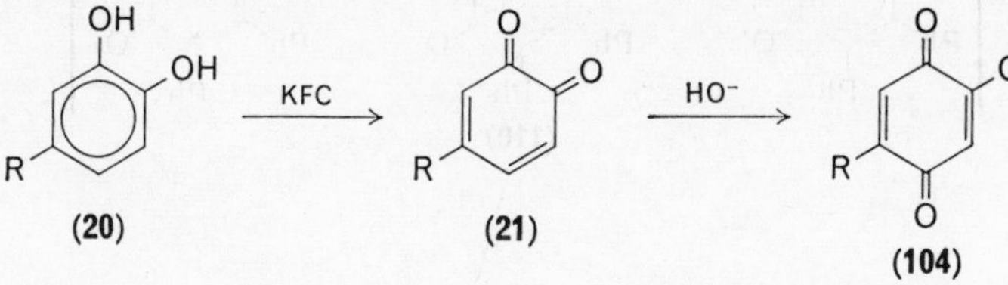

(20) —KFC→ (21) —HO⁻→ (104) (25)

The oxidation of orcinol (**105**) with alkaline ferricyanide proceeds as with monophenols (equation 26) and produces, via the intermediate radicals **106**, a mixture of C–C and C–O coupled dimeric (**107**) and polymeric products (**108**)[5, 6, 142].

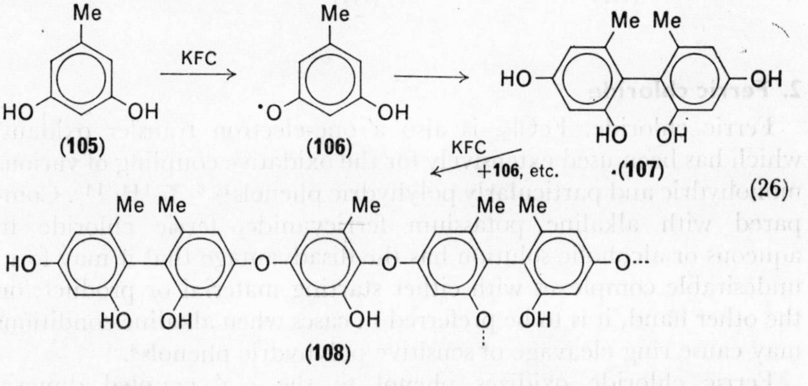

(105) —KFC→ (106) → (107) —KFC, +106, etc.→ (108) (26)

When treated with alkaline ferricyanide, 2,6-di-, 2,4,6-tri- and tetraphenylresorcinol, e.g., **109**, undergo ring contraction with evolution of carbon monoxide (equation 27) and afford phenyl substituted cyclopentadienones, such as **112**[143]. The intermediate formation of resonance-stabilized biradicals **110** and bicyclic diketones **111** has been proposed to explain the course of this reaction[143].

(109) —HO⁻→ ; (109) —KFC→ [↔ ↔] (110) (27) → (111) —−CO→ (112)

2. Ferric chloride

Ferric chloride, $FeCl_3$, is also a one-electron transfer oxidant, which has been used extensively for the oxidative coupling of various monohydric and particularly polyhydric phenols[1, 6, 8, 144, 145]. Compared with alkaline potassium ferricyanide, ferric chloride in aqueous or alcoholic solution has the disadvantage that it may form undesirable complexes with either starting material or product; on the other hand, it is to be preferred in cases when alkaline conditions may cause ring cleavage of sensitive polyhydric phenols[8].

Ferric chloride oxidizes phenol to the *o–o′* coupled dimeric 2,2′-dihydroxydiphenyl[8], whereas *p*-cresol (**32**, R = CH_3) is converted to the same products (**34**, **35** and **37**; R = CH_3) obtained when alkaline ferricyanide is used as oxidant (equation 10)[146].

1-Naphthol (**113**) is oxidized by ferric chloride (equation 28) to all three possible *ortho* and *para* C–C coupled dimers **114**, **115** and **116** [147]. 2-Naphthol (**117**, R = H) and 3-methoxy-2-naphthol (**117**, R = OCH_3) when treated with ferric chloride in acidic or neutral solution afford (equation 29) the corresponding 1,1′-bisnaphthols (**119**) in good yield[105, 133, 148]. Oxidation with alkaline potassium ferricyanide, on the other hand, produces mostly polymeric products[105, 149, 150], from which, in the case of 2-naphthol itself (**117**,

$$\text{(113)} \xrightarrow{FeCl_3} \text{(114)} + \text{(115)} + \text{(116)} \qquad (28)$$

R = H), a small amount of the hydroxynaphthyl ether **118** could be isolated[149, 150]. The dimeric dixydroxy compound **119** can be oxidized further by potassium ferricyanide[151], silver oxide[152] or aryloxy radicals[133] to furnish **121, 123** and other intermolecularly coupled products derived from radicals **120** and **122** [133]. The existence of these highly coloured aryloxy radicals (**120** and **122**) has been verified by e.s.r. measurements[133, 153].

Ferric chloride oxidation of 4-methoxyphenols usually proceeds with demethylation and formation of *p*-quinones[100, 106, 127, 154], as illustrated by equation (30)[106]. Except in some special cases[127], the oxidation of methoxyphenols with alkaline ferricyanide does not involve the loss of a methyl group from methoxy substituents (e.g., see reactions 19–21).

The property of ferric compounds (potassium ferricyanide and ferric chloride) to effect oxidative coupling of phenols has been successfully applied for the syntheses of various natural products[1, 6, 8, 139, 144]. For example, oxidation of quaternary laudanosoline methiodide (**128**) with ferric chloride gives a 62% yield (equation 31) of the glaucine **129** from the aporphine series[144, 155].

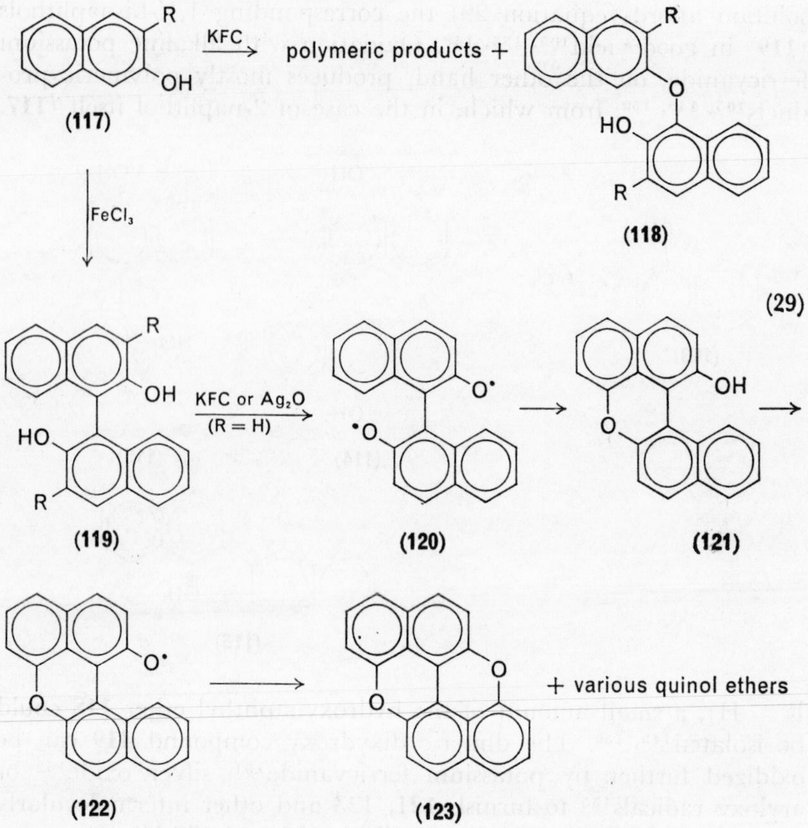

(117) —KFC→ polymeric products + (118)

(117) —$FeCl_3$→ (119)

(119) —KFC or Ag_2O (R = H)→ (120) → (121) →

(122) → (123) + various quinol ethers

(29)

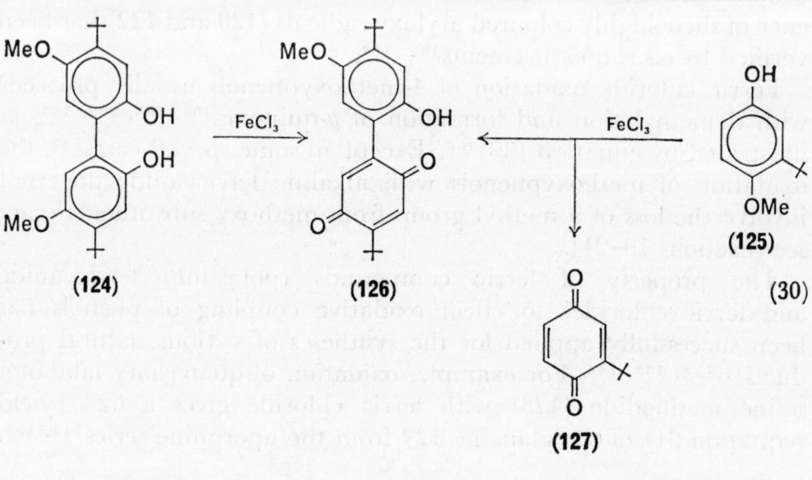

(124) —$FeCl_3$→ (126) ←$FeCl_3$— (125) —$FeCl_3$→ (127)

(30)

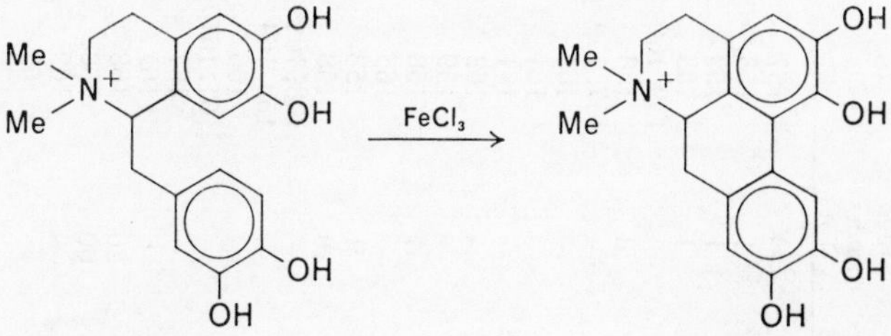

(128) $\xrightarrow{FeCl_3}$ (129) (31)

D. Oxidation by Tetravalent Lead

1. Lead tetraacetate

The extensive work on phenol and naphthol oxidations with lead tetraacetate $Pb(OAc)_4$, has been recently reviewed[156]. Depending on the nature, number and position of the substituents in the phenol, on the properties of the solvent, and on the ratio of oxidizing agent to substrate and dilution, different types of products may be obtained, e.g., **130–137** [38, 100, 105, 156–182]. Table 3 summarizes some of the results.

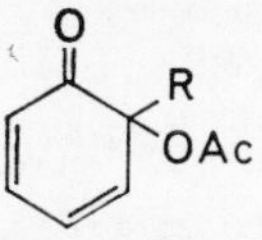

(130) 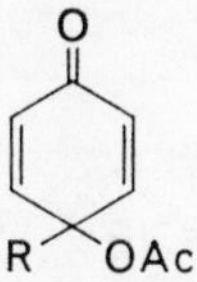 (131) quinol acetates

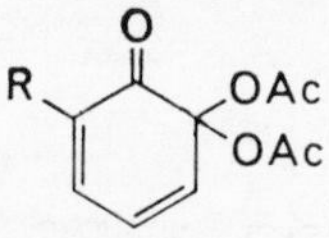

(132) *o*-quinone diacetate

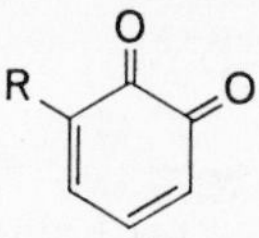

(133) 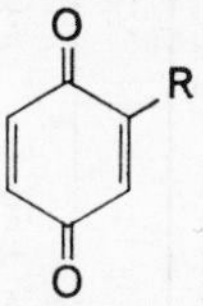 (134) quinones

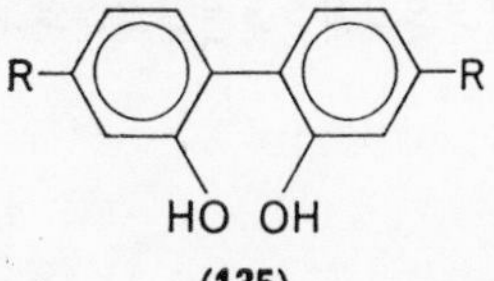

(135) 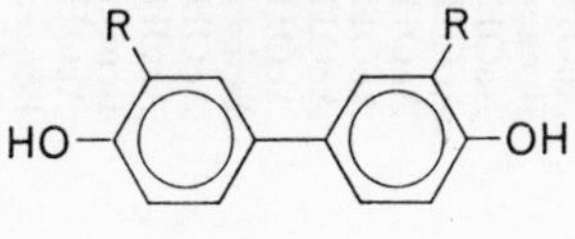 (136) (137) and/or diphenoquinones

ortho- and *para*-coupled dimeric products

The effect of the solvent on the distribution of products is considerable, acetic acid favouring the formation of quinol acetates (**130–132**) and nonpolar solvents such as benzene enhancing C–C coupling reactions which lead to the dimeric products **135–137**, particularly when no excess of lead tetraacetate is used[158, 182]. In

TABLE 3. Oxidation of phenols by lead tetraacetate.

Phenol	Solvent	Yields of products (in % of theoretical)					Refs.
		o-Quinol acetate	*o*-Quinone diacetate	*p*-Quinol acetate	Quinone	*o*- and *p*-coupled dimers	
Phenol	AcOH	—	4	—	—	—	157
	Benzene	—	—	—	—	1	158
2-Methyl	Ether	22	3	—	4	1	159
	Benzene	—	—	—	—	1	158
4-Methyl	AcOH	—	57	14	—	—	159, 160
	Benzene	—	—	—	—	8	158
3-Methyl	AcOH	—	0, 5	—	4	—	157, 159
2-Ethyl	AcOH	36	4	—	—	—	161
2-*n*-Propyl	AcOH	42	4	—	—	—	161
2-*t*-Butyl	AcOH	2	44	—	—	—	161
4-*t*-Butyl	AcOH	—	14	5	—	—	162
2,4-Dimethyl	AcOH	34	—	—	—	—	159
	Benzene	—	—	8	—	7	158
2,3-Dimethyl	AcOH	33	5	—	1	—	163
2,6-Dimethyl	AcOH	95	—	—	—	1	158
	Benzene	—	—	—	—	20	158
2,4,6-Trimethyl	AcOH	65	—	—	—	—	158, 164
	Benzene	71	—	—	—	—	158, 164
	Chloroform	92	—	—	—	+	166
2,4,6-Tri-*t*-butyl	AcOH	60	—	30	—	—	161, 164
2-Methoxy	AcOEt	—	—	+	57	—	167
2,6-Dimethoxy	AcOEt	—	—	—	78	—	167
2-Methyl-6-formyl	AcOH	5	—	—	—	20	168
2-Methyl-6-cyano	AcOH	—	—	—	—	100	168
2,4-Dimethyl-6-cyano	AcOH	23	—	—	—	—	169
2,4-Dimethyl-6-acetyl	Chloroform	65	—	—	—	—	169
2,4-Dimethyl-6-nitro	AcOEt	18	—	—	—	—	170
2-Methyl-5-bromo	Acetone	38	—	—	—	—	171
2,4-Dimethyl-5-bromo	AcOH	32	—	6	—	—	171
2,6-Dimethoxy-4-carboxy	AcOEt	—	—	—	57	—	167
2,4-Dimethyl-6-carbethoxy	AcOH	50	—	—	—	—	169

acetic acid as solvent *ortho*-substituted phenols usually give *o*-quinol acetates (**130**) as main products, whereas *p*-quinol acetates (**131**) rom phenols containing a 4-alkyl group are formed in lower yield. When one *ortho*-position is free *o*-quinol diacetates (**132**) may be obt... ned; these compounds are unstable but have been isolated in several cases[159–162]. If an electron-withdrawing group is present in one of the *o*-positions, acetoxylation takes place at the *o*-side with higher electron density, i.e. carrying an alkyl group or a hydrogen atom. *o*- and *p*-Quinones (**133** and **134**) are probably secondary products, arising from decomposition or hydrolysis of the initially formed unstable quinol diacetates (e.g., **132**)[105, 156, 180].

If one assumes that the first step of the lead tetraacetate oxidation of phenols involves the reversible formation of aryloxy-lead(IV)-acetates **138** (equation 32)[156, 183, 184], then two possibilities can be envisaged for the decomposition of these unstable intermediates:

$$\mathrm{ArOH + Pb(OAc)_4 \rightleftharpoons \underset{\textbf{(138)}}{ArOPb(OAc)_3} + AcOH} \qquad (32)$$

homolytic cleavage of the O–Pb bond (or one-electron transfer from O to Pb if this bond is ionic[184, 185]) leading to radical species **139** (equation 33)[156–158, 164, 165], or heterolytic cleavage of the O–Pb bond (or double-electron transfer if the bond is ionic[184, 185]) with formation of resonance-stabilized cationic aryloxy species **140** (equation 34)[156, 178, 179, 183]. Evidence, including e.s.r. spectroscopy, has been cited for both mechanisms[165, 179] and even the formation of coupling products (e.g., **135** and **136**) has been explained in terms of the ionic mechanism (34), as electrophilic substitution involving the intermediate phenoxonium ion (**140**) and the starting phenol (see also equation 24)[156]. On the basis of results obtained in the lead tetraacetate oxidation of alcohols, it appears that the radical mechanism (33) is more probable[184, 186]. Homolytic formation of various oxidation products is shown in scheme (35).

The lead tetraacetate acetoxylation of oestrone (**143**) affords (equation 36) about 20% of 10-acetoxy-1,4-oestradiene-3,7-dione (**144**), accompanied by a small amount of the *o*-quinone diacetate **145** [176–179].

Oxidative demethylation has been observed in the lead tetraacetate oxidation of 4-methoxyphenols[100, 180, 181], as illustrated by reactions (37)[180] and (38)[181], particularly when an excess of oxidant is employed or acetic acid is used as solvent.

The dihydric phenols catechol, hydroquinone and their derivatives

OPb(OAc)$_3$ R (138) → $^{\bullet}$Pb(OAc)$_3$ + [O$^{\bullet}$ R ⟷ O R] (139)

$^{\bullet}$Pb(OAc)$_3$ → Pb(OAc)$_2$ + AcO$^{\bullet}$ (33)

(138) → $^{-}$Pb(OAc)$_3$ + [O^{+} R ⟷ O R +] (140)

$^{-}$Pb(OAc)$_3$ → Pb(OAc)$_2$ + AcO^{-} (34)

139 → R, H, O, R, H, O (dimer) → 136 $\xrightarrow{-2H^{\bullet}}$ 137

139 $\xrightarrow{AcO^{\bullet}}$ 130

139 $\xrightarrow{AcO^{\bullet}}$ AcO, H, O, R → AcO, OH, R (141) $\xrightarrow{Pb(OAc)_4}$ AcO, O, R (35)

$\xrightarrow{AcO^{\bullet}}$ AcO, AcO, O, R (132) $\xrightarrow{H_2O}$ (133)

$\xrightarrow{AcO^{\bullet}}$ AcO, O, R, OAc (142)

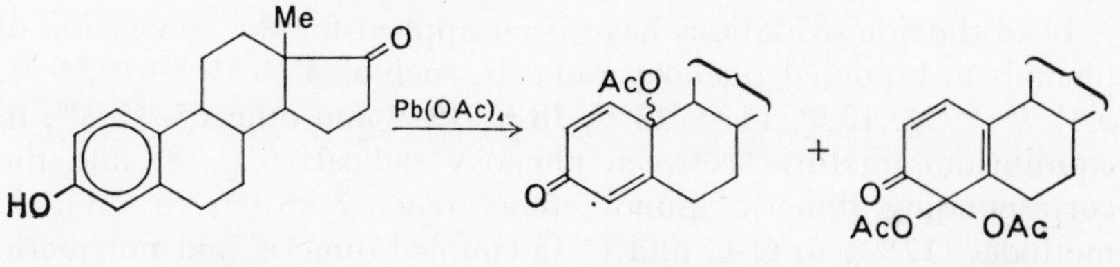

(143) (144) (145)

(36)

are rapidly and quantitatively oxidized by lead tetraacetate to the corresponding quinones[156], and even quinones with very high oxidation potentials can be conveniently prepared in this way[156, 187, 188].

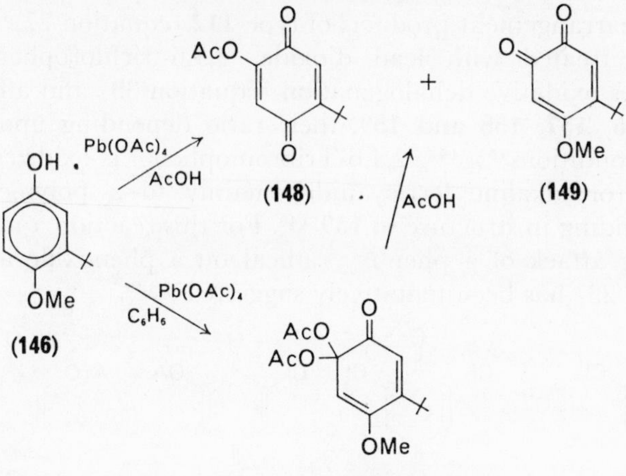

(146) (147) (148) (149)

(37)

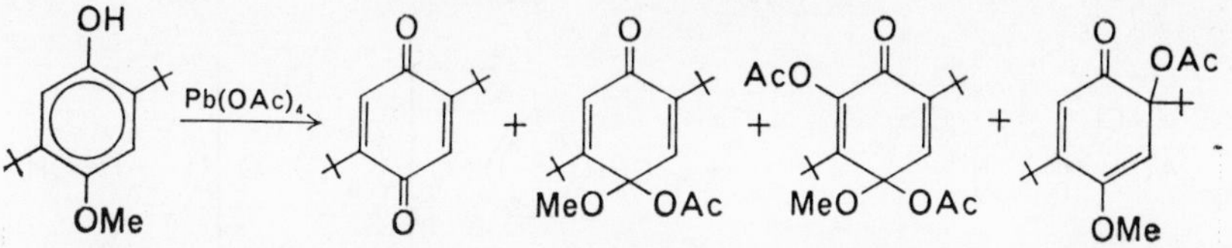

(150) (151) (152) (153) (154)

(38)

2. Lead dioxide

Lead dioxide, PbO_2, has similar oxidizing properties to alkaline potassium ferricyanide. It is used as suspension in organic solvents, preferably benzene, and its efficiency as oxidant frequently depends on the content of active oxygen and on the dilution[25, 189].

Lead dioxide oxidations have been applied for the conversion of phenols to hindered phenoxy radicals, such as **4** [22, 24, 25, 55, 60, 87], **5** [26, 28, 29, 32], **10** [40], **11** [41], **13** [43], **18** [55], **19** [57] and others[51, 87, 190]; to equilibrium mixtures between phenoxy radicals (e.g., **8**) and the corresponding dimeric quinol ethers (e.g., **7**)[34, 121]; to quinone methides (**17**)[18]; to C–C and C–O coupled dimeric and polymeric products **42** (equation 11)[97, 103], **46** (equation 12)[103], **68** (equation 18)[123], **69** (equation 18)[123], **81** (equation 19)[127], **95** (equation 20)[96, 97], **99** (equation 21)[94, 95], **107** and **108** (equation 26)[5, 6, 142]; to the C(methyl)–C(methyl) coupled dihydroxy compound **54** and the corresponding quinone **55**, arising from dimerization of the initially formed unstable quinone methide **59** (equations 14, 16)[22]; and to rearrangement products of type **112** (equation 27)[143].

When treated with lead dioxide, 2,4,6-trichlorophenol (**155**) undergoes oxidative dehalogenation (equation 39) and affords products **156**, **157**, **158** and **159**, their ratio depending upon experimental conditions[191, 192]. 2,4,6-Tribromophenol is oxidized by lead dioxide (or alkaline ferricyanide) mainly to a polymeric ether corresponding in structure to **159** [140]. For this reaction a mechanism involving attack of a phenoxy radical on a phenoxide anion (see equation 23) has been tentatively suggested[7, 140].

(155) —PbO_2→ **(156)** + **(157)** + **(158)** + **(159)** (39)

Ar = 2,4,6-trichlorophenyl

Oxidative lactonization involving carboxyl–phenol coupling (reaction 40) and leading to the formation of the spirolactone **161** was observed upon oxidation of 2-carboxy-4′-hydroxydiphenylether (**160**) with active lead dioxide[193]. This reaction has been successfully applied in the last step of the synthesis of geodoxin (**162**)[193] and the geodoxin analogue of griseofulvin (**163**)[194].

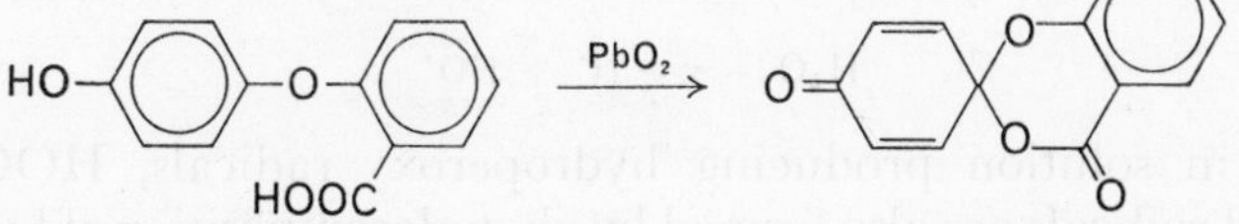

HO— —O— (HOOC) $\xrightarrow{PbO_2}$ O= —O— O (40)

(160) (161)

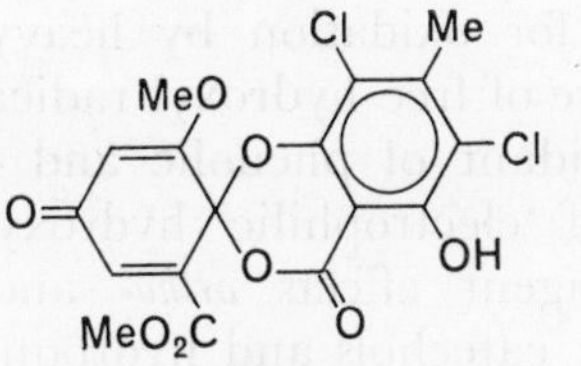

(162) 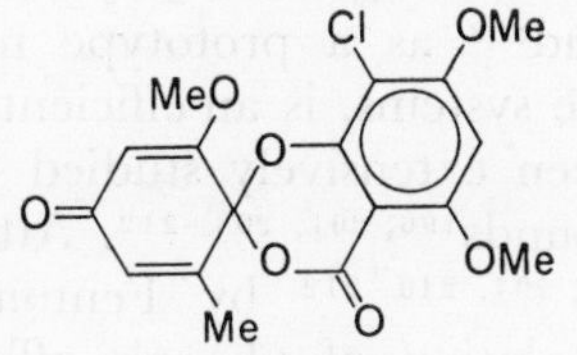 (163)

Active lead dioxide, obtained from lead tetraacetate and water, oxidizes dihydric phenols to quinones; for example, it has been used for the preparation (equation 41) of amphi-naphthoquinone (**165**) from 2,6-dihydroxynaphthalene (**164**)[189, 195].

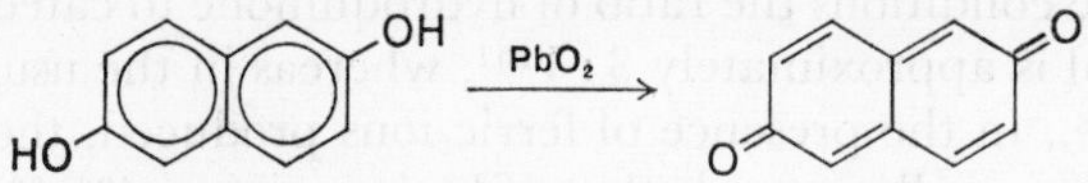

$\xrightarrow{PbO_2}$ (41)

(164) (165)

E. Oxidation by Peroxy Compounds

I. Hydrogen peroxide

Species resulting either from homolytic or heterolytic decomposition of hydrogen peroxide can attack and oxidize phenolic and other organic compounds.

Several different ways are possible for the generation of hydroxyl radicals ($HO^{\bullet}$) and hydroperoxy radicals ($HOO^{\bullet}$) derived from hydrogen peroxide[196, 197]. (a) Reduction of hydrogen peroxide by ferrous salts (Fenton's reagent) in aqueous solution produces hydroxyl radicals, according to the Heber–Weiss reaction (42)[197]; (b) molyb-

$$M^{n+} + H_2O_2 \longrightarrow M^{n+1} + HO^{\bullet} + HO^{-} \qquad (42)$$

dates as well as oxides and salts of other so called 'peracid formers', such as titanium, vanadium, tungsten, etc., in the presence of hydrogen peroxide, afford hydroxyl radicals[63a, 198–200]; (c) rupture of water molecules by ionizing radiations such as X-rays, γ-rays or neutrons furnishes hydroxyl radicals and hydrogen atoms (reaction

43)[201], whereby hydrogen atoms react immediately with oxygen

$$H_2O \rightsquigarrow H^{\bullet} + HO^{\bullet} \quad (43)$$

present in solution producing hydroperoxy radicals, $HOO^{\bullet}$; (d) hydroxyl radicals are also formed by photodecomposition of hydrogen peroxide in aqueous media[202].

Fenton's reagent (H_2O_2 + ferrous salts), which was regarded by Wieland[203] as a prototype model for oxidation by heavy-metal enzyme systems, is an efficient source of free hydroxyl radicals and has been extensively studied as oxidant of phenolic and related compounds[196, 201, 203–212]. Attack of 'electrophilic' hydroxyl radicals[199, 207, 210–212] by Fenton's reagent effects *ortho*- and *para*-hydroxylation of phenols affording catechols and hydroquinones. Secondary processes, including quinone formation, are also observed[201, 208], because the Fe^{3+} ions produced in reaction (42) form complexes with the phenols and the ferrous–ferric system enters into oxidation–reduction processes with the products[201]. However, the formation of quinones can be inhibited by the addition of ionic fluoride or pyrophosphate, which remove ferric salts as complexes. Under these conditions the ratio of hydroquinone to catechol formed from phenol is approximately 3 : 1[201], whereas in the usual Fenton's reaction, i.e., in the presence of ferric ions produced, the formation of catechol generally exceeds that of hydroquinone [196, 201, 204–208].

When the oxidation of phenols by Fenton's reagent is performed in an acidic medium, a change of product ratio is often observed and the reaction can even take a different course. Thus, the oxidation of *p*-cresol with ferrous ion–hydrogen peroxide in dilute sulphuric acid solution[209] does not yield the expected *o*-hydroxylation product but only dimeric compounds resulting from oxidative coupling and identical to those obtained by alkaline ferricyanide oxidation (equation 10)[78, 88–91], namely 2,2′-dihydroxy-5,5′-dimethylbiphenyl (**34**) and Pummerer's ketone (**37**).

The use of Ti^{3+} ions[63a, 199] and molybdate salts[198, 200] in the hydrogen peroxide oxidation of phenols has been reported. With hydrogen peroxide and ammonium molybdate the oxidation of 2-naphthol does not stop at the *o*-hydroxylation stage, but proceeds further (equation 44) to 4-(2′-hydroxy-1′-naphthyl)-1,2-naphthoquinone (**166**)[200]. This product was shown to arise from initial oxidation of a 2-naphthol molecule to the corresponding 1,2-quinone followed by combination with a 2-naphthyloxy radical.

Secondary reactions which often complicate the *o,p*-hydroxylation

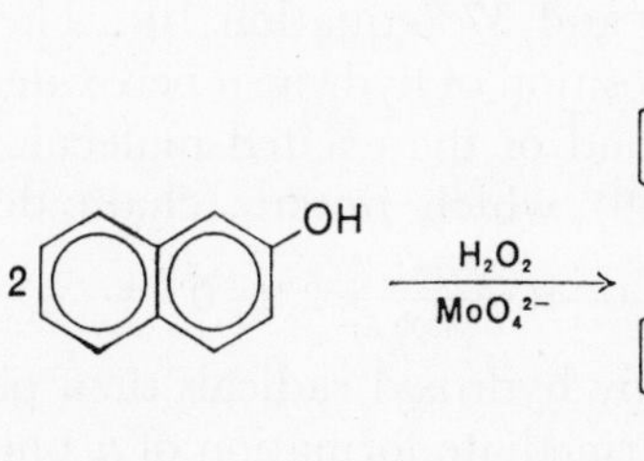

$$2\ \text{(2-naphthol)} \xrightarrow[MoO_4^{2-}]{H_2O_2} \text{(166)} \quad (44)$$

(166)

of phenols can mostly be avoided if instead of Fenton's reagent hydroxyl radicals are generated by the action of ionizing radiations on water (reaction 43), since in this case metal ions are absent and a much wider range of pH can be used. Moreover, Fenton's reagent can lead to a chain reaction which is often difficult to reproduce exactly, whereas when penetrating rays are used the amount of radicals is known and easily controlled[201]. As with Fenton's reagent, hydroxylation of phenol by X-ray irradiation of its aqueous solutions takes place exclusively in the *ortho*- and *para*-position, the hydroquinone–catechol ratio in the products being between 1·5 and 2 in neutral solution, and rising to higher values (4·0–4·7) in acidic or alkaline media[201]. Towards both extremes of the pH range quinones are formed, apparently not from, but in place of, the dihydroxybenzenes. In acidic solution a possible mode of formation of *o*-benzoquinone has been envisaged as involving an intermediate resulting from coupling of phenoxy and hydroperoxy radicals (equation 45)[201].

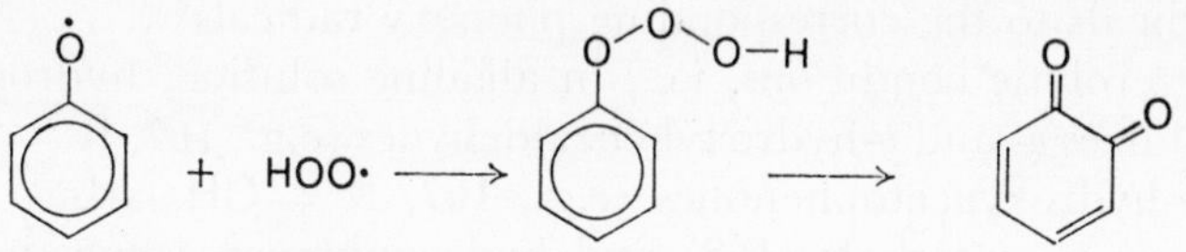

$$\text{PhO}^\bullet + HOO^\bullet \longrightarrow \text{(O–O–O–H adduct)} \longrightarrow \text{(o-benzoquinone)} + H_2O \quad (45)$$

The photolysis of hydrogen peroxide in the presence of phenols in aqueous solution can be conveniently applied for the preparative *o*- and *p*-hydroxylation of various phenols[202]. Irradiation with light at 2537 Å results in the formation of catechols and hydroquinones as the main products, *ortho*-hydroxylation being predominant. With *p*-carboxy- and *p*-methoxyphenols, hydroquinone was obtained (as a result of displacement of the *p*-substituent by a hydroxyl group), in addition to the usual catechol derivative. Using light over 2800 Å at pH 1 *p*-cresol did not undergo *o*-hydroxylation but was converted to a mixture of dimers **34** and **37** (equation 10). The primary reaction of the photodecomposition of hydrogen peroxide is considered as a fission of the O–O bond of the excited molecule to form two HO$^{\bullet}$ radicals (equation 46) which initiate chain decomposition.

$$\text{H—O—O—H} \xrightarrow[(<4000\ \text{Å})]{h\nu} 2\ \text{H—O}^{\bullet} \tag{46}$$

Hydroxylation of phenols by hydroxyl radicals then proceeds, most probably, through the intermediate formation of a phenoxy radical which combines with a second HO$^{\bullet}$ radical to give an *o*- or *p*-dihydroxy compound (equation 47)[202]. This mechanism would also

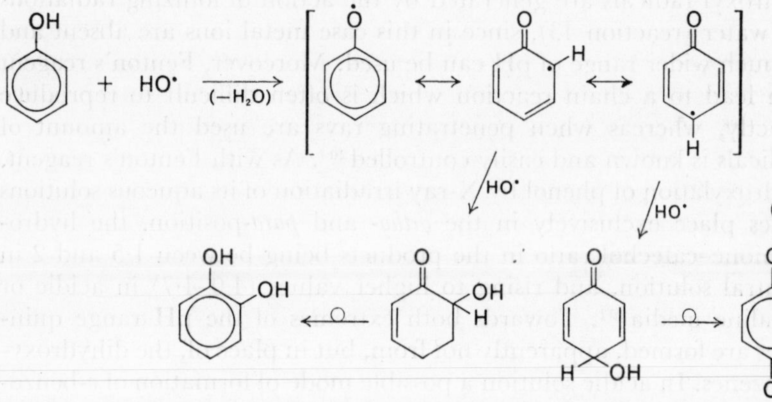

(47)

explain the formation of dimeric coupling products (from the phenoxy radicals). Furthermore, it is substantiated by e.s.r. studies, which have shown that phenol and *p*-cresol are converted by hydroxy radicals to the corresponding phenoxy radicals[63a].

Under heterolytic conditions, i.e., in alkaline solution, hydrogen peroxide oxidizes *o*- and *p*-hydroxybenzaldehydes (e.g., **167**, R = H) or *o*- and *p*-hydroxyacetophenones (e.g., **167**, R = CH_3) (but not the *m*-isomers) to catechols (**168**) and hydroquinones, respectively (reaction 48)[213–217]. This reaction, discovered by Dakin[213], which is formally, (equation 48), the substitution of an *ortho*- or *para*-formyl or acyl group by a hydroxyl group, was first regarded as being applicable mainly to hydroxybenzaldehydes, but was later successfully extended[214–216] to the generally more accessible *o*- and *p*-hydroxyacetophenones and other higher alkyl hydroxyaryl ketones. Difficulties resulting from the tendency of *o*-acylated phenols to form sparingly soluble chelated alkali salts when the Dakin reaction is performed in aqueous sodium or potassium hydroxide, may be

$$\text{(167)} \xrightarrow[\text{NaOH-H}_2\text{O}]{\text{H}_2\text{O}_2} \text{(168)} + \text{RCOOH} \qquad (48)$$

R = H, Me

overcome by using as the base tetramethylammonium hydroxide (or somewhat less satisfactorily benzyltrimethylammonium hydroxide), whose salt with the starting material is ionized and therefore readily soluble[216]. Thus, the yield of 3,4-dimethylcatechol (**170**) obtained in the Dakin reaction (equation 49) of 2-hydroxy-3,4-dimethylacetophenone (**169**) is only 2·5% when the base is potassium hydroxide, and over 25% when tetramethylammonium hydroxide is used instead[216].

$$\text{(169)} \xrightarrow[\text{HO}^-]{\text{H}_2\text{O}_2} \text{(170)} \qquad (49)$$

The Dakin reaction has been interpreted by a mechanism similar to that proposed for the Baeyer–Villiger oxidation and is formulated in scheme (**50**)[218].

$$\text{R-C(=}^{18}\text{O)-C}_6\text{H}_4\text{-O}^- + \text{HO-O}^- \longrightarrow \text{R-C(}^{18}\text{O}^-\text{)(O-OH)-C}_6\text{H}_4\text{-O}^- \longrightarrow \text{R-C(=O}^{18}\text{)-O-C}_6\text{H}_4\text{-O}^- + \text{HO}^- \qquad (50)$$

$$\longrightarrow \text{HO-C}_6\text{H}_4\text{-O}^- + \text{R-C(=O}^{18}\text{)-O}^-$$

2. Alkyl peroxides

The reactions of peroxy radicals, ROO•, and alkoxy radicals, RO•, with phenols are of particular interest in relation to autoxidation phenomena, since phenols act as inhibitors and destroy these radical species, which are active intermediates in autoxidation free radical chain processes[3, 7, 9].

A large kinetic isotope effect (k_H/k_D = 6·5–10·5 at room temperature) is observed when the hydroxyl hydrogen in phenol is replaced by deuterium. On the basis of the retardation of the reaction rate and the decrease of phenol inhibiting efficiency[219–222], Ingold[219–221] and Shelton[222] have suggested that the rate-determining step (*a*), in the reaction (51) of a phenol with peroxy radicals (X = RO) consists in the abstraction of a hydrogen atom from the hydroxyl group by a peroxy radical, possibly through a transition state of type **171a** [12], with formation of a phenoxy free radical (**172**), which subsequently reacts (*a'*) with a second peroxy radical. A similar scheme was proposed for the rate-determining step in the reaction between phenols and alkoxy radicals[115, 223].

$$\mathrm{ArOH + XO^{\bullet} \xrightarrow{(a)} [ArOH{\cdot}OX \longleftrightarrow \overset{+}{\overbrace{ArO{\cdot}H}}{:}\overset{-}{\overbrace{OX}} \longleftrightarrow ArO{\cdot}HOX]} \quad \textbf{(171a)}$$

(*b*) ↓ ... ↘ (−XOH) (51)

$$\left[\mathrm{XO{:}^{-}}\ \ \text{(phenol radical cation, OH, +)}\right] \textbf{(171b)} \xrightarrow[(b')]{\mathrm{XO^{\bullet}(-XOH)}} \text{products} \xleftarrow[(a')]{\mathrm{XO^{\bullet}}} \mathrm{ArO^{\bullet}}\ \textbf{(172)}$$

X = RO or R

On the other hand, from kinetic measurements it appears that the transition state of the rate-determining step involves one molecule of phenol and two peroxy radicals[224], and Coppinger[48], following the proposal of Hammond and Boozer[224], has therefore suggested that as the rate-determining step (*b*) a charge-transfer complex (**171b**) was initially formed between a peroxy radical and the phenol which reacts directly (*b'*) with a second peroxy radical to give products, without passing through a phenoxy free radical (**172**). Hyperconjugation in this complex (**171b**) presumably could account for the observed kinetic isotope effect[48].

The relative rates of hydrogen abstraction from mononuclear phenols by peroxy and alkoxy radicals, and therefore the relative inhibiting properties of phenols in autoxidation processes, have been found to give a remarkably good correlation with Hammett's σ values[219, 224–226] and particularly with Brown's electrophilic σ^+ constants[220, 223, 227]. The maximum rate, i.e., inhibiting efficiency, is achieved when the substituents of the phenol have the largest possible negative $\sum \sigma^+$ consistent with a minimum of steric protection afforded to the hydroxyl group. Thus mononuclear phenols which contain a *t*-butyl group in the 2-position, substituents with large negative σ^+ constants (electron-releasing groups) in 3-, 4- and 5-positions, and a vacant 6-position should be and are in fact autoxidation inhibitors with optimum efficiency.

The oxidation of 2,4,6-trialkyl phenols (**173**) with peroxy radicals (equation 52) generated by the cobalt-catalysed decomposition of hydroperoxides[114, 228] or by the thermal decomposition (at 40–60°) of compounds producing alkyl radicals such as α,α′-azobisisobutyronitrile, in the presence of oxygen (and oxidation initiators)[114, 224, 229], generally affords mainly 4-peroxycyclohexadienones **174**. The yields of the latter depend on the phenol, on the peroxy radical and on the experimental conditions, and are nearly quantitative when 2,6-di-*t*-butylphenols containing a 4-methyl or 4-*t*-butyl substituent and *t*-butyl peroxy radicals or a mixture of α-tetralyl peroxy and isobutyronitrile peroxy radicals are used[114, 228, 229].

OH ... R'OO• ... R'OO• ... R OOR'

(173) R = Me, *t*-Bu (174)

(52)

Under certain conditions 2,6-di-*t*-butyl-4-methylphenol (**173**, R = CH_3) undergoes oxidative C(methyl)–C(methyl) dimerization to products **54** and/or **55** (reaction 14), via the unstable *p*-quinone methide **59** (reaction 16), e.g., with triphenylmethylperoxy radicals[111, 114], or in the high-temperature air oxidation of cumene retarded by **173** (R = CH_3)[111, 230].

When mono- and di-alkyl phenols with an unsubstituted *ortho*- or *para*-position are oxidized by peroxy radicals a variety of products can be formed[111, 230, 231]. Thus, from the reaction of 2,4-di-*t*-butylphenol (**175**) with *t*-butoxy radicals (generated by the cobalt toluate catalysed decomposition of *t*-butyl hydroperoxide below 30°) products **176**–**182** shown in scheme (53) have been isolated[231]. These can be divided into three groups, depending on the type of radical coupling reaction by which they are produced or in which theirprecursors are formed: (*i*) phenoxy–peroxy 4C–O and 6C–O

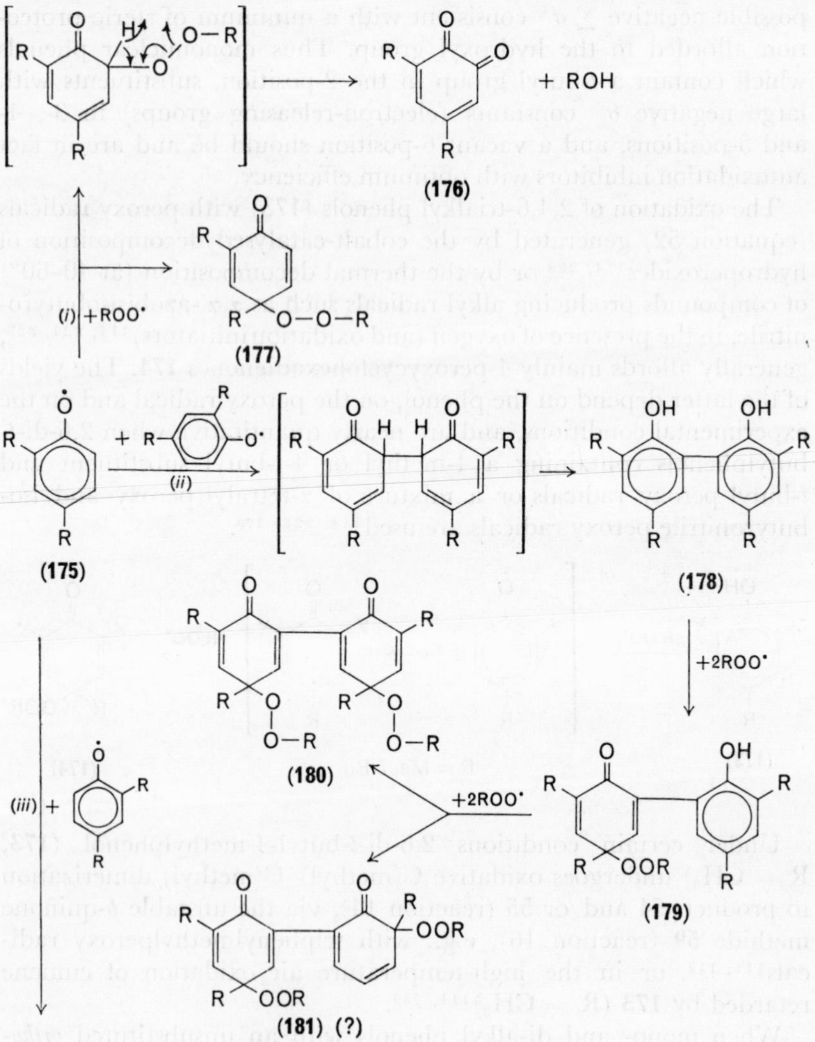

(176) + ROH

(*i*) +ROO˙

(177)

(175) + (*ii*)

(178)

+2ROO˙

(179)

+2ROO˙

(180)

(*iii*) +

(181) (?)

Scheme 53 (*cont. on next page*)

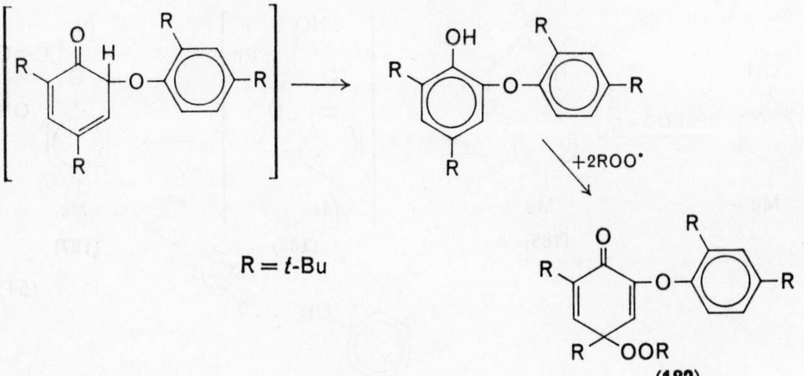

R = *t*-Bu

+2ROO$^{\bullet}$

(182)

Scheme 53 (*cont.*)

coupling; (*ii*) phenoxy–phenoxy 6C–6C coupling; and (*iii*) phenoxy–phenoxy 6C–O coupling.

The reaction of *t*-butoxy radicals, $(CH_3)_3CO^{\bullet}$, generated by thermal decomposition at 122° of di-*t*-butyl peroxide, with 2,6-di-*t*-butylphenol [**38**, R = R′ = $C(CH_3)_3$] affords the expected 4C–4C coupled dimers **41** and **42** (equation 11)[115], whereas *t*-butoxy radical oxidation of 2,6-di-*t*-butyl-4-methylphenol (**53**) proceeds with the initial formation of the *p*-quinone methide **59** (equation 16), followed by C(methyl)–C(methyl) dimerization to **54, 55** (equation 14) and **183** [115]. The absence of the 4-*t*-butoxycyclohexadienone **184** in this reaction is probably due to increased steric effects [compared to those present during the formation (equation 52) of the 4-peroxycyclohexadienone (**174**)] and to the instability of product **184** at 122°.

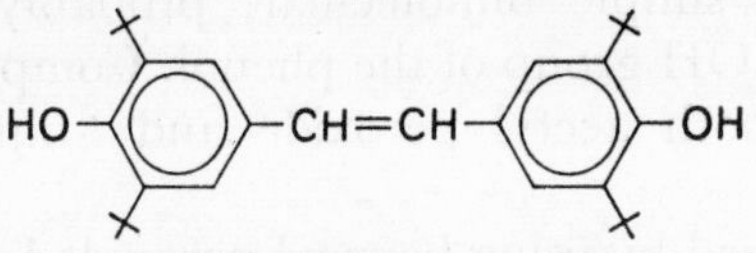

(183) 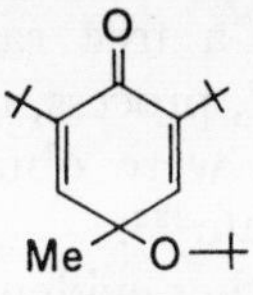 **(184)**

3. Acyl peroxides and peracids

When treated with benzoyl peroxide in refluxing chloroform, *p*-cresol is converted in 35% yield to 4-benzoyloxy-3-hydroxytoluene (**187**), and the same product is obtained, though in poorer yield (20%), from *m*-cresol (reaction 54)[232]. Other phenols with a free *ortho*-position behave similarly, i.e., the benzoyloxy group is introduced preferentially into a position adjacent to the hydroxyl

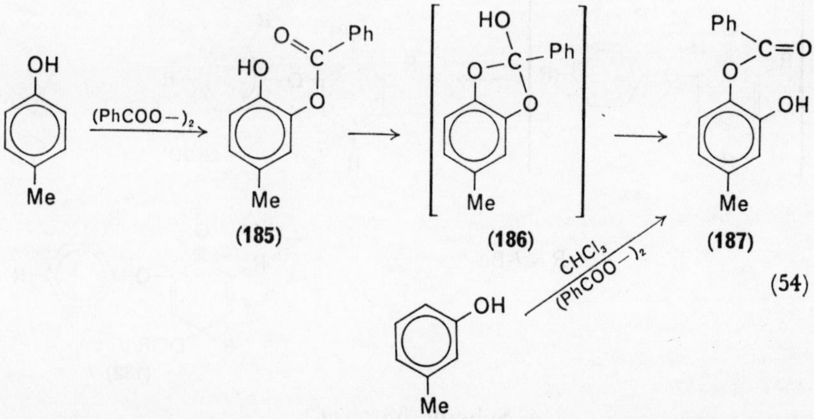

(54)

group[232]. However, with *p*-substituted phenols (such as *p*-cresol, 2,4-dimethylphenol and hydroquinone monomethyl ether), the benzoyl group of the initially formed product (**185**) usually undergoes rapid migration (**185**→**187**); this migration probably occurs by transesterification through an intermediate of type **186** [232]. When one of the *o*-positions in the phenol is occupied by a bulky substituent, as in 2-*t*-butyl-4-methoxyphenol, *ortho*-substitution by the benzoyloxy group is not followed by benzoyl migration[180].

Kinetic studies of this reaction by Walling and Hodgdon[233] have shown that radical traps (oxygen and iodine) have no effect upon rate or products, and that if the hydrogen of the phenolic OH group is replaced by deuterium the reaction velocity decreases ($k_H/k_D = 1{\cdot}32$); moreover, no carbon dioxide is evolved during the reaction. These and other results indicate that the phenol–benzoyl peroxide reaction is not a free radical, but a simple bimolecular, probably 'four centre', process, involving the OH group of the phenol. Comparable results were obtained with both acetyl peroxide and *t*-butylperbenzoate[233].

Further evidence was obtained by using benzoyl peroxide labelled with ^{18}O in the carbonyl groups[234]. Analysis of the product **190** (scheme 56) showed that about 87% of the excess ^{18}O was present in the carbonyl group of the benzoate substituent which had been introduced into the phenol. Therefore, benzoyloxy radicals could

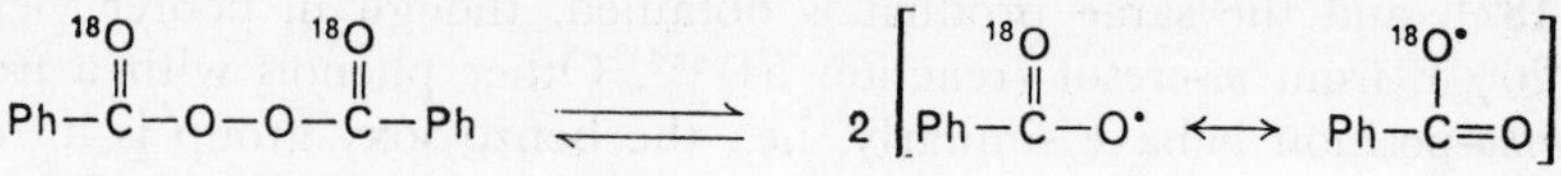

$$\mathrm{Ph{-}C(={}^{18}O){-}O{-}O{-}C(={}^{18}O){-}Ph} \rightleftharpoons 2\left[\mathrm{Ph{-}C(={}^{18}O){-}O^{\bullet}} \longleftrightarrow \mathrm{Ph{-}C({-}{}^{18}O^{\bullet}){=}O}\right] \quad (55)$$

not have been generated since the formation (equation 55) of such radicals would have led to product **190** in which half of ^{18}O was in the ester carbonyl and the other half in the phenolic hydroxyl group.

All these facts suggest that the reaction (scheme 56) involves a 'four centre' mechanism either concerted (path *a*)[7], or proceeding (path *b*) via an unstable perester (**188**) and ion-pair (**189**)[233, 234].

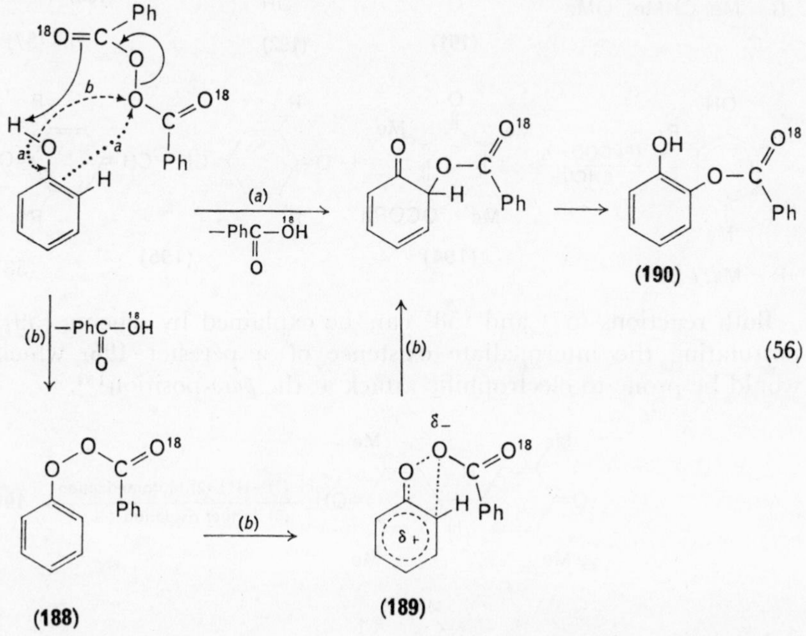

(56)

(**190**)

(**188**) (**189**)

When both *ortho*-positions, but not the *para*-position, of a phenol are occupied, oxidation by peroxides in refluxing chloroform or benzene (reaction 57) affords as major product (50–70%) the 3,3′,5,5′-tetrasubstituted diphenoquinone (**191**), accompanied by small amounts (up to 10%) of the corresponding 4,4′-dihydroxydiphenyl (**192**) and the *p*-benzoyloxy derivative of the starting phenol (**193**)[110, 233].

On the other hand, when mesitol, in which both *o*-positions and the *p*-position are blocked by methyl groups, is treated with benzoyl peroxide in refluxing chloroform, the reaction (58, R = CH_3) affords over 90% of 4-benzoyloxy-2,4,6-trimethylcyclohexa-2,5-dienone (**194**) and only traces of 3,3′,5,5′-tetramethylstilbenequinone (**195**, R = CH_3)[110].

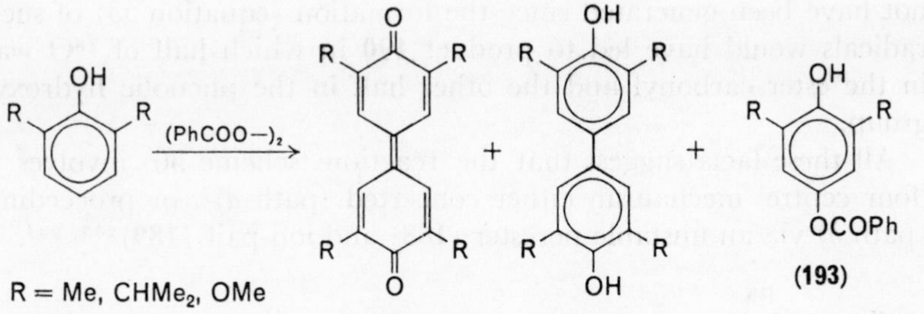

R = Me, CHMe$_2$, OMe

(PhCOO—)$_2$ → (191) + (192) + (193) (57)

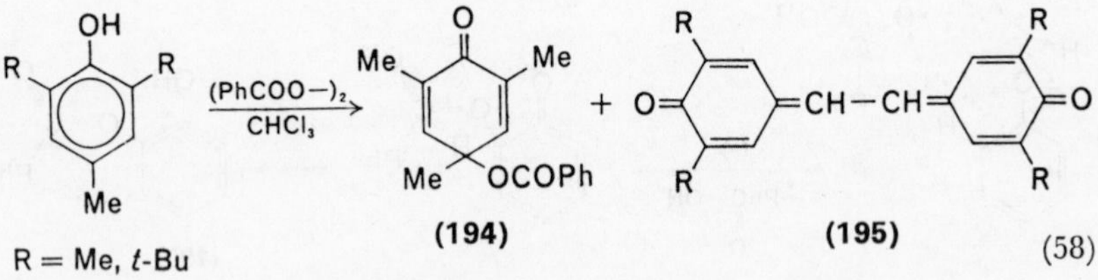

R = Me, *t*-Bu

(PhCOO—)$_2$, CHCl$_3$ → (194) + (195) (58)

Both reactions (57) and (58) can be explained by scheme (59), postulating the intermediate existence of a perester **196**, which would be prone to electrophilic attack at the *para*-position[184].

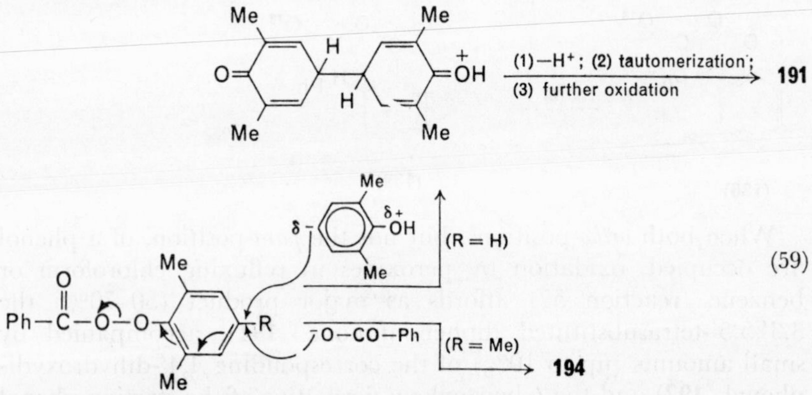

(1) —H$^+$; (2) tautomerization; (3) further oxidation → 191

(R = H)

(R = Me) → 194

(196) (59)

Bulky *t*-butyl groups in the *o*-positions retard considerably the reaction (58) of 2,6-di-*t*-butyl-4-methylphenol (R = *t*-butyl) with benzoyl peroxide; no dienone of type **194** is obtained (probably because of steric hindrance to formation of **196**), and the only products isolated in moderate yield are the stilbenequinone **195** (R = *t*-butyl) and the corresponding 4,4′-dihydroxydibenzyl[110, 112].

Treatment of the isomeric cresols and 2,4-dimethylphenol with acetyl peroxide in acetic acid at 62–77°, followed by acid hydrolysis of the phenol fraction, affords dihydric phenols corresponding to the introduction of the acetoxy group into the free *ortho*- and *para*-positions; moreover, derivatives of *o*- and *p*-hydroxyphenylacetic acid were found in the acid fraction, indicating that probably the radical $^{\bullet}CH_2COOH$ participates in the reaction[165, 235]. Acetyl peroxide in benzene reacts with 2,6-disubstituted phenols in the same way as benzoyl peroxide (reaction 57)[233]. When 2,4,6-trimethylphenol is treated with acetyl peroxide in acetic acid at 65° (reaction 60) it is converted in 80–90% yield to 3,5-dimethyl-4-hydroxybenzyl acetate (**199**), possibly by rearrangement of the initially formed, but not isolated, 4- and/or 2-acetoxy-2,4,6-trimethylcyclohexadienones **197** and **198** [166, 235].

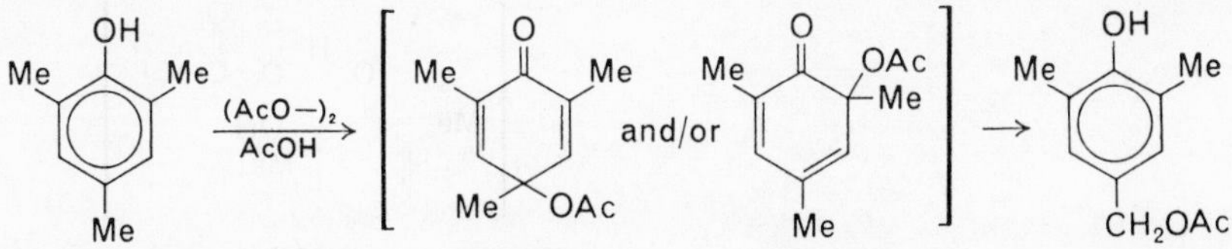

OH Me Me Me $\xrightarrow[AcOH]{(AcO-)_2}$ [O Me Me Me OAc (**197**) and/or O Me OAc Me Me (**198**)] → OH Me Me CH_2OAc (**199**)

(60)

Peracetic acid in acetic acid attacks preferentially the *ortho*-positions of phenol and *p*-substituted phenols, which are converted (reaction 61), probably through *o*-quinones, to the corresponding *cis-cis*-muconic acids (**200**) (phenol gives, in addition, some *p*-benzoquinone)[236, 237]. Since these acids easily undergo further reaction

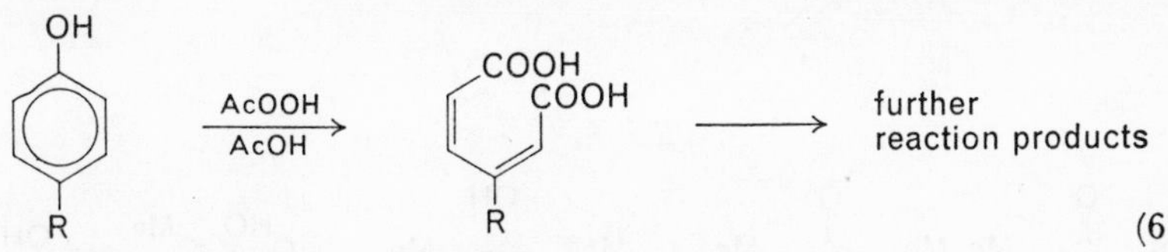

OH R $\xrightarrow[AcOH]{AcOOH}$ COOH COOH R (**200**) ⟶ further reaction products

R = H, Me, Cl, Br

(61)

(cyclization to lactones, hydroxylation, etc.) and the resulting products are difficult to separate, this oxidation is not of great preparative value.

However, when peracetic acid is used in a mixture of sulphuric acid and acetic acid[238], or when oxidations are performed with trifluoroperacetic acid in methylene chloride[239], 4-unsubstituted di- and tri-methylphenols, even when they contain a free *ortho*-position, are converted in yields up to 80% to the corresponding *p*-benzoquinones.

When trifluoroperacetic acid is generated slowly in situ, i.e., by slow addition of hydrogen peroxide to a solution of 2,6-dimethylphenol and trifluoroacetic acid in methylene chloride, the reaction (62)[240] affords as major product (42%) the Diels–Alder dimer **203** of 2,6-dimethyl-*o*-quinol (**202**), while 2,6-dimethyl-*p*-benzoquinone (**204**), which predominates under usual conditions (i.e., when hydrogen peroxide is added all at once)[239, 240], is obtained in only 27% yield. A cyclic hydrogen-bonded transition state **201** has been suggested to account for *o*-hydroxylation leading to the *o*-quinol **202** [240].

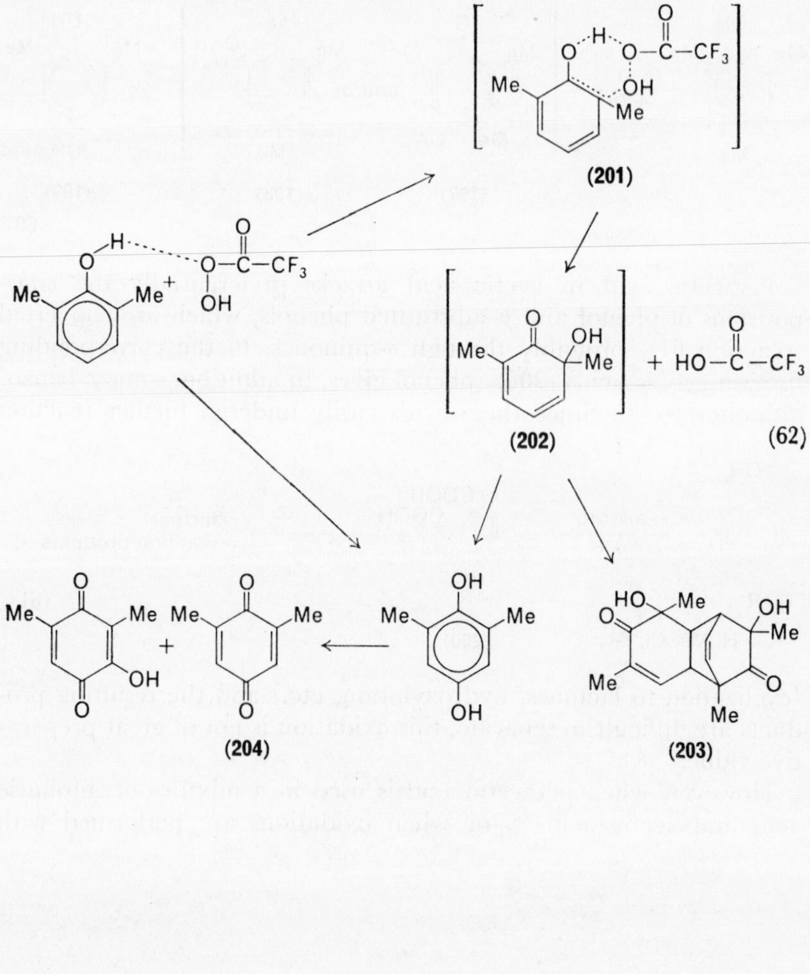

(201)

(202)

(62)

(204)

(203)

4. Persulphate oxidation

The oxidation of monohydric phenols to dihydric phenols by potassium or ammonium persulphate in cold aqueous alkali was discovered by Elbs[241]. When the *para*-position is free, *p*-hydroxylation (equation 63) gives hydroquinone derivatives (**206**); with *p*-substituted phenols reaction takes place at the *ortho*-position and derivatives of catechol are obtained, though usually in much lower yield[242, 243].

It was shown that a hydroxyphenyl alkali sulphate **205** is formed as an intermediate and is subsequently hydrolysed in acid solution to hydroquinone[242, 243].

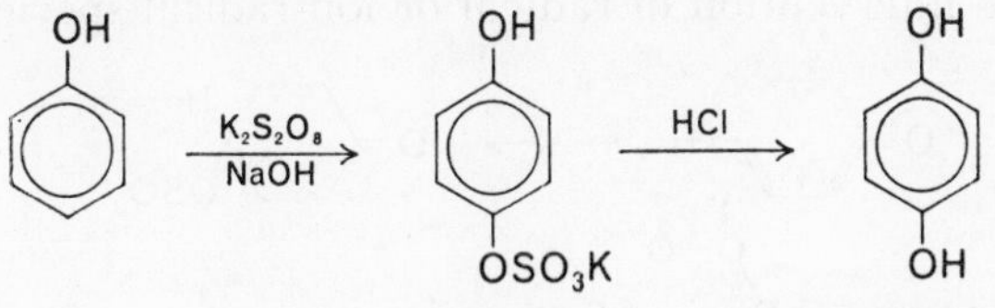

OH—C_6H_5 $\xrightarrow[NaOH]{K_2S_2O_8}$ HO—C_6H_4—OSO_3K (**205**) $\xrightarrow{HCl}$ HO—C_6H_4—OH (**206**) (63)

Baker and Brown[243] have pointed out that the direct introduction of the sulphate group *para* or *ortho* (but never *meta*) to the phenolic oxygen atom, suggests that in the Elbs persulphate oxidation is the resonance hybrid of the phenoxide ion undergoing attack (scheme 64), and that the substituting agent is a reactive sulphate ion-radical, $^{\bullet}OSO_3^-$, which although an anion is yet electrophilic in character. This ion-radical might be initially generated by interaction of a trace of a metal cation present as impurity in the persulphate salt

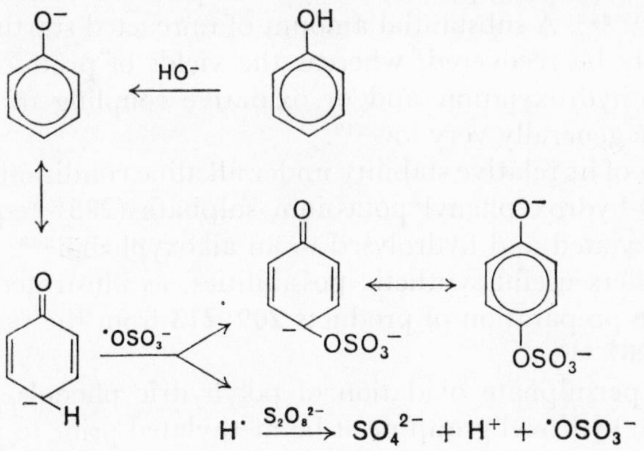

(64)

O⁻ (phenoxide) $\xleftarrow{HO^-}$ OH (phenol)

$\updownarrow$

O (cyclohexadienone, –H) $\xrightarrow{^{\bullet}OSO_3^-}$ O (cyclohexadienone, $^-$ OSO_3^-) $\longleftrightarrow$ O⁻ (OSO_3^-)

$H^{\bullet} \xrightarrow{S_2O_8^{2-}} SO_4^{2-} + H^+ + {}^{\bullet}OSO_3^-$

(such as a ferrous or silver ion) with a persulphate anion (equation 65).

$$Fe^{2+} + S_2O_8^{2-} \longrightarrow Fe^{3+} + SO_4^{2-} + {}^{\cdot}OSO_3^- \qquad (65)$$

The slow rate of the reaction can be correlated with the requirement in the above mechanism (64) that the sulphate anion-radical has to attack an anion. A further point in agreement with such a mechanism (64) is the fact that in general the yield of *para*-hydroxylation is increased by the presence of electron-attracting groups, by increasing substitution and by the effect of substituents on the activity of the position *para* to the hydroxyl group[242, 243].

Waters[7], however, considers the alkaline persulphate oxidation of phenols to be a completely heterolytic reaction (66), proceeding without the intervention of radical or ion-radical species.

$$^-O-C_6H_4-H + {}^-O_3S-O-O-SO_3^- \longrightarrow O{=}C_6H_4(H)(OSO_3^-) + SO_4^{2-} \qquad (66)$$

$$\downarrow$$

$$HSO_4^- + HO-C_6H_4-OH \xleftarrow[H_2O]{H^+} HO-C_6H_4-O-SO_3^-$$

(205)

A variety of mono- and poly-substituted monohydric phenols (in which the position *para* to the hydroxyl group is free) have been successfully used as substrates in the persulphate reaction, the yield of the corresponding *para*-hydroxylation products ranging from 18 to 50% [241–243]. A substantial amount of unreacted starting material can usually be recovered, whereas the yields of products resulting from *ortho*-hydroxylation and/or oxidative coupling of the phenol nuclei, are generally very low[242].

Because of its relative stability under alkaline conditions, the intermediate *p*-hydroxyphenyl potassium sulphate (**205**) (equation 63) can be alkylated and hydrolysed to an alkoxyphenol[243]. This modification offers useful synthetic possibilities, as illustrated (equation 67) by the preparation of products **209**–**213** from the same starting material **207** [243].

In the persulphate oxidation of polyhydric phenols, usually all except one hydroxyl group must be methylated prior to reaction, in

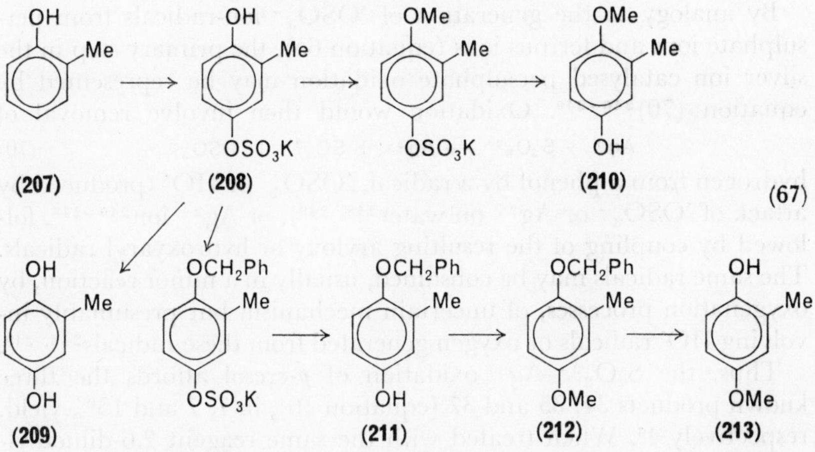

(207) (208) (210) (209) (211) (212) (213)

(67)

order to protect the molecule against general oxidation. A number of such partially alkylated phenols have been successfully oxidized to *para*-hydroxy derivatives[242]. 1-Naphthol and its derivatives are converted in good yields to the corresponding 4-hydroxy compounds, whereas 2-naphthols give poor yields of 1,2-dihydroxy products[244]. This is in agreement with the above mentioned observation that *ortho*-hydroxylation of *p*-substituted phenols proceeds in low yield. The persulphate oxidation converts coumarins to 6-hydroxycoumarins (equation 68)[242, 244], and 5-hydroxyflavones to the corresponding 5,8-dihydroxy compounds (equation 69)[242, 245].

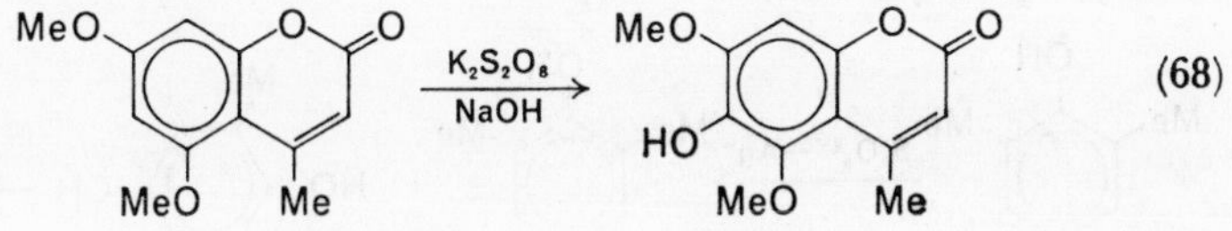

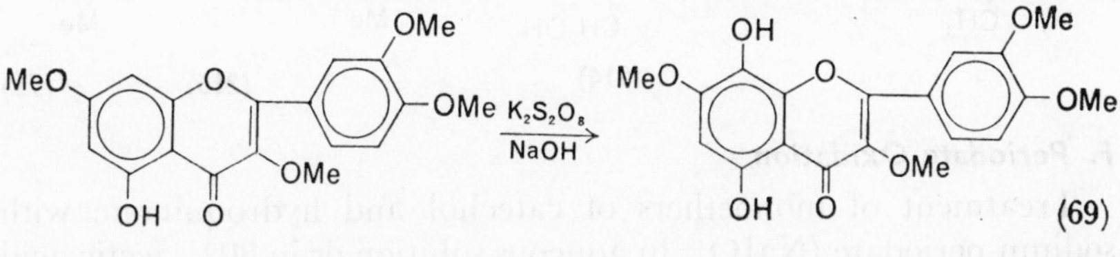

$K_2S_2O_8$, NaOH (68)

$K_2S_2O_8$, NaOH (69)

In neutral aqueous solution and in the presence of catalytic amounts of ferrous, ferric or silver ions, the reaction leads predominantly to oxidative coupling products[246, 247], similar to those described for oxidations with ferricyanide, ferric chloride or aqueous ferrous salts and hydrogen peroxide.

By analogy to the generation of $^{\bullet}OSO_3^-$ ion-radicals from persulphate ions and ferrous ions (equation 65), the primary step in the silver ion catalysed persulphate oxidation may be represented by equation (70)[246–249]. Oxidation would then involve removal of

$$Ag^+ + S_2O_8^{2-} \longrightarrow Ag^{2+} + SO_4^{2-} + {}^{\bullet}OSO_3^- \qquad (70)$$

hydrogen from a phenol by a radical, $^{\bullet}OSO_3^-$ or $HO^{\bullet}$ (produced by attack of $^{\bullet}OSO_3^-$ or Ag^{2+} on water[248, 249]), or Ag^{2+} ion[246–248], followed by coupling of the resulting aryloxy or hydroxyaryl radicals. The same radicals may be consumed, usually in a minor reaction, by oxygenation processes, of uncertain mechanism but presumably involving $HO^{\bullet}$ radicals or oxygen generated from these radicals[248, 249].

Thus, the $S_2O_8^{2-}$–Ag^+ oxidation of *p*-cresol affords the three known products **34**, **35** and **37** (equation 10), in 7, 7 and 15% yield, respectively[246]. When treated with the same reagent 2,6-dimethylphenol (**38**, R = R′ = CH_3) is converted in major part (about 60%) to the nuclear C–C coupling products **41** and **42** (equation 11), whereas nuclear *p*-oxygenation (either direct or involving *p*-hydroxylation followed by oxidation; see above) to 2,6-dimethyl-*p*-benzoquinone proceeds only in about 10% yield[247]. Under similar conditions, 2,4,6-trimethylphenol is attacked at the *para*-methyl group (equation 71), which, presumably via the corresponding benzyl radical, undergoes 22% of hydroxylation resulting in the formation of 4-hydroxy-3,5-dimethylbenzyl alcohol (**214**), and 13% of oxidative coupling and elimination to give 4,4′-dihydroxy-3,3′,5,5′-tetramethyldiphenylmethane (**215**)[247].

OH, Me, Me, CH_3 $\xrightarrow{S_2O_8^{2-}-Ag^+}$ OH, Me, Me, CH_2OH (**214**) + HO–(Me, Me)–CH_2–(Me, Me)–OH (**215**) (71)

F. Periodate Oxidation

Treatment of monoethers of catechol and hydroquinone with sodium periodate ($NaIO_4$) in aqueous solution or in 80% acetic acid leads mainly to oxidative removal of the ether substituent with formation of *ortho*- and *para*-benzoquinone, respectively, and the corresponding alcohol (reactions 72 and 73)[250–252]. Thus, guaiacol (equation 72, R = CH_3) is rapidly converted to *o*-benzoquinone, which can be isolated in about 65% yield, but with excess periodate is itself slowly oxidized to *cis-cis*-muconic acid (**216**)[252]. Catechol and hydroquinone (72 and 73, R = H) are also rapidly oxidized with periodate to the corresponding quinones[251, 252]. Sodium bismuthate, $NaBiO_3$, behaves in these reactions like sodium periodate[180, 252].

Resorcinol and its monomethyl ether are only slowly attacked by periodate, and so also is phenol itself[252, 253]. However, alkyl-substituted phenols are readily oxidized by sodium periodate to give mainly dimeric products[254, 255]. For example, 2,4-dimethylphenol is first converted (reaction 74) to 2,4-dimethyl-*p*-quinol (**217**), 2,4-dimethyl-*o*-quinol (**218**) and 3,5-dimethyl-*o*-quinone (**219**)[254]; of these products only the *p*-quinol **217** can be isolated, whereas the

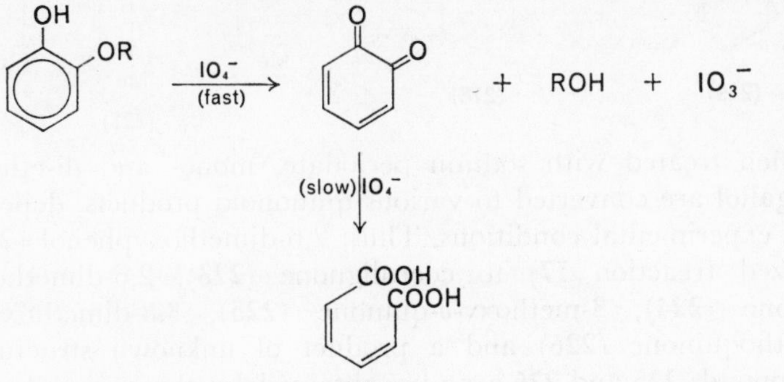

OH, OR → (IO_4^-, fast) → O, O + ROH + IO_3^- (72)

(slow) IO_4^- → COOH, COOH

(216)

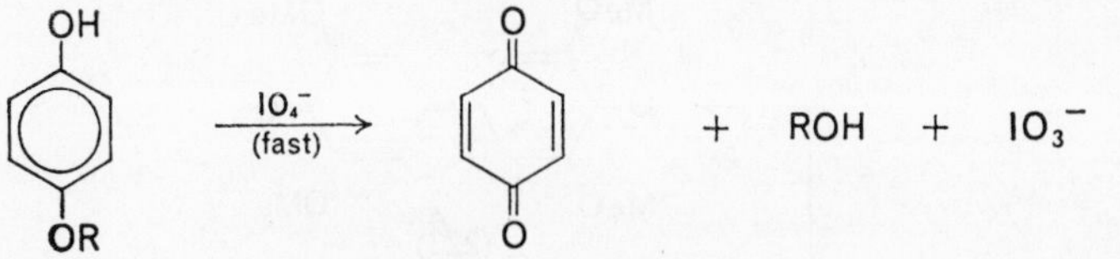

OH, OR → (IO_4^-, fast) → O, O + ROH + IO_3^- (73)

R = Me, Ph, $PhCH_2$, H

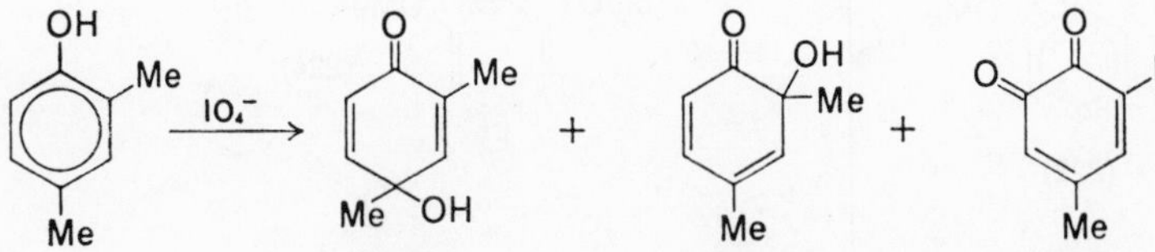

OH, Me, Me → IO_4^- → O, Me, Me OH (217) + O, OH, Me, Me (218) + O, O, Me, Me (219) (74)

o-quinol **218** rapidly undergoes two further reactions: dimerization by a Diels–Alder addition (equation 75) to give a 1,4-ethenonaphthalene derivative **220**, and a similar type of addition (equation 76) to the *o*-quinone **219** resulting in the formation of the adduct **221**.

Similar dimeric products have been reported for the periodate oxidation of 2,6-dimethyl- and 2,4,6-trimethyl-phenol[255].

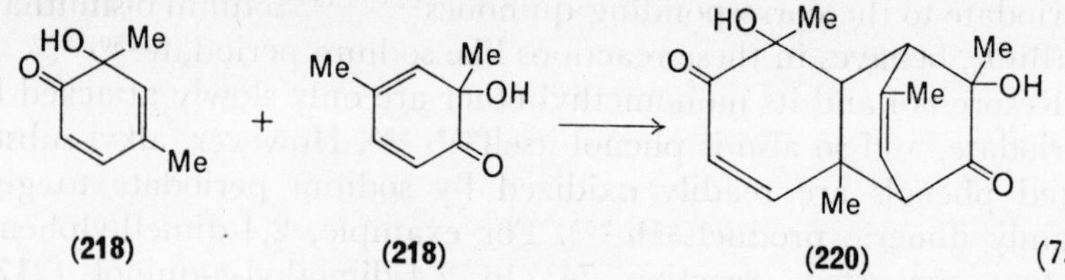

(218) + (218) → (220) (75)

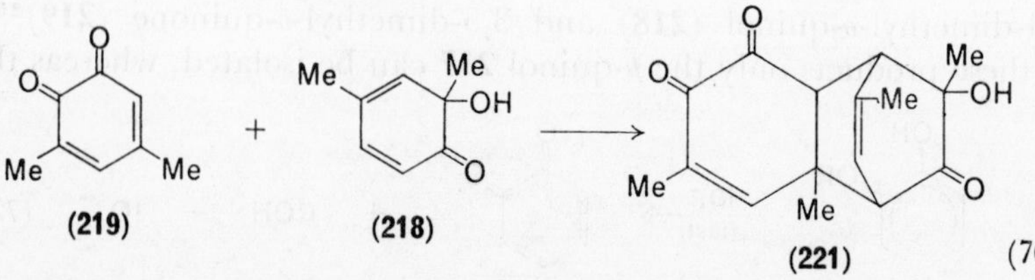

(219) + (218) → (221) (76)

When treated with sodium periodate, mono- and di-ethers of pyrogallol are converted to various quinonoid products, depending upon experimental conditions. Thus, 2,6-dimethoxyphenol (**222**) is oxidized (reaction 77) to coerulignone (**223**), 2,6-dimethoxy-*p*-quinone (**224**), 3-methoxy-*o*-quinone (**225**), 3,8-dimethoxy-1,2-naphthoquinone (**226**) and a product of unknown structure[256]. Compounds **225** and **226** are also obtained by the periodate oxid-

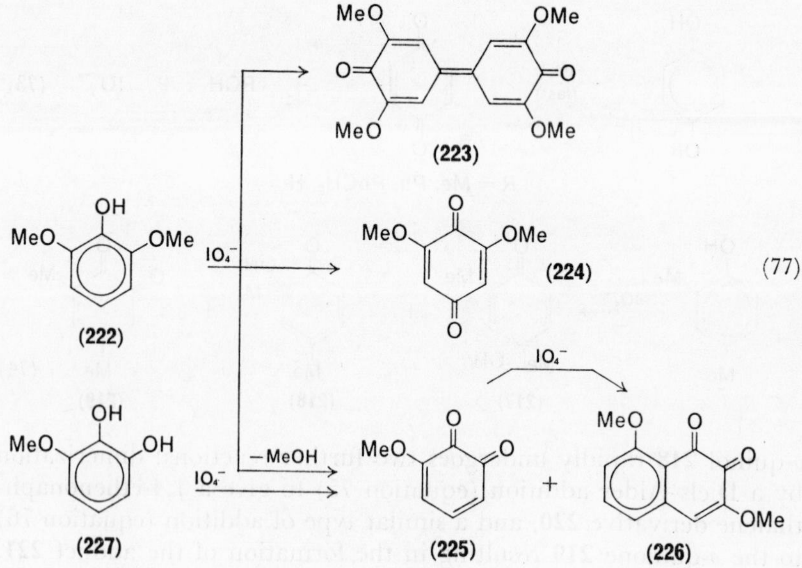

(222) $\xrightarrow{IO_4^-}$ (223) + (224) + (225) + (226) (77)

(227) $\xrightarrow[IO_4^-]{-MeOH}$ (225) $\xrightarrow{IO_4^-}$ (226)

ation of 3-methoxycatechol (**227**)[256]. The formation of the naphthoquinone **226**, which can be obtained by treating the quinone **225** with periodate[256], has been formulated as involving a Diels–Alder type addition of the *o*-quinone **225** to its hydrated form followed by periodate oxidation[257, 258].

The oxidative demethylation of the monomethyl ethers of catechol (reaction 72, R = CH_3) and hydroquinone (reaction 73, R = CH_3) by sodium periodate in ^{18}O–water affords labelled *o*- and *p*-benzoquinone, respectively (and methanol without ^{18}O), whereas unlabelled quinones are obtained from catechol or hydroquinone (R = H) under the same conditions[259]. On the basis of these results, Adler has suggested the schemes (78) and (79), which both involve as intermediates aryl esters of periodic acid (**228–231**)[259]. Cyclic diester (**231**) formation (pathway *a* in 79) would be possible only with catechol, whereas hydroquinone oxidation would have to proceed via ester **230** according to path *b*.

Recent kinetic work on the periodate oxidation at pH 1–4 of

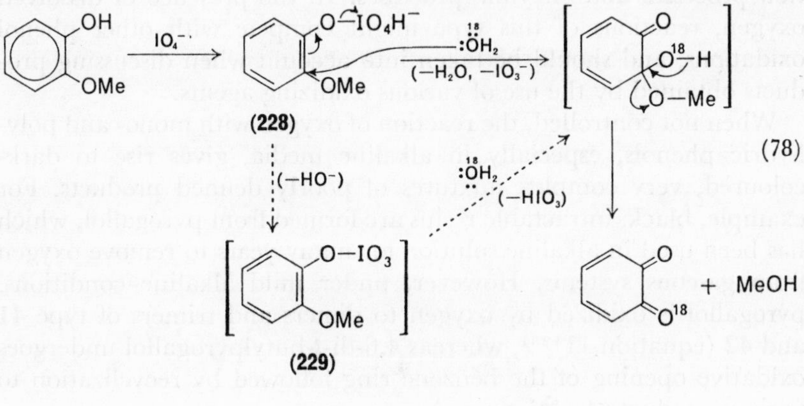

(78)

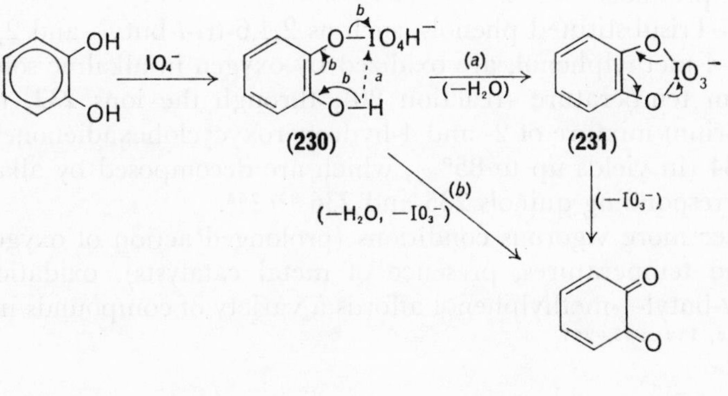

(79)

hydroquinone, its monomethyl ether and catechol monomethyl ether (guaiacol) to form the corresponding benzoquinones (reactions 78 and 79), has shown that the reaction is second-order but that there was no evidence for a detectable intermediate[260]; this suggests that if substrate–periodate complexes (possibly of type **228** and **230**) are intermediates, their formation rather than their decomposition to products is the rate-determining step. For the periodate oxidation of catechol, over a pH range 0–10, however, it was found that an intermediate is formed in a second-order reaction and that this intermediate (not isolated but discussed in terms of the cyclic diester **231** (equation 79) or a dissociable *o*-benzoquinone–iodate charge-transfer complex) then decomposes in a slower first-order reaction to products[260].

G. Oxidation by Molecular Oxygen

The reactions between atmospheric oxygen and phenols or their corresponding radicals are of special interest in relation to autoxidation processes and enzymic processes. In the presence of dissolved oxygen, reactions of this type might compete with other phenol oxidations, and should be taken into account when discussing products obtained by the use of various oxidizing agents.

When not controlled, the reaction of oxygen with mono- and polyhydric phenols, especially in alkaline media, gives rise to dark-coloured, very complex mixtures of poorly defined products. For example, black, intractable resins are formed from pyrogallol, which has been used in alkaline solution for many years to remove oxygen from gaseous systems. However, under mild alkaline conditions, pyrogallol is oxidized by oxygen to dimers and trimers of type **41** and **42** (equation 11)[261], whereas 4,6-di-*t*-butylpyrogallol undergoes oxidative opening of the benzene ring followed by recyclization to various products[262, 263].

2,4,6-Trisubstituted phenols, such as 2,4,6-tri-*t*-butyl- and 2,6-di-*t*-butyl-4-methylphenol, are oxidized by oxygen in alkaline solution at room temperature (reaction 80), through the ions **232**, to an equilibrium mixture of 2- and 4-hydroperoxycyclohexadienones **233** and **234** (in yields up to 85%), which are decomposed by alkali to the corresponding quinols **235** and **236** [92, 264].

Under more vigorous conditions (prolonged action of oxygen at elevated temperatures, presence of metal catalysts), oxidation of 2,6-di-*t*-butyl-4-methylphenol affords a variety of compounds in low yield[112, 113, 265–267].

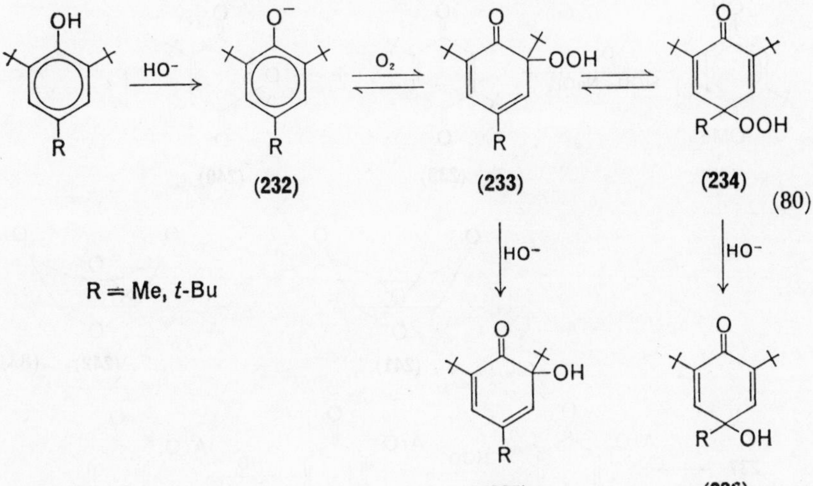

R = Me, *t*-Bu

(232) (233) (234) (80)

(235) (236)

If the *p*-position is unsubstituted, as in 2,6-di-*t*-butylphenol, the intermediate phenoxy radical reacts faster with itself than with oxygen[92, 93, 107, 268], and the diphenoquinone **42** [reaction 11, R = R′ = $C(CH_3)_3$] is produced in very good yield[92, 264].

When one or both *o*-positions in the starting 4-methoxyphenol are free, the course of the reaction with oxygen depends mainly on the number and position of the *t*-butyl substituents, as illustrated by reactions (81–83)[269]. It has been suggested[269] that the C–O coupling

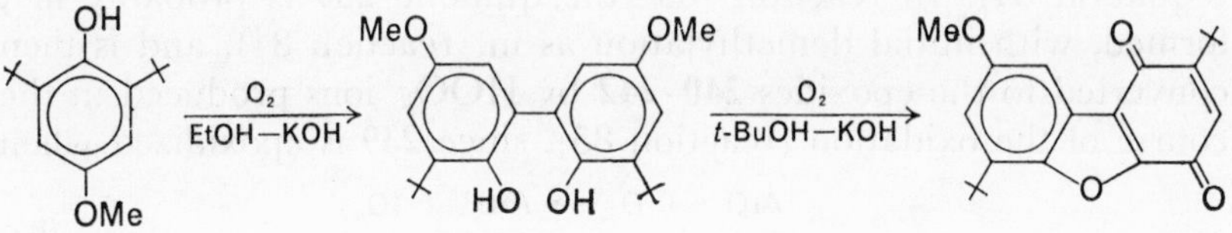

(81)

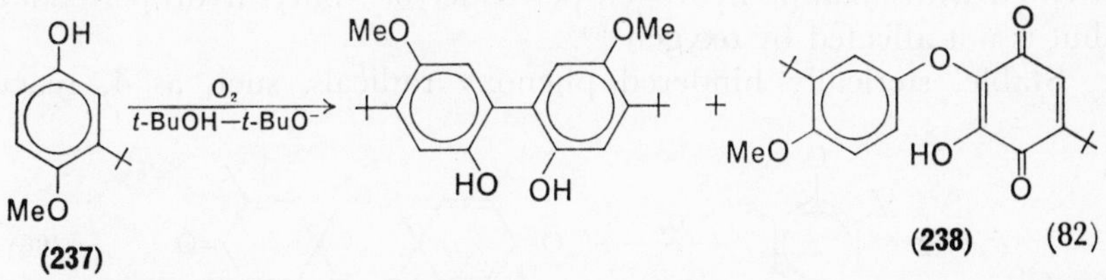

(237) (238) (82)

reaction in the formation (equation 82) of the phenoxy-quinone **238** precedes oxidative demethylation, which takes place as shown

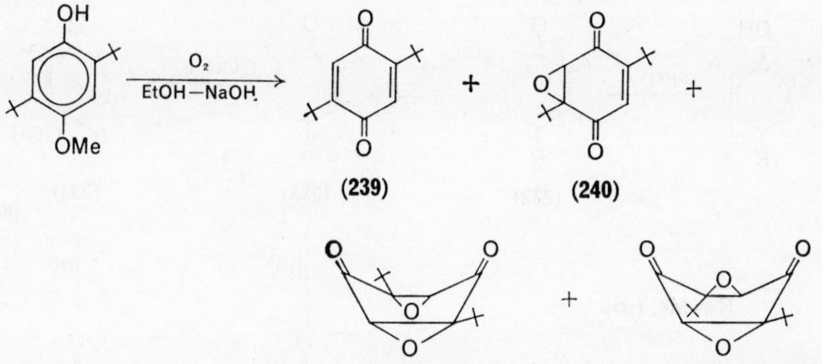

(83)

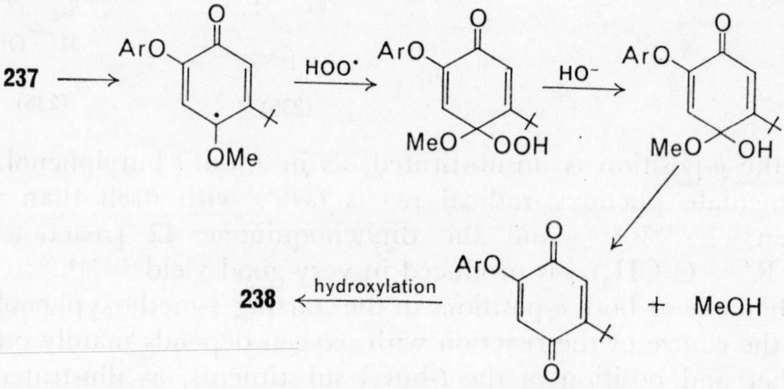

(84)

(equation 84). In reaction (83) the quinone **239** is probably first formed, with initial demethylation as in (reaction 84), and is then converted to the epoxides **240–242** by HOO^- ions produced in the course of the oxidation (reaction 85), since **239** is epoxidized when

$$\mathrm{ArO^- + O_2 \longrightarrow ArO^{\bullet} + (O_2)^{\overline{\bullet}}}$$
$$\mathrm{ArO^- + (O_2)^{\overline{\bullet}} + H^+ \longrightarrow ArO^{\bullet} + HOO^-} \qquad (85)$$

treated with alkaline hydrogen peroxide (or *t*-butyl hydroperoxide) but is not affected by oxygen[269].

Stable, sterically hindered phenoxy radicals, such as **4**, react

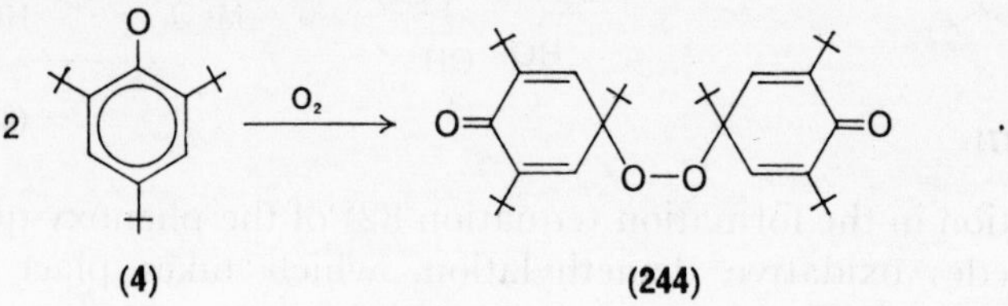

(86)

with oxygen (equation 86) to produce quinol peroxides, e.g., **244** [18, 22–24, 87, 190]. The reactivity of phenoxy radicals towards oxygen is particularly decreased by phenyl substitution, 2,4,6-triphenylphenoxy radical being remarkably stable to attack by oxygen[34–36, 52].

Air-oxidation (reactions 87 and 88) of 4-alkylcatechols (**245**) and 2-alkylhydroquinones (**248**) in alkaline media affords up to 75% of hydroxy-*p*-benzoquinones (**247** and **253**, respectively), which differ

(245) $\xrightarrow[HO^-]{O_2}$ **(246)** $\xrightarrow[HO^-]{O_2}$ **(247)** $\xleftarrow[HO^-]{O_2}$ (87)

R = H, Me, *t*-alkyl

in the position of the alkyl substituent[69, 70, 270, 271]. According to available evidence the mechanism of both reactions (87) and (88) is very probably the same[69, 70] and, as shown for the oxidation (equation 88) of hydroquinones **248**, proceeds successively through dianions **249** [72] and benzosemiquinone radicals of type **250** and **252**. These radicals have been detected by the e.s.r. technique[68–70]. Catechol itself (**245**, R = H) is oxidized by atmospheric oxygen to 2,5-dihydroxy-1,4-benzoquinone (**247**, R = OH)[272], but in the presence of dimethylformamide, dibenzo[1,4]dioxin-2,3-quinone (**254**) is also obtained[68].

Oxidation of alkyl-derivatives of resorcinol with oxygen in alkaline solution affords a variety of monomeric and dimeric products, depending on the number, position and bulk of the alkyl substituents[5, 6, 72, 142, 273]. For example, orcinol (**255**) is converted in about 50% yield (equation 89) to a mixture of the dimeric mono- and bis-(hydroxy-*p*-quinones) **256** and **257**, respectively[5, 6, 142]. Free resorcinol monoradicals are not formed as intermediates in this reaction (compare reaction 26), and the rate-determining step appears to be electrophilic attack of oxygen on the resorcinol monoanion.

Air-oxidation of catechols and resorcinols in the presence of ammonia also involves the substitution of a hydroxyl group by an amino group, which can itself then undergo oxidation[5, 6, 274, 275]. In this way resorcinol derivatives have been converted to orceine and litmus dyes[5, 6, 274].

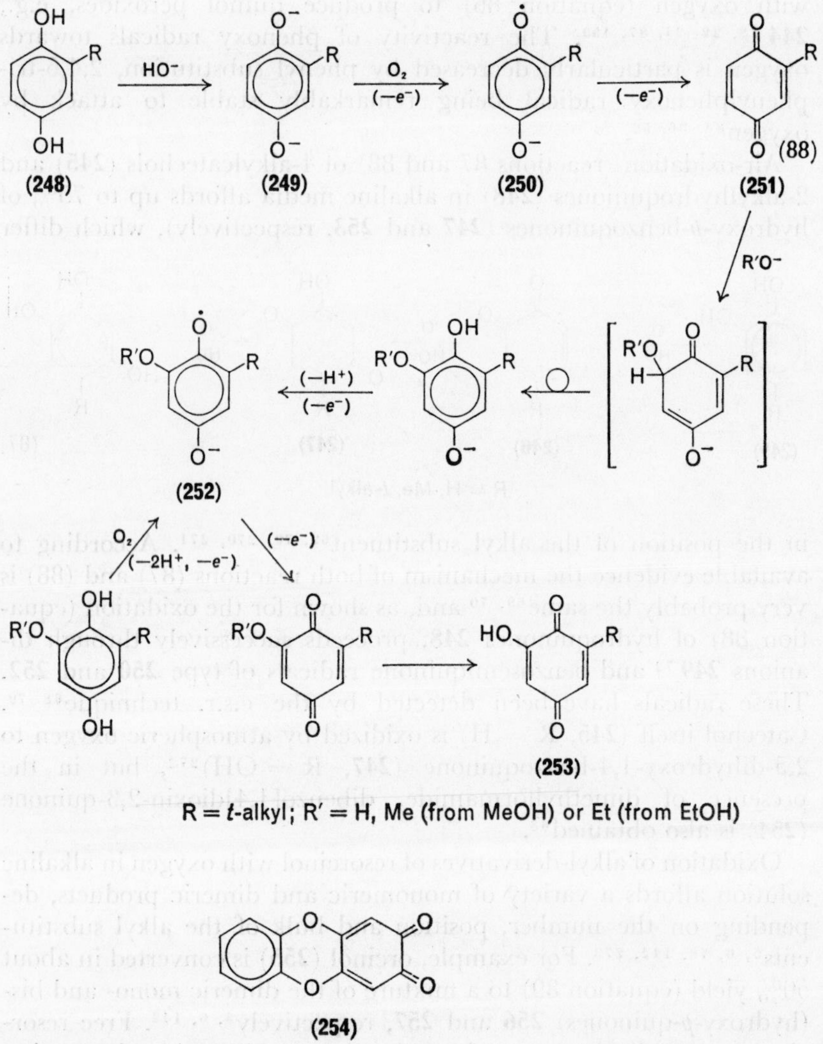

(248) $\xrightarrow{HO^-}$ (249) $\xrightarrow[(-e^-)]{O_2}$ (250) $\xrightarrow{(-e^-)}$ (251) (88)

(251) $\xrightarrow{R'O^-}$ [intermediate] → (252) $\xleftarrow[(-e^-)]{(-H^+)}$

(252): O_2 (−2H⁺, −e⁻); (−e⁻) → (253)

(253)

R = *t*-alkyl; R′ = H, Me (from MeOH) or Et (from EtOH)

(254)

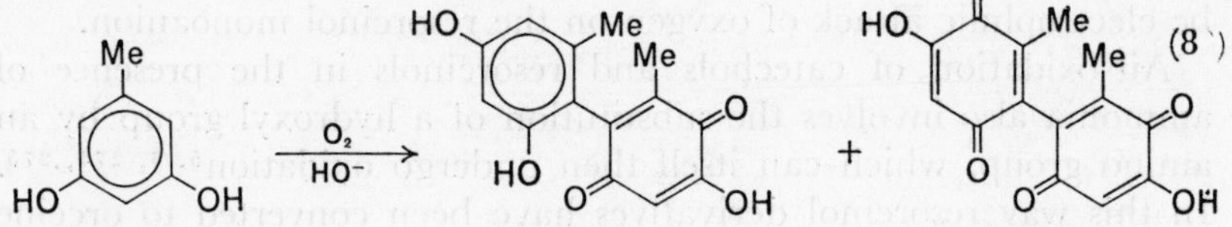

(255) $\xrightarrow[HO^-]{O_2}$ (256) + (257) (89)

When the oxidation of monohydric phenols by molecular oxygen is accomplished in the presence of cupric ions and a secondary amine such as morpholine, a rapid reaction takes place at room temperature affording amino-substituted *o*-quinones[276]. Thus, with morpholine as the amine ligand in the cupric salt–amine complex catalyst, 1- and 2-naphthol are converted (reaction 90) to 4-morpholino-1,2-naphthoquinone (**258**), whereas phenol affords (reaction 91) 4,5-dimorpholino-1,2-benzoquinone (**259**). In scheme (92)[276],

$$\text{1-naphthol} + R_2NH \xrightarrow[37\%]{O_2,\ Cu^{2+}} \textbf{(258)} \xleftarrow[84\%]{O_2,\ Cu^{2+}} \text{2-naphthol} + R_2NH \qquad (90)$$

$$\text{phenol} + 2\,R_2NH \xrightarrow[64\%]{O_2,\ Cu^{2+}} \textbf{(259)} \qquad (91)$$

$R_2NH = HN$ (morpholine)

the initially formed copper complex **260** would explain the exclusive *o*-hydroxylation of phenols. In the case of phenol itself, product **264** is further converted into the dimorpholino-derivative **259**. These reactions have been particularly studied in search of a homogenous catalytic system which would represent a model simulating the action of tyrosinase, since it is known that this group of copper containing enzymes catalyses the oxidation of phenols and catechols to *o*-quinone derivatives.

2,6-Dimethylphenol reacts with oxygen in the presence of a cuprous chloride–amine (usually pyridine) catalyst to yield (reaction 93*a*) a high molecular weight linear polyphenylene ether **265** (see also equation 18, **65**→**69**)[123, 277, 278]. *p*-Cresols behave similarly[279].

It was shown that in a series of 2,6-dialkyl substituted phenols, two different types of products tend to form, depending on the bulk of the substituents (equation 93). With larger groups, such as *t*-butyl,

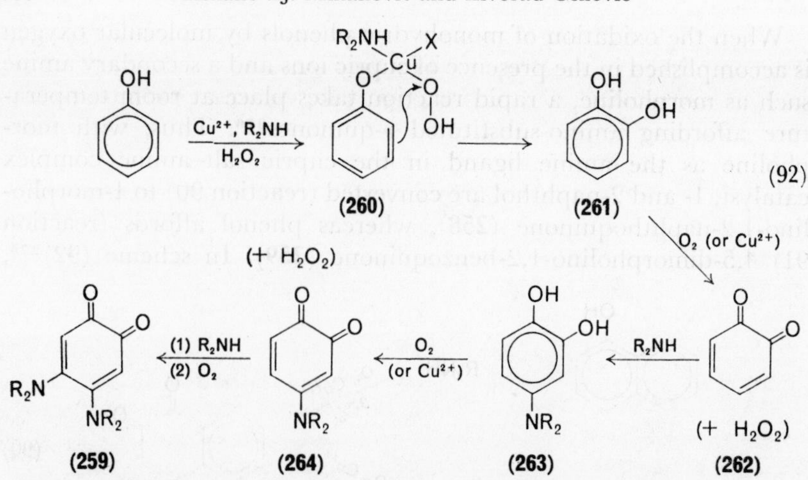

(92)

C–C coupling (93*b*) predominates and tetrasubstituted diphenoquinones **267** [e.g., R = $(CH_3)_3C$] are produced, via intermediate dihydroxybiphenyl derivatives **266** [277, 278, 280]. On the other hand, with smaller substituents, such as methyl, a facile C–O coupling (93*a*) can occur, resulting in poly(2,6-dialkyl-1,4-phenylene ethers) (**265**) (e.g., R = CH_3) of high molecular weight[277–279]. However, C–O and C–C coupling can be competitive reactions even in the oxidation of 2,6-dimethylphenol, their relative rates being very sensitive to catalyst concentration, ligand ratio in the cuprous chloride–amine catalyst, temperature and steric hindrance in the amine ligand

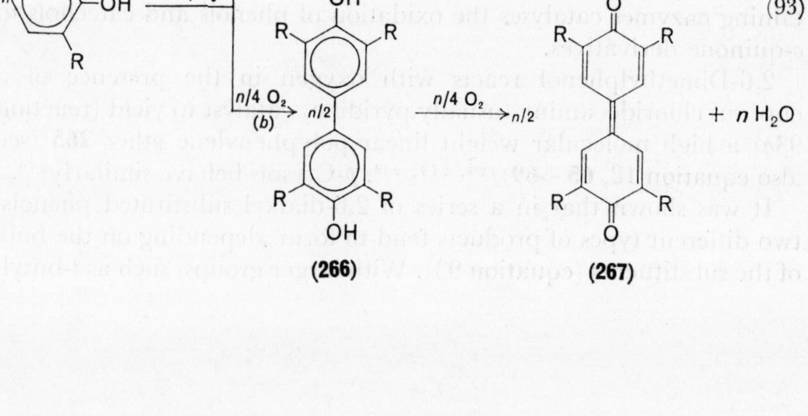

(93)

of the complex[281]. It is highly probable that both reactions (93*a* and 93*b*) proceed through radical intermediates, with initial C–O or C–C coupling of monomeric phenoxy radicals to dimeric products such as **268** (X = H, reaction 94)[123, 279, 282, 283]. However, since the increase in the degree of polymerization towards the end of the reaction is inconsistent with the addition of monomeric phenoxy radicals to the growing polymer chain, in the formation of **265** polymeric quinone ketals **269** have been postulated as intermediates (equation 94), which arise from combination of two aryloxy radicals **268** and which decompose to give redistributed polymeric molecules (**270**), in such a way that one phenol unit at a time is transferred from polymer chain to polymer chain[123, 283–285].

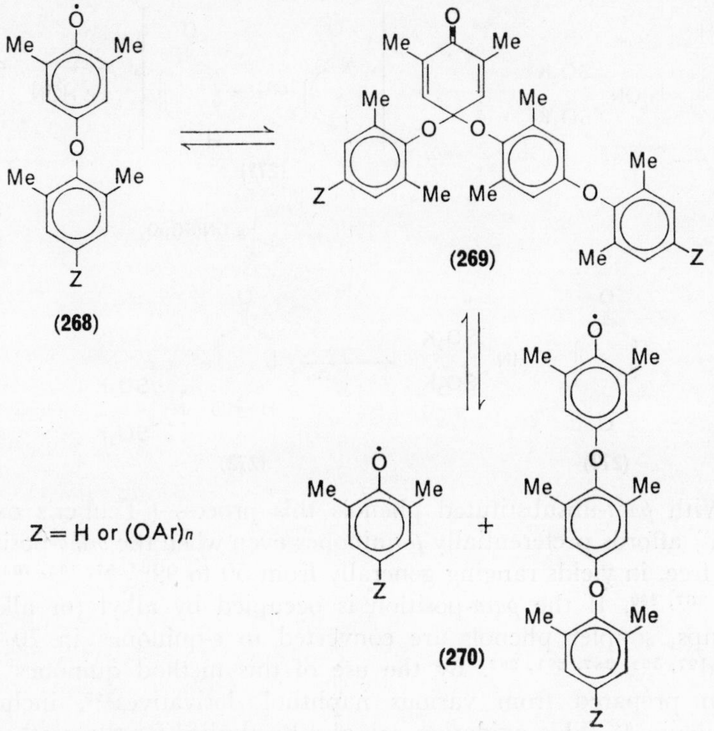

(268) ⇌ (269) ⇌ (270) (94)

Z = H or $(OAr)_n$

H. Miscellaneous Oxidations

1. Potassium nitrosodisulphonate (Fremy's salt)

Potassium nitrosodisulphonate, $ON(SO_3K)_2$, known as Fremy's salt, is one of the most efficient agents for the preparative oxidation of monohydric phenols to *o*- and *p*-quinones[286, 287]. This salt—a relatively unstable, deep yellow, solid dimer—is completely dissociated in water, forming a deep purple solution which is reasonably stable in the pH range 8–11, and contains the anion-radical $^{\bullet}ON(SO_3^-)_2$ [288]. Two moles of Fremy's salt are consumed per mole of phenol (reaction 95) and produce one mole each of potassium hydroxylamine-*N*,*N*-disulphonate and potassium imidodisulphonate, whereby the first mole generates from phenol a phenoxy radical (**271**) which then combines with the second mole to give an intermediate of the quinol type **272** ('quinitrol')[287]. The formation of such quinols has been demonstrated in particular cases[289], but they usually decompose rapidly into the quinone **273** and the potassium salt of imidodisulphonic acid.

OH, + ON(SO3K)2 ⟶ [(271)] + HON(SO3K)2; +ON(SO3K)2; (272) ⟶ (273) + HN(SO3K)2 (95)

With *para*-unsubstituted phenols this process ('Teuber's oxidation') affords preferentially *p*-quinones even when the *ortho*-positions are free, in yields ranging generally from 50 to 99% [97, 102, 104, 176, 286, 287, 290]. If the *para*-position is occupied by alkyl (or alkoxy) groups, simpler phenols are converted to *o*-quinones in 70–90% yield[97, 102, 287, 291, 292]. By the use of this method quinones have been prepared from various naphthol derivatives[290], including equilenin[293]. This oxidation can also be applied for the synthesis of more complicated, labile *o*-quinones[291] and hydroxy-*p*-quinones[294], provided that the redox potential of the quinone to be formed is not too high.

Oxidative dealkylation and dealkoxylation with formation of quinones has been observed in the reaction of Fremy's salt with

2,4,6-trisubstituted phenols (reaction 96)[105, 116, 292], and with 2,4,5-trisubstituted phenols (reaction 97)[127, 292] in which attack by the reagent at the free *o*-position is sterically hindered by a *meta*-substituent. When in 2,4,6-trialkyl phenols the *para*-blocking group is methyl;

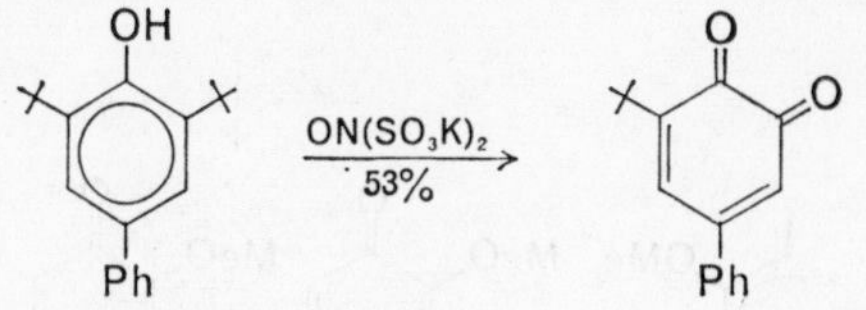

$$\text{ON(SO}_3\text{K)}_2,\ 53\% \qquad (96)$$

$$\text{ON(SO}_3\text{K)}_2 \qquad (97)$$

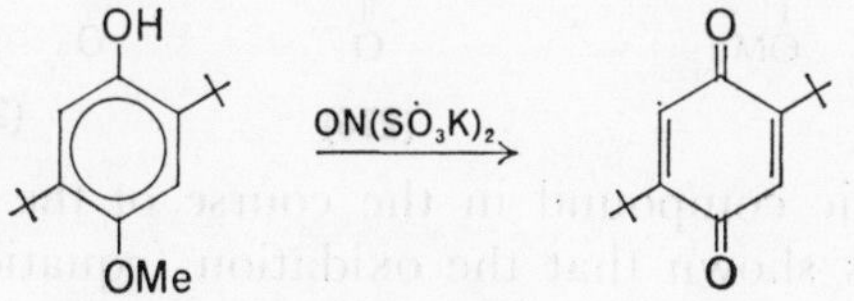

coupling products, e.g., **54**, **55** (equation 14) and **274**, arising from the intermediate *p*-quinone methide **59** (equation 16), are also formed[116].

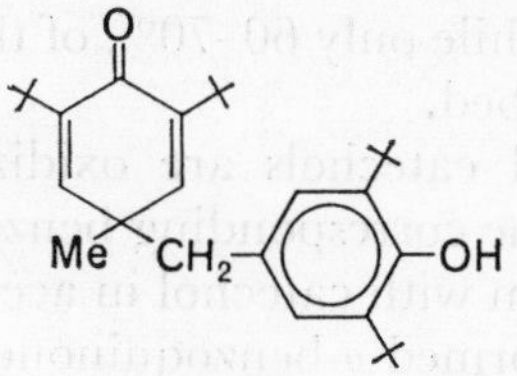

(274)

The use of other stable nitroxides instead of Fremy's salt in the oxidation of phenols does not show any particular advantage[295].

2. Silver oxide

A one-electron oxidant, silver oxide (usually in benzene or diethyl ether), converts phenols to phenoxy radicals[1, 6, 12, 24, 41]. These, depending on their stability, may undergo the characteristic C–C and/or C–O coupling reactions such as in equations (11)[93, 101, 102], (14)[110], (18)[102, 122, 123], (20)[91, 96–99, 129], (22)[111], (29)[152] and oxidation (equation 98)[191], involving both coupling and oxidative demethylation.

Since silver oxide undergoes a rather facile thermal, photolytic or metal-catalysed decomposition to silver metal and oxygen, it may happen that some of the oxygen from silver oxide is incorporated

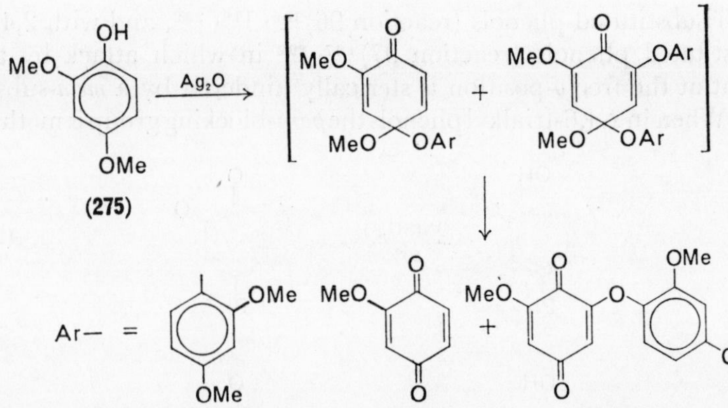

(98)

(99)

into the phenolic compound in the course of the reaction. Thus, Blanchard[93] has shown that the oxidation (equation 86) of 2,4,6-tri-*t*-butylphenol to bis(1,3,5-tri-*t*-butyl-2,5-cyclohexadien-4-on-1-yl) peroxide (**244**), via the corresponding phenoxy radical (**4**), with silver oxide in the presence of oxygen, requires the partial utilization of the oxygen from silver oxide, since 90–100% yields of peroxide (**244**) were obtained while only 60–70% of the theoretical amount of free oxygen was absorbed.

Hydroquinones and catechols are oxidized by silver oxide (in benzene or ether) to the corresponding benzoquinones[1, 6]. However, on running the reaction with catechol in acetone solution, two molecules of the initially formed *o*-benzoquinone undergo a Diels–Alder type addition (equation 99) to produce the yellow crystalline dimer **278** [296].

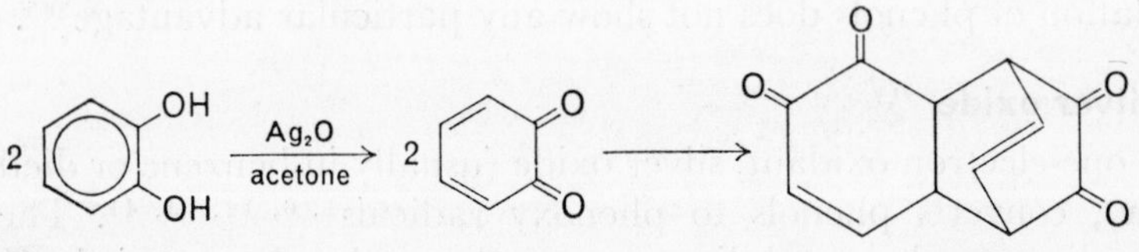

3. Halogens

When 2,4-dialkyl- and 2,4,6-trialkyl-phenols are treated with bromine in the cold in solvent systems containing a proton acceptor (e.g., $AcOH–H_2O$, $Et_2O–H_2O$–pyridine, CCl_4–pyridine, hexane–dioxan, etc.), 4-bromo-2,5-cyclohexadienones (*p*-quinobromides) (**279**) are obtained (reaction 100), in yields up to 98% [23, 25, 109, 297–302]. This

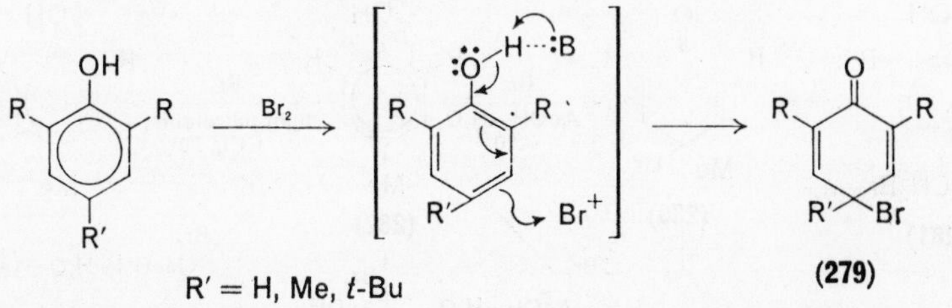

(100)

$R' = H, Me, t\text{-Bu}$

(279)

oxidation is considered to involve electrophilic attack of a bromonium ion (Br^+) at the 4-position of the phenol with elimination of the phenolic proton (reaction 100) [297, 299–301, 303]. 4-Bromo-4-methylcyclohexadienones (**280**) undergo a facile rearrangement (often on standing at room temperature) to 3,5-dialkyl-4-hydroxybenzyl bromides (**281**) (reaction 101)[109, 299, 301, 304]. According to e.s.r.

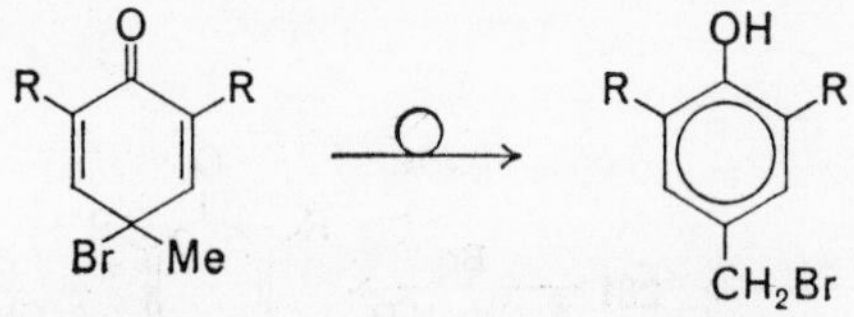

(101)

(280) (281)

measurements such a rearrangement in inert solvents, which is accelerated by u.v. light, proceeds homolytically through phenoxy radicals[109, 301, 302], but in the presence of traces of acids or bases and in polar solvents a heterolytic mechanism might be operative[109, 297, 299, 301, 305]. 2,4,6-Trialkyl-4-bromo-2,5-cyclohexadienones (**279**) when shaken with a metal (Hg, Ag, Cu, Zn, etc.) in an inert solvent under nitrogen can be converted, frequently in nearly quantitative yield, to the corresponding phenoxy radicals[12, 23, 25, 109].

By varying experimental conditions (solvent and temperature) the bromination of 2,6-dialkyl-4-methylphenols (**282**) can afford a variety of products in very good yields (reaction 102), most of which are derived from the initially formed 4-bromo-4-methylcyclohexadienones (**280**)[25, 297, 299, 301, 306]. Scheme (103) represents the possible reaction paths leading to products, most of which have been confirmed experimentally[297, 301, 306].

When 2,6-dialkylphenols containing a *para*-electron-withdrawing group (e.g., NO_2, CN, etc.) are oxidized by bromine in ether–water–pyridine, 2-bromo-3,5,cyclohexadienones (*o*-quinobromides) can be obtained in good yield[307].

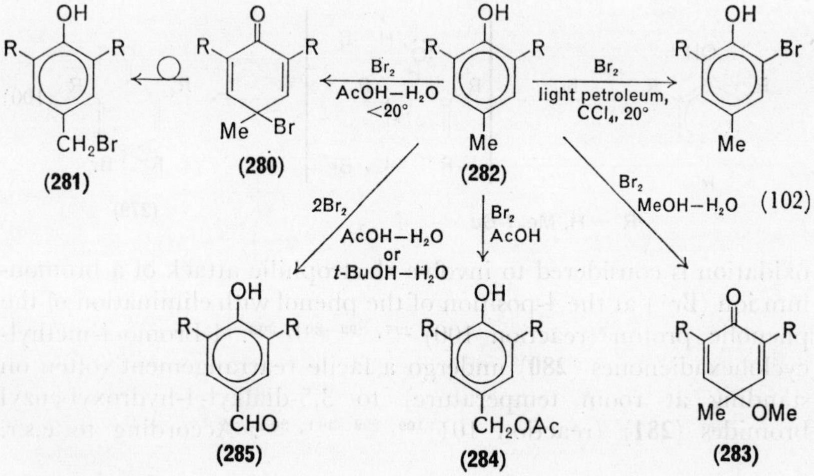

(102)

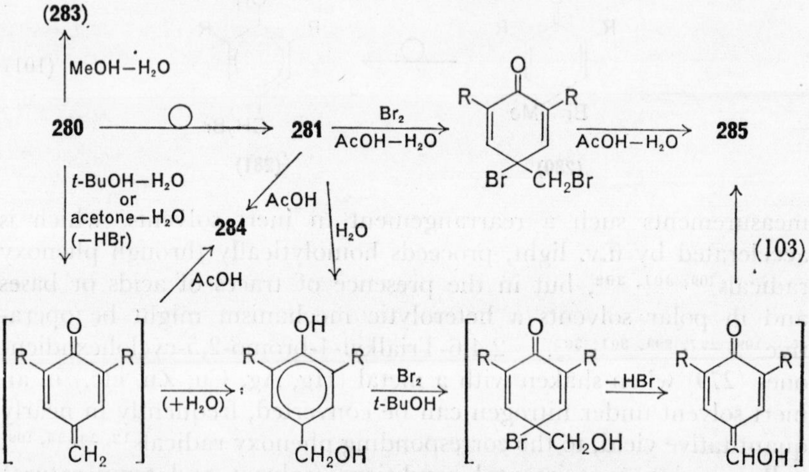

(103)

Oxidation of 2,4,6-trialkylphenols with chlorine or nitric acid in polar solvents at or below room temperature affords the corresponding 4-chloro- and 4-nitro-2,5-cyclohexadienones, respectively[23, 25, 190, 301, 303, 308, 309].

4. Perchloryl fluoride

Perchloryl fluoride, $FClO_3$, has not found much use in oxidations of phenolic compounds, since it offers no preparative advantages over other one-electron oxidizing agents. With 2,6-dimethylphenoxide anion (**287**) in toluene or dioxan at 0° it reacts exothermically

(a) by a one-electron oxidation–reduction process to give 3,3′,5,5′-tetramethyl-4,4′-diphenoquinone (**288**) and 2,6-dimethyl-*p*-benzoquinone (**289**), and (b) by nucleophilic displacement on fluorine leading to the Diels–Alder type fluoro-dimer **291** of the 2,4-cyclohexadienone intermediate **290** [310]. The neutral phenol **286** in dimethylformamide reacts slowly with perchloryl fluoride to give as additional and major product 2,6-dimethyl-4-chlorophenol (**292**), which probably arises from chlorination by intermediate species such as ClO_2 and HOCl [310] (scheme 104).

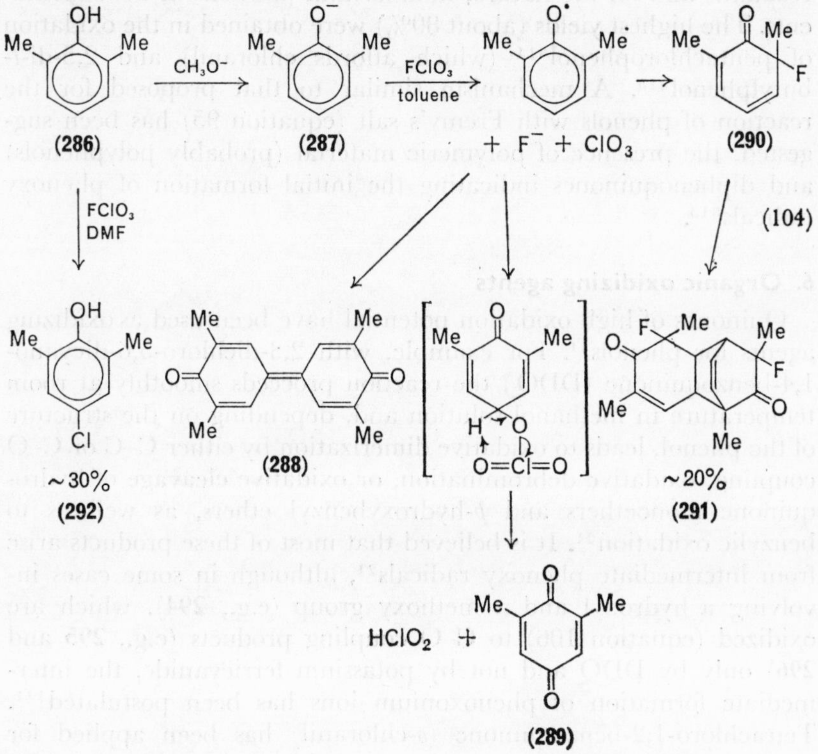

Me OH Me (286) —CH_3O^-→ (287) —$FClO_3$, toluene→ Me O• Me + F^- + ClO_3 → (290) Me O Me F

(286) —$FClO_3$, DMF→ (292) OH Me Me Cl ~30%

(288) Me Me O= =O Me Me

[Me O Me H O O=Cl=O] → $HClO_2$ + (289)

(291) ~20% F Me Me F O O Me Me

(104)

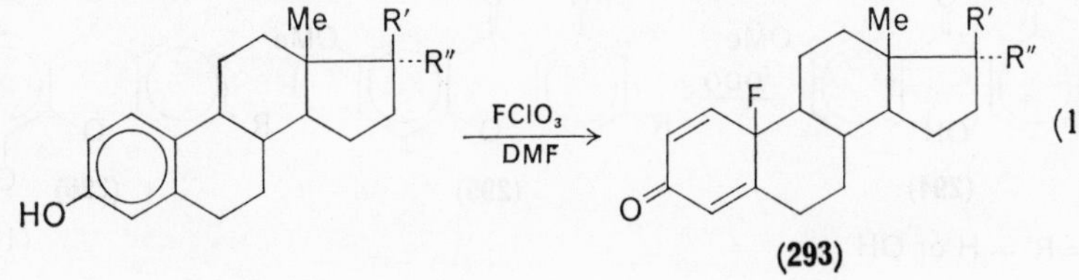

Me R′ ···R″ HO —$FClO_3$, DMF→ (293) Me R′ ···R″ F O (105)

Treatment of steroidal ring A phenols with perchloryl fluoride in dimethylformamide solution results in the *para*-introduction of fluorine with formation of 10β-fluoro-dienones **293** (reaction 105)[311].

5. Chromyl chloride

Chromyl chloride (CrO_2Cl_2) has been used for the oxidation of different types of organic compounds[312, 313], but its action on phenols has only recently been reported. *p*-Benzoquinones are the major reaction products, their yields depending on the ratio of reactants and on the nature, number and positions of the substituents. The highest yields (about 80%) were obtained in the oxidation of pentachlorophenol[314] (which affords chloranil) and 2,5-di-*t*-butylphenol[315]. A mechanism similar to that proposed for the reaction of phenols with Fremy's salt (equation 95) has been suggested, the presence of polymeric material (probably polyphenols) and diphenoquinones indicating the initial formation of phenoxy radicals[315].

6. Organic oxidizing agents

Quinones of high oxidation potential have been used as oxidizing agents for phenols[41]. For example, with 2,3-dichloro-5,6-dicyano-1,4-benzoquinone (DDQ) the reaction proceeds smoothly at room temperature in methanol solution and, depending on the structure of the phenol, leads to oxidative dimerization by either C–C or C–O coupling, oxidative debromination, or oxidative cleavage of hydroquinone monoethers and *p*-hydroxybenzyl ethers, as well as to benzylic oxidation[21]. It is believed that most of these products arise from intermediate phenoxy radicals[21], although in some cases involving a hydroxyl and a methoxy group (e.g., **294**), which are oxidized (equation 106) to C–O coupling products (e.g., **295** and **296**) only by DDQ and not by potassium ferricyanide, the intermediate formation of phenoxonium ions has been postulated[141]. Tetrachloro-1,2-benzoquinone (*o*-chloranil) has been applied for

R O OMe R' OH **(294)** —DDQ→ R O OMe R' O **(295)** + R O R' O OMe **(296)**

R = R' = H or OH (106)

the preparation of labile *o*-quinones from catechols and pyrogallols[6, 8, 316], and for the generation of aryloxy radicals from resorcinols[142].

Phenoxy radicals can also abstract the hydroxylic hydrogen from phenols to produce new aryloxy radicals[19, 38, 94, 121]. If the new radicals are stable, the position of the equilibrium will depend upon structural features (particularly upon oxidation potentials), concentration and solvent[34, 36, 51, 80, 87, 136, 190, 317]. By using two moles of a starting hindered phenoxy radical, e.g., **297**, per mole of phenol (**298**), mixed quinol ethers (**301**) are formed (equation 107), the less hindered phenol (**298**) reacting in the oxygen radical form (**300**)[12, 35, 37, 38, 55, 318–321].

(297) + (298) $\underset{\text{fast}}{\rightleftharpoons}$ (299) + (300)

(107)

[(297) ⟷ (297)] + (300) $\xrightarrow{\text{fast}}$ (301)

Several enzymatic systems and cell-free extracts of higher plants have been found to catalyse the oxidative coupling reactions of phenolic compounds[6, 8, 12].

7. Electrolytic oxidation

Electrochemical methods for the preparative scale oxidation of phenolic compounds have received far less attention than chemical procedures.

The phenoxy radicals produced by electrolytic oxidation of phenols in neutral or basic media may undergo, in addition to the usual C–C and C–O coupling processes, other reactions as well, such as hydroxylation and further oxidation to quinones[15]. An interesting case, related to coumarin biosynthesis, is the electrooxidation at a platinum anode of *p*-hydroxyphenylpropionic acids **302** (phloretic acids) (reaction 108), which leads, probably by way of radical coupling, to the dienone lactones **303** [322, 323]. When treated with

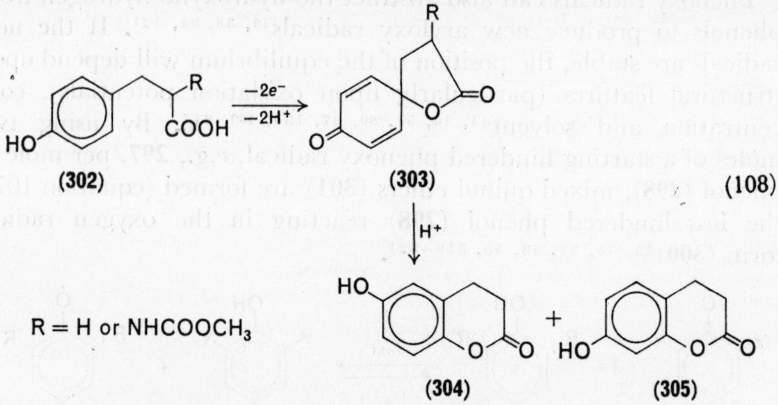

R = H or $NHCOOCH_3$

(302) $\xrightarrow[-2H^+]{2e^-}$ (303) (108)

$\downarrow H^+$

(304) + (305)

mineral acid the latter undergo rearrangement to a mixture of 6- and 7-hydroxydihydrocoumarins (**304, 305**) [323]. *p*-Hydroxy-*cis*- cinnamic acid behaves similarly[323].

In acid solution the anodic oxidation of polyarylsubstituted phenols consists, according to the slopes of the curves corresponding to the half-wave potentials, of two single electron transfers, the initially formed phenoxy radicals being further oxidized to cations (equation 109) [75]. When aqueous acetic acid containing sodium acetate was used as solvent, quinol acetates (**306**) were obtained in nearly quantitative yields[75].

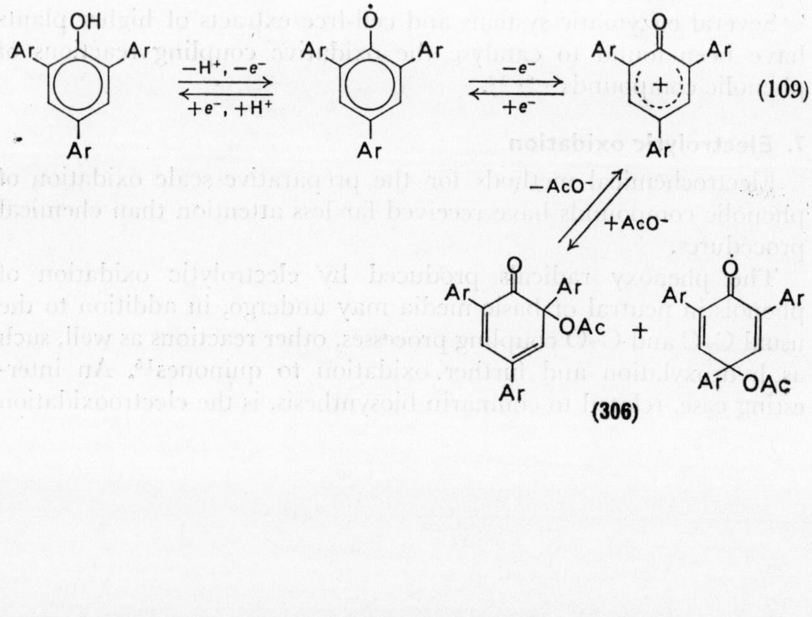

$\underset{+e^-, +H^+}{\overset{-H^+, -e^-}{\rightleftharpoons}}$ $\underset{+e^-}{\overset{-e^-}{\rightleftharpoons}}$ (109)

$-AcO^-$ $+AcO^-$

(306)

8. Other oxidants

The conversion of phenols to phenoxy radicals and further products can be effected by means of manganic ions (Mn^{3+})[324], ceric ions (Ce^{4+})[16, 17, 73, 100, 102, 106, 180, 325], activated manganese dioxide (MnO_2)[104, 123, 321], mercuric oxide (HgO)[49], cupric salts of carboxylic acids[326] and sodium bismuthate ($NaBiO_3$)[100, 101, 180, 252]. It is reported[327] that phenols are easily oxidized by vanadium (v) and cobalt (III) salts. Permanganate readily attacks phenols and, given sufficient oxidant, converts them mainly to carbon dioxide and water[328]. Halate ions (XO_3^-) can also oxidize phenolic compounds[104, 130]. For example, pyrogallol and 4-substituted pyrogallols (**307**) are converted (equation 110) by means of aqueous

(307) $\xrightarrow{IO_3^-}$ (308) (110)

(309) $\xleftarrow[-CO_2]{-2H}$

$\xrightarrow[-H^+,\ -e^-_{aq}]{h\nu\ (2537\ \text{Å})}$ $\xrightarrow{OH^-}$ (310) (111)

potassium or sodium iodate to purpurogallin and its 4′,7-disubstituted derivatives (**309**), probably through the initially formed *o*-quinone **308** [329–332].

Flash photolysis of phenols affords phenoxy radicals, which have been analysed by electron (u.v. and visible) and e.s.r. spectroscopy[12]. Various C–O and C–C coupled dimers and hydroxylated products were obtained (equation 111), the latter (**310**) being formed even when photolysis was performed under nitrogen[333].

III. REDUCTION OF PHENOLS

In general, the conversion of phenols to compounds of lower oxidation levels involves either hydrogenolysis of the phenolic hydroxyl group and/or hydrogenation of the aromatic ring[334]. Depending on the reaction conditions (nature and amount of catalyst, hydrogen pressure, temperature, solvent), the following reduction processes (equation 112) are possible: (i) hydrogenation of the benzene ring with retention of the hydroxyl group[334–336]; (ii) hydrogenation to alicyclic ketones[334, 337]; (iii) hydrogenolysis to aromatic hydrocarbons[334, 338, 339]; (iv) hydrogenolysis and hydrogenation to alicyclic hydrocarbons[334, 338–341]; (v) hydrodealkylation of alkyl-substituted phenols to lower homologues[334, 342].

OH, R; $3H_2$ (i); OH, R; H_2 (v); OH + RH

$2H_2$ (ii); H_2 (iii); (iv) $4H_2$ (112)

O, R; R + H_2O; R + H_2O

A. Hydrogenation and Hydrogenolysis of Phenols

When phenol undergoes hydrogenation, the predominant product may be cyclohexanol, benzene or cyclohexane, depending on the catalyst employed and the reaction. Under certain conditions and with specific catalysts, cyclohexanone may be isolated during the course of the reaction[334, 343–345]. According to kinetic evidence and product distribution in various reductions, it appears that the conversion of phenols to cyclohexanols and cyclohexanes proceeds according to a complex mechanism (scheme 113)[334, 341, 344–346], involving successive hydrogenations of the substrate adsorbed on the catalyst surface to give short-lived intermediate cyclohexadienols **311** and cyclohexenols **312**. Tautomerization of 1-cyclohexenol to

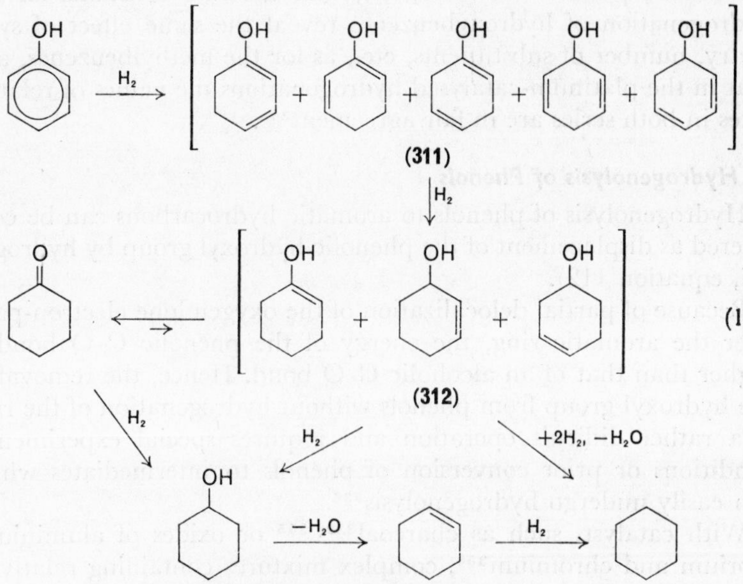

(113)

cyclohexanone would be expected to take place since the keto form is more stable than the enol form by about 18 kcal/mole, and this would account for the presence of ketones in the reaction mixture without requiring that cyclohexanones are intermediates directly involved in the formation of cyclohexanols[344, 345]. Since reduction of cyclohexanol to cyclohexane proceeds by loss of water, i.e., by C–O bond cleavage, the hydrogenative conversion of phenols to alicyclic hydrocarbons (cyclohexanes and isomerization products, such as methylcyclopentanes) usually must be performed at higher temperatures (over 200°) and in the presence of catalysts with dehydrating properties[334, 340, 341, 345].

Catalysts used for the hydrogenation of phenols to cyclohexanols and further hydrogenolysis to cyclohexanes (and isomeric alicyclic hydrocarbons) are platinum, palladium, rhodium[344, 345, 347, 348], and nickel on alumina[348, 349], oxides of nickel, molybdenum and wolfram[350], and mixtures of metal oxides and sulphides, such as $WS_2 + NiS + Al_2O_3$ [351, 352], $MoS_2 + WS_2$ or $MoO_3 + S$ [353].

The effect of the hydroxyl group on the rate of catalytic hydrogenation of the benzene ring was investigated by comparing reaction rates of phenol and dihydric and trihydric phenols with those of benzene and its alkyl derivatives[345, 354]. It was found that in general the kinetic picture is similar, i.e., that the rate constants for the hydrogenation of hydroxybenzenes reveal the same effect of symmetry, number of substituents, etc., as for the methylbenzenes, and that in the platinum-catalysed hydrogenations the values of relative rates in both series are in fair agreement[345].

B. *Hydrogenolysis of Phenols*

Hydrogenolysis of phenols to aromatic hydrocarbons can be considered as displacement of the phenolic hydroxyl group by hydrogen (iii, equation 112).

Because of partial delocalization of the oxygen lone electron-pairs over the aromatic ring, the energy of the phenolic C–O bond is higher than that of an alcoholic C–O bond. Hence, the removal of the hydroxyl group from phenols without hydrogenation of the ring is a rather difficult operation and requires special experimental conditions or prior conversion of phenols to intermediates which can easily undergo hydrogenolysis[334].

With catalysts such as charcoal[339, 355] or oxides of aluminium, thorium and chromium[356], complex mixtures containing relatively low yields of corresponding aromatic hydrocarbons are usually obtained. More effective catalysts for the hydrogenolysis of mixtures of phenols are molybdenum oxides (MoO_2 and MoO_3)[357], their activity being increased by the addition of small amounts of copper or chromium oxides[358], or sulphur compounds[359]. Freshly prepared molybdenum disulphide, MoS_2 (obtained by reduction of molybdenum trisulphide), is an efficient catalyst for the hydrogenolysis of phenol and *o*-cresol, the yield of aromatic hydrocarbon obtained at 25 atm being about 90%[360]. However, as the pressure increases more products hydrogenated in the ring are formed[360, 361]. In general, the amount of aromatic hydrocarbon (iii, equation 112) increases and that of saturated hydrocarbons (iv, equation 112) decreases by decreasing the hydrogen pressure and reaction temperature and by increasing the number of alkyl substituents in the starting phenol[334, 351, 360–362].

Hydrogenolysis of the hydroxyl group can be achieved by treating phenolic compounds with phosphorus trisulphide at high temperature (equation 114)[338], but only one fourth of phenol present is

$$8\,ArOH + P_2S_3 \longrightarrow 2\,ArH + 2\,(ArO)_3PO + 3\,H_2S \qquad (114)$$

reduced. Therefore, by adding phenol itself, as coreactant, to the phenolic compound which is to be reduced, this reaction can be successfully applied for the synthesis of various polycyclic aromatic hydrocarbons[363].

Hydrogenolysis of the phenolic hydroxyl group can be achieved under milder conditions and in better yield if the starting phenol is first converted (by treatment with *p*-tosyl chloride) to its toluene-*p*-sulphonate ester followed by reduction in the presence of Raney-nickel as catalyst (equation 115)[364] or (by treatment with diethyl

$$2\,Ar{-}O{-}SO_2C_6H_4CH_3\text{-}p \xrightarrow[\text{Ra-Ni}]{H_2} 2\,ArH + Ni(O{-}SO_2C_6H_4CH_3\text{-}p)_2 \qquad (115)$$

phosphite in carbon tetrachloride containing triethylamine) to the corresponding diethyl phosphate, which is then easily reduced by

$$ArOH + HOP(OEt)_2 + CCl_4 + NEt_3 \longrightarrow ArO{-}\overset{O}{\overset{\|}{P}}(OEt)_2 + CHCl_3 + NHEt_3Cl \qquad (116)$$

$$ArO{-}\overset{O}{\overset{\|}{P}}(OEt)_2 + Na + NH_3 \longrightarrow ArH + NaNH_2 + NaO{-}\overset{O}{\overset{\|}{P}}(OEt)_2$$

lithium or sodium in liquid ammonia (equation 116)[365]. In reaction (116) the yields of aromatic hydrocarbons from simple phenols vary from 60 to 90%, but the reaction is much less successful with dihydric phenols[365]. Because of mild experimental conditions in both steps, the reaction sequence (116) has been applied for the hydrogenative conversion of sensitive polycyclic phenolic compounds to the corresponding aromatic hydrocarbons[366].

IV. REFERENCES

1. Reviews: D. H. R. Barton and T. Cohen, in *Festschrift Arthur Stoll*, Birkhäuser, Basel, 1957, pp. 117–143; H. Erdtman and C. A. Wachtmeister, in *Festschrift Arthur Stoll*, Birkhäuser, Basel, 1957, pp. 144–165.
2. Review: B. S. Thyagarajan, *Chem. Rev.*, **58**, 439 (1958).
3. Review: W. A. Waters, *Progr. Org. Chem.*, **5**, 35–45 (1961).
4. Review: J. D. Loudon, *Progr. Org. Chem.*, **5**, 46 (1961).
5. Review: H. Beecken, U. v. Gizycki, E. M. Gottschalk, H. Krämer, D. Maassen, H.-G. Matthies, H. Musso, C. Rathjen and U. I. Záhorsky, *Angew. Chem.*, **73**, 665 (1961).
6. Review: H. Musso, *Angew. Chem.*, **75**, 965 (1963); *Angew. Chem., Int. Ed. Engl.*, **2**, 723 (1963).
7. W. A. Waters, *Mechanisms of Oxidation of Organic Compounds*, Methuen, London, 1964, pp. 132–149.

8. Review: A. I. Scott, *Quart. Rev.* (*London*), **19**, 1 (1965).
9. Review: K. U. Ingold, *Chem. Rev.*, **61**, 563 (1961).
10. Ref. 3, pp. 17–26.
11. Ref. 7, pp. 6–16 and 145–147.
12. Review: E. R. Altwicker, *Chem. Rev.*, **67**, 475 (1967).
13. Reviews: L. M. Strigun, L. S. Vartanyan and N. M. Emanuel, *Usp. Khim.*, **37**, 969 (1968); V. D. Pokhodenko, V. A. Khizhnii and V. A. Bidzilya, *Usp. Khim.*, **37**, 998 (1968).
14. Ref. 3, pp. 26–45.
15. Review: N. L. Weinberg and H. R. Weinberg, *Chem. Rev.*, **68**, 449 (1968).
16. T. J. Stone and W. A. Waters, *Proc. Chem. Soc.*, 253 (1962).
17. T. J. Stone and W. A. Waters, *J. Chem. Soc.*, 213 (1964).
18. C. D. Cook and B. E. Norcross, *J. Am. Chem. Soc.*, **78**, 3797 (1956).
19. E. Müller, R. Mayer, U. Heilmann and K. Scheffler, *Ann. Chem.*, **645**, 66 (1961).
20. R. H. Bauer and G. M. Coppinger, *Tetrahedron*, **19**, 1201 (1963).
21. H.-D. Becker, *J. Org. Chem.*, **30**, 982 (1965).
22. C. D. Cook, *J. Org. Chem.*, **18**, 261 (1953).
23. C. D. Cook and R. C. Woodworth, *J. Am. Chem. Soc.*, **75**, 6242 (1953).
24. E. Müller and K. Ley, *Chem. Ber.*, **87**, 922 (1954).
25. E. Müller, K. Ley and W. Kiedaisch, *Chem. Ber.*, **87**, 1605 (1954).
26. G. M. Coppinger, *J. Am. Chem. Soc.*, **79**, 501 (1957).
27. M. S. Kharasch and B. S. Joshi, *J. Org. Chem.*, **22**, 1435 (1957).
28. P. D. Bartlett and D. Rüchardt, *J. Am. Chem. Soc.*, **82**, 1756 (1960); P. D. Bartlett and T. Funahashi, *J. Am. Chem. Soc.*, **84**, 2596 (1962).
29. C. Besev, A. Lund and T. Vänngard, *Acta Chem. Scand.*, **17**, 2281 (1963).
30. K. Ley, E. Müller and K. Scheffler, *Angew. Chem.*, **70**, 74 (1958).
31. E. Müller, K. Ley, K. Scheffler and R. Mayer, *Chem. Ber.*, **91**, 2682 (1958).
32. C. Steelink and R. E. Hansen, *Tetrahedron Letters*, 105 (1966).
33. N. C. Yang and A. J. Castro, *J. Am. Chem. Soc.*, **82**, 6208 (1960); D. Kearns and S. Ehrenson, *J. Am. Chem. Soc.*, **84**, 739 (1962).
34. K. Dimroth, F. Kalk and G. Neubauer, *Chem. Ber.*, **90**, 2058 (1957); K. Dimroth, F. Kalk, R. Sell and K. Schlömer, *Ann. Chem.*, **624**, 51 (1959).
35. E. Müller, K. Ley and G. Schlechte, *Chem. Ber.*, **90**, 2660 (1957).
36. K. Dimroth, *Angew. Chem.*, **72**, 714 (1960).
37. K. Dimroth and A. Berndt, *Angew. Chem.*, **76**, 434 (1964); *Angew. Chem., Int. Ed. Engl.*, **3**, 385 (1964).
38. K. Dimroth, H. Perst, K. Schlömer, K. Worschech and K.-H. Müller, *Chem. Ber.*, **100**, 629 (1967).
39. K. Dimroth, W. Umbach and H. Thomas, *Chem. Ber.*, **100**, 132 (1967).
40. G. M. Coppinger, *Tetrahedron*, **18**, 61 (1962).
41. O. Neunhoeffer and P. Heitmann, *Chem. Ber.*, **96**, 1027 (1963).
42. E. Müller, H. Eggensperger and K. Scheffler, *Ann. Chem.*, **658**, 103 (1962).
43. E. Müller, H. B. Stegmann and K. Scheffler, *Ann. Chem.*, **645**, 79 (1961).
44. F. D. Greene, W. Adam and J. E. Cantrill, *J. Am. Chem. Soc.*, **83**, 3461 (1961); R. C. Lamb, P. W. Ayers and M. K. Toby, *J. Am. Chem. Soc.*, **85**, 3483 (1963); J. P. Lorand and P. D. Bartlett, *J. Am. Chem. Soc.*, **88**, 3294 (1966); C. Walling and Ž. Čekovič, *J. Am. Chem. Soc.*, **89**, 6681 (1967).
45. P. D. Bartlett and S. T. Purrington, *J. Am. Chem. Soc.*, **88**, 3303 (1966).

46. C. D. Cook and B. E. Norcross, *J. Am. Chem. Soc.*, **81**, 1176 (1959).
47. A. Hubele, H. Suhr and U. Heilmann, *Chem. Ber.*, **95**, 639 (1962).
48. G. M. Coppinger, *J. Am. Chem. Soc.*, **86**, 4385 (1964).
49. B. R. Loy, *J. Org. Chem.*, **31**, 2386 (1966).
50. Review: A. B. Turner, *Quart. Rev. (London)*, **18**, 347 (1964).
51. E. Müller, R. Mayer and K. Ley, *Angew. Chem.*, **70**, 73 (1958).
52. E. Müller, A. Schick, R. Mayer and K. Scheffler, *Chem. Ber.*, **93**, 2649 (1960).
53. Review: E. Müller, A. Rieker, K. Scheffler and A. Moosmayer, *Angew. Chem.*, **78**, 98 (1966); *Angew. Chem., Int. Ed. Engl.*, **5**, 6 (1966).
54. W. E. Wertz, C. F. Koelsch and L. Vivo, *J. Chem. Phys.*, **23**, 2194 (1955).
55. E. Müller, A. Rieker and K. Scheffler, *Ann. Chem.*, **645**, 92 (1961).
56. E. Müller, H. Eggensperger, A. Rieker, K. Scheffler, H.-D. Spanagel, H. B. Stegmann and B. Teissier, *Tetrahedron*, **21**, 227 (1965).
57. A. Rieker and K. Scheffler, *Ann. Chem.*, **689**, 78 (1965).
58. K. Dimroth, A. Berndt, F. Bär, R. Volland and A. Schweig, *Angew. Chem.*, **79**, 69 (1967); *Angew. Chem., Int. Ed. Engl.*, **6**, 34 (1967).
59. A. Rieker and K. Scheffler, *Tetrahedron Letters*, 1337 (1965).
60. A. Rieker, K. Scheffler and E. Müller, *Ann. Chem.*, **670**, 23 (1963). See also A. Rieker and P. Ziemek, *Z. Naturforsch.*, **20b**, 640 (1965); A. Rieker, *Z. Naturforsch.*, **21b**, 647 (1966).
61. K. Dimroth, F. Bär and A. Berndt, *Angew. Chem.*, **77**, 217 (1965); *Angew. Chem., Int. Ed. Engl.*, **4**, 240 (1965).
62. K. Dimroth, A. Berndt and R. Volland, *Chem. Ber.*, **99**, 3040 (1966).
63a. W. T. Dixon and R. O. C. Norman, *J. Chem. Soc.*, 4857 (1964);
63b. A. L. Buchachenko, *Stable Radicals*, Consultants Bureau, New York, 1965, Chapter III.
64. L. Michaelis, M. P. Schubert and S. Granick, *J. Am. Chem. Soc.*, **61**, 1981 (1939).
65. H. Diebler, M. Eigen and P. Matthies, *Z. Elektrochem. Ber. Bunsenges. Physik. Chem.*, **65**, 634 (1961); M. Eigen and P. Matthies, *Chem. Ber.*, **94**, 3309 (1961).
66. W. Flaig and J. C. Salfeld, *Naturwissenschaften*, **47**, 516 (1960).
67. Review: A. Carrington, *Quart. Rev. (London)*, **17**, 67 (1963).
68. F. R. Hewgill, T. J. Stone and W. A. Waters, *J. Chem. Soc.*, 408 (1964).
69. T. J. Stone and W. A. Waters, *J. Chem. Soc.*, 1488 (1965).
70. J. Pilař, I. Buben and J. Pospíšil, *Tetrahedron Letters*, 4203 (1968).
71. K. Ley and E. Müller, *Angew. Chem.*, **70**, 469 (1958).
72. H. Musso and H. Döpp, *Chem. Ber.*, **100**, 3627 (1967).
73. T. J. Stone and W. A. Waters, *J. Chem. Soc.*, 4302 (1964).
74. G. E. Panketh, *J. Appl. Chem. (London)*, **7**, 512 (1957).
75. F. W. Steuber and K. Dimroth, *Chem. Ber.*, **99**, 258 (1966).
76. K. Dimroth and K. J. Kraft, *Chem. Ber.*, **99**, 264 (1966).
77. L. F. Fieser, *J. Am. Chem. Soc.*, **52**, 4915, 5204 (1930).
78. C. G. Haynes, A. H. Turner and W. A. Waters, *J. Chem. Soc.*, 2823 (1956).
79. N. S. Hush, *J. Chem. Soc.*, 2375 (1953); T. Fueno, T. Ree and H. Eyring, *J. Phys. Chem.*, **63**, 1940 (1959).
80. C. D. Cook, C. B. Depatie and E. S. English, *J. Org. Chem.*, **24**, 1356 (1959).
81. H. Musso, K. Figge and D. Becker, *Chem. Ber.*, **94**, 1107 (1961).

82. C. Párkányi and R. Zahradník, *Collection Czech. Chem. Commun.*, **30**, 4287 (1965).
83. J. C. Suatoni, R. E. Snyder and R. O. Clark, *Anal. Chem.*, **33**, 1894 (1961).
84. H. N. Simpson, C. K. Hancock and E. A. Meyers, *J. Org. Chem.*, **30**, 2678 (1965).
85. P. J. Elving and A. F. Krivis, *Anal. Chem.*, **30**, 1645 (1958).
86. J. L. Bolland and P. ten Have, *Discussions Faraday Soc.*, **2**, 252 (1947); see also K. U. Ingold and J. A. Howard, *Nature (London)*, **195**, 280 (1962).
87. C. D. Cook, D. A. Kuhn and P. Fianu, *J. Am. Chem. Soc.*, **78**, 2002 (1956).
88. R. Pummerer, H. Puttfarcken and P. Schopflocher, *Ber.*, **58**, 1808 (1925); R. Pummerer, D. Melamed and H. Puttfarcken, *Ber.*, **55**, 3116 (1922).
89. D. H. R. Barton, A. M. Deflorin and O. E. Edwards, *J. Chem. Soc.*, 530 (1956); *Chem. Ind.*, 1039 (1955).
90. V. Arkley, F. M. Dean, A. Robertson and P. Sidisunthorn, *J. Chem. Soc.*, 2322 (1956).
91. D. F. Bowman and R. F. Hewgill, *Chem. Commun.*, 471 (1967).
92. M. S. Kharasch and B. S. Joshi, *J. Org. Chem.*, **22**, 1439 (1957).
93. H. S. Blanchard, *J. Org. Chem.*, **25**, 264 (1960).
94. E. Müller, R. Mayer, B. Narr, A. Rieker and K. Scheffler, *Ann. Chem.*, **645**, 25 (1961).
95. R. H. Rosenwald and J. A. Chenicek, *J. Am. Oil Chemists' Soc.*, **28**, 185 (1951); J. Baltes and F. Volbert, *Fette, Seifen, Anstrichmittel*, **57**, 660 (1955).
96. F. R. Hewgill, *J. Chem. Soc.*, 4987 (1962).
97. F. R. Hewgill and B. S. Middleton, *J. Chem. Soc.*, 2914 (1965).
98. F. R. Hewgill and D. G. Hewitt, *J. Chem. Soc.*, 3660 (1965).
99. D. F. Bowman, F. R. Hewgill and B. R. Kennedy, *J. Chem. Soc.* (**C**), 2274 (1966).
100. F. R. Hewgill and D. G. Hewitt, *J. Chem. Soc.* (**C**), 726 (1967).
101. C. J. R. Adderley and F. R. Hewgill, *J. Chem. Soc.* (**C**), 1434 (1968).
102. C. J. R. Adderley and F. R. Hewgill, *J. Chem. Soc.* (**C**), 1438 (1968).
103. C. D. Cook, E. S. English and B. J. Wilson, *J. Org. Chem.*, **23**, 755 (1958).
104. R. G. R. Bacon and A. R. Izzat, *J. Chem. Soc.* (**C**), 791 (1966).
105. F. R. Hewgill and B. S. Middleton, *J. Chem. Soc.* (**C**), 2316 (1967).
106. F. R. Hewgill and D. G. Hewitt, *J. Chem. Soc.* (**C**), 723 (1967), and references therein.
107. K. Ley, E. Müller, R. Mayer and K. Scheffler, *Chem. Ber.*, **91**, 2670 (1958).
108. C. D. Cook and N. D. Gilmour, *J. Org. Chem.*, **25**, 1429 (1960).
109. C. D. Cook, N. G. Nash and H. R. Flanagan, *J. Am. Chem. Soc.*, **77**, 1783 (1955).
110. S. L. Cosgrove and W. A. Waters, *J. Chem. Soc.*, 388 (1951).
111. R. F. Moore and W. A. Waters, *J. Chem. Soc.*, 243 (1954).
112. G. R. Yohe, D. R. Hill, J. E. Dunbar and F. M. Scheidt, *J. Am. Chem. Soc.*, **75**, 2688 (1953).
113. G. R. Yohe, J. E. Dunbar, R. L. Pedrotti, F. M. Scheidt, F. G. H. Lee and E. C. Smith, *J. Org. Chem.*, **21**, 1289 (1956).
114. A. F. Bickel and E. C. Kooyman, *J. Chem. Soc.*, 3211 (1953).
115. K. U. Ingold, *Can. J. Chem.*, **41**, 2807 (1963).
116. R. Magnusson, *Acta Chem. Scand.*, **18**, 759 (1964); **20**, 2211 (1966).
117. A. Rieker and H. Kessler, *Tetrahedron*, **24**, 5133 (1968).

118. E. Müller, R. Mayer, H.-D. Spanagel and K. Scheffler, *Ann. Chem.*, **645,** 53 (1961); E. Müller, H.-D. Spanagel and A. Rieker, *Ann. Chem.*, **681**, 141 (1965).
119. R. Pummerer, G. Schmidutz and H. Seifert, *Chem. Ber.*, **85,** 535 (1952); R. Pummerer and I. Veit, *Chem. Ber.*, **86,** 412 (1953).
120. K. Ley, E. Müller and G. Schlechte, *Chem. Ber.*, **90,** 1530 (1957).
121. E. Müller, K. Schurr and K. Scheffler, *Ann. Chem.*, **627,** 132 (1959).
122. B. O. Lindgren, *Acta Chem. Scand.*, **14,** 1203, 2089 (1960).
122a. C. C. Price, in *The Chemistry of the Ether Group* (Ed. S. Patai), John Wiley, New York, 1967, Chap. 11, pp. 517–522.
123. E. McNelis, *J. Org. Chem.*, **31,** 1255 (1966).
124. M. B. Neiman, Y. G. Mamedova, P. Blenke and A. L. Buchachenko, *Dokl. Akad. Nauk SSSR.*, **144,** 392 (1962).
125. W. R. Hatchard, R. G. Lipscomb and F. W. Stacey, *J. Am. Chem. Soc.*, **80,** 3636 (1958).
126. E. Müller and K. Ley, *Chemiker-Ztg.*, **80,** 618 (1956).
127. E. Müller, H. Kaufmann and A. Rieker, *Ann. Chem.*, **671,** 61 (1964).
128. F. R. Hewgill and B. R. Kennedy, *J. Chem. Soc.*, 2921 (1965).
129. F. R. Hewgill and D. G. Hewitt, *Tetrahedron Letters*, 3737 (1965).
130. F. R. Hewgill and B. R. Kennedy, *J. Chem. Soc.* (**C**), 362 (1966).
131. K. Fries and E. Brandes, *Ann. Chem.*, **542,** 48 (1939).
132. R. Pummerer and E. Cherbuliez, *Ber.*, **52,** 1392 (1919).
133. A. Rieker, N. Zeller, K. Schurr and E. Müller, *Ann. Chem.*, **697,** 1 (1966).
134. M. L. Khiedekel, A. L. Buchachenko, G. A. Razuvaev, L. V. Gorbunova and M. B. Neiman, *Dokl. Akad. Nauk SSSR*, **140,** 1096 (1962).
135. C. Walling and S. A. Buckler, *J. Am. Chem. Soc.*, **77,** 6032 (1955).
136. E. Müller, K. Ley and W. Schmidhuber, *Chem. Ber.*, **89,** 1738 (1956).
137. K. Scheffler, *Z. Anal. Chem.*, **181,** 456 (1960).
138. D. H. R. Barton, *Proc. Chem. Soc.*, 293 (1963), and references therein.
139. D. H. R. Barton, *Chemistry in Britain*, 330 (1967), and references therein.
140. G. D. Staffin and C. C. Price, *J. Am. Chem. Soc.*, **82,** 3632 (1960).
141. J. W. A. Findlay, P. Gupta and J. R. Lewis, *Chem. Commun.*, 206 (1969).
142. H. Musso, U. v. Gizycki, H. Krämer and H. Döpp, *Chem. Ber.*, **98,** 3952 (1965).
143. H. Güsten, G. Kirsch and D. Schulte-Frohlinde, *Tetrahedron*, **24,** 4393 (1968).
144. B. Franck, G. Blaschke and G. Schlingloff, *Angew. Chem.*, **75,** 957 (1963); *Angew. Chem., Int. Ed. Engl.*, **3,** 192 (1964), and references therein.
145. B. Franck and G. Blaschke, *Ann. Chem.*, **695,** 144 (1966).
146. K. Bowden and C. H. Reece, *J. Chem. Soc.*, 2249 (1950).
147. J. D. Edwards and J. L. Cashaw, *J. Am. Chem. Soc.*, **76,** 6141 (1954).
148. R. Pummerer, E. Prell and A. Rieche, *Ber.*, **59,** 2159 (1926).
149. R. Pummerer, *Ber.*, **52,** 1403 (1919).
150. R. Pummerer and E. Cherbuliez, *Ber.*, **52,** 1414 (1919).
151. R. Pummerer and A. Rieche, *Ber.*, **59,** 2161 (1926).
152. R. Pummerer and R. Frankfurter, *Ber.*, **47,** 1472 (1914).
153. A. Rieche, B. Elschner and M. Landbeck, *Angew. Chem.*, **72,** 385 (1960).
154. T. Posternak, W. Alcalay, R. Luzzati and A. Tardent, *Helv. Chim. Acta*, **31,** 525 (1948).

155. B. Franck and G. Schlingloff, *Ann. Chem.*, **659,** 123 (1962).
156. Review: R. Criegee, in *Oxidation in Organic Chemistry* (Ed. K. Wiberg), Part A, Academic Press, New York, 1965, pp. 288–292.
157. W. Metlesics, E. Schinzel, H. Vilcsek and F. Wessely, *Monatsh.*, **88,** 1069 (1957).
158. G. W. K. Cavill, E. R. Cole, P. T. Gilham and D. J. McHugh, *J. Chem. Soc.*, 2785 (1954).
159. F. Wessely and F. Sinwel, *Monatsh.*, **81,** 1055 (1950).
160. F. Wessely, J. Swoboda and V. Guth, *Monatsh.*, **95,** 649 (1964).
161. F. Takacs, *Monatsh.*, **95,** 961 (1964).
162. F. Wessely, J. Kotlan and F. Sinwel, *Monatsh.*, **83,** 902 (1952).
163. F. Wessely, J. Kotlan and W. Metlesics, *Monatsh.*, **85,** 69 (1954).
164. G. N. Bogdanov and V. V. Ershov, *Izv. Akad. Nauk SSSR, Otd. Khim. Nauk*, 2145 (1962).
165. G. N. Bogdanov, M. S. Postnikova and N. M. Emanuel, *Izv. Akad. Nauk SSSR, Otd. Khim. Nauk*, 173 (1963).
166. F. Wessely and E. Schinzel, *Monatsh.*, **84,** 425 (1953).
167. F. Wessely and J. Kotlan, *Monatsh.*, **84,** 291 (1953).
168. F. Wessely, E. Zbiral and H. Sturm, *Chem. Ber.*, **93,** 2840 (1960).
169. E. Zbiral, F. Wessely and H. Sturm, *Monatsh.*, **93,** 15 (1962).
170. G. Kunesch and F. Wessely, *Monatsh.*, **96,** 1291 (1965).
171. F. Wessely, E. Zbiral and J. Jörg, *Monatsh.*, **94,** 227 (1963).
172. E. Zbiral, F. Wessely and J. Jörg, *Monatsh.*, **92,** 654 (1961).
173. L. J. Smith and H. H. Hoehn, *J. Am. Chem. Soc.*, **61,** 2619 (1939).
174. A. Ebnöther, T. M. Meijer and H. Schmid, *Helv. Chim. Acta*, **35,** 910 (1952); H. Schmid and M. Burger, *Helv. Chim. Acta*, **35,** 928 (1952).
175. R. R. Holmes, J. Conrady, J. Guthrie and R. McKay, *J. Am. Chem. Soc.*, **76,** 2400 (1954).
176. A. M. Gold and E. Schwenk, *J. Am. Chem. Soc.*, **80,** 5683 (1958).
177. E. Hecker, *Naturwissenschaften*, **46,** 514 (1959); E. Hecker and E. Walk, *Chem. Ber.*, **93,** 2928 (1960).
178. E. Hecker, *Chem. Ber.*, **92,** 1386 (1959).
179. E. Hecker and R. Lattrell, *Ann. Chem.*, **662,** 48 (1963).
180. F. R. Hewgill, B. R. Kennedy and D. Kilpin, *J. Chem. Soc.*, 2904 (1965).
181. F. R. Hewgill and S. L. Lee, *J. Chem. Soc.* (**C**), 1556 (1968).
182. H. E. Barron, G. W. K. Cavill, E. R. Cole, P. T. Gilham and D. H. Solomon, *Chem. Ind.*, 76 (1954).
183. R. Criegee, *Angew. Chem.*, **70,** 173 (1958).
184. Review: K. Heusler and J. Kalvoda, *Angew. Chem.*, **76,** 518 (1964); *Angew. Chem., Int. Ed. Engl.*, **3,** 525 (1964).
185. K. Heusler, *Chimia*, **21,** 557 (1967); K. Heusler, H. Labhart and H. Loeliger, *Tetrahedron Letters*, 2847 (1965).
186. M. Lj. Mihailović, M. Jakovljević, V. Trifunović, R. Vukov and Ž. Čeković, *Tetrahedron*, **24,** 6959 (1968); M. Lj. Mihailović, Ž. Čeković, V. Andrejević, R. Matić and D. Jeremić, *Tetrahedron*, **24,** 4947 (1968), and references therein.
187. O. Dimroth, O. Friedemann and H. Kämmerer, *Ber.*, **53,** 481 (1920); O. Dimroth and V. Hilcken, *Ber.*, **54,** 3050 (1921); K. Zahn and P. Ochwat, *Ann. Chem.*, **462,** 72 (1928).
188. K. H. König, W. Schulze and G. Möller, *Chem. Ber.*, **93,** 554 (1960).

189. R. Kuhn and I. Hammer, *Chem. Ber.*, **83,** 413 (1950).
190. E. Müller and K. Ley, *Chem. Ber.*, **88,** 601 (1955).
191. W. H. Hunter and A. A. Levine, *J. Am. Chem. Soc.*, **48,** 1608 (1926); W. H. Hunter and M. Morse, *J. Am. Chem. Soc.*, **48,** 1615 (1926).
192. M. Hedayatullah and L. Denivelle, *Compt. Rend.*, **254,** 2369 (1962).
193. C. H. Hassal and J. R. Lewis, *J. Chem. Soc.*, 2312 (1961); J. R. Lewis, *Chem. Ind.*, 159 (1962).
194. J. R. Lewis and J. A. Vickers, *Chem. Ind.*, 779 (1963).
195. R. Willstätter and J. Parnas, *Ber.*, **40,** 1406 (1907).
196. Review: O. C. Dermer and M. T. Edmison, *Chem. Rev.*, **57,** 77 (1957).
197. Review: N. Uri, *Chem. Rev.*, **50,** 375 (1950).
198. I. D. Raacke-Fels, G. H. Wang, R. K. Robins and B. E. Christensen, *J. Org. Chem.*, **15,** 627 (1950).
199. C. R. E. Jefcoate and R. O. C. Norman, *J. Chem. Soc.* (**B**), 48 (1968).
200. A. R. Bader, *J. Am. Chem. Soc.*, **73,** 3731 (1951).
201. G. Stein and J. Weiss, *J. Chem. Soc.*, 3265 (1951).
202. K. Omura and T. Matsuura, *Tetrahedron*, **24,** 3475 (1968).
203. H. Wieland and W. Franke, *Ann. Chem.*, **451,** 1 (1927); **475,** 1 (1929).
204. M. Martinon, *Bull. Soc. Chim. France*, [2] **43,** 155 (1885).
205. O. Y. Magidson, E. Y. Porozovska and N. E. Seligsohn, *Trans. Sci. Chem.-Pharm. Inst. (Moscow)*, **6,** 23 (1923) [*Chem. Abstr.*, **22,** 3884 (1928)]; O. Y. Magidson and N. A. Preobrazhenskii, *Trans. Sci. Chem.-Pharm. Inst. (Moscow)*, No. 16, 65 (1926) [*Chem. Abstr.*, **23,** 1630 (1929)].
206. H. Goldhammer, *Biochem. Z.*, **189,** 81 (1927).
207. H. Wheland, *J. Am. Chem. Soc.*, **64,** 900 (1942).
208. J. H. Mertz and W. A. Waters, *J. Chem. Soc.*, 2427 (1949).
209. S. L. Cosgrove and W. A. Waters, *J. Chem. Soc.*, 1726 (1951).
210. G. H. Williams, *Homolytic Aromatic Substitution*, Pergamon Press, New York, 1960, pp. 110.
211. R. O. C. Norman and G. K. Radda, *Proc. Chem. Soc.*, 138 (1962).
212. G. A. Hamilton and J. P. Friedman, *J. Am. Chem. Soc.*, **85,** 1008 (1963).
213. H. D. Dakin, *Am. Chem. J.*, **42,** 477 (1909); *Org. Synth.*, Coll. Vol. 1 (2nd ed.), 149 (1946); A. R. Surrey, *Org. Synth.*, Coll. Vol. 3, 759 (1955).
214. W. Baker, E. H. T. Jukes and C. A. Subrahmanyam, *J. Chem. Soc.*, 1681 (1934).
215. W. Baker, *J. Chem. Soc.*, 662 (1941).
216. W. Baker, H. F. Bondy, J. Gumb and D. Miles, *J. Chem. Soc.*, 1615 (1953).
217. A. v. Wacek and H. O. Eppinger, *Ber.*, **73,** 644 (1940); A. v. Wacek and A. v. Bézard, *Ber.*, **74,** 845 (1941).
218. C. A. Bunton, in J. O. Edwards, *Peroxide Reaction Mechanisms*, Interscience, New York, 1962, pp. 14–15; J. Hine, *Physical Organic Chemistry*, 2nd ed., McGraw-Hill Book Co., New York, 1962, p. 341.
219. J. A. Howard and K. U. Ingold, *Can. J. Chem.*, **40,** 1851 (1962).
220. J. A. Howard and K. U. Ingold, *Can. J. Chem.*, **41,** 1744 (1963).
221. J. A. Howard and K. U. Ingold, *Can. J. Chem.*, **42,** 2324 (1964).
222. J. R. Shelton and D. W. Vincent, *J. Am. Chem. Soc.*, **85,** 2433 (1963).
223. K. U. Ingold, *Can. J. Chem.*, **41,** 2816 (1963).
224. C. E. Boozer, G. S. Hammond, C. E. Hamilton and J. N. Sen, *J. Am. Chem. Soc.*, **77,** 3233 (1955).

225. K. U. Ingold and D. R. Taylor, *Can. J. Chem.*, **39,** 471 (1961); K. U. Ingold, *Can. J. Chem.*, **40,** 111 (1962).
226. C. Walling, *Free Radicals in Solution,* John Wiley & Sons, New York, 1957, pp. 397–466.
227. J. A. Howard and K. U. Ingold, *Can. J. Chem.*, **41,** 2800 (1963).
228. T. W. Campbell and G. M. Coppinger, *J. Am. Chem. Soc.*, **74,** 1469 (1952).
229. E. C. Horswill and K. U. Ingold, *Can. J. Chem.*, **44,** 263 (1966).
230. M. E. Hey and W. A. Waters, *J. Chem. Soc.*, 2753 (1955).
231. E. C. Horswill and K. U. Ingold, *Can. J. Chem.*, **44,** 269 (1966).
232. S. L. Cosgrove and W. A. Waters, *J. Chem. Soc.*, 3189 (1949).
233. C. Walling and R. B. Hodgdon Jr., *J. Am. Chem. Soc.*, **80,** 228 (1958).
234. D. B. Denney and D. Z. Denney, *J. Am. Chem. Soc.*, **82,** 1389 (1960).
235. F. Wessely and E. Schinzel, *Monatsh.*, **84,** 969 (1953).
236. J. Böeseken and R. Engelberts, *Proc. Acad. Sci. Amsterdam,* **34,** 1292 (1931) [*Chem. Abstr.*, **26,** 2970 (1932)]; J. Böeseken, *Proc. Acad. Sci. Amsterdam,* **35,** 750 (1932) [*Chem. Abstr.*, **27,** 1332 (1933)].
237. J. Böeseken and C. F. Metz, *Rec. Trav. Chim.*, **54,** 345 (1935).
238. D. Bryce-Smith and A. Gilbert, *J. Chem. Soc.*, 873 (1964).
239. R. D. Chambers, P. Goggin and W. K. R. Musgrave, *J. Chem. Soc.*, 1804 (1959).
240. J. D. McClure, *J. Org. Chem.*, **28,** 69 (1963).
241. K. Elbs, *J. Prakt. Chem.*, **48,** 179 (1893).
242. Review: S. M. Sethna, *Chem. Rev.*, **49,** 91 (1951).
243. W. Baker and N. C. Brown, *J. Chem. Soc.*, 2303 (1948), and references therein.
244. R. B. Desai and S. Sethna, *J. Indian Chem. Soc.*, **28,** 213 (1951).
245. T. R. Seshardi, *Experientia,* Suppl. 2, 258 (1955).
246. R. G. R. Bacon, R. Grime and D. J. Munro, *J. Chem. Soc.*, 2275 (1954).
247. R. G. R. Bacon and D. J. Munro, *J. Chem. Soc.*, 1339 (1960).
248. R. G. R. Bacon and J. R. Doggart, *J. Chem. Soc.*, 1332 (1960).
249. C. E. H. Bawn and D. Margerison, *Trans. Faraday Soc.*, **51,** 925 (1955).
250. D. E. Pennington and D. M. Ritter, *J. Am. Chem. Soc.*, **68,** 1391 (1946); **69,** 187 (1947).
251. E. Adler and S. Hernestam, *Acta Chem. Scand.*, **9,** 319 (1955).
252. E. Adler and R. Magnusson, *Acta Chem. Scand.*, **13,** 505 (1959).
253. J. P. Feiser, M. A. Smith and B. R. Willeford, *J. Org. Chem.*, **24,** 90 (1959).
254. E. Adler, L. Junghahn, U. Lindberg, B. Berggren and G. Westin, *Acta Chem. Scand.*, **14,** 1261 (1960); E. Adler, *Angew. Chem.*, **69,** 272 (1957).
255. E. Adler, J. Dahlén and G. Westin, *Acta Chem. Scand.*, **14,** 1580 (1960).
256. E. Adler, R. Magnusson, B. Berggren and H. Thomelines, *Acta Chem. Scand.*, **14,** 515 (1960); E. Adler, *Angew. Chem.*, **71,** 580 (1959).
257. E. Adler and B. Berggren, *Acta Chem. Scand.*, **14,** 529 (1960).
258. E. Adler, R. Magnusson and B. Berggren, *Acta Chem. Scand.*, **14,** 539 (1960).
259. E. Adler, I. Falkehag and B. Smith, *Acta Chem. Scand.*, **16,** 529 (1962).
260. E. T. Kaiser and S. W. Weidman, *J. Am. Chem. Soc.*, **86,** 4354 (1964); *Tetrahedron Letters,* 497 (1965); S. W. Weidman and E. T. Kaiser, *J. Am. Chem. Soc.*, **88,** 5820 (1966).
261. C. Harries, *Ber.*, **35,** 2957, (1902); H. Erdtman, *Ann. Chem.*, **513,** 240 (1934); M. Nierenstein, *J. Chem. Soc.*, **107,** 1217 (1915).

262. T. W. Campbell and G. M. Coppinger, *J. Am. Chem. Soc.*, **73,** 2708 (1951).
263. T. W. Campbell, *J. Am. Chem. Soc.*, **73,** 4190 (1951).
264. H. R. Gersmann and A. F. Bickel, *J. Chem. Soc.*, 2711 (1959); 2356 (1962).
265. J. I. Wasson and W. M. Smith, *Ind. Eng. Chem.*, **45,** 197 (1953).
266. G. R. Yohe, J. E. Dunbar, M. W. Lansford, R. L. Pedrotti, F. M. Scheidt, F. G. Lee and E. C. Smith, *J. Org. Chem.*, **24,** 1251 (1959).
267. J. K. Becconsall, S. Clough and F. Scott, *Trans. Faraday Soc.*, **56,** 459 (1960).
268. K. Ley, *Angew. Chem.*, **70,** 74 (1958).
269. F. R. Hewgill and S. L. Lee, *J. Chem. Soc.* (**C**), 1549 (1968).
270. J. Pospíšil and V. Ettel, *Collection Czech. Chem. Commun.*, **24,** 729 (1959).
271. I. Buben and J. Pospíšil, *Tetrahedron Letters*, 5123 (1967).
272. V. Ettel and J. Pospíšil, *Collection Czech. Chem. Commun*, **22,** 1613, 1624 (1957).
273. H. Musso, U. I. Záhorsky, D. Maassen and I. Seeger, *Chem. Ber.*, **96,** 1579 (1963); H. Musso, U. v. Gizycki, U. I. Záhorsky and D. Bormann, *Ann. Chem.*, **676,** 10 (1964); H. Musso and D. Bormann, *Chem. Ber.*, **98,** 2774 (1965); H. Musso and D. Maassen, *Ann. Chem.*, **689,** 93 (1965).
274. H. Musso and U. I. Záhorsky, *Chem. Ber.*, **98,** 3964 (1965).
275. K. Ley, *Angew. Chem.*, **74,** 871 (1962); *Angew. Chem., Int. Ed. Engl.*, **1,** 591 (1962).
276. W. Brackman and E. Havinga, *Rec. Trav. Chim.*, **74,** 937, 1021, 1070, 1100, 1107 (1955); E. Talman, Ph.D. Thesis, Universität Leiden, Holland (1961).
277. A. S. Hay, H. S. Blanchard, G. F. Endres and J. W. Eustance, *J. Am. Chem. Soc.*, **81,** 6335 (1959).
278. A. S. Hay, *J. Polymer Sci.*, **58,** 581 (1962).
279. Y. Ogata and T. Morimoto, *Tetrahedron*, **21,** 2791 (1965).
280. E. Ochiai, *Tetrahedron*, **20,** 1831 (1964).
281. G. F. Endres, A. S. Hay and J. W. Eustance, *J. Org. Chem.*, **28,** 1300 (1963).
282. G. F. Endres and J. Kwiatek, *J. Polymer Sci.*, **58,** 593 (1962).
283. W. J. Mijs, O. E. van Lohuizen, J. Bussink and L. Vollbracht, *Tetrahedron*, **23,** 2253 (1967).
284. G. D. Cooper, H. S. Blanchard, G. F. Endres and H. Finkbeiner, *J. Am. Chem. Soc.*, **87,** 3996 (1965).
285. D. A. Bolon, *J. Org. Chem.*, **32,** 1584 (1967).
286. H.-J. Teuber and G. Jellinek, *Chem. Ber.*, **85,** 95 (1952).
287. H.-J. Teuber and W. Rau, *Chem. Ber.*, **86,** 1036 (1953).
288. G. D. Allen and W. A. Waters, *J. Chem. Soc.*, 1132 (1956).
289. H.-J. Teuber and N. Gotz, *Chem. Ber.*, **89,** 2654 (1956); H.-J. Teuber and G. Thaler, *Chem. Ber.*, **92,** 667 (1959).
290. H.-J. Teuber and N. Gotz, *Chem. Ber.*, **87,** 1236 (1954).
291. H.-J. Teuber and G. Staiger, *Chem. Ber.*, **88,** 802 (1955).
292. E. Müller, F. Günter and A. Rieker, *Z. Naturforsch.*, **18,** 1002 (1963).
293. H.-J. Teuber, *Chem. Ber.*, **86,** 1495 (1953).
294. H. Musso, *Chem. Ber.*, **91,** 349 (1958); H. Musso and H. Beecken, *Chem. Ber.*, **92,** 1416 (1959); H. Musso and H.-G. Matthies, *Chem. Ber.*, **94,** 356 (1961).
295. A. R. Forrester and R. H. Thomson, *J. Chem. Soc.* (**C**), 1844 (1966).
296. J. Harley-Mason and A. H. Laird, *J. Chem. Soc.*, 1718 (1958).
297. G. M. Coppinger and T. W. Campbell. *J. Am. Chem. Soc.*, **75,** 734 (1963).
298. A. A. Volodkin and V. V. Ershov, *Izv. Akad. Nauk SSSR, Otd. Khim. Nauk*, 1108 (1962).

299. V. V. Ershov and A. A. Volodkin, *Izv. Akad. Nauk SSSR, Otd. Khim. Nauk*, 2015 (1962).
300. A. A. Volodkin and V. V. Ershov, *Izv. Akad. Nauk SSSR, Otd. Khim. Nauk*, 2022 (1962).
301. Review: V. V. Ershov, A. A. Volodkin and G. N. Bogdanov, *Usp. Khim.*, **32**, 154 (1963).
302. V. D. Pokhodenko and N. Kalibabchuk, *Zh. Organ. Khim.*, **2**, 1397 (1966).
303. E. Grovenstein Jr. and U. V. Henderson Jr., *J. Am. Chem. Soc.*, **78**, 569 (1956).
304. A. A. Volodkin and V. V. Ershov, *Izv. Akad. Nauk SSSR, Otd. Khim. Nauk*, 1292 (1962).
305. V. V. Ershov and A. A. Volodkin, *Izv. Akad. Nauk SSSR, Otd. Khim. Nauk*, 2026 (1962).
306. V. V. Ershov, A. A. Volodkin, G. A. Nikiforov and K. M. Dyumaev, *Izv. Akad. Nauk SSSR, Otd. Khim. Nauk*, 1389 (1962).
307. V. V. Ershov and A. A. Volodkin, *Izv. Akad. Nauk SSSR, Otd. Khim. Nauk*, 893 (1963).
308. V. V. Ershov and G. A. Zlobina, *Izv. Akad. Nauk SSSR, Otd. Khim. Nauk*, 1667 (1963).
309. K. Ley and E. Müller, *Chem. Ber.*, **89**, 1402 (1956).
310. A. S. Kende and P. MacGregor, *J. Am. Chem. Soc.*, **83**, 4197 (1961).
311. J. S. Mills, *J. Am. Chem. Soc.*, **81**, 5515 (1959); J. S. Mills, J. Barrera, E. Olivares and H. García, *J. Am. Chem. Soc.*, **82**, 5882 (1960).
312. Review: W. H. Hartford and M. Darrin, *Chem. Rev.*, **58**, 1 (1958).
313. Review: K. B. Wiberg, in *Oxidation in Organic Chemistry* (Ed. K. B. Wiberg), Part A, Academic Press, New York, 1965, pp. 69–184.
314. J. A. Strickson and C. A. Brooks, *Tetrahedron*, **23**, 2817 (1967).
315. J. A. Strickson and M. Leigh, *Tetrahedron*, **24**, 5145 (1968).
316. L. Horner and W. Dürkheimer, *Z. Naturforsch.*, **14b**, 741 (1959).
317. E. Müller, K. Ley, A. Rieker, R. Mayer and K. Scheffler, *Chem. Ber.*, **92**, 2278 (1959).
318. E. Müller, K. Ley and G. Schlechte, *Angew. Chem.*, **69**, 204 (1957).
319. H. J. Cahnmann and T. Matsuura, *J. Am. Chem. Soc.*, **82**, 2050, 2055 (1960); T. Matsuura and A. Nishinaga, *J. Org. Chem.*, **27**, 3072 (1962).
320. E. Müller, A. Rieker and A. Schick, *Ann. Chem.*, **673**, 40 (1964).
321. H.-D. Becker, *J. Org. Chem.*, **29**, 3068 (1964).
322. H. Iwasaki, L. A. Cohen and B. Witkop, *J. Am. Chem. Soc.*, **85**, 3701 (1963).
323. A. I. Scott, P. A. Dodson, F. McCapra and M. B. Meyers, *J. Am. Chem. Soc.*, **85**, 3702 (1963).
324. R. Van Helden and E. C. Kooyman, *Rec. Trav. Chim.*, **80**, 57 (1961).
325. C. Steelink, *J. Am. Chem. Soc.*, **87**, 2056 (1965).
326. W. W. Kaeding, *J. Org. Chem.*, **28**, 1063 (1963).
327. Review: W. A. Waters and J. S. Littler, in *Oxidation in Organic Chemistry* (Ed. K. B. Wiberg), Part A, Academic Press, New York, 1965, pp. 237–238.
328. Review: R. Stewart, in *Oxidation in Organic Chemistry* (Ed. K. B. Wiberg), Part A, Academic Press, New York, 1965, pp. 59–60.
329. A. Critchlow, R. D. Haworth and P. L. Pauson, *J. Chem. Soc.*, 1318 (1951); W. Crow and R. D. Haworth, *J. Chem. Soc.*, 1325 (1951).
330. J. C. Salfeld, *Angew. Chem.*, **69**, 723 (1957); *Chem. Ber.*, **93**, 737 (1960).

331. L. Horner and W. Dürckheimer, *Z. Naturforsch.*, **14b**, 744 (1959); *Chem. Ber.*, **95**, 1206, 1219 (1962); L. Horner, W. Dürckheimer, K. H. Weber and K. Dolling, *Chem. Ber.*, **97**, 312 (1964).
332. A. Critchlow, E. Haslam, R. D. Haworth, P. B. Tinker and N. M. Waldron, *Tetrahedron*, **23**, 2829 (1967), and references therein.
333. H.-I. Joschek and I. Miller, *J. Am. Chem. Soc.*, **88**, 3269, 3273 (1966), and references therein; H.-I. Joschek and L. I. Grossweiner, *J. Am. Chem. Soc.*, **88**, 3261 (1966).
334. Review: N. I. Shuĭkin and L. A. Erivanskaya, *Usp. Khim.*, **29**, 648 (1960).
335. V. N. Ipatieff, *J. Russ. Phys.-Chem. Soc.*, **38**, 89 (1906); **39**, 693 (1907); A. E. Osterberg and E. C. Kendall, *J. Am. Chem. Soc.*, **42**, 2616 (1920); L. Palfray, *Bull. Soc. Chim. France*, [5] **7**, 401, 407 (1940).
336. A. K. Macbeth and J. A. Mills, *J. Chem. Soc.*, 709 (1945); L. M. Jackman, A. K. Macbeth and J. A. Mills, *J. Chem. Soc.*, 1717 (1949); G. Vavon and P. Anziani, *Bull. Soc. Chim. France*, [4] **41**, 1638 (1927); G. Vavon and A. Callier, *Bull. Soc. Chim. France*, [4] **41**, 357, 677 (1927).
337. G. G. Joris and J. Vitrone Jr., *U.S. Pat. 2, 829, 166* (1958) [*Chem. Abstr.*, **52**, 14671 (1958)]; A. L. Barney and H. B. Hass, *Ind. Eng. Chem.*, **36**, 85 (1944).
338. A. Geuther, *Ann. Chem.*, **221**, 55 (1883).
339. W. Smith, *J. Chem. Soc. Ind.*, **9**, 445 (1890).
340. S. Andô, *J. Fuel Soc. Japan*, **12**, 62 (1933); [*Chem. Abstr.*, **27**, 3702 (1933)].
341. V. Ipatieff, *J. Am. Chem. Soc.*, **55**, 3696 (1933).
342. P. H. Given, *J. Appl. Chem.*, **7**, 172 (1957); J. Pigman, E. D. Bel and M. B. Neuworth, *J. Am. Chem. Soc.*, **76**, 6169 (1954).
343. G. Vavon and A. L. Berton, *Bull. Soc. Chim. France*, [4] **37**, 296 (1925).
344. R. J. Wicker, *J. Chem. Soc.*, 3299 (1957).
345. H. A. Smith and B. L. Stump, *J. Am. Chem. Soc.*, **83**, 2739 (1961).
346. Y. Takagi, S. Nishimura, K. Taya and K. Hirota, *J. Catalysis*, **8**, 100 (1967).
347. P. N. Rylander and N. Himelstein, *Engelhard Ind., Tech. Bull.*, **5**, 43 (1964); [*Chem. Abstr.*, **62**, 3968 (1965)].
348. M. Kraus, K. Kochloefl, L. Beranek and V. Bazant, *Proc. Intern. Congr. Catalysis, 3rd, Amsterdam*, **1**, 577 (1964) [*Chem. Abstr.*, **63**, 12998 (1965)]; N. I. Shuĭkin, A. E. Viktorova and N. I. Cherkasin, *Vestnik Mosk. Univ.*, **11**, No. 6, *Ser. Fiz.-Mat. i Estestven. Nauk*, No. 4, 57 (1956) [*Chem. Abstr.*, **51**, 7321 (1957)]; A. Skita and W. Faust, *Ber.*, **72**, 1127 (1939).
349. N. I. Shuĭkin, E. A. Viktorova, I. E. Pokrovska and A. I. Afanaseva, *Vestnik Mosk. Univ.*, **12**, *Ser. Mat. Mekh., Astron. Fiz., Khim.*, No. 2, 157 (1957); [*Chem. Abstr.*, **57**, 299 (1958)].
350. S. Andô, *J. Soc. Chem. Ind., Japan*, **40**, 83 (1937) [*Chem. Abstr.*, **31**, 6851 (1937)]; *J. Soc. Chem. Ind., Japan*, **42**, 27 (1939) [*Chem. Abstr.*, **33**, 4406 (1939)].
351. K. A. Alekseeva and B. L. Moldovskiĭ, *Khim. i Tekhnol. Topliv i Masel*, **4**, 43 (1959); [*Chem. Abstr.*, **53**, 10104 (1959)].
352. G. Guenther, *Chem. Tech. (Berlin)*, **13**, 720 (1961); [*Chem. Abstr.*, **57**, 4954 (1962)].
353. N. I. Shuĭkin and L. A. Erivanskaya, *Neftekhimiya*, **4**, 431 (1964) [*Chem. Abstr.*, **61**, 6927 (1964)]; S. Andô, *J. Soc. Chem. Ind., Japan*, **43**, 328, 355 (1940) [*Chem. Abstr.*, **35**, 1770, 3980 (1941)]; *J. Soc. Chem. Ind., Japan*, **41**, 413 (1938) [*Chem. Abstr.*, **33**, 6807 (1939)].

354. G. Gilman and G. Cohn, *Advances in Catalysis*, Vol. IX, Academic Press, New York, 1957, p. 736.
355. T. A. Antonova and V. E. Rakovskii, *Tr. Kalininsk. Torf. Inst.*, 29 (1960); [*Chem. Abstr.*, **57**, 2516 (1962)].
356. A. Kling and D. Florentin, *Compt. Rend.*, **182**, 389, 526 (1926); **193**, 859, 1023 (1931).
357. H. Tropsch, *Fuel*, **11**, 61 (1932) [*Chem. Abstr.*, **26**, 3493 (1932)]; L. Woodward and A. T. Glover, *Trans. Faraday Soc.*, **44**, 608 (1948); H. E. Newall, *Fuel Res. Tech.*, *Paper*, No. 48, 55 (1938) [*Chem. Abstr.*, **32**, 9448 (1938)].
358. T. Bahr and A. J. Patrick, *Brennstoff-Chem.*, **14**, 161 (1933); [*Chem. Abstr.*, **27**, 4906 (1933)].
359. J. Varga and I. Makray, *Brennstoff-Chem.*, **17**, 81 (1936); [*Chem. Abstr.*, **30**, 7821 (1936)].
360. B. L. Moldavskii and S. E. Levshitz, *J. Gen. Chem. USSR*, **3**, 603 (1933); [*Chem. Abstr.*, **28**, 2693 (1934)].
361. V. N. Ipatieff and J. Orlov, *Compt. Rend.*, **181**, 793 (1925).
362. Review: C. M. Cawley, *Research*, **1**, 553 (1948).
363. W. N. Moulton and C. G. Wade, *J. Org. Chem.*, **26**, 2528 (1961).
364. G. W. Kenner and M. A. Murray, *J. Chem. Soc.*, S 178 (1949).
365. G. W. Kenner and N. R. Williams, *J. Chem. Soc.*, 522 (1955).
366. S. W. Pelletier and D. M. Locke, *J. Org. Chem.*, **23**, 131 (1958).

CHAPTER 11

Displacement of hydroxyl groups

GEOFFREY W. BROWN

Sir John Cass College, London, England

I. INTRODUCTION

Alcohols are among the most easily obtained reagents of organic chemistry. For this reason the overall conversion

$$ROH \longrightarrow RX$$

where R is any alkyl, aryl, allyl, benzyl, propargyl, vinyl or acyl group, where X is (typically) halide, hydride, azide, alkyl or amine, and where the reaction proceeds with rupture of the carbon–oxygen bond, is of great importance. Through it, a wide range of organic derivatives is available from readily accessible (including naturally occurring) starting materials, and furthermore, many of the fundamental mechanisms of organic reactions have been elucidated using similar displacements.

This chapter seeks to provide a survey of both the synthetic and mechanistic importance of replacing hydroxyl groups by other functional groups, although lack of space precludes an exhaustive coverage of these topics.

Displacement reactions involving the initial formation of an ester, ROY, even if the material is isolable, are included where the ester is usually prepared as an intermediate in the synthesis of RX (equation 1). Thus, reagents such as **1**, prepared from alcohols as intermediates in their conversion to iodides[1], are included, whereas displacements of *p*-toluenesulphonate (tosylate), acetate, and similar derivatives of alcohols will not be covered in detail.

$$ROH \longrightarrow [ROY] \longrightarrow RX \qquad (1)$$

$$C_6H_4O_2PCl \xrightarrow{ROH} C_6H_4O_2POR \xrightarrow{I_2} RI$$

(1)

Displacement of hydroxyl groups by different groups will be dealt with in sequence, and different types of hydroxyl function will be discussed separately where applicable under each heading. The hydroxyl group in carboxylic and sulphonic acids will not be considered in detail.

The commonly available literature has been searched through the greater part of 1968. Many of the references quoted, particularly for reactions of classical importance, are of recent applications of the technique, and are not necessarily included for any other reason. Undoubtedly some contributions to our understanding of the problem have been overlooked.

II. DISPLACEMENT BY HALOGEN

A. Displacement by Iodine

I. Direct halogenation

(a) The classical method for converting alcohols to alkyl iodides[2] involves heating the alcohol with iodine in the presence of red phosphorus (equation 2). Like other iodinations using phosphorus-

$$6\,ROH + 2\,P + 3\,I_2 \longrightarrow 6\,RI + 2\,H_3PO_3 \tag{2}$$

containing reagents, the reaction proceeds through an intermediate ester which is decomposed by hydriodic acid liberated in the formation of the ester (equation 3). The analogous mechanism of reactions

$$PI_3 \xrightarrow[ROH]{} P(OR)_3 \xrightarrow[HI]{} RI \tag{3}$$

leading to alkyl chlorides[3] and bromides[3] is discussed in detail later: products of greater optical purity are formed from phosphorus trichloride and tribromide than from phosphorus triiodide. The intermediate phosphite esters are isolable in the absence of free acids.

(b) Methanol is converted rapidly and quantitatively to methyl iodide using iodine in the presence of diborane[4] (equation 4),

$$MeOH \xrightarrow[LiBH_4]{I_2} MeI \tag{4}$$

whereas the same alcohol requires an excess of red phosphorus mixed with the yellow allotrope[5] when the classical method is used for the preparation of methyl iodide.

2. The use of hydriodic acid

α-Glycols have been converted to vicinal diiodides[6] in high yield (equation 5) under very mild conditions, although a more typical

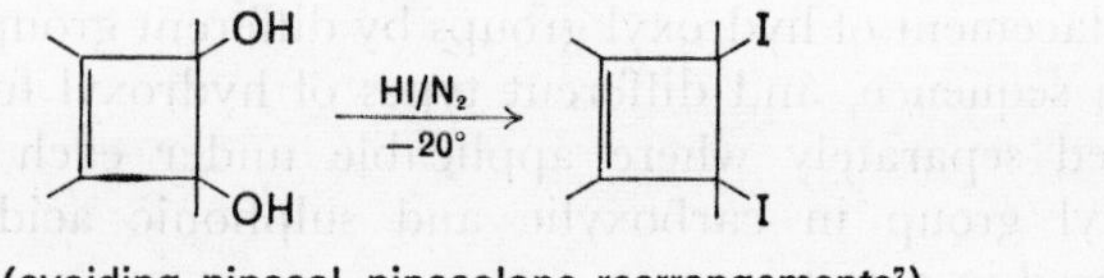

(5)

(avoiding pinacol–pinacolone rearrangements[7])

example of the displacement uses refluxing concentrated acid[8]

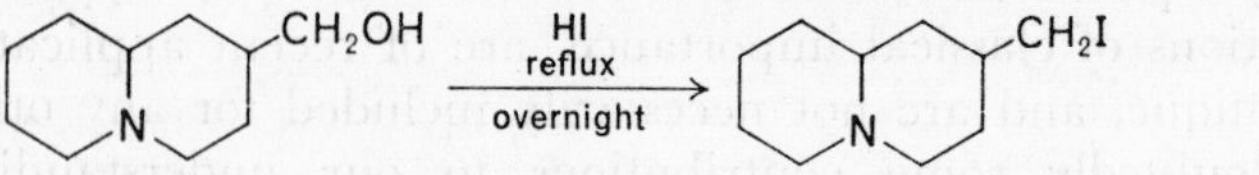

(6)

(equation 6) or the in situ generation of the acid[9] (equation 7). The

$$HO(CH_2)_6OH \xrightarrow[95\%\ H_3PO_4]{KI} I(CH_2)_6I \qquad (7)$$

generally accepted mechanism of the reaction involves displacement of water from the protonated alcohol (equation 8) by either an S_N1

$$ROH \xrightarrow{H^+} [R\overset{+}{O}H_2] \xrightarrow{I^-} RI + H_2O \qquad (8)$$

or S_N2 process, depending on the stability of the carbonium ion which is generated in an S_N1 ionization, but a recent suggestion[10], that all substitutions at saturated carbon proceed by a single mechanism accommodating these extremes, must be considered too. Hydriodic acid is more acidic than the other halogen acids, and iodide ion is a better nucleophile than the other halide ions, so that alkyl iodides are formed more readily than the other halides under corresponding conditions.

In common with all acid-catalysed reactions of alcohols, the production of alkyl iodides by this method is accompanied by rearrangements in unsaturated alcohols, although dehydration reactions are not troublesome. But hydriodic acid is a reducing agent, and can convert alkyl iodides to alkanes. This reduction was used by Cope in the structure-determination of the macrolide antifungal antibiotic fungichromin (**2**)[11], and in determining the carbon skeleton (**3**) of the aglycone of the related rimocidin[12].

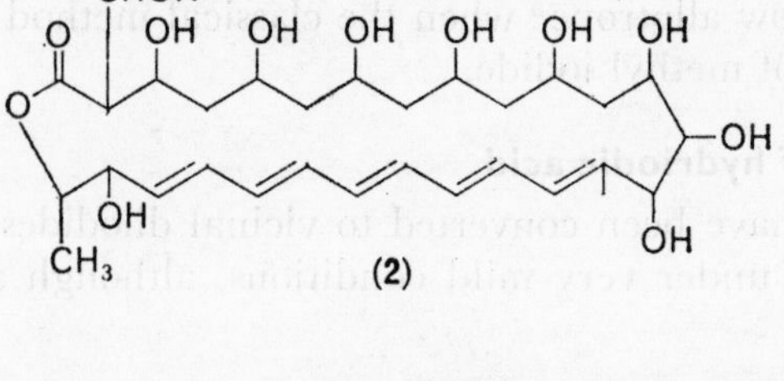

(2)

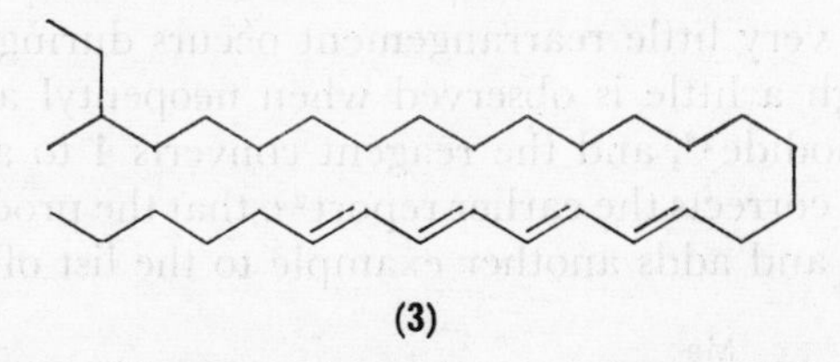

(3)

3. Phosphonium salts and derivatives

(a) Work by Arbusov developed the reaction which now bears his name: phosphonium salts derived from alkyl halides and various organic phosphites (equation 9) disproportionate, on heating, to give (equation 10) a phosphonate, and a different alkyl halide, derived from the phosphite[13].

$$RBr + P(OR')_3 \longrightarrow (R'O)_3\overset{+}{P}R\ Br^- \tag{9}$$

$$(R'O)_3\overset{+}{P}R\ Br^- \longrightarrow R'Br + RPO(OR')_2 \tag{10}$$

The reaction has recently regained prominence because of controversy over the reaction of phosphines with halo-ketones[14] and because the phosphonate by-products from the Arbusov reaction are used in a modification[15] of the Wittig olefin-forming reaction.

Triisopropyl phosphite gives a high yield of isopropyl iodide when heated with methyl iodide[16] (equation 11). In general the conversion of an alcohol to the iodide is effected in two stages (equation 12); an acidic reaction medium is avoided during the complete sequence, and rearrangements do not occur.

$$[(CH_3)_2CHO]_3P \xrightarrow[\Delta]{MeI} (CH_3)_2CHI \tag{11}$$

$$ROH \xrightarrow[base]{PCl_3} P(OR)_3 \xrightarrow[\Delta]{R'I} RI \tag{12}$$

(b) The method was modified by Rydon and Landor[17], and developed by their co-workers. Phosphite methiodides give good yields of iodides when treated with alcohols (equation 13) and the only serious disadvantage in the method is in separating involatile, base-sensitive iodides from the phenol by-product. Where this separation is easy the method works well, and it has been used on a wide variety of alcohols, including primary[17], secondary[17], tertiary[17], propargylic[18] and other unsaturated systems[17], glycols[17], hydroxy acids[17], sugars (both as free carbohydrate[19, 20] and in nucleosides[21, 22]) and cholesterol[17].

$$(PhO)_3\overset{+}{P}Me\ I^- + ROH \longrightarrow RI + PhOH + (PhO)_2POMe \tag{13}$$

In general, very little rearrangement occurs during the displacement, although a little is observed when neopentyl alcohol is converted to the iodide[23], and the reagent converts **4** to a mixture of **5** and **6** [24]. This corrects the earlier report[25], that the products obtained were **6** and **7**, and adds another example to the list of S_N2 reactions

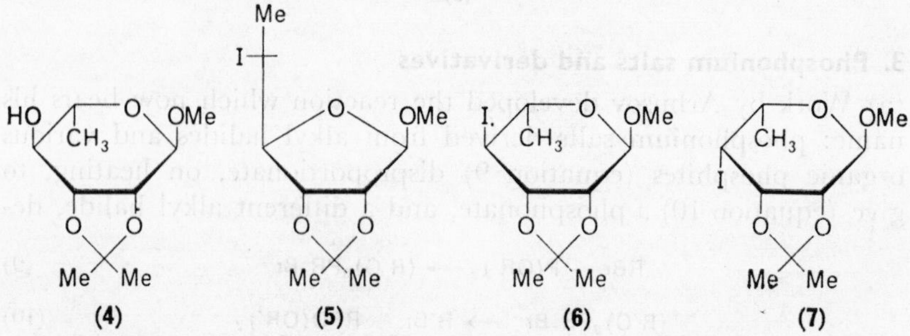

(4) (5) (6) (7)

which involve rearrangements at C-4 [26, 27].

Propargyl alcohols give iodoallenes[18] in dimethylformamide solution, but unrearranged products in methylene chloride. The gross mechanism (equation 14) for displacements using phosphite methiodides involves either expulsion of phenol by the alcohol, followed by nucleophilic attack by iodide ion, or, more probably (with highly electrophilic triphenoxy alkyl salts) attack by iodide ion on an intermediate pentavalent phosphorus derivative.

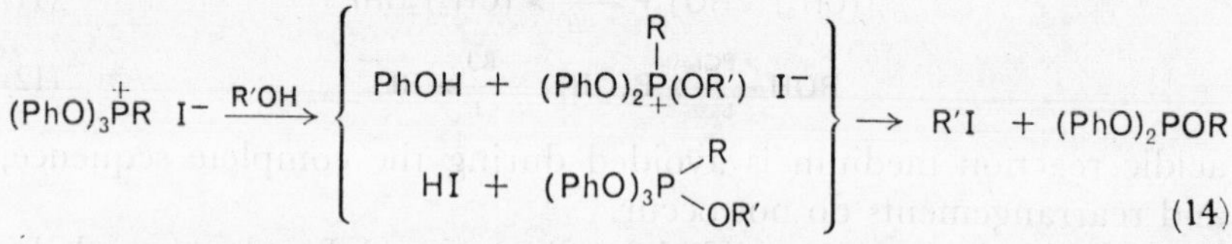

$$(PhO)_3\overset{+}{P}R\ I^- \xrightarrow{R'OH} \left\{ \begin{array}{l} PhOH + (PhO)_2\overset{+}{P}(R)(OR')\ I^- \\ HI + (PhO)_3P(R)(OR') \end{array} \right\} \rightarrow R'I + (PhO)_2POR \qquad (14)$$

(c) The very similar phosphite diiodides $(RO)_3PI_2$ have also been used in this conversion, as have the corresponding dibromides and dichlorides for the preparation of alkyl (and aryl) bromides and chlorides. The diiodide has been used much less often than the other dihalides, and it has been shown in one case[18] that it gives poorer yields than the methiodide.

An early example of the interaction of iodine with phosphites was given by Forsman and Lipkin[28] (equation 15) and was developed by Corey[1] (equation 16) to give what seems to be the method of choice for the preparation of iodides without side-reactions.

The advantages of phosphine dihalide reagents are discussed in the

$$(PhO)_2POC_6H_{11} \xrightarrow{I_2} C_6H_{11}I \quad (15)$$

$$\text{(o-phenylene)}O_2POR \xrightarrow{I_2} RI \quad (16)$$

sections on bromination and chlorination. The stereochemical consequences of using the diiodo-,* dibromo- and dichloro-reagents are similar, and these also will be discussed later.

Modifications of the methiodide reaction have been reported, using halogen acids or metal halides instead of methyl iodide[17]. In general the yields are much less satisfactory, although *n*-butyl chloride can be formed in 63% yield using ammonium chloride as halogen source. The halogen acid-catalysed decomposition of phosphite esters is dealt with in section II. B.3.b.

4. Indirect displacements

The *p*-toluenesulphonate esters of alcohols readily undergo displacement by iodide ions in polar anhydrous solvents (equation 17). Dimethylformamide is rarely used, but acetic anhydride[29], acetone[30], dimethyl sulphoxide[31] and methyl ethyl ketone[32] have often been used.

$$ROTs \xrightarrow{I^-} RI \quad (17)$$

Again, the displacement proceeds by S_N1 or S_N2 mechanisms and this is reflected in the observed optical purity of the products. The most polar solvents favour S_N1 reactions.

1-Apocamphanol has been converted to the corresponding iodide by irradiating the oxalic acid ester in the presence of iodine and mercuric oxide[33].

5. The use of inorganic iodides

Aluminium iodide, generated in situ, has been used to convert cholesterol into cholesteryl iodide[34].

B. Displacement by Bromine and Chlorine

The general methods for preparing bromo- and chloro-compounds from hydroxyl groups are so similar that they will be discussed together. Nearly a hundred categories of the conversion of alcohols to chloro-compounds are discussed in Houben–Weyl[35].

* Unpublished work using Ph_3PI_2 is referred to by Wiley and co-workers, *J. Am. Chem. Soc.*, **86**, 964 (1964).

1. Direct halogenation

(a) Chlorination of alcohols, both photochemically and under ordinary conditions, often causes oxidation[36] (equation 18) and

$$HO(CH_2)_4SH \xrightarrow{Cl_2} Cl(CH_2)_4SO_2Cl \qquad (18)$$

affords no worthwhile preparative route to chloro-compounds[37].

(b) 1-Hydroxyadamantane gives a high yield of 1-bromoadamantane (**8**) when heated in refluxing anhydrous bromine[38] (equation 19). Since the adamantyl cation is known[39] to be formed much more

$$\text{1-hydroxyadamantane (OH)} \xrightarrow{Br_2} \text{1-bromoadamantane (Br)} \qquad (19)$$

(**8**)

easily than other bridgehead ions, and bromine is a Lewis acid, the mechanism of the displacement probably involves the cation as an intermediate.

2. The use of halogen acids

The use of hydrochloric and hydrobromic acids in the preparation of halo-compounds from alcohols has long been a commonplace, and examples of these reactions are legion[40].

The observed reactivity gradations HI > HBr > HCl > HF for the acids, and tertiary > secondary > primary for the alcohols (the former the result both of acidity and anion nucleophilicity), and the use of hydrochloric acid in the presence of zinc chloride for distinguishing primary, secondary and tertiary alcohols (Lucas' test[41]) complete the classical scope of the methods. Phenols are not attacked by the halogen acids HBr and HCl.

The acids cannot be used on acid-sensitive alcohols. Dehydration to alkenes, and rearrangement of cyclopropyl-[42], allyl-[43, 44], propargyl-[45, 46] and, indeed, alkylcarbonium[47] ions in general, occur frequently. One of the earliest reported examples[48] of neighbouring group participation occurs during such a displacement (equation 20)

$$EtSCHMeCH_2OH \xrightarrow{HCl} \text{(episulfonium ion: } \overset{+}{S}\text{–Et, Me)} \xrightarrow{Cl^-} EtSCH_2CHClMe \qquad (20)$$

and other examples of rearrangements observed are given in equations (21–23).

HOCH₂–(tetrahydrothiophene)–CH₂OH —HCl[49]→ Cl–(thiane)–CH₂Cl (21)

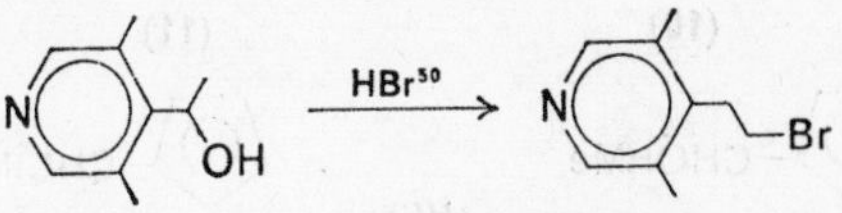

(pyridyl carbinol) —HBr[50]→ (pyridyl)–CH₂–Br (22)

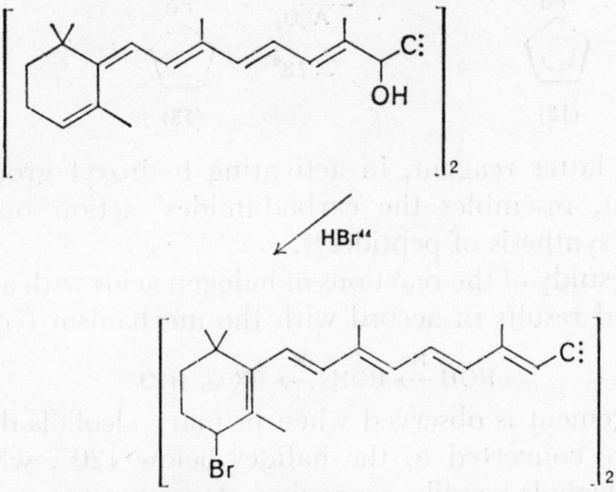

[... C: (OH)]₂ —HBr[44]→ [... C: (Br)]₂ (23)

In the presence of cuprous chloride and calcium chloride, hydrochloric acid converted the propargyl alcohol (**9**) into the unrearranged chloro-compound in high yield[51] at 0°.

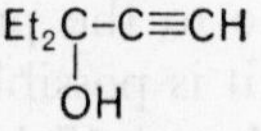

$$Et_2C(OH){-}C{\equiv}CH$$

(**9**)

The rearrangement of alkylcarbonium ions in the course of reactions of alcohols with hydrochloric acid is catalysed by zinc chloride and by sulphuric acid[47, 52], but the results cast no doubt on the validity of Lucas' test since they were obtained at 100°. Zinc bromide and hydrobromic acid convert 1-(bicyclo[2,2,1]heptyl)-methanol (**10**) to 1-bromobicyclo[2,2,2]octane (**11**)[53].

Other catalysts used in conjunction with hydrochloric acid for the conversion of alcohols to chloro-compounds include alumina[54] (equation 24)—the reaction on **12** is carried out at low temperatures to render the highly reactive product **13** isolable—and methyl cyanide[55] and trichloromethyl cyanide[56, 57] as solvents (section II.

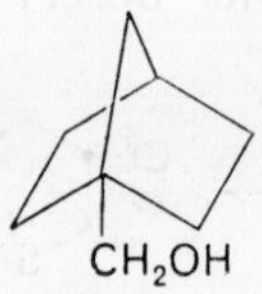

(10) 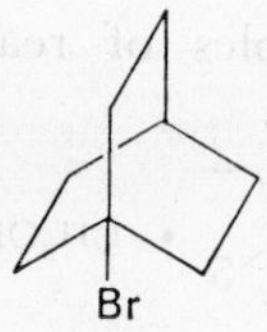 (11)

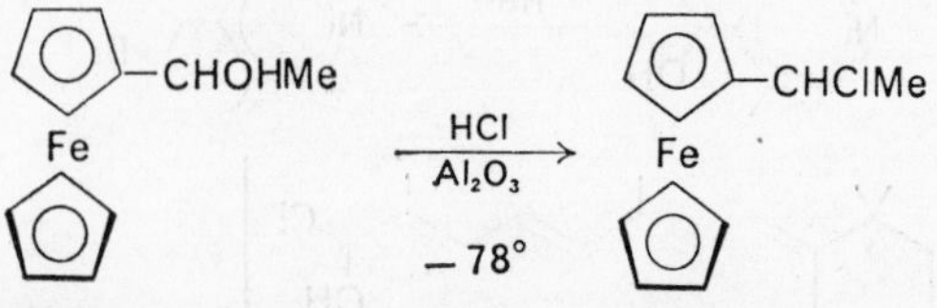

$$\text{(12)} \xrightarrow[\text{Al}_2\text{O}_3,\ -78^\circ]{\text{HCl}} \text{(13)} \tag{24}$$

B.5.c). The latter reagent, in activating hydroxyl groups towards displacement, resembles the carbodiimides' action on carboxylic acids in the synthesis of peptides[58].

A general study of the reactions of halogen acids with alcohols[47, 52] has produced results in accord with the mechanism (equation 25).

$$\text{ROH} \xrightarrow{\text{H}^+} \text{R}\overset{+}{\text{O}}\text{H}_2 \xrightarrow{\text{X}^-} \text{RX} + \text{H}_2\text{O} \tag{25}$$

No rearrangement is observed when primary alcohols derived from *n*-alkanes are converted to the halides below 120°, whereas some secondary alcohols readily rearrange at room temperature. These rearrangements occur not by a dehydration–addition mechanism, but through a carbonium ion mechanism catalysed by zinc chloride and by concentrated sulphuric acid.

Another report is germane to the problem of the mechanism of these and similar reactions: it is possible to distinguish between the diastereoisomeric alcohols **14** and **15** by n.m.r. spectroscopy, using

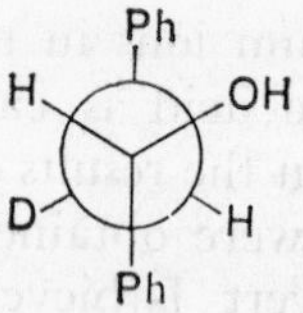

(14) 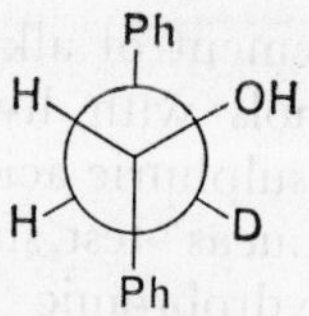 (15)

the magnitude of the proton–proton coupling in the two forms. Conversion of the alcohols to their chloro-derivatives was then shown[59] to involve almost complete racemization using lithium chloride in hydrochloric acid, but retention of configuration with thionyl chloride (S_Ni mechanism[60]).

3. Phosphorus-containing reagents

a. Derivatives of P^V. The use of pentavalent phosphorus derivatives for halogenations, at first confined to the pentahalides and to the oxychloride, has been extended widely, and now includes reagents such as Ph_2PCl_3 [61], $PhPOCl_2$ [62], **16** [63], and the phosphine or phosphite adducts with halogens, **17**.

(i) Phosphorus pentabromide converts certain alcohols to bromides more cleanly than does the tribromide[64]. Phosphorus pentahalides react with certain phenols[65] (but not with phenol esters and ethers[66]) and with alcohols[66] to give halo-derivatives, with carboxylic acids to give acyl halides[67], and with amides to give haloimines[68] (equations 26–29). A phosphate ester has been isolated from the reaction of phosphorus pentachloride with a large excess of β-naphthol (equation 30). The ester decomposed to 2-chloronaphthalene only when

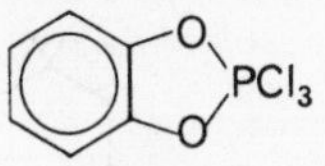

$C_6H_4O_2PCl_3$ **(16)** R_3PX_2 **(17)**

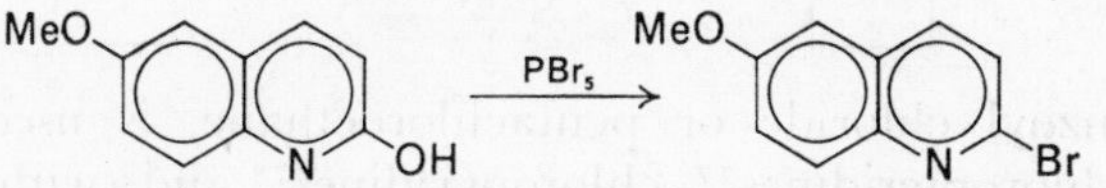

6-MeO-quinolin-2-ol $\xrightarrow{PBr_5}$ 6-MeO-2-bromoquinoline (26)

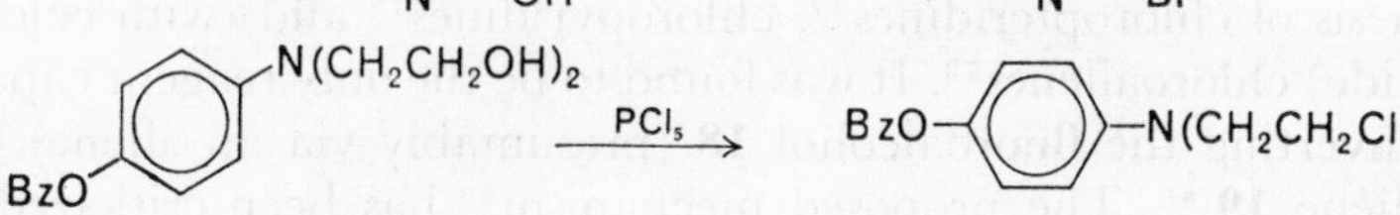

$BzO{-}C_6H_4{-}N(CH_2CH_2OH)_2 \xrightarrow{PCl_5} BzO{-}C_6H_4{-}N(CH_2CH_2Cl)_2$ (27)

$p\text{-}O_2NC_6H_4CO_2H \xrightarrow{PCl_5} p\text{-}O_2NC_6H_4COCl$ (28)

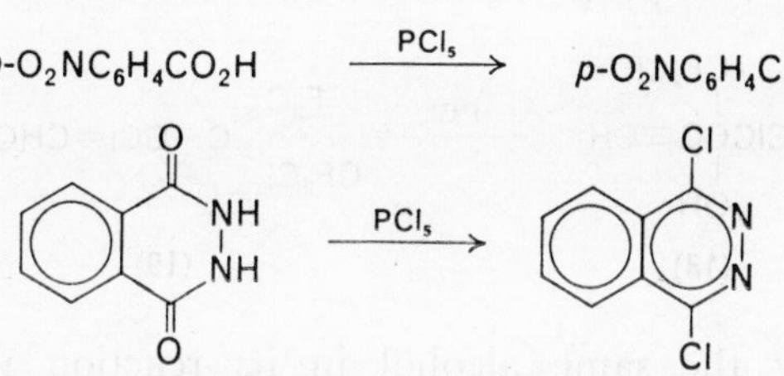

phthalhydrazide $\xrightarrow{PCl_5}$ 1,4-dichlorophthalazine (29)

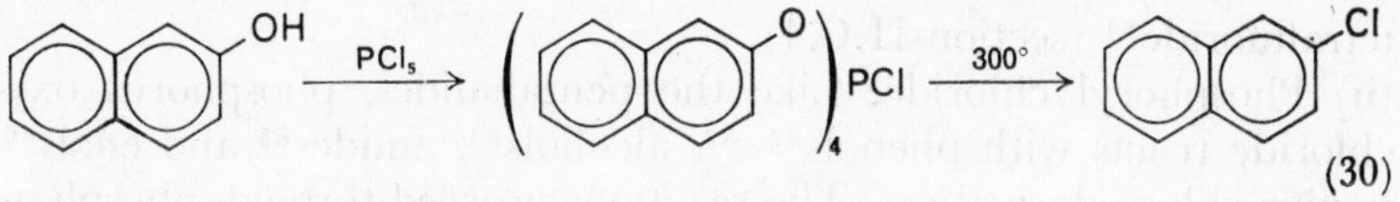

2-naphthol $\xrightarrow{PCl_5}$ $(C_{10}H_7O)_4PCl \xrightarrow{300°}$ 2-chloronaphthalene (30)

heated to 300° [69]. Tetraphenoxyphosphonium chloride has been reported to give no chlorobenzene[70]. The reagent can bring about Beckmann rearrangements in oximes, giving chloroazines[71] (equation 31) or analogous products (equation 32)[72] and converts secondary alcohols to vicinal dihalides in dry chloroform[73].

$$PhC(=NOH)\text{—}C(=NOH)Ph \xrightarrow{PCl_5} PhCCl{:}N{\cdot}N{:}CClPh \quad (31)$$

$$\text{(HON)(Ph)-substituted imidazole, 2-Ph} \longrightarrow \text{2-Cl-4,6-diphenyl-1,3,5-triazine} \quad (32)$$

A comparison of the action of phosphorus pentachloride, phosphine dihalides (section II.B.3a.v) and thionyl chloride is made in section II.B.4: particularly significant differences occur with homoallylic alcohols[74], and these have been exploited in the steroid field[75].

Like oxalyl chloride (section II.B.6), phosphorus pentachloride converts keto-enamines to chlorovinyl immonium salts[76] (equation 33, cf equation 29) and, in the presence of phosphorus oxy-

$$\text{5,5-dimethyl-3-pyrrolidino-cyclohex-2-enone} \longrightarrow \text{3-chloro-5,5-dimethylcyclohex-2-enylidene pyrrolidinium} \quad (33)$$

chloride, benzoyl chloride or pentachloroethane, is used in the synthesis of chloropteridines[77], chloropyridines[78] and (with calcium chloride) chloroallenes[79]. It was found to be the only reagent capable of converting the fluoroalcohol **18** (presumably via an allene) into the diene **19** [80]. The proposed mechanism[80] has been criticized by

$$F_2ClCC(CF_2Cl)(OH)C{\equiv}CH \ \textbf{(18)} \xrightarrow{PCl_5} (F_2C{=})(CF_2Cl)C\text{—}CCl{=}CHCl \ \textbf{(19)}$$

others studying the same alcohol in its reaction with sulphur tetrafluoride[81] (section II.C.4).

(ii) Phosphoryl chloride. Like the pentahalides, phosphorus oxychloride reacts with phenols[82, 83], alcohols[84], amides[85] and enols[86], to give chloro-derivatives. The reactions proceed through phosphate esters which are decomposed by halide ion, and give hydrochloric acid and phosphoric acid as by-products. (The reagent is a vigorous dehydrating agent[87], particularly in the presence of pyridine, though perhaps less so than thionyl chloride under the same conditions[88]. Both reagents catalyse the esterification of carboxylic acids[89], through intermediate mixed anhydrides.) Phosphorus oxychloride was used in the presence of lithium chloride in the conversion of **20** to **21** [86].

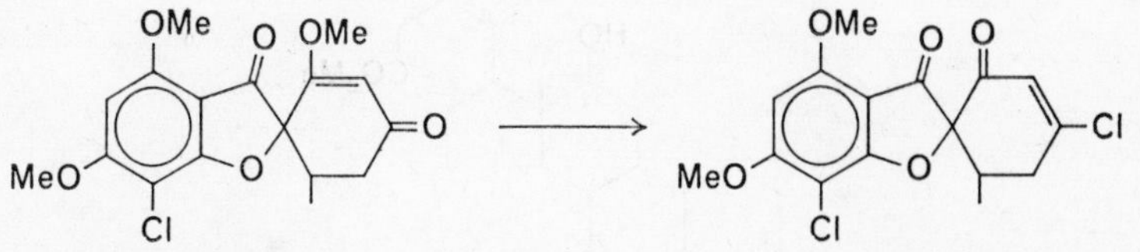

(**20**) → (**21**)

(iii) The imide **22** was converted to **23** in high yield using a large excess of $PhPOCl_2$ [62] at 160°: because of the involatility of the reagent no sealed tube was necessary, and the method was shown to be preferable to that using phosphorus oxychloride.

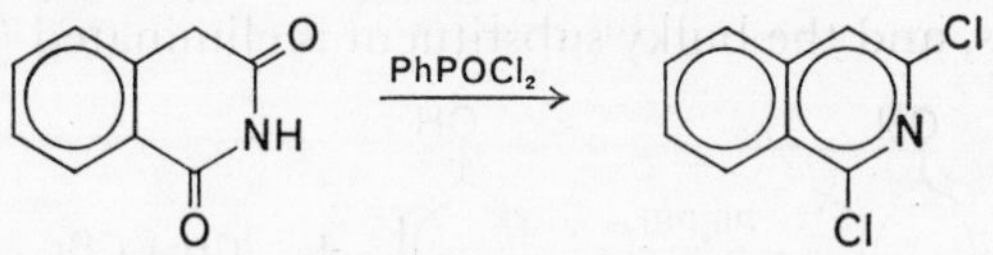

(**22**) —$PhPOCl_2$→ (**23**)

(iv) The cyclic phosphorotrihalidates **16**, developed by Gross[63] have been used to prepare vinyl chlorides from ketones, and other uses of the bromo-reagents have been reviewed[90].

(v) As with the conversion of alcohols to iodides, the reagents most suited to the mild displacement of hydroxyl by chloride or bromide are the dihalophosphoranes and analogous compounds prepared by the interaction of phosphines with other halogen sources, notably carbon tetrahalides.

(α) Triphenylphosphine dibromide is readily prepared from bromine and the phosphine, and is available commercially. Its reaction with alcohols[91] and with phenols[61, 92] is outlined in equation (34): the

$$Ph_3PBr_2 + ROH \longrightarrow RBr + Ph_3PO + HBr \tag{34}$$

conversion occurs at low temperatures, no addition occurs with unsaturated alcohols, and the major by-product, triphenylphosphine oxide, is neutral and unable to cause subsequent side-reactions.

Carboxylic acids yield acid chlorides with Ph_3PCl_2 [75]. However, amides[75] and oximes[75] may be dehydrated, the latter may undergo Beckmann rearrangement[93], and ethers[94] cleaved with the reagent, so that some thought must be given to whether a particular hydroxyl group can be displaced without changing the rest of the molecule.

Hydroxy ketones in the steroid field can probably not be converted to pure bromoketones, as dehydrogenation occurs[75] adjacent to carbonyl groups, but the ester **24** can be brominated selectively[75]

(24)

with a relative reactivity 3 > 7 > 12, the same as that found for acetylation.

A limitation of the method has been found with phenols substituted in the *ortho* position with *tert*-butyl groups[95]. No halogenation of the nucleus occurs, and the bulky substituent is eliminated (equation 35)

$$\xrightarrow{Ph_3PBr_2} \quad + \quad (CH_3)_3CBr \qquad (35)$$

by a mechanism thought to involve a betaine intermediate. Smaller substituents are unaffected, and indeed the reaction of phenols, naphthols and heterocyclic hydroxy-compounds with triphenylphosphine dibromide provides a very useful and efficient route to aryl halides. When 1-bromo-2-naphthol is heated to 200° with triphenylphosphine[61] it is converted into 2-bromonaphthalene, an overall displacement of hydroxyl by bromine. Similarly, *p*-bromophenol gives bromobenzene: in view of the result with the naphthol derivative this must be also displacement of hydroxyl by bromine (with rearrangement) and not by hydrogen.

Direct bimolecular displacements from an aromatic nucleus only occur when the reaction is aided by an electron-poor nucleus (as in the displacement of halide ion from 2,4-dinitrohalobenzenes) or when the displaced group is very stable. Displacements are possible using the phosphine dibromide because triphenylphosphine oxide is a good leaving group. Enol-phosphonium salts behave similarly in some reactions[96].

In solution the structure of the chlorine reagent has been shown[97] to involve an equilibrium (equation 36) and haloform-solvated

$$Ph_3PCl_2 \rightleftharpoons Ph_3\overset{+}{P}Cl\ Cl^- \rightleftharpoons Ph_3\overset{+}{P}Cl\ Ph_3\overset{-}{P}Cl_3 \qquad (36)$$

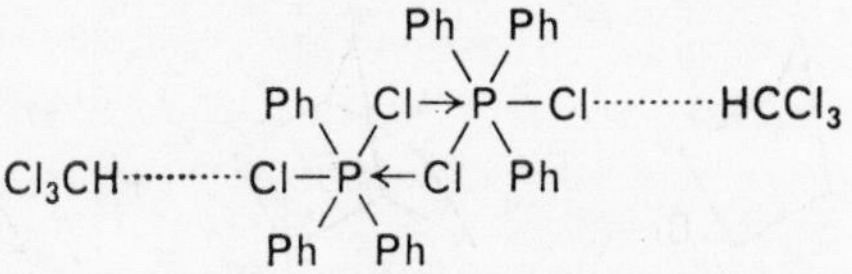

$$
\begin{array}{c}
\mathrm{Ph\ \ Ph} \\
\mathrm{Ph\quad Cl \rightarrow P - Cl \cdots\cdots HCCl_3} \\
\mathrm{Cl_3CH \cdots\cdots Cl - P \leftarrow Cl\quad Ph} \\
\mathrm{Ph\ \ Ph}
\end{array}
$$

(25)

dimers, for which structure **25** [98] has been suggested, are isolable solids, stable in the absence of moisture. The structures of many other pentacoordinate chlorinated phosphorus compounds are similar in showing equilibria between covalent and ionic species[99], and nucleophilic attack by the alcohol seems the most reasonable first step in the conversion to halo-compounds using triphenylphosphine dihalides (equation 37).

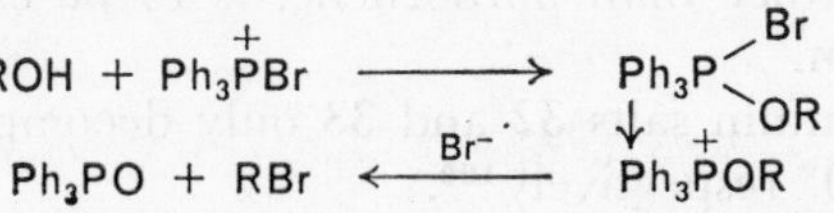

$$
\begin{array}{ccc}
\mathrm{ROH + Ph_3\overset{+}{P}Br} & \longrightarrow & \mathrm{Ph_3P{<}^{Br}_{OR}} \\
 & & \downarrow \\
\mathrm{Ph_3PO + RBr} & \xleftarrow{\ Br^-\ } & \mathrm{Ph_3\overset{+}{P}OR}
\end{array}
\qquad (37)
$$

Preference for decomposition of the alkoxy-phosphonium salt by an S_N2 mechanism is shown in many cases.

The isomeric norbornanols **26**–**29** were treated with triphenylphosphine dibromide, with the formation of the phosphonium salts **30**–**33** [100].

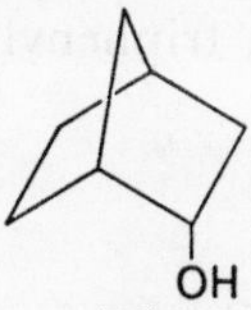

OH **(26)** OH **(27)** OH **(28)** OH **(29)**

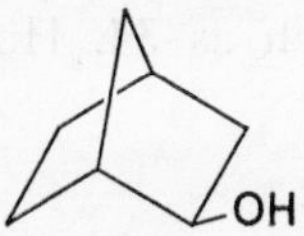

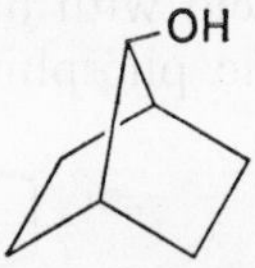

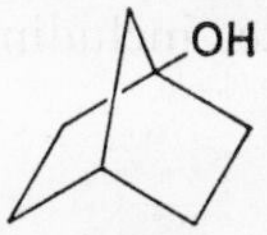

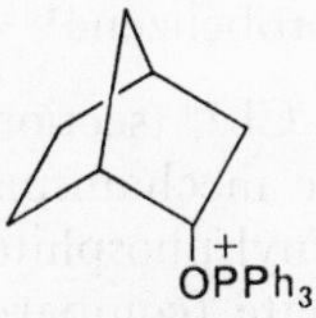

$\overset{+}{O}PPh_3$ **(30)** $\overset{+}{O}PPh_3$ **(31)** $\overset{+}{O}PPh_3$ **(32)** $\overset{+}{O}PPh_3$ **(33)**

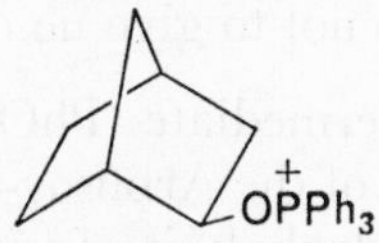

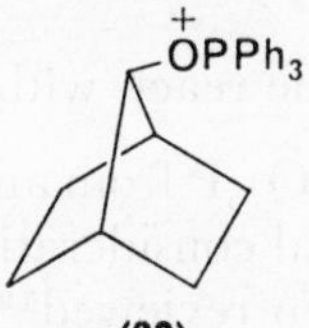

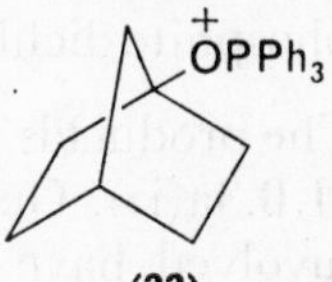

The *endo*-alcohol **26** was converted smoothly to the *exo*-bromide with no loss of optical activity, classical evidence for an S_N2 mechanism. The phosphonium salt **30** is isolable, and only decomposes when heated above 115°. No analogous displacement is possible on **31** however, which decomposes at room temperature to give a

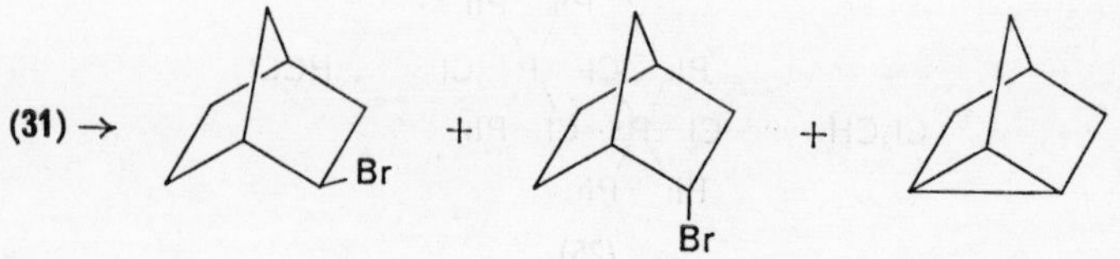

$$(31) \rightarrow \text{[exo-2-bromonorbornane]} + \text{[endo-2-bromonorbornane]} + \text{[nortricyclene]} \quad (38)$$

complex mixture of products (equation 38) the composition of which is very solvent dependent.

In triglyme the relative ratio of ion-pair-derived products (racemic *exo*-bromide and nortricyclene) to S_N2-derived product (optically active *endo*-bromide) is about 7:1; in dimethylformamide it is nearly 100:1. An $E1$ mechanism for the decomposition is made more likely by the isolation of nortricyclene, since Kwart[101] has shown that this product, rather than norbornene, is to be expected from the norbornyl cation.

The phosphonium salts **32** and **33** only decompose when heated to 170° and 200° respectively[100].

Analogous adducts between phosphites and halogen are also effective in converting alcohols to halo-compounds[102]. Many dihalotrialkyl phosphites decompose spontaneously to phosphorohalidates $(RO)_2POCl$ and alkyl halide[103], but stable adducts are also known[104]. An early report claimed that allyl alcohol was best substituted using butyl phosphite and bromine, whereas benzyl alcohols react in high yield with halogen and any phosphine or phosphite, including cyclic phosphites such as **34**. However, triphenyl

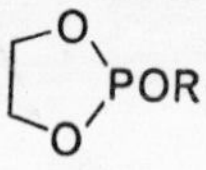

(34)

phosphite dichloride reacts with phenol to give no chlorobenzene[105]. The product is $(PhO)_5P$ from an intermediate $(PhO)_4\overset{+}{P}\overset{-}{Cl}$ [70] (section II.B.3a(i)). General considerations of the Arbusov-type mechanisms involved have been reviewed[106]: alcoholysis of triphenyl phosphite dibromide gives phenol[107], and not triphenyl phosphate (compare equation 37) as by-product in the formation of alkyl bromides.

(β) Intermediates similar to those of equation (36) are involved in the oxidation of phosphites to phosphates in the presence of alcohols and carbon tetrachloride[108, 109], a process which serves to convert alcohols to alkyl chlorides (equation 39). No acidic by-products are

$$P(OR)_3 + CCl_4 + R'OH \longrightarrow (RO)_3PO + R'Cl + CHCl_3 \quad (39)$$

formed and the reaction proceeds at moderate temperatures: phosphines may be used instead of phosphites.

Different mechanisms have been suggested for the phosphine and phosphite reactions however, the more nucleophilic phosphines attacking halogen, and phosphites carbon.

Thus, carbon tetrachloride and trialkyl phosphites react with alcohols by the scheme outlined in equation (40): a tetraalkoxyphosphonium salt is formed by displacement of the trichloromethyl anion from the initially formed salt **35**. (A radical process for the

$$CCl_4 + P(OR)_3 \longrightarrow \underset{(35)}{(RO)_3\overset{+}{P}CCl_3\overset{-}{Cl}} \xrightarrow{R'OH} (RO)_3\overset{+}{P}OR'\overset{-}{Cl} + CHCl_3,$$

$$(RO)_3\overset{+}{P}OR' \xrightarrow[S_N2]{Cl^-} (RO)_3PO + R'Cl \qquad (40)$$

interaction (in the absence of alcohols) of phosphites with carbon tetrachloride[110], chloroform[111] and bromoform[112] has also been suggested.) The method suffers from the disadvantage that two alkyl halides can be formed, one from the phosphite and one from the alcohol. The ratio of products observed will depend on the relative susceptibilities of R and R′ to nucleophilic attack.

No such ambiguity exists when phosphines are used instead of phosphites, and a different mechanism[113] (equation 41) applies.

$$CCl_4 + PPh_3 \longrightarrow Ph_3\overset{+}{P}Cl\ \overset{-}{C}Cl_3 \xrightarrow{EtOH} Ph_3\overset{+}{P}OEt\ Cl^-$$

$$Ph_3\overset{+}{P}Cl\ \overset{-}{C}Cl_3 \downarrow \quad \underset{(36)}{Ph_3\overset{+}{P}CCl_3\ Cl^-} \qquad Ph_3\overset{+}{P}OEt\ Cl^- \downarrow \quad EtCl + Ph_3PO \qquad (41)$$

Triphenylphosphine and carbon tetrachloride apparently provide the ylid $Ph_3P{=}CCl_2$ also[114]—ketones undergo the Wittig reaction under these conditions. The phosphite intermediate **35** is known to decompose in the absence of an alcohol to alkyl chloride and the phosphonate, **37**, which, heated in alcohols, gives no further alkyl

$$\underset{(37)}{(RO)_2POCCl_3}$$

halide[115]. Tri(*n*-octyl)phosphine behaves in the same way as triphenylphosphine with carbon tetrachloride and alcohols, using the halide as solvent, and with carbon tetrabromide and alcohols, when only two equivalents of halide and phosphine are used[116].

Another report[117] on the mechanism of the similar reaction between alcohols and carbon tetrachloride in the presence of tris(dimethylamino)phosphine favours the mechanism of equation (41), as expected for a very nucleophilic reagent. (Aminophosphines are frequently used in place of alkyl or arylphosphines because the basic groups greatly simplify the removal of phosphine oxide from the product mixture[118].)

Evidence for the proposed mechanism (equation 42) comes from three results[117]. The phosphonium salt **38** is not attacked by water,

$$(Me_2N)_3P + CCl_4 \longrightarrow (Me_2N)_3\overset{+}{P}Cl\ \bar{C}Cl_3 \xrightarrow{ROH} (Me_2N)_3\overset{+}{P}ORCl^- + CHCl_3$$

$$(Me_2N)_3\overset{+}{P}Cl\ \bar{C}Cl_3 \longrightarrow (Me_2N)_3\overset{+}{P}CCl_3\ Cl^-\ \textbf{(38)}$$

$$(Me_2N)_3\overset{+}{P}ORCl^-\ \textbf{(39)} \longrightarrow RCl + (Me_2N)_3PO \qquad (42)$$

and, therefore, not by alcohol in all probability. This was confirmed by adding alcohol to the mixture of phosphine and tetrahalide only after the interaction was complete. No alkyl halide was obtained. The phosphine, tetrahalide and alcohol, mixed in ether at low temperature, gave two layers, a lower oily one, presumably **39**, and an etherial one containing chloroform but no alkyl halide. The alkyl halide was liberated by a slow displacement step as the reagents were warmed.

The stereochemical consequences of the displacement should be configurational inversion. This has been found to be so[119, 120], but examples of reactions with racemization[119] are known, and retention of configuration[121] occurs with alcohols in which S_N2 reactions are prevented by any factor. In this, the reagent resembles the penta-

(40) $\xrightarrow[CCl_4]{PPh_3}$ **(41)**

halides, and the phosphine halides. Thus, the conversion of **40** to **41** occurs with retention of configuration[121].

Halogen sources other than carbon tetrachloride and tetrabromide have been used: bromotrichloromethane[110], chloroform[122], bromocyanacetamide[123] and hexachlorocyclopentadiene[124] have all been employed successfully, but thionyl chloride, sulphuryl chloride, cupric chloride and hexachloroethane proved less efficient[102].

The mechanisms involved are probably similar[125], although Trippett has shown[126] that the phosphonium salts derived from bromocyanacetamide are mixtures of keto-and enol-phosphonium salts (equation 43). Others have reported[127] that, at least with α-haloketones, only the enol-phosphonium salts react with alcohols.

$$Ph_3P + BrCHCNCONH_2 \longrightarrow Ph_3\overset{+}{P}CH(CN)CONH_2 + Ph_3\overset{+}{P}OC(NH_2){=}CHCN \quad (43)$$

Modifications of the phosphite methiodide method for preparing alkyl iodides have been used to synthesize alkyl chlorides (benzyl chloride)[107] and alkyl bromides (benzyl bromide)[107].

Few direct comparisons between the methods of using phosphine dibromide, phosphine in the presence of carbon tetrahalide, and phosphite in the presence of alkyl halide, have been reported. All are mild reactions, and rearrangement products are of only minor importance. Sugar-protecting groups are not attacked by the reagents, and whereas the phosphite methiodides give phosphonates instead of halides with vicinal diols, triphenylphosphine in carbon tetrachloride gives dihalides[128].

General reviews of the nucleophilic attack by phosphites and phosphines on alkyl halides, and of the reactions of derivatives of tetravalent phosphorus, have appeared recently[13, 129].

b. Derivatives of P^{III}. The standard use[130] of the phosphorus trihalides in the synthesis of alkyl halides from alcohols needs little discussion.

The mechanism of the reaction has been studied most recently by Gerrard and co-workers[3] who showed that both the formation and decomposition of the intermediate phosphite ester proceed (equation 44) in a stepwise manner.

$$PX_3 \xrightarrow{ROH} (RO)PX_2 \xrightarrow{ROH} (RO)_2PX \xrightarrow{ROH} (RO)_3P$$

$$(RO)_3P \xrightarrow{HX} RX + (RO)_2POH \xrightarrow{HX} RX + (RO)P(OH)_2 \xrightarrow{HX} RX + H_3PO_3 \quad (44)$$

Optical purity of the alkyl halide produced is high in each stage of the decomposition, and is lowest in the third stage when S_N1 reactions become increasingly important as the stability of the phosphite ion increases (equation 45). The earlier stages of the de-

$$(RO)P(OH)_2 \rightleftharpoons R^+ + H_2PO_3^- \quad (45)$$

composition involve S_N2 attack on the protonated phosphite (equation 46). Overall, the optical purity achieved using these

$$(RO)_3\overset{+}{P}H \xrightarrow{X^-} RX + (RO)_2P(OH) \quad (46)$$

phosphorus trihalides is better than when hydrochloric acid or phosphorus and iodine are used to effect the same reactions.

Attempts to prepare the propargyl bromide from alcohol **42** using

$$\textbf{(42)} \xrightarrow{PBr_3} \textbf{(43)}$$

phosphorus tribromide gave only **43** [131], and rearrangements in propargylic systems are common with this reagent, particularly with tertiary alcohols (equations 47 [132] and 48 [133]) and probably no tertiary propargyl aryl halides are known. It has been shown[134] that all earlier claims for such halides have given instead the haloallenes.

$$Me_2C(OH)C{\equiv}CH \xrightarrow{PCl_3} Me_2CClC{\equiv}CH\ (47\%) + CH_2{=}CMe{-}CH{=}CHCl\ (23\%) \quad (47)$$

$$\xrightarrow[0^\circ]{PBr_3} \quad (48)$$

A curious difference in the reactivity of enantiomers towards phosphorus tribromide in the bridged-diphenyl series has been reported by Mislow[135]. Racemic **44** reacts with the reagent in the solid state to give **45** as expected, whereas the (+) form disproportionates under the same conditions to optically active **46** and **47**. The allylic alcohol **48** reacts with phosphorus tribromide and pyridine in petroleum ether to give the unrearranged bromide[136].

4. Sulphur-containing reagents

Of the many sulphur-containing reagents used to convert hydroxy compounds to halides[137], only thionyl chloride (and bromide) have enjoyed wide application. Thionyl chloride reacts with alcohols to

$H_2C{-}S{-}CH_2$; HOH_2C CH_2OH (44) $H_2C{-}S{-}CH_2$; BrH_2C CH_2Br (45) $H_2C{-}S{-}CH_2$; $H_2C{-}S{-}CH_2$ (46)

BrH_2C CH_2Br; BrH_2C CH_2Br (47)

CH_2OH (48)

form intermediate esters (equation 49) but the stereochemical path

$$ROH + SOCl_2 \longrightarrow ROSOCl + HCl \qquad (49)$$

to the final products (and the nature of the products) is determined by the reaction conditions employed. Equimolar proportions of alcohol, chloride and pyridine give the product of S_N2 attack by chloride ion on the chlorosulphite ester (equation 50) with inversion

$$ROSOCl \xrightarrow{Cl^-} RCl + SO_2 + Cl^- \qquad (50)$$

of the alkyl group configuration.

Two moles of alcohol and of pyridine to one of chloride give an alkyl sulphite in good yield (equation 51).

$$ROSOCl \xrightarrow{ROH} ROSOOR \qquad (51)$$

In the absence of any base, the S_Ni mechanism operates, and an alkyl chloride of the same configuration as the alcohol is obtained (equation 52), probably through an ion-pair formed by loss of SO_2 from the chlorosulphite.

$$R{-}O{-}S(=O){-}Cl \longrightarrow RCl + SO_2 \qquad (52)$$

A similar difference occurs when allylic alcohols react with thionyl chloride with or without added base, giving either S_N2 attack (equation 53) or S_N2' (equation 54). Propargyl alcohols behave

$$CH_3CH{=}CHCH_2OH \xrightarrow[\text{base}]{SOCl_2} CH_3CH{=}CHCH_2Cl \quad (53)$$

$$CH_3CH{=}CHCH_2OH \xrightarrow[\text{no base}]{SOCl_2} CH_3CHClCH{=}CH_2 \quad (54)$$

similarly[138], undergoing direct substitution (equation 55) or rearrangement[139].

$$RC{\equiv}CCH_2OH \xrightarrow{SOCl_2} RC{\equiv}CCH_2Cl \quad (55)$$

Thionyl chloride is the standard reagent for converting carboxylic acids to acid chlorides[140]. It also displaces hydroxyl groups from tropolones[141] (equation 56) and, in dimethylformamide solution, from highly acidic phenols[142] (equation 57). The importance of the

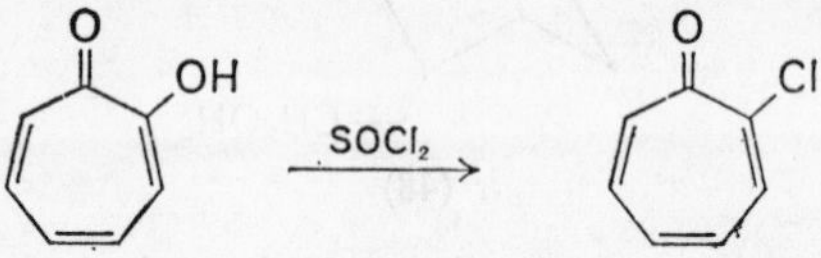

(56)

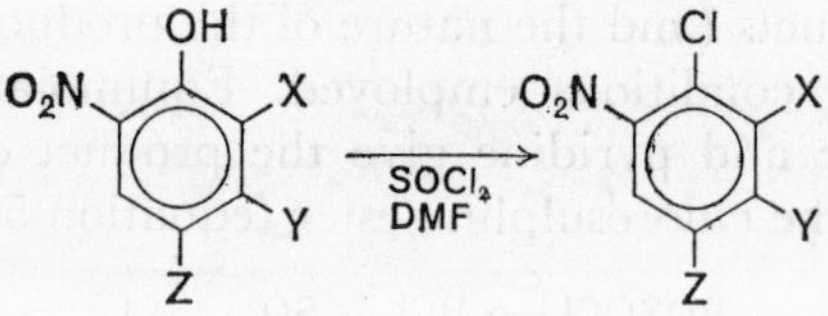

(57)

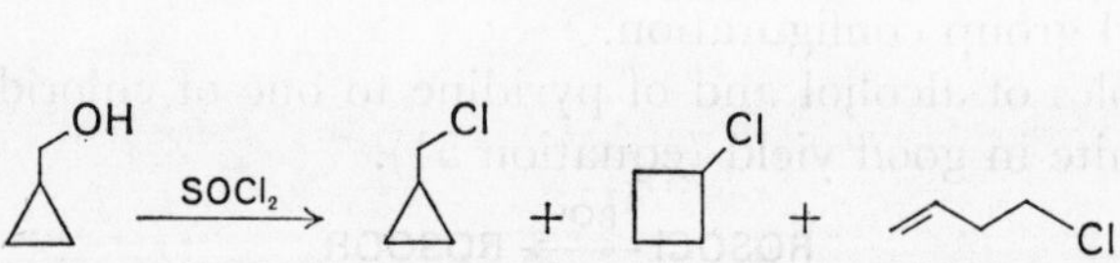

(58)

solvent in this reaction is discussed in the following section.

The reagent converts amides to chloroimines[143] and nitroalcohols to nitrochlorides[144, 145] but does not attack phenol esters[146]. In common with other chlorinating reagents it gives rise to rearrangements in cyclopropylcarbinyl systems[147] (equation 58) and in the attempted conversion of aryl and other hindered tertiary propargylic alcohols to halides[139, 148]. Alcohol **49** was converted into the chlorocompound **50** using two moles of thionyl chloride to one of pyri-

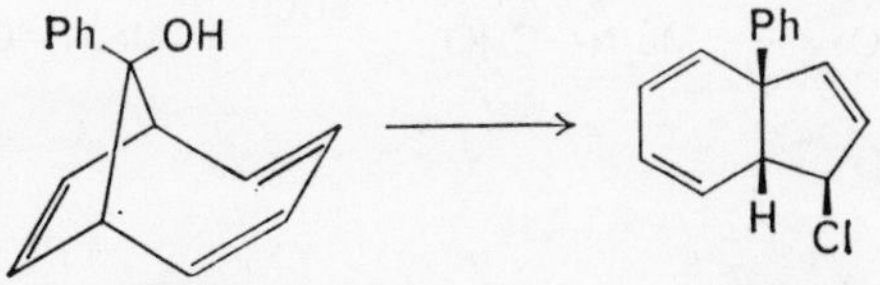

(49) → (50)

dine[149], and **51** into **52** but **53** into **54** [150, 151], an interesting example of ring-size effects in bicyclic systems.

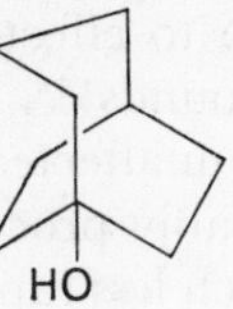

(51) 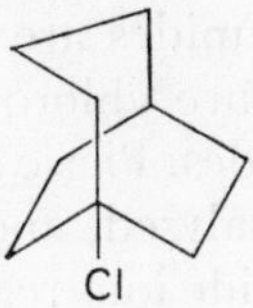 (52) 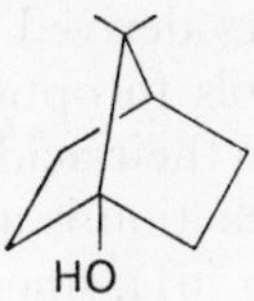 (53) 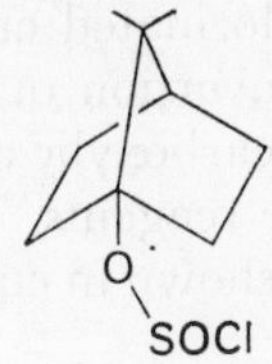 (54)

Comparisons have been made of the relative usefulness of thionyl chloride and hydrochloric acid[52], and of phosphorus pentachloride and thionyl chloride[152], in converting alcohols to chloro-compounds. In general, however, since the reagents often give rise to different stereochemical results, the choice is not always available.

Shoppee and co-workers have made a particularly extensive study[153] of the 5α-cholestane skeleton and have rationalized the variation in products observed from changing the hydroxyl group position in the molecule.

In general, phosphorus pentachloride is more likely than thionyl chloride to give products from inversion of configuration since the intermediate chlorophosphate ester is more easily ionized than the corresponding chlorosulphite.

5. Nitrogen-containing reagents

(a) The active intermediates in many displacement reactions (equation 59) which are carried out in dimethylformamide solu-

$$ROH + SOCl_2 \xrightarrow{DMF} RCl + SO_2 + HCl \qquad (59)$$

tion[142, 154] are nitrogen-containing salts. A mechanism (equation 60) predicting retention of optical activity but inversion of configuration through an S_N2 reaction is reasonable.

(b) A recent review[155] of polychloroamine compounds dealt only with their preparations and their intramolecular reactions. But

$$\mathrm{Me_2NCHO} \longleftrightarrow \mathrm{Me_2\overset{+}{N}{=}CH\bar{O}} \xrightarrow{\mathrm{SOCl_2}} \mathrm{Me_2\overset{+}{N}{=}CHCl\ Cl^-} \xrightarrow{\mathrm{ROH}} \mathrm{Me_2NCHCl(\overset{+}{H}OR)\ Cl^-} \xrightarrow{S_N2} \mathrm{Me_2NCHCl(OH) + RCl} \longrightarrow \mathrm{Me_2NCHO} \qquad (60)$$

chlorinated enamines derived from amides are known to effect the conversion of alcohols to optically pure chloro-compounds[156], and of carboxylic acids to their acid chlorides. Picric acid is unaffected by the reagents. The reaction is acid-catalysed, and probably proceeds as shown in equation (61), since alkoxide ions react much less rapidly

$$\mathrm{Cl_2C{=}CClNEt_2} \longleftrightarrow \mathrm{Cl_2\bar{C}CCl{=}\overset{+}{N}Et_2} \xrightarrow{\mathrm{H^+}} \mathrm{Cl_2CHCCl{=}\overset{+}{N}Et_2} \xrightarrow{\mathrm{ROH}} \mathrm{Cl_2CHCClNEt_2(\overset{+}{R}OH)} \longrightarrow \mathrm{Cl_2CHCClNEt_2(OR)} \xrightarrow{\mathrm{Cl^-}} \mathrm{Cl_2CHCONEt_2 + Cl^- + RCl} \qquad (61)$$

with the vinyl amine than alcohols do, and, as expected for an S_N2 reaction, the alkyl halide obtained is of the opposite configuration to that of the alcohol.

(Similar halogenations are known to occur using chlorinated vinyl ethers as the reagents, for example **55** [157] and **56** [158].)

$$\underset{\mathbf{(55)}}{\mathrm{CH_2{=}C(OEt)Cl}} \qquad \underset{\mathbf{(56)}}{\mathrm{CHCl{=}C(OEt)Cl}}$$

(c) Alkyl cyanides give salts with alcohols (equation 62 [55]). These are susceptible to S_N2 reactions, giving alkyl halides of configuration

$$\mathrm{RC{\equiv}N} \xrightarrow[\mathrm{H^+}]{\mathrm{R'OH}} \mathrm{R{-}C(OR'){=}\overset{+}{N}H_2} \longleftrightarrow \mathrm{R{-}C({=}\overset{+}{O}R')NH_2} \xrightarrow{\mathrm{Cl^-}} \mathrm{R'Cl + RCONH_2} \qquad (62)$$

opposite to that found in the alcohol. The effect is similar to that of carbodiimides in activating carboxyl groups towards amide formation in peptide syntheses[58] (equation 63).

$$RN{=}C{=}NR \xrightarrow{R'CO_2H} RNHC(OCOR'){=}NR \xrightarrow{R''NH_2} RNHCONHR + R''NHCOR' \quad (63)$$

6. Organic acid chlorides

Many acid chlorides react with alcohols to form esters, which may either decompose spontaneously to give alkyl chlorides, or which can be converted to them in a subsequent step.

Oxalyl chloride forms chloroxalate esters with alcohols at room temperature (equation 64). The esters decompose when heated

$$ROH + (COCl)_2 \longrightarrow ROCOCOCl + HCl \quad (64)$$

$$ROCOCOCl \xrightarrow[125°]{pyridine} RCl + CO + CO_2 \quad (65)$$

above 100° in pyridine[159]. Analogous chloroformates also decompose to give alkyl chlorides[160]—a survey of the breakdown of chloroesters in general appears in standard texts on physical organic chemistry.

A similar (but multistep) reaction has been used to convert 1-apocamphanol to the iodide (equation 66) by irradiating the oxalate in carbon tetrachloride in the presence of iodine and mercuric oxide[33] (section II.A.4) based on earlier work on the de-

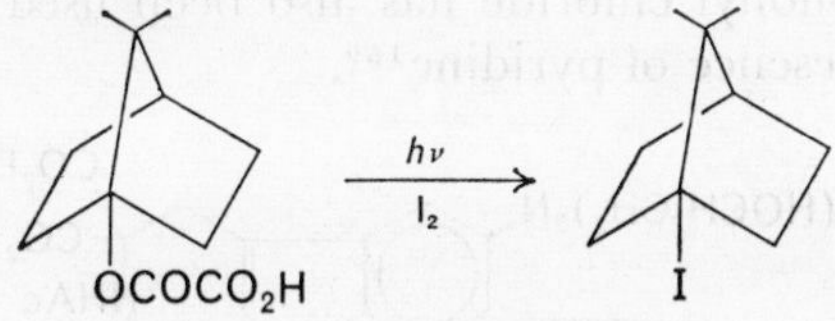

OCOCO₂H $\xrightarrow[I_2]{h\nu}$ I (66)

composition of hypoiodites by Barton[161].

Oxalyl chloride[162], like phosgene[163], converts certain carbonyl

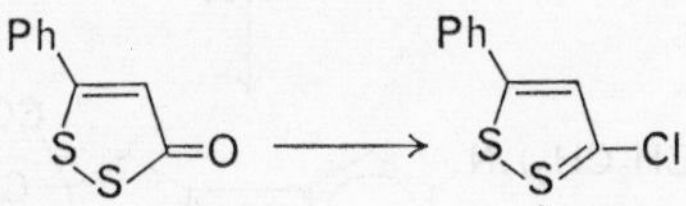

Ph ... S S =O ⟶ Ph ... S S⁺ —Cl (67)

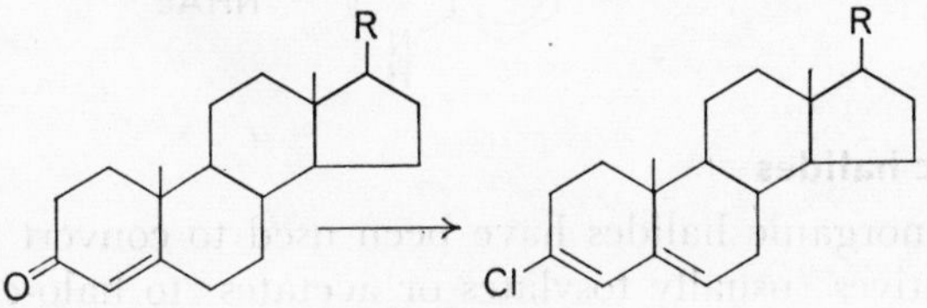

R ... O ⟶ R ... Cl (68)

groups to chlorine-substituted salts (e.g., equation 67) and, in the presence of oxalic acid, to vinyl chlorides[164, 165] (e.g., equation 68).

α-Acetoxy acid chlorides effect the conversion shown in equation (69)[166].

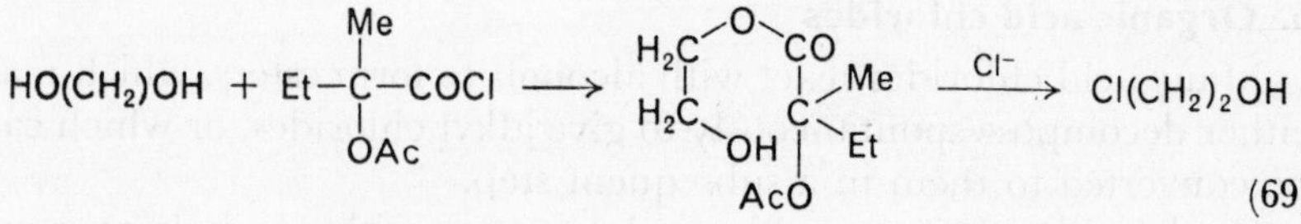

$$HO(CH_2)OH + Et{-}C(Me)(OAc){-}COCl \longrightarrow \text{[cyclic adduct: } H_2C, H_2C, O, CO, C(Me)(Et), OH, AcO] \xrightarrow{Cl^-} Cl(CH_2)_2OH \quad (69)$$

Acetyl bromide has been used to prepare bromo-compounds from alcohols, but products from acetylation alone, and from dehydration, are common, and the reagent is little used[167]. Thus, carbohydrate hemiacetals are converted to (acetylated) bromoethers:

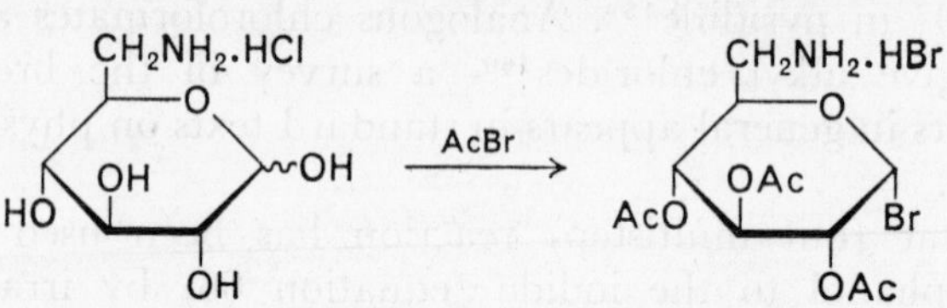

$CH_2NH_2 \cdot HCl$ (sugar with HO, OH, OH, ~OH) $\xrightarrow{AcBr}$ $CH_2NH_2 \cdot HBr$ (sugar with AcO, OAc, OAc, Br)

Halogen acids in acetic acid react with the protected sugars to effect the same result[168].

Methanesulphonyl chloride has also been used as a chlorinating agent in the presence of pyridine[169].

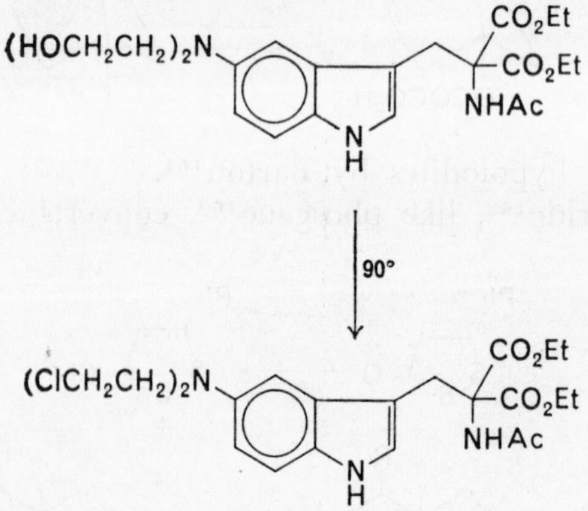

$(HOCH_2CH_2)_2N$–(indol-5-yl)–3-$CH_2C(CO_2Et)_2NHAc$

$\downarrow$ 90°

$(ClCH_2CH_2)_2N$–(indol-5-yl)–3-$CH_2C(CO_2Et)_2NHAc$

7. Inorganic halides

Various inorganic halides have been used to convert alcohols or their derivatives (usually tosylates or acetates) to halo-compounds.

Among these are sodium bromide[170], lithium chloride[171] and bromide[172], calcium chloride in the presence of hydrochloric acid[173, 174], zinc chloride with hydrochloric acid[175] or with dichloromethyl methyl ether[176], magnesium bromide[177], boron and aluminium trichlorides[178, 179], and titanium tetrachloride[180, 181] and tetrabromide[182]. Polar solvents such as dimethylformamide and dimethyl sulphoxide (with which tosylates react reversibly) and acetone are usually used for these displacements. Treatment of a cyclopropylcarbinol with lithium chloride and hydrochloric acid gave ring-opened products[183].

Tosylate displacements in particular provide one of the best methods of converting secondary alcohols to bromides[170] without rearrangement, although inversion of configuration occurs.

Pyridine hydrochloride has been used on steroid tosylates for the preparation of chloro compounds[184].

Phenol esters are inert to hydrobromic acid in acetic anhydride (equation 70 [185]) whereas benzylic ones are cleaved, and similar

OAc OAc Ph CH_2OAc $\xrightarrow{HBr}$ OAc OAc Ph CH_2Br (70)

differentiations are apparent (equations 71, 72) in the preparation of γ-pyrone derivatives, both with free hydroxyl groups and with derivatives.

HO O CH_2OH OH O $\xrightarrow{HBr^{186}}$ HO O CH_2Br OH O (71)

$MeSO_2OH_2C$ O OSO_2Me O $\xrightarrow{HBr^{187}}$ BrH_2C O OSO_2Me O (72)

C. Displacement by Fluorine

1. Direct halogenation

Alcohols do not react with fluorine to give alkyl fluorides. In general, the other halogens can be used directly or indirectly to bring about substitutive halogenation of alcohols under certain conditions

(1-adamantanol with liquid bromine[38], alcohols with iodine in the presence of borohydride[4] or phosphorus[2], alcohols[36] with chlorine), although oxidations occur if labile functional groups are also present in the molecule.

2. Hydrofluoric acid

Despite the reversibility of the reaction (equation 73) and the conclusion[188] that the reaction is of little synthetic value, hydro-

$$ROH + HF \rightleftharpoons RF + H_2O \qquad (73)$$

fluoric acid has been used quite widely for preparing fluoro compounds (equations 74 [189], 75 [190] and 76 [191]). The low acidity of

$$Ph_2C(OH)CH_2F \xrightarrow[\text{1 minute}, -78°]{HF} Ph_2CFCH_2F \qquad (74)$$

hydrofluoric acid compared with the other halogen acids, and the low nucleophilicity of fluoride ion are here overcome. Very variable

$$R_3SiOH \longrightarrow R_3SiF \qquad (75)$$

yields were obtained with the silanols, (equation 75), from 7% when R = Et, to 100% when R = Ph, but the reaction, using hydrofluoric acid in acetone, has not been used widely for carbon hydroxy compounds.

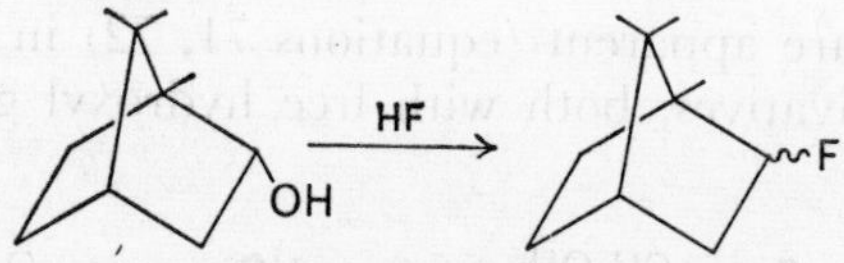

(76)

Alcohols add to diphenylcyanamide, Ph_2NCN, under the influence of potassium butoxide, to give intermediates[192], analogous to those used in section II.B.5c, which react with hydrofluoric acid to give alkyl fluorides.

3. Fluoramines

The fluoramine $FCHClCF_2NEt_2$ has been used to good effect[193] in the example shown (equation 77) and in converting hydroxyamino

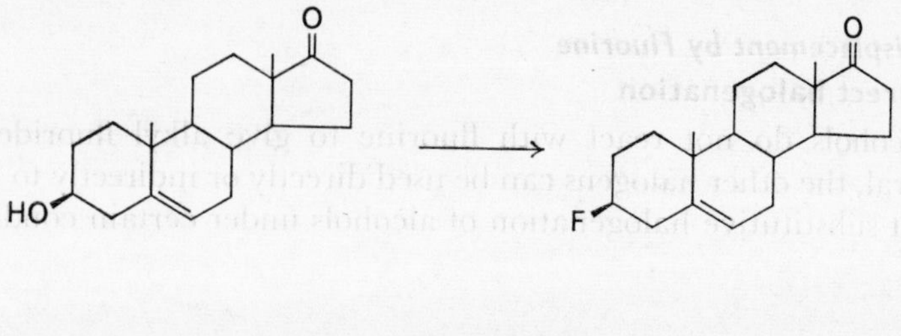

(77)

acids to fluoroamino acids[194], but products resulting from the rearrangement of the intermediate carbonium ion are also found in many cases[195].

4. Sulphur tetrafluoride

A great many oxygen-containing compounds, including alcohols, are attacked by sulphur tetrafluoride[196, 197]. This toxic gas is perhaps the most efficient reagent for converting alcohols[198] to fluoro compounds, but its great reactivity can be a disadvantage. Carbonyl groups are converted to *gem*-difluorides, and carboxylic acid derivatives to trifluoromethyl groups. Ethers, alkenes and alkynes are not attacked, however.

The conversions are catalysed by Lewis acids (*L*), and the mechanism is therefore probably as shown (equation 78). With propargylic

$$\mathrm{ROH} \quad \overset{\delta+}{\mathrm{SF_4}}\cdots\cdots\overset{\delta-}{L} \longrightarrow \mathrm{R\overset{H}{\underset{+}{O}}SF_3}\cdots\cdots L \quad \mathrm{F^-}$$

$$\mathrm{R\overset{H}{\underset{+}{O}}SF_3}\cdots\cdots L \quad \mathrm{F^-} \longrightarrow \mathrm{R{-}O{-}SF_2{-}F} \quad \mathrm{F^-} \tag{78}$$

$$\longrightarrow \mathrm{RF + SOF_2 + F^-}$$

alcohols[81] (equation 79) a mechanism similar to that invoked by

$$\mathrm{(F_3C)_2C(OH)C{\equiv}CH} \xrightarrow{\mathrm{SF_4}} \mathrm{(F_3C)_2C{-}C{\equiv}CH} \;(\text{cyclic O–SF}_2\text{–F intermediate}) \tag{79}$$

$$\rightarrow \mathrm{(F_3C)_2CFC{\equiv}CH + (F_3C)_2C{=}C{=}CHF}$$

Landor[199] for the rearrangement of sulphinate esters of propargylic alcohols explains the product formation. A similar mechanism was proposed by the same[81] authors for the interaction of PCl_5 with the alcohol[80].

Sulphur tetrafluoride substitutes the hydroxyl group of tropolones by fluorine[196] in benzene solution at 60°, (equation 80) conditions considerably milder than those usually used (ca 150° in a pressure vessel). This very ready reaction is the consequence of the acidity of the enol and also of the increased susceptibility of the intermediate

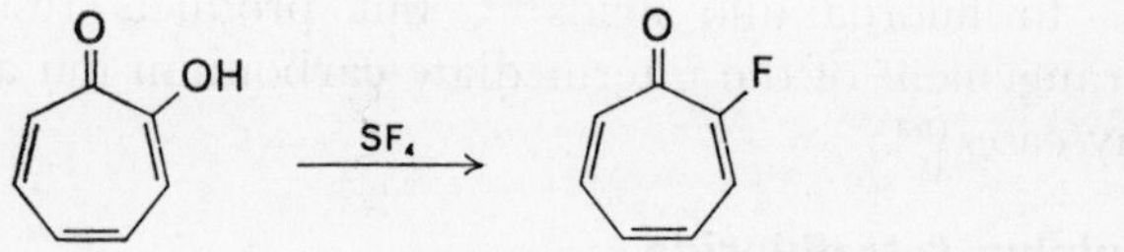

(80)

to attack by fluoride ion due to the presence of the carbonyl group (equation 81). The analogous reaction with simple phenols does not

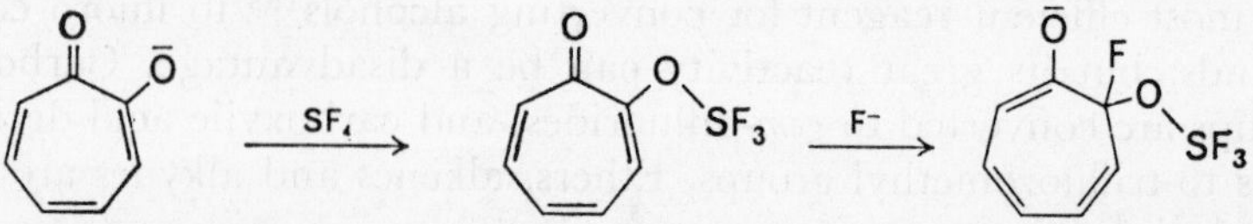

(81)

proceed, a consequence of the resistance of the electron-rich ring in the phenyl ether intermediate to nucleophilic attack (equation 82)

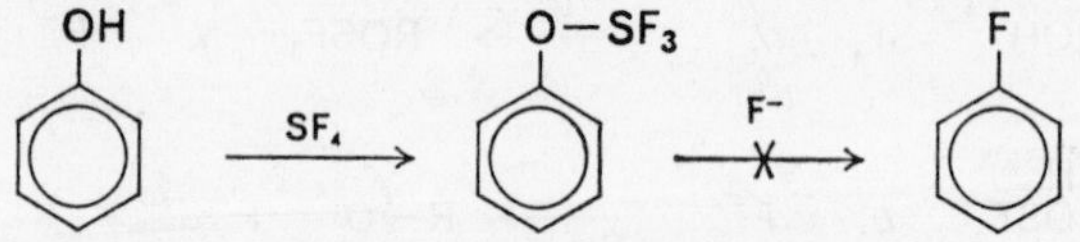

(82)

However, hydroxyquinones are attacked by the reagent[196] (equation 83) and borate esters[200] of alcohols also.

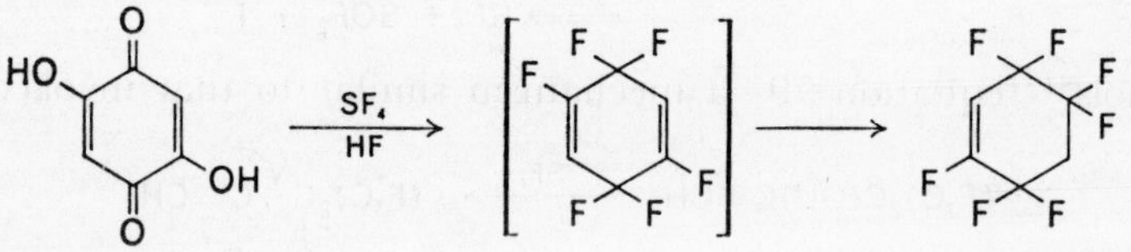

(83)

A recent review[201] has briefly summarized other uses of SF_4 in organic chemistry.

Phenylsulphur trifluoride reacts in many ways as SF_4 [201], but is a milder reagent.

5. Tosylate displacements

Fluoride ion can displace tosyl ion, from sugar derivatives for example[202], and in other molecules also.

6. Fluoroformate decomposition

Carbonyl fluoride reacts with alcohols to form fluoroformates, and the decomposition of these esters to alkyl fluorides has been described[203].

III. DISPLACEMENT BY NITROGEN

A. *Displacement by NH_2*

1. Bucherer reaction. The reversible conversion of naphthols to naphthylamines in the presence of aqueous sodium bisulphite and ammonia, the Bucherer reaction[204] (equation 84), has also been

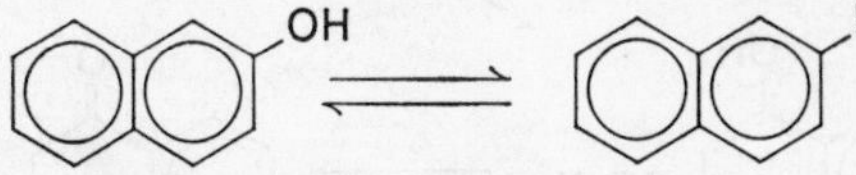

(84)

applied[205] to some phenols, to more highly condensed hydroxy-aromatic compounds, and to some heterocyclic systems.

The mechanism of the reaction, for long assumed[205] to involve addition of ammonia to the bisulphite addition compound of the keto-form of the naphthol (equation 85) has been shown erroneous,

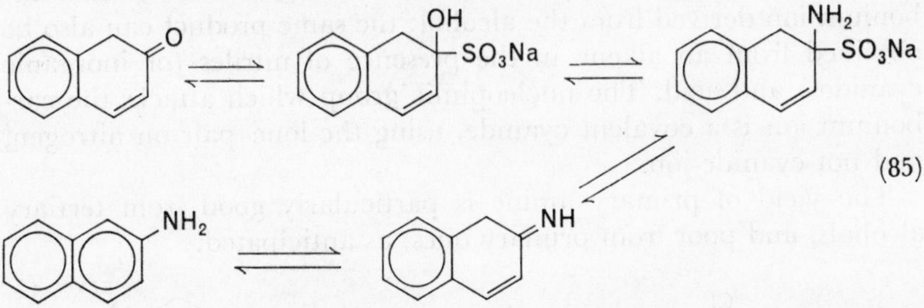

(85)

and recent reviews have corrected it[204, 206]. The older formulation of the intermediate as the bisulphite addition product of a ketone was untenable in the light of i.r. evidence[207]. This clearly showed the presence of a ketone, which forms an oxime and a semicarbazone in the usual way. The intermediate may be regarded as the product of 1,4 addition to an α,β-unsaturated ketone (equation 86) or, more

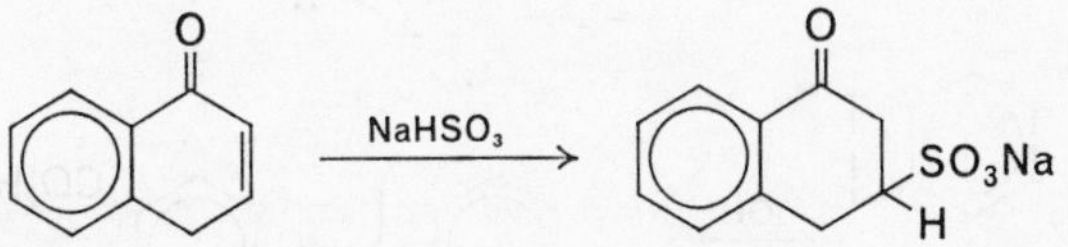

$\xrightarrow{NaHSO_3}$ (86)

fully, as the product of the scheme outlined in equation (87). The conversion of the ketone, via an imine, to naphthylamine derivatives then is unexceptional.

2. Ritter reaction[208]. A general method for converting alcohols, rather than phenols (section III.A.1), to amines is that due to Ritter

$$\text{1-naphthol} \longleftrightarrow \text{(keto form, }\overset{+}{O}H\text{, carbanion)} \xrightarrow{H^+} \overset{+}{O}H \text{ cation} \xrightarrow{SO_3Na^-} \text{enol}-SO_3Na \rightleftharpoons \text{ketone}-SO_3Na \qquad (87)$$

(equation 88). A review of its application in the synthesis of hetero-

$$ROH \xrightarrow[R'CN]{H_2SO_4} RNHCOR' \longrightarrow RNH_2 \qquad (88)$$

cyclic compounds has appeared[209]. The reaction involves the carbonium ion derived from the alcohol: the same product can also be obtained from an alkene in the presence of nitriles (or inorganic cyanides) and acid. The nucleophilic group which attacks the carbonium ion is a covalent cyanide, using the lone pair on nitrogen, and not cyanide ion.

The yield of primary amine is particularly good from tertiary alcohols, and poor from primary ones, as anticipated.

$$\text{4-chloro-2-phenylquinazoline} + \text{oestrone phenoxide} \longrightarrow \text{4-aryloxy-2-phenylquinazoline} \xrightarrow{\Delta} \text{N-aryl-2-phenylquinazolin-4-one} \xrightarrow{OH^-} H_2N\text{-aryl} + \text{anthranilic acid } (CO_2H, NH_2) + PhCO_2H \qquad (89)$$

3. A particularly neat synthetic route to amines from phenols, though not a direct displacement mechanistically, has been reported by Scherrer[210], and applied to oestrone[211] (equation 89).

Another, patented[212] reaction in which the overall conversion is of a phenol to an aniline derivative is shown in equation (90) although it is very unlikely that the mechanism involves any direct displacement of hydroxyl by amine group.

$$HO-C_6H_4-CMe_2-C_6H_4-OH \xrightarrow{PhNH_2 \cdot HCl} H_2N-C_6H_4-CMe_2-C_6H_4-NH_2 \quad (90)$$

4. Tosylates can be converted to azides (section III.C.2) and the azides reduced to amines to provide a useful synthetic route from alcohols to amines[213].

B. Displacement by NR_2

The Bucherer reaction can be extended to prepare secondary amines[205] and hydrazines[205] from phenols. Examples of these reactions are included in review articles.

C. Displacement by Azide

1. Direct displacements of hydroxyl by azide are uncommon, but carbonium ions derived from alcohols are attacked by azide ion (compare the Ritter reaction, section III.A.2) to give organic azides[214] (equation 91).

$$Ph_3COH \xrightarrow[H_2SO_4]{NaN_3} Ph_3CN_3 \quad (91)$$

2. Tosylate displacement by inorganic azide ion provides the principal route to azides from alcohols[215, 216]. The stereochemistry of the product obtained in the displacement of a secondary alcohol depends markedly on the solvent polarity, as anticipated from the inevitable change in mechanism from S_N2 to S_N1 in more polar solvents.
3. Epoxides[217] and halohydrins[218] are attacked by sodium azide and acid, giving hydroxyazides.

IV. DISPLACEMENT BY HYDROGEN

A. Hydrogenolysis

It has long been known that benzylic alcohols[219] (and, more recently, homobenzylic alcohols)[220] are reduced to the parent hydrocarbons by catalytic or chemical hydrogenolysis (equation 92).

$$\text{Ar CH}_2\text{CH}_2\text{OH} \xrightarrow{[\text{H}]} \text{Ar CH}_2\text{CH}_3 \tag{92}$$

Cleavage of the carbon–oxygen bond occurs even more rapidly when ester derivatives of alcohols are used, and McQuillin and co-workers have shown[221] that in a series of conversions (equation 93) the rate of hydrogenolysis parallels the stability of X^-, the ease with which

$$\text{R}-\overset{\text{Ph}}{\underset{\text{X}}{\text{C}}}-\text{CO}_2\text{Et} \xrightarrow{[\text{H}]} \text{R}-\overset{\text{Ph}}{\underset{\text{H}}{\text{C}}}-\text{CO}_2\text{Et} \tag{93}$$

X^- is eliminated, $OH^- < OCOCH_3^- < OCOCF_3^-$.

The hydrogenolysis of secondary alcohols is characterized by a high degree of stereospecificity[222], different catalysts (and different leaving groups) determining whether retention or inversion of configuration occurs.

Garbisch[223] has discussed the mechanism in terms of a modified Horiuti–Polanyi model[224]. He concludes that whereas homobenzylic hydroxyl groups are lost as water in a β-elimination process involving a benzylic hydrogen (no carbon–oxygen bond cleavage is observed in the absence of such hydrogens) and the final product is therefore produced from an alkene, benzylic hydrogenolysis proceeds by a different and more complex mechanism.

Hydrogenolysis of phenols to benzene derivatives using palladium on carbon can be effected through the derivative **57** [225] (compare equation 89).

ArO–C(=N–N=N–N(Ph)–) (tetrazole ring: ArO attached to C; N—N, N—N, N-Ph)

(57)

B. Reduction

The apparent reduction observed when *p*-bromophenol is heated with triphenylphosphine[61] (section II.B.3a) (equation 94) is seen

$$\text{4-BrC}_6\text{H}_4\text{OH} \xrightarrow[\text{Ph}_3\text{P}]{200^\circ} \text{C}_6\text{H}_5\text{Br} \qquad (94)$$

instead as a displacement of hydroxyl by bromine in the light of a similar reaction (equation 95) with 1-bromo-2-naphthol, but several

$$\text{1-bromo-2-naphthol} \xrightarrow{\text{Ph}_3\text{P}} \text{2-bromonaphthalene} \qquad (95)$$

methods for replacing hydroxyl groups other than by hydrogenolysis are known. Iodides (most conveniently prepared from tosylates) are reduced by zinc and hydrochloric acid in methanol[226] or by bisulphite in dioxan[227].

Triphenylcarbinol is reduced to triphenylmethane by 98% formic acid at 20° [228], 2-naphthol to naphthalene by phosphorus trisulphide in phenol at 400° [229], and the alcohol **58** to **59** by lithium aluminium hydride in the presence of aluminium chloride[230].

(3-thienyl)–CHOH–(3-thienyl) **(58)** (3-thienyl)–CH_2–(3-thienyl) **(59)**

V. DISPLACEMENT BY OXYGEN

A. *Exchange with Labelled Hydroxyl Groups*[231] *(equation 96)*

This topic is discussed in another chapter in this volume.

$$\text{PhOH} + \text{H}^+ \xrightarrow{\text{H}_2\text{O}^*} \text{Ph}\overset{*}{\text{O}}\text{H} \qquad (96)$$

B. *Exchange with Alkoxy Groups*[232] *(equation 97)*

$$\text{PhOH} + \text{H}^+ \xrightarrow{\text{ROH}} \text{PhOR} \qquad (97)$$

For a discussion of ether formation from alcohols, see an earlier volume in this series[233].

VI. DISPLACEMENT BY SULPHUR

Paralleling the acid-catalysed exchange of hydroxyl for alkoxy groups in the phenols are the reported conversions (equations 98, 99)

$$\text{2-naphthol (OH)} \xrightarrow[\text{H}^+]{\text{HSCH}_2\text{CO}_2\text{H}} \text{2-naphthyl-SCH}_2\text{CO}_2\text{H} \qquad (98)$$

of β-naphthol to a sulphur ether[234], and of 3-hydroxy-3-methylbut-

$$\text{Me}_2\text{COHC}\equiv\text{CH} \xrightarrow{\text{PhCH}_2\text{SH}} [\text{PhCH}_2\text{SCMe}_2\text{C}\equiv\text{CH}] \longrightarrow \text{PhCH}_2\text{SCMe}_2\text{COMe} \qquad (99)$$

(61) **(60)**

1-yne to the ketone **60** via **61** [235]. Phosphorus pentasulphide converts

$$\text{1,8-naphthalene-(CH(OH)Ph)}_2 \xrightarrow{\text{P}_2\text{S}_5} \text{1,8-naphthalene-(CHPh-S-CHPh)}$$

HOHC CHOH, Ph Ph **(62)** → Ph S Ph **(63)**

the diol **62** to the sulphide **63** [236], and thiols and thiocyanates can also be used in the displacement of tosylate groups (equations 100, 101).

$$\text{Me}_2\text{C}(\text{OCH}_2)_2\text{C}(\text{CH}_2\text{OTs})_2 \xrightarrow{\text{HSCH}_2\text{CO}_2\text{K}^{237}} \text{Me}_2\text{C}(\text{OCH}_2)_2\text{C}(\text{CH}_2\text{SCH}_2\text{CO}_2\text{K})(\text{CH}_2\text{OH}) \qquad (100)$$

$$\text{MeCH(OTs)CH(OTs)Me} \xrightarrow[\text{DMSO}]{\text{KSCN}^{238}} \text{MeCH(SCN)CH(SCN)Me} \qquad (101)$$

VII. DISPLACEMENT BY CARBON

The multitudinous reactions in which the carbonium ion derived from an alcohol attacks another carbon atom, as for example in alkene polymerizations, cannot be discussed here. Examples of nucleophilic displacement reactions by alcohol-derived carbanions are included however, in section VII.B.

A. Displacement by Cyanide

High yields of cyanoallenes are obtained from propargyl alcohols in the presence of hydrobromic acid, cuprous cyanide and potassium cyanide. There is no evidence that the acetylenic cyanide is an intermediate in this reaction, and no bromoallene is formed[239].

Inorganic cyanides displace tosyl groups from alcohol derivatives[240], usually in dimethylformamide or dimethyl sulphoxide solution, in which the tosylates exist in the equilibrium[241] shown in equation (102). (Displacements by inorganic ions in these polar

$$ROTs + Me_2SO \rightleftharpoons Me_2\overset{+}{S}OR\ \overset{-}{O}Ts \qquad (102)$$

solvents also seem aided by preferential solvation of the cation[242].)

B. *Displacement by Alkyl Groups*

A great many examples are known of reactions in which carbanions react with alcohols and diols to give alkylated products[243, 244]. The general reaction is represented in equation (103).

$$\text{fluorene} \xrightarrow[\text{MeOH}]{\text{KOH}} \text{9-methylfluorene (Me)} \qquad (103)$$

Grignard reagents are sources of potential carbanions, but react with alcohols to remove the acidic proton, usually the basis of the Zerewitinov determination of active hydrogen[245]. Allyl alcohols can, however, be alkylated by treating ester derivatives such as **64** with Grignard reagents[246] (equation 104).

$$CH_2{=}CHCH_2OCO{-}C_6H_2(Me)_3 \xrightarrow{\text{RMgX}} CH_2{=}CHCH_2R \qquad (104)$$

(64)

VIII. SURVEY OF DISPLACEMENTS BY HALOGEN ON SPECIFIC HYDROXYL FUNCTIONS

A. *Displacements on Propargyl Alcohols*

The alcohols are very susceptible to rearrangements during substitution, giving allene derivatives under a wide variety of conditions, often without the intermediacy of propargyl halides. Applications of the rearrangements to the synthesis of allenes have been recently reviewed by Taylor[247].

Reagents which are used on less labile systems to obviate rearrangements work well on propargyl systems generally, although no aryl tertiary propargyl halides have been prepared[134]. In addition to triphenyl phosphite methiodide at low temperature in methylene

TABLE 1. Displacements on propargyl alcohols.

Alcohol	Reagent	Product	Yield %	Ref.
A.				
$CH{\equiv}CCH_2OH$	$(PhO)_3PBr_2$/pyridine	$CH{\equiv}CCH_2Br$	72	248
$CH{\equiv}CCH_2OH$	$(PhO)_3PMeI/CH_2Cl_2$	$CH{\equiv}CCH_2I$	90	18
$CH_3C{\equiv}CCH_2OH$	$SOCl_2$/pyridine	$CH_3C{\equiv}CCH_2Cl$	81	249
$CH{\equiv}CCHMeOH$	$(PhO)_3PMeI/CH_2Cl_2$	$CH{\equiv}CCHMeI$	43	18
$CH{\equiv}CCHMeOH$	$(PhO)_3PBr_2$/pyridine	$CH{\equiv}CCHMeBr$	72	45
$CH{\equiv}CC(OH)Et_2$	$HCl/CuCl/CaCl_2/Cu/Zn$	$CH{\equiv}CCClEt_2$	83	51
$CH{\equiv}CC(OH)Me_2$	PCl_3	$CH{\equiv}CCClMe_2$	47	132
		$+CHCl{=}CHCMe{=}CH_2$	23	
$CH{\equiv}CC(OH)Me_2$	SF_4	$CH{\equiv}CCFMe_2$		81
B. see also References 80, 131, 133				
$CH{\equiv}CC(OH)MeBu\text{-}t$	$HBr/CuBr/Cu$	$CHBr{=}C{=}CMeBu\text{-}t$		45, 46
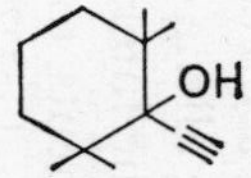 OH	$SOCl_2$/pyridine	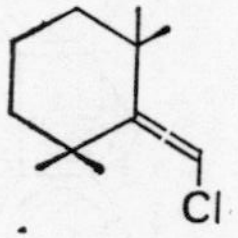 Cl	83	139
$t\text{-}BuC{\equiv}CC(OH)ArBu\text{-}t$	PBr_3	$t\text{-}BuCBr{=}C{=}CArBu\text{-}t$		134
$t\text{-}BuC{\equiv}CC(OH)ArBu\text{-}t$	$SOCl_2$	$t\text{-}BuCCl{=}C{=}CArBu\text{-}t$		134
$PhC{\equiv}CC(OH)PhAr$	$SOCl_2$	$PhCCl{=}C{=}CPhAr$		250
$HC{\equiv}CC(OH)(CF_3)_2$	SF_4	$CHF{=}C{=}C(CF_3)_2$		81
		$+HC{\equiv}CCF(CF_3)_2$		

chloride[18] and triphenyl phosphite dibromide in the presence of pyridine[248], however, thionyl chloride has been used on primary propargyl alcohols[138, 249], and hydrochloric acid[51], phosphorus trichloride[132] and sulphur tetrafluoride[81] on aliphatic tertiary ones without causing extensive rearrangement. Under different conditions, mineral acid[45, 46], phosphorus tribromide[134], thionyl chloride[134, 148, 250] and sulphur tetrafluoride[81] have all been reported to cause major rearrangements in propargyl alcohols. Propargyl iodides rearrange to iodoallenes in dimethylformamide[18].

B. Displacements on Allylic, Homallylic and Allenic Alcohols

The problems encountered in studying substitutions of allylic alcohols[251] are mainly those of determining whether the mechanism involves double bond participation.

From the synthetic aspect, as with homoallylic systems and, indeed, with simple alcohols, different reagents can be used to give stereochemically or isomerically different products. The different reactions effected by PCl_5 and $SOCl_2$ (7β-cholestanol gives 55% 7α-chlorocholestane with $PCl_5/CaCO_3$ but 59% 7β-chlorocholestane with $SOCl_2$ [152]) have been referred to in section II.B.4. Retention of configuration is found with many reagents on homoallylic alcohols: PCl_5, PX_3, $(PhO)_3PX_2$ and Ph_3PBr_2 all react in this way[74, 75]. An important factor encouraging S_Ni reactions in 3-hydroxy-5-ene steroids is the inability of the rigid molecule to assume the necessary planar transition state configuration for S_N2 reaction. (See also equation 77.)

Allenic alcohols also can be substituted directly by some reagents[79], but rearrange to 1,3-diene derivatives with others[79, 252].

See also References 6, 17, 18, 43, 44, 107, 136, 149, 193 and 248.

C. Reactions of Cyclopropanols[253] and of Cyclopropylcarbinols

No reactions are known in which direct substitution of cyclopropanols occurs; both with the free alcohols and with tosylate derivatives[254], ring opening[255] invariably takes place preferentially.

Cyclopropylcarbinols are also very susceptible to ring opening reactions, although the three-membered ring survives substitution at the adjacent carbon atom to some extent[147]. Earlier speculations on the intermediacy of the tricyclobutonium ion in these reactions are no longer tenable[256]. Halogen acids effect ring opening to give homoallylic halides in good yields[42, 183].

D. Displacements on Phenols (see also section II.B.3.a)

Phenols readily form esters with many of the reagents used for converting alcohols to alkyl halides, but these fail to undergo nucleophilic substitution in a subsequent, halide-forming step, except in a few cases. Phosphorus pentachloride gives poor conversions of phenols to aryl halides because of side reactions: tetraaryloxyphosphonium halides in general are stable at quite high temperatures[69] although their decomposition can be carried out at 140° using the method due to Rydon[70]. The same method allows for the preparation of aryl bromides and iodides by exchanging the chloride ion in the phosphonium salts.

The use of triphenylphosphine dihalides in place of phosphorus pentahalides is preferred because of the easier reaction resulting from making triphenylphosphine oxide the leaving group, and the only disadvantage of the reagent seems to be encountered with very hindered phenols[95] (equation 35).

The hydroxyl group in tropolones can be substituted smoothly with sulphur tetrafluoride or with thionyl chloride[141]. Very acidic phenols react with thionyl chloride in dimethylformamide[142].

The relative inertia of phenolic hydroxyl groups towards displacement enables chemical differentiations to be made between them and, for example, benzylic hydroxyl groups. This has been exploited many times (section II.B.7 and Reference 257).

IX. REFERENCES

1. H. Gross, S. Katzwinkel and J. Gloede, *Chem. Ber.*, **99**, 2631 (1966) and E. J. Corey and J. E. Anderson, *J. Org. Chem.*, **32**, 4160 (1967).
2. *Organic Syntheses*, Coll. Vol. 2, John Wiley & Sons Inc., New York, 1943, p. 322.
3. E. J. Coulson, W. Gerrard and H. R. Hudson, *J. Chem. Soc.*, 2364 (1965).
4. G. F. Freeguard and L. H. Long, *Chem. Ind.* (*London*), 1582 (1964).
5. Ref. 2, p. 399.
6. R. Criegee, H. Kristinsson, D. Seebach and F. Zanker, *Chem. Ber.*, **98**, 2331 (1965).
7. C. J. Collins and J. F. Eastham, in *The Chemistry of the Carbonyl Group* (Ed. S. Patai), Interscience, New York, 1963, p. 762.
8. N. J. Leonard, K. Conrow and R. W. Fulmer, *J. Org. Chem.*, **22**, 1445 (1957).
9. H. Stone and H. Shechter, *J. Org. Chem.*, **15**, 491 (1950).
10. R. A. Sneen and J. W. Larsen, *J. Am. Chem. Soc.*, **91**, 362 (1969).
11. A. C. Cope, R. K. Bly, E. P. Burrows, O. J. Ceder, E. Ciganek, B. T. Gillis, R. F. Porter and H. E. Johnson, *J. Am. Chem. Soc.*, **84**, 2170 (1962).

12. A. C. Cope, E. P. Burrows, M. E. Derieg, S. Moon and W.-D. Wirth, *J. Am. Chem. Soc.*, **87**, 5452 (1965).
13. See, for example, R. F. Hudson, *Structure and Mechanism in Organo-phosphorus Chemistry*, Academic Press, London, 1965, p. 135.
14. F. W. Lichtenthaler, *Chem. Rev.*, **61**, 607 (1961).
15. W. S. Wadsworth and W. D. Emmons, *J. Am. Chem. Soc.*, **83**, 1733 (1961).
16. A. H. Ford-Moore and B. J. Perry, *Organic Syntheses*, Coll. Vol. 4, John Wiley & Sons Inc., New York, 1963, p. 325.
17. S. R. Landauer and H. N. Rydon, *J. Chem. Soc.*, 2224 (1953).
18. C. S. L. Baker, P. D. Landor, S. R. Landor and A. N. Patel, *J. Chem. Soc.*, 4348 (1965).
19. N. K. Kochetkov and A. I. Usov, *Tetrahedron Letters*, 973 (1963).
20. N. K. Kochetkov and A. I. Usov, *Izvest. Akad. Nauk SSSR*, 475 (1964); [*Chem. Abstr.*, **60**, 15952g (1964)].
21. J. P. H. Verheyden and J. G. Moffatt, *J. Am. Chem. Soc.*, **88**, 5684 (1966).
22. J. P. H. Verheyden and J. G. Moffatt, *J. Am. Chem. Soc.*, **86**, 2093 (1964).
23. N. Kornblum and D. C. Iffland, *J. Am. Chem. Soc.*, **77**, 6653 (1955).
24. K. Kefurt, J. Jary and Z. Samek, *J. Chem. Soc.* (**D**), 213 (1969).
25. N. K. Kochetkov and A. I. Usov, *Tetrahedron Letters*, 519 (1963).
26. C. L. Stevens, R. P. Glinski, K. G. Taylor, P. Blumbergs and F. Sirokman, *J. Am. Chem. Soc.*, **88**, 2073 (1966).
27. S. Hanessian, *Chem. Commun.*, 796 (1966).
28. J. P. Forsman and D. P. Lipkin, *J. Am. Chem. Soc.*, **75**, 3145 (1953).
29. A. Roedig, in *Houben—Weyl, Methoden der Organischen Chemie*, Vol. 5, Pt. 4, George Thieme Verlag, Stuttgart, 1960, p. 627.
30. J. M. Sugihara, D. L. Schmidt, V. D. Calbi and S. M. Dorrence, *J. Org. Chem.*, **28**, 1406 (1963).
31. L. F. Fieser and M. Fieser, *Reagents for Organic Synthesis*, John Wiley & Sons Inc., New York, 1967, p. 296.
32. F. C. Uhle, *J. Am. Chem. Soc.*, **83**, 1460 (1961).
33. A. Goosen, *J. Chem. Soc.* (**D**), 145 (1969).
34. J. Broome, B. R. Brown and G. H. R. Summers, *J. Chem. Soc.*, 2071 (1957).
35. R. Stroh and W. Hahn, in *Houben–Weyl, Methoden der Organischen Chemie*, Vol. 5, Pt. 3, Georg Thieme Verlag, Stuttgart, 1962, p. 503.
36. E. J. Goethals and M. Verzele, *Bull. Soc. Chim. Belges*, **74**, 21 (1965).
37. G. Sosnowsky, in *Free Radical Reactions in Preparative Organic Chemistry*, Macmillan Co., New York, 1964, p. 376.
38. M. R. Peterson and G. H. Wahl, *Chem. Commun.*, 1552 (1968).
39. P. v. R. Schleyer, R. C. Fort, W. E. Watts, M. B. Comisarow and G. A. Olah, *J. Am. Chem. Soc.*, **86**, 4195 (1964).
40. Ref. 31, p. 450.
41. H. J. Lucas, *J. Am. Chem. Soc.*, **52**, 802 (1930).
42. M. Julia, S. Julia and J. Amaudric du Chaffaut, *Bull. Soc. Chim. France*, 1735 (1960).
43. J. D. Surmatis and A. Ofner, *J. Org. Chem.*, **26**, 1171 (1961).
44. O. Isler, H. Lindlar, M. Montavon, R. Ruegg and P. Zeller, *Helv. Chim. Acta*, **39**, 449 (1956).
45. D. K. Black, S. R. Landor, A. N. Patel and P. F. Whiter, *Tetrahedron Letters*, 483 (1963).

46. S. R. Landor, A. N. Patel, P. F. Whiter and P. M. Greaves, *J. Chem. Soc.* (**C**), 1223 (1966).
47. W. Gerrard and H. R. Hudson, *J. Chem. Soc.*, 2310 (1964).
48. R. C. Fuson, C. C. Price and D. M. Burness, *J. Org. Chem.*, **11**, 475 (1946).
49. J. V. Cerny and J. Hora, *Collection Czech. Chem. Commun.*, **25**, 711 (1960).
50. J. P. Kutney and T. Tabata, *Can. J. Chem.*, **41**, 695 (1963).
51. G. F. Hennion and A. P. Boisselle, *J. Org. Chem.*, **26**, 725 (1961).
52. W. Gerrard and H. R. Hudson, *J. Chem. Soc.*, 1059 (1963).
53. W. P. Whelan, quoted in K. Wiberg and B. R. Lowry, *J. Am. Chem. Soc.*, **85**, 3188 (1963).
54. R. A. Benkeser and W. P. Fitzgerald, *J. Org. Chem.*, **26**, 4179 (1961).
55. C. L. Stevens, D. Morrow and J. Lawson, *J. Am. Chem. Soc.*, **77**, 2341 (1955).
56. W. Steinkopf, *Ber.*, **41**, 2541 (1908).
57. F. Cramer and H. J. Baldauf, *Chem. Ber.*, **92**, 370 (1959).
58. N. L. Albertson, in *Organic Reactions*, **12**, John Wiley & Sons Inc., New York, 1962, p. 205.
59. C. A. Kingsbury and W. B. Thornton, *J. Am. Chem. Soc.*, **88**, 3159 (1966).
60. E. S. Gould, *Mechanism and Structure in Organic Chemistry*, Holt, Rinehart & Winston, New York, 1959, p. 294.
61. H. Hoffmann, L. Horner, H. G. Wippel and D. Michael, *Chem. Ber.*, **95**, 523 (1962).
62. M. M. Robison, *J. Am. Chem. Soc.*, **80**, 5481 (1958).
63. H. Gross and J. Gloede, *Chem. Ber.*, **96**, 1387 (1963).
64. E. L. Eliel and R. G. Haber, *J. Org. Chem.*, **24**, 143 (1959).
65. C. E. Kaslow and M. M. Marsh, *J. Org. Chem.*, **12**, 456 (1947).
66. M. H. Benn, A. M. Creighton, L. N. Owen and G. R. White, *J. Chem. Soc.*, 2365 (1961).
67. Ref. 31, p. 866.
68. A. Hirsch and D. Orphanos, *Can. J. Chem.*, **43**, 2708 (1965).
69. O. M. Nefedov, Y. L. Levkov and A. D. Petrov, *Dokl. Akad. Nauk SSSR*, **133**, 855 (1960); *Chem. Abstr.*, **54**, 24567d (1960).
70. D. G. Coe, H. N. Rydon and B. L. Tonge, *J. Chem. Soc.*, 323 (1957).
71. G. Ponzio, *Gazz. Chim. Ital.*, **62**, 1025 (1932).
72. M. Ruccia, *Gazz. Chim. Ital.*, **89**, 1670 (1959).
73. H. L. Goering and F. H. McCarron, *J. Am. Chem. Soc.*, **78**, 2270 (1956).
74. C. W. Shoppee, *Chemistry of Steroids*, Butterworths, London, 2nd ed., 1964, p. 50.
75. D. Levy and R. Stevenson, *J. Org. Chem.*, **30**, 3469 (1965).
76. G. H. Alt and A. J. Speziale, *J. Org. Chem.*, **29**, 794 (1964).
77. A. Albert and J. Clark, *J. Chem. Soc.*, 1666 (1964).
78. H. M. Wuest, J. A. Bigot, Th. J. de Boer, B. van der Wal and J. P. Wibaut, *Rec. Trav. Chim.*, **78**, 226 (1959).
79. M. Bertrand and J. Le Gras, *Compt. Rend.*, **261**, 474 (1965).
80. H. E. Simmons and D. W. Wiley, *J. Am. Chem. Soc.*, **82**, 2288 (1960).
81. R. E. A. Dear and E. G. Gilbert, *J. Org. Chem.*, **33**, 819 (1968).
82. M. Warman and V. I. Siele, *J. Org. Chem.*, **26**, 2997 (1961).
83. N. Yamaoka and K. Aso, *J. Org. Chem.*, **27**, 1462 (1962).
84. R. S. Tipson, *J. Org. Chem.*, **27**, 1449 (1962).
85. D. Harrison, J. T. Ralph and A. C. B. Smith, *J. Chem. Soc.*, 2930 (1963).

86. L. Stephenson, T. Walker, W. K. Warburton and G. B. Webb, *J. Chem. Soc.*, 1282 (1962).
87. L. H. Sarett, *J. Am. Chem. Soc.*, **70,** 1454 (1948).
88. Ref. 31, p. 879.
89. Ref. 31, pp. 877, 1160.
90. H. Gross and U. Karsch, *J. Prakt. Chem.*, **29,** 315 (1965).
91. G. A. Wiley, R. L. Hershkowitz, B. M. Rein and B. C. Chung, *J. Am. Chem. Soc.*, **86,** 964 (1964).
92. J. P. Schaeffer and J. Higgins, *J. Org. Chem.*, **32**, 1607 (1967).
93. M. Ohno and I. Sakai, *Tetrahedron Letters*, 4541 (1965).
94. A. G. Anderson and F. J. Freenor, *J. Am. Chem. Soc.*, **86**, 5037 (1964).
95. D. G. Lee, *Chem. Commun.*, 1554 (1968).
96. A. J. Speziale and R. D. Partos, *J. Am. Chem. Soc.*, **85**, 3312 (1963).
97. D. B. Denney, D. Z. Denney and B. C. Chang, *J. Am. Chem. Soc.*, **90,** 6332 (1968).
98. G. G. Arzoumanidis, *J. Chem. Soc.* (**D**), 217 (1969).
99. Refs. quoted in Ref. 97.
100. J. P. Schaeffer and D. S. Weinberg, *J. Org. Chem.*, **30**, 2635, 2639 (1965).
101. H. Kwart, T. Takeshita and J. L. Nyce, *J. Am. Chem. Soc.*, **86,** 2606 (1964).
102. A. W. Frank and C. F. Baranauckas, *J. Org. Chem.*, **31,** 872 (1966).
103. H. McCombie, B. C. Saunders and G. J. Stacey, *J. Chem. Soc.*, 380 (1945).
104. H. N. Rydon and B. L. Tonge, *J. Chem. Soc.*, 4682 (1957).
105. L. Anschütz and F. Wenger, *Ann. Chem.*, **482,** 25 (1930).
106. Ref. 13, ch. 7.
107. D. G. Coe, S. R. Landauer and H. N. Rydon, *J. Chem. Soc.*, 2281 (1954); compare Ref. 106, p. 217.
108. P. C. Crofts and I. M. Downie, *J. Chem. Soc.*, 2559 (1963).
109. A. J. Burn and J. I. G. Cadogan, *Chem. Ind.* (*London*), 736 (1963).
110. A. J. Burn and J. I. G. Cadogan, *J. Chem. Soc.*, 5788 (1963).
111. C. E. Griffin, *Abstracts of 135th A.C.S. Meeting*, 1959, p. 690.
112. F. Ramirez and N. McKelvie, *J. Am. Chem. Soc.*, **79**, 5829 (1957).
113. B. Miller, in *Topics in Phosphorus Chemistry*, Vol. 2 (Ed. M. Grayson and E. J. Griffith), John Wiley & Sons Inc., New York, 1965, p. 133.
114. R. Rabinowitz and R. Marcus, *J. Am. Chem. Soc.*, **84**, 1312 (1962).
115. T. Kamai and Z. S. Egorova, *Zh. Obshch. Khim.*, **16**, 1521 (1946).
116. J. Hooz and S. S. H. Gilani, *Can. J. Chem.*, **46,** 86 (1968).
117. I. M. Downie, J. B. Lee and M. F. S. Matough, *Chem. Commun.*, 1350 (1968).
118. H. Oediger and K. Eiter, *Ann. Chem.*, **682,** 58 (1965).
119. R. G. Weiss and E. I. Snyder, *Chem. Commun.*, 1358 (1968).
120. J. B. Lee and I. M. Downie, *Tetrahedron*, **23,** 359 (1967).
121. J. B. Lee and T. J. Nolan, *Tetrahedron*, **23,** 2789 (1967).
122. A. J. Burn, J. I. G. Cadogan and P. J. Bunyan, *J. Chem. Soc.*, 4369 (1964).
123. T. Mukaiyama, C. Mitsunobu and T. Obata, *J. Org. Chem.*, **30,** 101 (1965).
124. H. von Brachel, *Ger. Pat. 1,103,328* (1961); *Chem. Abstr.*, **56,** 7176 (1962).
125. For hexachlorocyclopentadiene, see V. Mark, *Tetrahedron Letters*, 295 (1961).
126. S. Trippett, *J. Chem. Soc.*, 2337 (1962).
127. I. J. Borowitz and R. Virkhaus, *J. Am. Chem. Soc.*, **85,** 2183 (1963).
128. J. B. Lee and T. J. Nolan, *Can. J. Chem.*, **44,** 1331 (1966).

129. A. J. Kirby and S. G. Warren, *The Organic Chemistry of Phosphorus*, Elsevier, Amsterdam, 1967.
130. Ref. 31, p. 873.
131. C. C. Leznoff and F. Sondheimer, *J. Am. Chem. Soc.*, **90**, 731 (1968).
132. E. D. Bergmann and D. Herrman, *J. Am. Chem. Soc.*, **73**, 4014 (1951).
133. H. Tani and F. Toda, *Bull. Chem. Soc. Japan*, **37**, 470 (1964).
134. T. L. Jacobs and D. M. Fenton, *J. Org. Chem.*, **30**, 1808 (1965).
135. K. Mislow and M. A. W. Glass, *J. Am. Chem. Soc.*, **83**, 2780 (1961).
136. P. R. Bai, B. B. Ghatge and S. C. Bhattacharyya, *Tetrahedron*, **22**, 907 (1966).
137. Ref. 35, p. 857 ff.
138. M. S. Newman and J. H. Wotiz, *J. Am. Chem. Soc.*, **71**, 1292 (1949).
139. Y. R. Bhatia, P. D. Landor and S. R. Landor, *J. Chem. Soc.*, 24 (1959).
140. Ref. 31, p. 1158.
141. R. B. Johns, A. W. Johnson and M. Tisler, *J. Chem. Soc.*, 4605 (1954).
142. I. Matsumoto, *Yakugaku Zasshi*, **85**, 544 (1965); *Chem. Abstr.*, **63**, 6898g (1965).
143. R. Huisgen, J. Sauer and M. Seidel, *Chem. Ber.*, **93**, 2885 (1960).
144. A. Dornov and A. Muller, *Chem. Ber.*, **93**, 41 (1960).
145. F. Borgardt, A. K. Seeler and P. Noble, *J. Org. Chem.*, **31**, 2806 (1966).
146. A. Butenandt, E. Biekert, M. Däuble and K. H. Köhrmann, *Chem. Ber.*, **92**, 2172 (1959).
147. J. D. Roberts and R. H. Mazur, *J. Am. Chem. Soc.*, **73**, 2509 (1951).
148. T. L. Jacobs, C. Hall, D. A. Babbe and P. Prempree, *J. Org. Chem.*, **32**, 2283 (1967).
149. A. S. Kende and T. L. Bogard, *Tetrahedron Letters*, 3383 (1967).
150. U. Schöllkopf, *Angew. Chem.*, **72**, 147 (1960).
151. S. Winstein, *Experientia*, Supp. 2, 137 (1955).
152. R. J. Cremlyn and C. W. Shoppee, *J. Chem. Soc.*, 3794 (1954).
153. C. W. Shoppee, R. E. Lack, S. C. Sharma and L. R. Smith, *J. Chem. Soc.* (**C**), 1155 (1967).
154. M. Ikehara, H. Uno and F. Ishikawa, *Chem. Pharm. Bull.* (*Tokyo*), **12**, 267 (1964).
155. H. Holdschmidt, E. Degener, H.-G. Schmelzer, H. Tarnov and W. Zecher, *Angew. Chem.*, *Inter. Ed.*, **7**, 856 (1968).
156. A. J. Speziale and R. C. Freeman, *J. Am. Chem. Soc.*, **82**, 903, 909 (1960).
157. Th. R. Rix and J. F. Arens, *Proc. Koninkel. Ned. Akad. Wetenschap.*, **56B**, 368, 372 (1953); *Chem. Abstr.*, **44**, 2300 (1955).
158. I. A. Smith, *J. Chem. Soc.*, 1099 (1927).
159. S. J. Rhoads and R. E. Michel, *J. Am. Chem. Soc.*, **85**, 585 (1963).
160. W. Gerrard and F. Schild, *Chem. Ind.* (*London*), 1232 (1954).
161. D. H. R. Barton, H. P. Faro, E. P. Serebryakov and N. F. Woolsey, *J. Chem. Soc.*, 2438 (1965).
162. J. Faust and R. Mayer, *Angew. Chem.*, **75**, 573 (1963).
163. H. Eilingsfeld, M. Seefelder and H. Weidinger, *Chem. Ber.*, **96**, 2671 (1963).
164. J. A. Ross and M. D. Martz, *J. Org. Chem.*, **29**, 2784 (1964).
165. R. Deghenghi and R. Gaudry, *Can. J. Chem.*, **40**, 818 (1962).
166. A. R. Mattocks, *J. Chem. Soc.*, 1918, 4840 (1964).
167. A. Roedig in Ref. 35, p. 407.
168. Y. Ito, S. Koto and S. Umezawa, *Bull. Chem. Soc. Japan*, **35**, 1618 (1962);

L. Zervas and S. Konstas, *Chem. Ber.*, **93**, 435 (1960); W. Weidmann and H. K. Zimmerman, *Chem. Ber.*, **92**, 1523 (1959); M. Nys and J. P. Verheijden, *Bull. Soc. Chim. Belges*, **69**, 57 (1960).
169. J. De Graw and L. Goodman, *J. Org. Chem.*, **27**, 1395 (1962).
170. J. Cason and J. S. Correia, *J. Org. Chem.*, **26**, 3645 (1961).
171. K. W. Buck and A. B. Foster, *J. Chem. Soc.*, 2217 (1963).
172. J. W. Cornforth, R. H. Cornforth and K. K. Mathew, *J. Chem. Soc.*, 2539 (1959).
173. L. Skattebol, *Tetrahedron*, **21**, 1357 (1965).
174. G. F. Hennion, J. J. Sheehan and D. E. Maloney, *J. Am. Chem. Soc.*, **72**, 3542 (1950).
175. M. Lora-Tamayo, R. Madroñero and G. G. Muñoz, *Chem. Ber.*, **93**, 289 (1960).
176. H. Gross and I. Farkas, *Chem. Ber.*, **93**, 95 (1960).
177. W. J. Baumann and H. K. Mangold, *J. Lipid Res.*, **7**, 568 (1966).
178. W. Gerrard, H. R. Hudson and W. S. Murphy, *J. Chem. Soc.*, 2314 (1964).
179. S. Umezawa, S. Koto and Y. Ito, *Bull. Chem. Soc. Japan*, **36**, 183 (1963); A. Momose, K. Kamei and Y. Nitta, *Chem. Pharm. Bull.*, **14**, 199 (1966).
180. E. Pacsu, *Ber.*, **61**, 1508 (1929).
181. C. D. Hurd and R. D. Kimbrugh, *J. Am. Chem. Soc.*, **83**, 236 (1961).
182. P. A. Finan and C. D. Warren, *J. Chem. Soc.*, 3089 (1962).
183. R. Ginsig and A. D. Cross, *J. Org. Chem.*, **31**, 1761 (1966).
184. R. T. Blickenstaff, *J. Am. Chem. Soc.*, **82**, 3673 (1960).
185. D. L. Fields, J. B. Miller and D. D. Reynolds, *J. Org. Chem.*, **29**, 2640 (1964).
186. Y. Kawase and C. Numata, *Bull. Chem. Soc. Japan*, **35**, 1366 (1962).
187. J. H. Looker, T. T. Okamoto, E. R. Magnuson, D. L. Shaneyfelt and R. J. Prokop, *J. Org. Chem.*, **27**, 4349 (1962).
188. E. Forche in Ref. 35, p. 141.
189. J. Bornstein, *International Symposium on Fluorine Chemistry*, Birmingham, July 1959.
190. C. Eaborn, *J. Chem. Soc.*, 2846 (1952).
191. M. Hanack, H. Eggensperger and R. Hähnle, *Ann. Chem.*, **652**, 96 (1962).
192. J. H. Amin, J. Newton and F. L. M. Pattison, *Can. J. Chem.*, **43**, 3173 (1965).
193. D. E. Ayer, *Tetrahedron Letters*, 1065 (1962).
194. A. Cohen and E. D. Bergmann, *Tetrahedron*, **22**, 3545 (1966).
195. L. H. Knox, E. Velarde, S. Berger, I. Delfin, R. Grezemkovsky and A. D. Cross, *J. Org. Chem.*, **30**, 4160 (1965).
196. E. Forche in Ref. 35, p. 85 ff.
197. W. C. Smith, *Angew. Chem.*, *Inter. Ed.*, **1**, 467 (1962).
198. D. C. England, *U.S. Pat. 3, 236, 894* (1966).
199. S. R. Landor, *Chem. Soc. (London) Spec. Publ.*, **19**, 164 (1965).
200. A. Dornow and M. Siebrecht, *Chem. Ber*, **95**, 763 (1962).
201. C. M. Sharts, *J. Chem. Educ.*, **45**, 185 (1968).
202. E. R. Blakely, *Biochem. Prep.*, **7**, 39 (1960).
203. W. A. Sheppard, *J. Org. Chem.*, **29**, 1 (1964), and S. Nakanishi, T. C. Myers and E. V. Jensen, *J. Am. Chem. Soc.*, **77**, 3099 (1955).
204. E. H. White and D. J. Woodcock, in *The Chemistry of the Amino Group* (Ed. S. Patai), Interscience, New York, 1968, p. 486.

205. N. L. Drake, in *Organic Reactions*, Vol. 1, John Wiley & Sons Inc., New York, 1942, p. 105.
206. H. Seeboth, *Angew. Chem., Inter. Ed.*, **6,** 307 (1967).
207. A. Rieche and H. Seeboth, *Ann. Chem.*, **638,** 43, 76 (1960).
208. J. J. Ritter and J. Kalish, *J. Am. Chem. Soc.*, **70,** 4048 (1948).
209. F. Johnson and R. Madroñero, *Advan. Het. Chem.*, **6,** 96 (1966).
210. R. A. Scherrer, *Abstracts of the 145th A.C.S. Meeting*, Sept. 1963, p. 33Q.
211. D. F. Morrow and R. M. Hofer, *J. Med. Chem.*, **9,** 249 (1966).
212. H. Krimm, H. Ruppert and H. Schnell, *F. Pat. 1,398,652* (*Cl. C O 7c*), May 7, 1965; *Chem. Abstr.*, **63,** P 13152g.
213. R. Goutarel, A. Cave, L. Tan and M. Lebœuf, *Bull. Soc. Chem. France*, 646 (1962).
214. C. L. Arcus and R. J. Mesley, *Chem. Ind.* (*London*), 701 (1951).
215. D. N. Jones, *Chem. Ind.* (*London*), 179 (1962).
216. L. A. Freiberg, *J. Org. Chem.*, **30,** 2476 (1965).
217. C. A. van der Werf, R. Y. Heisler and W. E. McEwen, *J. Am. Chem. Soc.*, **76,** 1231 (1954).
218. P. A. Leveune and A. Schormüller, *J. Biol. Chem.*, **105,** 547 (1934).
219. W. H. Hartung and R. Simonoff, in *Organic Reactions*, Vol. 7, John Wiley & Sons Inc., New York, 1953, p. 263.
220. T. W. Greenlee and W. A. Bonner, *J. Am. Chem. Soc.*, **81,** 4303 (1959).
221. A. M. Khan, F. J. McQuillin and I. Jardine, *J. Chem. Soc.* (**C**), 136 (1967).
222. A. M. Khan, F. J. McQuillin and I. Jardine, *Tetrahedron Letters*, 2649 (1966).
223. E. W. Garbisch, L. Schreader and J. J. Frankel, *J. Am. Chem. Soc.*, **89,** 4233 (1967).
224. S. Siegel, *Advan. Catalysis*, **16,** 123 (1966).
225. W. J. Musliner and J. W. Gates, *J. Am. Chem. Soc.*, **88,** 4271 (1966).
226. G. D. Meakins and R. Swindells, *J. Chem. Soc.*, 1044 (1959).
227. D. Rosenthal, P. Grabowich, E. F. Sabo and J. Fried, *J. Am. Chem. Soc.*, **85,** 3971 (1963).
228. R. Grinter and S. F. Mason, *Trans. Faraday Soc.*, **60,** 889 (1964).
229. W. N. Moulton and C. G. Wade, *J. Org. Chem.*, **26,** 2528 (1961).
230. A. Kraak, A. K. Wiersema, P. Jordens and H. Wynberg, *Tetrahedron*, **24,** 3381 (1968).
231. S. Oae, R. Kiritani and W. Tagaki, *Bull. Chem. Soc. Japan*, **39,** 1961 (1966).
232. S. Oae and R. Kiritani, *Bull. Chem. Soc. Japan*, **39,** 611 (1966).
233. H. Feuer and J. Hooz, in *The Chemistry of the Ether Linkage* (Ed. S. Patai), Interscience, New York, 1967, p. 468.
234. G. M. Furman, J. H. Thelin, D. W. Hein and W. B. Hardy, *J. Am. Chem. Soc.*, **82,** 1450 (1960).
235. G. W. Stacey, B. F. Barnett and P. L. Strong, *J. Org. Chem.*, **30,** 592 (1965).
236. R. H. Schlessinger and A. G. Schultz, *J. Am. Chem. Soc.*, **90,** 1676 (1968).
237. P. Bladon and L. N. Owen, *J. Chem. Soc.*, 585 (1950).
238. E. J. Corey and R. B. Mitra, *J. Am. Chem. Soc.*, **84,** 2938 (1962).
239. P. M. Greaves, S. R. Landor and D. R. J. Laws, *Chem. Commun.*, 321 (1965).
240. A. C. Cope and A. S. Mehta, *J. Am. Chem. Soc.*, **86,** 5626 (1964).
241. S. G. Smith and S. Winstein, *Tetrahedron*, **3,** 317 (1958).
242. H. E. Zaugg, B. W. Horrom and S. Borgwardt, *J. Am. Chem. Soc.*, **82,** 2895 (1960).

243. I. D. Ruben and E. I. Becker, *J. Org. Chem.*, **22,** 1623 (1957).
244. H. E. Fritz, D. W. Peck, M. A. Eccles and K. E. Atkins, *J. Org. Chem.*, **30,** 2540 (1965).
245. F. T. Weiss, in *Treatise on Analytical Chemistry* (Ed. I. M. Kolthoff and P. J. Elving), Interscience, **13,** 1966, p. 37.
246. G. M. C. Higgins, B. Saville and M. B. Evans, *J. Chem. Soc.*, 702 (1965).
247. D. R. Taylor, *Chem. Rev.*, **67,** 317 (1967).
248. D. K. Black, S. R. Landor, A. N. Patel and P. F. Whiter, *J. Chem. Soc.* **(C),** 2260 (1967).
249. M. G. Ettlinger and J. E. Hodgkins, *J. Am. Chem. Soc.*, 77 1831 (1955).
250. P. D. Landor and S. R. Landor, *Proc. Chem. Soc.*, 77 (1962).
251. P. B. D. de la Mare, in *Molecular Rearrangements* (Ed. P. de Mayo), Interscience, New York, 1963, Chap. 2.
252. M. Bertrand and J. Le Gras, *Compt. Rend.*, **257,** 456 (1963).
253. C. H. DePuy, *Chem. Soc. (London) Spec. Pub.*, **19,** 163 (1965); *Acc. Chem. Res.*, **1,** 33 (1968).
254. U. Schöllkopf, *Angew. Chem., Inter. Ed.*, **7,** 588 (1968).
255. S. Sarel, J. Yovell and M. Sarel-Imber, *Angew. Chem., Inter. Ed.*, **7,** 577 (1968).
256. R. Breslow in Ref. 251, Chap. 4.
257. E. Ziegler, *Monatsh.*, **79,** 146 (1948).